# The Selected Works of
# Roderick S. C. Wong
## Volume 4

# The Selected Works of
# Roderick S. C. Wong

## Volume 4

Editors

**Dan Dai**
City University of Hong Kong, Hong Kong

**Ding-Xuan Zhou**
The University of Sydney, Australia

World Scientific

NEW JERSEY • LONDON • SINGAPORE • BEIJING • SHANGHAI • HONG KONG • TAIPEI • CHENNAI • TOKYO

*Published by*

World Scientific Publishing Co. Pte. Ltd.
5 Toh Tuck Link, Singapore 596224
*USA office:* 27 Warren Street, Suite 401-402, Hackensack, NJ 07601
*UK office:* 57 Shelton Street, Covent Garden, London WC2H 9HE

**British Library Cataloguing-in-Publication Data**
A catalogue record for this book is available from the British Library.

**THE SELECTED WORKS OF RODERICK S C WONG**
**Volume 4**



ISBN 978-981-12-9085-5 (hardcover)
ISBN 978-981-12-9086-2 (ebook for institutions)
ISBN 978-981-12-9087-9 (ebook for individuals)

For any available supplementary material, please visit
https://www.worldscientific.com/worldscibooks/10.1142/13778#t=suppl

# List of Publications by Roderick S. C. Wong from Volumes 1–3

## Volume 1

1. The Asymptotic Behaviour of $\mu(z, \beta, \alpha)$, *Canad. J. Math.*, **21**(1969), 1013–1023 (with M. Wyman).

2. A Generalization of Watson's Lemma, *Canad. J. Math.*, **24**(1972), 185–208 (with M. Wyman).

3. Linear Equations in Infinite Matrices, *Linear Algebra and Appl.*, **7**(1973), 53–62 (with P. N. Shivakumar).

4. Asymptotic Solutions of Linear Volterra Integral Equations with Singular Kernels, *Trans. Amer. Math. Soc.*, **189**(1974), 185–200 (with J. S. W. Wong).

5. On Infinite Systems of Linear Differential Equations, *Canad. J. Math.*, **27**(1975), 691–703 (with J. P. McClure).

6. Error Bounds for Asymptotic Expansions of Hankel Transforms, *SIAM J. Math. Anal.*, **7**(1976), 799–808.

7. Explicit Error Terms for Asymptotic Expansions of Stieltjes Transforms, *J. Inst. Math. Appl.*, **22**(1978), 129–145 (with J. P. McClure).

8. Explicit Error Terms for Asymptotic Expansions of Mellin Convolutions, *J. Math. Anal. Appl.*, **72**(1979), 740–756.

9. Asymptotic Expansion of Multiple Fourier Transforms, *SIAM J. Math. Anal.*, **10**(1979), 1095–1104 (with P. N. Shivakumar).

10. Exact Remainders for Asymptotic Expansions of Fractional Integrals, *J. Inst. Math. Appl.*, **23**(1979), 139–147 (with J. P. McClure).

11. Asymptotic Expansion of the Hilbert Transform, *SIAM J. Math. Anal.*, **11**(1980), 92–99.

12. Error Bounds for Asymptotic Expansions of Integrals, *SIAM Rev.*, **22**(1980), 401–435.

13. Distributional Derivation of an Asymptotic Expansion, *Proc. Amer. Math. Soc.*, **80**(1980), 266–270.

14. On a Method of Asymptotic Evaluation of Multiple Integrals, *Math. Comp.*, **37**(1981), 509–521 (with J. P. McClure).

15. Asymptotic Expansion of the Lebesgue Constants Associated with Polynomial Interpolation, *Math. Comp.*, **39**(1982), 195–200 (with P. N. Shivakumar).

16. Quadrature Formulas for Oscillatory Integral Transforms, *Numer. Math.*, **39**(1982), 351–360.

17. Generalized Mellin Convolutions and Their Asymptotic Expansions, *Canad. J. Math.*, **36**(1984), 924–960 (with J. P. McClure).

18. A Uniform Asymptotic Expansion of the Jacobi Polynomials with Error Bounds, *Canad. J. Math.*, **37**(1985), 979–1007 (with C. L. Frenzen).

19. Asymptotic Expansion of a Multiple Integral, *SIAM J. Math. Anal.*, **18**(1987), 1630–1637 (with J. P. McClure).

20. Asymptotic Expansion of a Double Integral with a Curve of Stationary Points, *IMA J. Appl. Math.*, **38**(1987), 49–59 (with J. P. McClure).

21. Szegö's Conjecture on Lebesgue Constants for Legendre Series, *Pacific J. Math.*, **135**(1988), 157–188 (with C. K. Qu).

22. Uniform Asymptotic Expansions of Laguerre Polynomials, *SIAM J. Math. Anal.*, **19**(1988), 1232–1248 (with C. L. Frenzen).

23. Transformation to Canonical Form for Uniform Asymptotic Expansions, *J. Math. Anal. Appl.*, **149**(1990), 210–219 (with C. K. Qu).

24. Multidimensional Stationary Phase Approximation: Boundary Stationary Point, *J. Comput. Appl. Math.*, **30**(1990), 213–225 (with J. P. McClure).

25. Two-Dimensional Stationary Phase Approximation: Stationary Point at a Corner, *SIAM J. Math. Anal.*, **22**(1991), 500–523 (with J. P. McClure).

26. Asymptotic Expansions for Second-Order Linear Difference Equations, *J. Comput. Appl. Math.*, **41**(1992), 65–94 (with H. Li).

27. Asymptotic Expansions for Second-Order Linear Difference Equations, II, *Stud. Appl. Math.*, **87**(1992), 289–324 (with H. Li).

28. Asymptotic Behaviour of the Fundamental Solution to $\partial u/\partial t = -(-\Delta)^m u$, *Proc. Roy. Soc. London Ser. A*, **441**(1993), 423–432 (with X. Li).

29. A Bernstein-Type Inequality for the Jacobi Polynomial, *Proc. Amer. Math. Soc.*, **121**(1994), 703–709 (with Y. Chow and L. Gatteschi).

30. Error Bounds for Asymptotic Expansions of Laplace Convolutions, *SIAM J. Math. Anal.*, **25**(1994), 1537–1553 (with X. Li).

**Volume 2**

31. Asymptotic Behaviour of the Pollaczek Polynomials and Their Zeros, *Stud. Appl. Math.*, **96**(1996), 307–338 (with R. Bo).

32. Justification of the Stationary Phase Approximation in Time-Domain Asymptotics, *Proc. Roy. Soc. London Ser. A*, **453**(1997), 1019–1031 (with J. P. McClure).

33. Asymptotic Expansions of the Generalized Bessel Polynomials, *J. Comput. Appl. Math.*, **85**(1997), 87–112 (with J.-M. Zhang).

34. Uniform Asymptotic Expansions for Meixner Polynomials, *Constr. Approx.*, **14**(1998), 113–150 (with X.-S. Jin).

35. "Best Possible" Upper and Lower Bounds for the Zeros of the Bessel Function $J_\nu(x)$, *Trans. Amer. Math. Soc.*, **351**(1999), 2833–2859 (with C. K. Qu).

36. Justification of a Perturbation Approximation of the Klein–Gordon Equation, *Stud. Appl. Math.*, **102**(1999), 375–417 (with X.-H. Jiang).

37. Smoothing of Stokes's Discontinuity for the Generalized Bessel Function. II, *Proc. Roy. Soc. London Ser. A*, **455**(1999), 3065–3084 (with Y.-Q. Zhao).

38. Uniform Asymptotic Expansions of a Double Integral: Coalescence of Two Stationary Points, *Proc. Roy. Soc. London Ser. A*, **456**(2000), 407–431 (with W.-Y. Qiu).

39. Uniform Asymptotic Formula for Orthogonal Polynomials with Exponential Weight, *SIAM J. Math. Anal.*, **31**(2000), 992–1029 (with W.-Y. Qiu).

40. On the Asymptotics of the Meixner–Pollaczek Polynomials and Their Zeros, *Constr. Approx.*, **17**(2001), 59–90 (with X. Li).

41. Gevrey Asymptotics and Stieltjes Transforms of Algebraically Decaying Functions, *Proc. Roy. Soc. London Ser. A*, **458**(2002), 625–644 (with Y.-Q. Zhao).

42. Exponential Asymptotics of the Mittag–Leffler Function, *Constr. Approx.*, **18**(2002), 355–385 (with Y.-Q. Zhao).

43. On the Ackerberg–O'Malley Resonance, *Stud. Appl. Math.*, **110** (2003), 157–179 (with H.-P. Yang).

44. Asymptotic Expansions for Second-Order Linear Difference Equations with a Turning Point, *Numer. Math.*, **94**(2003), 147–194 (with Z. Wang).

45. On a Two-Point Boundary-Value Problem with Spurious Solutions, *Stud. Appl. Math.*, **111**(2003), 377–408 (with C.-H. Ou).

46. Shooting Method for Nonlinear Singularly Perturbed Boundary-Value Problems, *Stud. Appl. Math.*, **112**(2004), 161–200 (with C.-H. Ou).

## Volume 3

47. Asymptotic Expansion of the Krawtchouk Polynomials and Their Zeros, *Comput. Methods Funct. Theory*, **4**(1)(2004), 189–226 (with W.-Y. Qiu).

48. On a Uniform Treatment of Darboux's Method, *Constr. Approx.*, **21**(2005), 225–255 (with Y.-Q. Zhao).

49. Linear Difference Equations with Transition Points, *Math. Comp.*, **74**(2005), 629–653 (with Z. Wang).

50. Uniform Asymptotics for Jacobi Polynomials with Varying Large Negative Parameters — A Riemann–Hilbert Approach, *Trans. Amer. Math. Soc.*, **358**(2006), 2663–2694 (with W.-J. Zhang).

51. Uniform Asymptotics of the Stieltjes–Wigert Polynomials via the Riemann–Hilbert Approach, *J. Math. Pures Appl.*, **85**(5)(2006), 698–718 (with Z. Wang).

52. A Singularly Perturbed Boundary-Value Problem Arising in Phase Transitions, *European J. Appl. Math.*, **17**(6)(2006), 705–733 (with Y. Zhao).

53. On the Number of Solutions to Carrier's Problem, *Stud. Appl. Math.*, **120**(3)(2008), 213–245 (with Y. Zhao).

54. Asymptotic Expansions for Riemann–Hilbert Problems, *Anal. Appl. (Singap.)*, **6**(2008), 269–298 (with W.-Y. Qiu).

55. On the Connection Formulas of the Third Painlevé Transcendent, *Discrete Contin. Dyn. Syst.*, **23**(2009), 541–560 (with H. Y. Zhang).

56. Hyperasymptotic Expansions of the Modified Bessel Function of the Third Kind of Purely Imaginary Order, *Asymptot. Anal.*, **63**(2009), 101–123 (with W. Shi).

57. Global Asymptotics for Polynomials Orthogonal with Exponential Quartic Weight, *Asymptot. Anal.*, **64**(2009), 125–154 (with L. Zhang).

58. The Riemann–Hilbert Approach to Global Asymptotics of Discrete Orthogonal Polynomials with Infinite Nodes, *Anal. Appl. (Singap.)*, **8**(2010), 247–286 (with C.-H. Ou).

59. Global Asymptotics of the Meixner Polynomials, *Asymptot. Anal.*, **75**(2011), 211–231 (with X.-S. Wang).

60. Asymptotics of Orthogonal Polynomials via Recurrence Relations, *Anal. Appl. (Singap.)*, **10**(2)(2012), 215–235 (with X.-S. Wang).

61. Uniform Asymptotic Expansions for the Discrete Chebyshev Polynomials, *Stud. Appl. Math.*, **128**(2012), 337–384 (with J. H. Pan).

62. Global Asymptotics of the Hahn Polynomials, *Anal. Appl. (Singap.)*, **11**(3) (2013), 1350018 (47 pages) (with Y. Lin).

63. Global Asymptotics of Stieltjes–Wigert Polynomials, *Anal. Appl. (Singap.)*, **11**(5)(2013), 1350028 (12 pages) (with Y.-T. Li).

# Contents

List of Publications by Roderick S. C. Wong from Volumes 1–3 v

Preface xv

**Volume 4**

64. Uniform Asymptotic Expansion of the Jacobi Polynomials in a Complex Domain, *Proc. R. Soc. Lond. Ser. A Math. Phys. Eng. Sci.*, **460**(2004), 2569–2586 (with Y.-Q. Zhao). 1513

65. Uniform Asymptotics for Orthogonal Polynomials with Exponential Weights — the Riemann–Hilbert Approach, *Stud. Appl. Math.*, **115**(2005), 139–155 (with Z. Wang). 1531

66. Bessel-Type Asymptotic Expansions via the Riemann–Hilbert Approach, *Proc. R. Soc. Lond. Ser. A Math. Phys. Eng. Sci.*, **461**(2005), 2839–2856 (with Z. Wang). 1548

67. Asymptotic Analysis of a Perturbed Periodic Solution for the KdV Equation, *Stud. Appl. Math.*, **116**(2006), 21–33 (with X. H. Jiang). 1566

68. Asymptotic Solutions of a Fourth Order Differential Equation, *Stud. Appl. Math.*, **118**(2007), 133–152 (with H. Y. Zhang). 1579

69. Uniform Asymptotic Expansions of the Tricomi–Carlitz Polynomials, *Proc. Amer. Math. Soc.*, **138**(2010), 2513–2519 (with K. F. Lee). 1599

70. Uniform Treatment of Darboux's Method and the Heisenberg Polynomials, *Proc. Amer. Math. Soc.*, **141**(2013), 2683–2691 (with S.-Y. Liu and Y.-Q. Zhao). 1606

71. Global Asymptotics of the Discrete Chebyshev Polynomials, *Asymptot. Anal.*, **82**(2013), 39–64 (with Y. Lin). 1615

72. Global Asymptotics for Meixner–Pollaczek Polynomials with a Varying Parameter, *Stud. Appl. Math.*, **130**(2013), 345–392 (with J. Wang and W.-Y. Qiu). 1641

73. Asymptotics of the Discrete Chebyshev Polynomials, *Stud. Appl. Math.*, **132**(2014), 13–49 (with J. H. Pan). 1689

74. Asymptotic Expansion of the Tricomi–Carlitz Polynomials and Their Zeros, *J. Comput. Appl. Math.*, **265**(2014), 220–242 (with K. F. Lee). 1726

75. A Novel Generalized Solution Expansion for the Lagerstrom Model, *Asymptot. Anal.*, **88**(2014), 111–123 (with C.-H. Ou). 1749

76. Asymptotic Expansion of the Modified Lommel Polynomials $h_{n,\nu}(x)$ and Their Zeros, *Proc. Amer. Math. Soc.*, **142**(2014), 3953–3964 (with K. F. Lee). 1762

77. Global Asymptotics of the Szegő–Askey Polynomials, *Anal. Appl. (Singap.)*, **12**(2014), 727–746 (with Y. Lin). 1774

78. Asymptotics of Racah Polynomials with Varying Parameters, *J. Math. Anal. Appl.*, **436**(2016), 1149–1164 (with X.-S. Wang). 1794

79. Special Functions, Integral Equations and a Riemann–Hilbert Problem, *Proc. Amer. Math. Soc.*, **144**(2016), 4367–4380 (with Y.-Q. Zhao). 1810

80. Asymptotics of Pseudo-Jacobi Polynomials with Varying Parameters, *Stud. Appl. Math.*, **139**(2017), 179–217 (with Z. Song). 1824

81. Asymptotics of Racah Polynomials with Fixed Parameters, *Proc. Amer. Math. Soc.*, **146**(2018), 1083–1096 (with X.-S. Wang). 1863

82. Asymptotics of Orthogonal Polynomials, *Int. J. Numer. Anal. Model.*, **15**(2018), 193–212. 1877

83. Global Asymptotics of Orthogonal Polynomials Associated with a Generalized Freud Weight, *Chinese Ann. Math. Ser. B*, **39**(2018), 553–596 (with Z.-T. Wen and S.-X. Xu). 1897

84. Asymptotics of the Associated Pollaczek Polynomials, *Proc. Amer. Math. Soc.*, **147**(2019), 2583–2597 (with M.-J. Luo). 1941

85. Uniform Asymptotics and Zeros of the Associated Pollaczek Polynomials, *Stud. Appl. Math.*, **145**(2020), 625–646 (with X.-M. Huang). 1956

86. Recent Advances in Asymptotic Analysis, *Anal. Appl. (Singap.)*, **20**(2022), 1103–1146 (with Y.-Q. Zhao). 1978

87. Error Bounds for the Asymptotic Expansions of the Hermite Polynomials, *Proc. Roy. Soc. Edinburgh Sect. A*, **153**(2023), 417–440 (with W. Shi, G. Nemes and X.-S. Wang). 2022

Permissions 2047

# Preface

This volume serves as a continuation of *The Selected Works of Roderick S. C. Wong, Volumes 1–3*, which was published around 10 years ago. Within this new volume, a selection of papers published between 2004 and 2023 is included. Despite being in the later stages of his academic career, Wong remains actively engaged in various aspects. He continues to supervise students, mentor young scholars, conduct research, publish papers, and author books. His energy and passion for mathematical research have not diminished with age, reflecting the same vigor he possessed as a young mathematician over half a century ago.

The present volume, alongside the preceding three volumes, not only showcases Wong's extensive expertise and profound insights, but also serves as a tribute to his remarkable achievements. This collection of 87 papers provides a comprehensive and valuable literature on asymptotics and related fields. It is poised to make significant and enduring contributions to the field of analysis and its practical applications.

The final volume comprises 24 papers, which explore a wide range of topics including asymptotic analysis, perturbation methods, special functions and orthogonal polynomials, integral equations, and ordinary differential equations. These papers not only enjoy widespread appreciation but have also played a pivotal role in shaping and influencing numerous research areas within the field of applied mathematics.

Now, let us provide a brief overview of Wong's contributions in the field of asymptotic analysis.

## Asymptotic Approximations of Integrals and Differential Equations

Wong is a renowned expert in the field of asymptotic analysis, particularly focusing on the asymptotic approximation of integrals. His influential monograph, titled *Asymptotic Approximation of Integrals*, was first published in 1989 by Academic Press and later reprinted in 2001 by the Society for Industrial and Applied Mathematics (SIAM) as part of its Classics in Applied Mathematics series. Despite being published over 30 years ago, it remains a definitive reference in this research area.

Following the publication of his book, Wong continued to make significant contributions to the theory of asymptotic approximations of integrals. For instance, in collaboration with his colleagues, he extended the classical Darboux's method and established a uniform treatment, as described in [48]*. In this work, they derived simpler forms of uniform asymptotic expansions when two or more algebraic singularities on the circle of convergence coalesce as a certain parameter approaches a critical value. Building upon this idea, they obtained a uniform asymptotic expansion for the Heisenberg polynomials in [17], which involved the Kummer function and its derivatives.

In addition to studying the asymptotic approximation of single integrals, Wong also investigated double integrals. By employing a delicate steepest descent analysis, Wong and J. H. Pan derived uniform asymptotic expansions for discrete Chebyshev polynomials in [61] and [73]. Furthermore, Wong and his collaborators improved the classical asymptotic expansions for Hermite polynomials by obtaining explicit and computable error bounds; see [87].

Wong's research contributions extend beyond integrals to the realm of differential equations. In addition to the papers included in the previous volumes, he and X. H. Jiang applied the energy method to derive asymptotics for perturbed periodic solutions of the Korteweg–de Vries (KdV) equation in [67]. Moreover, he and Z. Song employed Olver's asymptotic method for differential equations with large parameters to study pseudo-Jacobi polynomials with varying parameters in [80].

**Difference Equation Method**

The development of the asymptotic theory for difference equations can be traced back to the early 20th century. However, progress in this field came to a halt in the 1930s and remained stagnant until the beginning of the 21st century. An intriguing observation was made by Frank Olver during a lecture given at a conference in honor of his 80th birthday. He expressed that "*In my view, Birkhoff and Trjitzinsky (1933) set back all research into the asymptotic solutions of difference equations for most of the 20th century*".

While a turning-point theory for second-order differential equations was satisfactorily developed during the 1930s–1960s by Langer (1931), Cherry (1949), Olver (1956), and others, progress in establishing a corresponding theory for difference equations has been limited. However, a breakthrough in turning-point theory for difference equations was achieved by Wong and his student Z. Wang in a series of papers [44, 49]. Subsequently, Wong's other students, such as L.-H. Cao and Y.-T. Li (*Anal. Appl. (Singap.)*, (2014)), and X.-S. Wang and Wong [60], completed the entire theory.

This breakthrough included the establishment of the classification for

*Numbered references without letters refer to the papers included in this book and the previous three volumes.

transition points in difference equations and the derivation of Airy-type and Bessel-type asymptotic expansions, depending on the nature of the transition points. The determination of coefficients for two linearly independent asymptotic solutions was also achieved. Currently, this method has evolved into a powerful approach for deriving asymptotics of various special functions from their recurrence relations. This is particularly significant for orthogonal polynomials, which all satisfy a three-term recurrence relation. Notable examples include certain non-classical orthogonal polynomials in [69, 76] and Racah polynomials in [78, 81], which are listed at the top level of the Askey scheme.

**Riemann–Hilbert Approach**

Near the end of the 20th century, a significant breakthrough was made in establishing a connection between orthogonal polynomials and matrix-valued Riemann–Hilbert problems. This breakthrough was accomplished by A. S. Fokas, A. R. Its, and A. V. Kitaev in their work in 1992. Given a weight function, the corresponding orthogonal polynomials are associated with a $2 \times 2$ matrix-valued Riemann–Hilbert problem: find an analytic function in $\mathbb{C} \setminus \Sigma$ based on its information on the boundary curve $\Sigma$. Around the same time, P. Deift and X. Zhou introduced a new asymptotic method to determine the large-time behavior of solutions to the MKdV equation in 1993. Recognizing the crucial connection between these two developments, Deift, Zhou, and their colleagues successfully extended their asymptotic method to investigate asymptotics of orthogonal polynomials. They also rigorously prove the universality results for random unitary ensembles. Contour deformation played a pivotal role in their asymptotic analysis, transforming an oscillatory Riemann–Hilbert problem into one with jumps close to the identity matrix. Due to the similarity with the classical steepest descent method used in studying asymptotic expansions of integrals, their method became known as the *Deift–Zhou nonlinear steepest descent method* for Riemann–Hilbert problems.

Wong made significant contributions in this field through his work on *global asymptotics*. In classical Riemann–Hilbert analysis, most results are local, providing asymptotic information in different regions of the complex plane. These results sufficed for establishing universality results in random matrix theory, such as obtaining the sine kernel in the bulk, the Airy kernel near the soft edge, and the Bessel kernel near the hard edge. However, from Wong's perspective as an expert in asymptotic analysis, such kind of results do not seem to be necessary. He aimed to extend the validity of results obtained in one region (often in terms of special functions) to other regions. Wong and his coauthors endeavored to introduce a global treatment to the Riemann–Hilbert approach, leading to the successful derivation of global asymptotics in much broader regions; examples of their work can be found in [71, 72, 77].

Polynomials orthogonal with respect to the Freud weight $\exp(-|x|^\beta)$, where $\beta > 0$, have attracted significant research interests. In the literature, T. Kriecherbauer and K. T.-R. McLaughlin (1999) successfully applied the Riemann–Hilbert approach to obtain asymptotics for $\beta \geq 1$. However, when $0 < \beta < 1$, the problem was only recently solved by Wong and Y.-Q. Zhao in [79]. They demonstrated that the dominant approximant does not involve the usual special functions such as Airy or Bessel functions; instead, it relies on two solutions to a pair of decoupled scalar linear integral equations.

Editors

**Dan Dai**
*City University of Hong Kong*
*Hong Kong*

**Ding-Xuan Zhou**
*The University of Sydney*
*Australia*

10.1098/rspa.2004.1296

THE ROYAL SOCIETY

# Uniform asymptotic expansion of the Jacobi polynomials in a complex domain

BY R. WONG[1] AND YU-QIU ZHAO[2]

[1]*Department of Mathematics, City University of Hong Kong, Tat Chee Avenue, Kowloon, Hong Kong* (mawong@cityu.edu.hk)

[2]*Mathematics Department, ZhongShan University, GuangZhou 510275, People's Republic of China*

*Received 2 July 2003; revised 18 December 2003; accepted 10 February 2004; published online 2 June 2004*

An asymptotic formula is found that links the behaviour of the Jacobi polynomial $P_n^{(\alpha,\beta)}(z)$ in the interval of orthogonality $[-1, 1]$ with that outside the interval. The two infinite series involved in this formula are shown to be exponentially improved asymptotic expansions. The method used in this paper can also be adopted in other cases of orthogonal polynomials such as Hermite and Laguerre.

**Keywords: Jacobi polynomials; uniform asymptotic expansion; complex domain; exponential improvement**

## 1. Introduction

Let $\alpha$ and $\beta$ be arbitrary real numbers. For an arbitrary real or complex number $x$ outside of the closed interval $[-1, 1]$, the Jacobi polynomials have the asymptotic formula

$$P_n^{(\alpha,\beta)}(x) \cong (x-1)^{-\alpha/2}(x+1)^{-\beta/2}\{(x+1)^{1/2}+(x-1)^{1/2}\}^{\alpha+\beta} \times (2\pi n)^{-1/2}(x^2-1)^{-1/4}\{x+(x^2-1)^{1/2}\}^{n+(1/2)}, \quad (1.1)$$

as $n \to \infty$, where $(x^2-1)^{-1/4}$, $(x^2-1)^{1/2}$ and $\{x+(x^2+1)^{1/2}\}^{n+(1/2)}$ are all real and positive if $x$ is real and greater than 1. Furthermore, this formula holds uniformly in the exterior of an arbitrary closed curve which encloses the interval $[-1, 1]$ in the sense that the ratio tends uniformly to 1 (see Szegö 1975, p. 196). When $x$ is real and lies in the interval $(-1, 1)$, we have

$$P_n^{(\alpha,\beta)}(\cos\theta) = (\pi n)^{-1/2}(\sin\tfrac{1}{2}\theta)^{-\alpha-(1/2)}(\cos\tfrac{1}{2}\theta)^{-\beta-(1/2)}\cos(N\theta+\gamma) + O(n^{-3/2}), \quad (1.2)$$

where $x = \cos\theta$, $0 < \theta < \pi$, $N = n + \frac{1}{2}(\alpha+\beta+1)$ and $\gamma = -(\alpha+\frac{1}{2})\frac{1}{2}\pi$. The bound for the error term holds uniformly in the interval $\varepsilon \leqslant \theta \leqslant \pi - \varepsilon$ (see Szegö 1975, p. 196).

In (1.1), if we write $x = \cos\theta$ with $\theta$ being complex, then the asymptotic approximant there can be expressed as

$$\tfrac{1}{2}(\pi n)^{-1/2}(\sin\tfrac{1}{2}\theta)^{-\alpha-(1/2)}(\cos\tfrac{1}{2}\theta)^{-\beta-(1/2)}\mathrm{e}^{\mathrm{i}(N\theta+\gamma)}, \quad (1.3)$$

with $N$ and $\gamma$ being the same as in (1.2). (Note that $z = \cos\theta$ maps the strip $0 < \operatorname{Re}\theta < \pi$ conformally onto the whole $z$-plane cut along the real axis from $-\infty$ to $-1$ and from $+1$ to $+\infty$.) When $\theta$ approaches the interval $(0, \pi)$ on the real axis, one readily sees that the real part of the expression in (1.3) is half of the approximant in (1.2). This fact was observed recently by Y. Chen (1999, personal communication), although it was known earlier to Heine (1878, p. 175) and Szegö (1975, p. 196, theorem 8.21.9). Chen also pointed out that this fact is true for the Hermite and the Laguerre polynomials, and asked whether this phenomenon is universal, i.e. whether the real part of an asymptotic approximant for orthogonal polynomials in the complex plane is always half of the corresponding asymptotic approximant in the interval of orthogonality on the real axis. The answer to Chen's question is affirmative, at least for orthogonal polynomials in the Szegö class, and this can be verified by using the two asymptotic formulae given in Szegö (1975, p. 297). It is interesting to note that Szegö seemed to know that this phenomenon is true even for orthogonal polynomials not in his class, as he had explicitly used this fact to derive the asymptotic formula of the Pollaczek polynomial $P_n(x; a, b)$ when $x$ is in the interval $(-1, 1)$ (see (Szegö 1975, p. 395, § 5 $a$). (One of the referees made the following observation, and we believe it should be included: the Jacobi polynomial $P_n^{(\alpha,\beta)}(x)$ can be defined in terms of the hypergeometric function and, in doing so, the value of $n$ no longer needs to be an non-negative integer. However, the more general function so obtained now may be multivalued, and may have singularities. To define a real-valued function on the cut of this more general function, one often takes the average of the limiting values of the function from above and below the cut, as is done in the case of the Legendre function $P_\nu^\mu(x)$ on the cut $(-1, 1)$. This provides a heuristic explanation of why the discontinuity factor $\frac{1}{2}$ shows up in (1.3).)

Chen's observation has prompted us to consider two related problems. The first one is to find an asymptotic formula that includes both the exponential and oscillatory approximations in (1.1) and (1.2). This problem is analogous to the one associated with the turning-point theory for differential equations. The desired approximation in that case involves the Airy function (see Olver 1974, chapter 11; Bender & Orszag 1978, chapter 10, § 10.4). In our case, we claim that the desired formula is provided by the asymptotic expansion

$$(\sin\tfrac{1}{2}\theta)^\alpha(\cos\tfrac{1}{2}\theta)^\beta P_n^{(\alpha,\beta)}(\cos\theta) = -J_\alpha(N\theta)\sum_{k=0}^{p-1}\frac{c_k(\theta)}{N^k} - J_{\alpha+1}(N\theta)\sum_{k=0}^{p-1}\frac{d_k(\theta)}{N^k} + \delta_p(N,\theta), \quad (1.4)$$

where $N$ has the same meaning as in (1.2). In Wong & Zhao (2003), we have shown that the coefficients $c_k(\theta)$ and $d_k(\theta)/\theta$, $k = 0, 1, 2, \dots$, are all analytic functions in $|\theta| \leqslant \frac{1}{2}\pi$. Furthermore, we have established (1.4) as an identity with an exact remainder for $\theta \in [0, \frac{1}{2}\pi]$, and proved that for $\alpha > -1$, $\beta$ real and $\theta \in [0, \frac{1}{2}\pi]$, the remainder $\delta_p(N, \theta)$ satisfies

$$|\delta_p(N,\theta)| \leqslant \frac{\Lambda_p}{N^p}\{|J_\alpha(N\theta)| + |J_{\alpha+1}(N\theta)|\}, \quad (1.5)$$

where $\Lambda_p$ is a positive constant independent of $\theta$ and $N$. Dividing both sides of (1.4) by $(\frac{1}{2}\theta)^\alpha$, one readily sees that, apart from the remainder, all other terms are analytic functions of $\theta$. Thus, to establish our claim, we only need to extend the domain of analyticity in $\theta$ so that its image $x = \cos\theta$ contains a neighbourhood of the interval $(-1, 1]$. By analytic continuation, equation (1.4) will continue to hold in the larger domain, but we have to show that the inequality in (1.5) remains valid in this extended domain in the complex $x$-plane.

The second problem deals with a more recent topic, namely, exponential asymptotics (Wong 2001). Here, we wish to demonstrate that equation (1.4) actually provides an exponentially improved asymptotic expansion (Olver 1991) for the Jacobi polynomials. To do this, we need to show that the bounds for the coefficients $c_k(\theta)$ and $d_k(\theta)/\theta$, $k = 0, 1, 2, \ldots$, and for the remainder $\delta_p(N, \theta)$, are all in the form of 'factorial divided by power' (Berry & Howls 1993), and that when the series in (1.4) is optimally truncated, the remainder is exponentially small.

A slightly different type of uniform expansion has been given by Dunster (1999) for the Jacobi polynomial $P_n^{(\alpha,\beta)}(x)$, which holds for unbounded complex $x$, except at the points near $x = 1$, and for $q$ taking any value in the range $0 \leqslant q \leqslant An$, $A$ being any fixed positive constant. Dunster's result was obtained by applying an existing asymptotic theory of a coalescing turning point and a double pole in the complex plane.

Our approach using analytic continuation and exponential improvement can be adopted in other cases of orthogonal polynomials such as Hermite and Laguerre. Although in most cases asymptotic methods for differential equations give stronger results than the corresponding ones for integrals, the result in this paper presents one example in which the integral approach has an advantage over the differential-equation theory. By this we mean that it is easier to see the validity of the extension of the uniform asymptotic expansion for orthogonal polynomials in the interval of orthogonality to regions in the complex plane containing this interval, and that the uniform expansion provides a connection between the oscillatory and exponential behaviour inside and outside the interval of orthogonality. Furthermore, the availability of an explicit expression for the remainder term makes it simpler to derive the sharper error estimates in the form of factorial divided by power, which in turn leads to the analysis of exponential asymptotics.

## 2. Summary of the results in Wong & Zhao (2003)

For $x$ in the interval $[-1, 1]$ we may write $x = \cos\theta$ with $\theta \in [0, \pi]$. In view of the reflection formula (Szegö 1975, p. 59)

$$P_n^{(\alpha,\beta)}(x) = (-1)^n P_n^{(\alpha,\beta)}(-x), \tag{2.1}$$

we may without loss of generality assume that $x$ belongs to the interval $[0, 1]$ or, equivalently, $\theta \in [0, \frac{1}{2}\pi]$. From Rodrigues's formula (Szegö 1975, p. 67) one readily obtains the integral representation

$$P_n^{(\alpha,\beta)}(x) = 2^{(\alpha+\beta+1)/2}(1-x)^{-\alpha}(1+x)^{-\beta}\frac{1}{2\pi\mathrm{i}}\int_C g(z)\mathrm{e}^{Nf(z,x)}\,\mathrm{d}z, \tag{2.2}$$

where

$$N = n + \tfrac{1}{2}(\alpha+\beta+1), \tag{2.3}$$

$$f(z, x) = \ln\frac{(z+x+1)(z+x-1)}{2z} \tag{2.4}$$

and

$$g(z) = \frac{(1-z-x)^{\alpha}}{(z+x-1)^{(\alpha+\beta+1)/2}} \frac{(z+x+1)^{(-\alpha+\beta-1)/2}}{z^{(-\alpha-\beta+1)/2}}. \tag{2.5}$$

The integration path $C$ in (2.2) is a simple closed contour encircling the point $z=0$, but not the points $z=\pm1-x$. For real $x$, the $z$-branches are chosen so that

$$\arg(z+1+x) \in [-\pi,\pi], \quad \arg(z+x-1) \in [0,2\pi], \quad \arg z \in [0,2\pi] \tag{2.6}$$

and

$$(1-z-x)^{\alpha} = \mathrm{e}^{-\alpha\pi\mathrm{i}}(z+x-1)^{\alpha}. \tag{2.7}$$

The functions $f(z,x)$ and $g(z)$ are thus well defined.

It is easily seen that $f(z,x)$ has two saddle points

$$z_{\pm} = \pm\mathrm{i}\sqrt{1-x^2} = \pm\mathrm{i}\sin\theta, \tag{2.8}$$

where $\theta \in [0,\pi)$ for $x \in (-1,1]$. The saddle points coalesce with each other when $\theta \to 0$.

To ensure that the result in Wong & Zhang (1996) is genuinely a uniform asymptotic expansion, Wong & Zhao (2003) have introduced the canonical transformation

$$f(z,x) = Z = \varphi(s,\theta), \qquad f(z_{\pm},x) = \varphi(s_{\pm},\theta), \tag{2.9}$$

where $Z$ is an intermediate variable,

$$\varphi(s,\theta) = \tfrac{1}{2}\theta\left(s-\frac{1}{s}\right) \tag{2.10}$$

and

$$s_{\pm} = \pm\mathrm{i} \tag{2.11}$$

are the saddle points of $\varphi(s,\theta)$. It is further demonstrated in Wong & Zhao (2003) that (2.9) defines a one-to-one analytic mapping $z \leftrightarrow s$ between a certain $z$-domain $D_z$ and $s$-domain $D_s$. These domains can be described, via the intermediate variable, as $|\mathrm{Im}\, Z| < \pi$, and they contain the entire paths of steepest descent.

Under the transformation (2.9), equation (2.2) now becomes

$$(\sin\tfrac{1}{2}\theta)^{\alpha}(\cos\tfrac{1}{2}\theta)^{\beta}P_n^{(\alpha,\beta)}(\cos\theta) = \frac{1}{2\pi\mathrm{i}}\int_L h_0(s,\theta)\mathrm{e}^{(N\theta/2)(s-(1/s))}\frac{\mathrm{d}s}{s^{\alpha+1}}, \tag{2.12}$$

where $L$ is the image of the contour $C$,

$$h_0(s,\theta) = c(\theta)g(z)s^{\alpha+1}\frac{\mathrm{d}z}{\mathrm{d}s}, \tag{2.13}$$

$g(z)$ is given in (2.5) and

$$c(\theta) = 2^{-(\alpha+\beta-1)/2}(\sin\tfrac{1}{2}\theta)^{-\alpha}(\cos\tfrac{1}{2}\theta)^{-\beta}. \tag{2.14}$$

The complex $s$-plane is cut so that in (2.13) we can specify

$$\arg s \in [-\pi,\pi]. \tag{2.15}$$

It is shown in Wong & Zhao (2003) that $h_0(s,\theta)$ is analytic in the above-mentioned $s$-domain $D_s$, namely,

$$\left|\operatorname{Im}\left\{\tfrac{1}{2}\theta\left(s-\frac{1}{s}\right)\right\}\right| < \pi. \tag{2.16}$$

It is also demonstrated that for $s$ close to $\pm\mathrm{i}$, $h_0(s,\theta)$ is an analytic function of $\theta$. More precisely, we have

$$h_0(s,\theta) = -\sum_{k=0}^{\infty} \xi_k(s)\theta^k, \quad |\theta| \leqslant \tfrac{1}{2}\pi + \varepsilon, \tag{2.17}$$

for fixed $s$ in $|s-\mathrm{i}| \leqslant \frac{1}{2}$ or $|s+\mathrm{i}| \leqslant \frac{1}{2}$, where each $\xi_k(s)$ is an analytic function of $s$ in $|s-\mathrm{i}| \leqslant \frac{1}{2}$ and $|s+\mathrm{i}| \leqslant \frac{1}{2}$, and $\xi_0(s) \equiv 1$. Furthermore, we have

$$|\xi_k(s)| \leqslant \frac{\tilde{M}}{(\frac{1}{2}\pi+\varepsilon)^k}, \quad k = 0, 1, 2, \ldots, \tag{2.18}$$

for $|s \pm \mathrm{i}| \leqslant \frac{1}{2}$. The argument in Wong & Zhao (2003, § 3) leading to (2.17) and (2.18) can actually be improved to read that for $\varepsilon \in (0, \pi)$, there exists $\delta > 0$ such that

$$h_0(s,\theta) = -\sum_{k=0}^{\infty} \xi_k(s)\theta^k, \quad |\theta| \leqslant \pi - \varepsilon, \tag{2.19}$$

and

$$|\xi_k(s)| \leqslant \frac{\tilde{M}}{(\pi-\varepsilon)^k}, \quad k = 0, 1, 2, \ldots, \tag{2.20}$$

for fixed $s$ in $|s-\mathrm{i}| \leqslant \delta$ or $|s+\mathrm{i}| \leqslant \delta$, where $\delta \in (0, \delta_\varepsilon)$ and

$$\delta_\varepsilon = \frac{-2(2\pi-\varepsilon) + 2\sqrt{(2\pi-\varepsilon)^2 + (\pi-\varepsilon)\varepsilon}}{\pi-\varepsilon} \sim \frac{\varepsilon}{2\pi} \quad \text{as } \varepsilon \to 0^+.$$

Each $\xi_k(s)$ is an analytic function in $|s-\mathrm{i}| \leqslant \delta$ and $|s+\mathrm{i}| \leqslant \delta$.

By repeated application of integration by parts, we obtain in Wong & Zhao (2003)

$$\begin{aligned}(\sin\tfrac{1}{2}\theta)^\alpha(\cos\tfrac{1}{2}\theta)^\beta P_n^{(\alpha,\beta)}(\cos\theta) = {} & -J_\alpha(N\theta)\sum_{k=0}^{p-1}\frac{c_k(\theta)}{N^k} \\ & - J_{\alpha+1}(N\theta)\sum_{k=0}^{p-1}\frac{d_k(\theta)}{N^k} + \delta_p(N,\theta),\end{aligned} \tag{2.21}$$

where

$$h_k(s,\theta) = c_k(\theta) + \frac{d_k(\theta)}{s} + \left(1+\frac{1}{s^2}\right)g_k(s,\theta), \quad k=0,1,2,\ldots, \tag{2.22}$$

$$h_{k+1}(s,\theta) = -\frac{2}{\theta}s^{\alpha+1}\frac{\mathrm{d}}{\mathrm{d}s}\{s^{-\alpha-1}g_k(s,\theta)\} = \frac{2}{\theta}\left\{\frac{\alpha+1}{s} - \frac{\mathrm{d}}{\mathrm{d}s}\right\}g_k(s,\theta), \quad k=0,1,2,\ldots, \tag{2.23}$$

and

$$\delta_p(N,\theta) = \frac{1}{N^p}\frac{1}{2\pi \mathrm{i}}\int_L h_p(s,\theta)\mathrm{e}^{(N\theta/2)(s-(1/s))}\frac{\mathrm{d}s}{s^{\alpha+1}}. \tag{2.24}$$

In deriving (2.21), we have made use of the integral representation of the Bessel function

$$J_\nu(z) = \frac{1}{2\pi\mathrm{i}}\int_{-\infty}^{(0^+)} s^{-\nu-1}\exp\left\{\tfrac{1}{2}z\left(s-\frac{1}{s}\right)\right\}\mathrm{d}s. \tag{2.25}$$

The main result in Wong & Zhao (2003) is the uniform asymptotic expansion (2.21) for $\theta \in [0, \frac{1}{2}\pi]$. With the improvement in (2.19) we can in fact slightly extend the interval of validity from $[0, \frac{1}{2}\pi]$ to $[0, \pi-\varepsilon]$. Indeed, we can show that the coefficients $c_k(\theta)$ and $d_k(\theta)/\theta$ in (2.21) have the bounds

$$|c_k(\theta)| \leqslant N_k \quad \text{and} \quad \frac{|d_k(\theta)|}{\theta} \leqslant N_k \tag{2.26}$$

for $\theta \in [0, \pi-\varepsilon]$ and $k = 0, 1, 2, \ldots$, where $N_k$ is a positive constant independent of $\theta$ (see a statement following (3.16) below). Moreover, considering $\theta$ as a complex variable, $c_k(\theta)$ and $d_k(\theta)/\theta$ are analytic functions in $|\theta| < \pi$. For $\alpha > -1$, $\beta$ real and $\theta \in [0, \pi-\varepsilon]$, the remainder $\delta_p(N,\theta)$ satisfies

$$|\delta_p(N,\theta)| \leqslant \frac{\Lambda_p}{N^p}\{|J_\alpha(N\theta)| + |J_{\alpha+1}(N\theta)|\} \tag{2.27}$$

for any $p \geqslant 1$, where $\Lambda_p$ is a positive constant independent of $\theta$ and $N$.

## 3. Extension to a complex domain

Since the result in (2.21) is actually an equality, it is natural to first extend all terms on both sides of the equation analytically into the complex plane. This process can continue until a singularity is reached. In this section, we show how to extend the validity of (2.21) from the interval $[\cos(\pi-\varepsilon), 1]$ on the real axis of the complex $x$-plane to a domain containing $[0, 1]$. To this end, we rewrite (2.21) as

$$\begin{aligned}\left(\frac{\sin(\theta/2)}{\theta/2}\right)^\alpha &(\cos\tfrac{1}{2}\theta)^\beta P_n^{(\alpha,\beta)}(\cos\theta)\\ &= -\{(\tfrac{1}{2}\theta)^{-\alpha}J_\alpha(N\theta)\}\sum_{k=0}^{p-1}\frac{c_k(\theta)}{N^k} - \{(\tfrac{1}{2}\theta)^{-\alpha-1}J_{\alpha+1}(N\theta)\}\sum_{k=0}^{p-1}\frac{\theta d_k(\theta)/2}{N^k} + R_p(N,\theta),\end{aligned} \tag{3.1}$$

where

$$R_p(N,\theta) = (\tfrac{1}{2}\theta)^{-\alpha}\delta_p(N,\theta). \tag{3.2}$$

Taking into account the well-known result

$$J_\nu(\tau) = (\tfrac{1}{2}\tau)^\nu\sum_{k=0}^{\infty}\frac{(-1/4\tau^2)^k}{k!\,\Gamma(\nu+k+1)} \tag{3.3}$$

for complex $\tau$, we readily see that both $(\frac{1}{2}\theta)^{-\alpha}J_\alpha(N\theta)$ and $(\frac{1}{2}\theta)^{-\alpha-1}J_{\alpha+1}(N\theta)$ are even and entire functions of $\theta$. Thus

$$(\tfrac{1}{2}\theta)^{-\nu}J_\nu(N\theta) = (-\tfrac{1}{2}\theta)^{-\nu}J_\nu(-N\theta) \tag{3.4}$$

for $\nu = \alpha$ and $\alpha + 1$.

As pointed out in the statement following (2.26), the coefficients $c_k(\theta)$ and $d_k(\theta)/\theta$ are analytic functions in $|\theta| < \pi$. We now proceed to show that for $\theta \in [0, \pi)$, we have

$$c_k(\theta) = c_k(-\theta) \quad \text{and} \quad \frac{d_k(\theta)}{\theta} = \frac{d_k(-\theta)}{-\theta}. \tag{3.5}$$

To do this we first give a brief account of the procedure used to establish the analyticity of $c_k(\theta)$ and $d_k(\theta)/\theta$ in a neighbourhood of the origin in the $\theta$-plane. The reader is referred to Wong & Zhao (2003) for full details. The discussion in that paper can be divided into several stages.

In the first stage, we adopt a method introduced by Olde Daalhuis & Temme (1994). The basic idea is to find Cauchy-type integral representations for the coefficients $c_k(\theta)$ and $d_k(\theta)$, and for the function $h_p(N, \theta)$ in the remainder, in terms of a class of rational functions. Define

$$A_0(s, \theta) = \frac{s}{s^2 + 1}, \qquad B_0(s, \theta) = -\frac{1}{s^2 + 1}, \tag{3.6}$$

and

$$A_{k+1}(s, \theta) = \frac{2}{\theta}\left(1 + \frac{1}{s^2}\right)^{-1} \left\{\frac{\alpha + 1}{s} + \frac{\mathrm{d}}{\mathrm{d}s}\right\} A_k(s, \theta), \tag{3.7}$$

$$B_{k+1}(s, \theta) = \frac{2}{\theta}\left(1 + \frac{1}{s^2}\right)^{-1} \left\{\frac{\alpha + 1}{s} + \frac{\mathrm{d}}{\mathrm{d}s}\right\} B_k(s, \theta) \tag{3.8}$$

for $k = 0, 1, 2, \ldots$. In terms of these rational functions, we have the following representations:

$$c_k(\theta) = \frac{1}{2\pi \mathrm{i}} \int_{C_\mathrm{L} \cup C_\mathrm{S}} h_0(s, \theta) A_k(s, \theta) \, \mathrm{d}s \tag{3.9}$$

and

$$d_k(\theta) = \frac{1}{2\pi \mathrm{i}} \int_{C_\mathrm{L} \cup C_\mathrm{S}} h_0(s, \theta) B_k(s, \theta) \, \mathrm{d}s, \tag{3.10}$$

$k = 0, 1, 2, \ldots$, where $C_\mathrm{S}$ and $C_\mathrm{L}$ are specific simple closed curves such that

$$\delta_0 \theta < |s| < \Delta_0 \theta, \quad s \in C_\mathrm{S},$$

and

$$\frac{\delta_0}{\theta} < |s| < \frac{\Delta_0}{\theta}, \quad s \in C_\mathrm{L},$$

$\delta_0$ and $\Delta_0$ being positive constants independent of $\theta$ and $s$. Furthermore, $C_\mathrm{L}$ encircles both i and $-$i, while $C_\mathrm{S}$ excludes both of them.

In the second stage, we have shown in Wong & Zhao (2003) that

$$|h_0(s, \theta)| \leqslant M_0, \quad s \in C_\mathrm{S} \cup C_\mathrm{L}, \tag{3.11}$$

where $M_0$ is a positive number independent of $\theta$ and $s$. In fact, this estimate is valid in regions whose images are bounded subsets of $|\mathrm{Im}\, Z| < \pi$. By induction, it can be verified that

$$A_k(s,\theta) = \left(\frac{1}{\theta}\right)^k \left(\frac{1}{s}\right)^{k+1} \left(\frac{s^2}{s^2+1}\right)^{k+1} \sum_{l=0}^{k} c_{k,l} \left(\frac{s^2}{s^2+1}\right)^l \tag{3.12}$$

and

$$B_k(s,\theta) = \left(\frac{1}{\theta}\right)^k \left(\frac{1}{s}\right)^{k+2} \left(\frac{s^2}{s^2+1}\right)^{k+1} \sum_{l=0}^{k} d_{k,l} \left(\frac{s^2}{s^2+1}\right)^l \tag{3.13}$$

for $k = 0, 1, 2, \ldots$, where $c_{k,l}$ and $d_{k,l}$ are constants independent of $\theta$ and $s$. Consequently, we have

$$|A_k(s,\theta)| \leqslant M_k\theta, \qquad |B_k(s,\theta)| \leqslant M_k \tag{3.14}$$

for $s \in C_{\mathrm{S}}$, and

$$|A_k(s,\theta)| \leqslant M_k\theta, \qquad |B_k(s,\theta)| \leqslant M_k\theta^2 \tag{3.15}$$

for $s \in C_{\mathrm{L}}$, where again $M_k$ is a positive constant independent of $\theta$ and $s$ (cf. lemma 2 in Wong & Zhao 2003). We also note that

$$\int_{C_{\mathrm{L}}} |\mathrm{d}s| \leqslant \frac{M}{\theta} \quad \text{and} \quad \int_{C_{\mathrm{S}}} |\mathrm{d}s| \leqslant M\theta, \tag{3.16}$$

where $M$ is a positive constant. The estimates in (2.26) are consequences of (3.9)–(3.11) and (3.14)–(3.16).

Since $h_0(s,\theta)$ is analytic in the domain $D_s$ for $\theta \in (0, \pi - \varepsilon]$, the integration path $C_{\mathrm{L}} \cup C_{\mathrm{S}}$ in (3.9) and (3.10) can be deformed into two circles $|s \pm \mathrm{i}| = \delta$. Furthermore, (2.19) and (2.20) hold along these two circles. In view of (3.12) and (3.13), we can write

$$A_k(s,\theta) = A_k^*(s)\theta^{-k} \quad \text{and} \quad B_k(s,\theta) = B_k^*(s)\theta^{-k} \tag{3.17}$$

with

$$|A_k^*(s)| \leqslant A_k \quad \text{and} \quad |B_k^*(s)| \leqslant B_k \quad \text{for } |s \pm \mathrm{i}| = \delta, \tag{3.18}$$

$A_k$ and $B_k$ being constants. Coupling (2.19) and (3.17), we obtain

$$h_0(s,\theta)A_k(s,\theta) = -\theta^{-k} \sum_{l=0}^{\infty} \zeta_l(s)\theta^l, \quad \theta \in (0, \pi - \varepsilon], \tag{3.19}$$

where $\zeta_l(s) := \xi_l(s)A_k^*(s)$ is analytic (and hence continuous) on $|s \pm \mathrm{i}| = \delta$ and

$$|\zeta_l(s)| \leqslant \frac{A_k\tilde{M}}{(\pi - \varepsilon)^l}.$$

For fixed $\theta \in (0, \pi - \varepsilon]$, the infinite series on the right-hand side of (3.19) is uniformly convergent on the circles $|s \pm \mathrm{i}| = \delta$. Termwise integration of (3.19) with respect to $s$ over $|s \pm \mathrm{i}| = \delta$ then gives

$$c_k(\theta) = \frac{1}{\theta^k} \sum_{l=0}^{\infty} \gamma_l \theta^l, \quad \theta \in (0, \pi - \varepsilon], \tag{3.20}$$

with

$$|\gamma_l| = \left|\frac{1}{2\pi i}\int_{|s\pm i|=\delta} \zeta_l(s)\,ds\right| \leqslant \frac{2\delta A_k \tilde{M}}{(\pi-\varepsilon)^l}, \tag{3.21}$$

$l = 0, 1, 2, \ldots$. From (3.20) and (3.21), it is easily seen that $c_k(\theta)$ can be extended to an analytic function in $0 < |\theta| < \pi - \varepsilon$ with a possible pole at $\theta = 0$. The boundedness of $c_k(\theta)$ shown in (2.26) then implies that

$$\gamma_0 = \gamma_1 = \cdots = \gamma_{k-1} = 0,$$

that is, $c_k(\theta)$ is in fact analytic in $|\theta| < \pi - \varepsilon$. In a similar manner, we can establish the analyticity of $d_k(\theta)/\theta$ in $|\theta| < \pi - \varepsilon$. Since $\varepsilon$ is an arbitrary small number, we conclude that $c_k(\theta)$ and $d_k(\theta)/\theta$ are analytic in $|\theta| < \pi$.

Now we are in a position to justify (3.5). Indeed, from (2.5), (2.9) and (2.13) we have

$$h_0(s,\theta) = \tfrac{1}{2}\theta e^{-\alpha\pi i} c(\theta) \frac{s^2+1}{z^2+\sin^2\theta} h_0^*(s,\theta), \tag{3.22}$$

where $c(\theta)$ is given by (2.14) and

$$h_0^*(s,\theta) = (z+x-1)^{(\alpha-\beta+1)/2}(z+x+1)^{(-\alpha+\beta+1)/2} z^{(\alpha+\beta+1)/2} s^{\alpha+1}. \tag{3.23}$$

From (3.22) and (2.9), one readily sees that

$$h_0(s,\theta) = h_0(-s,-\theta). \tag{3.24}$$

Let us temporarily restrict $\theta$ to $(0,\pi)$. From (3.12) and (3.13), we have

$$A_k(s,\theta) = -A_k(-s,-\theta) \quad \text{and} \quad B_k(s,\theta) = B_k(-s,-\theta). \tag{3.25}$$

In view of the fact that both $C_L$ and $C_S$ are symmetrical with respect to the origin, the identities in (3.5) now follow from (3.9), (3.10), (3.24) and (3.25). As a byproduct of (3.5), we also obtain

$$c_k(\theta) = \sum_{l=0}^{\infty} \gamma_{2l}\theta^{2l}, \quad d_k(\theta) = \sum_{l=0}^{\infty} \delta_{2l+1}\theta^{2l+1}, \quad |\theta| < \pi. \tag{3.26}$$

We observe that, except possibly for the error term, all terms in equation (3.1) have been extended to analytic functions in $|\theta| < \pi$. Hence, naturally the error term $R_p(N,\theta)$ in (3.2) can also be analytically continued to $|\theta| < \pi$. Furthermore, from (3.4) and (3.5) we have

$$R_p(N,\theta) = R_p(N,-\theta) \tag{3.27}$$

for $\theta \in [0,\pi)$.

In the remaining part of this section, we shall show that for complex $\theta$ the remainder $R_p(N,\theta)$ possesses an integral representation similar to (2.24). Clearly, we can extend the function $\varphi(s,\theta)$ in (2.10) to complex $\theta$ in $|\theta| < \pi$, and the function $f(x,z)$ in (2.4) to complex $x$ with $x = \cos\theta = \frac{1}{2}(e^{i\theta} + e^{-i\theta})$. However, the integration path $L$ in (2.24) needs some modification, otherwise the integral may not even converge when $\theta$ becomes complex. In § 4, we shall also demonstrate that equation (2.9) still defines a one-to-one mapping $z \leftrightarrow s$ between a $z$-domain and an $s$-domain. In terms

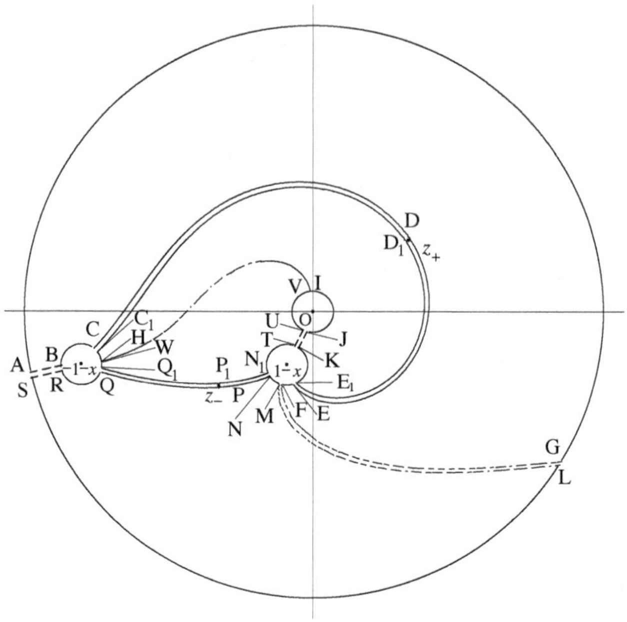


Figure 1. The $z$-plane.

of the intermediate variable $Z$, both domains can be described as $|\mathrm{Im}\, Z| < \pi$. These domains contain the contours

$$L(\theta) : \arg\left[-\tfrac{1}{2}\theta\left(s - \frac{1}{s}\right) \pm \theta\mathrm{i}\right] = \arg[Z(\pm\mathrm{i}) - Z(s)] = 0 \tag{3.28}$$

for complex $\theta$ in $|\theta| < \pi$, and $h_p(s,\theta)$ is analytic in this $s$-domain. Replacing the integration path $L$ in (2.24) by $L(\theta)$ with $\theta$ extended to the complex, we have

$$R_p(N,\theta) = (\tfrac{1}{2}\theta)^{-\alpha}\frac{1}{N^p}\frac{1}{2\pi\mathrm{i}}\int_{L(\theta)} h_p(s,\theta)\mathrm{e}^{(N\theta/2)(s-(1/s))}\frac{\mathrm{d}s}{s^{\alpha+1}} \tag{3.29}$$

(see figure 3 for an illustration of $L(\theta)$).

It is easily verified that the image of the set $\{\theta : \theta = \rho\mathrm{e}^{\mathrm{i}\phi},\ 0 < \rho < \pi,\ \phi \in [-\pi, 0]\}$ under the mapping $x = \cos\theta$ is a domain in the complex $x$-plane with an edge along the interval $(-1, 1]$ on the real axis. In view of the symmetry shown in (3.4), (3.5) and (3.27), by analytic continuation we can conclude that equation (3.1) holds for $x$ in a complex domain containing the interval $(-1, 1]$ on the real line. It is also clear that $\phi \in [-\frac{1}{2}\pi, 0]$ corresponds to $x$ in the upper half of this domain and $\phi \in [-\pi, -\frac{1}{2}\pi]$ corresponds to the lower half of the domain.

## 4. Estimates

When $\theta$ is complex, the branch points of $f(z,x)$ are still at $1 - \cos\theta, -1 - \cos\theta$ and 0, but now the points $\pm 1 - \cos\theta$ are no longer on the real line. However, the

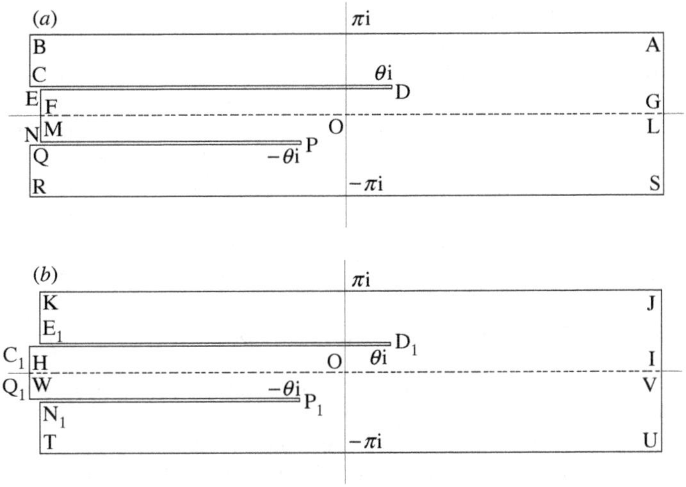


Figure 2. The $Z$-plane.

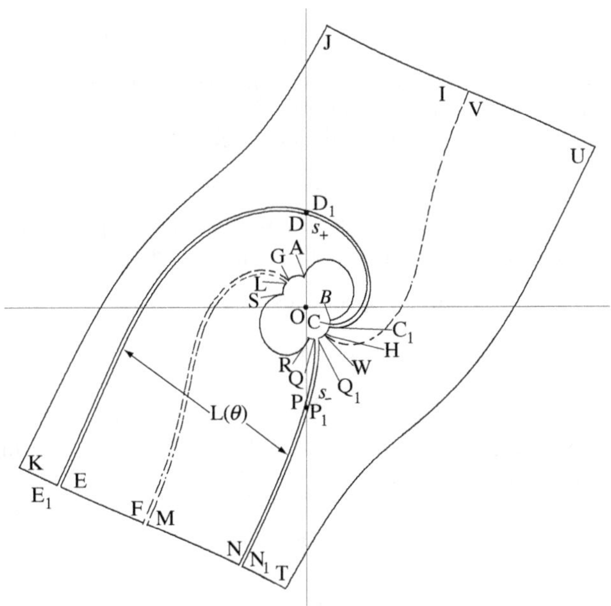


Figure 3. The $s$-plane.

one-to-one and analytic nature of the mapping $z \leftrightarrow s$, defined by (2.9), can again be established via the intermediate variable $Z$, by using the boundary correspondence as illustrated in figures 1–3, for $x = \cos\theta$, $|\theta| < \pi$ and $\arg\theta \in [-\pi, 0]$. The function $z(s)$ is therefore analytic in the domain $D_s$ given in (2.16).

Since the only possible singularities of $h_0(s,\theta)$ are at the branch points $z=0$, $-1-x$, $1-x$ and $s=0$, $h_0(s,\theta)$ is analytic in the domains corresponding to the upper and lower parts of the regions shown in figure 2. From (2.13) it is readily verified that the values of $h_0(s,\theta)$ on the curves $\overline{\mathrm{FG}}$ and $\overline{\mathrm{HI}}$ are identical to those along $\overline{\mathrm{ML}}$ and $\overline{\mathrm{WV}}$, respectively, thus justifying the analyticity of $h_0(s,\theta)$ in $D_s$. In the verification, attention has been paid to the choice of cuts: the cut for $z=-1-x$ is along $\overline{\mathrm{AB}}$; the cut for $z=0$ is along a curve from the origin to infinity, portions of which are labelled $\overline{\mathrm{JK}}$ and $\overline{\mathrm{FG}}$; the cut for $z=1-x$ is along $\overline{\mathrm{FG}}$ (see figure 1). The cut for $s=0$ in the $s$-plane is along $\overline{\mathrm{FG}}$ (see figure 3).

As in Wong & Zhao (2003), we let $C_{\mathrm{L}}$ and $C_{\mathrm{S}}$ denote, respectively, the simple closed curves

$$\overline{\mathrm{UV(I)JKE_1}},\quad \overline{\mathrm{(E)F(M)N(N_1)TU}}\quad\text{and}\quad \overline{\mathrm{GABC(C_1)H(W)Q_1(Q)RSL(G)}}$$

(see figure 3). The corresponding curves in the $Z$-plane can be described as $\operatorname{Im} Z=\pm\pi, -M\leqslant \operatorname{Re} Z\leqslant M$ and $\operatorname{Re} Z=\pm M, -\pi\leqslant\operatorname{Im} Z\leqslant\pi$.

To estimate $h_0(s,\theta)$, we first consider the unbounded $Z$-domain: $|\operatorname{Im} Z|<\pi$ and $\operatorname{Re} Z<-M$, where $M$ is large but fixed; this domain lies to the left of the curve $\overline{\mathrm{KE_1(E)F(M)N(N_1)T}}$ (cf. figure 2*a*, *b*). The corresponding $z$-domain is a small neighbourhood of $z=1-x$ (cf. figure 1). Hence, there exist constants $\lambda$ and $\Lambda$ such that

$$\lambda\leqslant|z+x+1|\leqslant\Lambda,\qquad \lambda|\theta|^2\leqslant|z|\leqslant\Lambda|\theta|^2. \tag{4.1}$$

The validity of the last two inequalities can be seen from the approximation $z\sim 1-x\sim\theta^2/2$. The reader is referred to Wong & Zhao (2003) for details. Taking the real part of both sides of the first equality in (2.9), we obtain from (4.1)

$$\lambda|\theta|^2\mathrm{e}^{\operatorname{Re} Z}\leqslant|z+x-1|\leqslant\Lambda|\theta|^2\mathrm{e}^{\operatorname{Re} Z}. \tag{4.2}$$

Here we have used $\lambda$ and $\Lambda$ as generic symbols for constants, whose values may change from place to place. Thus, from (2.5) it follows that

$$|g(z)|\leqslant C|\theta|^{2\alpha-2}\mathrm{e}^{(\alpha-\beta-1)\operatorname{Re} Z/2}, \tag{4.3}$$

where $C$ is another generic constant. From (2.9), we have $s=1/\theta(Z+\sqrt{Z^2+\theta})\sim 2Z/\theta$ and hence

$$|s^{\alpha+1}|\leqslant C|\theta|^{-\alpha-1}|Z|^{\alpha+1}. \tag{4.4}$$

Since $1+1/s^2$ is bounded and $|z^2+\sin^2\theta|\sim|(1-x)^2+\sin^2\theta|\sim\theta^2$ in this domain, we also have, from (2.9), (4.2) and (4.3),

$$\left|\frac{\mathrm{d}z}{\mathrm{d}s}\right|\leqslant C|\theta|^3\mathrm{e}^{\operatorname{Re} Z} \tag{4.5}$$

(see Wong & Zhao 2003, eqn (3.3)). A combination of (2.13), (2.14) and (4.2)–(4.4) gives

$$|h_0(s,\theta)|\leqslant C|Z|^{\alpha+1}\mathrm{e}^{(\alpha-\beta+1)\operatorname{Re} Z/2} \tag{4.6}$$

for $Z$ on the left side of the curve $\overline{\mathrm{KE_1(E)F(M)N(N_1)T}}$, i.e. for $|\operatorname{Im} Z|\leqslant\pi$ and $\operatorname{Re} Z<-M$.

Similarly, we have

$$|h_0(s,\theta)|\leqslant C|Z|^{-\alpha-1}\mathrm{e}^{-(\alpha-\beta-1)\operatorname{Re} Z/2} \tag{4.7}$$

for $Z$ on the left side of $\overline{\mathrm{BC(C_1)H(W)Q_1(Q)R}}$, except that in this case $s = 1/\theta(Z - \sqrt{Z^2+\theta^2}) \sim -\theta/2Z$. For $s$ bounded by $C_{\mathrm{L}}$ and $C_{\mathrm{S}}$, i.e. $|\mathrm{Im}\, Z| \leqslant \pi$ and $|\,\mathrm{Re}\, Z| \leqslant M$, the analyticity of $h_0(s,\theta)$ gives

$$|h_0(s,\theta)| \leqslant C. \tag{4.8}$$

The constant $C$ in (4.8) is independent of $s$ and $\theta$.

Summing up the above results we conclude that, for complex $\theta$, we have the same analyticity and boundedness properties as in the case when $\theta$ is real. Hence, for $0 < |\theta| \leqslant \pi - \varepsilon$ and $\arg z \in [-\pi, 0]$, the estimates in (2.26) continue to hold, and a procedure parallel to that used in Wong & Zhao (2003) shows that the integral representation (3.29) for the remainder term $R_p(N,\theta)$ in (3.1) satisfies

$$|R_p(N,\theta)| \leqslant \frac{\Lambda_p}{N^p}\{|\theta^{-\alpha} J_\alpha(N\theta)| + |\theta^{-\alpha} J_{\alpha+1}(N\theta)|\}, \tag{4.9}$$

where $\Lambda_p$ is a constant independent of $\theta$ and $N$. Therefore, (3.1) is genuinely a uniform asymptotic expansion in a complex $x$-domain containing the interval $(-1+\varepsilon, 1]$ on the real line.

It can be shown that the first two coefficients in (3.1) are given by $c_0(\theta) = -\sqrt{\theta/\sin\theta}$ and $d_0(\theta) = 0$ (see Wong 2001). Hence, with $p = 1$ in (3.1), we get from (4.9)

$$P_n^{(\alpha,\beta)}(\cos\theta) = (\sin\tfrac{1}{2}\theta)^{-\alpha}(\cos\tfrac{1}{2}\theta)^{-\beta}\sqrt{\frac{\theta}{\sin\theta}} J_\alpha(N\theta) + \delta_1(N,\theta) \tag{4.10}$$

with

$$|\delta_1(N,\theta)| \leqslant \frac{\Lambda_1'}{N}\{|J_\alpha(N\theta)| + |J_{\alpha+1}(N\theta)|\}, \tag{4.11}$$

$\Lambda_1' = 2^\alpha \Lambda_1$ being a constant. This result holds uniformly for $|\theta| < \pi$ and $\arg\theta \in [-\pi, 0]$, or equivalently for $x$ in a domain containing the interval $(-1, 1]$ on the real line. From well-known asymptotic formulae of the Bessel function (see Szegö (1975, p. 15) and equation (5.24) below), it is readily seen that the result (4.9), (4.10) includes both (1.2) and (1.3).

## 5. Exponential improvement

To demonstrate that equation (3.1) actually provides an exponentially improved asymptotic expansion, we need to show that the error term $R_p(N,\theta)$ is exponentially small when the two series in (3.1) are optimally truncated (see, for example, Berry & Howls 1990; Olver 1991; Wong & Zhao 1999). Since the basic ideas have already been given in Wong & Zhao (2003) and in the previous sections and, since the details of analysis are very complicated, we will limit ourselves to giving just a brief outline of the argument.

First, we extend the domain $D_s$ given in (2.16) to

$$\tilde{D}_s : \left|\mathrm{Im}\left\{\tfrac{1}{2}\theta\left(s - \frac{1}{s}\right)\right\}\right| < \mathrm{Im}\{(2\pi - \theta)\mathrm{i}\}, \tag{5.1}$$

where $\theta$ is a complex parameter satisfying $|\theta| < \pi$ and $\arg\theta \in (-\pi, 0]$, and show that the transformation $z \leftrightarrow s$ defined in (2.9) is one-to-one and analytic in this extended domain. The extension can again be illustrated via the intermediate variable $Z$. Basically, the $z$-domain is mapped twofold onto the corresponding $Z$-domains, which are in turn mapped onto the single $s$-domain $\tilde{D}_s$. In some circumstances, we may need to extend $D_s$ to a domain that is even larger than $\tilde{D}_s$ given in (5.1). In the $Z$-plane this domain may be described as having an upper boundary connecting the points $-\infty + \mathrm{i}\,\mathrm{Im}\{(2\pi - \theta)\mathrm{i}\}$, $(2\pi - \theta)\mathrm{i}$, $\mathrm{Re}\{(2\pi - \theta)\mathrm{i}\} + \mathrm{i}\,\mathrm{Im}\{(2\pi + \theta)\mathrm{i}\}$ and $+\infty + \mathrm{i}\,\mathrm{Im}\{(2\pi + \theta)\mathrm{i}\}$, and a lower boundary symmetrical with respect to the origin. The one-to-one analyticity of the mapping $z \leftrightarrow s$ can again be established in this $s$-domain.

Next, we show by induction that the rational functions introduced in (3.6)–(3.8) have integral representations:

$$\begin{aligned} A_k(s,\theta) &= \left(\frac{2}{\theta}\right)^k s^{-\alpha-1}\left[\frac{1}{1+s^{-2}}\frac{\partial}{\partial s}\right]^k\left(\frac{s}{s^2+1}s^{\alpha+1}\right) \\ &= \left(\frac{2}{\theta}\right)^k s^{-\alpha-1}\frac{k!}{2\pi\mathrm{i}}\int_{\{s\}}\frac{1+u^{-2}}{[(u-u^{-1})-(s-s^{-1})]^{k+1}}\left(\frac{u}{u^2+1}u^{\alpha+1}\right)\,du \end{aligned} \tag{5.2}$$

and

$$\begin{aligned} B_k(s,\theta) &= \left(\frac{2}{\theta}\right)^k s^{-\alpha-1}\left[\frac{1}{1+s^{-2}}\frac{\partial}{\partial s}\right]^k\left(-\frac{1}{s^2+1}s^{\alpha+1}\right) \\ &= \left(\frac{2}{\theta}\right)^k s^{-\alpha-1}\frac{k!}{2\pi\mathrm{i}}\int_{\{s\}}\frac{1+u^{-2}}{[(u-u^{-1})-(s-s^{-1})]^{k+1}}\left(\frac{-1}{u^2+1}u^{\alpha+1}\right)\,du, \end{aligned} \tag{5.3}$$

where $\{s\}$ denotes a closed curve in the $u$-plane encircling $u = s$ but not $u = \pm\mathrm{i}$ and 0. (This result was kindly communicated to us by A. B. Olde Daalhuis (2002, personal communication).)

To specify the integration path $\{s\}$, we again use the intermediate variable $Z$ in (2.9). Thus, $Z(u) = \frac{1}{2}\theta(u - 1/u)$ and

$$u = \frac{1}{\theta}(Z \pm \sqrt{Z^2+\theta^2}), \tag{5.4}$$

where

$$\sqrt{Z^2+\theta^2} = |Z^2+\theta^2|^{1/2}\mathrm{e}^{\mathrm{i}[\arg(Z-\mathrm{i}\theta)+\arg(Z+\mathrm{i}\theta)]}.$$

If we restrict $\arg(Z - \mathrm{i}\theta)$ and $\arg(Z + \mathrm{i}\theta)$ to their principal branch $[-\pi, \pi)$, then we have to take the '+' sign in (5.4) for the region $\mathrm{IJKE_1D_1C_1HI}$, and the '−' sign for the region GABCDEFG (cf. figure 2*a*, *b*). This can easily be verified by comparing the upper half of the imaginary axis in the $u$-plane with the ray $\arg Z(u) = \arg(\mathrm{i}\theta)$ in the $Z$-plane.

In terms of $Z$, equation (5.2) can be written as

$$A_k(s,\theta) = s^{-\alpha-1}\frac{k!}{2\pi\mathrm{i}}\int_{\{Z(u)\}}\frac{f_A(u)}{[Z(u)-Z(s)]^{k+1}}\,\mathrm{d}Z(u), \tag{5.5}$$

where

$$f_A(u) = \frac{u}{u^2+1}u^{\alpha+1}. \tag{5.6}$$

With $u = (Z + \sqrt{Z^2 + \theta^2})/\theta$, (5.6) becomes

$$f_A(u) = \tfrac{1}{2}\theta^{-\alpha}(Z^2 + \theta^2)^{-1/2}\{Z + \sqrt{Z^2 + \theta^2}\}^{\alpha+1} = \tfrac{1}{2}\theta(Z^2 + \theta^2)^{-1/2}u^{\alpha+1}. \quad (5.7)$$

There are two obvious branch points at $Z = \pm i\theta$, and no other singularities except at $\infty$.

Let $r(\theta) = \min\{|2\pi - 2\theta|, 2\pi\}$. If $\theta = \rho e^{i\phi}$, then we have explicitly

$$r(\theta) = \begin{cases} 2\sqrt{\pi^2 - 2\pi\rho\cos\phi + \rho^2}, & \phi \in \left(-\arccos\left(\dfrac{\rho}{2\pi}\right), 0\right], \\ 2\pi, & \phi \in \left[-\tfrac{1}{2}\pi, -\arccos\left(\dfrac{\rho}{2\pi}\right)\right]. \end{cases} \quad (5.8)$$

Deform the contours $C_{\mathrm{L}}$ and $C_{\mathrm{S}}$ in (3.9) and (3.10) so that their images in the $Z$-plane do not pass through the points $(\pm 2\pi \pm \theta)$i and, for $s$ in the deformed curves $C_{\mathrm{L}}$ and $C_{\mathrm{S}}$, the distance between $Z(s)$ and all branch points and saddle points are greater than or equal to $r(\theta) - \sigma$, where $\sigma$ is a small positive number. For $s \in C_{\mathrm{S}} \cup C_{\mathrm{L}}$, we specify the contour $\{Z(u)\}$ in (5.5) to be the circle

$$|Z(u) - Z(s)| = r_1,$$

where we write $r_1 = r(\theta) - 2\sigma$ for short. Estimating the integral in (5.5), we obtain

$$|A_k(s,\theta)| \leqslant C|\theta|\frac{k!}{r_1^k} \quad (5.9)$$

for $s \in C_{\mathrm{S}} \cup C_{\mathrm{L}}$ with $C_{\mathrm{S}}$ and $C_{\mathrm{L}}$ being the deformed curves described above. A combination of (3.9), (3.11), (3.16) and (5.9) gives

$$|c_k(\theta)| \leqslant C\frac{k!}{(r(\theta) - 2\sigma)^k}, \quad (5.10)$$

where $C$ is a generic constant. In a similar manner, we obtain

$$\left|\frac{d_k(\theta)}{\theta}\right| \leqslant C\frac{k!}{(r(\theta) - 2\sigma)^k}. \quad (5.11)$$

Finally, we estimate the remainder $R_p(N,\theta)$ given in (3.2). In Wong & Zhao (2003) we have expressed the function $h_p(s,\theta)$ in (2.24) in terms of the rational functions $Q_k(u,s,\theta)$ defined inductively by

$$Q_0(u,s,\theta) = \frac{1}{u - s} \quad (5.12)$$

and

$$Q_{k+1}(u,s,\theta) = \frac{2}{\theta}\left(1 + \frac{1}{u^2}\right)^{-1}\left\{\frac{\alpha+1}{u} + \frac{\mathrm{d}}{\mathrm{d}u}\right\}Q_k(u,s,\theta) \quad (5.13)$$

for $k = 0, 1, 2, \ldots$. Analogous to (5.2) and (5.3), A. B. Olde Daalhuis (2002, personal communication) has suggested the representation

$$\begin{aligned} Q_k(u,s,\theta) &= \left(\frac{2}{\theta}\right)^k u^{-\alpha-1}\left\{\frac{1}{1+u^{-2}}\frac{\partial}{\partial u}\right]^k\left(\frac{u^{\alpha+1}}{u - s}\right) \\ &= \left(\frac{2}{\theta}\right)^k u^{-\alpha-1}\frac{k!}{2\pi i}\int_{\{u\}}\frac{(1+w^{-2})w^{\alpha+1}}{[(w - w^{-1}) - (u - u^{-1})]^{k+1}(w - s)}\,\mathrm{d}w, \end{aligned} \quad (5.14)$$

where, as before, $\{u\}$ denotes a simple closed curve encircling $w = u$ but not any other singularities of the integrand. In terms of the intermediate variable $Z$, (5.14) can be written as

$$Q_k(u, s, \theta) = u^{-\alpha-1}\frac{k!}{2\pi \mathrm{i}}\int_{\{Z(u)\}}\frac{f_Q(w)}{[Z(w)-Z(u)]^{k+1}}\,\mathrm{d}Z(w), \tag{5.15}$$

where $f_Q(w) = w^{\alpha+1}/(w-s)$ and $\{Z(u)\}$ is a simple closed curve encircling $Z(u)$ but not any other singularities.

Again, as in (5.9), we choose the contour $\{Z(u)\}$ in (5.15) to be the circle $|Z(w) - Z(u)| = r(\theta) - 2\sigma$. An easy estimation gives

$$|Q_k(u, s, \theta)| \leqslant C\frac{k!}{r_1^k}|\theta| \tag{5.16}$$

for $|Z(u) - Z(s)| = r(\theta)$ and $s \in L_{\mathrm{L}}(\theta)$, where $L_{\mathrm{L}}(\theta)$ denotes the part of the contour $L(\theta)$ whose image in the $Z$-plane lies on the left side of $\overline{\mathrm{KE_1(E)F(M)N(N_1)T}}$ (see figure 2*a*). Similarly, we have

$$|Q_k(u, s, \theta)| \leqslant C\frac{k!}{r_1^k}|\theta|^{-1} \tag{5.17}$$

for $|Z(u) - Z(s)| = r(\theta)$ and $s \in L_{\mathrm{S}}(\theta)$, where $L_{\mathrm{S}}(\theta)$ denotes the part of $L(\theta)$ whose image in the $Z$-plane lies to the left of $\overline{\mathrm{BC(C_1)}H(W)Q(\mathrm{Q_1})R}$.

Equation (5.5) in Wong & Zhao (2003) gives an explicit expression for $h_k(s, \theta)$ in terms of integrals involving $h_0(u, \theta)$ and $Q_1(u, s, \theta), \dots Q_k(u, s, \theta)$ on a simple closed contour $C_E$ in the $u$-plane, which encloses $u = s$, but not $u = 0$ and the saddle points $u = s_{\pm}$ (see fig. 4 in Wong & Zhao (2003)). Using the improved bounds for $c_l(\theta)$ and $d_l(\theta)/\theta$ in (5.10), (5.11) and the estimates in (5.16), (5.17), one can show that

$$|h_k(s, \theta)| \leqslant C|Z|^{c_1}\mathrm{e}^{c_2 \operatorname{Re} Z}\frac{k!}{(r(\theta) - 2\sigma)^k} \tag{5.18}$$

for $s \in L_{\mathrm{L}}(\theta) \cup L_{\mathrm{S}}(\theta)$, where $C$, $c_1$ and $c_2$ are appropriately chosen constants depending only on $\alpha$ and $\beta$.

To estimate the integral in (3.29) we first note that the parts of the contour $L(\theta)$, denoted by $L_{\mathrm{L}}(\theta)$ and $L_{\mathrm{S}}(\theta)$, either tend to infinity or approach the origin. For $s \in L_{\mathrm{L}}(\theta)$ we have from (5.18)

$$\left|\int_{L_{\mathrm{L}}(\theta)} h_p(s, \theta)\mathrm{e}^{(N\theta/2)(s-(1/s))}\frac{\mathrm{d}s}{s^{\alpha+1}}\right| \leqslant C\frac{p!\,|\theta|^{\alpha}}{(r(\theta) - 2\sigma)^p}\mathrm{e}^{-MN/2}, \tag{5.19}$$

where $C$ is a constant independent of $p$. The contour $L(\theta)\backslash L_{\mathrm{L}}(\theta)$ consists of two steepest descent paths; one is the curve $\overline{\mathrm{E_1D_1C_1}}$, and the other is $\overline{\mathrm{Q_1P_1N_1}}$, with $\mathrm{C_1}$ and $\mathrm{Q_1}$ both approaching the origin in the $s$-plane (see figure 3). On account of (5.18), one can show that the integral (5.19) on $L_{\mathrm{S}}(\theta)$ can be replaced by the same integral over $\overline{\mathrm{C_1H(W)Q_1}}$. Hence the integration path $L(\theta)\backslash L_{\mathrm{L}}(\theta)$ can be deformed into the curve $\overline{\mathrm{N_1P_1Q_1H(W)C_1D_1E_1}}$. We further deform the curve so that its image in the $Z$-plane has at least a distance of $r(\theta) - \sigma$ away from the points $(\pm 2\pi \pm \theta)\mathrm{i}$, and a distance of $\sigma|\theta|/\pi$ from $\pm\theta\mathrm{i}$. Let this final curve be denoted by $L_{\mathrm{D}}(\theta)$. For $s \in L_{\mathrm{D}}(\theta)$, we then show that

$$|h_p(s, \theta)| \leqslant C\frac{p!}{(r(\theta) - 2\sigma)^p}. \tag{5.20}$$

Consider the case when $N\theta$ is bounded, say $|N\theta| \leqslant C$. For $s \in L_{\mathrm{D}}(\theta)$, $|s| \geqslant C_0 > 0$, and hence $(N\theta/2)1/s$ is bounded by $C/2C_0$. Making the change of variable $t = N\theta s/2$, and using the fact that $\mathrm{Re}\, t$ is bounded from above, we get

$$\left| \int_{L(\theta)\setminus L_{\mathrm{L}}(\theta)} h_p(s,\theta) \mathrm{e}^{(N\theta/2)(s-(1/s))} \frac{\mathrm{d}s}{s^{\alpha+1}} \right| \leqslant C \frac{p!}{(r(\theta)-2\sigma)^p} |N\theta|^{\alpha}. \tag{5.21}$$

When $N\theta$ is unbounded, i.e. $|N\theta| \to \infty$, one can use the ordinary method of steepest descent to conclude that

$$\left| \int_{L(\theta)\setminus L_{\mathrm{L}}(\theta)} h_p(s,\theta) \mathrm{e}^{(N\theta/2)(s-(1/s))} \frac{\mathrm{d}s}{s^{\alpha+1}} \right| \leqslant C \frac{p!}{(r(\theta)-2\sigma)^p} \frac{1}{\sqrt{|N\theta|}}; \tag{5.22}$$

the factor $p!/(r(\theta)-2\sigma)^p$ comes from the bound for $h_p(\pm\mathrm{i},\theta)$, which is derived in a manner similar to that leading to (5.10) and (5.11). The constant $C$ is independent of $\theta$, since this parameter is absorbed in the large asymptotic variable $N\theta$.

Clearly, the results in (5.21) and (5.22) agree with that in (5.19). In view of the asymptotic behaviour of the Bessel function $J_\alpha(\cdot)$ near infinity and near the origin, we can combine (5.19), (5.21) and (5.22) to give

$$|R_p(N,\theta)| \leqslant C \frac{p!}{[N(r(\theta)-2\sigma)]^p} \{|\theta^{-\alpha} J_\alpha(N\theta)| + |\theta^{-\alpha} J_{\alpha+1}(N\theta)|\}. \tag{5.23}$$

As mentioned previously, the letter $C$ in the above equations is used as a generic symbol for constants whose value may change from time to time.

Recall that

$$J_\nu(\tau) \sim (2\pi\tau)^{-1/2} \left\{ \mathrm{e}^{\mathrm{i}\chi} \sum_{k=0}^{\infty} \frac{(-\mathrm{i})^k a_k}{\tau^k} + \mathrm{e}^{-\mathrm{i}\chi} \sum_{k=0}^{\infty} \frac{\mathrm{i}^k a_k}{\tau^k} \right\} \tag{5.24}$$

as $|\tau| \to \infty$ in $|\arg \tau| < \pi$, where $\chi = \tau - (\frac{1}{2}\nu + \frac{1}{4})\pi, a_0 = 1$ and

$$a_k = \frac{(1^2 - 4\nu^2)(3^2 - 4\nu^2) \ldots\ldots ((2k-1)^2 - 4\nu^2)}{8^k k!}, \tag{5.25}$$

$k = 1, 2, \ldots$ (see, for example, Abramowitz & Stegun 1964, p. 364). For $\theta = \rho \mathrm{e}^{\mathrm{i}\phi}$ and $\phi \in (-\pi, 0), \mathrm{e}^{\mathrm{i}\chi}$ is of the magnitude $O(\mathrm{e}^{N\rho(-\sin\phi)})$, while $\mathrm{e}^{-\mathrm{i}\chi}$ is $O(\mathrm{e}^{N\rho\sin\phi})$. Thus, the first term inside the curly brackets in (5.24) dominates the second. Furthermore, for any finite $p$, the estimate in (5.23) for the remainder $R_p(N,\theta)$ is of higher asymptotic order than the exponentially small subdominant term in (5.24). Since the subdominant term cannot be observed, if one replaces the Bessel functions in (3.1) by their asymptotic expansions obtained from (5.24), then one still sees a discontinuous change from the expansion resulting from (3.1) to the approximation in (1.2), as $\theta$ approaches the interval $(-\pi, 0)$ on the real line. To see the subdominant term, we need to truncate the two series in (3.1) at an optimal place

$$p \sim N(r(\theta) - 2\sigma) \tag{5.26}$$

(cf. Berry & Howls 1990; Olver 1991; Wong & Zhao 1999). With $p$ so chosen, (5.23) yields

$$|R_p(N,\theta)| \leqslant C\sqrt{N} \mathrm{e}^{-N(r(\theta)-2\sigma)} \{|\theta^{-\alpha} J_\alpha(N\theta)| + |\theta^{-\alpha} J_{\alpha+1}(N\theta)|\}. \tag{5.27}$$

Compared with (4.9), it is evident that (5.27) is an exponential improvement.

The research of R.W. was partly supported by grants from the Research Grant Council of Hong Kong. The research of Y.-Q.Z. was partly supported by the National Natural Science Foundation of China, GuangDong Natural Science Foundation and a fund from the Advanced Research Centre of ZhongShan University.

## References

Abramowitz, M. & Stegun, I. A. (eds) 1964 *Handbook of mathematical functions.* National Bureau of Standards, Applied Mathematics Series, vol. 55. Washington, DC: National Bureau of Standards.

Bender, C. M. & Orszag, S. A. 1978 *Advanced mathematical methods for scientists and engineers.* McGraw-Hill.

Berry, M. V. & Howls, C. J. 1990 Hyperasymptotics. *Proc. R. Soc. Lond.* A **430**, 653–667.

Berry, M. V. & Howls, C. J. 1993 Unfolding the high orders of asymptotic expansions with coalescing saddles: singularity theory, crossover and duality. *Proc. R. Soc. Lond.* A **433**, 107–126.

Dunster, T. M. 1999 Uniform asymptotic approximations for the Jacobi and ultraspherical polynomials, and related functions. *Meth. Applic. Analysis* **6**, 281–316.

Heine, E. 1878 *Handbuch der Kugelfunctionen*, 2nd edn, vol. 1. Berlin: G. Reimer.

Olde Daalhuis, A. B. & Temme, N. M. 1994 Uniform Airy-type expansions of integrals. *SIAM J. Math. Analysis* **25**, 304–321.

Olver, F. W. J. 1974 *Asymptotics and special functions.* Academic. (Reprinted 1997 Natick, MA: A. K. Peters.)

Olver, F. W. J. 1991 Uniform, exponentially improved, asymptotic expansions for the generalized exponential integral. *SIAM J. Math. Analysis* **22**, 1460–1474.

Szegö, G. 1975 *Orthogonal polynomials*, 4th edn. Colloquium Publications, vol. 23. Providence, RI: American Mathematical Society.

Wong, R. 2001 Exponential asymptotics. In *Special functions 2000: current perspective and future directions* (ed. J. Bustoz, M. E. H. Ismail & S. K. S. Suslov). NATO Science Series II, vol. 30, pp. 505–518. Deventer: Kluwer.

Wong, R. & Zhang, J.-M. 1996 A uniform asymptotic expansion for the Jacobi polynomials with explicit remainder. *Applic. Analysis* **61**, 17–29.

Wong, R. & Zhao, Y.-Q. 1999 Smoothing of Stokes discontinuity for the generalized Bessel function. II. *Proc. R. Soc. Lond.* A **455**, 3065–3084.

Wong, R. & Zhao, Y.-Q. 2003 Estimates for the error term in a uniform asymptotic expansion of the Jacobi polynomials. *Analysis Appl.* **1**, 213–241.

# Uniform Asymptotics for Orthogonal Polynomials with Exponential Weights—the Riemann–Hilbert Approach

*By Z. Wang and R. Wong*

Let $Q(x) = q_{2m}x^{2m} + q_{2m-1}x^{2m-1} + \cdots$ be a polynomial of degree $2m$ with $q_{2m} > 0$, and let $\{\pi_n(x)\}_{n\geq 1}$ be the sequence of monic polynomials orthogonal with respect to the weight $w(x) = e^{-Q(x)}$ on $\mathbb{R}$. Furthermore, let $\alpha_n$ and $\beta_n$ denote the Mhaskar–Rakhmanov–Saff (MRS) numbers associated with $Q(x)$. By using the Riemann–Hilbert approach, an asymptotic expansion is constructed for $\pi_n(c_n z + d_n)$, which holds uniformly for all $z$ bounded away from $(-\infty, -1)$, where $c_n = \frac{1}{2}(\beta_n - \alpha_n)$ and $d_n = \frac{1}{2}(\beta_n + \alpha_n)$.

## 1. Introduction

Because the publication of the two papers by Nevai [8, 9] in 1980 on the asymptotic behavior of the orthogonal polynomials associated with the weight function $\exp(-x^4)$, there has been a considerable amount of interest in the asymptotics of the polynomials orthogonal with respect to the more general exponential weight $w(x) = e^{-Q(x)}$, where $Q(x)$ is a polynomial of even degree and with positive leading coefficient; see, e.g., [2, 4, 6, 11, 13]. Here, we are concerned with only [4], which deals with the most general case of $Q(x)$. In the other references, only special cases are considered, although the results there are stronger and more complete.

Address for correspondence: R. Wong, Department of Mathematics, City University of Hong Kong, Tat Chee Avenue, Kowloon, Hong Kong; e-mail: mawong@cityu.edu.hk

*STUDIES IN APPLIED MATHEMATICS* 115:139–155 

Published by Blackwell Publishing, 350 Main Street, Malden, MA 02148, USA, and 9600 Garsington Road, Oxford, OX4 2DQ, UK.

To describe the main result in [4], we first write

$$Q(x) = \sum_{k=0}^{2m} q_k x^k, \quad q_{2m} > 0, \quad m > 0. \tag{1}$$

Denote by

$$\pi_n(x) = \pi_n(x, Q) = x^n + \cdots \tag{2}$$

the monic polynomials orthogonal with respect to the weight

$$w(x) = e^{-Q(x)} \tag{3}$$

on the real line $\mathbb{R}$, and let $p_n(x) = \gamma_n \pi_n(x)$ be the normalized $n$th orthogonal polynomials, i.e.,

$$\int_{-\infty}^{\infty} p_n(x) p_m(x) e^{-Q(x)}\, dx = \delta_{n,m}, \quad n, m \in \mathbb{N}. \tag{4}$$

The statement of the result given by Deift et al. involves the Mhaskar–Rakhmanov–Saff (MRS) numbers $\alpha_n$ and $\beta_n$; see [7, 12]. These numbers can be determined from the equations

$$\frac{1}{2\pi} \int_{\alpha_n}^{\beta_n} \frac{Q'(t)(t - \alpha_n)}{\sqrt{(\beta_n - t)(t - \alpha_n)}}\, dt = n, \tag{5}$$

$$\frac{1}{2\pi} \int_{\alpha_n}^{\beta_n} \frac{Q'(t)(\beta_n - t)}{\sqrt{(\beta_n - t)(t - \alpha_n)}}\, dt = -n. \tag{6}$$

The existence of the MRS numbers for sufficiently large $n$ has been established in [4], including their convergent series representations in powers of $n^{-1/2m}$. Define

$$c_n := \frac{\beta_n - \alpha_n}{2}, \quad d_n := \frac{\beta_n + \alpha_n}{2}. \tag{7}$$

These numbers provide the width and the midpoint of the interval $[\alpha_n, \beta_n]$, and can be represented by the power series

$$c_n = n^{1/2m} \sum_{l=0}^{\infty} c^{(l)} n^{-l/2m}, \quad c^{(0)} = (q_{2m} m A_m)^{-1/2m}, \quad c^{(1)} = 0, \tag{8}$$

$$d_n = \sum_{l=0}^{\infty} d^{(l)} n^{-l/2m}, \quad d^{(0)} = -\frac{q_{2m-1}}{2m q_{2m}}, \tag{9}$$

where

$$A_m := \prod_{j=1}^{m} \frac{2j - 1}{2j}, \quad m > 0. \tag{10}$$

The coefficients $c^{(l)}$ and $d^{(l)}$, $l = 1, 2, \ldots$, can be computed explicitly.

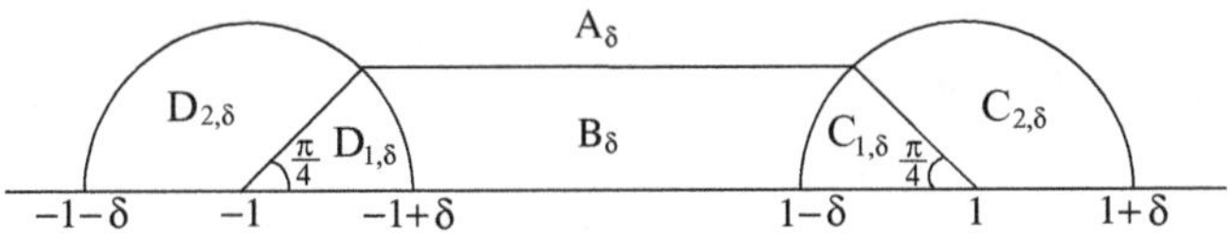


Figure 1. Different asymptotic regions for $p_n(c_nz + d_n)$ in $\bar{C}_+$.

Because $p_n(z) = \overline{p_n(\bar{z})}$, one may restrict $z$ to the closed upper half-plane $\bar{C}_+$. In [4], six asymptotic approximations/expansions are given for $z$ in six different regions $A_\delta$, $B_\delta$, $C_{1,\delta}$, $C_{2,\delta}$, $D_{1,\delta}$, and $D_{2,\delta}$, where $\delta$ is a small positive parameter; see Figure 1. The purpose of this paper is to show that the asymptotic expansions in the two regions $C_{1,\delta}$ and $C_{2,\delta}$ are identical, and that the common expansion holds for all $z$ (strictly) bounded away from the cut along the negative real axis from $-\infty$ to $-1$. Using a similar argument, a corresponding expansion holds for all $z$ bounded away from the infinite half-line $[1, \infty)$.

## 2. Statement of the result

As in [4], we first normalize the interval $[\alpha_n, \beta_n]$ to be $[-1, 1]$. Then, we make the linear change of variable

$$z \mapsto \lambda_n(z) = c_n z + d_n, \tag{11}$$

which takes the interval $[-1, 1]$ onto $[\alpha_n, \beta_n]$. Define

$$V_n(z) := \frac{1}{n} Q(\lambda_n(z)). \tag{12}$$

Clearly, $V_n$ is again a polynomial of degree $2m$ whose leading coefficient is asymptotic to $(mA_m)^{-1} > 0$ and all other coefficients tend to 0 as $n \to \infty$. Furthermore, we define the function

$$h_n(z) = \frac{1}{2\pi i} \int_{\Gamma_z} \frac{V_n'(\zeta)}{\sqrt{\zeta^2 - 1}} \frac{d\zeta}{\zeta - z}, \tag{13}$$

where $\Gamma_z$ is any simple, closed, and positively oriented contour containing $[-1, 1]$ and $\{z\}$. In [4, p. 1513], it has been shown that

$$h_n(z) = \sum_{k=0}^{2m-2} h_{n,k} z^k, \quad h_{n,k} = \sum_{l=0}^{\infty} h_k^{(l)} n^{-l/2m}, \tag{14}$$

$$h_{2k}^{(0)} = 2\frac{A_{m-k-1}}{A_m}, \quad h_{2k+1}^{(0)} = 0, \quad O \le k \le m-1, \tag{15}$$

where $A_m$ is given in (10).

To state our main theorem, we need a few more notations. From (14), it is readily seen that the constant

$$l_n := \frac{1}{\pi}\int_{-1}^{1}\sqrt{1-t^2}h_n(t)\log|t|\,dt - V_n(0) \tag{16}$$

can also be expanded in power series of $n^{-1/2m}$

$$l_n = \sum_{k=0}^{\infty} l^{(k)} n^{-k/2m}, \quad l^{(0)} = -\frac{1}{m} - 2\log 2; \tag{17}$$

see [4, p. 1497]. Put

$$\psi_n(z) := \frac{1}{2\pi}(1-z)^{1/2}(1+z)^{1/2}h_n(z), \quad z \notin (-\infty,-1]\cup[1,\infty), \tag{18}$$

and

$$\xi_n(z) := -2\pi i\int_1^z \psi_n(\zeta)\,d\zeta, \quad z \in \mathbb{C}\backslash((-\infty,-1)\cup(1,\infty)). \tag{19}$$

Define $\varphi_n : \mathbb{C}\backslash\mathbb{R} \to \mathbb{C}$ by

$$\varphi_n(z) := \begin{cases} -\frac{1}{2}\xi_n(z) & \text{for } \operatorname{Im} z > 0, \\ \frac{1}{2}\xi_n(z) & \text{for } \operatorname{Im} z < 0. \end{cases} \tag{20}$$

From (19), it is easily verified that

$$(\varphi_n)_+(x) = (\varphi_n)_-(x) \quad \text{for } x > 1 \tag{21}$$

and

$$(\varphi_n)_+(x) = (\varphi_n)_-(x) - 2\pi i \quad \text{for } x < -1; \tag{22}$$

see [4, p. 1521, line 6 from the bottom]. Hence, the function $e^{n\varphi_n(z)}$ can be analytically continued to $\mathbb{C}\backslash[-1,1]$. Let

$$\zeta_n(z) := \left[\frac{3}{2}\varphi_n(z)\right]^{2/3}. \tag{23}$$

Because $(\zeta_n)_+(x) = (\zeta_n)_-(x)$ for $x \in (-1,1)$, $\zeta_n(z)$ can be analytically continued to $\mathbb{C}\backslash(-\infty,-1]$.

THEOREM 1. *With $c_n$, $d_n$, $l_n$, $V_n(z)$, and $\zeta_n(z)$ defined as above, we have*

$$\pi_n(c_n z + d_n) = \sqrt{\pi}\,c_n^n e^{nl_n/2+nV_n(z)/2} \times\left\{n^{1/6}\operatorname{Ai}(n^{2/3}\zeta_n)\mathbf{A}(z,n) - n^{-1/6}\operatorname{Ai}'(n^{2/3}\zeta_n)\mathbf{B}(z,n)\right\}, \tag{24}$$

*where* $\mathbf{A}(z, n)$ *and* $\mathbf{B}(z, n)$ *are analytic functions of* $z$ *in* $\mathbb{C}\backslash(-\infty, -1]$. *Furthermore, for* $z$ *bounded away from* $(-\infty, -1]$, *the asymptotic expansions*

$$\mathbf{A}(z, n) \sim \frac{\zeta_n^{1/4}(z)}{a(z)} \left[ 1 + \sum_{k=2m}^{\infty} \frac{A_k(z)}{n^{k/2m}} \right], \tag{25}$$

$$\mathbf{B}(z, n) \sim \frac{a(z)}{\zeta_n^{1/4}(z)} \left[ 1 + \sum_{k=2m}^{\infty} \frac{B_k(z)}{n^{k/2m}} \right] \tag{26}$$

*hold uniformly, where*

$$a(z) = \frac{(z-1)^{1/4}}{(z+1)^{1/4}} \tag{27}$$

*and the coefficient functions* $A_k(z)$ *and* $B_k(z)$ *are analytic functions in* $\mathbb{C}\backslash(-\infty, -1]$.

*Remark* 1. The function $f(z) = f_n(z)$ in [4, eq. (7.18)] is related to our $\zeta_n(z)$ in (23) via

$$f_n(z) = n^{2/3}\zeta_n(z).$$

Furthermore, from equation (7.18) in [4], it can be shown that $\zeta_n(z)$ behaves like a constant multiple of $(z-1)$ near $z = 1$. Hence, the function $\zeta_n^{1/4}(z)/a(z)$ in (25) is analytic and nonvanishing in a neighborhood of $z = 1$.

*Remark* 2. Note that

$$\pi_n(z) = \frac{1}{\gamma_n} p_n(z) \sim \sqrt{\pi}\, c_n^{n+\frac{1}{2}} e^{nl_n/2} p_n(z);$$

see [4, p. 1497]. Thus, our result agrees with that given by Deift et al. [4].

## 3. Riemann–Hilbert Problems

In [5], Fokas et al. gave an important theorem that connects orthogonal polynomials to a Riemann–Hilbert problem (RHP). To state their result, we begin with the following RHP for $Y$. Let $Y(z)$ be a $2 \times 2$ matrix function satisfying

$(Y_a)$ $Y(z)$ is analytic in $\mathbb{C}\backslash\mathbb{R}$,

$(Y_b)$ for $x \in \mathbb{R}$,

$$Y_+(x) = Y_-(x) \begin{pmatrix} 1 & w(x) \\ 0 & 1 \end{pmatrix}, \tag{28}$$

$(Y_c)$ as $|z| \to \infty$,

$$Y(z)\begin{pmatrix} z^{-n} & 0 \\ 0 & z^n \end{pmatrix} = I + O\left(\frac{1}{|z|}\right), \tag{29}$$

where $w(x)$ is a positive weight function on $\mathbb{R}$. Fokas et al. [5] showed that a unique solution to the above problem is given by

$$Y(z) = \begin{pmatrix} \pi_n(z) & C[\pi_n w](z) \\ -2\pi i \gamma_{n-1}^2 \pi_{n-1}(z) & -2\pi i \gamma_{n-1}^2 C[\pi_{n-1} w](z) \end{pmatrix} \tag{30}$$

for $z \in \mathbb{C}\backslash\mathbb{R}$, where $\pi_n(x)$ is the monic polynomials orthogonal with respect to $w(x)$, $\gamma_n$ is the leading coefficient of the normalized orthogonal polynomial $p_n(x)$, i.e., $p_n(x) = \gamma_n \pi_n(x)$, and $C[f](z)$ denotes the Cauchy transform of $f$ defined by

$$C[f](z) := \frac{1}{2\pi i} \int_{\mathbb{R}} \frac{f(\zeta)}{\zeta - z} d\zeta, \quad z \in \mathbb{C}\backslash\mathbb{R}. \tag{31}$$

In the present case, $w(x)$ is the weight function given in (3).

Our next step is to apply the innovative method of steepest descent introduced by Deift and Zhou [3]. This method consists of a sequence of transformations

$$Y \to U \to T \to S, \tag{32}$$

which ultimately leads to a RHP with a constant jump matrix. The transformation $Y \to U$ is a rescaling, and the transformation $U \to T$ is a normalization. The transformation $T \to S$ involves contour deformations so that the jump matrix of the resulting RHP has exponentially small off-diagonal entries.

Recall the sequences $\{c_n\}$ and $\{d_n\}$ defined in (7), and the linear change of variable $z \mapsto \lambda_n(z)$ given in (11). Define the matrix

$$U(z) := \begin{pmatrix} c_n^{-n} & 0 \\ 0 & c_n^n \end{pmatrix} Y(\lambda_n(z)). \tag{33}$$

It is easily verified that $U$ satisfies the RHP:

$(U_a)$ $U(z)$ is analytic in $\mathbb{C}\backslash\mathbb{R}$,

$(U_b)$ for $x \in \mathbb{R}$,

$$U_+(x) = U_-(x)\begin{pmatrix} 1 & e^{-nV_n(x)} \\ 0 & 1 \end{pmatrix}, \tag{34}$$

$(U_c)$ as $|z| \to \infty$,

$$U(z)\begin{pmatrix} z^{-n} & 0 \\ 0 & z^n \end{pmatrix} = I + O\left(\frac{1}{|z|}\right), \tag{35}$$

where $nV_n(z) = Q(\lambda_n(z))$; see (12).

To define the transformation $U \to T$, we first recall the Pauli matrix $\sigma_3$ defined by

$$\sigma_3 = \begin{pmatrix} 1 & 0 \\ 0 & -1 \end{pmatrix}. \tag{36}$$

Now, set

$$\begin{aligned} T(z) &:= e^{-\frac{1}{2}nl_n\sigma_3} U(z) e^{-n(g(z)-\frac{1}{2}l_n)\sigma_3} \\ &= \begin{pmatrix} e^{-\frac{1}{2}nl_n} & 0 \\ 0 & e^{\frac{1}{2}nl_n} \end{pmatrix} U(z) \begin{pmatrix} e^{-n(g(z)-\frac{1}{2}l_n)} & 0 \\ 0 & e^{n(g(z)-\frac{1}{2}l_n)} \end{pmatrix}, \end{aligned} \tag{37}$$

where $l_n$ is the constant given in (16) and $g(z)$ is to be determined. First, we require $g(z)$ to be analytic so that $T(z)$ solves a RHP on the same contour as $U(z)$. Next, to normalize the RHP for $T$ at infinity, we also require that

$$g(z) \sim \log z \quad \text{as } |z| \to \infty. \tag{38}$$

Simple calculation shows that $T$ is the unique solution of the RHP:

$(T_a)$ $T$ is analytic in $\mathbb{C}\backslash\mathbb{R}$,

$(T_b)$ for $x \in \mathbb{R}$,

$$T_+(x) = T_-(x) \begin{pmatrix} e^{-n(g_+-g_-)} & e^{n(g_++g_-V_n-l_n)} \\ 0 & e^{n(g_+-g_-)} \end{pmatrix}, \tag{39}$$

$(T_c)$ as $|z| \to \infty$,

$$T(z) = I + O\left(\frac{1}{|z|}\right). \tag{40}$$

The so-called $g$-function is further required to satisfy the following conditions:

$$g_+(x) + g_-(x) - V_n(x) - l_n = 0, \quad x \in [-1, 1], \tag{41}$$

$$g_+(x) - g_-(x) \text{ is purely imaginary}, \quad x \in [-1, 1], \tag{42}$$

$$g_+(x) + g_-(x) - V_n(x) - l_n < 0, \quad x \in \mathbb{R}\backslash[-1, 1], \tag{43}$$

$$e^{g_+(x)-g_-(x)} = 1, \quad x \in \mathbb{R}\backslash[-1, 1]. \tag{44}$$

Deift et al. [4] called these conditions the *phase conditions* for $g$, and showed that the required $g$-function is given by

$$g_n(z) := \int_{-1}^{1} \log(z-x)\psi_n(x)\,dx, \quad z \in \mathbb{C}\backslash(-\infty, 1], \tag{45}$$

where $\log z = \log|z| + i \arg z$ with $|\arg z| < \pi$,

$$\psi_n(z) := \frac{1}{2\pi}(1-z)^{1/2}(1+z)^{1/2}h_n(z), \quad z \notin (-\infty, -1] \cup [1, \infty), \tag{46}$$

and $h_n(z)$ is the polynomial given in (13) and (14); see [4, eqs. (4.16), (5.23), and (5.33)]. From (45), it can be shown that

$$(g_n)_+(x) - (g_n)_-(x) = \xi_n(x), \quad x \in (-1, 1), \tag{47}$$

where $\xi_n(z)$ is defined in (19), and that

$$(g_n)_+(x) - (g_n)_-(x) = 2\pi i, \quad x \in (-\infty, -1). \tag{48}$$

see [4, eq. (5.37)]. Note that the $g$-function in (37) actually depends on $n$; cf. (45). However, for convenience, this dependence is sometimes suppressed in our later discussion.

We are now ready to define the transformation $T \to S$. Let $\Sigma_S$ denote the contour $\Sigma_S = \cup_{i=1}^{5} \Sigma_i$ shown in Figure 2, and set

$$S(z) := T(z) \quad \text{for } z \text{ outside the lens-shaped region,} \tag{49}$$

$$S(z) := T(z) \begin{pmatrix} 1 & 0 \\ -e^{-n\xi_n(z)} & 1 \end{pmatrix} \quad \text{in the upper lens region,} \tag{50}$$

$$S(z) := T(z) \begin{pmatrix} 1 & 0 \\ e^{n\xi_n(z)} & 1 \end{pmatrix} \quad \text{in the lower lens region.} \tag{51}$$

Furthermore, we define the jump matrix

$$J_S(z) = \begin{pmatrix} 1 & 0 \\ e^{-n\xi_n(z)} & 1 \end{pmatrix}, \quad z \in \Sigma_1, \tag{52}$$

$$J_S(x) = \begin{pmatrix} 0 & 1 \\ -1 & 0 \end{pmatrix}, \quad x \in \Sigma_2, \tag{53}$$

$$J_S(z) = \begin{pmatrix} 1 & 0 \\ e^{n\xi_n(z)} & 1 \end{pmatrix}, \quad z \in \Sigma_3, \tag{54}$$

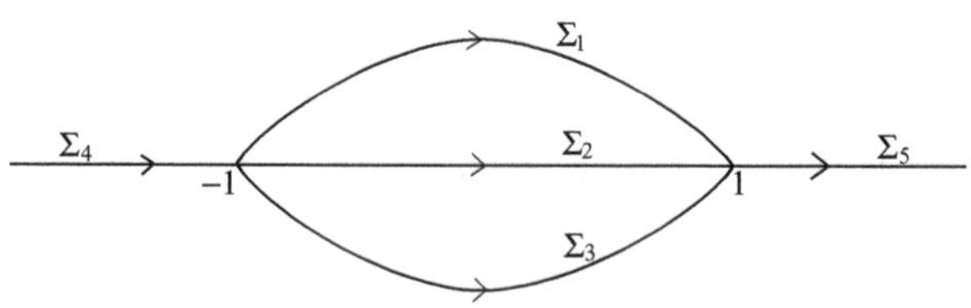


Figure 2. Contour $\Sigma_S$.

and

$$J_S(x) = \begin{pmatrix} 1 & e^{n(g_+(x)+g_-(x)-V_n(x)-l_n)} \\ 0 & 1 \end{pmatrix}, \qquad x \in \Sigma_4 \cup \Sigma_5. \tag{55}$$

It is readily verified that $S$ satisfies the following conditions:

$(S_a)$ $S(z)$ is analytic in $\mathbb{C}\backslash\Sigma_S$,

$(S_b)$ for $z \in \Sigma_S$

$$S_+(z) = S_-(z)J_s(z). \tag{56}$$

$(S_c)$ as $|z| \to \infty$,

$$S(z) = I + O\left(\frac{1}{|z|}\right). \tag{57}$$

Note that for $z \in \Sigma_S/\Sigma_2$, the jump matrix $J_S(z)$ tends to the identity matrix as $n \to \infty$; see the signs of $\operatorname{Re}\xi_n(z)$ given below. Also, note that for $x \in \Sigma_2$, $J_S(x)$ is the constant matrix given in (53). Thus, it is natural to suggest that for large $n$, the solution of the RHP for $S$ may behave asymptotically like the solution of the following RHP for $N$:

$(N_a)$ $N(z)$ is analytic in $\mathbb{C}\backslash[-1, 1]$,

$(N_b)$ for $x \in (-1, 1)$,

$$N_+(x) = N_-(x)\begin{pmatrix} 0 & 1 \\ -1 & 0 \end{pmatrix}, \tag{58}$$

$(N_c)$ as $|z| \to \infty$,

$$N(z) = I + O\left(\frac{1}{|z|}\right). \tag{59}$$

This problem can be solved explicitly, and its solution is

$$\begin{aligned} N(z) &= \frac{1}{2}\begin{pmatrix} a(z)+a(z)^{-1} & i(a(z)^{-1}-a(z)) \\ i(a(z)-a(z)^{-1}) & a(z)+a(z)^{-1} \end{pmatrix} \\ &= \frac{1}{2}\begin{pmatrix} 1 & -1 \\ -i & -i \end{pmatrix}\begin{pmatrix} a(z)^{-1} & 0 \\ 0 & a(z) \end{pmatrix}\begin{pmatrix} 1 & i \\ -1 & i \end{pmatrix}, \end{aligned} \tag{60}$$

where $a(z)$ is the analytic function in $\mathbb{C}\backslash[-1, 1]$ defined by (27); cf. [4, p. 1520]. Because

$$\operatorname{Re}\xi_n(z) > 0 \quad \text{in the upper lens region}$$

and

$$\operatorname{Re}\xi_n(z) < 0 \quad \text{in the lower lens region}$$

(see [4, p. 1508]), we have $T \sim S$ as $n \to \infty$ on account of (49)–(51). Furthermore, because $S \sim N$ as $n \to \infty$, it follow from (60) that as $n \to \infty$,

$$T(z) \sim \frac{1}{2}\begin{pmatrix} 1 & -1 \\ -i & -i \end{pmatrix}\begin{pmatrix} a(z)^{-1} & 0 \\ 0 & a(z) \end{pmatrix}\begin{pmatrix} 1 & i \\ -1 & i \end{pmatrix} \tag{61}$$

for $z \in \mathbb{C}\backslash\mathbb{R}$. By virtue of (37), we also have

$$U(z) \sim e^{\frac{1}{2}nl_n\sigma_3}\begin{pmatrix} 1 & -1 \\ -i & -i \end{pmatrix}\begin{pmatrix} a(z)^{-1} & 0 \\ 0 & a(z) \end{pmatrix}\begin{pmatrix} \frac{1}{2} & \frac{1}{2}i \\ -\frac{1}{2} & \frac{1}{2}i \end{pmatrix} e^{n(g_n(z)-\frac{1}{2}l_n)\sigma_3}. \tag{62}$$

The argument in this paragraph is only formal (heuristic); we give it here just as a motivation for the asymptotic formula, which we establish rigorously.

By (20), (41), and (47), we have

$$(g_n(x) + \varphi_n(x))_+ = (g_n(x) + \varphi_n(x))_- = \frac{1}{2}V_n(x) + \frac{1}{2}l_n$$

for $x \in (-1, 1)$. Hence, by analytic continuation

$$g_n(z) + \varphi_n(z) = \frac{1}{2}V_n(z) + \frac{1}{2}l_n \quad \text{for } z \in \mathbb{C}\backslash(-\infty, -1]. \tag{63}$$

Because the right–hand side is a polynomial, (63) in fact holds for all $z$. In view of this, (62) can be written as

$$\begin{aligned} e^{-\frac{1}{2}nl_n\sigma_3}U(z) \sim{}& \begin{pmatrix} 1 & -1 \\ -i & -i \end{pmatrix}\begin{pmatrix} a(z)^{-1} & 0 \\ 0 & a(z) \end{pmatrix} \\ & \begin{pmatrix} \frac{1}{2} & \frac{1}{2}i \\ -\frac{1}{2} & \frac{1}{2}i \end{pmatrix}\begin{pmatrix} e^{-n\varphi_n} & 0 \\ 0 & e^{n\varphi_n} \end{pmatrix}\begin{pmatrix} e^{\frac{1}{2}nV_n} & 0 \\ 0 & e^{-\frac{1}{2}nV_n} \end{pmatrix}. \end{aligned} \tag{64}$$

Now, recall the well-known asymptotic formulas

$$\mathrm{Ai}(z) \sim \frac{1}{2\sqrt{\pi}}z^{-1/4}e^{-\frac{2}{3}z^{3/2}}, \quad \mathrm{Ai}'(z) \sim -\frac{1}{2\sqrt{\pi}}z^{1/4}e^{-\frac{2}{3}z^{3/2}},$$

as $|z| \to \infty$ in $|\arg z| < \pi$, and

$$\mathrm{Bi}(z) \sim \frac{1}{\sqrt{\pi}}z^{-1/4}e^{\frac{2}{3}z^{3/2}}, \quad \mathrm{Bi}'(z) \sim \frac{1}{\sqrt{\pi}}z^{1/4}e^{\frac{2}{3}z^{3/2}},$$

as $|z| \to \infty$ in $|\arg z| < \frac{\pi}{3}$. Using these formulas, we have from (64)

$$e^{-\frac{1}{2}nl_n\sigma_3}U(z) \sim \sqrt{\pi}\begin{pmatrix}1 & -1\\ -i & -i\end{pmatrix}\begin{pmatrix}a(z)^{-1} & 0\\ 0 & a(z)\end{pmatrix}\begin{pmatrix}\left(n^{2/3}\zeta_n\right)^{1/4} & 0\\ 0 & \left(n^{2/3}\zeta_n\right)^{-1/4}\end{pmatrix}$$
$$\times\begin{pmatrix}\mathrm{Ai}\left(n^{2/3}\zeta_n\right) & \mathrm{Bi}\left(n^{2/3}\zeta_n\right)\\ \mathrm{Ai}'\left(n^{2/3}\zeta_n\right) & \mathrm{Bi}'\left(n^{2/3}\zeta_n\right)\end{pmatrix}\begin{pmatrix}1 & c\\ 0 & \frac{1}{2}i\end{pmatrix}\begin{pmatrix}e^{\frac{1}{2}nV_n} & 0\\ 0 & e^{-\frac{1}{2}nV_n}\end{pmatrix} \tag{65}$$

for $z \in \mathbb{C}\backslash\mathbb{R}$, where $c = \frac{1}{2}$ for $\operatorname{Im} z > 0$ and $c = -\frac{1}{2}$ for $\operatorname{Im} z < 0$. Note that $\zeta_n^{1/4}(z)/a(z)$ can be analytically continued to $\mathbb{C}\backslash(-\infty, -1]$; see Remark 1 in Section 2. Define the matrix $P(z)$ by

$$P(z) := \sqrt{\pi}\begin{pmatrix}1 & -1\\ -i & -i\end{pmatrix}\begin{pmatrix}n^{1/6}\zeta_n^{1/4}/a(z) & 0\\ 0 & a(z)/\left(n^{1/6}\zeta_n^{1/4}\right)\end{pmatrix}$$
$$\times\begin{pmatrix}\mathrm{Ai}\left(n^{2/3}\zeta_n\right) & \mathrm{Bi}\left(n^{2/3}\zeta_n\right)\\ \mathrm{Ai}'\left(n^{2/3}\zeta_n\right) & \mathrm{Bi}'\left(n^{2/3}\zeta_n\right)\end{pmatrix}\begin{pmatrix}1 & \frac{1}{2}\\ 0 & \frac{1}{2}i\end{pmatrix} \tag{66a}$$

for Im $z > 0$, and

$$P(z) := \sqrt{\pi}\begin{pmatrix}1 & -1\\ -i & -i\end{pmatrix}\begin{pmatrix}n^{1/6}\zeta_n^{1/4}/a(z) & 0\\ 0 & a(z)/\left(n^{1/6}\zeta_n^{1/4}\right)\end{pmatrix}$$
$$\times\begin{pmatrix}\mathrm{Ai}\left(n^{2/3}\zeta_n\right) & \mathrm{Bi}\left(n^{2/3}\zeta_n\right)\\ \mathrm{Ai}'\left(n^{2/3}\zeta_n\right) & \mathrm{Bi}'\left(n^{2/3}\zeta_n\right)\end{pmatrix}\begin{pmatrix}1 & -\frac{1}{2}\\ 0 & \frac{1}{2}i\end{pmatrix} \tag{66b}$$

for Im $z < 0$. Formula (65) then becomes

$$U(z) \sim e^{\frac{1}{2}nl_n\sigma_3}P(z)\begin{pmatrix}e^{\frac{1}{2}nV_n} & 0\\ 0 & e^{-\frac{1}{2}nV_n}\end{pmatrix} \tag{67}$$

for $z \in \mathbb{C}\backslash\mathbb{R}$. Put

$$\tilde{U}(z) := P(z)\begin{pmatrix}e^{\frac{1}{2}nV_n} & 0\\ 0 & e^{-\frac{1}{2}nV_n}\end{pmatrix}. \tag{68}$$

We show that the matrix $\tilde{U}(z)$ defined in (68) is the leading term in an asymptotic expansion of $Y(\lambda_n(z))$ multiplied by a simple diagonal matrix independent of $z$, which holds uniformly for all $z \in \mathbb{C}$ bounded away from the interval $(-\infty, -1]$.

### 4. Riemann-Hilbert Problem for $\tilde{U}(z)$

In this section, we show that the matrix $\tilde{U}(z)$ defined in (68) is a solution of the following RHP:

$(\tilde{U}_a)$ $\tilde{U}(z)$ is analytic in $\mathbb{C}\backslash\mathbb{R}$,
$(\tilde{U}_b)$ $\tilde{U}(z)$ satisfies the jump conditions

$$\tilde{U}_+(x) = \tilde{U}_-(x)\begin{pmatrix}1 & e^{-nV_n(x)}\\ 0 & 1\end{pmatrix}, \quad x \in (-1,\infty), \tag{69}$$

and

$$\tilde{U}_+(x) = M(x)\tilde{U}_-(x)\begin{pmatrix}1 & e^{-nV_n(x)}\\ 0 & 1\end{pmatrix}, \quad x \in (-\infty,-1), \tag{70}$$

where the jump matrix $M(x)$ has the uniform asymptotic expansion

$$M(x) \sim I + \sum_{k=2m}^{\infty} M_k(x)n^{-k/2m} \tag{71}$$

for $x \in (-\infty,-1]$, $\|M_k(x)\|$ is bounded by $|x|^{-2m}$ for $x \le -1$ and $k = 2m$, $2m+1,\ldots,$ and $\|\cdot\|$ denotes a matrix norm,
$(\tilde{U}_c)$ as $z \to \infty$,

$$e^{\frac{1}{2}nl_n\sigma_3}\tilde{U}(z)\begin{pmatrix}z^{-n} & 0\\ 0 & z^n\end{pmatrix} = I + O\left(\frac{1}{|z|}\right). \tag{72}$$

We first establish the jump condition (69). From (66a) and (66b), it is easy to show that

$$P_+(x) = P_-(x)\begin{pmatrix}1 & 1\\ 0 & 1\end{pmatrix} \tag{73}$$

for $x \in (-1,\infty)$. Hence,

$$\tilde{U}_-^{-1}(x)\tilde{U}_+(x) = \begin{pmatrix}1 & e^{-nV_n}\\ 0 & 1\end{pmatrix};$$

i.e., (69) holds. Next, we verify the jump condition (70). Note that (48) is equivalent to

$$\int_{-1}^{1} \psi_n(y)\,dy = 1.$$

By a combination of (18)–(20) and (23), we have

$$\frac{2}{3}\left(\zeta_n^{3/2}\right)_\pm(x) = \mp\pi i + \frac{1}{2}\int_x^{-1}\sqrt{t^2-1}\,h_n(t)\,dt \tag{74}$$

for $x < -1$. Put

$$M(x) := \tilde{U}_+(x) \begin{pmatrix} 1 & -e^{-nV_n} \\ 0 & 1 \end{pmatrix} \tilde{U}_-^{-1}(x); \tag{75}$$

cf. (70). Inserting (66a) and (66b) in (75) gives

$$M(x) = \pi \begin{pmatrix} a(x)^{-1} & -a(x) \\ -ia(x)^{-1} & -ia(x) \end{pmatrix} \begin{pmatrix} m_{11} & m_{12} \\ m_{21} & m_{22} \end{pmatrix} \times \begin{pmatrix} \frac{1}{2}a(x) & \frac{i}{2}a(x) \\ -\frac{1}{2}a(x)^{-1} & \frac{i}{2}a(x)^{-1} \end{pmatrix}, \tag{76}$$

where

$$\begin{aligned} m_{11} &= \zeta_+^{1/4}\zeta_-^{-1/4}\left[\mathrm{Ai}(n^{2/3}\zeta_+)\mathrm{Bi}'(n^{2/3}\zeta_-) - \mathrm{Bi}(n^{2/3}\zeta_+)\mathrm{Ai}'(n^{2/3}\zeta_-)\right], \\ m_{12} &= -n^{1/3}\zeta_+^{1/4}\zeta_-^{1/4}\left[\mathrm{Ai}(n^{2/3}\zeta_+)\mathrm{Bi}(n^{2/3}\zeta_-) - \mathrm{Bi}(n^{2/3}\zeta_+)\mathrm{Ai}(n^{2/3}\zeta_-)\right], \\ m_{21} &= n^{-1/3}\zeta_+^{-1/4}\zeta_-^{-1/4}\left[\mathrm{Ai}'(n^{2/3}\zeta_+)\mathrm{Bi}'(n^{2/3}\zeta_-) - \mathrm{Bi}'(n^{2/3}\zeta_+)\mathrm{Ai}'(n^{2/3}\zeta_-)\right], \\ m_{22} &= -\zeta_+^{-1/4}\zeta_-^{1/4}\left[\mathrm{Ai}'(n^{2/3}\zeta_+)\mathrm{Bi}(n^{2/3}\zeta_-) - \mathrm{Bi}'(n^{2/3}\zeta_+)\mathrm{Ai}(n^{2/3}\zeta_-)\right]. \end{aligned} \tag{77}$$

For convenience, in (77) we have dropped the dependence of $\zeta$ on $n$, i.e., $\zeta(z) = \zeta_n(z)$. Now we recall the well-known asymptotic expansions [10, pp. 392–393]

$$\begin{aligned} \mathrm{Ai}(z) &\sim \frac{z^{-1/4}}{2\sqrt{\pi}}e^{-\eta}\sum_{s=0}^{\infty}(-1)^s\frac{u_s}{\eta^s}, & \mathrm{Ai}'(z) &\sim -\frac{z^{1/4}}{2\sqrt{\pi}}e^{-\eta}\sum_{s=0}^{\infty}(-1)^s\frac{v_s}{\eta^s}, \\ \mathrm{Bi}(z) &\sim \frac{z^{-1/4}}{\sqrt{\pi}}e^{\eta}\sum_{s=0}^{\infty}\frac{u_s}{\eta^s}, & \mathrm{Bi}'(z) &\sim \frac{z^{1/4}}{\sqrt{\pi}}e^{\eta}\sum_{s=0}^{\infty}\frac{v_s}{\eta^s}, \end{aligned} \tag{78}$$

where $\eta = (2/3)z^{3/2}$, and $u_s$, $v_s$ are constants with $u_0 = v_0 = 1$. Coupling (76) and (74), it is easily verified that the asymptotic expansion in (71) holds.

## 5. Asymptotic Expansion of $U(z)$

Let $R(z) := e^{-\frac{1}{2}nl_n\sigma_3}U(z)\tilde{U}^{-1}(z)$. From (34) and (69), we observe that $U(z)$ and $\tilde{U}(z)$ have the same jump matrix on $(-1, \infty)$. Hence, $R_+(x) = R_-(x)$ for $x \in (-1, 1)$, and $R(z)$ can be analytically extended to $\mathbb{C}\backslash(-\infty, -1)$. On account of (70) and (71), it is readily shown that $R(z)$ is a solution of the RHP:

($R_a$) $R(z)$ is analytic for $z \in \mathbb{C}\backslash(-\infty, -1)$,

($R_b$) for $x \in (-\infty, -1)$,

$$R_+(x) = R_-(x)[I + \Delta(x)], \tag{79}$$

where

$$I + \Delta(x) := M^{-1}(x) \sim I + \sum_{s=2m}^{\infty} \frac{\Delta_s(x)}{n^{s/2m}} \qquad \text{as } n \to \infty, \tag{80}$$

$(R_c)$ as $z \to \infty$,

$$R(z) = I + O\left(\frac{1}{|z|}\right), \qquad z \notin (-\infty, -1). \tag{81}$$

LEMMA 1. *There exists a sequence of matrices $\{R_k(z)\}$ with the property that for every $p \geq 2m$, there is a constant $C_p > 0$ such that*

$$\left\| R(z) - I - \sum_{k=2m}^{p} \frac{R_k(z)}{n^{k/2m}} \right\| \leq \frac{C_p}{|z| n^{(p+1)/2m}} \tag{82}$$

*for all $z$ strictly bounded away from $(-\infty, -1)$, where $\|\cdot\|$ denotes any matrix norm.*

*Proof*: From (79), we have $R_+ = R_- + R_-\Delta$ (or, equivalently, $(R - I)_+ = (R - I)_- + R_-\Delta$). Hence, by the Plemeltj formula [1, p. 590]

$$R(z) = I + \frac{1}{2\pi i} \int_{-\infty}^{-1} \frac{R_-(\zeta)\Delta(\zeta)}{\zeta - z} d\zeta, \quad z \in \mathbb{C}\backslash(-\infty, -1), \tag{83}$$

for all $z$ strictly bounded away from $(-\infty, -1)$. In (83), we now insert the formal series

$$R(z) = I + \sum_{k=2m}^{\infty} \frac{R_k(z)}{n^{k/2m}}$$

and the asymptotic expansion in (80). Upon collecting terms with like powers of $1/n$, we obtain the recursive formula

$$R_k(z) = \frac{1}{2\pi i} \int_{-\infty}^{-1} \sum_{j=1}^{k} (R_{k-j})_-(\zeta)\Delta_j(\zeta) \frac{d\zeta}{\zeta - z}, \quad z \in \mathbb{C}\backslash(-\infty, -1), \tag{84}$$

$k = 1, 2, \ldots$, where $\Delta_1(z) = \cdots = \Delta_{2m-1}(z) = 0$ and $(R_0)_-(\zeta) = I$. (From (84), it is also clear that $R_1(z) = \cdots = R_{2m-1}(z) = 0$.) By induction, it can be easily verified that for $z \in \mathbb{C}\backslash\mathbb{R}$,

$$R_k(z) = O\left(\frac{1}{|z|}\right) \quad \text{as } z \to \infty. \tag{85}$$

To establish (82), we put

$$E_p(z) := R(z) - I - \sum_{k=2m}^{p} \frac{R_k(z)}{n^{k/2m}}. \tag{86}$$

Using (79) and (84), it can be shown that

$$(E_p)_+(x) = (E_p)_-(x)[I + \Delta(x)] + F_p(x) \tag{87}$$

for $x \in (-\infty, -1)$, where

$$F_p(x) := \left[ I + \sum_{k=2m}^{p} \frac{(R_k)_-(x)}{n^{k/2m}} \right] \Delta(x) - \sum_{k=2m}^{p} \frac{1}{n^{k/2m}} \left[ \sum_{j=1}^{k} (R_{k-j})_-(x) \Delta_j(x) \right]. \tag{88}$$

By the Plemelj formula,

$$E_p(z) = \frac{1}{2\pi i} \int_{-\infty}^{-1} \frac{(E_p)_-(\zeta)\Delta(\zeta)}{\zeta - z} d\zeta + \frac{1}{2\pi i} \int_{-\infty}^{-1} \frac{F_p(\zeta)}{\zeta - z} d\zeta. \tag{89}$$

Now define the sequence $\{E_p^{(l)}(z) : l = 0, 1, 2, \ldots\}$ successively by $E_p^{(0)}(z) = 0$ and

$$E_p^{(l)}(z) = \frac{1}{2\pi i} \int_{-\infty}^{-1} \frac{\left(E_p^{(l-1)}\right)_-(\zeta)\Delta(\zeta)}{\zeta - z} d\zeta + \frac{1}{2\pi i} \int_{-\infty}^{-1} \frac{F_p(\zeta)}{\zeta - z} d\zeta \tag{90}$$

for $l = 1, 2, \ldots$. Note that we can rewrite (88) as

$$\begin{aligned} F_p(x) = & \left[ I + \sum_{k=2m}^{p} \frac{(R_k)_-(x)}{n^{k/2m}} \right] \left[ \Delta(x) - \sum_{k=2m}^{p} \frac{\Delta_k(x)}{n^{k/2m}} \right] \\ & + \sum_{k=p+1}^{2p} \frac{1}{n^{k/2m}} \left[ \sum_{j=k-p}^{p} (R_{k-j})_-(x) \Delta_j(x) \right], \end{aligned} \tag{91}$$

and that the matrix norm of $F_p(x)$ is bounded by $(1 + |x|)^{-1} n^{-(p+1)/2m}$, as $n \to \infty$, uniformly for $x \in (-\infty, -1)$. Writing

$$E_p^{(l)}(z) = \sum_{i=0}^{l-1} [E_p^{(i+1)}(z) - E_p^{(i)}(z)], \tag{92}$$

and applying the usual argument used in the method of successive approximation, one can show that the limit

$$\lim_{l \to \infty} E_p^{(l)}(z) = E_p(z) \tag{93}$$

is the unique solution of Equation (89). Furthermore,

$$\| E_p(z) \| \le \frac{C_p}{1 + |z|} n^{-(p+1)/2m} \tag{94}$$

for all $z$ bounded away from $(-\infty, -1)$. This completes the proof of Lemma 1. ∎

Recall that $U(z) = e^{\frac{1}{2}nl_n\sigma_3}R(z)\tilde{U}(z)$ and

$$Y(\lambda_n(z)) = \begin{pmatrix} c_n^n & 0 \\ 0 & c_n^{-n} \end{pmatrix} U(z);$$

see (33). Let $R_{ij}(z)$, $i$, $j = 1, 2$, denote the entries in the matrix $R(z)$. Then, from (67) we have

$$Y(\lambda_n(z)) = \begin{pmatrix} c_n^n e^{\frac{1}{2}nl_n} & 0 \\ 0 & c_n^{-n}e^{-\frac{1}{2}nl_n} \end{pmatrix} \begin{pmatrix} R_{11} & R_{12} \\ R_{21} & R_{22} \end{pmatrix}$$

$$\times P(z) \begin{pmatrix} e^{\frac{1}{2}nV_n} & 0 \\ 0 & e^{-\frac{1}{2}nV_n} \end{pmatrix}. \tag{95}$$

Simple computation shows that the entries in the first column of the matrix $P(z)$, given in (66), are

$$\begin{pmatrix} \sqrt{\pi}\left[n^{1/6}\zeta^{1/4}a^{-1}(z)\text{Ai} - n^{-1/6}\zeta^{-1/4}a(z)\text{Ai}'\right] \\ -i\sqrt{\pi}\left[n^{1/6}\zeta^{1/4}a^{-1}(z)\text{Ai} + n^{1/6}\zeta^{-1/4}a(z)\text{Ai}'\right] \end{pmatrix}.$$

Equation (95) yields, in particular,

$$\pi_n(c_nz + d_n) = \sqrt{\pi}c_n^n e^{\frac{1}{2}nl_n + \frac{1}{2}nV_n}\left\{n^{1/6}\text{Ai}\left(n^{2/3}\zeta_n\right)\mathbf{A}(z, n)\right.$$

$$\left. -n^{-1/6}\text{Ai}'\left(n^{2/3}\zeta_n\right)\mathbf{B}(z, n)\right\}, \tag{96}$$

where

$$\mathbf{A}(z, n) = \zeta_n^{1/4}(z)a^{-1}(z)[R_{11}(z) - iR_{12}(z)] \tag{97}$$

and

$$\mathbf{B}(z, n) = \zeta_n^{-1/4}(z)a(z)[R_{11}(z) + iR_{12}(z)]. \tag{98}$$

Note that $\mathbf{A}(z, n)$ and $\mathbf{B}(z, n)$ are analytic functions of $z$ in $\mathbb{C}\backslash(-\infty, -1]$, and by (86) and (94) they have the asymptotic expansions given in (25) and (26). Furthermore, because $\text{Ai}(n^{2/3}\zeta_n)$ and $\text{Ai}'(n^{2/3}\zeta_n)$ are also analytic in $\mathbb{C}\backslash(-\infty, -1]$, by analytic continuation the result (96) holds for $z \in \mathbb{C}\backslash(-\infty, -1]$. This completes the proof of Theorem 1 stated in Section 2.

## References

1. M. J. Ablowitz and A. S. Fokas, *Complex Variables: Introduction and Applications*, Cambridge University Press, Cambridge, 1997.
2. B. Rui and R. Wong, A uniform asymptotic formula for orthogonal polynomials associated with, $\exp(-x^4)$, *J. Approx. Theory* 98:146–166 (1999).

3. P. DEIFT and X. ZHOU, A steepest descent method for oscillatory Riemann–Hilbert problems. Asymptotics for the MKdV equation, *Ann. Math.* 137:295–368 (1993).

4. P. DEIFT, T. KRIECHERBAUER, K. T.-R. MCLAUGHLIN, S. VENAKIDES, and X. ZHOU, Strong asymptotics of orthogonal polynomials with respect to exponential weights, *Commun. Pure Appl. Math.* 52:1491–1552 (1999).

5. A. S. FOKAS, A. R. ITS, and A. V. KITAEV, The isomonodromy approach to matrix models in 2D quantum gravity, *Commun. Math. Phys.* 147:395–430 (1992).

6. D. S. LUBINSKY, Strong asymptotics for extremal errors and polynomials associated with Erdös-type weights, *Pitman Research Notes in Mathematical Series*, Longman, Essex, England, 1989.

7. H. N. MHASKAR and E. B. SAFF, Extremal problems for polynomials with exponential weights, *Trans. Am. Math. Soc.* 285:203–234 (1984).

8. P. NEVAI, Orthogonal polynomials associated with $\exp(-x^4)$, in "Second Edmonton Conference on Approximation Theory," *Can. Math. Soc. Conf. Proc.* 3:263–285 (1983).

9. P. NEVAI, Asymptotics for orthogonal polynomials associated with $\exp(-x^4)$, *SIAM J. Math. Anal.* 15:1177–1187 (1984).

10. F. W. J. OLVER, *Asymptotics and Special Functions*, Academic Press, New York, 1974. Reprinted by A. K. Peters, Wellesley, 1997.

11. W.-Y. QIU and R. WONG, Uniform asymptotic formula for orthogonal polynomials with exponential weight, *SIAM J. Math. Anal.* 31:992–1029 (2000).

12. E. A. RAKHMANOV, Asymptotic properties of orthogonal polynomials on the real axis (Russian), *Math. Sb. (N.S.)* 119 (161) 163–203, 303. (1982). *English translation in Math. USSR-Sb.* 47:155–193 (1984).

13. R. C. SHEEN, Asymptotics for orthogonal polynomials associated with $\exp(-x^6)$, *J. Approx. Theory* 50:232–293 (1987).

CITY UNIVERSITY OF HONG KONG

(Received January 31, 2005)

PROCEEDINGS OF THE ROYAL SOCIETY A

*Proc. R. Soc. A* (2005) **461**, 2839–2856
doi:10.1098/rspa.2005.1467
*Published online* 1 August 2005

# Bessel-type asymptotic expansions via the Riemann–Hilbert approach

By Z. Wang and R. Wong

*Department of Mathematics, City University of Hong Kong, Tat Chee Avenue, Kowloon, Hong Kong*
(mawong@cityu.edu.hk)

A Bessel-type asymptotic expansion is established for the monic polynomials $\pi_n(x)$ that are orthogonal with respect to the modified Jacobi weight $w(x) = (1-x)^\alpha(1+x)^\beta h(x)$, $x \in (-1,1)$, where $\alpha$, $\beta > -1$ and $h(x)$ is real analytic and strictly positive on $[-1,1]$. This expansion holds uniformly in a region containing the neighbourhood of the critical value $x=1$. This result complements the two recent results obtained by Kuijlaars and his co-workers, one for $x$ bounded away from $(-1,1)$ and the other for $x$ in $(-1+\delta, 1-\delta)$, $\delta > 0$. Our method is also based on the Riemann–Hilbert approach.

**Keywords: uniform asymptotic expansion; orthogonal polynomials; modified Jacobi weight; Riemann–Hilbert problems**

## 1. Introduction

We consider monic polynomials $\pi_n(x)$ that are orthogonal on $[-1,1]$ with respect to the modified Jacobi weight

$$w(x) = (1-x)^\alpha(1+x)^\beta h(x), \quad x \in (-1,1), \tag{1.1}$$

where $\alpha$, $\beta > -1$ and $h(x)$ is real analytic and strictly positive on $[-1,1]$. These polynomials occur in the study of the modified Jacobi unitary ensemble, which is a probability measure on the space of $n \times n$ Hermitian matrices with all eigenvalues in $(-1,1)$. When the size of the matrices tends to infinity, the distribution of the eigenvalues is related to the asymptotic behaviour of these polynomials (see some of the references given in Kuijlaars & Vanlessen 2002; Kuijlaars *et al.* 2004).

Since the weight function $w(x)$ satisfies the Szegő condition

$$\int_{-1}^{1} \frac{\log w(x)}{\sqrt{1-x^2}} \mathrm{d}x > -\infty,$$

by using Szegő's theorem (Szegő 1975, p. 297) we have

$$\lim_{n\to\infty} \frac{2^n \pi_n(z)}{\varphi^n(z)} = \frac{D_\infty}{D(z)} \frac{\varphi^{1/2}(z)}{\sqrt{2}(z^2-1)^{1/4}} \tag{1.2}$$

*Received* 21 May 2004
*Accepted* 7 February 2005

uniformly for $z$ in compact subsets of $\mathbb{C}\backslash[-1,1]$, where $D_\infty = \lim_{z\to\infty} D(z)$,

$$\varphi(z) = z + \sqrt{z^2-1}, \quad z \in \mathbb{C}\backslash[-1,1], \tag{1.3}$$

and

$$D(z) = \exp\left( -\frac{(z^2-1)^{1/2}}{2\pi} \int_{-1}^{1} \frac{\log w(t)}{\sqrt{1-t^2}} \frac{\mathrm{d}t}{t-z} \right); \tag{1.4}$$

see also Kuijlaars *et al.* (2004). In (1.3) and (1.4), we take the branch of $(z^2-1)^{1/2}$ which is analytic in $\mathbb{C}\backslash[-1,1]$ and which behaves like $z$ as $z\to\infty$. The function $\varphi(z)$ in (1.3) is the familiar Joukowski or aerofoil map that maps the exterior of $[-1,1]$ conformally onto the exterior of the unit disc, and the function $D(z)$ is the so-called Szegő function associated with the weight $w(x)$; see Szegő (1975, p. 277).

Note that $h$ is real analytic on $[-1,1]$; hence it can be analytically extended to a region $U_\delta := \{z \in \mathbb{C}; |\varphi(z)| \leq 1+\delta\}$, $\delta > 0$, containing the interval $[-1,1]$. By using this analytic extension and the Riemann–Hilbert approach, Kuijlaars *et al.* (2004) obtained an asymptotic expansion for the polynomials $\pi_n(x)$, holding uniformly in compact subsets of $\mathbb{C}\backslash[-1,1]$. In another paper, Kuijlaars & Vanlessen (2002) presented a uniform asymptotic expansion for $\pi_n(x)$ in the interval $(-1+\varepsilon, 1-\varepsilon)$, $\varepsilon>0$, which agrees with the following result given by Szegő (1975, p. 298, theorem 12.1.6 and footnote 59)

$$\pi_n(x) = \frac{\sqrt{2} D_\infty}{2^n w^{1/2}(x)(1-x^2)^{1/4}} [\cos(n \arccos x + \gamma(x)) + o(1)], \tag{1.5}$$

where

$$D_\infty = \exp\left( \frac{1}{2\pi} \int_{-1}^{1} \frac{\log w(t)}{\sqrt{1-t^2}} \mathrm{d}t \right), \tag{1.6}$$

and

$$\gamma(x) = \frac{(1-x^2)^{1/2}}{2\pi} \mathrm{PV} \int_{-1}^{1} \frac{\log[\sqrt{1-t^2}\, w(t)]}{\sqrt{1-t^2}} \frac{\mathrm{d}t}{t-x}. \tag{1.7}$$

Here the integral is taken in the Cauchy principal-value sense. However, the results are not valid in the neighbourhoods of the critical values $x = \pm 1$.

The objective of this paper is to derive a uniform asymptotic expansion for the orthogonal polynomials $\pi_n(z)$, which holds uniformly in a region containing the interval $(-1+\delta, 1]$, $\delta > 0$. Our approach is also based on the nonlinear steepest descent method for Riemann–Hilbert problems (RHPs), introduced by Deift & Zhou (1993). As in Kuijlaars *et al.* (2004), we start with a RHP for a matrix $Y$, and make a sequence of transformations

$$Y \to T \to S.$$

The purpose of the first transformation is to 'normalize' the behaviour of the matrix $T$ at $z = \infty$, and that of the second transformation is to deform the contour so that the off-diagonal entries of the jump matrix of $S$, when $z$ is away from the interval $(-1,1)$, are exponentially small. From $S$, we guess an asymptotic formula

$$S \sim N \quad \text{as } n \to \infty.$$

Based on $N$, we construct a RHP for $\tilde{Y}$, which is the leading term of a uniform asymptotic expansion for $Y$. The main result of this paper is stated below.

**Theorem 1.1.** *Let* $U_\delta := \{z \in \mathbb{C}; |\varphi(z)| \leq 1+\delta\}$, $\zeta^{1/2}(z) := \log(z+\sqrt{z^2-1})$, $W(z) := (z-1)^{\alpha/2+1/4}(z+1)^{\beta/2+1/4}h^{1/2}(z)$ *for* $z \in U_\delta \backslash (-\infty,-1]$ *and* $\tilde{D}(z)$ *and* $\tilde{D}_\infty$ *be given as in (2.7) and (2.8). Then, as* $n \to \infty$, *the monic orthogonal polynomial* $\pi_n(z)$ *has an asymptotic expansion of the form*

$$\pi_n(z) \sim \sqrt{2\pi n \zeta^{1/2} \tilde{D}_\infty / 2^n}\, W^{-1/2}(z)$$
$$\times \left\{ I_\alpha(n\zeta^{1/2}) \sum_{k=0}^{\infty} \frac{A_k(z)}{n^k} + \sqrt{z^2-1}\, I_{\alpha+1}(n\zeta^{1/2}) \sum_{k=0}^{\infty} \frac{B_k(z)}{n^k} \right\}, \tag{1.8}$$

*where* $I_\alpha$ *is the modified Bessel function and the expansion holds uniformly for* $z \in U_\delta$ *bounded away from* $(-\infty,-1]$. *The coefficients* $A_k(z)$ *and* $B_k(z)$ *are analytic functions in* $z \in U_\delta \backslash (-\infty,-1]$; *the leading coefficients are given explicitly by*

$$A_0(z) = \frac{1}{2}(W/\tilde{D} + \tilde{D}/W), \quad B_0(z) = \frac{1}{2\sqrt{z^2-1}}(W/\tilde{D} - \tilde{D}/W).$$

## 2. The Riemann–Hilbert problem

Denote by $\{p_n(x)\}$ the sequence of orthonormal polynomials with respect to the weight function $w(x) = (1-x)^\alpha(1+x)^\beta h(x)$ on $(-1,1)$, i.e.,

$$p_n(x) = \gamma_n x^n + \cdots, \quad \gamma_n > 0,$$

$$\int_{-1}^{1} p_n(x) p_m(x) w(x) \mathrm{d}x = \delta_{n,m}.$$

Consider the following RHP for a $2\times2$ matrix-valued function $Y(z)$.

**RHP for *Y*:**

(i) $Y(z)$ is analytic in $\mathbb{C}\backslash[-1,1]$;

(ii) let $Y_+(x)$ and $Y_-(x)$ denote the limiting values of $Y(z)$ as $z$ tends to $x \in (-1,1)$ from above and below, respectively; $Y_+(x)$ and $Y_-(x)$ are continuous on $(-1,1)$, and

$$Y_+(x) = Y_-(x)\begin{pmatrix} 1 & w(x) \\ 0 & 1 \end{pmatrix}, \quad x \in (-1,1); \tag{2.1}$$

(iii) as $z \to \infty$,

$$Y(z) = \left(I + O\left(\frac{1}{z}\right)\right)\begin{pmatrix} z^n & 0 \\ 0 & z^{-n} \end{pmatrix}; \tag{2.2}$$

(iv) near $z=1$,

$$Y(z) = \begin{cases} O\begin{pmatrix} 1 & |z-1|^\alpha \\ 1 & |z-1|^\alpha \end{pmatrix}, & \text{if } \alpha < 0, \\ O\begin{pmatrix} 1 & \log|z-1| \\ 1 & \log|z-1| \end{pmatrix}, & \text{if } \alpha = 0, \\ O\begin{pmatrix} 1 & 1 \\ 1 & 1 \end{pmatrix}, & \text{if } \alpha > 0; \end{cases} \tag{2.3}$$

(v) near $z=-1$,

$$Y(z) = \begin{cases} O\begin{pmatrix} 1 & |z+1|^\beta \\ 1 & |z+1|^\beta \end{pmatrix}, & \text{if } \beta < 0, \\ O\begin{pmatrix} 1 & \log|z+1| \\ 1 & \log|z+1| \end{pmatrix}, & \text{if } \beta = 0, \\ O\begin{pmatrix} 1 & 1 \\ 1 & 1 \end{pmatrix}, & \text{if } \beta > 0. \end{cases} \tag{2.4}$$

The unique solution to the above RHP for $Y$ is given by

$$Y(z) = \begin{pmatrix} \pi_n(z) & C[\pi_n w](z) \\ -2\pi i \gamma_{n-1}^2 \pi_{n-1}(z) & -2\pi i \gamma_{n-1}^2 C[\pi_{n-1} w](z) \end{pmatrix}, \quad z \in \mathbb{C}\backslash[-1,1], \tag{2.5}$$

where $\pi_n(x) = p_n(x)/\gamma_n = x^n + \cdots$ is the $n$th monic orthogonal polynomial with respect to the weight $w(x)$ and

$$C[f](z) = \frac{1}{2\pi i}\int_{-1}^{1} \frac{f(\xi)}{\xi - z} d\xi, \quad z \in \mathbb{C}\backslash[-1,1],$$

is the Cauchy transformation of $f$; see Fokas *et al.* (1991) and Kuijlaars *et al.* (2004).

To find the asymptotic behaviour of $Y(z)$, we first formulate an equivalent RHP which is normalized at infinity. Its jump matrix has rapidly oscillating entries. Let $\sigma_3$ be the Pauli matrix

$$\sigma_3 = \begin{pmatrix} 1 & 0 \\ 0 & -1 \end{pmatrix}$$

so that

$$a^{\sigma_3} = \begin{pmatrix} a & 0 \\ 0 & a^{-1} \end{pmatrix}.$$

Define

$$T(z) = (2^n/\tilde{D}_\infty)^{\sigma_3} Y(z)(\varphi^n/\tilde{D}(z))^{-\sigma_3}$$

$$= \begin{pmatrix} 2^n/\tilde{D}_\infty & 0 \\ 0 & 2^{-n}\tilde{D}_\infty \end{pmatrix} Y(z) \begin{pmatrix} \varphi^{-n}\tilde{D}(z) & 0 \\ 0 & \varphi^n/\tilde{D}(z) \end{pmatrix}, \quad z \in \mathbb{C}\backslash[-1,1], \tag{2.6}$$

where

$$\tilde{D}(z) = \exp\left( -\frac{(z^2-1)^{1/2}}{2\pi} \int_{-1}^{1} \frac{\log[\sqrt{1-t^2}\,w(t)]}{\sqrt{1-t^2}} \frac{\mathrm{d}t}{t-z} \right), \quad z \in \mathbb{C}\backslash[-1,1], \tag{2.7}$$

and

$$\tilde{D}_\infty = \lim_{z\to\infty} \tilde{D}(z) = \exp\left( \frac{1}{2\pi} \int_{-1}^{1} \frac{\log[\sqrt{1-t^2}\,w(t)]}{\sqrt{1-t^2}} \mathrm{d}t \right). \tag{2.8}$$

Recall that $\varphi(z) = z + (z^2-1)^{1/2}$ is the conformal map from $\mathbb{C}\backslash[-1,1]$ onto the exterior of the unit disc. For convenience, we have defined the Szegő function $\tilde{D}(z)$ in (2.7) slightly differently from the one given in (1.4). Note that $T(z)$ is analytic in $\mathbb{C}\backslash[-1,1]$. Using (2.1) and (2.6), it is readily verified that $T(z)$ satisfies the following jump condition:

$$T_+(x) = T_-(x)(\varphi_-(x)^n/\tilde{D}_-(x))^{\sigma_3} \begin{pmatrix} 1 & w(x) \\ 0 & 1 \end{pmatrix} (\varphi_+(x)^n/\tilde{D}_+(x))^{-\sigma_3}.$$

Since $\tilde{D}_+(x)\tilde{D}_-(x) = \sqrt{1-x^2}\,w(x)$, it follows that

$$T_+(x) = T_-(x) \begin{pmatrix} \dfrac{\tilde{D}_+(x)}{\tilde{D}_-(x)} \varphi_+(x)^{-2n} & \dfrac{1}{\sqrt{1-x^2}} \\ 0 & \dfrac{\tilde{D}_-(x)}{\tilde{D}_+(x)} \varphi_-(x)^{-2n} \end{pmatrix} \tag{2.9}$$

for $x \in (-1,1)$. Combining (2.2), (2.6) and the fact that $\varphi(z) = 2z + O(1/z)$ as $z \to \infty$, we have $T(z) = I + O(1/z)$ as $z \to \infty$. Thus, the transformation $Y \to T$ gives the RHP that follows.

**RHP for *T*:**

(i) $T(z)$ is analytic in $\mathbb{C}\backslash[-1,1]$;

(ii) $T(z)$ satisfies the following jump relation

$$T_+(x) = T_-(x) \begin{pmatrix} \dfrac{\tilde{D}_+(x)}{\tilde{D}_-(x)} \varphi_+(x)^{-2n} & \dfrac{1}{\sqrt{1-x^2}} \\ 0 & \dfrac{\tilde{D}_-(x)}{\tilde{D}_+(x)} \varphi_-(x)^{-2n} \end{pmatrix},$$

$$x \in (-1,1); \tag{2.10}$$

(iii) $T(z)$ has the following behaviour

$$T(z) = I + O\left(\frac{1}{z}\right), \quad z \to \infty; \tag{2.11}$$

(iv) near $z=1$,

$$T(z) = \begin{cases} O\begin{pmatrix} |z-1|^{1/4} & |z-1|^{-1/4}\log|z-1| \\ |z-1|^{1/4} & |z-1|^{-1/4}\log|z-1| \end{pmatrix}, & \text{if } \alpha = 0, \\ O\begin{pmatrix} |z-1|^{\alpha/2+1/4} & |z-1|^{-\alpha/2-1/4} \\ |z-1|^{\alpha/2+1/4} & |z-1|^{-\alpha/2-1/4} \end{pmatrix}, & \text{if } \alpha \neq 0; \end{cases} \tag{2.12}$$

(v) near $z=-1$,

$$T(z) = \begin{cases} O\begin{pmatrix} |z+1|^{1/4} & |z+1|^{-1/4}\log|z+1| \\ |z+1|^{1/4} & |z+1|^{-1/4}\log|z+1| \end{pmatrix}, & \text{if } \beta = 0, \\ O\begin{pmatrix} |z+1|^{\beta/2+1/4} & |z+1|^{-\beta/2-1/4} \\ |z+1|^{\beta/2+1/4} & |z+1|^{-\beta/2-1/4} \end{pmatrix}, & \text{if } \beta \neq 0. \end{cases} \tag{2.13}$$

Using the fact that $\varphi_+(x)\varphi_-(x)=1$ for $x\in(-1,1)$, the jump matrix in (2.10) can be factorized as a product of three matrices, and we have

$$\begin{pmatrix} \frac{\tilde{D}_+(x)}{\tilde{D}_-(x)}\varphi_+(x)^{-2n} & \frac{1}{\sqrt{1-x^2}} \\ 0 & \frac{\tilde{D}_-(x)}{\tilde{D}_+(x)}\varphi_-(x)^{-2n} \end{pmatrix}$$
$$= \begin{pmatrix} 1 & 0 \\ \frac{\tilde{D}_-^2(x)}{w(x)}\varphi_-(x)^{-2n} & 1 \end{pmatrix} \begin{pmatrix} 0 & \frac{1}{\sqrt{1-x^2}} \\ -\sqrt{1-x^2} & 0 \end{pmatrix} \begin{pmatrix} 1 & 0 \\ \frac{\tilde{D}_+^2(x)}{w(x)}\varphi_+(x)^{-2n} & 1 \end{pmatrix}. \tag{2.14}$$

Note that the weight function $w$ has an analytic extension to $U_\delta\backslash((-\infty,-1]\cup[1,\infty))$, also denoted by $w$, given by

$$w(z) = (1-z)^\alpha(1+z)^\beta h(z).$$

We transform the RHP for $T$ into a RHP with jumps on a lens-shaped contour $\Sigma = \Sigma_1 \cup \Sigma_2 \cup \Sigma_3$, as shown in figure 1. Here $\Sigma_1$ and $\Sigma_3$ are the upper and lower lips of the lens, respectively, and $\Sigma_2=[-1,1]$. The lens is contained in $U_\delta$.

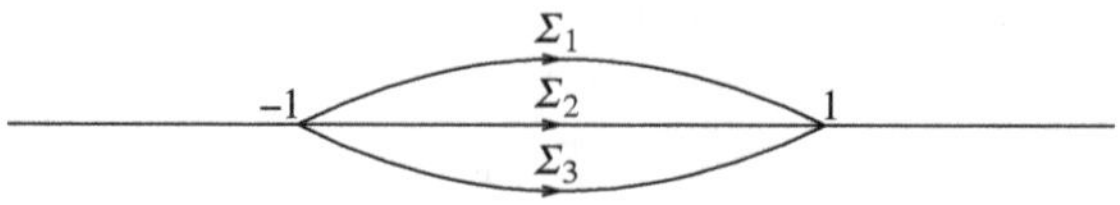


Figure 1. Contour of $\Sigma$.

Let $T$ be given as in (2.6), and define the matrix-valued function $S$ on $\mathbb{C}\backslash\Sigma$ by

$$S(z)=\begin{cases} T(z) & \text{for } z \text{ outside the lens,}\\ T(z)\begin{pmatrix} 1 & 0\\ -\dfrac{\tilde{D}^2(z)}{w(z)}\varphi(z)^{-2n} & 1\end{pmatrix} & \text{for } z \text{ in the upper part of the lens,}\\ T(z)\begin{pmatrix} 1 & 0\\ \dfrac{\tilde{D}^2(z)}{w(z)}\varphi(z)^{-2n} & 1\end{pmatrix} & \text{for } z \text{ in the lower part of the lens.}\end{cases} \tag{2.15}$$

The purpose of this transformation is to simplify the jump matrix for $T$. Indeed, the jump matrix for the RHP of $S$ given below behaves like the identity matrix when $z$ lies on $\Sigma_1\cup\Sigma_3$; see the penultimate paragraph in §1. Straightforward calculation shows that $S$ is a solution of the following RHP.

**RHP for $S$:**

(i) $S(z)$ is analytic in $\mathbb{C}\backslash\Sigma$;

(ii) $S(z)$ satisfies the jump conditions

$$S_+(x)=S_-(x)\begin{pmatrix} 1 & 0\\ \dfrac{\tilde{D}^2(z)}{w(z)}\varphi(z)^{-2n} & 1\end{pmatrix} \quad \text{for } x\in\Sigma_1^{\mathrm{o}}\cup\Sigma_3^{\mathrm{o}}, \tag{2.16}$$

$$S_+(x)=S_-(x)\begin{pmatrix} 0 & \dfrac{1}{\sqrt{1-x^2}}\\ -\sqrt{1-x^2} & 0\end{pmatrix} \quad \text{for } x\in(-1,1), \tag{2.17}$$

where $\Sigma_i^{\mathrm{o}}$, $i=1, 2, 3$, denote the curve $\Sigma_i$ with the two endpoints being removed;

(iii) $S(z)$ has the following behaviour

$$S(z)=I+O\left(\frac{1}{z}\right),\quad z\to\infty; \tag{2.18}$$

(iv) near $z=1$,

$$S(z)=\begin{cases} O\begin{pmatrix} |z-1|^{1/4}\log|z-1| & |z-1|^{-1/4}\log|z-1|\\ |z-1|^{1/4}\log|z-1| & |z-1|^{-1/4}\log|z-1|\end{pmatrix} & \text{if } \alpha=0,\\ O\begin{pmatrix} |z-1|^{\alpha/2}+|z-1|^{-\alpha/2} & |z-1|^{-\alpha/2-1/4}\\ |z-1|^{\alpha/2}+|z-1|^{-\alpha/2} & |z-1|^{-\alpha/2-1/4}\end{pmatrix} & \text{if } \alpha\neq 0;\end{cases} \tag{2.19}$$

(v) near $z=-1$,

$$S(z) = \begin{cases} O\begin{pmatrix} |z+1|^{1/4}\log|z+1| & |z+1|^{-1/4}\log|z+1| \\ |z+1|^{1/4}\log|z+1| & |z+1|^{-1/4}\log|z+1| \end{pmatrix} & \text{if } \beta = 0, \\ O\begin{pmatrix} |z+1|^{\beta/2} + |z+1|^{-\beta/2} & |z+1|^{-\beta/2-1/4} \\ |z+1|^{\beta/2} + |z+1|^{-\beta/2} & |z+1|^{-\beta/2-1/4} \end{pmatrix} & \text{if } \beta \neq 0. \end{cases} \tag{2.20}$$

Note that $\varphi$ maps $\mathbb{C}\backslash[-1,1]$ onto $\{z : |z| > 1\}$. Hence, we have $\varphi^{-2n}(z) \to 0$ as $n \to \infty$ for $z \notin [-1,1]$, and the jump matrix in (2.16) must tend to the identity matrix $I$ as $n \to \infty$. It is, therefore, reasonable to expect that

$$S(z) \sim N(z) \quad \text{as } n \to \infty, \tag{2.21}$$

where $N(z)$ is the solution of the RHP:

(i) $N(z)$ is analytic in $\mathbb{C}\backslash[-1,1]$;
(ii) $N(z)$ satisfies the jump condition

$$N_+(x) = N_-(x)\begin{pmatrix} 0 & \dfrac{1}{\sqrt{1-x^2}} \\ -\sqrt{1-x^2} & 0 \end{pmatrix} \quad \text{for } x \in (-1,1);$$

(iii) $N(z)$ has the behaviour

$$N(z) = I + O\left(\frac{1}{z}\right), \quad z \to \infty.$$

It is not difficult to find that

$$N(z) = \begin{pmatrix} 1 & \dfrac{i}{\sqrt{z^2-1}} \\ \dfrac{-i}{2\varphi} & \dfrac{\varphi}{2\sqrt{z^2-1}} \end{pmatrix}. \tag{2.22}$$

From (2.6) and (2.15), it then follows that

$$\begin{pmatrix} 2^n/\tilde{D}_\infty & 0 \\ 0 & 2^{-n}\tilde{D}_\infty \end{pmatrix} Y(z) \begin{pmatrix} \tilde{D}(z) & 0 \\ 0 & 1/\tilde{D}(z) \end{pmatrix}$$

behaves asymptotically like

$$\begin{pmatrix} \varphi^n & \dfrac{i}{\sqrt{z^2-1}}\varphi^{-n} \\ -\dfrac{i}{2}\varphi^{n-1} & \dfrac{1}{2\sqrt{z^2-1}}\varphi^{-(n-1)} \end{pmatrix} \tag{2.23}$$

for $z\in\mathbb{C}\backslash[-1,1]$. If we let $\zeta^{1/2}(z)=\log\varphi(z)$ for $z\in\mathbb{C}\backslash(-\infty,1]$, then $\zeta(z)=[\log\varphi(z)]^2$ for $z\in\mathbb{C}\backslash(-\infty,-1]$. Furthermore, $\zeta_\pm^{1/2}(x)=\pm\mathrm{i}\arccos x$ and $\zeta_\pm(x)=-(\arccos x)^2$ for $x\in[-1,1]$.

Since

$$\varphi^n(z)=\mathrm{e}^{n\zeta^{1/2}(z)},$$

from the well-known formulae

$$I_\alpha(z)\sim\frac{\mathrm{e}^z}{\sqrt{2\pi z}}\quad\text{as}\quad z\to\infty\quad\text{in}\quad -\frac{\pi}{2}<\arg z<\frac{3\pi}{2}$$

and

$$K_\alpha(z)\sim\sqrt{\frac{\pi}{2z}}\mathrm{e}^{-z}\quad\text{as}\quad z\to\infty\quad\text{in}\quad |\arg z|<\frac{3\pi}{2},$$

we conclude that the matrix in (2.23) is asymptotically equal to

$$\sqrt{2\pi n\zeta^{1/2}}\begin{pmatrix} I_\alpha(n\zeta^{1/2}) & \dfrac{\mathrm{i}}{\pi\sqrt{z^2-1}}K_\alpha(n\zeta^{1/2}) \\ -\dfrac{\mathrm{i}}{2}\left(1-\dfrac{1}{n}\right)^{1/2}I_\alpha((n-1)\zeta^{1/2}) & \dfrac{1}{2\pi\sqrt{z^2-1}}\left(1-\dfrac{1}{n}\right)^{1/2}K_\alpha((n-1)\zeta^{1/2})\end{pmatrix}.$$

We use the modified Bessel functions, instead of the simpler exponential form given in (2.23), in order to make the final approximation also valid in the neighbourhood of the endpoint $z=1$. By lemma 1 in Wang & Wong (2005), the last matrix is in turn asymptotically equal to

$$P:=\sqrt{2\pi n\zeta^{1/2}}\begin{pmatrix}1 & 0\\ -\dfrac{\mathrm{i}}{2}z & 1\end{pmatrix}\begin{pmatrix} I_\alpha(n\zeta^{1/2}) & \dfrac{\mathrm{i}}{\pi\sqrt{z^2-1}}K_\alpha(n\zeta^{1/2}) \\ \dfrac{\mathrm{i}}{2}\sqrt{z^2-1}I_{\alpha+1}(n\zeta^{1/2}) & \dfrac{1}{2\pi}K_{\alpha+1}(n\zeta^{1/2})\end{pmatrix}. \tag{2.24}$$

We shall show that $P$ is indeed the leading term in an asymptotic expansion of $Y$ which holds uniformly for all $z\in U_\delta$ bounded away from the interval $(-\infty,-1]$.

## 3. Riemann–Hilbert problem for $\tilde{Y}$

Recall that $h$ can be analytically extended to $U_\delta:=\{z\in\mathbb{C};|\varphi(z)|\le 1+\delta\}$ for some $\delta>0$. Hence, the function

$$W(z)=(z-1)^{\alpha/2+1/4}(z+1)^{\beta/2+1/4}h^{1/2}(z) \tag{3.1}$$

is defined and analytic for $z\in U_\delta\backslash(-\infty,1]$. Similarly, the function

$$\tilde{W}(z)=(1-z)^{\alpha/2+1/4}(-1-z)^{\beta/2+1/4}h^{1/2}(z) \tag{3.2}$$

is defined and analytic for $z\in U_\delta\backslash[-1,\infty)$.

Let $\Gamma=\Gamma_1\cup\Gamma_2\cup\Gamma_3$ denote the contour shown in figure 2, where $\Gamma_3=\{z\in\mathbb{C};|z+1|=\varepsilon\}$ for some small $\varepsilon>0$, $\Gamma_2=\{z\in\mathbb{R};-1\le z\le 1\}$, and $\Gamma_1$ consists of

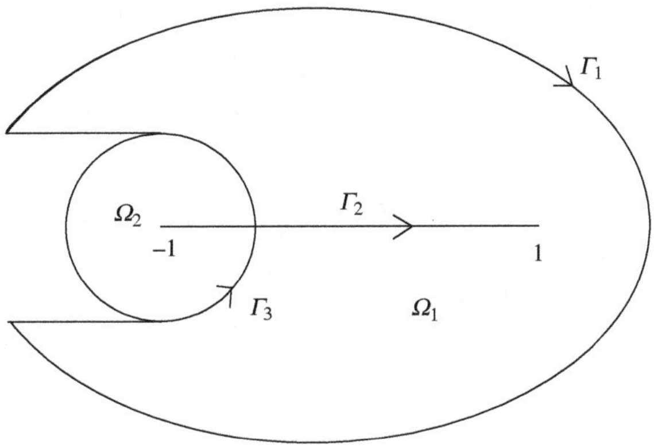


Figure 2. Contour of $\Gamma$.

part of the boundary $\partial U_\delta$ and two parallel line segments. The contour $\Gamma$ divides the complex plane $\mathbb{C}$ into three regions. Let $\Omega_1$ denote the region bounded between $\Gamma_1$ and $\Gamma_3$, and let $\Omega_2$ denote the region inside the circle $\Gamma_3$. (Both $\Omega_1$ and $\Omega_2$ are slit along the straight line $\Gamma_2$.)

For $z \in \mathbb{C}\backslash(\bar{\Omega}_1 \cup \bar{\Omega}_2)$, i.e. for $z$ outside $\Gamma_1 \cup \Gamma_3$, we define the matrix-valued function

$$\tilde{Y}(z) := \begin{pmatrix} \tilde{D}_\infty/2^n & 0 \\ 0 & 2^n/\tilde{D}_\infty \end{pmatrix} \begin{pmatrix} \varphi^n & \dfrac{\mathrm{i}}{\sqrt{z^2-1}}\varphi^{-n} \\ -\dfrac{\mathrm{i}}{2}\varphi^{n-1} & \dfrac{1}{2\sqrt{z^2-1}}\varphi^{-(n-1)} \end{pmatrix} \begin{pmatrix} 1/\tilde{D}(z) & 0 \\ 0 & \tilde{D}(z) \end{pmatrix}. \tag{3.3}$$

For $z \in \Omega_1$, we define

$$\tilde{Y}(z) := \begin{pmatrix} \tilde{D}_\infty/2^n & 0 \\ 0 & 2^n/\tilde{D}_\infty \end{pmatrix} P_\alpha(z) \begin{pmatrix} 1/W(z) & 0 \\ 0 & W(z) \end{pmatrix}, \tag{3.4}$$

where

$$P_\alpha(z) := \sqrt{2\pi n \zeta^{1/2}} \begin{pmatrix} 1 & 0 \\ -\dfrac{\mathrm{i}}{2}z & 1 \end{pmatrix} Q_\alpha(z) \begin{pmatrix} I_\alpha(n\zeta^{1/2}) & \dfrac{\mathrm{i}}{\pi\sqrt{z^2-1}} K_\alpha(n\zeta^{1/2}) \\ \dfrac{\mathrm{i}}{2}\sqrt{z^2-1}\, I_{\alpha+1}(n\zeta^{1/2}) & \dfrac{1}{2\pi} K_{\alpha+1}(n\zeta^{1/2}) \end{pmatrix} \tag{3.5}$$

and

$$Q_\alpha(z) := \begin{pmatrix} \dfrac{1}{2}\left(\dfrac{W}{\tilde{D}} + \dfrac{\tilde{D}}{W}\right) & \dfrac{\mathrm{i}}{\sqrt{z^2-1}}\left(\dfrac{\tilde{D}}{W} - \dfrac{W}{\tilde{D}}\right) \\ -\dfrac{\mathrm{i}}{4}\sqrt{z^2-1}\left(\dfrac{\tilde{D}}{W} - \dfrac{W}{\tilde{D}}\right) & \dfrac{1}{2}\left(\dfrac{W}{\tilde{D}} + \dfrac{\tilde{D}}{W}\right) \end{pmatrix}. \tag{3.6}$$

Put $\tilde{\zeta}(z) := \log^2(-\varphi)$, $z \in \mathbb{C}\backslash[-1,\infty)$. For $z \in \Omega_2$, we define

$$\tilde{Y}(z) := \begin{pmatrix} \tilde{D}_\infty/2^n & 0 \\ 0 & 2^n/\tilde{D}_\infty \end{pmatrix} P_\beta(z) \begin{pmatrix} 1/\tilde{W}(z) & 0 \\ 0 & \tilde{W}(z) \end{pmatrix}, \tag{3.7}$$

where

$$P_\beta(z) := \sqrt{2\pi n \tilde{\zeta}^{1/2}} \begin{pmatrix} 1 & 0 \\ -\frac{\mathrm{i}}{2}z & 1 \end{pmatrix} Q_\beta(z) \begin{pmatrix} I_\beta(n\tilde{\zeta}^{1/2}) & \frac{\mathrm{i}}{\pi\sqrt{z^2-1}} I_{\beta+1}(n\tilde{\zeta}^{1/2}) \\ \frac{\mathrm{i}}{2}\sqrt{z^2-1}\, I_{\beta+1}(n\tilde{\zeta}^{1/2}) & \frac{1}{2\pi} K_{\beta+1}(n\tilde{\zeta}^{1/2}) \end{pmatrix} \tag{3.8}$$

and

$$Q_\beta(z) := \begin{pmatrix} \frac{1}{2}\left(\frac{\tilde{W}}{\tilde{D}} + \frac{\tilde{D}}{\tilde{W}}\right) & \frac{\mathrm{i}}{\sqrt{z^2-1}}\left(\frac{\tilde{D}}{\tilde{W}} - \frac{\tilde{W}}{\tilde{D}}\right) \\ -\frac{\mathrm{i}}{4}\sqrt{z^2-1}\left(\frac{\tilde{D}}{\tilde{W}} - \frac{\tilde{W}}{\tilde{D}}\right) & \frac{1}{2}\left(\frac{\tilde{W}}{\tilde{D}} + \frac{\tilde{D}}{\tilde{W}}\right) \end{pmatrix}. \tag{3.9}$$

Note that for $z \in U_\delta \backslash (-\infty, 1]$ we have

$$\frac{W(z)}{\tilde{D}(z)} = \varphi(z)^{\alpha/2+\beta/2+1/2} \exp\left( \frac{(z^2-1)^{1/2}}{4\pi\mathrm{i}} \int_{\partial U_\delta} \frac{\log h(\zeta)}{\sqrt{\zeta^2-1}} \frac{\mathrm{d}\zeta}{\zeta - z} \right), \tag{3.10}$$

and for $z \in U_\delta \backslash [-1, \infty)$ we have

$$\frac{\tilde{W}(z)}{\tilde{D}(z)} = (-\varphi(z))^{\alpha/2+\beta/2+1/2} \exp\left( \frac{(z^2-1)^{1/2}}{4\pi\mathrm{i}} \int_{\partial U_\delta} \frac{\log h(\zeta)}{\sqrt{\zeta^2-1}} \frac{\mathrm{d}\zeta}{\zeta - z} \right); \tag{3.11}$$

see Kuijlaars *et al.* (2004, Lemma 6.4). The matrices $Q_\alpha(z)$ and $Q_\beta(z)$ are both analytic in $U_\delta$.

The matrix $\tilde{Y}(z)$, so defined, satisfies the following RHP.

**RHP for $\tilde{Y}$:**

(i) $\tilde{Y}(z)$ is analytic in $\mathbb{C}\backslash\Gamma$;

(ii) $\tilde{Y}(z)$ has the jump conditions

$$\tilde{Y}_+(x) = \tilde{Y}_-(x) \begin{pmatrix} 1 & w(x) \\ 0 & 1 \end{pmatrix}, \quad x \in (-1, 1), \tag{3.12}$$

$$\tilde{Y}_+(z) = M(z)\tilde{Y}_-(z), \quad z \in \Gamma_1 \cup \Gamma_3, \tag{3.13}$$

where the jump matrix $M(z)$ has a uniform asymptotic expansion

$$(2^n/\tilde{D}_\infty)^{\sigma_3} M(z) (\tilde{D}_\infty/2^n)^{\sigma_3} \sim I + \sum_{k=1}^{\infty} \frac{M_k(z)}{n^k} \tag{3.14}$$

for $z \in \Gamma_1 \cup \Gamma_3$; for each $k=1,2,\ldots$, $\|M_k(z)\|$ is bounded for $z \in \Gamma_1 \cup \Gamma_3$ and $\|\cdot\|$ can be any matrix norm; in (3.13) we have put the jump matrix on the left of $\tilde{Y}_-$, instead of on the right, in order to have the jump matrix of $Y(z)\tilde{Y}^{-1}(z)$ to appear on the right (see §4);

(iii) as $z \to \infty$,

$$\tilde{Y}(z) = \left(I + O\left(\frac{1}{z}\right)\right)\begin{pmatrix} z^n & 0 \\ 0 & z^{-n} \end{pmatrix}; \tag{3.15}$$

(iv) near $z=1$,

$$\tilde{Y}(z) = \begin{cases} O\begin{pmatrix} 1 & 1 \\ 1 & 1 \end{pmatrix}, & \text{if } \alpha \neq 0, \\ O\begin{pmatrix} 1 & \log|z-1| \\ 1 & \log|z-1| \end{pmatrix}, & \text{if } \alpha = 0; \end{cases} \tag{3.16}$$

(v) near $z=-1$,

$$\tilde{Y}(z) = \begin{cases} O\begin{pmatrix} 1 & 1 \\ 1 & 1 \end{pmatrix}, & \text{if } \beta \neq 0, \\ O\begin{pmatrix} 1 & \log|z+1| \\ 1 & \log|z+1| \end{pmatrix}, & \text{if } \beta = 0. \end{cases} \tag{3.17}$$

We first establish the jump condition (3.12). Note that $(z-1)^{-\alpha/2}I_\alpha[n\zeta^{1/2}(z)]$ and $(z-1)^{\alpha/2}K_\alpha[n\zeta^{1/2}(z)]$ are analytic for $z \in \mathbb{C}\backslash(-\infty,-1]$. Hence, it is easy to show that

$$(P_\alpha(x))_+ = (P_\alpha(x))_- \begin{pmatrix} e^{(\alpha+(1/2))\pi i} & \dfrac{1}{\sqrt{1-x^2}} \\ 0 & e^{-(\alpha+(1/2))\pi i} \end{pmatrix} \tag{3.18}$$

for $-1+\varepsilon < x < 1$, and

$$(P_\beta(x))_+ = (P_\beta(x))_- \begin{pmatrix} e^{-(\beta+(1/2))\pi i} & \dfrac{1}{\sqrt{1-x^2}} \\ 0 & e^{(\beta+(1/2))\pi i} \end{pmatrix} \tag{3.19}$$

for $-1 < x < -1+\varepsilon$. Furthermore,

$$W_+(x) = e^{(\alpha/2+1/4)\pi i}(1-x)^{\alpha/2+1/4}(1+x)^{\beta/2+1/4}h^{1/2}(x),$$
$$W_-(x) = e^{-(\alpha/2+1/4)\pi i}(1-x)^{\alpha/2+1/4}(1+x)^{\beta/2+1/4}h^{1/2}(x)$$

for $-1+\varepsilon < x < 1$, and

$$\tilde{W}_+(x) = e^{-(\beta/2+1/4)\pi i}(1-x)^{\alpha/2+1/4}(1+x)^{\beta/2+1/4}h^{1/2}(x),$$
$$\tilde{W}_-(x) = e^{(\beta/2+1/4)\pi i}(1-x)^{\alpha/2+1/4}(1+x)^{\beta/2+1/4}h^{1/2}(x)$$

for $-1<x<-1+\varepsilon$. Hence,

$$\tilde{Y}_-^{-1}(x)\,\tilde{Y}_+(x) = \begin{pmatrix} 1 & w(x) \\ 0 & 1 \end{pmatrix}.$$

Next, we verify the jump conditions (3.13) and (3.14). For $z\in\Gamma_1$, we have, from (3.5) and (3.6),

$$P_\alpha(z)\begin{pmatrix} \dfrac{\tilde{D}}{W} & 0 \\ 0 & \dfrac{W}{\tilde{D}} \end{pmatrix} = \begin{pmatrix} p_{11} & p_{12} \\ p_{21} & p_{22} \end{pmatrix}, \tag{3.20}$$

where

$$p_{11} = \frac{1}{2}\sqrt{2\pi n\zeta^{1/2}}\left[\left(1+\frac{\tilde{D}^2}{W^2}\right)I_\alpha(n\zeta^{1/2}) + \left(1-\frac{\tilde{D}^2}{W^2}\right)I_{\alpha+1}(n\zeta^{1/2})\right], \tag{3.21}$$

$$p_{12} = \frac{\mathrm{i}\sqrt{2\pi n\zeta^{1/2}}}{2\pi\sqrt{z^2-1}}\left[\left(1+\frac{W^2}{\tilde{D}^2}\right)K_\alpha(n\zeta^{1/2}) + \left(1-\frac{W^2}{\tilde{D}^2}\right)K_{\alpha+1}(n\zeta^{1/2})\right], \tag{3.22}$$

$$p_{21} = -\frac{\mathrm{i}}{4}\sqrt{2\pi n\zeta^{1/2}}\left[\left(\frac{1}{\varphi}+\frac{\varphi\tilde{D}^2}{W^2}\right)I_\alpha(n\zeta^{1/2}) + \left(\frac{1}{\varphi}-\frac{\varphi\tilde{D}^2}{W^2}\right)I_{\alpha+1}(n\zeta^{1/2})\right], \tag{3.23}$$

$$p_{22} = \frac{\sqrt{2\pi n\zeta^{1/2}}}{4\pi\sqrt{z^2-1}}\left[\left(\varphi+\frac{W^2}{\varphi\tilde{D}^2}\right)K_\alpha(n\zeta^{1/2}) + \left(\varphi-\frac{W^2}{\varphi\tilde{D}^2}\right)K_{\alpha+1}(n\zeta^{1/2})\right]. \tag{3.24}$$

Also, from (3.3) and (3.4) we have

$$\begin{aligned}(2^n/\tilde{D}_\infty)^{\sigma_3}M(z)(\tilde{D}_\infty/2^n)^{\sigma_3} &= \begin{pmatrix} \varphi^n & \dfrac{\mathrm{i}}{\sqrt{z^2-1}}\varphi^{-n} \\ -\dfrac{\mathrm{i}}{2}\varphi^{n-1} & \dfrac{1}{2\sqrt{z^2-1}}\varphi^{-(n-1)} \end{pmatrix}\begin{pmatrix} \dfrac{W}{\tilde{D}} & 0 \\ 0 & \dfrac{\tilde{D}}{W} \end{pmatrix}P_\alpha^{-1}(z) \\ &:= \begin{pmatrix} m_{11} & m_{12} \\ m_{21} & m_{22} \end{pmatrix},\end{aligned} \tag{3.25}$$

where

$$m_{11} = p_{22}\varphi^n - \frac{\mathrm{i}}{\sqrt{z^2-1}} p_{21}\varphi^{-n},$$

$$m_{12} = -p_{12}\varphi^n + \frac{\mathrm{i}}{\sqrt{z^2-1}} p_{11}\varphi^{-n},$$

$$m_{21} = -\frac{\mathrm{i}}{2} p_{22}\varphi^{n-1} - \frac{1}{2\sqrt{z^2-1}} p_{21}\varphi^{-n+1},$$

$$m_{22} = \frac{\mathrm{i}}{2} p_{12}\varphi^{n-1} + \frac{1}{2\sqrt{z^2-1}} p_{11}\varphi^{-n+1}.$$

Here we have made use of the identity

$$I_\alpha(z)K_{\alpha+1}(z) + K_\alpha(z)I_{\alpha+1}(z) = \frac{1}{z}.$$

From the well-known asymptotic expansions

$$K_\nu(z) \sim \sqrt{\frac{\pi}{2z}}\mathrm{e}^{-z}\sum_{n=0}^{\infty}\frac{(\nu,n)}{(2z)^n}, \quad |\arg z| < \frac{3\pi}{2},$$

and

$$I_\nu(z) \sim \frac{\mathrm{e}^z}{\sqrt{2\pi z}}\sum_{n=0}^{\infty}\frac{(-1)^n(\nu,n)}{(2z)^n} + \frac{\mathrm{e}^{-z+(\nu+1/2)\pi\mathrm{i}}}{\sqrt{2\pi z}}\sum_{n=0}^{\infty}\frac{(\nu,n)}{(2z)^n}, \quad -\frac{\pi}{2} < \arg z < \frac{3\pi}{2},$$

we can get (3.14) for $z \in \Gamma_1$. In a similar manner, we can demonstrate the jump condition (3.13)-(3.14) for $z \in \Gamma_3$.

## 4. Asymptotic expansion of $Y(z)$

Let $R(z) := (2^n/\tilde{D}_\infty)^{\sigma_3}Y(z)\tilde{Y}^{-1}(z)(\tilde{D}_\infty/2^n)^{\sigma_3}$. Note that $Y(z)$ and $\tilde{Y}(z)$ have the same jump matrix in $(-1,1)$; see (2.1) and (3.12). Hence, $R_+(x) = R_-(x)$ for $x \in (-1,1)$. Since the singularities of $R(z)$ at $z = \pm 1$ are removable, $R(z)$ can be analytically extended to $\mathbb{C}\backslash\Sigma_R$, where $\Sigma_R := \Gamma_1 \cup \Gamma_3$. On account of (3.13) and (3.14), $R$ is a solution of the following RHP.

**RHP for $R$:**

(i) $R(z)$ is analytic for $z \in \mathbb{C}\backslash\Sigma_R$;

(ii) $R(z)$ satisfies the jump condition

$$R_+(z) = R_-(z)(I + \Delta(z)) \text{ for } z \in \Sigma_R, \tag{4.1}$$

where

$$I + \Delta(z) = (2^n/\tilde{D}_\infty)^{\sigma_3}[M(z)]^{-1}(\tilde{D}_\infty/2^n)^{\sigma_3} \sim I + \sum_{s=1}^{\infty}\frac{\Delta_s(z)}{n^s} \text{ for } z \in \Sigma_R; \tag{4.2}$$

(iii) as $z \to \infty$,

$$R(z) = I + O\left(\frac{1}{z}\right). \tag{4.3}$$

As in theorem 7.10 of Deift *et al.* (1999), one can obtain from (4.1) an asymptotic expansion of the form

$$R(z) \sim I + \sum_{k=1}^{\infty} \frac{R_k(z)}{n^k} \tag{4.4}$$

as $n \to \infty$, which holds uniformly for $z \in \mathbb{C} \backslash \Sigma_R$. Indeed, we have the following result.

**Lemma 4.1.** *For every p, there exists a constant $C_p > 0$ such that*

$$\| R(z) - I - \sum_{k=1}^{p} \frac{R_k(z)}{n^k} \| \le \frac{C_p}{(1+|z|)n^{p+1}} \tag{4.5}$$

*for any z bounded away from $\Sigma_R$. Here, $\| \cdot \|$ denotes any matrix norm.*

*Proof.* Since $R_+ = R_- + R_- \varDelta$, by the Plemelj formula we have

$$R(z) = I + \frac{1}{2\pi \mathrm{i}} \int_{\Sigma_R} \frac{R_-(s)\varDelta(s)}{s-z} \mathrm{d}s \quad \text{for} \quad z \in \mathbb{C} \backslash \Sigma_R. \tag{4.6}$$

Inserting (4.2) and (4.4) in (4.6), and collecting powers of $1/n$, we obtain the recursive formula

$$R_k(z) = \frac{1}{2\pi \mathrm{i}} \int_{\Sigma_R} \sum_{j=1}^{k} (R_{k-j})_-(s) \varDelta_j(s) \frac{\mathrm{d}s}{s-z} \quad \text{for } z \in \mathbb{C} \backslash \Sigma_R, \tag{4.7}$$

$k = 1, 2, \ldots$, where $R_0 = I$. It is easy to show that

$$R_k(z) = O\left(\frac{1}{|z|}\right), \quad \text{as } z \to \infty, \quad \text{for } z \in \mathbb{C} \backslash \Sigma_R.$$

Let

$$E_p(z) = R(z) - I - \sum_{k=1}^{p} \frac{R_k(z)}{n^k}.$$

From (4.1) and (4.7), we have

$$(E_p)_+(s) = (E_p)_-(s)(I + \varDelta(s)) + F_p(s) \tag{4.8}$$

for $s \in \mathbb{C} \backslash \Sigma_R$, where

$$F_p(s) := \left(I + \sum_{k=1}^{p} \frac{(R_k)_-(s)}{n^k}\right) \varDelta(s) - \sum_{k=1}^{p} \frac{1}{n^k} \left(\sum_{j=1}^{k} (R_{k-j})_-(s) \varDelta_j(s)\right). \tag{4.9}$$

By the Plemelj formula,

$$E_p(z) = \frac{1}{2\pi \mathrm{i}} \int_{\Sigma_R} \frac{(E_p)_- \Delta}{s-z} \mathrm{d}s + \frac{1}{2\pi \mathrm{i}} \int_{\Sigma_R} \frac{F_p(s)}{s-z} \mathrm{d}s. \tag{4.10}$$

Starting with $E_p^{(0)}(z)=0$, and

$$E_p^{(l)}(z) = \frac{1}{2\pi \mathrm{i}} \int_{\Sigma_R} \frac{(E_p^{(l-1)})_- \Delta}{s-z} \mathrm{d}s + \frac{1}{2\pi \mathrm{i}} \int_{\Sigma_R} \frac{F_p(s)}{s-z} \mathrm{d}s \tag{4.11}$$

for $l=1,2,\ldots$, we can prove that the sequence $\{E_p^{(l)}(z)\}_{l=1}^{\infty}$ converges to a solution of equation (4.10) as $l \to \infty$, i.e.

$$\lim_{l\to\infty} E_p^{(l)}(z) = \tilde{E}_p(z) \quad \text{for } z \in \mathbb{C} \backslash \Sigma_R, \tag{4.12}$$

and $\tilde{E}_p(z)$ is a solution of equation (4.10). Moreover,

$$F_p(s) = \left(I + \sum_{k=1}^{p} \frac{(R_k)_-(s)}{n^k}\right)\left(\Delta(s) - \sum_{k=1}^{p} \frac{\Delta_k(s)}{n^k}\right) + \sum_{k=p+1}^{2p} \left[\frac{1}{n^k}\left(\sum_{j=k-p}^{p} (R_{k-j})_-(s)\Delta_j(s)\right)\right], \tag{4.13}$$

the matrix norm of which is bounded by $1/(1+|s|)n^{p+1}$ as $n \to \infty$, uniformly for $s \in \Sigma_R$. Since

$$\tilde{E}_p^{(l)}(z) = \sum_{k=0}^{l-1} [\tilde{E}_p^{(k+1)}(z) - \tilde{E}_p^{(k)}(z)],$$

by applying a usual argument in successive approximation, it can be shown that there exists a constant $C_p > 0$ such that

$$\|\tilde{E}_p(z)\| \le \frac{C_p}{(1+|z|)n^{p+1}} \tag{4.14}$$

for any $z$ bounded away from $\Sigma_R$. Let

$$\tilde{R}_p(z) := \tilde{E}_p(z) + I + \sum_{k=1}^{p} \frac{R_k(z)}{n^k}.$$

Note that $\tilde{R}_p(z)$ is a solution of the RHP for $R$; thus, $\tilde{R}_p(z)\tilde{Y}(z)$ can be analytically extended to $\mathbb{C} \backslash [-1,1]$, and satisfies the conditions of the RHP for $Y$. By uniqueness, $Y(z) = \tilde{R}_p(z)\tilde{Y}(z)$ or, equivalently, we have $R(z) = \tilde{R}_p(z)$ and $E_p(z) = \tilde{E}_p(z)$. This completes the proof of the lemma. ∎

By the lemma, we have

$$Y(z) \sim (\tilde{D}_\infty/2^n)^{\sigma_3}\left(I+\sum_{k=1}^{\infty}\frac{R_k(z)}{n^k}\right)(2^n/\tilde{D}_\infty)^{\sigma_3}\tilde{Y}(z), \tag{4.15}$$

uniformly for any $z$ bounded away from $\Sigma_R$. Let $R_{ij}(z)$, $i$, $j=1,2$, denote the elements in the matrix $R(z)$. Then, in particular, we have

$$\begin{aligned}\pi_n(z) = {}& \sqrt{2\pi n\zeta^{1/2}}\,\tilde{D}_\infty/2^n\, W^{-1/2}(z)\{I_\alpha(n\zeta^{1/2})A(z)\\ &+\sqrt{z^2-1}\,I_{\alpha+1}(n\zeta^{1/2})B(z)\}\end{aligned} \tag{4.16}$$

for $z\in\Omega_1$, where

$$\begin{aligned}A(z) = {}& \frac{R_{11}(z)}{2}\left(\frac{W}{\tilde{D}}+\frac{\tilde{D}}{W}\right)\\ &+\frac{\mathrm{i}R_{12}(z)}{2}\left[\sqrt{z^2-1}\left(\frac{W}{\tilde{D}}-\frac{\tilde{D}}{W}\right)-\frac{z}{2}\left(\frac{W}{\tilde{D}}+\frac{\tilde{D}}{W}\right)\right],\end{aligned} \tag{4.17}$$

and

$$\begin{aligned}B(z) = {}& \frac{R_{11}(z)}{2\sqrt{z^2-1}}\left(\frac{W}{\tilde{D}}-\frac{\tilde{D}}{W}\right)\\ &+\frac{\mathrm{i}R_{12}(z)}{2}\left[\frac{z}{2}\left(\frac{W}{\tilde{D}}+\frac{\tilde{D}}{W}\right)-\frac{z}{\sqrt{z^2-1}}\left(\frac{W}{\tilde{D}}-\frac{\tilde{D}}{W}\right)\right].\end{aligned} \tag{4.18}$$

Here, $A(z)$ and $B(z)$ are analytic for $z\in U_\delta\backslash(-\infty,-1]$, and have uniform asymptotic expansions of the form

$$A(z)\sim\sum_{k=0}^{\infty}\frac{A_k(z)}{n^k},\quad B(z)\sim\sum_{k=0}^{\infty}\frac{B_k(z)}{n^k}, \tag{4.19}$$

with

$$A_0(z)=\frac{1}{2}(W/\tilde{D}+\tilde{D}/W),\quad B_0(z)=\frac{1}{2\sqrt{z^2-1}}(W/\tilde{D}-\tilde{D}/W)$$

for any $z\in U_\delta$ bounded away from $(-\infty,-1]$.

Note that $\zeta^{1/4}W^{-1/2}(z)I_\alpha(n\zeta^{1/2})$ and $\zeta^{1/4}W^{-1/2}(z)\sqrt{z^2-1}\,I_{\alpha+1}(n\zeta^{1/2})$ are analytic functions for $z\in U_\delta\backslash(-\infty,-1]$. By analytic continuation, (4.16) holds for $z\in U_\delta$ and $z$ bounded away from $(-\infty,-1]$. In particular, when $z=x=\cos\theta$, $\theta\in(0,\pi-\delta)$, we have $\zeta^{1/2}(x\pm 0\mathrm{i})=\pm\theta\mathrm{i}$. Since $I_\alpha(\mathrm{i}x)=\mathrm{e}^{\alpha\pi\mathrm{i}/2}J_\alpha(x)$ for $x\in\mathbb{R}$, it follows that

$$
\begin{aligned}
&w^{1/2}(\cos\theta)\left(\frac{\sin\theta}{\theta}\right)^{1/2}\pi_n(\cos\theta) \\
&\quad\sim \sqrt{2\pi}\tilde{D}_\infty/2^n\left[n^{1/2}J_\alpha(n\theta)\sum_{s=0}^{\infty}\frac{A_s(-\theta^2)}{n^s} - n^{1/2}\sin\theta J_{\alpha+1}(n\theta)\sum_{s=0}^{\infty}\frac{B_s(-\theta^2)}{n^s}\right],
\end{aligned} \tag{4.20}
$$

which agrees with the solution given in Wang & Wong (2005). Here $J_\alpha$ is the Bessel function of the first kind.

We would like to thank Professor Arno Kuijlaars for sending us a copy of Kuijlaars *et al.* (2004) prior to its publication. We are also grateful to the referees for making some very helpful suggestions. In particular, we thank one of the referees for pointing out a mathematical problem that was not addressed satisfactorily in the earlier version of this paper. The work of the second author is partially supported by the Research Grant Council of Hong Kong under project 9040980.

## References

Deift, P. & Zhou, X. 1993 A steepest descent method for oscillatory Riemann–Hilbert problems, applications for the MKdV equation. *Ann. Math.* **137**, 295–368.

Deift, P., Kriecherbauer, T., McLaughlin, K. T.-R., Venakides, S. & Zhou, X. 1999 Strong asymptotics of orthogonal polynomials with respect to exponential weights. *Commun. Pure Appl. Math.* **52**, 1491–1552.

Fokas, A. S., Its, A. R. & Kitaev, A. V. 1991 Discrete Painlevé equations and their appearance in quantum gravity. *Commun. Math. Phys.* **142**, 313–344.

Kuijlaars, A. B. J. & Vanlessen, M. 2002 Universality for eigenvalue correlations from the modified Jacobi unitary ensemble. *Int. Math. Res. Notices* **2002**, 1575–1600. (doi:10.1155/s1073792802203116.)

Kuijlaars, A. B. J., McLaughlin, K. T.-R., Van Assche, W. & Vanlessen, M. 2004 The Riemann–Hilbert approach to strong asymptotics for orthogonal polynomials on $[-1, 1]$. *Adv. Math.* **188**, 337–398.

Szegő, G. 1975 *Orthogonal polynomials*, vol. 23, 4th edn. Providence, RI: Colloquium Publications, American Mathematical Society.

Wang, Z. & Wong, R. 2005 Linear difference equations with transition points. *Math. Comp.* **74**, 629–653.

# Asymptotic Analysis of a Perturbed Periodic Solution for the KdV Equation

*By X. H. Jiang and R. Wong*

We consider the solution of the Korteweg–de Vries (KdV) equation

$$v_t + vv_x + \beta v_{xxx} = 0, \quad -\infty < x < \infty, \quad t \geq 0,$$

with periodic initial value

$$v(x, 0) = C + \varepsilon(\mu + A \sin kx), \quad -\infty < x < \infty,$$

where $C$, $A$, $k$, $\mu$, and $\beta$ are constants. The solution is shown to be uniformly bounded for all small $\varepsilon$, and a formal expansion is constructed for the solution via the method of multiple scales. By using the energy method, we show that for any given number $T > 0$, the difference between the true solution $v(x, t; \varepsilon)$ and the $N$th partial sum of the asymptotic series is bounded by $\varepsilon^{N+1}$ multiplied by a constant depending on $T$ and $N$, for all $-\infty < x < \infty$, $0 \leq t \leq T/\varepsilon$, and $0 \leq \varepsilon \leq \varepsilon_0$.

## 1. Introduction

The Korteweg–de Vries (KdV) equation is a typical nonlinear wave equation. It was first obtained by Korteweg and de Vries in 1895 in the study of long water waves in a rectangular canal. Since then, many papers have been published

Address for correspondence: R. Wong, Department of Mathematics, City University of Hong Kong, Tat Chee Avenue, Kowloon, Hong Kong; e-mail: rscwong@cityu.edu.hk

*STUDIES IN APPLIED MATHEMATICS* 116:21–33 

Published by Blackwell Publishing, 350 Main Street, Malden, MA 02148, USA, and 9600 Garsington Road, Oxford, OX4 2DQ, UK.

in the study of mechanics, hydrodynamics, etc. A very important result for KdV equation was given by Miura et al. [1] who discovered that there are infinitely many conservation laws for this equation. In another paper [2], they introduced the famous inverse-scattering method to find the exact solutions of the KdV equation. This method is very powerful and has since been used widely. Later on, Lax [3] used the mathematical machinery in the theory of operator algebra to show why the method of inverse-scattering works and how to construct the conservation laws. At the same time, much attention has been paid to the periodic solution of the KdV equation; see, e.g., Bourgain [4], Lax [5], and Zabusky and Kruskal [6].

In this paper, we consider a perturbed periodic initial-value problem for KdV equation

$$\begin{cases} \dfrac{\partial v}{\partial t} + v\dfrac{\partial v}{\partial x} + \beta\dfrac{\partial^3 v}{\partial x^3} = 0, \\ v(x,0) = C + \varepsilon f(x), \end{cases}$$

where $f(x)$ is a $2\pi$-periodic function, $\varepsilon$ is a small positive parameter, and $C$, $\beta$ are constants. The problem describes the evolution of the perturbation around a constant state $C$. For the existence and uniqueness of the solution of this problem, we refer to [4] and [7]. For simplicity, we assume $C = 1$ and consider only the simplest case $f(x) = \mu + A\sin kx$, where $\mu$, $A$, and $k$ are constants. A change of variable $v = 1 + \varepsilon u$ gives the following equivalent perturbed equation:

$$\frac{\partial u}{\partial t} + \frac{\partial u}{\partial x} + \varepsilon u\frac{\partial u}{\partial x} + \beta\frac{\partial^3 u}{\partial x^3} = 0 \tag{1}$$

with initial data

$$u(x,0) = \mu + A\sin kx. \tag{2}$$

Our aim is to obtain an asymptotic expansion for the solution as $\varepsilon \to 0$. As is well known, if one starts with a regular expansion

$$u(x,t,\varepsilon) = \sum_{n=0}^{\infty} u_n(x,t)\varepsilon^n,$$

secular terms will appear in $u_1(x,t)$ and consequently the expansion is valid only for $\varepsilon t \ll 1$. To enlarge the time interval, as suggested in Kevorkian and Cole [8] and in Holms [9], we use the method of two-timing to construct an expansion for the solution of the problem (1)–(2), and use rigorous analysis to show that the expansion is indeed asymptotic. The main result of this paper is stated below.

THEOREM 1. *The solution of Equation* (1) *with initial conditions* (2) *exists globally and is uniformly bounded for all* $t > 0$. *Furthermore, for any given* $T > 0$ *and any integer* $N \geq 1$, *there is a constant* $M(T, N, \varepsilon_0) > 0$ *such that*

$$\left| u(x,t;\varepsilon) - \sum_{n=0}^{N-1} u_n(x,t,\varepsilon t)\varepsilon^n \right| \leq M(T,N,\varepsilon_0)\varepsilon^N \tag{3}$$

*for* $0 \leq t \leq T/\varepsilon$ *and* $0 \leq \varepsilon \leq \varepsilon_0$, *where* $\varepsilon_0$ *is a small constant,*

$$u_0(x,t,\varepsilon t) = \mu + A\sin(kx - \omega t - \mu k t \varepsilon), \tag{4}$$

$\omega = k - \beta k^3$, *and* $u_n(x, t, \varepsilon t)$ $(n \geq 1)$ *can be determined successively as in Section* 2.

The present paper is organized as follows. In Section 2, we show that all coefficient functions $u_n(x, t, \varepsilon t)$ are bounded for all $x \in (-\infty, \infty)$ and $0 < t \leq T/\varepsilon$. In Section 3, based on the conservation laws for KdV equation, we use the energy method to obtain a uniform bound of the solution $u(x, t; \varepsilon)$ for all $-\infty < x < \infty$ and $t \geq 0$. The final section is devoted to estimate the error term associated with the asymptotic expansion.

## 2. Construction of $u_n(x, t, \varepsilon t)$ in (3)

To construct $u_n(x, t, \varepsilon t)$, we introduce the following two-time scales:

$$s = t \quad \text{and} \quad \tau = \varepsilon t.$$

Clearly,

$$\frac{\partial}{\partial t} = \frac{\partial}{\partial s} + \varepsilon\frac{\partial}{\partial \tau}.$$

In terms of the new variables, the solution can be written as $u(x, t; \varepsilon) = u(x, s, \tau; \varepsilon)$. Assume that we have a formal series expansion

$$u(x,s,\tau;\varepsilon) = \sum_{n=0}^{\infty} u_n(x,s,\tau)\varepsilon^n. \tag{5}$$

With $u_{-1} \equiv 0$, the following relation holds:

$$\frac{\partial u}{\partial t} = \sum_{n=0}^{\infty} \varepsilon^n \left( \frac{\partial u_n}{\partial s} + \frac{\partial u_{n-1}}{\partial \tau} \right). \tag{6}$$

Substituting (5) and (6) into (1)–(2), and comparing the coefficients of like powers of $\varepsilon$, we obtain

$$\begin{cases} \dfrac{\partial u_0}{\partial s} + \dfrac{\partial u_0}{\partial x} + \beta \dfrac{\partial^3 u_0}{\partial x^3} = 0, \\ u_0(x, 0, 0) = \mu + A \sin kx; \end{cases} \tag{7}$$

$$\begin{cases} \dfrac{\partial u_1}{\partial s} + \dfrac{\partial u_1}{\partial x} + \beta \dfrac{\partial^3 u_1}{\partial x^3} = -\dfrac{\partial u_0}{\partial \tau} - u_0 \dfrac{\partial u_0}{\partial x}, \\ u_1(x, 0, 0) = 0; \end{cases} \tag{8}$$

and for $n \geq 2$

$$\begin{cases} \dfrac{\partial u_n}{\partial s} + \dfrac{\partial u_n}{\partial x} + \beta \dfrac{\partial^3 u_n}{\partial x^3} = -\dfrac{\partial u_{n-1}}{\partial \tau} - \left( u_0 \dfrac{\partial u_{n-1}}{\partial x} + \cdots + u_{n-1} \dfrac{\partial u_0}{\partial x} \right), \\ u_n(x, 0, 0) = 0. \end{cases} \tag{9}$$

The solution of (7) is given by

$$u_0(x, s, \tau) = \mu_0(\tau) + A_0(\tau) \sin[kx - \omega s + \varphi_0(\tau)], \tag{10}$$

where $\omega = k - \beta k^3$, $A_0(0) = A$, and $\varphi_0(0) = 0$. Inserting (10) into (8) gives

$$\frac{\partial u_1}{\partial s} + \frac{\partial u_1}{\partial x} + \beta \frac{\partial^3 u_1}{\partial x^3} = b(x, s, \tau), \tag{11}$$

where

$$\begin{aligned} b(x, s, \tau) = & -\mu_0'(\tau) - A_0'(\tau) \sin[kx - \omega s + \varphi_0(\tau)] \\ & - [k\mu_0(\tau)A_0(\tau) + A_0(\tau)\varphi_0'(\tau)] \cos[kx - \omega s + \varphi_0(\tau)] \\ & - \frac{k}{2} A_0^2(\tau) \sin[2kx - 2\omega s + 2\varphi_0(\tau)]. \end{aligned} \tag{12}$$

To avoid secular terms occurring in $u_1(x, s, \tau)$, we must eliminate all resonance terms on the right-hand side of (11). Hence, we set

$$\mu_0'(\tau) = 0, \quad A_0'(\tau) = 0, \tag{13}$$

$$k\mu_0(\tau)A_0(\tau) + A_0(\tau)\varphi_0'(\tau) = 0. \tag{14}$$

From the initial conditions, we obtain

$$\mu_0(\tau) = \mu, \quad A_0(\tau) = A, \quad \varphi_0(\tau) = -k\mu\tau. \tag{15}$$

This gives

$$u_0(x, s, \tau) = \mu + A \sin(kx - \omega s - k\mu\tau). \tag{16}$$

Using (13)–(15), (8) becomes

$$\begin{cases} \dfrac{\partial u_1}{\partial s} + \dfrac{\partial u_1}{\partial x} + \beta \dfrac{\partial^3 u_1}{\partial x^3} = -\dfrac{1}{2} k A^2 \sin(2kx - 2\omega s - 2k\mu\tau), \\ u_1(x, 0, 0) = 0. \end{cases} \tag{17}$$

The solution of (17) is given by

$$\begin{aligned} u_1(x, s, \tau) = & \Big[ A_1(\tau) \cos 2\gamma s + B_1(\tau) \sin 2\gamma s \\ & - \frac{kA^2\gamma}{4(\omega^2 - \gamma^2)} \sin(2\omega s + 2k\mu\tau) \Big] \sin 2kx \\ & + \Big[ B_1(\tau) \cos 2\gamma s - A_1(\tau) \sin 2\gamma s \\ & + \frac{kA^2(\omega^2 - \omega\gamma - \gamma^2)}{4\gamma(\omega^2 - \gamma^2)} \cos(2\omega s + 2k\mu\tau) \Big] \cos 2kx, \end{aligned} \tag{18}$$

where $\gamma = k - 4\beta k^3$, $A_1(0) = 0$, and

$$B_1(0) = -\frac{kA^2(\omega^2 - \omega\gamma - \gamma^2)}{4\gamma(\omega^2 - \gamma^2)}. \tag{19}$$

The functions $A_1(\tau)$ and $B_1(\tau)$ are to be chosen so that there will be no secular terms in $u_2$. We assume that

$$\begin{aligned} u_2(x, s, \tau) = & H_1(s, \tau) \sin kx + H_2(s, \tau) \cos kx + H_3(s, \tau) \sin 2kx \\ & + H_4(s, \tau) \cos 2kx + H_5(s, \tau) \sin 3kx + H_6(s, \tau) \cos 3kx. \end{aligned} \tag{20}$$

Substituting (16), (18), and (20) into (9) for $n = 2$ and comparing the coefficients of $\sin 2kx$ and $\cos 2kx$, we obtain

$$\begin{cases} \dfrac{\partial H_3}{\partial s} - 2\gamma H_4 = G_1(s, \tau), \\ \dfrac{\partial H_4}{\partial s} + 2\gamma H_3 = G_2(s, \tau), \end{cases} \tag{21}$$

where

$$\begin{aligned} G_1(s, \tau) = & \left( -\frac{\partial A_1}{\partial \tau} + 2k\mu B_1 \right) \cos 2\gamma s - \left( \frac{\partial B_1}{\partial \tau} + 2k\mu A_1 \right) \sin 2\gamma s \\ & + \frac{k^2 A^2 \mu\gamma}{2(\omega^2 - \gamma^2)} \cos(2\omega s + 2k\mu\tau) \\ & - \frac{k^2 A^2 \mu^2 (\omega^2 - \omega\gamma - \gamma^2)}{2\gamma(\omega^2 - \gamma^2)} \sin(2\omega s + 2k\mu\tau) \end{aligned} \tag{22}$$

and

$$G_2(s,\tau) = -\left(\frac{\partial B_1}{\partial\tau} + 2k\mu A_1\right)\cos 2\gamma s + \left(\frac{\partial A_1}{\partial\tau} - 2k\mu B_1\right)\sin 2\gamma s$$
$$+ \frac{k^2A^2\mu(\omega^2 - \omega\gamma - \gamma^2)}{2\gamma(\omega^2-\gamma^2)}\cos(2\omega s + 2k\mu\tau)$$
$$+ \frac{k^2A^2\mu\gamma}{2(\omega^2-\gamma^2)}\sin(2\omega s + 2k\mu\tau). \tag{23}$$

The eigenvalues of system (21) are $\pm 2\gamma i$. Hence, to avoid secular term occurring in $H_3$ and $H_4$, we set

$$\begin{cases} \dfrac{\partial A_1}{\partial\tau} - 2k\mu B_1 = 0, \\ \dfrac{\partial B_1}{\partial\tau} + 2k\mu A_1 = 0. \end{cases} \tag{24}$$

With the initial conditions $A_1(0) = 0$ and (19), we have

$$\begin{cases} A_1(\tau) = -\dfrac{kA^2(\omega^2-\omega\gamma-\gamma^2)}{4\gamma(\omega^2-\gamma^2)}\sin 2k\mu\tau, \\ B_1(\tau) = -\dfrac{kA^2(\omega^2-\omega\gamma-\gamma^2)}{4\gamma(\omega^2-\gamma^2)}\cos 2k\mu\tau. \end{cases} \tag{25}$$

Substituting (25) into (18), we obtain

$$u_1(x,s,\tau) = -\frac{kA^2(\omega^2-\omega\gamma-\gamma^2)}{4\gamma(\omega^2-\gamma^2)}$$
$$\times [\sin(2\gamma s + 2k\mu\tau) + \gamma^2\sin(2\omega s + 2k\mu\tau)]\sin 2kx$$
$$-\frac{kA^2(\omega^2-\omega\gamma-\gamma^2)}{4\gamma(\omega^2-\gamma^2)}$$
$$\times [\cos(2\gamma s + 2k\mu\tau) - \cos(2\omega s + 2k\mu\tau)]\cos 2kx. \tag{26}$$

The above process can be continued, and the coefficients $u_n(x, s, \tau)$ in (5) are all linear combinations of sin or cos functions of $x$, $s$, $\tau$. As in (7) and (8), we choose $\{u_1, \ldots, u_{n-1}\}$ so that there are no secular terms in $u_n(x, s, \tau)$. The unspecified functions in $u_{n-1}$, i.e., those correspond to $\mu_0(\tau)$, $A_0(\tau)$ in $u_0$ and $A_1(\tau)$, $B_1(\tau)$ in $u_1$, can be uniquely determined by requiring not to have secular terms in $u_n$. To see this, we return to Equation (9). The right-hand side of this equation is

$$-\frac{\partial u_{n-1}}{\partial\tau} - \left(u_0\frac{\partial u_{n-1}}{\partial x} + \cdots + u_{n-1}\frac{\partial u_0}{\partial x}\right).$$

To require $u_n$ not to have secular terms, we can see, as in (24), that any unspecified function in $u_{n-1}$ must satisfy a first-order linear ordinary differential equation with smooth coefficients. These equations can be solved uniquely, and the solutions are $C^\infty$-functions.

## 3. Uniform bound for the solution of (1)–(2)

In this section, our main aim is to obtain a uniform bound for the solution of (1)–(2) for small $\varepsilon$, say $0 < \varepsilon \leq \varepsilon_0$. The method used here can be also applied to estimate the remainder of the first few terms of the asymptotic solution; see Section 4.

For any fixed $t$, according to Sjöberg's result [7], the solution $u(x, t; \varepsilon)$ of (1)–(2) is a $2\pi/k$–periodic function. Integrating (1) from 0 to $2\pi/k$ gives

(i)
$$\frac{d}{dt}\int_0^{2\pi/k} u(x,t)\,dx = 0,$$

i.e.,

$$\int_0^{2\pi/k} u(x,t)\,dx = \int_0^{2\pi/k} (\mu + A\sin kx)\,dx = \frac{2\pi}{k}\mu. \tag{27}$$

As shown by Miura et al. [1], there are infinitely many conservation laws for (1), and Equation (27) is just the first one. Similarly, we can obtain

(ii)
$$\frac{d}{dt}\int_0^{2\pi/k} u^2(x,t)\,dx = 0,$$

that is

$$\int_0^{2\pi/k} u^2(x,t)\,dx = \int_0^{2\pi/k} (\mu + A\sin kx)^2\,dx. \tag{28}$$

(iii)
$$\frac{d}{dt}\int_0^{2\pi/k}\left[\left(\frac{\partial u}{\partial x}\right)^2 - \frac{\varepsilon}{3\beta}u^3\right]dx = 0,$$

$$\int_0^{2\pi/k}\left[\left(\frac{\partial u}{\partial x}\right)^2 - \frac{\varepsilon}{3\beta}u^3\right]dx = \int_0^{2\pi/k}\left[(kA\cos kx)^2 - \frac{\varepsilon}{3\beta}(\mu + A\sin kx)^3\right]dx. \tag{29}$$

Equation (29) can be obtained as follows. Multiplying (1) by $\partial^2 u/\partial x^2$ and integrating it over $[0, 2\pi/k]$, we get

$$\frac{d}{dt}\int_0^{2\pi/k}\left(\frac{\partial u}{\partial x}\right)^2 dx + \varepsilon\int_0^{2\pi/k}\left(\frac{\partial u}{\partial x}\right)^3 dx = 0. \tag{30}$$

Multiplying (1) by $3u^2$ and integrating it over $[0, 2\pi/k]$, one obtains

$$\frac{d}{dt}\int_0^{2\pi/k} u^3\,dx + 3\beta\int_0^{2\pi/k}\left(\frac{\partial u}{\partial x}\right)^3 dx = 0. \tag{31}$$

Coupling (30) and (31) gives (29).

In a similar manner, multiplying (1) by

$$2\frac{\partial^4 u}{\partial x^4} + \frac{5\varepsilon^2}{9\beta^2}u^3 + \frac{5\varepsilon}{3\beta}\left(\frac{\partial u}{\partial x}\right)^2 + \frac{10\varepsilon}{3\beta}u\frac{\partial^2 u}{\partial x^2},$$

integrating it with respect to $x$ from 0 to $2\pi/k$, and making use of an integration by parts, we have

(iv) $$\frac{d}{dt}\int_0^{2\pi/k}\left[\left(\frac{\partial^2 u}{\partial x^2}\right)^2 + \frac{5\varepsilon^2}{36\beta^2}u^4 - \frac{5\varepsilon}{3\beta}u\left(\frac{\partial u}{\partial x}\right)^2\right]dx = 0.$$

Therefore,

$$\begin{aligned}\int_0^{2\pi/k}&\left[\left(\frac{\partial^2 u}{\partial x^2}\right)^2 + \frac{5\varepsilon^2}{36\beta^2}u^4 - \frac{5\varepsilon}{3\beta}u\left(\frac{\partial u}{\partial x}\right)^2\right]dx\\ &= \int_0^{2\pi/k}\left[(k^2 A\sin kx)^2 + \frac{5\varepsilon^2}{36\beta^2}(\mu + A\sin kx)^4\right.\\ &\qquad\left. - \frac{5\varepsilon}{3\beta}(\mu + A\sin kx)(kA\cos kx)^2\right]dx. \end{aligned}\tag{32}$$

This is the fourth conservation law of Equation (1).

Now we can state the main theorem in this section.

THEOREM 2. *Suppose that* $u(x, t; \varepsilon)$ *is the solution to* (1)–(2). *Then there exist constants* $C_0(\varepsilon_0)$ *and* $C_1(\varepsilon_0)$ *such that*

$$|u(x, t; \varepsilon)| \le C_0(\varepsilon_0), \qquad \left|\frac{\partial u(x, t; \varepsilon)}{\partial x}\right| \le C_1(\varepsilon_0) \tag{33}$$

*for all* $x \in \mathbb{R}$, $t \ge 0$, *and* $0 < \varepsilon \le \varepsilon_0$.

*Proof*: For any fixed $t$, suppose that

$$M(t) = \max_{x\in\mathbb{R}}|u(x, t)| = \max_{x\in[0,2\pi/k]}|u(x, t)|. \tag{34}$$

Then there exists one point $x_0 = x_0(t) \in [0, 2\pi/k]$ such that $|u(x_0, t)| = M(t)$, and

$$M^2(t) = u^2(x_0, t) = u^2(x, t) + \int_x^{x_0} 2u \frac{\partial u}{\partial x} \, dx. \tag{35}$$

From this, it follows that

$$M^2(t) \le u^2(x, t) + \int_{x_0 - 2\pi/k}^{x_0} \left[ u^2 + \left( \frac{\partial u}{\partial x} \right)^2 \right] dx \tag{36}$$

for all $x_0 - 2\pi/k \le x \le x_0$. Integrating it with respect to $x$ from $x_0 - 2\pi/k$ to $x_0$, we obtain

$$\begin{aligned} M^2(t) &\le \left( \frac{k}{2\pi} + 1 \right) \int_{x_0 - 2\pi/k}^{x_0} u^2 \, dx + \int_{x_0 - 2\pi/k}^{x_0} \left( \frac{\partial u}{\partial x} \right)^2 dx \\ &= \left( \frac{k}{2\pi} + 1 \right) \int_0^{2\pi/k} u^2 \, dx + \int_0^{2\pi/k} \left( \frac{\partial u}{\partial x} \right)^2 dx. \end{aligned} \tag{37}$$

Note that

$$|u(x, t)|^3 \le M(t) u^2(x, t),$$

from which it follows that

$$\left| \frac{\varepsilon}{3\beta} \int_0^{2\pi/k} u^3(x, t) \, dx \right| \le \frac{1}{2} M^2(t) + \frac{1}{2} \left[ \frac{\varepsilon}{3\beta} \int_0^{2\pi/k} u^2(x, t) \, dx \right]^2. \tag{38}$$

Combining (28), (29), (37), and (38) gives

$$M(t) \le C_0(\varepsilon_0) \tag{39}$$

for all $t \ge 0$ and $0 < \varepsilon \le \varepsilon_0$, where

$$\begin{aligned} C_0(\varepsilon_0) := \Bigg\{ \int_0^{2\pi/k} \Bigg[ &\frac{k + 2\pi}{\pi} (\mu + A \sin kx)^2 \\ &+ 2(kA \cos kx)^2 + \frac{2\varepsilon_0}{|\beta|} |\mu + A \sin kx|^3 \Bigg] dx \\ &+ \left[ \frac{\varepsilon_0}{3\beta} \int_0^{2\pi/k} (\mu + A \sin kx)^2 \, dx \right]^2 \Bigg\}^{\frac{1}{2}}. \end{aligned} \tag{40}$$

Thus the first inequality in (33) is proved. Furthermore, together with (29) and (32), we can also obtain uniform bounds for $\|\partial u/\partial x\|_{L^2}$ and $\|\partial^2 u/\partial x^2\|_{L^2}$. These in turn will give the second inequality in (33). ■

## 4. Proof of Theorem 1

We only need to establish the inequality in (3). Suppose that we have constructed $u_0, u_1, \ldots, u_n$ $(n \geq 1)$. Then we can write

$$u(x,t;\varepsilon) = \sum_{j=0}^{n} \varepsilon^j u_j(x,t,\varepsilon t) + \varepsilon^n R_n. \tag{41}$$

Substituting (41) into (1)–(2), we obtain

$$\begin{aligned}
&\varepsilon^n \left[ \frac{\partial R_n}{\partial t} + \frac{\partial R_n}{\partial x} + \beta \frac{\partial^3 R_n}{\partial x^3} + \frac{\partial}{\partial x}\left( \varepsilon R_n \sum_{j=0}^{n} \varepsilon^j u_j \right) + \varepsilon^{n+1} R_n \frac{\partial R_n}{\partial x} \right] \\
&\quad = \varepsilon^{n+1} f_n(x,t,\varepsilon t;\varepsilon)
\end{aligned} \tag{42}$$

and

$$\varepsilon^n R_n(x,0;\varepsilon) = 0, \tag{43}$$

where

$$\begin{aligned}
\varepsilon^{n+1} f_n(x,t,\varepsilon t;\varepsilon) = &- \sum_{j=0}^{n} \varepsilon^j \left( \frac{\partial u_j}{\partial t} + \frac{\partial u_j}{\partial x} + \beta \frac{\partial^3 u_j}{\partial x^3} \right) \\
&- \varepsilon \left( \sum_{j=0}^{n} \varepsilon^j u_j \right) \left( \sum_{j=0}^{n} \varepsilon^j \frac{\partial u_j}{\partial x} \right).
\end{aligned} \tag{44}$$

Equations (42) and (43) can be rewritten as

$$\frac{\partial R_n}{\partial t} + \frac{\partial R_n}{\partial x} + \beta \frac{\partial^3 R_n}{\partial x^3} + \frac{\partial}{\partial x}\left( \varepsilon R_n \sum_{j=0}^{n} \varepsilon^j u_j \right) + \varepsilon^{n+1} R_n \frac{\partial R_n}{\partial x} = \varepsilon f_n(x,t,\varepsilon t;\varepsilon) \tag{45}$$

and

$$R_n(x,0;\varepsilon) = 0. \tag{46}$$

Now define $u_{-1} \equiv 0$ and $u_j \equiv 0$ for all $j \geq n+1$. As in Section 2, let $s = t$ and $\tau = \varepsilon t$. By using (1) and (9), it can be shown that

$$f_n(x,t,\varepsilon t;\varepsilon) = -\frac{\partial u_n}{\partial \tau} - \sum_{j=0}^{n} \varepsilon^j \left( \sum_{p=0}^{n-j} u_{n-p} \frac{\partial u_{j+p}}{\partial x} \right). \tag{47}$$

Because each $u_k$, $k = 0, 1, \ldots, n$, is a linear combination of sin or cos functions of $x$, $t$, and $t\varepsilon$, it is easily seen that $f_n(x,s,\tau;\varepsilon)$ is at most a linear combination of products of such functions. Hence for any $t > 0$ and $0 < \varepsilon < \varepsilon_0$, there exists a constant $M_1(n, \varepsilon_0)$ such that

$$\left|\sum_{j=0}^{n} \varepsilon^j \frac{\partial u_j}{\partial x}\right| \leq M_1(n, \varepsilon_0), \tag{48}$$

$$\left|\sum_{j=0}^{n} \varepsilon^j \frac{\partial^3 u_j}{\partial x^3}\right| \leq M_1(n, \varepsilon_0), \tag{49}$$

and

$$\int_0^{2\pi/k} \left[f_n^2 + \left(\frac{\partial f_n}{\partial x}\right)^2\right] dx \leq M_1(n, \varepsilon_0). \tag{50}$$

Now we turn to the estimation of the remainder $R_n$ in (41) by using the method of energy. Note that $R_n$, $\partial R_n/\partial x$, and $\partial^2 R_n/\partial x^2$ are also $2\pi/k$-periodic functions. Multiplying (45) by $2R_n - 2\partial^2 R_n/\partial x^2$ and integrating it with respect to $x$ from 0 to $2\pi/k$, we obtain

$$\begin{aligned}
&\frac{d}{dt}\int_0^{2\pi/k} \left[R_n^2 + \left(\frac{\partial R_n}{\partial x}\right)^2\right] dx + \varepsilon \int_0^{2\pi/k} \left[\left(\sum_{j=0}^{n} \varepsilon^j \frac{\partial u_j}{\partial x}\right)\right. \\
&\left. - \left(\sum_{j=0}^{n} \varepsilon^j \frac{\partial^3 u_j}{\partial x^3}\right)\right] R_n^2\, dx + \varepsilon \int_0^{2\pi/k} \left[3\left(\sum_{j=0}^{n} \varepsilon^j \frac{\partial u_j}{\partial x}\right) + \varepsilon^n \frac{\partial R_n}{\partial x}\right] \left(\frac{\partial R_n}{\partial x}\right)^2 dx \\
&= 2\varepsilon \int_0^{2\pi/k} \left[f_n R_n + \frac{\partial f_n}{\partial x}\frac{\partial R_n}{\partial x}\right] dx.
\end{aligned} \tag{51}$$

Note that

$$\left|3\left(\sum_{j=0}^{n} \varepsilon^j \frac{\partial u_j}{\partial x}\right) + \varepsilon^n \frac{\partial R_n}{\partial x}\right| = \left|2\left(\sum_{j=0}^{n} \varepsilon^j \frac{\partial u_j}{\partial x}\right) + \frac{\partial u}{\partial x}\right| \leq 2M_1(n, \varepsilon_0) + C_1(\varepsilon_0)$$

(see (33) and (48)), and the right-hand side of the Equation (51) is bounded by

$$\varepsilon \int_0^{2\pi/k} \left[f_n^2 + \left(\frac{\partial f_n}{\partial x}\right)^2\right] dx + \varepsilon \int_0^{2\pi/k} \left[R_n^2 + \left(\frac{\partial R_n}{\partial x}\right)^2\right] dx.$$

Hence

$$\frac{d}{dt}\int_0^{2\pi/k} \left[R_n^2 + \left(\frac{\partial R_n}{\partial x}\right)^2\right] dx \leq \varepsilon M_2(n, \varepsilon_0)\left\{1 + \int_0^{2\pi/k} \left[R_n^2 + \left(\frac{\partial R_n}{\partial x}\right)^2\right] dx\right\}, \tag{52}$$

where

$$M_2(n,\varepsilon_0) = 2M_1(n,\varepsilon_0) + C_1(\varepsilon_0) + 1.$$

By Gronwall's inequality, we have

$$\int_0^{2\pi/k} \left[ R_n^2 + \left( \frac{\partial R_n}{\partial x} \right)^2 \right] dx \le e^{\varepsilon t M_2(n,\varepsilon_0)} - 1 \le e^{T M_2(n,\varepsilon_0)} \tag{53}$$

for $\varepsilon t \le T$. Note that for any $|x - y| \le 2\pi/k$,

$$\begin{aligned} R_n^2(x,t;\varepsilon) - R_n^2(y,t;\varepsilon) &= \int_y^x 2R_n(\xi,t;\varepsilon)\, \frac{\partial R_n}{\partial \xi}(\xi,t,\varepsilon)\, d\xi \\ &\le \int_0^{2\pi/k} \left[ R_n^2 + \left( \frac{\partial R_n}{\partial \xi} \right)^2 \right] d\xi. \end{aligned} \tag{54}$$

Integrating it with respect to $y$ from $x$ to $x + 2\pi/k$, we obtain from (53) that

$$\begin{aligned} R_n(x,t;\varepsilon) &\le \left\{ \frac{k+2\pi}{2\pi} \int_0^{2\pi/k} \left[ R_n^2 + \left( \frac{\partial R_n}{\partial y} \right)^2 \right] dy \right\}^{\frac{1}{2}} \\ &\le \sqrt{\frac{k+2\pi}{2\pi}}\, e^{\frac{1}{2} T M_2(n,\varepsilon_0)} \end{aligned} \tag{55}$$

for all $-\infty < x < +\infty$, $0 \le \varepsilon \le \varepsilon_0$, and $0 \le \varepsilon t \le T$.

Because $u_n$ is bounded by some constant $C(n,\varepsilon_0)$ for $0 \le \varepsilon \le \varepsilon_0$, coupling (41) and (55) gives

$$\begin{aligned} \left| u(x,t;\varepsilon) - \sum_{j=0}^{n-1} \varepsilon^j u_j \right| &\le \varepsilon^n |u_n| + \varepsilon^n \sqrt{\frac{k+2\pi}{2\pi}}\, e^{\frac{1}{2} T M_2(n,\varepsilon_0)} \\ &\le \varepsilon^n M(T,n,\varepsilon_0), \end{aligned} \tag{56}$$

where

$$M(T,n,\varepsilon_0) := C(n,\varepsilon_0) + \sqrt{\frac{k+2\pi}{2\pi}}\, e^{\frac{1}{2} T M_2(n,\varepsilon_0)}.$$

This completes the proof of Theorem 1.

## References

1. R. M. Miura, C. S. Gardner, and M. D. Kruskal, The Korteweg-de Vries equation and generalizations, II. Existence of conservation laws and constants of motion, *J. Math. Phys.* 9:1204–1209 (1968).

2. C. S. Gardner, J. M. Greene, M. D. Kruskal, and R. M. Miura, Method for solving the Korteweg-de Vries equation, *Phys. Rev. Lett.* 19:1095–1097 (1967).
3. P. D. Lax, Integrals of nonlinear equations of evolution and solitary waves, *Comm. Pure Appl. Math.* 21:467–490 (1968).
4. J. Bourgain, On the Cauchy problem for periodic KdV-type equations, *J. Fourier Anal. Appl.* Special Issue:17–86 (1995).
5. P. D. Lax, Periodic solutions of the Korteweg-de Vries equation, *Comm. Pure Appl. Math.* 28:141–188 (1975).
6. N. J. Zabusky and M. D. Kruskal, Interaction of solitons in a collisionless plasma and the recurrence of initial states, *Phys. Rev. Lett.* 15:240–243 (1965).
7. A. Sjöberg, On the Korteweg-de Vries equation: Existence and uniqueness, *J. Math. Anal. Appl.* 29:569–579 (1970).
8. J. Kevorkian and J. D. Cole, *Multiple Scale and Singular Perturbation Methods*, Springer, New York, 1996.
9. M. H. Holms, *Introduction to Perturbation Methods*, Springer, New York, 1995.

Beijing University of Chemical Technology
City University of Hong Kong

(Received March 28, 2005)

# Asymptotic Solutions of a Fourth Order Differential Equation

*By R. Wong and H. Y. Zhang*

In this paper, we derive uniform asymptotic expansions of solutions to the fourth order differential equation

$$y^{(4)} + \lambda^2(xy'' + y) = 0,$$

where $x$ is a real variable and $\lambda$ is a large positive parameter. The solutions of this differential equation can be expressed in the form of contour integrals, and uniform asymptotic expansions are derived by using the cubic transformation introduced by Chester, Friedman, and Ursell in 1957 and the integration-by-part technique suggested by Bleistein in 1966. There are two advantages to this approach: (i) the coefficients in the expansion are defined recursively, and (ii) the remainder is given explicitly. Moreover, by using a recent method of Olde Daalhuis and Temme, we extend the validity of the uniform asymptotic expansions to include all real values of $x$.

## 1. Introduction

A major contribution to the study of the stability of viscous flows was made by Orr (1907) and Sommerfeld (1908), who derived the celebrated equation

$$(i\alpha R)^{-1}(\varphi^{(4)} - 2\alpha^2\varphi'' + \alpha^4\varphi) = (U - c)(\varphi'' - \alpha^2\varphi) - U''\varphi, \tag{1}$$

Address for correspondence: R. Wong, Department of Mathematics, City University of Hong Kong, 83 Tat Chee Avenue, Kowloon, Hong Kong; e-mail: rscwong@cityu.edu.hk.



Published by Blackwell Publishing, 350 Main Street, Malden, MA 02148, USA, and 9600 Garsington Road, Oxford, OX4 2DQ, UK.

which is now named after them. All the quantities involved in this equation are dimensionless; $U(\varsigma)$ is the basic velocity distribution, $R$ is the Reynolds number, $\alpha$, $c$ are constants describing the nature of the disturbance, and $\varphi(\varsigma)e^{i\alpha(x-ct)}$ is the stream function. Reynolds was the first to point out that when $R$ is large, for flows without an inflection point, viscosity may be the cause of instability. Such instability is called "weak" instability. Asymptotic approximation is therefore of importance in dealing with the Orr–Sommerfeld equation (see Reid [1]).

In attempting to develop an asymptotic theory for Equation (1), it would be natural to suppose that the relevant comparison equation should be of fourth order. This was the approach first used by Wasow [2–4] and later in greater generality by Lin and Rabenstein [5, 6]. Wasow [3] used

$$y^{(4)} + \lambda^2(xy'' + y) = 0 \tag{2}$$

as a comparison equation, and investigated the asymptotic behavior of its solutions as $\lambda \to +\infty$ by using the method of steepest descent. But, the intervals of validity of Wasow's expansions do not cover the entire real axis. One of his approximations involves a function (integral), which is not one of the known special functions of mathematical physics. Later, Rabenstein showed that the comparison equation must be of the form

$$\varepsilon^3 u^{(4)} - (xu'' + \alpha u' + \beta u) = 0, \tag{3}$$

where $\varepsilon$ is a small parameter and $\alpha$, $\beta$ are constants. However, his results are similar to those of Wasow. In this paper, we shall restudy the four linearly independent solutions of Equation (2), and derive asymptotic expansions for these solutions that are uniformly valid for $x$ in the whole real line. (This problem was considered extremely difficult by Reid [1, p. 382].) Our expansions involve only known special functions such as Airy functions and Bessel functions. Our approach is to use the cubic transformation introduced by Chester, Friedman, and Ursell [7] and the sequence of rational functions suggested by Olde Daalhuis and Temme [8]. These techniques were, of course, not available at the time of investigation by Wasow and Rabenstein. Our method can also be used to derive similar results for the four solutions of Rabenstein's equation (3). For simplicity, we shall just take $\alpha = 0$, $\beta = 1$ in Equation (3), and restrict ourselves to Wasow's equation (2).

The material in this paper is arranged as follows. In Section 2, we give integral representations to solutions of the linear differential equation. Then we summarize the key steps in the Chester–Friedman–Ursell method in Section 3, which includes the reduction of a contour integral into a canonical form by using a cubic transformation. In Section 4, we prove the cubic transformation in Section 3 is one-to-one and analytic in a domain containing the path of integration. The Bleistain technique of repeated integration by parts is reviewed in Section 5, which will lead to Airy-type expansions with explicit remainders.

In Section 6, we present the uniform asymptotic expansions, and compute the leading term coefficients. The last section is devoted to estimate the error terms in all these expansions.

## 2. Integral representations

By using Laplace transformation, we can express solutions to differential equation (2) in the form

$$y(x) = \int_L \tau^{-2} \exp\left(\tau x - \tau^{-1} + \frac{\lambda^{-2}\tau^3}{3}\right) \mathrm{d}\tau, \tag{4}$$

where the contour $L$ is either a closed curve or satisfies

$$\exp\left(\tau x - \tau^{-1} + \frac{\lambda^{-2}\tau^3}{3}\right) = 0 \tag{5}$$

at both endpoints of the curve. To describe $L$, we note that Re $(\tau^3) < 0$ for $\tau$ in one of the three sectors $T_k$, $k = 1, 2, 3$, defined by

$$T_k : -\frac{\pi}{2} + (k-1)\frac{2\pi}{3} < \arg \tau < -\frac{\pi}{6} + (k-1)\frac{2\pi}{3}; \tag{6}$$

see the shaded regions in Figure 1. Thus, we choose $L$ to begin and end at infinity and asymptotically approach the bisectors of two different sectors $T_k$, so that Equation (5) holds and the integral in Equation (4) is convergent. To fix

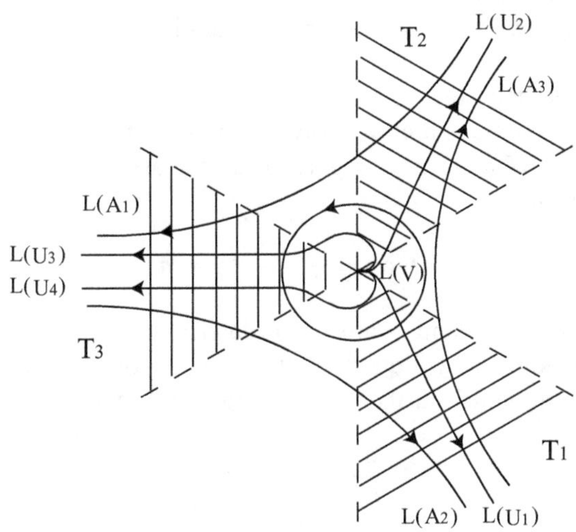


Figure 1. The eight integral paths in the $\tau$-plane.

this idea, let $L(A_1)$, $L(A_2)$, $L(A_3)$ be the three curves that lie entirely in a half-plane bounded by a dash line through the origin (see Figure 1). Let these three nonequivalent paths be combined to form a circular contour $L(V)$ around the origin described in the positive direction (see, again, Figure 1). Three more solutions $U_1$, $U_2$, and $U_3$ can be defined by choosing paths of integration $L(U_1)$, $L(U_2)$, $L(U_3)$ to start from zero with direction $\arg\tau = 0$, so that $\mathrm{Re}(\tau^{-1}) > 0$, and to end at infinity in the sectors $T_1$, $T_2$, and $T_3$, respectively. The last solution $U_4$ can be obtained by reflecting $L(U_3)$ with respect to the real axis (see Figure 1). The eight solutions are linked by the following identities:

$$A_1 + A_2 + A_3 = V, U_2 - U_1 = A_3, U_3 - U_2 = A_1, U_1 - U_4 = A_2. \tag{7}$$

From these identities, we also get $U_1 - U_3 = A_2 - V$, $U_4 - U_2 = A_1 - V$, and $U_3 - U_4 = V$ immediately.

## 3. Method of Chester, Friedman, and Ursell

Before constructing asymptotic expansions for the integral (4), let us first consider the more standard integral

$$I(\lambda; \alpha) = \int_C g(t) e^{-\lambda f(t;\alpha)}\, dt, \tag{8}$$

where $f$ and $g$ are analytic functions of the complex variable $t$ and $f$ is also an analytic function of the parameter $\alpha$. The position of the saddle points varies with $\alpha$, and we assume that there exists a critical value of $\alpha$, say $\alpha = \alpha_0$, such that for $\alpha \neq \alpha_0$ there are two distinct saddle points $t_+$ and $t_-$ of multiplicity 1, but when $\alpha = \alpha_0$ these two points coincide and give a single saddle point $t_0$ of multiplicity 2.

In view of the fact that the simplest phase function that exhibits two coalescing saddle points is a cubic polynomial, Chester, Friedman, and Ursell (1957) introduced what is now regarded as a classic method, a change of variable of the form

$$f(t, \alpha) = \frac{1}{3}u^3 - \zeta u + \eta, \tag{9}$$

where the coefficients $\zeta$ and $\eta$ are to be determined. In order to ensure the mapping $t \leftrightarrow u$ defined by Equation (9) is one-to-one and analytic, neither $dt/du$ nor $du/dt$ can vanish in the relevant regions. Since

$$\frac{dt}{du} = \frac{u^2 - \zeta}{f_t(t; \alpha)} \tag{10}$$

and $f_t(t; \alpha) = 0$ when $t = t_+$ and $t_-$, we must make $t_+$ and $t_-$ correspond to $\zeta^{1/2}$ and $-\zeta^{1/2}$, respectively.

The transformation (9) was shown by Chester, Friedman, and Ursell to be one-to-one and analytic for all $t$, $\alpha$ in a neighborhood of $t_0$, $\alpha_0$. More precisely, there is a theorem in [9], which says that with the above values of $\zeta$ and $\eta$, the transformation (9) has exactly one branch $u = u(t, \alpha)$ which can be expanded into a power series in $t$, with coefficients which are continuous in $\alpha$ for $\alpha$ near $\alpha_0$. On this branch the points $t = t_{\pm}$ correspond to $u = \pm\zeta^{1/2}$, respectively. Furthermore, for $\alpha$ near $\alpha_0$, the correspondence $t \leftrightarrow u$ is one-to-one.

The above theorem is only a local result, but not a global one. For our integral (4), we can prove that the transformation (9) is in fact one-to-one and analytic in a domain containing the entire paths of integration.

## 4. Study of the cubic transformation

In order to apply the method described in the last section, we shall first make a slight change of variables. Let $x = 4\xi - \frac{2}{\lambda}$ and $\tau = -\lambda t$, so that (4) becomes,

$$y(\xi) = \int_{L'} -\lambda^{-1} t^{-2} e^{-\lambda f(\xi, t)}\, dt, \tag{11}$$

where

$$f(\xi, t) = \frac{t^3}{3} + \left(4\xi - \frac{2}{\lambda}\right) t - \lambda^{-2} t^{-1}, \tag{12}$$

$L'$ is the image of contour $L$. Corresponding to the contours $L(A_k)$, $k = 1, 2, 3$, and $L(U_q)$, $q = 1, 2, 3, 4$, we now have the new contours $L'(A_k)$, $k = 1, 2, 3$ and $L'(U_q)$, $q = 1, 2, 3, 4$. To use the method in the last section, we must prove that the mapping from $t$ to $u$ under the cubic transformation (9) is one-to-one and analytic in a domain containing the paths of integration. The one-to-one and analytic properties will be established under two separate cases $-\infty < x < 2/\lambda$ and $-2/\lambda < x < \infty$. Note that these two intervals overlap, and together cover the entire real axis. We will give a detailed study of the transformation (9) for the contours $L'(U_1)$, $L'(U_2)$ and $L'(A_3)$ in the case $-\infty < x < 2/\lambda$ in this section as an illustration. Similar argument works for $L'(A_1)$, $L'(A_2)$ in the case $-\infty < x < 2/\lambda$ and $L'(A_1)$, $L'(A_2)$, $L'(U_q)$ ($q = 1, 2, 3, 4$) in the case $-2/\lambda < x < \infty$. The asymptotic behavior of the solutions $U_3$ and $U_4$ in the first case, and that of the solution $A_3$ in the second case can be derived directly from the identities in (7).

It is easily seen that the function $f(\xi, t)$ in (12) has four symmetrically located saddle points. Since the contours $L'(U_1)$, $L'(U_2)$, and $L'(A_3)$ are all in the left half $t$-plane, only the saddle points in the left half $t$-plane will be considered; they are

$$\begin{aligned} t_+ &= -\sqrt{\lambda^{-1} - \xi} - \sqrt{-\xi}, \\ t_- &= -\sqrt{\lambda^{-1} - \xi} + \sqrt{-\xi}, \end{aligned} \tag{13}$$

for $\xi < 0$ and

$$\begin{aligned} t_+ &= -\sqrt{\lambda^{-1} - \xi} + i\sqrt{\xi}, \\ t_- &= -\sqrt{\lambda^{-1} - \xi} - i\sqrt{\xi}, \end{aligned} \tag{14}$$

for $0 < \xi < \frac{1}{\lambda}$.

As $\xi \to 0$, $t_+$ and $t_-$ coincide at $t_0 = -\lambda^{-1/2}$. This is exactly the situation described in the last section. Therefore we shall use the cubic transformation given in Equation (9) with $(t, \alpha)$ replaced by $(t, \xi)$, and set

$$\frac{t^3}{3} + \left(4\xi - \frac{2}{\lambda}\right) t - \lambda^{-2} t^{-1} = \frac{1}{3} u^3 - \zeta u + \eta. \tag{15}$$

Differentiating (15) with respect to $u$, we find that we must make the correspondence $u = \zeta^{1/2}$ with $t = t_+$, and $u = -\zeta^{1/2}$ with $t = t_-$. This gives

$$\begin{aligned} \zeta^{\frac{3}{2}} &= -4(-\xi)^{\frac{3}{2}}, \\ \eta &= \frac{8}{3}\left(\frac{1}{\lambda} - \xi\right)^{\frac{3}{2}}. \end{aligned} \tag{16}$$

As usual, the one-to-one and analytic nature of the transformation $t \to u$ is best seen by introducing two intermediate variables $Z$ and $\varphi$ defined by

$$\begin{aligned} Z &= \frac{t^3}{3} + \left(4\xi - \frac{2}{\lambda}\right) t - \lambda^{-2} t^{-1} - \eta, \\ -\frac{2}{3}\zeta^{\frac{3}{2}} \sin\varphi &= Z. \end{aligned} \tag{17}$$

In the following analysis, we shall temporarily assume $\xi < 0$. The second quadrant of the $t$-plane is shown in Figure 2; the saddle point $t = t_+$, $t = t_-$, and the singularity $t = 0$ are excluded by small indentations and the quadrant is closed by a large circular arc at $\infty$; the curve CE is the upper-half of the contour $L'(A_3)$.

By using the definition of $Z$ and $\varphi$ in (17), we can obtain the corresponding figures in the $Z$-plane and $\varphi$-plane, respectively (see Figures 3 and 4).

The cubic equation (15) for $u$ can be solved explicitly in terms of trigonometric functions. When $\xi \neq 0$, the three solutions are given by

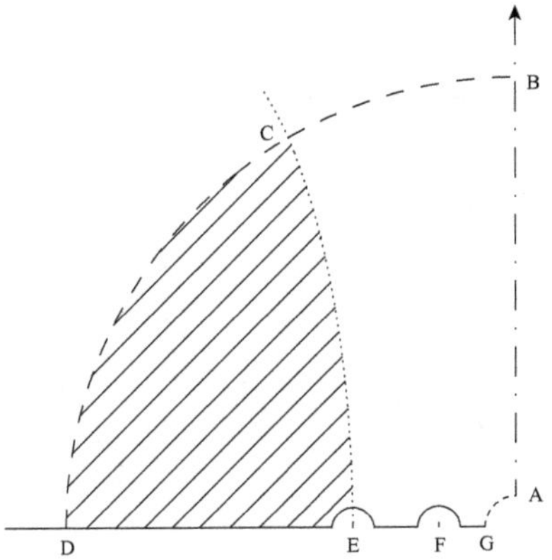


Figure 2. $t$-plane ($-\infty < \xi < 0$).

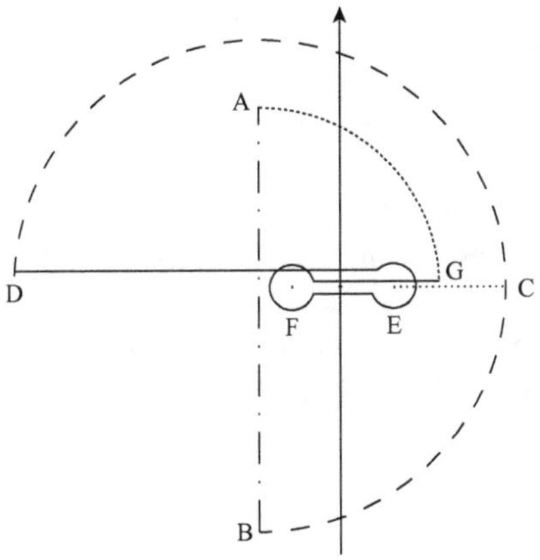


Figure 3. $Z$-plane ($-\infty < \xi < 0$).

$$
\begin{aligned}
u_1(t,\xi) &= -2\zeta^{1/2}\sin\left(\frac{\pi}{3}+\frac{\varphi}{3}\right),\\
u_2(t,\xi) &= 2\zeta^{1/2}\sin\frac{\varphi}{3},\\
u_3(t,\xi) &= -2\zeta^{1/2}\sin\left(-\frac{\pi}{3}+\frac{\varphi}{3}\right),
\end{aligned}
\tag{18}
$$

where

$$
\varphi = \arcsin\left(\frac{-3Z}{2\zeta^{\frac{3}{2}}}\right). \tag{19}
$$

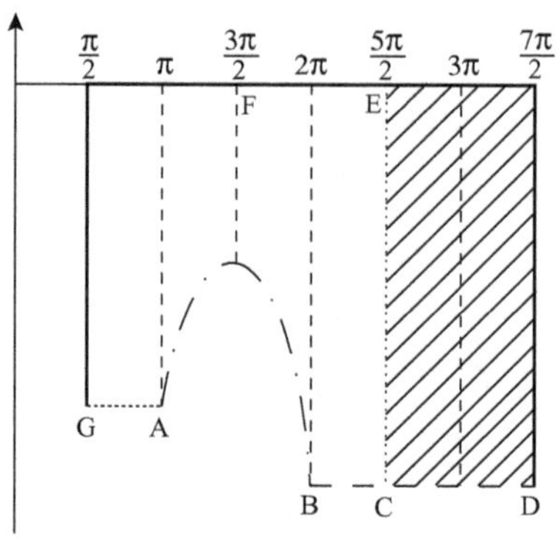


Figure 4. $\varphi$-plane ($-\infty < \xi < 0$).

When $\xi = 0$, we have

$$u(t,0) = \left[t^3 - 6t\lambda^{-1} - 3\lambda^{-2}t^{-1} - 8\lambda^{-\frac{3}{2}}\right]^{\frac{1}{3}}, \tag{20}$$

and again, there are three branches.

Since $t = t_+$ corresponds to $u = \zeta^{1/2}$, from (15) and (19) it follows that $\sin\varphi = 1$ when $t = t_+$. Similarly, we have $\sin\varphi = -1$ when $t = t_-$. Therefore, $u_1(t, \xi)$ is the desired solution of (15), and, from here on, we set $u(t, \xi) \equiv u_1(t, \xi)$, and obtain the corresponding figure in the $u$-plane (see Figure 5).

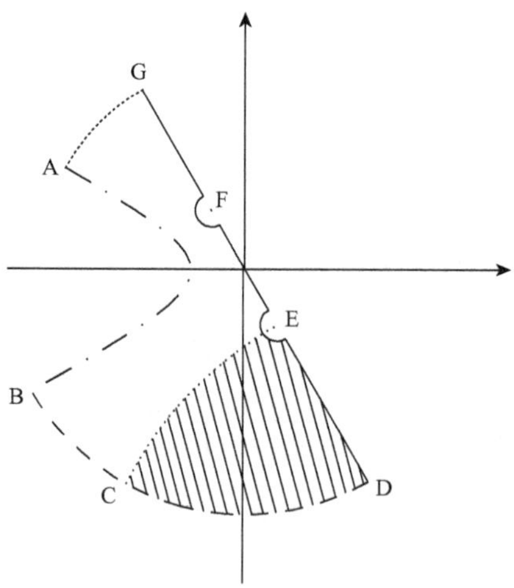


Figure 5. $u$-plane ($-\infty < \xi < 0$).

As $t$ moves once along the boundary of the region ABCDEFGA in Figure 2, its image point $Z$ also moves once along the corresponding boundary in Figure 3. The line segments DE, EF, and FG in Figure 3 are considered as distinct parts of the boundary. By the same argument, as $u$ describes once the closed curved ABCDEFGA in Figure 5, $\varphi$ and $Z$ describe the corresponding curve in Figure 4 and 3, respectively. Hence, the mapping $t \leftrightarrow u$ is one-to-one on the boundary curve ABCDEFGA, and consequently, by Theorem 4.5 in Markushevich [10, p. 118], in the interior of this region. The parametric equation of the curve CE in Figure 5 is given by

$$\sqrt{3}rs - s^2 + \frac{3}{2}|\zeta| = 0, \quad s < 0, \tag{21}$$

where $r = \text{Re}\, u$ and $s = \text{Im}\, u$.

By the same procedure, we can get the properties of the mapping $t \leftrightarrow u$ in the third quadrant, and obtain the image in the $u$-plane just from Figure 5 by reflection with respect to the curve DEFG.

By the one-to-one and analytic properties of the mapping from the $t$-plane to the $u$-plane, we can get the corresponding contour $\tilde{L}(A_3)$, $\tilde{L}(U_1)$, $\tilde{L}(U_2)$ in the $u$-plane (see Figure 6), which are the images of $L'(A_3)$, $L'(U_1)$, $L'(U_2)$ in the $t$-plane.

Note that the mapping properties mentioned above are established under the condition $\xi < 0$. The validity of these properties for $0 < \xi < 1/\lambda$ can be established in a similar manner. Figure 7 shows the $u$-plane and contours $\tilde{L}(A_3)$, $\tilde{L}(U_1)$, and $\tilde{L}(U_2)$ in the case $0 < \xi < 1/\lambda$. By continuity, the mapping from $t$ to $u$ is also one-to-one and analytic for $\xi = 0$.

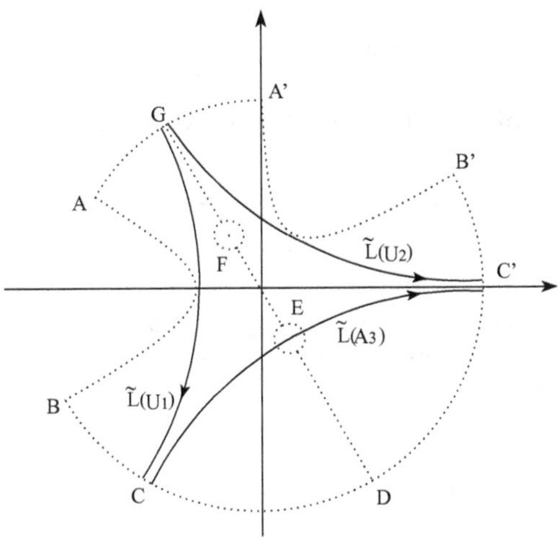


Figure 6. $u$-plane $(-\infty < \xi < 0)$.

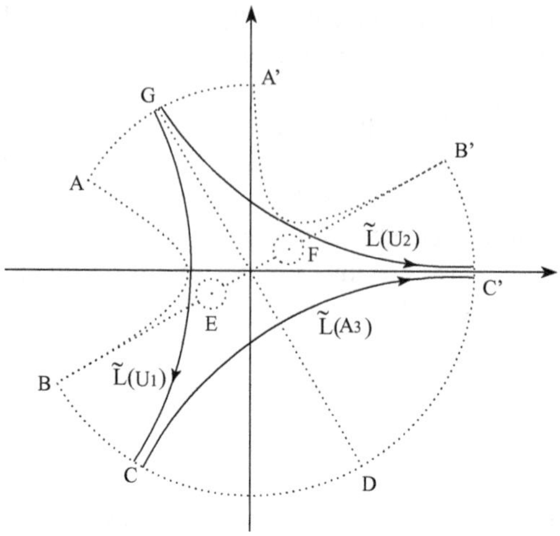


Figure 7. $u$-plane $(0 < \xi < \frac{1}{\lambda})$.

## 5. Bleistein's sequence

We now return to the integral (11). With the transformation (15), this integral becomes

$$y(\xi) = -e^{-\lambda\eta} \int_{\tilde{L}} h_0(u) e^{-\lambda(u^3/3 - \zeta u)}\, du, \tag{22}$$

where

$$h_0(u) = \lambda^{-1} t^{-2} \frac{dt}{du} \tag{23}$$

and $\tilde{L}$ is the image of $L'$ under the cubic transformation. From (10), we have

$$h_0(u) = \frac{u^2 - \zeta}{t^4 + \left(4\xi - \dfrac{2}{\lambda}\right) t^2 + \lambda^{-2}}. \tag{24}$$

The method of obtaining an Airy-type expansion, based on repeated integration-by-parts, was introduced by Bleistein [11]; see also [12, Section 9.2] and [9, Section 7.5].

First, we write

$$h_0(u) = a_0 + b_0 u + (u^2 - \zeta) g_0(u). \tag{25}$$

Since the last term vanishes at $u = \pm\zeta^{1/2}$, the coefficients $a_0$ and $b_0$ are easily determined to be

$$a_0 = \frac{1}{2}\left[h_0(\zeta^{\frac{1}{2}}) + h_0(-\zeta^{\frac{1}{2}})\right],$$
$$b_0 = \frac{1}{2\zeta^{\frac{1}{2}}}\left[h_0(\zeta^{\frac{1}{2}}) - h_0(-\zeta^{\frac{1}{2}})\right]. \tag{26}$$

Clearly,

$$\lim_{\zeta^{1/2}\to 0} a_0 = h_0(0), \qquad \lim_{\zeta^{1/2}\to 0} b_0 = h_0'(0). \tag{27}$$

Thus, $b_0$ is an analytic function of $\zeta^{1/2}$ with a removable singularity at the origin. With $a_0$ and $b_0$ so chosen, it is easily seen that the function

$$g_0(u) = \frac{h_0(u) - a_0 - b_0 u}{u^2 - \zeta} \tag{28}$$

is analytic in $u$ and has removable singularities at $u = \pm\,\zeta^{1/2}$. Indeed, we have

$$\lim_{u\to\pm\zeta^{1/2}} g_0(u) = \pm\frac{1}{2\zeta^{1/2}}\left[h_0'(\pm\zeta^{1/2}) - b_0\right]. \tag{29}$$

Recall that the Airy function Ai($z$) has the integral representation

$$\mathrm{Ai}(z) = \frac{1}{2\pi i}\int_{\gamma_0} e^{\frac{1}{3}u^3 - zu}\,du, \tag{30}$$

where $\gamma_0$ is shown in Figure 8. Inserting (25) into (22) gives

$$-e^{\lambda\eta}y(\xi) = \frac{a_0}{\lambda^{1/3}}Q(\lambda^{2/3}\zeta) + \frac{b_0}{\lambda^{2/3}}Q'(\lambda^{2/3}\zeta) + I_1(\lambda,\xi), \tag{31}$$

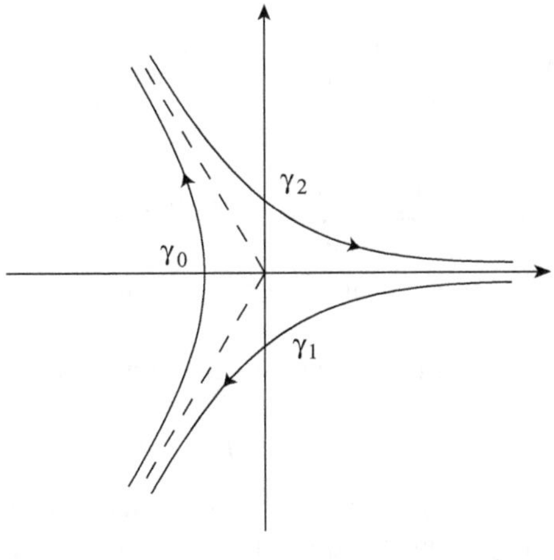


Figure 8. Three contours of the integral $Q(\lambda)$.

where

$$I_1(\lambda, \xi) = \int_{\tilde{L}} (u^2 - \zeta) e^{-\lambda(\frac{1}{3}u^3 - \zeta u)} g_0(u)\, du \tag{32}$$

and

$$Q(\lambda) = \int_{\tilde{L}} \mathrm{e}^{-\frac{v^3}{3} + \lambda v} \mathrm{d}v, \tag{33}$$

$\tilde{L}$ being one of the contours $\gamma_0$, $\gamma_1$, and $\gamma_2$ shown in Figure 8. If $\tilde{L}$ is the contour $\gamma_j, j = 0, 1, 2$, then it can be easily verified that

$$Q(\lambda) = 2\pi i \omega^j \operatorname{Ai}(\lambda \omega^j), \tag{34}$$

with $\omega$ being the cube root $\mathrm{e}^{2\pi i/3}$ of unity.

An integration by parts shows that the integral (32) can be expressed as

$$I_1(\lambda, \xi) = \frac{1}{\lambda} \int_{\tilde{L}} e^{-\lambda(\frac{1}{3}u^3 - \zeta u)} h_1(u)\, du, \tag{35}$$

where

$$h_1(u) = \frac{\mathrm{d}}{\mathrm{d}u} g_0(u). \tag{36}$$

Since $h_1(u)$ is analytic in the same domain as $h_0(u)$, and the integral in Equation (35) is exactly of the same form as that in Equation (32), the above process can be repeated. Thus we define inductively

$$\begin{aligned} h_m(u) &= a_m + b_m(u) + (u^2 - \zeta) g_m(u), \\ h_{m+1}(u) &= \frac{\mathrm{d}}{\mathrm{d}u} g_m(u), \quad m = 0, 1, 2, 3, \ldots. \end{aligned} \tag{37}$$

This leads to

$$-e^{\lambda\eta} y(\xi) = Q(\lambda^{2/3}\zeta) \sum_{m=0}^{n-1} \frac{a_m}{\lambda^{m+1/3}} + Q'(\lambda^{2/3}\zeta) \sum_{m=0}^{n-1} \frac{b_m}{\lambda^{m+2/3}} + I_n(\lambda; \xi), \tag{38}$$

where

$$I_n(\lambda; \xi) = \frac{1}{\lambda^n} \int_{\tilde{L}} e^{-\lambda(\frac{1}{3}u^3 - \zeta u)} h_n(u)\, \mathrm{d}u, \tag{39}$$

and $h_n(u)$ has the same domain of analyticity as $h_0(u)$.

## 6. Uniform asymptotic expansions

### *6.1. Solution $V(x, \lambda)$*

As the integration path $L(V)$ is a simple closed curve encircling the point $\tau = 0$ and does not pass any singular point of the integrand, it is easy to derive the uniform asymptotic expansion of the solution $V(x, \lambda)$, which is given by

$$V(x,\lambda)=\int_{L(V)}\tau^{-2}\exp\left(\tau x-\tau^{-1}+\frac{\lambda^{-2}\tau^{3}}{3}\right)\mathrm{d}\tau. \tag{40}$$

Since the Taylor series

$$\exp\left(\frac{\lambda^{-2}\tau^{3}}{3}\right)=\sum_{n=0}^{\infty}\frac{(\lambda^{-2}\tau^{3})^{n}}{3^{n}n!} \tag{41}$$

converges uniformly for $\tau$ in any bounded domain and $|\lambda|\geq|\lambda_0|>0$, termwise integration gives

$$\begin{aligned}V(x,\lambda)&=\int_{L(V)}\tau^{-2}\exp(\tau x-\tau^{-1})\sum_{n=0}^{\infty}\frac{(\lambda^{-2}\tau^{3})^{n}}{3^{n}n!}\,\mathrm{d}\tau\\&=\sum_{n=0}^{\infty}\frac{\lambda^{-2n}}{3^{n}n!}\int_{L(V)}\exp(\tau x-\tau^{-1})\tau^{3n-2}\,\mathrm{d}\tau.\end{aligned} \tag{42}$$

Substituting $t=\tau\sqrt{x}$, we obtain

$$V(x,\lambda)=2\pi i\sum_{n=0}^{N-1}\frac{(-1)^{3n-1}}{3^{n}n!}\sqrt{x}^{1-3n}J_{3n-1}(2\sqrt{x})\lambda^{-2n}+\varepsilon_N(x,\lambda), \tag{43}$$

where $J_{3n-1}\left(2\sqrt{x}\right)$ is the Bessel function and

$$\varepsilon_N(x,\lambda)=\sum_{n=N}^{\infty}\frac{\lambda^{-2n}}{3^{n}n!}\int_{L(V)}(\sqrt{x})^{1-3n}\exp\left(t\sqrt{x}-\frac{\sqrt{x}}{t}\right)t^{3n-2}\,\mathrm{d}t. \tag{44}$$

Note that the first term of (43), i.e.,

$$-2\pi i\sqrt{x}J_{-1}(2\sqrt{x})=2\pi i\sqrt{x}J_{1}(2\sqrt{x}), \tag{45}$$

is a solution of the limiting differential equation $xy''+y=0$ and an entire function.

### *6.2. Solutions $A_k(x,\lambda)$, $k=1, 2, 3$ and $U_q(x,\lambda)$, $q=1, 2, 3, 4$*

In view of the cubic transformation considered in Section 4, all these solutions $A_k(x,\lambda)$, $k=1, 2, 3$, and $U_q(x,\lambda)$, $q=1, 2, 3, 4$ have asymptotic expansions of Airy-type. We shall determine the function $Q(\lambda^{2/3}\zeta)$ in the general form of Airy-type expansion given in Equation (38). According to the cubic transformation studied in Section 4, we have

$$U_1:\ Q(\lambda^{2/3}\zeta)=-2\pi i\,\mathrm{Ai}(\lambda^{2/3}\zeta), \tag{46}$$

$$U_2:\ Q(\lambda^{2/3}\zeta)=2\pi i e^{i4\pi/3}\mathrm{Ai}(\lambda^{2/3}\zeta e^{i4\pi/3}), \tag{47}$$

$$A_3:\ Q(\lambda^{2/3}\zeta)=-2\pi i e^{i2\pi/3}\mathrm{Ai}(\lambda^{2/3}\zeta e^{i2\pi/3}), \tag{48}$$

where $\zeta^{\frac{3}{2}}=-4(-\xi)^{\frac{3}{2}}$; cf. Figures 6, 7, and 8.

We conclude this section with an evaluation of the leading coefficient $a_0$ and $b_0$ in (38). When $\xi < 0$, from (26) it follows that

$$a_0 = \frac{e^{-\frac{7}{6}\pi i}2^{-\frac{5}{3}}}{(1-\xi\lambda)^{\frac{1}{4}}}\left[i(\sqrt{1-\xi\lambda}-\sqrt{-\xi\lambda})^{\frac{3}{2}}+(\sqrt{1-\xi\lambda}+\sqrt{-\xi\lambda})^{\frac{3}{2}}\right], \tag{49}$$

$$b_0 = \frac{e^{-\frac{5}{6}\pi i}2^{-\frac{7}{3}}}{\sqrt{-\xi}(1-\xi\lambda)^{\frac{1}{4}}}\left[i(\sqrt{1-\xi\lambda}-\sqrt{-\xi\lambda})^{\frac{3}{2}}-(\sqrt{1-\xi\lambda}+\sqrt{-\xi\lambda})^{\frac{3}{2}}\right]. \tag{50}$$

A corresponding formula exists for the case $0 < \xi < \frac{1}{\lambda}$, and can be formally obtained from (49) and (50) by simply replacing $\sqrt{-\xi}$ by $-i\sqrt{\xi}$. The case of critical value $\xi = 0$ can be obtained directly from (27), and we have $a_0 = 4^{-\frac{1}{3}}$, $b_0 = 0$. Similar results can be given for the solutions $A_1$, $A_2$ in the case $-\infty < x < 2/\lambda$ and $A_1$, $A_2$, $U_q$ $(q = 1, 2, 3, 4)$ in the case $-2/\lambda < x < \infty$. The asymptotic behavior of the solutions $U_3$ and $U_4$ in the first case, and that of the solution $A_3$ in the second case can be derived directly from the identities in (7).

## 7. Estimation of error terms

### *7.1. Error terms associated with solution $V(x, \lambda)$*

To show that the expansions we have derived are indeed asymptotic, we must estimate the error terms associated with these expansions. The error term in the asymptotic expansion of solution $V(x, \lambda)$ is given in (44):

$$\begin{aligned}\varepsilon_N(x,\lambda) &= \sum_{n=N}^{\infty}\frac{\lambda^{-2n}}{3^n n!}\int_{L(V)}(\sqrt{x})^{1-3n}\exp\left(t\sqrt{x}-\frac{\sqrt{x}}{t}\right)t^{3n-2}\,dt \\ &= 2\pi i\sum_{n=N}^{\infty}\frac{(-1)^{3n-1}}{3^n n!}\sqrt{x}^{1-3n}J_{3n-1}(2\sqrt{x})\lambda^{-2n}.\end{aligned} \tag{51}$$

When $x < 0$, by using the ascending series of Bessel functions given in [13, p. 57], it can be shown that $(\sqrt{x})^{-\nu}J_\nu(2\sqrt{x})$ is decreasing with respect to $\nu$. Thus, we have

$$|\varepsilon_N(x,\lambda)| \le M_1\lambda^{-2N}\left|\sqrt{x}^{1-3N}J_{3N-1}(2\sqrt{x})\right|, \tag{52}$$

where $M_1$ is a constant independent of $x$ and $\lambda$.

When $x \ge 0$, using the bounds for Bessel functions given in [13, p. 59, Ex. 9.6], we can obtain

$$|\varepsilon_N(x,\lambda)| \le M_2\lambda^{-2N}, \tag{53}$$

where $M_2$ is also a constant independent of $x$ and $\lambda$.

A combination of (52) and (53) yields

$$|\varepsilon_N(x,\lambda)| \leq M\lambda^{-2N}\left(\left|\sqrt{x}^{1-3N}J_{3N-1}\left(2\sqrt{x}\right)\right|+1\right), \tag{54}$$

where $M = \max(M_1, M_2)$. This implies that (43) is an asymptotic expansion uniform for all real $x$.

### *7.2. Error terms in the Airy-type expansions for finite x*

To estimate $\varepsilon_n$ in the Airy-type expansions for solutions $A_k(x,\lambda)$ ($k = 1, 2, 3$) and $U_q(x,\lambda)$ ($q = 1, 2, 3, 4$), we employ the following auxiliary functions

$$\tilde{\mathrm{Ai}}(z) := \left\{E_1^2(z)\left|\mathrm{Ai}^2(\omega^2 z)\right| + E_{-1}^2(z)\left|\mathrm{Ai}^2(\omega z)\right|\right\}^{1/2} \tag{55}$$

and

$$\tilde{\tilde{\mathrm{Ai}}}(z) := \left\{E_1^2(z)\left|\mathrm{Ai}'^2(\omega^2 z)\right| + E_{-1}^2(z)\left|\mathrm{Ai}'^2(\omega z)\right|\right\}^{1/2}, \tag{56}$$

where $\omega = e^{2\pi i/3}$ and $E_1(z)$, $E_{-1}(z)$ are the weight functions given by

$$E_j(z) = \left|\exp\left(\frac{2}{3}z^{3/2}\right)\right| \quad (j = \pm 1), \tag{57}$$

the branch of the fractional power being chosen in such a way that $E_1(z) \geq 1$ in $\frac{1}{3}\pi \leq \mathrm{ph}\, z \leq \pi$ and $E_1(z) \leq 1$ in $-\pi \leq \mathrm{ph}\, z \leq \frac{1}{3}\pi$; $E_{-1}(z) \geq 1$ in $-\pi \leq \mathrm{ph}\, z \leq -\frac{1}{3}\pi$ and $E_{-1}(z) \leq 1$ in $-\frac{1}{3}\pi \leq \mathrm{ph}\, z \leq \pi$.

Furthermore, we define

$$\tilde{a}_p = \begin{cases} 1 & \text{if} - 2/\lambda < x < 2/\lambda \\ |a_p| & \text{if} - N < x < -2/\lambda \end{cases} \tag{58}$$

and

$$\tilde{b}_p = \begin{cases} 1 & \text{if} - 2/\lambda < x < 2/\lambda \\ |b_p| & \text{if} - N < x < -2/\lambda, \end{cases} \tag{59}$$

where $a_p$ and $b_p$ are the coefficients in the expansion (38), $N$ is a large positive number. $\tilde{\mathrm{Ai}}(z)$ and $\tilde{\tilde{\mathrm{Ai}}}(z)$ mimic the behavior of Ai $(z)$ and $\mathrm{Ai}'(z)$, respectively. Moreover, $\tilde{a}_p$ and $\tilde{b}_p$ have the same behavior as $a_p$ and $b_p$, respectively. For large values of $\lambda$ and for $-N < x < 2/\lambda$, the error bound of solution $U_1$ satisfies the estimate

$$|\varepsilon_n| \leq \frac{N_n}{\lambda^{n+1/3}}|\tilde{a}_n||\tilde{\mathrm{Ai}}(\lambda^{2/3}\zeta)| + \frac{M_n}{\lambda^{n+2/3}}|\tilde{b}_n||\tilde{\tilde{\mathrm{Ai}}}(\lambda^{2/3}\zeta)|, \tag{60}$$

$N_n$ and $M_n$ being some positive constants independent of $\lambda$ and $x$. The same argument works for all other solutions with Airy-type expansions.

### 7.3. *Error terms in the Airy-type expansions for unbounded x*

We can extend the validity of the Airy-type uniform asymptotic expansions to include all real values of $x$. Our method makes use of the rational functions introduced by Olde Daalhuis and Temme [8], and their Cauchy-type integral representation for the remainder in the expansion. A new estimate for the remainder will be given in this subsection. Since the estimations of error terms in all Airy-type expansions in our problem are similar, we only consider the solution $A_3$ under the condition $-\infty < x < 2/\lambda$ as an illustration.

7.3.1. *A class of rational functions.* In [8], Olde Daalhuis and Temme introduced the sequence of rational functions

$$R_0(w, u, \zeta) = \frac{1}{w - u}, \tag{61}$$

$$R_{n+1}(w, u, \zeta) = \frac{-1}{w^2 - \zeta} \frac{\mathrm{d}}{\mathrm{d}w} R_n(w, u, \zeta), \quad n = 0, 1, 2, \ldots, \tag{62}$$

where $w, u, \zeta \in \mathbb{C}$, $w \neq u$, and $w^2 \neq \zeta$. Then they showed that the function $h_n(u)$ defined by the recursive formula (37) has the representation

$$h_n(u) = \frac{1}{2\pi i} \int_C R_n(w, u, \zeta) h_0(w) \, \mathrm{d}w, \tag{63}$$

where $C$ is any simple closed contour in the domain of analyticity of $h_0(u)$, enclosing the points $u$ and $\pm\zeta^{\frac{1}{2}}$.

Furthermore, by using induction on $n$, they gave the representation

$$R_n(w, u, \zeta) = \sum_{i=0}^{n-1} \sum_{j=0}^{k_{n,i}} \frac{C_{ij} w^{i-j}}{(w - u)^{n+1-i-j}(w^2 - \zeta)^{n+i}}, \quad n = 1, 2, \ldots, \tag{64}$$

where $k_{n,i} = \min(i, n - 1 - i)$ and $C_{ij}$ do not depend on $w$, $u$ and $\zeta$. Note that the rational functions defined in (61) and (62) are independent of the function $h_0(u)$; hence, the representation (63) can be considered as the analogue of the Cauchy integral for the remainder of a Taylor series. Furthermore, they established the following two results:

(i) Let $u \in \mathbb{C}$ be such that $|u - \zeta^{\frac{1}{2}}| = O(|\zeta|^{\frac{1}{2}})$ as $|\zeta| \to \infty$, and let $\Gamma$ be a simple closed contour that encircles $\zeta^{\frac{1}{2}}$ and $u$. Then for $n = 1, 2, \ldots,$

$$\frac{1}{2\pi i} \int_\Gamma R_n(w, u, \zeta) \, \mathrm{d}w = O(|\zeta|^{-\frac{3}{2}n}) \tag{65}$$

as $|\zeta| \to \infty$.

(ii) Let $\zeta \in \mathbb{C}$ and $\Omega(\zeta) = \{(w, u) \in \mathbb{C}^2 : |w - \zeta^{\frac{1}{2}}| = \rho(\zeta), |u - \zeta^{\frac{1}{2}}| \leq \frac{1}{2}\rho(\zeta)\}$, where $\rho(\zeta) = |\zeta|^{\frac{1}{2}\theta}$ as $|\zeta| \to \infty$ and $-\frac{1}{2} < \theta \leq 1$. Then there are numbers $A_n$ independent of $\zeta$ such that

$$\sup_{(w,u)\in\Omega(\zeta)} |R_n(w, u, \zeta)| \leq A_n|\zeta|^{\frac{-(1+2\theta)n-\theta}{2}} \quad \text{as } |\zeta| \to \infty. \tag{66}$$

Since $h_0(u)$ has no singularity in the domain that concerns us, we can choose $\theta$ as large as possible (as indicated in [8, p. 310]). Thus, let $\theta = 1$; i.e., $\rho(\zeta) = |\zeta|^{\frac{1}{2}}$, and we have

$$\sup_{(w,u)\in\Omega(\zeta)} |R_n(w, u, \zeta)| \leq A_n|\zeta|^{\frac{-3n-1}{2}} \quad \text{as } |\zeta| \to \infty. \tag{67}$$

7.3.2. *Estimation of $h_n(u)$.* Our next task is to give an estimate for the function $h_0(u)$ in (24), when $u$ lies in the disks $|u \pm \zeta^{\frac{1}{2}}| \leq \rho(\zeta)$. Put

$$\tilde{h}_n(\zeta^{\frac{1}{2}}) := \sup\{|h_n(u)| : |u \pm \zeta^{\frac{1}{2}}| \leq \frac{1}{2}\rho(\zeta)\} \tag{68}$$

and

$$h_n^*(\zeta^{\frac{1}{2}}) := \sup\{|h_n(u)| : |u \pm \zeta^{\frac{1}{2}}| \leq \rho(\zeta)\}. \tag{69}$$

By the same argument used in [8, p. 310], we can show that

$$\tilde{h}_n(\zeta^{\frac{1}{2}}) \leq C_n|\zeta|^{-\frac{3n}{2}} h_0^*(\zeta^{\frac{1}{2}}) \quad \text{as } |\zeta| \to \infty, \tag{70}$$

where $C_n$ does not depend on $\zeta$. Notice that $h_0(u)$ is analytic on $|u \pm \zeta^{\frac{1}{2}}| < \frac{3}{2}\rho(\zeta)$, thus $h_0^*(\zeta^{\frac{1}{2}})$ is finite.

7.3.3. *Estimation of $\varepsilon_n$.* To estimate the remainder term in Equations (38) and (39), we first split the contour $\tilde{L}(A_3)$ into two parts $\Gamma'$ and $\Gamma''$. In the case $-\infty < x < 2/\lambda$ and $|x|$ is large, we take $\Gamma' = \{u \in \tilde{L}(A_3) : |u - \zeta^{\frac{1}{2}}| \leq \frac{1}{2}\rho(\zeta)\}$ and define $\Gamma'' = \tilde{L}(A_3) - \Gamma'$. The corresponding integrals are denoted by

$$\varepsilon_n' = \lambda^{-n} \int_{\Gamma'} e^{-\lambda(\frac{1}{3}u^3-\zeta u)} h_n(u)\, du, \tag{71}$$

$$\varepsilon_n'' = \lambda^{-n} \int_{\Gamma''} e^{-\lambda(\frac{1}{3}u^3-\zeta u)} h_n(u)\, du. \tag{72}$$

In this case, we have from (68), (69) and (70)

$$|\varepsilon_n'| \leq C_n\lambda^{-n-\frac{1}{3}}|\zeta|^{-\frac{3}{2}n} h_0^*(\zeta^{\frac{1}{2}})\tilde{\mathrm{A}}\mathrm{i}(\lambda^{\frac{2}{3}}\zeta e^{\frac{2\pi i}{3}}), \tag{73}$$

where the constant $C_n$ does not depend on $\zeta$ and $\lambda$, and the function $\tilde{\mathrm{A}}\mathrm{i}$ is the modulus function defined in (55). In [8], Olde Daalhuis and Temme have already shown that $\varepsilon_n''$ is exponentially small compared to the estimate of $\varepsilon_n'$ as $\lambda \to \infty$ uniformly with respect to large $|\zeta|$, provided that $h_0(u)$ satisfies the following conditions:

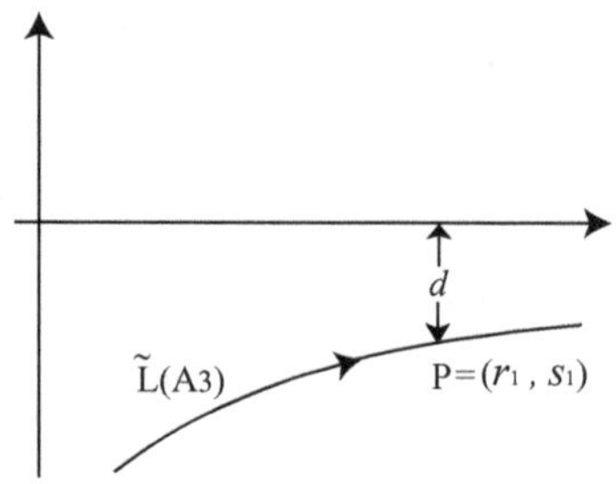


Figure 9. Enlarged part of contour $\tilde{L}(A_3) - \infty < x < 2/\lambda$.

(i) $h_0(u)$ is analytic in a neighborhood $\Omega_0(\zeta^{\frac{1}{2}})$ of $\Gamma''$, such that for every $u \in \Gamma''$ there exists a disc with center $u$ and radius

$$R(u, \zeta^{\frac{1}{2}}, p, q, k) = k\left|u^{-q} e^{p(\frac{1}{3}u^3 - \zeta u + \frac{2}{3}\zeta^{\frac{3}{2}})}\right| \tag{74}$$

contained in $\Omega_0(\zeta^{\frac{1}{2}})$, where $k > 0$ and $p, q \geq 0$ do not depend on $\zeta$ and $u$.

(ii) There exist constants $\sigma \geq 0$ and $C_0 > 0$ such that

$$|h_0(u)| \leq C_0 h_0^*(\zeta^{\frac{1}{2}})\left|e^{-\sigma(\frac{1}{3}u^3 - \zeta^2 u + \frac{2}{3}\zeta^3)}\right| \tag{75}$$

for all $u \in \Omega_0(\zeta^{\frac{1}{2}}) \cup \tilde{L}(A_3)$.

To establish the existence of the domain $\Omega_0(\zeta^{\frac{1}{2}})$ described in the first condition, let us enlarge part of the contour $\tilde{L}(A_3)$ in Figure 9. Since we need the final estimation to be valid for all the solutions with Airy-type expansions, we choose radius $R$ less than $d$ in Figure 9. With $u = r + si$, the equation of the contour $\tilde{L}(A_3)$ is given by

$$\sqrt{3}rs - s^2 + \frac{3}{2}|\zeta| = 0, \quad s < 0. \tag{76}$$

Let $P = (r_1, s_1)$ be a point on $\Gamma''$, from Equation (76) we have

$$d = |s_1| = \frac{\sqrt{3r_1^2 + 6|\zeta|} - \sqrt{3}r_1}{2}. \tag{77}$$

Thus, we can choose $k = 1$, $p = 0$, $q = 2$ in (74), i.e.,

$$R(u, \zeta^{\frac{1}{2}}, p, q, k) = |u|^{-2}, \tag{78}$$

so that the ratio of $d$ and $R$ at the point $P$ is

$$\begin{aligned} \frac{d}{R} &= \frac{\sqrt{3r_1^2 + 6|\zeta|} - \sqrt{3}r_1}{2|u_1|^{-2}} \\ &= \frac{3|\zeta|\left(r_1^2 + s_1^2\right)}{\sqrt{3r_1^2 + 6|\zeta|} + \sqrt{3}r_1}. \end{aligned} \tag{79}$$

It is easy to prove that $d/R$ is always larger than 1 on $\Gamma''$ for large values of $|\zeta|$, thus establishing condition (i).

Now, consider condition (ii). By using Equation (76), it is easy to verify that when $u \in \Gamma''$,

$$\left|e^{-\sigma(\frac{1}{3}u^3 - \zeta^2 u + \frac{2}{3}\zeta^{\frac{3}{2}})}\right| = \left|e^{\sigma\frac{2}{3}(r-\zeta^{\frac{1}{2}})(2r+\zeta^{\frac{1}{2}})^2}\right|, \tag{80}$$

where $r = \operatorname{Re} u$. The above quantity could be exponentially large for $u \in \Gamma''$, if $\sigma$ is positive.

Since $h_0(u)$ is analytic in $\Omega_0(\zeta^{\frac{1}{2}})$, by (24) we have

$$h_0(u) = \frac{u^2 - \zeta}{t^4 + (4\xi - \frac{2}{\lambda})t^2 + \lambda^{-2}} = O(u^2) \tag{81}$$

for $u \in \Gamma''$. It is now clear that there exist constants $C_0 > 0$ and $\sigma \geq 0$ such that

$$|h_0(u)| \leq C_0 h_0^*(\zeta^{\frac{1}{2}})\left|e^{-\sigma(\frac{1}{3}u^3 - \zeta u + \frac{2}{3}\zeta^{\frac{3}{2}})}\right| \tag{82}$$

for $u \in \Gamma''$. The fact that (82) holds for $u \in \Gamma'$ is obvious, thus establishing condition (ii).

With conditions (i) and (ii) being satisfied, we can now use the following result proved in [8, (A.5)]:

$$|\varepsilon_n''| \leq \tilde{C}_n \lambda^{-n-\frac{4}{3}} e^{-\delta(\lambda - 2n - 2 - \sigma)|\zeta|^{\frac{3}{2}}} h_0^*(\zeta^{\frac{1}{2}}) \tilde{\mathrm{A}}\mathrm{i}\left(\lambda^{\frac{2}{3}}\zeta e^{\frac{2\pi i}{3}}\right), \tag{83}$$

where $\tilde{C}_n$, $\delta$ are positive constants not depending on $\zeta$ and $\lambda$. The two estimates (73) and (83) together give

$$|\varepsilon_n| \leq C_n'(1 + |\zeta|)^{-\frac{3}{2}n} h_0^*(\zeta^{\frac{1}{2}}) \lambda^{-n-\frac{1}{3}} \tilde{\mathrm{A}}\mathrm{i}\left(\lambda^{\frac{2}{3}}\zeta e^{\frac{2\pi i}{3}}\right) \tag{84}$$

with $C_n'$ being a positive constant again not depending on $\zeta$ and $\lambda$. This establishes that the asymptotic expansion of solution $A_3$ indeed holds uniformly with respect to $-\infty < x < 2/\lambda$.

## References

1. W. H. Reid, On the asymptotic theory of the Orr–Sommerfeld problem, in *Asymptotic and computational analysis: Conference in honor of Frank W.J. Olver's 65th birthday* (R. Wong, Ed.), pp. 379–391. New York, M. Dekker, 1990.

2. W. Wasow, The complex asymptotic theory of a fourth order differential equation of hydrodynamics, *Ann. of Math.* 49:852–871 (1948).
3. W. Wasow, A study of the solutions for the differential equation $y^{(4)} + \lambda^2(xy'' + y) = 0$ for large values of $\lambda$, *Ann. of Math.* 52:350–361 (1950).
4. W. Wasow, Asymptotic solution of the differential equation of hydrodynamic stability in a domain containing a transition point, *Ann. of Math.* 58:222–252 (1953).
5. C. C. Lin and A. L. Rabenstein, On the asymptotic solutions of a class of ordinary differential equations of the fourth order: I. Existence of regular formal solutions, *Trans. Am. Math. Soc.* 94:24–57 (1960).
6. A. L. Rabenstein, Asymptotic solutions of $u^{(iv)} + \lambda^2 (zu'' + \alpha u' + \beta u) = 0$ for large $|\lambda|$, *Arch. Rat. Mech. Anal.* 1:418–435 (1958).
7. C. Chester, B. Friedman, and F. Ursell, An extension of the method of steepest descents, *Proc. Camb. Phil. Soc.*, 53:599–611 (1957).
8. A. B. Olde Daalhuis and N. M. Temme, Uniform Airy-type expansions of integrals, *SIAM J. Math. Anal.* 25:304–321 (1994).
9. R. Wong, *Asymptotic Approximations of Integrals*, Academic Press, Boston, 1989. (Reprinted by SIAM, Philadelphia, 2001.)
10. A. I. Markushevich, *Theory of Functions of a Complex Variable*, 2, Chelsea, New York, 1977.
11. N. Bleistein, Uniform asymptotic expansions of integrals with many nearby stationary points near algebraic singularities, *J. Math. Mech.* 17:533–559 (1967).
12. N. Bleistein and R. A. Handelsman, *Asymptotic Expansions of Integrals*, Dover, Holt, Rinehart and Winston, New York, 1986.
13. F. W. J. Olver, *Asymptotics and Special Functions*, Academic Press, New York, 1974. (Reprinted by A. K. Peters, Ltd., Wellesley, 1997.)

City University of Hong Kong

(Received August 24, 2006)

PROCEEDINGS OF THE
AMERICAN MATHEMATICAL SOCIETY
Volume 138, Number 7, July 2010, Pages 2513–2519
S 0002-9939(10)10301-3
Article electronically published on March 4, 2010

# UNIFORM ASYMPTOTIC EXPANSIONS OF THE TRICOMI-CARLITZ POLYNOMIALS

K. F. LEE AND R. WONG

(Communicated by Walter Van Assche)

*Dedicated to the Lord*

ABSTRACT. The Tricomi-Carlitz polynomials satisfy the second-order linear difference equation

$$(n+1)f_{n+1}^{(\alpha)}(x) - (n+\alpha)\,x\,f_n^{(\alpha)}(x) + f_{n-1}^{(\alpha)}(x) = 0, \qquad n \geq 1,$$

with initial values $f_0^{(\alpha)}(x) = 1$ and $f_1^{(\alpha)}(x) = \alpha x$, where $x$ is a real variable and $\alpha$ is a positive parameter. An asymptotic expansion is derived for these polynomials by using the turning-point theory for three-term recurrence relations developed by Wang and Wong [Numer. Math. **91**(2002) and **94**(2003)]. The result holds uniformly in regions containing the critical values $x = \pm 2/\sqrt{\nu}$, where $\nu = n + 2\alpha - 1/2$.

## 1. INTRODUCTION

The Tricomi polynomials are defined by

$$t_n^{(\alpha)}(x) = \sum_{k=0}^{n} (-1)^k \binom{x-\alpha}{k} \frac{x^{n-k}}{(n-k)!}, \qquad n = 0, 1, 2, \ldots, \tag{1.1}$$

which satisfy the recurrence relation

$$(n+1)t_{n+1}^{(\alpha)}(x) - (n+\alpha)t_n^{(\alpha)}(x) + x\,t_{n-1}^{(\alpha)}(x) = 0, \qquad n \geq 1, \tag{1.2}$$

with initial values $t_0^{(\alpha)}(x) = 1$, $t_1^{(\alpha)}(x) = \alpha$. Tricomi [7] observed that $\{t_n^{(\alpha)}(x)\}$ is not a system of orthogonal polynomials, since the recurrence relation (1.2) fails to have the required form [6, p.135]. However, Carlitz [1] discovered that if one sets

$$f_n^{(\alpha)}(x) = x^n t_n^{(\alpha)}(x^{-2}), \tag{1.3}$$

then $f_n^{(\alpha)}(x)$ satisfies

$$(n+1)f_{n+1}^{(\alpha)}(x) - (n+\alpha)\,x\,f_n^{(\alpha)}(x) + f_{n-1}^{(\alpha)}(x) = 0, \qquad n \geq 1, \tag{1.4}$$

Received by the editors June 23, 2009, and, in revised form, November 6, 2009.
2010 *Mathematics Subject Classification.* Primary 41A60, 39A10; Secondary 33C45.
*Key words and phrases.* Tricomi-Carlitz polynomials, uniform asymptotic expansion, difference equation.

with initial values $f_0^{(\alpha)}(x) = 1$, $f_1^{(\alpha)}(x) = \alpha x$. Furthermore, he gave the orthogonal relation

$$\int_{-\infty}^{\infty} f_m^{(\alpha)}(x)\, f_n^{(\alpha)}(x)\, d\psi^{(\alpha)}(x) = \frac{2\,e^{\alpha}}{(n+\alpha)n!}\,\delta_{mn}, \tag{1.5}$$

where $\psi^{(\alpha)}(x)$ is the step function whose jumps are

(1.6)

$$d\psi^{(\alpha)}(x) = \frac{(k+\alpha)^{k-1}e^{-k}}{k!} \quad \text{at} \quad x = x_k = \pm(k+\alpha)^{-1/2}, \qquad k = 0, 1, 2, \ldots.$$

The generating function

$$\exp\left\{\frac{w}{x} + \frac{1-\alpha x^2}{x^2}\log(1-wx)\right\} = \sum_{n=0}^{\infty} f_n^{(\alpha)}(x)\, w^n \tag{1.7}$$

can be derived from the recurrence relation (1.4) when $x \neq 0$. The series in (1.7) is convergent for $|wx| < 1$.

Asymptotic behavior of $f_n^{(\alpha)}(x)$ has been investigated by Goh and Wimp [2, 3]. In their first paper, they used an elementary approach to show that

$$\frac{f_n^{(\alpha)}(y/\sqrt{\alpha})}{\alpha^{-n/2}y^n n^{-\alpha(1-y^2)/y^2-1}} \to \frac{e^{\alpha/y^2}}{\Gamma(-\alpha(1-y^2)/y^2)}, \qquad \text{as} \quad n \to \infty, \tag{1.8}$$

uniformly for all $y$ in a compact set $K \subseteq \mathbb{C}\backslash[-1,1]$. (There is a minor error in the statement of their result; the factor $\alpha^{-n/2}$ in the denominator on the left-hand side of (1.8) is missing in their equation (23) in [2].) The validity of (1.8) can be verified by a direct application of Darboux's method [10, p. 116]. Goh and Wimp also observed that all zeros of $f_n^{(\alpha)}(y/\sqrt{\alpha})$ lie in the interval $[-1,1]$. In their second paper, they used a saddle-point method to study the asymptotics of $f_n^{(\alpha)}(z/\sqrt{n})$ for $z$ in neighborhoods of $z = \pm i/2$. Note that the scales in their two papers are different; in [2] the scale is $y/\sqrt{\alpha}$, whereas in [3] the scale is $z/\sqrt{n}$. The behavior of $f_n^{(\alpha)}(x)$ as $\alpha \to \infty$ has been studied by López and Temme [4]. Their result is expressed in terms of Hermite polynomials.

The purpose of this paper is to present an asymptotic expansion for $f_n^{(\alpha)}(t/\sqrt{\nu})$, which holds uniformly for $t$ in $[0,\infty)$, where $\nu = n + 2\alpha - 1/2$. In view of the reflection formula $f_n^{(\alpha)}(-x) = (-1)^n f_n^{(\alpha)}(x)$, a corresponding result can be given for $t$ in $(-\infty, 0]$. Note that our result is not covered in the two papers of Goh and Wimp; in fact, it complements those in [2] and [3]. We also point out that the Tricomi-Carlitz polynomials do not satisfy a second-order differential equation; hence the powerful tools developed for differential equations (see, e.g., [5]) are not applicable. Our approach is to use a turning-point theory recently introduced by Wang and Wong [8, 9] for three-term recurrence relations.

## 2. Difference equation

Returning to (1.4), we write

$$f_{n+1}^{(\alpha)}(x) - \frac{n+\alpha}{n+1}\, x\, f_n^{(\alpha)}(x) + \frac{1}{n+1} f_{n-1}^{(\alpha)}(x) = 0 \tag{2.1}$$

and introduce the sequence $\{K_n\}$ defined by

$$(n+1)K_{n+1} = K_{n-1} \tag{2.2}$$

with $K_0 = 1$ and $K_1 = \sqrt{2/\pi}$. Induction shows that

$$K_n = \frac{1}{2^{n/2}\Gamma\left(\frac{1}{2}n+1\right)}. \tag{2.3}$$

With the notation

$$F_n^{(\alpha)}(x) := \frac{f_n^{(\alpha)}(x)}{K_n}, \tag{2.4}$$

equation (2.1) can be put in the canonical form considered in [9],

$$F_{n+1}^{(\alpha)}(x) - (A_n x + B_n)F_n^{(\alpha)}(x) + F_{n-1}^{(\alpha)}(x) = 0 \tag{2.5}$$

with

$$A_n = \frac{n+\alpha}{n+1}\frac{K_n}{K_{n+1}} = \frac{n+\alpha}{\sqrt{2}}\frac{\Gamma\left(\frac{1}{2}(n+1)\right)}{\Gamma\left(\frac{1}{2}n+1\right)} \tag{2.6}$$

and $B_n = 0$. To find an asymptotic expansion for $A_n$, we recall the well-known result [10, p. 47]

$$\frac{\Gamma\left(\frac{1}{2}n+\frac{1}{2}\right)}{\Gamma\left(\frac{1}{2}n+1\right)} \sim \sqrt{\frac{2}{n}}\left[1 - \frac{1}{4n} + \frac{1}{32n^2} + \cdots\right].$$

Thus,

$$A_n \sim n^{1/2}\left[1 + (\alpha - \frac{1}{4})\frac{1}{n} + (\frac{1}{32} - \frac{\alpha}{4})\frac{1}{n^2} + \cdots\right]. \tag{2.7}$$

In terms of the notation

$$A_n \sim n^{-\theta}\sum_{s=0}^{\infty}\frac{\alpha_s}{n^s} \quad \text{and} \quad B_n \sim \sum_{s=0}^{\infty}\frac{\beta_s}{n^s}, \tag{2.8}$$

used in [9], we have

$$\theta = -\frac{1}{2}, \quad \alpha_0 = 1, \quad \alpha_1 = \alpha - \frac{1}{4}, \quad \alpha_2 = (\frac{1}{32} - \frac{\alpha}{4}), \cdots \tag{2.9}$$

and $\beta_0 = \beta_1 = \beta_2 = \cdots = 0$. If these expansions are recast in the form

$$A_n \sim \nu^{-\theta}\sum_{s=0}^{\infty}\frac{\alpha'_s}{\nu^s} \quad \text{and} \quad B_n \sim \sum_{s=0}^{\infty}\frac{\beta'_s}{\nu^s}, \tag{2.10}$$

where $\nu = n + \tau_0$ and $\tau_0$ is some fixed real number to be determined, it is easily found that

$$\alpha'_0 = 1, \quad \alpha'_1 = (\alpha - \frac{1}{4}) - \frac{\tau_0}{2}, \cdots \tag{2.11}$$

and $\beta'_0 = \beta'_1 = \beta'_2 = \cdots = 0$. To apply the result in [9], we first choose $\tau_0$ so that $\alpha'_1 = 0$. From (2.11), it is obvious that the choice is

$$\tau_0 = 2\alpha - \frac{1}{2}. \tag{2.12}$$

According to equation (2.4) in [9], the characteristic equation is

$$\lambda^2 - t\lambda + 1 = 0, \tag{2.13}$$

where $t$ is the rescaled variable $x = \nu^{-\frac{1}{2}}t$. The two roots of this equation are

$$\lambda_{\pm} = \frac{1}{2}(t \pm \sqrt{t^2 - 4}). \tag{2.14}$$

The points $t_\pm = \pm 2$ where the two roots coincide are called the *turning points* of equation (2.5). In view of the symmetry relation $F_n^{(\alpha)}(-x) = (-1)^n F_n^{(\alpha)}(x)$, we may restrict ourselves just to the case $0 < t < \infty$.

We now define the function $\zeta(t)$ introduced in [9, (4.10)]. With $t_+ = 2$, $\theta = -\frac{1}{2}$, $\alpha_0' = 1$ and $\beta_0' = 0$, this function is given by

$$\frac{2}{3}[\zeta(t)]^{3/2} := \log \frac{t+\sqrt{t^2-4}}{2} - \frac{1}{t^2}\int_2^t \frac{s^2}{\sqrt{s^2-4}}\,ds, \qquad t \geq 2,$$

and

$$\frac{2}{3}[-\zeta(t)]^{3/2} := t^{-2}\int_t^2 \frac{s^2}{\sqrt{4-s^2}}\,ds - \cos^{-1}\frac{t}{2}, \qquad 0 < t < 2.$$

By direct calculation, one obtains

(2.15)
$$\frac{2}{3}[\zeta(t)]^{3/2} = \log \frac{t+\sqrt{t^2-4}}{2} - \frac{1}{t^2}\left[\frac{1}{2}t\sqrt{t^2-4} + 2\log|t+\sqrt{t^2-4}| - 2\ln 2\right],$$

for $t \geq 2$, and

$$\frac{2}{3}[-\zeta(t)]^{3/2} = \frac{1}{t^2}\left[-2\sin^{-1}\frac{t}{2} + \pi + \frac{t}{2}\sqrt{4-t^2}\right] - \cos^{-1}\frac{t}{2}, \tag{2.16}$$

for $0 < t < 2$. We also define the functions $H_0(\zeta)$ and $\Phi(\zeta)$ introduced in [9, (4.19) and (4.28)]. In the present situation, these functions are given by

$$H_0(\zeta) = -\sqrt{\frac{t^2-4}{4\zeta}} \qquad \text{and} \qquad \Phi(\zeta) = 0, \tag{2.17}$$

where $\zeta$ is the function defined in (2.15) and (2.16). Note that in our special case, $\alpha_1' = \beta_1' = 0$; hence, according to the definition of $\Phi(\zeta)$ given in (4.28) of [9], the second equation in (2.17) holds for $0 < t < \infty$, instead of $t \geq \delta$, $0 < \delta < 2$.

With this preliminary work done, we can now apply the main result in [9] to conclude that there are constants $C_1(x)$ and $C_2(x)$ such that the polynomials $F_n^{(\alpha)}(x)$ in (2.4) can be expressed as

$$F_n^{(\alpha)}(x) = C_1(x)P_n(x) + C_2(x)Q_n(x), \tag{2.18}$$

where, with $x = \nu^{-\frac{1}{2}}t$, $P_n(x)$ and $Q_n(x)$ have the asymptotic expansions

$$P_n(\nu^{-\frac{1}{2}}t) \sim \left(\frac{4\zeta}{t^2-4}\right)^{\frac{1}{4}}\left[\mathrm{Ai}\left(\nu^{\frac{2}{3}}\zeta\right)\sum_{s=0}^{\infty}\frac{\widetilde{A}_s(\zeta)}{\nu^{s-\frac{1}{6}}} + \mathrm{Ai}'\left(\nu^{\frac{2}{3}}\zeta\right)\sum_{s=0}^{\infty}\frac{\widetilde{B}_s(\zeta)}{\nu^{s+\frac{1}{6}}}\right] \tag{2.19}$$

and

$$Q_n(\nu^{-\frac{1}{2}}t) \sim \left(\frac{4\zeta}{t^2-4}\right)^{\frac{1}{4}}\left[\mathrm{Bi}\left(\nu^{\frac{2}{3}}\zeta\right)\sum_{s=0}^{\infty}\frac{\widetilde{A}_s(\zeta)}{\nu^{s-\frac{1}{6}}} + \mathrm{Bi}'\left(\nu^{\frac{2}{3}}\zeta\right)\sum_{s=0}^{\infty}\frac{\widetilde{B}_s(\zeta)}{\nu^{s+\frac{1}{6}}}\right]. \tag{2.20}$$

In (2.19) and (2.20), $\mathrm{Ai}(\cdot)$ and $\mathrm{Bi}(\cdot)$ are the Airy functions, the leading coefficients are given by

$$\widetilde{A}_0(\zeta) = 1 \qquad \text{and} \qquad \widetilde{B}_0(\zeta) = 0, \tag{2.21}$$

and the expansions hold uniformly for $0 \leq t < \infty$.

## 3. Determination of $C_1(x)$ and $C_2(x)$

First we examine the behavior of $f_n^{(\alpha)}(x)$ as $n \to \infty$. By using the classical steepest decent method, one can show that to leading order, we have

$$
\begin{aligned}
f_n^{(\alpha)}(\frac{t}{\sqrt{\nu}}) \sim & \frac{\nu^{-\frac{n}{2}} e^{\nu\left(\frac{\lambda_+}{t}+(\frac{2}{t^2}-1)\log\lambda_+\right)}}{\sqrt{2\pi\nu}(t^2-4)^{\frac{1}{4}}}\cos(\pi\alpha-\frac{\nu\pi}{t^2}) \\
& + \frac{\nu^{-\frac{n}{2}} e^{\nu\left(\frac{\lambda_-}{t}+(\frac{2}{t^2}-1)\log\lambda_-\right)}}{\sqrt{2\pi\nu}(t^2-4)^{\frac{1}{4}}} 2\sin(\pi\alpha-\frac{\nu\pi}{t^2}),
\end{aligned} \tag{3.1}
$$

when $t > 2$;

(3.2)

$$
\begin{aligned}
f_n^{(\alpha)}(\frac{t}{\sqrt{\nu}}) \sim \frac{\sqrt{2}\nu^{-\frac{n}{2}} e^{\frac{\nu}{2}}}{\sqrt{\pi\nu}(4-t^2)^{\frac{1}{4}}} \sin\Bigg[\nu\left(\frac{1}{2t}\sqrt{4-t^2}+\frac{\pi}{t^2}-\frac{2}{t^2}\sin^{-1}\frac{t}{2}-\cos^{-1}\frac{t}{2}\right) \\
+\pi\alpha-\frac{\nu\pi}{t^2}+\frac{\pi}{4}\Bigg],
\end{aligned}
$$

when $0 < t < 2$ and

$$
f_n^{(\alpha)}(\frac{2}{\sqrt{\nu}}) \sim \nu^{-\frac{n}{2}} e^{\frac{\nu}{2}} 3^{\frac{1}{3}} \nu^{-\frac{1}{3}} \left(\frac{1}{3\Gamma(\frac{2}{3})}\cos(\pi\alpha-\frac{\nu\pi}{4}) + \frac{\sqrt{3}}{3\Gamma(\frac{2}{3})}\sin(\pi\alpha-\frac{\nu\pi}{4})\right), \tag{3.3}
$$

when $t = 2$. Next we recall the well-known asymptotic formulas

$$
\mathrm{Ai}(\eta) \sim \frac{\eta^{-\frac{1}{4}}}{2\sqrt{\pi}}\exp(-\frac{2}{3}\eta^{\frac{3}{2}}) \tag{3.4}
$$

and

$$
\mathrm{Bi}(\eta) \sim \frac{\eta^{-\frac{1}{4}}}{\sqrt{\pi}}\exp(\frac{2}{3}\eta^{\frac{3}{2}}) \tag{3.5}
$$

as $\eta \to +\infty$; see [5, p. 392]. Furthermore, since $\nu = n + 2\alpha - 1/2$, it is readily seen from (3.4) and (3.5) that

$$
\frac{P_n(\nu^{-\frac{1}{2}}t)}{2^{\frac{n}{2}}\Gamma(\frac{n}{2}+1)} \sim \frac{\nu^{-\frac{n}{2}}\nu^{-\frac{1}{2}}}{\sqrt{2}\,\pi(t^2-4)^{\frac{1}{4}}} e^{\nu\left(\frac{\lambda_+}{t}+(\frac{2}{t^2}-1)\log\lambda_+\right)},
$$

$$
\frac{Q_n(\nu^{-\frac{1}{2}}t)}{2^{\frac{n}{2}}\Gamma(\frac{n}{2}+1)} \sim \frac{\sqrt{2}\nu^{-\frac{n}{2}}\nu^{-\frac{1}{2}}}{\pi(t^2-4)^{\frac{1}{4}}} e^{\nu\left(\frac{\lambda_-}{t}+(\frac{2}{t^2}-1)\log\lambda_-\right)}.
$$

Comparing the two sides of (2.18), one concludes that

$$
C_1(x) = \sqrt{\pi}\cos(\pi\alpha-\frac{\pi}{x^2}), \qquad C_2(x) = \sqrt{\pi}\sin(\pi\alpha-\frac{\pi}{x^2}). \tag{3.6}
$$

Note that in obtaining (3.6), any one of the three formulas (3.1), (3.2) or (3.3) could have been used. In summary, we have from (2.4), (2.18), (2.19) and (2.20)

$$
\begin{aligned}
f_n^{(\alpha)}(\frac{t}{\sqrt{\nu}}) = & \frac{\sqrt{\pi}\cos(\pi\alpha-\frac{\nu\pi}{t^2})}{2^{\frac{1}{2}n}\Gamma(\frac{1}{2}n+1)}\left(\frac{4\zeta}{t^2-4}\right)^{\frac{1}{4}}\nu^{\frac{1}{6}}\left[\mathrm{Ai}\left(\nu^{\frac{2}{3}}\zeta\right)+O(\frac{1}{\nu})\right] \\
& + \frac{\sqrt{\pi}\sin(\pi\alpha-\frac{\nu\pi}{t^2})}{2^{\frac{1}{2}n}\Gamma(\frac{1}{2}n+1)}\left(\frac{4\zeta}{t^2-4}\right)^{\frac{1}{4}}\nu^{\frac{1}{6}}\left[\mathrm{Bi}\left(\nu^{\frac{2}{3}}\zeta\right)+O(\frac{1}{\nu})\right],
\end{aligned} \tag{3.7}
$$

where $x = \nu^{-\frac{1}{2}}t$ and $\nu = n + 2\alpha - 1/2$. This result holds uniformly for $0 \le t < \infty$. As a check on the validity of (3.7), we note that

$$f_{2n}^{(\alpha)}(0) = \frac{(-1)^n}{2^n n!}, \quad f_{2n+1}^{(\alpha)}(0) = 0, \quad n = 0, 1, 2, \ldots. \tag{3.8}$$

From (3.7), we obtain

$$f_n^{(\alpha)}(0) \sim \frac{e^{n/2}}{\sqrt{\pi} n^{(n+1)/2}} \sin(-\frac{n\pi}{2} + \frac{\pi}{2}). \tag{3.9}$$

In view of Stirling's formula, (3.9) agrees with (3.8).

## 4. Numerical results

The expansion (3.7) is particularly useful near $t = 2$, where the two characteristic roots in (2.14) coincide. As an illustration, we take $\alpha = 1.9$ and $n = 100$. Table 1 provides exact and approximate values of $2^{n/2}\Gamma(n/2+1)f_n^{(\alpha)}(t/\sqrt{\nu})$. The last column of the table shows the percentage error of the approximations.

TABLE 1. Numerical Results

| t | Exact values | Approximate values | Error |
|---|---|---|---|
| 1.0 | $-7.46841313\times10^{-1}$ | $-7.37169075\times10^{-1}$ | 0.0130 |
| 1.2 | $6.30560302\times10^{-1}$ | $6.34644034\times10^{-1}$ | 0.0065 |
| 1.4 | $8.46491439\times10^{-1}$ | $8.31094179\times10^{-1}$ | 0.0182 |
| 1.6 | -1.30620957 | -1.29091632 | 0.0117 |
| 1.8 | 1.51877275 | 1.49426911 | 0.0161 |
| 2.0 | 1.90931070 | 1.89864389 | 0.0056 |
| 2.2 | $3.26791053\times10^{2}$ | $3.16621347\times10^{2}$ | 0.0311 |
| 2.4 | $-9.21372780\times10^{4}$ | $-8.83780892\times10^{4}$ | 0.0408 |
| 2.6 | $3.94171949\times10^{9}$ | $3.74229636\times10^{9}$ | 0.0506 |
| 2.8 | $1.54972056\times10^{13}$ | $1.45539824\times10^{13}$ | 0.0609 |
| 3.0 | $7.21884326\times10^{16}$ | $6.70106781\times10^{16}$ | 0.0717 |

## References


1. L. Carlitz, On some polynomials of Tricomi, *Boll. Un. Mat. Ital.*, **13**, 58-64, 1958. MR0103303 (21:2078)
2. W.M.Y. Goh and J. Wimp, On the asymptotics of the Tricomi-Carlitz polynomials and their zero distribution. I, *SIAM J. Math. Anal.*, **25**, 420-428, 1994. MR1266567 (95b:42025)
3. W.M.Y. Goh and J. Wimp, The zero distribution of the Tricomi-Carlitz polynomials. Approximation theory and applications, *Comput. Math. Appl.*, **33**, 119-127, 1997. MR1442066 (99e:33006)
4. J.L. López and N.M. Temme, Approximation of orthogonal polynomials in terms of Hermite polynomials. Dedicated to Richard A. Askey on the occasion of his 65th birthday, Part II, *Methods Appl. Anal.*, **6**, 131–146, 1999. MR1803886 (2001m:33013)
5. F.W.J. Olver, *Asymptotics and Special Functions*, Academic Press, New York, 1974. (Reprinted by A. K. Peters, Ltd., Wellesley, 1997.) MR0435697 (55:8655)
6. G. Szegö, *Orthogonal Polynomials*, 4th edition, *Amer. Math. Soc. Colloq. Publ.*, 23, Amer. Math. Soc., Providence, RI, 1975. MR0372517 (51:8724)
7. F.G. Tricomi, A class of non-orthogonal polynomials related to those of Laguerre, *J. Analyse Math.*, **13**, 209–231, 1951. MR0051351 (14:466e)
8. Z. Wang and R. Wong, Uniform asymptotic expansion of $J_\nu(\nu a)$ via a difference equation, *Numer. Math.*, **91**, 147–193, 2002. MR1896091 (2003g:33008)

9. Z. Wang and R. Wong, Asymptotic expansions for second-order linear difference equations with a turning point, *Numer. Math.*, **94**, 147–194, 2003. MR1971216 (2004c:39012)
10. R. Wong, *Asymptotic Approximations of Integrals*, Academic Press, Boston, 1989. MR1016818 (90j:41061)

Department of Mathematics, City University of Hong Kong, 83 Tat Chee Avenue, Kowloon, Hong Kong
*E-mail address*: charleslkf8571@gmail.com

Department of Mathematics, City University of Hong Kong, 83 Tat Chee Avenue, Kowloon, Hong Kong
*E-mail address*: rscwong@cityu.edu.hk

PROCEEDINGS OF THE
AMERICAN MATHEMATICAL SOCIETY
Volume 141, Number 8, August 2013, Pages 2683–2691
S 0002-9939(2013)11587-X
Article electronically published on March 29, 2013

# UNIFORM TREATMENT OF DARBOUX'S METHOD AND THE HEISENBERG POLYNOMIALS

SAI-YU LIU, R. WONG, AND YU-QIU ZHAO

(Communicated by Walter Van Assche)

ABSTRACT. We show that the set of Heisenberg polynomials furnishes a simple non-trivial example in the uniform treatment of Darboux's method.

## 1. THE HEISENBERG POLYNOMIALS

The Heisenberg polynomials are polynomials in $z$ and $\bar{z}$, defined explicitly as

$$C_n^{(\alpha,\beta)}(z) = \sum_{j=0}^{n} \frac{(\alpha)_j(\beta)_{n-j}}{j!(n-j)!} \bar{z}^j z^{n-j}, \quad n = 0, 1, \cdots, \tag{1.1}$$

where $\alpha$ and $\beta$ are real constants, and $(\gamma)_k$ is the Pochhammer symbol defined by $(\gamma)_0 = 1$ and $(\gamma)_k = \gamma(\gamma+1)\cdots(\gamma+k-1)$. The representation can be readily derived from the generating function

$$(1-w\bar{z})^{-\alpha}(1-wz)^{-\beta} = \sum_{n=0}^{\infty} C_n^{(\alpha,\beta)}(z) w^n, \qquad |wz| < 1. \tag{1.2}$$

The notation $C_n^{(\alpha,\beta)}(z)$ was used by Gasper in [8] in the sense of (1.1) and (1.2), and yet the term *Heisenberg polynomials* was first used by Dunkl in [2].

The study of Heisenberg polynomials may be traced back to the investigation of the harmonic polynomials on Heisenberg group $H_N$. The group $H_N$ consists of $(N+1)$-tuples in $\mathbb{C}^N \times \mathbb{R}$, $N = 1, 2, 3 \cdots$, with the group operation

$$(\zeta, s) \cdot (z, t) = (\zeta + z, s + t + 2\,\mathrm{Im}\langle \zeta, z \rangle), \tag{1.3}$$

where $\zeta, z \in \mathbb{C}^N$, $s, t \in \mathbb{R}$, and $\langle \zeta, \xi \rangle = \sum_{j=1}^{N} \zeta_j \bar{\xi}_j$; cf. Dunkl [2] and Greiner [9].

The Heisenberg group $H_N$ arose in various branches of mathematics, such as harmonic analysis [10, 16] and partial differential operators and equations [6, 7]. The reason may partially be that the results in Abelian harmonic analysis could be adapted to Heisenberg groups.

The so-called $L_{2\gamma}$-harmonic polynomials are the polynomial solutions $p = p(z_1, \bar{z}_1, \cdots z_N, \bar{z}_N, t)$ of $L_{2\gamma} p = 0$ in $H_N$, where the differential operator $L_{2\gamma}$ is

Received by the editors August 18, 2011 and, in revised form, October 22, 2011.
2010 *Mathematics Subject Classification.* Primary 41A60; Secondary 33C15.
*Key words and phrases.* Heisenberg polynomials, Darboux's method, uniform asymptotic expansion, confluent hypergeometric function, Bessel function.
The research of the third author was supported in part by the National Natural Science Foundation of China under grant numbers 10471154 and 10871212.

defined by

$$L_{2\gamma} = -\frac{1}{2}\sum_{j=1}^{N}(Z_j\overline{Z}_j - \overline{Z}_jZ_j) + 2i\gamma\frac{\partial}{\partial t} \tag{1.4}$$

and

$$Z_j = \frac{\partial}{\partial z_j} + i\bar{z}_j\frac{\partial}{\partial t}, \ \overline{Z}_j = \frac{\partial}{\partial \bar{z}_j} - i\bar{z}_j\frac{\partial}{\partial t},$$

$\gamma$ being a constant. In particular, $L_0$ in $H_1$ is a sub-elliptic Laplacian on $\mathbb{R}^3$; cf. [6, 7].

Greiner [9] initiated the study of $L_{2\gamma}$-spherical harmonic polynomials. It is later shown by Dunkl (cf. [2]) that every such polynomial can be expressed in terms of $C_n^{(\alpha,\beta)}$.

A few facts are known about the Heisenberg polynomials. For example, the representation (1.1) of $C_n^{(\alpha,\beta)}$ can be interpreted as a Gauss hypergeometric function. The polynomials $C_n^{(\alpha,\beta)}(z)$ reduce to the Legendre polynomials on the unit circle when $\alpha = \beta = \frac{1}{2}$. The reader is referred to [2] for more properties of these polynomials.

We mention several relevant investigations on the Heisenberg polynomials. In 1981, Gasper [8] indicated that $\{C_n^{(\alpha,\beta)}(e^{i\theta})\}$ furnishes a system of orthogonal functions on $[0, 2\pi]$, with weight function $w^{(\alpha,\beta)}(\theta) = \left(1 - e^{-2i\theta}\right)^{\alpha}\left(1 - e^{2i\theta}\right)^{\beta}$, where $\alpha$, $\beta$, $\alpha + \beta > -1$.

In 1984, Dunkl [2] obtained a bi-orthogonality relation for these functions in $L^2(T, |\sin\theta|^{\alpha+\beta}d\theta)$, where $\alpha > 0$, $\beta > 0$, and $T = \{e^{i\theta}| - \pi < \theta \le \pi\}$. A 1986 paper [3] of Dunkl addressed the density problem for $\left\{C_n^{(\alpha,\beta)}(e^{i\theta})\right\}$ in $L_2$.

In 1986, Temme [15] considered two sets of bi-orthogonal polynomials, $P_n(z;\alpha,\beta)$ and $Q_n(z;\alpha,\beta)$, closely related to the Heisenberg polynomials, where

$$P_n(z;\alpha,\beta) = F(-n, \alpha+\beta+1; 2\alpha+1; 1-z) \text{ and } Q_n(z;\alpha,\beta) = P_n(z;\alpha,-\beta),$$

with $F(a, b; c; z)$ being the Gauss hypergeometric function; cf. [13, p. 384]. The connection with the Heisenberg polynomials is

$$P_n(e^{i\theta};\alpha,\beta) = \frac{n!}{(2\alpha+1)_n}e^{in\theta/2}C_n^{(\alpha-\beta,\alpha+\beta+1)}(e^{i\theta/2}). \tag{1.5}$$

The two sets $\{P_n\}$ and $\{Q_n\}$ are bi-orthogonal on the unit circle with respect to the weight function $(1 - e^{i\theta})^{\alpha+\beta}(1 - e^{-i\theta})^{\alpha-\beta}$. Based on an integral representation of the Gauss hypergeometric function, uniform asymptotic expansions for $P_n(z;\alpha,\beta)$ were obtained in [15], including error bounds. These expansions are in terms of confluent hypergeometric functions, and show that all points on the unit circle in the $z$-plane are transition points in the sense that the polynomials have different asymptotic behavior inside and outside the unit disk. To ensure the convergence of the integral representation, extra restrictions on the parameters, namely $\alpha+\beta > -1$ and $\alpha - \beta \ge 0$, are required in [15].

It is readily seen that the Heisenberg polynomials are homogeneous functions in $z$ and $\bar{z}$, which implies that $C_n^{(\alpha,\beta)}(\rho e^{i\theta}) = \rho^n C_n^{(\alpha,\beta)}(e^{i\theta})$. Hence, to study the behavior of the polynomials, it is natural to consider the polynomials on the unit circle. The generating function (1.2) now takes the form

$$(1.6)\qquad \left(1-ze^{-i\theta}\right)^{-\alpha}\left(1-ze^{i\theta}\right)^{-\beta}=\sum_{n=0}^{\infty}C_n^{(\alpha,\beta)}(e^{i\theta})z^n$$

for $|z|<1$ and $\theta\in[0,\pi]$.

In what follows, we will show that the set of Heisenberg polynomials on the unit circle provides a simple non-trivial example in the uniform treatment of Darboux's method given in [18].

## 2. Uniform treatment of Darboux's method

Darboux was the first to consider the asymptotic behavior of the coefficients $a_n$ in the Maclaurin expansion

$$(2.1)\qquad F(z)=\sum_{n=0}^{\infty}a_nz^n,$$

where $F(w)$ has only a finite number of singularities on its circle of convergence, all of which are algebraic in nature. A method was introduced to obtain the asymptotic expansion for $a_n$ as $n\to\infty$. The method, named after Darboux, indicates that the contribution to the expansion comes from the singularities on the circle of convergence; cf., e.g., Olver [12] and Wong [17] for the classical Darboux method.

If the singularities are allowed to move around on the circle of convergence, Darboux's method will continue to work provided that their essential configuration remains the same (although their relative positions vary). However, this method breaks down when two or more singularities coalesce with each other. A uniform treatment was later given by Fields [5], in which he considered generating functions of the type

$$(2.2)\qquad F(z,\theta)=(1-z)^{-\lambda}\left[\left(e^{i\theta}-z\right)\left(e^{-i\theta}-z\right)\right]^{-\Delta}f(z,\theta)=\sum_{n=0}^{\infty}a_n(\theta)z^n$$

for $|z|<1$, where $\lambda$ and $\Delta$ are bounded quantities, the branches are chosen such that both $(1-z)^{-\lambda}$ and $\left[\left(e^{i\theta}-z\right)\left(e^{-i\theta}-z\right)\right]^{-\Delta}$ reduce to 1 at $z=0$, and $f(z,\theta)$ is analytic in $|z|<e^{\eta}$ with $\eta>0$ not depending on $\theta$. The results in [5] were regarded as being too complicated to apply; see, e.g., [4, p. 167].

The problem of uniform treatment was revisited by Wong and Zhao [18], in which they first addressed a special case of (2.2) with $\lambda=0$, namely

$$(2.3)\qquad F(z,\theta)=\left[\left(e^{i\theta}-z\right)\left(e^{-i\theta}-z\right)\right]^{-\alpha}f(z,\theta)=\sum_{n=0}^{\infty}a_n(\theta)z^n.$$

Uniform asymptotic expansion was derived in terms of the Bessel functions $J_{\alpha\pm1/2}(n\theta)$, valid for $\theta\in[0,\pi-\varepsilon]$, $\varepsilon>0$. The general case considered in [18] deals with many coalescing singularities,

$$(2.4)\qquad F(z,\theta)=\left\{\prod_{k=1}^{q}(z_k(\theta)-z)^{-\alpha_k}\right\}f(z,\theta)=\sum_{n=0}^{\infty}a_n(\theta)z^n,$$

where $|z_k(\theta)|=1$, $q\geq2$, $\alpha_k$ are constants, and $f(z,\theta)$ is the same function as in (2.2) and (2.3). In (2.4), all the branch points $z_k(\theta)$ approach 1 as $\theta$ tends to 0; more precisely, $z_k(\theta)=e^{i\theta s_k(\theta)}$ with $s_k(0)=-iz_k'(0)$ not vanishing. A uniform asymptotic expansion for the general case was presented in [18] in terms of a set of special functions $T_l$, $l = 1, 2, \cdots, q$.

In a follow-up paper, Bai and Zhao [1] extended the essential ideas in [18] to study the classical Jacobi polynomials.

As mentioned in the last section and shown in (1.6), the set of Heisenberg polynomials serves as a good example of the uniform Darboux treatment, which is simple (only two singularities) and non-trivial (the exponents $\alpha$ and $\beta$ may be different). To derive the asymptotic expansion of $C_n^{(\alpha,\beta)}(e^{i\theta})$, we apply Theorem 2 of [18]. To this end, we define

$$T_l(x) = \frac{1}{2\pi i}\int_{\Gamma_0} s^{l-1}(s-i)^{-\beta}(s+i)^{-\alpha}e^{xs}ds,\ l=1,2, \tag{2.5}$$

where $\Gamma_0$ is a Hankel-type loop which starts and ends at $-\infty$ and encircles $s = \pm i$ in the positive sense; cf. [18, (4.7)]. We further introduce an auxiliary function

$$h_0(s,\theta) = e^{i\theta(\alpha-\beta)}\left[\frac{e^{i\theta}-e^{-\theta s}}{(s+i)\theta}\right]^{-\alpha}\left[\frac{e^{-i\theta}-e^{-\theta s}}{(s-i)\theta}\right]^{-\beta} \tag{2.6}$$

and a sequence of functions $\{h_k(s,\theta)\}_{k=0}^{\infty}$ defined inductively by

(2.7)

$$h_k = \alpha_k(\theta)+s\beta_k(\theta)+(s^2+1)g_k(s,\theta),\ h_{k+1}=\frac{s^2+1}{\theta}\left\{\frac{\alpha-1}{s+i}+\frac{\beta-1}{s-i}-\frac{d}{ds}\right\}g_k(s,\theta)$$

for $k = 0, 1, \cdots$. The coefficients $\alpha_k(\theta)$ and $\beta_k(\theta)$ are determined by requiring all $h_k(s,\theta)$ and $g_k(s,\theta)$ to be analytic in $\mathcal{D} = \{s : \mathrm{Re}\, s \geq -\frac{\eta}{\theta}, |s \pm i| < \frac{2\pi}{\theta}\}$. The uniform asymptotic expansion is given in the following theorem:

**Theorem 1.** *For $\theta \in [0, \pi - \delta]$ with arbitrary $\delta > 0$, we have*

$$C_n^{(\alpha,\beta)}(e^{i\theta}) = \theta^{1-\alpha-\beta}T_1(n\theta)\sum_{k=0}^{m-1}\frac{\alpha_k(\theta)}{n^k} + \theta^{1-\alpha-\beta}T_2(n\theta)\sum_{k=0}^{m-1}\frac{\beta_k(\theta)}{n^k} + \varepsilon_{\theta,m}, \tag{2.8}$$

*where $|\alpha_k(\theta)| \leq M_k$, $|\frac{\beta_k(\theta)}{\theta}| \leq M_k$, for $k = 0, 1, 2, \cdots$, and*

$$|\varepsilon_{\theta,m}| \leq M_m\theta^{1-\alpha-\beta}n^{-m}\{|T_1(n\theta)| + |T_2(n\theta)|\}, \tag{2.9}$$

*for $m = 1, 2, \cdots$. The positive constants $M_k$, $k = 1, 2, \cdots$, are independent of $\theta$ for $\theta \in [0, \pi - \delta]$. The coefficients $\alpha_k(\theta)$ and $\beta_k(\theta)$ are given iteratively in* (2.7), *with*

$$\alpha_0=\frac{e^{i\theta\alpha}}{2}\left(\frac{\sin\theta}{\theta}\right)^{-\alpha}+\frac{e^{-i\theta\beta}}{2}\left(\frac{\sin\theta}{\theta}\right)^{-\beta},\ \beta_0=\frac{e^{i\theta\alpha}}{2i}\left(\frac{\sin\theta}{\theta}\right)^{-\alpha}-\frac{e^{-i\theta\beta}}{2i}\left(\frac{\sin\theta}{\theta}\right)^{-\beta}.$$

## 3. Special functions $T_1$ and $T_2$

By expanding the slowly varying factor in the integrand of (2.5) in a uniformly convergent power series of $1/s$ and integrating term by term, we obtain

$$T_1(x) = x^{\alpha+\beta-1}\sum_{k=0}^{\infty}\sum_{l=0}^{k}\frac{i^k(-1)^l(\alpha)_l(\beta)_{k-l}}{l!(k-l)!\Gamma(\alpha+\beta+k)}x^k, \tag{3.1}$$

where $x^{\alpha+\beta-1}$ is positive for real positive $x$. It is easily seen that $x^{-\alpha-\beta+1}T_1(x)$ is an entire function. From (2.5), it is also readily verified that $T_2(x) = T_1'(x)$ in the cut plane $x \in \mathbb{C}\backslash(-\infty, 0]$.

The behavior of $T_1(x)$ at the origin is exhibited by the leading terms in (3.1). The asymptotic behavior of $T_l(x)$ as $x \to +\infty$ has already been given in [18, (4.45)]. Moreover, with $a = 2-\alpha-\beta$ and $b = \beta-\alpha$, $T_1(x)$ satisfies the differential equation

$$xT_1'' + aT_1' + (x - bi)T_1 = 0; \tag{3.2}$$

cf. [18, (4.52)]. This equation is of Laplace type in the sense that the coefficients are linear in $x$. We proceed to show that $T_1(x)$ is connected to the confluent hypergeometric function via the following change of variables:

$$T_1(x) = x^{1-a}e^{ix}y(x), \; z = -2ix. \tag{3.3}$$

Substituting (3.3) in (3.2) leads to the Kummer equation

$$z\frac{d^2y}{dz^2} + [(2-a) - z]\frac{dy}{dz} - \frac{2-a-b}{2}y = 0. \tag{3.4}$$

Taking into account the first two terms of the infinite series in (3.1), we have

$$T_1(x) = \frac{1}{\Gamma(\alpha+\beta)}x^{\alpha+\beta-1}e^{ix}M(\alpha, \alpha+\beta, -2ix), \tag{3.5}$$

where $M$ is the Kummer function; cf. [13, p. 322]. (One may also derive this result directly from (2.5) and a Laplace inversion integral of this Kummer function [13, p. 327, (13.4.13)].) Accordingly, from $T_2(x) = T_1'(x)$ one has

$$\begin{aligned} T_2(x) =& \frac{(\alpha+\beta-1+ix)x^{\alpha+\beta-2}e^{ix}}{\Gamma(\alpha+\beta)}M(\alpha, \alpha+\beta, -2ix) \\ &- \frac{2ix^{\alpha+\beta-1}e^{ix}}{\Gamma(\alpha+\beta)}M'(\alpha, \alpha+\beta, -2ix), \end{aligned} \tag{3.6}$$

where $M'(\gamma, \delta, z) = \frac{d}{dz}M(\gamma, \delta, z)$. Thus, substituting (3.5) and (3.6) in (2.8), we obtain an asymptotic expansion of the Heisenberg polynomials in terms of the Kummer function as follows:

**Theorem 2.** *Assume that $\alpha$ and $\beta$ are real and fixed and that $z = \rho e^{i\theta}$ with $\rho > 0$ and $\theta$ real. Then we have the compound asymptotic expansion (§4.2 of Chapter 4 in [12])*

$$\begin{aligned} C_n^{(\alpha,\beta)}(z) \sim n^{\alpha+\beta-1}z^n \Bigg\{ & M(\alpha, \alpha+\beta, -2in\theta)\sum_{k=0}^{\infty}\frac{c_k(\theta)}{n^k} \\ & + M'(\alpha, \alpha+\beta, -2in\theta)\sum_{k=0}^{\infty}\frac{d_k(\theta)}{n^k}\Bigg\} \end{aligned} \tag{3.7}$$

*as $n \to \infty$, uniformly with respect to $\rho \in (0, \infty)$ and $\theta \in [0, \pi-\delta]$, where $\delta$ is an arbitrary constant such that $0 < \delta \le \pi$. Here the coefficients are given by*

$$c_k(\theta) = \frac{\alpha_k(\theta) + i\beta_k(\theta)}{\Gamma(\alpha+\beta)} + \frac{\beta_k(\theta)/\theta}{\Gamma(\alpha+\beta-1)}, \; d_k(\theta) = \frac{-2i\beta_k(\theta)}{\Gamma(\alpha+\beta)} \tag{3.8}$$

*for $k = 0, 1, 2, \cdots$, $\alpha_k(\theta)$ and $\beta_k(\theta)$ being defined as in Section 2. In particular,*

$$c_0(\theta) = \frac{e^{i\theta\alpha}}{\Gamma(\alpha+\beta)}\left(\frac{\sin\theta}{\theta}\right)^{-\alpha}, \; d_0(\theta) = \frac{e^{-i\theta\beta}}{\Gamma(\alpha+\beta)}\left(\frac{\sin\theta}{\theta}\right)^{-\beta} - \frac{e^{i\theta\alpha}}{\Gamma(\alpha+\beta)}\left(\frac{\sin\theta}{\theta}\right)^{-\alpha}.$$

When the parameters $\alpha$ and $\beta$ in the above theorem are non-positive integers, the coefficients $c_k$ and $d_k$ in (3.7) all vanish; see (3.8). However, the asymptotic relation remains valid since the polynomials also vanish for large values of $n$; i.e., it is a trivial result.

For completeness, we write, by using (2.5) and (3.2), the asymptotic expansion in the special case $\alpha = \beta$; see also [18]. In this case, we have

$$T_1(x) = \frac{\sqrt{\pi}}{\Gamma(\alpha)}\left(\frac{x}{2}\right)^{\alpha-1/2} J_{\alpha-1/2}(x),\ T_2(x) = \frac{\sqrt{\pi}}{\Gamma(\alpha)}\left(\frac{x}{2}\right)^{\alpha-1/2} J_{\alpha-3/2}(x), \tag{3.9}$$

where $J_\nu(x)$ is the Bessel function of the first kind; cf. [13, p. 217]. The expansion in Theorems 1 and 2 now takes the form:

**Corollary 1.** *Assume that $\alpha$ is real and fixed and that $z = \rho e^{i\theta}$ with $\rho > 0$ and $\theta$ real. Then we have the compound asymptotic expansion*

$$C_n^{(\alpha,\alpha)}(z) \sim \rho^n \frac{\sqrt{\pi}}{\Gamma(\alpha)}\left(\frac{n}{2\theta}\right)^{\alpha-1/2}\left\{J_{\alpha-1/2}(n\theta)\sum_{k=0}^{\infty}\frac{\alpha_k(\theta)}{n^k} + J_{\alpha-3/2}(n\theta)\sum_{k=0}^{\infty}\frac{\beta_k(\theta)}{n^k}\right\} \tag{3.10}$$

*as $n \to \infty$, uniformly with respect to $\rho \in (0, \infty)$ and $\theta \in [0, \pi - \delta]$, where $\delta$ is an arbitrary constant such that $0 < \delta \le \pi$. Here $\alpha_k(\theta)$ and $\beta_k(\theta)$ are given in (2.7), and the leading coefficients are*

$$\alpha_0(\theta) = \cos(\alpha\theta)\left(\frac{\sin\theta}{\theta}\right)^{-\alpha} \quad \text{and} \quad \beta_0(\theta) = \sin(\alpha\theta)\left(\frac{\sin\theta}{\theta}\right)^{-\alpha}. \tag{3.11}$$

## 4. Asymptotic zeros

We complete this note by giving some remarks on the zeros of the Heisenberg polynomials. By (2.8), these zeros are approximated by the zeros of $T_1(n\theta)$. Thus, we shall later turn to consider the zeros of $T_1$. But, as a consistency check on our result, we first consider the special case $\alpha = \beta > -1/2$. In this case, the Heisenberg polynomials on the unit circle are related to the Gegenbauer polynomials via the identity

$$C_n^{(\alpha,\alpha)}(e^{i\theta}) = C_n^{(\alpha)}(\cos\theta),\ \theta \in [0, \pi], \tag{4.1}$$

which can be seen by comparing the generating functions in (1.6) and in [13, p. 448]. In this specific case, $C_n^{(\alpha,\alpha)}(e^{i\theta})$ is known to have $n$ distinct zeros in the interval $0 < \theta < \pi$.

The asymptotic behavior of the extreme zeros can be obtained from (3.10) and (3.11). Indeed, when $n\theta$ is bounded, the terms in the curly brackets in (3.10) can be written as $H := (\alpha_0 + \alpha_1/n)\, J_{\alpha-1/2}(n\theta) + \beta_0 J_{\alpha-3/2}(n\theta) + O(n^{-2})$; here, we have used the fact that $\beta_k(\theta) = O(\theta) = O(1/n)$ for $k = 0, 1, 2, \cdots$, as mentioned in Theorem 1. The quantity $H$ can be further expressed as

$$H = \Omega\left[J_{\alpha-1/2}(n\theta) + \Delta J'_{\alpha-1/2}(n\theta) + O(n^{-2})\right] = \Omega\left[J_{\alpha-1/2}(n\theta + \Delta) + O(n^{-2})\right], \tag{4.2}$$

where $\Omega = \alpha_0 + \{\alpha_1 + (\alpha - \frac{1}{2})\beta_0/\theta\}n^{-1}$ and $\Delta = \beta_0/\Omega$ both depend on $\theta$ and $n$. Note that $\Omega = \alpha_0 + O(1/n) = 1 + O(1/n)$ and $\Delta = \{n\theta\Omega^{-1}\beta_0/\theta\}/n = O(1/n)$ as

$n \to \infty$. In (4.2), we have used the Taylor expansion of $J_{\alpha-1/2}(x)$ at $x = n\theta$ and the formula

$$J_{\alpha-3/2}(x) = J'_{\alpha-1/2}(x) + \frac{2\alpha - 1}{2x} J_{\alpha-1/2}(x);$$

cf. [13, p. 222]. Hence, $C_n^{(\alpha,\alpha)}(z) = 0$ amounts to $H = 0$, which implies

$$J_{\alpha-1/2}(n\theta + \Delta) = O(n^{-2}).$$

From (3.11), it is readily seen that $\Delta = \alpha\theta + O(1/n^2)$. Denoting by $j_{\alpha-1/2,k}$ the $k$th positive zero of $J_{\alpha-1/2}(x)$, we have the asymptotic approximation

$$\theta_{n,k} = \frac{j_{\alpha-1/2,k}}{n+\alpha} + O(n^{-3}) \tag{4.3}$$

for the $k$th zero of $C_n^{(\alpha,\alpha)}(e^{i\theta})$ arranged in increasing order, where $k = 1, 2, \cdots$ are fixed and $\alpha > -1/2$. In deriving (4.3), we have made use of the fact that $J'_{\alpha-1/2}$ does not vanish in a neighborhood of $j_{\alpha-1/2,k}$.

If we relax the restriction that $k$ be fixed while still requiring $\theta = O(1/n)$ and $n \gg k \gg 1$, then formula (4.3) would continue to hold, with only a weaker error estimate. For instance, making use of the fact that

$$j_{\alpha-1/2,k} \sim (k + (\alpha - 1)/2)\,\pi$$

for $\alpha > 1/2$ and $k \gg 1$ (cf. [13, p. 236, (10.21.19)]) and substituting this into (4.3) gives

$$\theta_{n,k} \sim \frac{(\alpha-1)/2 + k}{n+\alpha}\pi \tag{4.4}$$

for $\alpha > 1/2$ and $n \gg k \gg 1$. The formula in (4.4) agrees with that in [14, p. 139, (6.6.3)] for $1/2 < \alpha < 1$.

On the other hand, taking the two leading terms in (3.10)-(3.11) and making use of the asymptotic approximation for the Bessel function [13, p. 223, (10.7.8)],

$$J_\nu(x) = \sqrt{2/(\pi x)}\cos(x - \nu\pi/2 - \pi/4) + O(x^{-1}) \text{ as } x \to +\infty, \tag{4.5}$$

we have from (3.10)

$$C_n^{(\alpha,\alpha)}(z) = \rho^n \frac{\theta^{-\alpha}}{\Gamma(\alpha)} \left(\frac{n}{2}\right)^{\alpha-1} \left(\frac{\sin\theta}{\theta}\right)^{-\alpha} \left\{\cos\left((n+\alpha)\theta - \frac{\alpha\pi}{2}\right) + O\left(\frac{1}{n\theta}\right)\right\} \tag{4.6}$$

for $n\theta \gg 1$ and $\theta \in (0, \pi - \delta]$. It is readily seen that this asymptotic approximation agrees well with (4.4).

It is worth mentioning that if $\theta = \theta_{n,\nu}$ is a zero of $C_n^{(\alpha,\alpha)}(e^{i\theta})$, then all points on the ray $\arg z = \theta_{n,\nu}$ are zeros of $C_n^{(\alpha,\alpha)}(z)$. This fact follows readily from the homogeneity of the Heisenberg polynomials, namely, $C_n^{(\alpha,\beta)}(\rho e^{i\theta}) = \rho^n C_n^{(\alpha,\beta)}(e^{i\theta})$ for $\rho > 0$ and $\theta \in (-\pi, \pi]$.

Now, we turn to the asymptotic zeros of the special function $T_1(x)$, whose asymptotic properties were investigated in [18]. It is of interest to consider the zeros of $T_1(x)$ for $x > 0$ when $\alpha \neq \beta$. Without loss of generality, we assume that $\beta > \alpha$.

**Proposition 1.** *The function $T_1(x)$ possesses no positive zeros when $\alpha + \beta > 0$.*

*Proof.* We use an argument similar to that given in Ince [11, p. 512]. Write $T_1(x) = x^{\frac{1}{2}(\alpha+\beta)-1}I(x)$. Equation (3.2) gives

$$I''(x) + [g_1(x) + ig_2(x)]I(x) = 0, \tag{4.7}$$

where the functions

$$g_1(x) = 1 - \frac{a(a-2)}{4x^2} \text{ and } g_2(x) = -\frac{b}{x}$$

are real and $g_2(x) < 0$ for $x > 0$. From (4.7) we have

$$(\bar{I}I')' = |I'|^2 + \bar{I}I'' = \left\{|I'|^2 - g_1|I|^2\right\} - i\left\{g_2(x)|I|^2\right\}.$$

Integrating over an arbitrary interval $[x_1, x_2]$ and taking the imaginary part, one obtains

$$\operatorname{Im}\left[\overline{I(x)}I'(x)\right]_{x_1}^{x_2} + \int_{x_1}^{x_2} g_2(x)|I(x)|^2 dx = 0.$$

From (3.1), it is straightforward to verify that when $\alpha+\beta>0$, $\lim_{x\to 0^+}\operatorname{Im}\left[\overline{I(x)}I'(x)\right] = 0$. Hence, for $x > 0$,

$$\operatorname{Im}\left[\overline{I(x)}I'(x)\right] = -\int_0^x g_2(s)|I(s)|^2 ds > 0,$$

from which it follows that $I(x)$, and consequently $T_1(x)$, has no zero in the interval $(0, \infty)$. □

In view of Theorem 2, we conclude that for large values of $n$ there are no zeros of $C_n^{(\alpha,\beta)}(z)$ in the complex $z$-plane when $\alpha \neq \beta$ and $\alpha + \beta > 0$. In our discussion of asymptotic zeros of $C_n^{(\alpha,\beta)}(z)$, we have not addressed the cases when $\alpha = \beta \leq -1/2$ and when $\alpha \neq \beta$ with $\alpha + \beta \leq 0$.

## References

[1] X.-X. Bai and Y.-Q. Zhao, A uniform asymptotic expansion for Jacobi polynomials via uniform treatment of Darboux's method, *J. Approx. Theory*, **148**(2007), 1-11. MR2356572 (2009c:33017)

[2] C. F. Dunkl, The Poisson kernel for Heisenberg polynomials on the disk, *Math. Z.*, **187**(1984), 527-547. MR760053 (85m:42027)

[3] C. F. Dunkl, Boundary value problems for harmonic functions on the Heisenberg group, *Canad. J. Math.*, **38**(1986), 478-512. MR833580 (87f:22006)

[4] A. Erdélyi, Uniform asymptotic expansion of integrals, 149-168, in *Analytic Methods in Mathematical Physics*, R. P. Gilbert, R. G. Newton, eds., Gordon and Breach, New York, 1970. MR0336181 (49:957)

[5] J. L. Fields, A uniform treatment of Darboux's method, *Arch. Rational Mech. Anal.*, **27**(1967), 289-305. MR0220984 (36:4036)

[6] G. B. Folland, A fundamental solution for a subelliptic operator, *Bull. Amer. Math. Soc.*, **79**(1973), 373-376. MR0315267 (47:3816)

[7] G. B. Folland and E. M. Stein, Estimates for the $\bar{\partial}_b$ complex and analysis on the Heisenberg group, *Comm. Pure Appl. Math.*, **27**(1974), 429-522. MR0367477 (51:3719)

[8] G. Gasper, Orthogonality of certain functions with respect to complex-valued weights, *Canad. J. Math.*, **33**(1981), 1261-1270. MR638380 (83a:33014)

[9] P. C. Greiner, Spherical harmonics on the Heisenberg group, *Canad. Math. Bull.*, **23**(1980), 383-396. MR602590 (82e:43009)

[10] R. Howe, On the role of the Heisenberg group in harmonic analysis, *Bull. Amer. Math. Soc.*, **3**(1980), no. 2, 821–843. MR0578375 (81h:22010)

[11] E. L. Ince, *Ordinary Differential Equations*, Dover, New York, 1956. MR0010757 (6:65f)

[12] F. W. J. Olver, *Asymptotics and Special Functions*, Academic Press, New York, 1974. MR0435697 (55:8655)

[13] F. W. J. Olver, D. W. Lozier, R. F. Boisvert and C. W. Clark, *NIST Handbook of Mathematical Functions*, Cambridge University Press, 2010. MR2723248 (2012a:33001)

[14] G. Szegő, *Orthogonal Polynomials*, 4th ed., American Mathematical Society, Providence, Rhode Island, 1975. MR0372517 (51:8724)

[15] N. M. Temme, Uniform asymptotic expansion for a class of polynomials biorthogonal on the unit circle, *Constr. Approx.*, **2**(1986), 369-376. MR892162 (88e:42047)

[16] S. Thangavelu, *Harmonic Analysis on the Heisenberg Group*, Birkhäuser, Boston, 1998. MR1633042 (99h:43001)

[17] R. Wong, *Asymptotic Approximations of Integrals*, Academic Press, Boston, 1989. MR1016818 (90j:41061)

[18] R. Wong and Y.-Q. Zhao, On a uniform treatment of Darboux's method, *Constr. Approx.*, **21**(2005), 225-255. MR2107940 (2005h:41064)

Department of Mathematics, ZhongShan University, GuangZhou 510275, People's Republic of China

Department of Mathematics, City University of Hong Kong, Tat Chee Avenue, Kowloon, Hong Kong

Department of Mathematics, ZhongShan University, GuangZhou 510275, People's Republic of China

*E-mail address*: `stszyq@mail.sysu.edu.cn`

Asymptotic Analysis 82 (2013) 39–64
DOI 10.3233/ASY-2012-1135
IOS Press



# Global asymptotics of the discrete Chebyshev polynomials

Y. Lin [a,b] and R. Wong [b,*]
[a] *Department of Mathematics, Zhongshan University, Guangzhou, China*
[b] *Department of Mathematics, City University of Hong Kong, Kowloon, Hong Kong*

**Abstract.** In this paper, we study the asymptotics of the discrete Chebyshev polynomials $t_n(z, N)$ as the degree grows to infinity. Global asymptotic formulas are obtained as $n \to \infty$, when the ratio of the parameters $n/N = c$ is a constant in the interval $(0, 1)$. Our method is based on a modified version of the Riemann–Hilbert approach first introduced by Deift and Zhou.

Keywords: global asymptotics, discrete Chebyshev polynomials, Riemann–Hilbert problems, Airy functions

## 1. Introduction

The discrete Chebyshev polynomials were introduced by Chebyshev in the study of least-squares data fitting. They are explicitly given by

$$t_n(z, N) = n!\Delta^n \left\{ \binom{z}{n} \binom{z-N}{n} \right\}, \quad n = 0, 1, \dots, N-1, \tag{1.1}$$

where $\Delta$ is the forward difference operator with unit spacing on the variable $z$; see [14], formula (2.8.1). These polynomials are orthogonal on the discrete set $\{0, 1, \dots, N-1\}$ with respect to the weight function $w(k) = 1$, $k = 0, 1, \dots, N-1$. More precisely, we have

$$\sum_{k=0}^{N-1} t_n(k, N)t_m(k, N) = \frac{(N+n)!}{(2n+1)(N-n-1)!}\delta_{n,m}, \quad n, m = 0, 1, \dots, N-1. \tag{1.2}$$

Like all other classical orthogonal polynomials, these polynomials also satisfy a three-term recurrence relation. Indeed, we have

$$(n+1)t_{n+1}(z, N) = 2(2n+1)\left(z - \frac{1}{2}(N-1)\right)t_n(z, N) - n\left(N^2 - n^2\right)t_{n-1}(z, N); \tag{1.3}$$

see [6].

*Corresponding author: R. Wong, Department of Mathematics, City University of Hong Kong, Tat Chee Avenue, Kowloon, Hong Kong. E-mail: mclbj@cityu.edu.hk.

The discrete Chebyshev polynomials are also known as a special case of the Hahn polynomials [2], p. 174,

$$Q_n(z;\alpha,\beta,N) := {}_3F_2(-n,-z,n+\alpha+\beta+1;-N,\alpha+1;1); \tag{1.4}$$

see also [8]. Recall the weight function of the Hahn polynomials

$$w(z;\alpha,\beta,N) := \frac{(\alpha+1)_z}{z!}\frac{(\beta+1)_{N-1-z}}{(N-1-z)!}, \quad z=0,1,\ldots,N-1. \tag{1.5}$$

Note that by taking $\alpha = \beta = 0$, the weight function becomes 1. In this case, the Hahn polynomials reduce to the discrete Chebyshev polynomials except for a constant factor.

For further properties and applications, we refer to Szegö [14] and Hildebrand [7]. Like other discrete orthogonal polynomials, these polynomials do not satisfy a second-order linear differential equation. Hence, the asymptotic theory for differential equations cannot be applied. By using the steepest descent method for double integrals, Pan and Wong [12] have recently derived infinite asymptotic expansions for the discrete Chebyshev polynomials when the variable $z$ is real and the ratio $n/N$ lies in the interval (0, 1); their results involve a confluent hypergeometric function and its derivative. For more information about the integral methods, we refer to Wong [17]. On the other hand, Baik et al. [1] have studied the asymptotics of discrete orthogonal polynomials with respect to a general weight function by using the Riemann–Hilbert approach. The results in [1] are very general, but it is difficult to use them to write out explicit formulas for specific polynomials. In the case of the Hahn polynomials (1.4), only the situation when both parameters $\alpha$ and $\beta$ are large has been considered. More precisely, the authors of [1] considered the case when $\alpha = NA$ and $\beta = NB$, where $A$ and $B$ are fixed positive numbers, thus excluding the special case of the discrete Chebyshev polynomials (i.e., $\alpha = \beta = 0$). Moreover, the results in [1] are more local in nature; that is, more formulas are needed to describe the behavior of the orthogonal polynomials in the complex plane.

In this paper, we investigate the asymptotics of the discrete Chebyshev polynomials as $n \to \infty$, when the ratio of the parameters $n/N$ is a constant $c \in (0, 1)$. Global asymptotic formulas are obtained in the complex $z$-plane when $n$ goes to infinity. Our approach is based on a modified version of the steepest-descent method for Riemann–Hilbert problems introduced by Deift and Zhou in [4]; see [3,11,15]. More precisely, we derive two Airy-type asymptotic formulas for the discrete Chebyshev polynomials. The regions of validity of these formulas overlap, and together they cover the whole complex plane.

The presentation of this paper is arranged as follows: In Section 2, we first present the basic Riemann–Hilbert problem associated with the discrete Chebyshev polynomials. Then, we transform the basic Riemann–Hilbert problem into a continuous one, which is similar to that in [3,15]. In Section 3, we introduce some auxiliary functions, known as the $g$-function and the $\phi$-function. Moreover, we define a function $D(z)$, which is analogous to the function first used in [15] to remove the jumps of the continuous Riemann–Hilbert problem near the endpoints of the interval of orthogonality. In Section 4, we construct our parametrix by using Airy functions and their derivatives, and prove that this parametrix is asymptotically equal to the solution of the Riemann–Hilbert problem associated with the discrete Chebyshev polynomials formulated in Section 2. Our main result is given in Section 5. In Section 6, we compare our formulas with those given by Pan and Wong [12].

## 2. Riemann–Hilbert problem

We start with the fundamental Riemann–Hilbert problem (RHP) for the discrete orthogonal polynomials. Note that the leading coefficient of $t_n(z,N)$ is $(2n)!/n!^2$. Hence, the monic discrete Chebyshev polynomials are given by

$$\pi_{N,n}(z) := \frac{n!^2}{(2n)!} t_n(z,N). \tag{2.1}$$

Clearly, they satisfy the new orthogonality relation

$$\sum_{k=0}^{N-1} \pi_{N,n}(k)\pi_{N,m}(k) = \delta_{n,m}/\gamma_{N,n}^2, \tag{2.2}$$

where

$$\gamma_{N,n}^2 = \frac{(2n+1)\Gamma(N-n)\Gamma^2(2n+1)}{\Gamma(N+n+1)\Gamma^4(n+1)}.$$

Consider the following RHP for a $2 \times 2$ matrix-value function $Y(z)$:

$(Y_a)$ $Y(z)$ is analytic for $z \in \mathbb{C} \setminus \mathbb{N}$;

$(Y_b)$ at each $z = k \in \mathbb{N}$, the first column of $Y$ is analytic and the second column of $Y$ has a simple pole where the residue is

$$\operatorname*{Res}_{z=k} Y(z) = \lim_{z \to k} Y(z) \begin{pmatrix} 0 & 1 \\ 0 & 0 \end{pmatrix};$$

$(Y_c)$ as $z \to \infty$,

$$Y(z) = \left(I + \mathrm{O}\left(\frac{1}{z}\right)\right) \begin{pmatrix} z^n & 0 \\ 0 & z^{-n} \end{pmatrix}.$$

By a well-known theorem of Fokas et al. [5] (see also [1]), we have the following theorem.

**Theorem 2.1.** *The unique solution to the above RHP is given by*

$$Y(z) := \begin{pmatrix} \pi_{N,n}(z) & \sum_{k=0}^{N-1} \frac{\pi_{N,n}(k)}{z-k} \\ \gamma_{N,n-1}^2 \pi_{N,n-1}(z) & \sum_{k=0}^{N-1} \frac{\gamma_{N,n-1}^2 \pi_{N,n-1}(k)}{z-k} \end{pmatrix}. \tag{2.3}$$

Let $X_N$ denote the set defined by

$$X_N := \{x_{N,k}\}_{k=0}^{N-1}, \quad \text{where } x_{N,k} := \frac{k+1/2}{N}.$$

The $x_{N,k}$'s are called *nodes* and they all lie in the interval (0, 1). Following Baik et al. [1], we introduce the first transformation

$$H(z) := \begin{pmatrix} N^{-n} & 0 \\ 0 & N^{n} \end{pmatrix} Y(Nz - 1/2) \begin{pmatrix} \prod_{j=0}^{N-1}(z - x_{N,j})^{-1} & 0 \\ 0 & \prod_{j=0}^{N-1}(z - x_{N,j}) \end{pmatrix}$$
$$= N^{-n\sigma_3} Y(Nz - 1/2) \left[ \prod_{j=0}^{N-1}(z - x_{N,j}) \right]^{-\sigma_3}, \tag{2.4}$$

where $\sigma_3 := \left(\begin{smallmatrix} 1 & 0 \\ 0 & -1 \end{smallmatrix}\right)$ is a Pauli matrix.

A straightforward calculation shows that $H(z)$ is a solution of the RHP:

$(H_a)$ $H(z)$ is analytic for $z \in \mathbb{C} \setminus X_N$;

$(H_b)$ at each $x_{N,k} \in X_N$, the second column of $H(z)$ is analytic and the first column of $H(z)$ has a simple pole where the residue is

$$\operatorname*{Res}_{z=x_{N,k}} H(z) = \lim_{z \to x_{N,k}} H(z) \begin{pmatrix} 0 & 0 \\ \prod_{\substack{j=0 \\ j \neq k}}^{N-1}(z - x_{N,j})^{-2} & 0 \end{pmatrix};$$

$(H_c)$ as $z \to \infty$,

$$H(z) = \left(I + O\left(\frac{1}{z}\right)\right) \begin{pmatrix} z^{n-N} & 0 \\ 0 & z^{-n+N} \end{pmatrix}.$$

We next consider the second transformation which will remove the poles as well as transform the discrete RHP into a continuous one. Let $\delta > 0$ be a sufficiently small number, and we define

$$R(z) := H(z) \begin{pmatrix} 1 & 0 \\ \frac{\mp \mathrm{i} e^{\pm N\pi \mathrm{i} z} \cos(N\pi z)}{N\pi \prod_{j=0}^{N-1}(z - x_{N,j})^2} & 1 \end{pmatrix} \tag{2.5}$$

for $z \in \Omega_\pm$, and

$$R(z) := H(z) \tag{2.6}$$

for $z \notin \Omega_\pm$, where the domains $\Omega_\pm$ are shown in Fig. 1. Let $\Sigma_+$ be the boundary of $\Omega_+$ in the upper half-plane, and $\Sigma_-$ be the mirror image of $\Sigma_+$ in the lower half-plane. For the contour $\Sigma = (0, 1) \cup \Sigma_\pm$, see also Fig. 1.

**Lemma 2.2.** *For each $x_{N,k} \in X_N$, the singularity of $R(z)$ at $x_{N,k}$ is removable*; *i.e.*, $\operatorname{Res}_{z=x_{N,k}} R(z) = 0$.

**Proof.** For $z \in \Omega_\pm$, it follows from (2.5) that

$$R_{11}(z) = H_{11}(z) + H_{12}(z) \frac{\mp \mathrm{i} e^{\pm N\pi \mathrm{i} z} \cos(N\pi z)}{N\pi \prod_{j=0}^{N-1}(z - x_{N,j})^2}, \qquad R_{12}(z) = H_{12}(z). \tag{2.7}$$

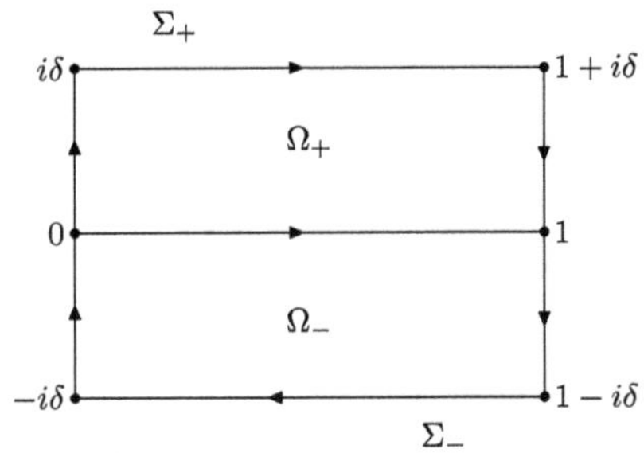


Fig. 1. The domains $\Omega_\pm$ and the contour $\Sigma$.

Since $H_{12}(z)$ is analytic by $(H_b)$, the residue of $R_{12}(z)$ at $x_{N,k}$ is zero. From $(H_b)$, we observe that the residue of $H_{11}(z)$ at $z = x_{N,k}$ is

$$\operatorname*{Res}_{z=x_{N,k}} H_{11}(z) = H_{12}(x_{N,k}) \prod_{\substack{j=0\\ j\neq k}}^{N-1} (x_{N,k} - x_{N,j})^{-2}. \tag{2.8}$$

On the other hand, it is readily seen that

$$\operatorname*{Res}_{z=x_{N,k}} \frac{\mp \mathrm{i} e^{\pm N\pi \mathrm{i} z}\cos(N\pi z)}{N\pi \prod_{j=0}^{N-1}(z - x_{N,j})^2} = -\prod_{\substack{j=0\\ j\neq k}}^{N-1} (x_{N,k} - x_{N,j})^{-2}. \tag{2.9}$$

Thus, applying (2.8) and (2.9) to (2.7) shows that the residue of $R_{11}(z)$ at $z = x_{N,k}$ is zero. The entries in the second row of the matrix $R(z)$ can be studied similarly. This completes the proof of the lemma. □

Note that this transformation makes $R_+(x)$ and $R_-(x)$ continuous on the interval (0, 1). Therefore, $R(z)$ is a solution of the following RHP:

$(R_a)$ $R(z)$ is analytic for $z \in \mathbb{C} \setminus \Sigma$;
$(R_b)$ the jump conditions on the contour $\Sigma$: for $x \in (0, 1)$,

$$R_+(x) = R_-(x) \begin{pmatrix} 1 & 0 \\ r(x) & 1 \end{pmatrix}, \tag{2.10}$$

where

$$r(x) = \frac{4\cos^2(N\pi x)}{W \prod_{j=0}^{N-1}(x - x_{N,j})^2}; \tag{2.11}$$

for $z \in \Sigma_\pm$,

$$R_+(z) = R_-(z) \begin{pmatrix} 1 & 0 \\ \tilde{r}_\pm(z) & 1 \end{pmatrix}, \tag{2.12}$$

where

$$\tilde{r}_{\pm}(z) = \frac{\mp 2\mathrm{e}^{\pm N\pi \mathrm{i} z}\cos(N\pi z)}{W\prod_{j=0}^{N-1}(z - x_{N,j})^2}; \tag{2.13}$$

$(R_c)$ as $z \to \infty$,

$$R(z) = \left(I + \mathrm{O}\left(\frac{1}{z}\right)\right)\begin{pmatrix} z^{n-N} & 0 \\ 0 & z^{-n+N} \end{pmatrix}. \tag{2.14}$$

For simplicity in the following section, here we have introduced the notation $W := 2N\pi\mathrm{i}$.

## 3. The auxiliary functions

For our subsequent analysis, we need some auxiliary functions. Consider the monic orthogonal polynomials $p_{N,n}(z) := N^{-n}\pi_{N,n}(Nz - 1/2)$. From (1.3) and (2.1), we have

$$p_{N,n+1}(z) = (z - a_n)p_{N,n}(z) - b_n p_{N,n-1}(z), \tag{3.1}$$

where

$$a_n = \frac{1}{2}, \qquad b_n = \frac{1 - (n/N)^2}{16 - 4/n^2}.$$

Using the method introduced by Kuijlaars and Van Assche [9], we can derive the equilibrium measure corresponding to the discrete Chebyshev polynomials in our case. Recall $c := n/N$, and let

$$a := \frac{1}{2} - \frac{1}{2}\sqrt{1 - c^2}, \qquad b := \frac{1}{2} + \frac{1}{2}\sqrt{1 - c^2}. \tag{3.2}$$

The density function is given by

$$\mu(x) = \begin{cases} \dfrac{2}{\pi c}\arcsin\left(\dfrac{c}{2\sqrt{x - x^2}}\right), & x \in [a, b], \\ \dfrac{1}{c}, & x \in [0, a] \cup [b, 1]. \end{cases} \tag{3.3}$$

As usual, we now define the auxiliary $g$-function and $\phi$-function.

**Definition 3.1.** The $g$-function is the logarithmic potential defined by

$$g(z) := \int_0^1 \log(z - s)\mu(s)\,\mathrm{d}s, \quad z \in \mathbb{C} \setminus (-\infty, 1], \tag{3.4}$$

and the so-called $\phi$-function is defined by

$$\phi(z) := \frac{l}{2} - g(z) \tag{3.5}$$

for $z \in \mathbb{C} \setminus (-\infty, 1]$, where $l := 2\int_0^1 \log|b - s|\mu(s)\,\mathrm{d}s$ is called the Lagrange multiplier.

An explicit formula for the $g$-function is given in (6.25) in Section 6. On account of (3.3), the derivative of $g(z)$ is given by

$$g'(z) = \frac{2}{c}\log\left(\frac{z + \sqrt{(z-a)(z-b)} + c/2}{(z-1) + \sqrt{(z-a)(z-b)} - c/2}\right) + \frac{1}{c}\log\left(\frac{z-1}{z}\right). \tag{3.6}$$

Introduce the $\phi^*$-function

$$\phi^*(z) := \int_b^z \left(-g'(s) \mp \frac{1}{c}\pi\mathrm{i}\right)\mathrm{d}s = \phi(z) \pm \frac{1}{c}\pi\mathrm{i}(1 - z) \tag{3.7}$$

for $z \in \mathbb{C}_\pm$. It is readily seen that

$$\phi^*(z) = -\int_b^z \left(\frac{2}{c}\log\left(\frac{s + \sqrt{(s-a)(s-b)} + c/2}{1 - s - \sqrt{(s-a)(s-b)} + c/2}\right) + \frac{1}{c}\log\left(\frac{1-s}{s}\right)\right)\mathrm{d}s \tag{3.8}$$

for $z \in \mathbb{C}\backslash(-\infty, b] \cup [1, \infty)$.

Similarly, we also set

$$\tilde{\phi}(z) := \int_a^z \left(-g'(s) \mp \frac{1}{c}\pi\mathrm{i}\right)\mathrm{d}s = \phi \pm \pi\mathrm{i}\left(1 - \frac{1}{c}z\right) \tag{3.9}$$

for $z \in \mathbb{C}_\pm$. Note that for $x \in (0, a)$, we have from (3.4)

$$g_\pm(x) = \int_0^1 \log|x - s|\mu(s)\,\mathrm{d}s \pm \pi\mathrm{i}\int_x^1 \mu(s)\,\mathrm{d}s.$$

Thus, from (3.9) and (3.5), it follows that $\tilde{\phi}_+(x) = \tilde{\phi}_-(x)$ for $x \in (0, a)$; that is, $\tilde{\phi}(z)$ can be analytically continued to the interval $(0, a)$.

The functions $\tilde{\phi}(z)$ and $\phi^*(z)$ play an important role in our argument, and the following are some of their properties.

**Proposition 3.2.** *For $x \in (a, b)$, we have*

$$\phi^*_+(x) + \phi^*_-(x) = 0, \qquad \tilde{\phi}_+(x) + \tilde{\phi}_-(x) = 0. \tag{3.10}$$

*Furthermore, we have*

$$\phi^*(x) < 0 \quad \text{for } x \in (b, 1) \quad \text{and} \quad \tilde{\phi}(x) < 0 \quad \text{for } x \in (0, a). \tag{3.11}$$

*For any $x \in (b, 1)$ and sufficiently small $\delta > 0$, we have*

$$\operatorname{Re}\phi^*(x \pm \mathrm{i}\delta) = \phi^*(x) + \mathrm{O}(\delta^2), \qquad \operatorname{Re}\phi(x \pm \mathrm{i}\delta) = \phi^*(x) - \frac{1}{c}\pi\delta + \mathrm{O}(\delta^2). \tag{3.12}$$

*For any $x \in (0, a)$ and sufficiently small $\delta > 0$, we have*

$$\operatorname{Re}\tilde{\phi}(x \pm \mathrm{i}\delta) = \tilde{\phi}(x) + \mathrm{O}(\delta^2), \qquad \operatorname{Re}\phi(x \pm \mathrm{i}\delta) = \tilde{\phi}(x) - \frac{1}{c}\pi\delta + \mathrm{O}(\delta^2). \tag{3.13}$$

**Proof.** Since

$$\phi^*_\pm(x) = -\int_b^x \left(\frac{2}{c}\log\left(\frac{s \pm \mathrm{i}\sqrt{(s-a)(b-s)} + c/2}{1 - s \mp \mathrm{i}\sqrt{(s-a)(b-s)} + c/2}\right) + \frac{1}{c}\log\left(\frac{1-s}{s}\right)\right)\mathrm{d}s \tag{3.14}$$

for $x \in (a, b)$, it is easy to verify that

$$\phi^*_+(x) + \phi^*_-(x) = -\int_b^x \left(\frac{2}{c}\log\left(\frac{s + sc}{(1-s)(1+c)}\right) + \frac{2}{c}\log\left(\frac{1-s}{s}\right)\right)\mathrm{d}s = 0. \tag{3.15}$$

Similarly, we also have $\tilde{\phi}_+(x) + \tilde{\phi}_-(x) = 0$ for $x \in (a, b)$, thus proving (3.10). Together with (3.5) and (3.7), this implies that $\int_0^1 \log|x - s|\mu(s)\,\mathrm{d}s \equiv l/2$ for $x \in (a, b)$.

For any small $\varepsilon > 0$ and $z \in U(b, \varepsilon) := \{z \in \mathbb{C}\colon |z - b| < \varepsilon\}$, we have from (3.8)

$$\begin{aligned}\phi^*(z) &= -\int_b^z \left[\frac{2}{c}\left(\frac{\sqrt{(b-a)(s-b)}}{b + c/2} + \frac{\sqrt{(b-a)(s-b)}}{1 - b + c/2}\right) + \mathrm{O}(\varepsilon)\right]\mathrm{d}s \\ &= -\frac{8}{3c^2}\left(1 - c^2\right)^{1/4}(z - b)^{3/2} + \mathrm{O}(\varepsilon^2).\end{aligned} \tag{3.16}$$

Here, we have used the fact that $a + b = 1$, $b - a = (1 - c^2)^{1/2}$ and $ab = c^2/4$; see (3.2). By straightforward calculation, it is readily seen from (3.8) that

$$\begin{aligned}\phi^{*\prime}(x) &= -\frac{2}{c}\log\left(\frac{x + \sqrt{(x-a)(x-b)} + c/2}{1 - x - \sqrt{(x-a)(x-b)} + c/2}\right) - \frac{1}{c}\log\left(\frac{1-x}{x}\right) \\ &= -\frac{1}{c}\log\left[\left(\frac{x + \sqrt{(x-a)(x-b)} + c/2}{1 - x - \sqrt{(x-a)(x-b)} + c/2}\right)^2\left(\frac{1-x}{x}\right)\right] < 0\end{aligned} \tag{3.17}$$

for $b < x < 1$. Thus, we obtain $\phi^*(x) < \phi(b + \varepsilon) < 0$.

In the same manner, if $z \in U(a, \varepsilon) := \{z \in \mathbb{C}\colon |z - a| < \varepsilon\}$, then

$$\begin{aligned}\tilde{\phi}(z) &= \int_a^z \left[\frac{2}{c}\left(\frac{\sqrt{(a-s)(b-a)}}{a + c/2} + \frac{\sqrt{(a-s)(b-a)}}{1 - a + c/2}\right) + \mathrm{O}(\varepsilon)\right]\mathrm{d}s \\ &= -\frac{8}{3c^2}\left(1 - c^2\right)^{1/4}(a - z)^{3/2} + \mathrm{O}(\varepsilon^2)\end{aligned} \tag{3.18}$$

and

$$\tilde{\phi}'(x) = -\frac{2}{c}\log\left(\frac{x-\sqrt{(a-x)(b-x)}+c/2}{1-x+\sqrt{(a-x)(b-x)}+c/2}\right) - \frac{1}{c}\log\left(\frac{1-x}{x}\right)$$
$$= -\frac{1}{c}\log\left[\left(\frac{x+\sqrt{(x-a)(x-b)}+c/2}{1-x-\sqrt{(x-a)(x-b)}+c/2}\right)^2\left(\frac{1-x}{x}\right)\right] > 0 \tag{3.19}$$

for $0 < x < a$. Consequently, $\tilde{\phi}(x) < \tilde{\phi}(a-\varepsilon) < 0$ for $0 < x < a$, thus proving (3.11).

For any $x \in (b, 1)$ and sufficiently small $\delta > 0$, we have from the Taylor expansion

$$\operatorname{Re}\phi^*(x \pm \mathrm{i}\delta) = \operatorname{Re}\phi^*(x) \mp \delta \operatorname{Im}\phi^{*\prime}(x) + \mathrm{O}(\delta^2) = \operatorname{Re}\phi^*(x) + \mathrm{O}(\delta^2). \tag{3.20}$$

The last equality follows from the fact that $\phi^{*\prime}(x)$ is real; see (3.17). Coupling this with (3.7) gives

$$\operatorname{Re}\phi(x \pm \mathrm{i}\delta) = \operatorname{Re}\phi^*(x) - \frac{1}{c}\pi\delta + \mathrm{O}(\delta^2); \tag{3.21}$$

thus (3.12) holds.

Similarly, for any $x \in (0, a)$ and sufficiently small $\delta > 0$, we have

$$\operatorname{Re}\tilde{\phi}(x \pm \mathrm{i}\delta) = \operatorname{Re}\tilde{\phi}(x) \mp \delta \operatorname{Im}\tilde{\phi}'(x) + \mathrm{O}(\delta^2) = \operatorname{Re}\tilde{\phi}(x) + \mathrm{O}(\delta^2). \tag{3.22}$$

From (3.9), it follows that

$$\operatorname{Re}\phi(x \pm \mathrm{i}\delta) = \operatorname{Re}\tilde{\phi}(x) - \frac{1}{c}\pi\delta + \mathrm{O}(\delta^2). \tag{3.23}$$

This ends the proof of (3.13). □

From the definition in (3.9), it is easy to see that $\tilde{\phi}(a) = 0$ and $\tilde{\phi}_+(b) = (1 - \frac{1}{c})\pi\mathrm{i}$. Furthermore, from (3.8) it follows that $\phi^*(b) = 0$, and from (3.7) and (3.9) we have $\phi^*_+(a) = (\frac{1}{c} - 1)\pi\mathrm{i}$. Let us now give a brief outline of the argument used in establishing the following mapping properties of the $\phi$-function.

**Proposition 3.3.** *The images of the upper half of the $z$-plane under the mappings $\phi^*$, $\tilde{\phi}$ and $\phi$ are depicted in Fig.* 2.

**Proof.** For $x \in (b, 1)$, $\phi^*(x)$ is real and negative in view of (3.11). Furthermore, (3.17) implies that $\phi^*(x)$ is a monotonically decreasing function in $(b, 1)$. Since $\phi^*(b) = 0$, $\phi^*(1)$ is negative. For $x \in (a, b)$, we have from (3.7), (3.5) and (3.4)

$$\phi^*_+(x) = \frac{l}{2} - \int_0^1 \log|x-s|\mu(s)\,\mathrm{d}s + \pi\mathrm{i}\int_x^1\left(\frac{1}{c} - \mu(s)\right)\mathrm{d}s.$$

On account of a statement following (3.15), the first two terms on the right-hand side cancel. Hence, we obtain

$$\phi^*_+(x) = \pi\mathrm{i}\int_x^1\left(\frac{1}{c} - \mu(s)\right)\mathrm{d}s$$

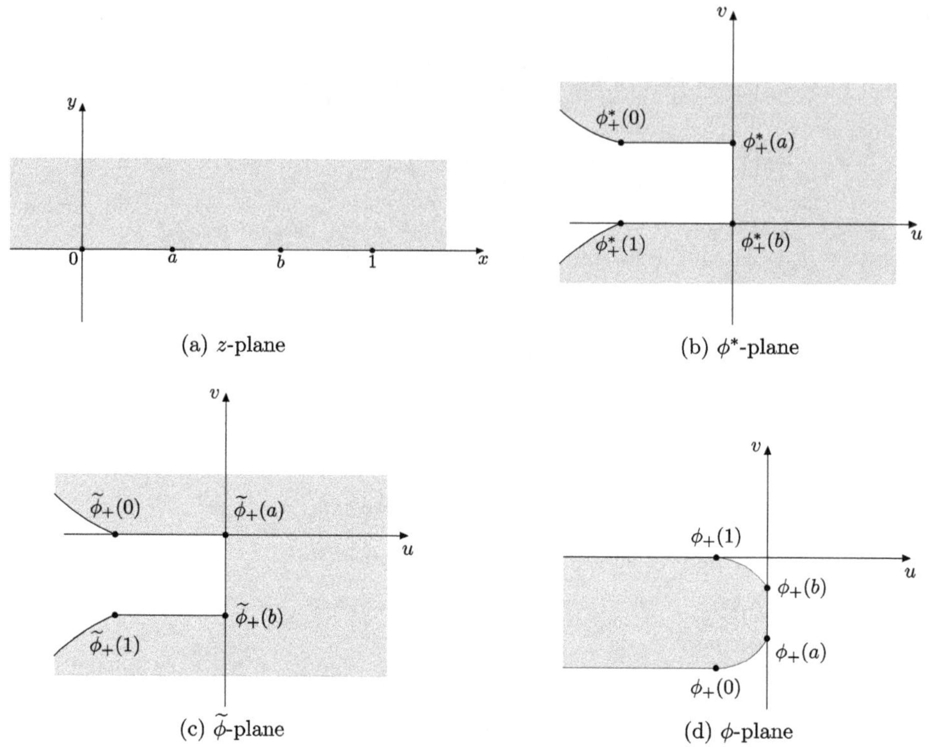


Fig. 2. The upper half-plane under the transformations $\phi$, $\tilde\phi$ and $\phi^*$.

for $x \in (a,b)$. By (3.3), $\frac{1}{c} > \mu(s)$ for $s \in (0,1)$. Thus, $\operatorname{Re}\phi_+^*(x) = 0$ and $\operatorname{Im}\phi_+^*(x) > 0$ for $x \in (a,b)$. Furthermore, since $(\operatorname{Im}\phi_+^*(x))' = -\pi(\frac{1}{c} - \mu(x)) < 0$, $\operatorname{Im}\phi_+^*(x)$ is monotonically decreasing in $(a,b)$. By a similar argument, it can be shown that for $x \in (1,+\infty)$, $\operatorname{Im}\phi_+^*(x)$ is negative and decreasing; also $\operatorname{Re}\phi_+^*(x) < \operatorname{Re}\phi_+^*(1)$ and $\operatorname{Re}\phi_+^*(x)$ is decreasing. Again, by a similar argument, it can be shown that for $x \in (0,a)$, $\operatorname{Re}\phi_+^*(0) = \operatorname{Re}\phi_+^*(1)$, $\operatorname{Im}\phi_+^*(0) = \operatorname{Im}\phi_+^*(a) = \pi(\frac{1}{c} - 1) > 0$ and $\operatorname{Im}\phi_+^*(x) = \pi(\frac{1}{c} - 1) = \operatorname{Im}\phi_+^*(a)$. Finally, one can show that for $x \in (-\infty,0)$, $\operatorname{Im}\phi_+^*(x) > \pi(\frac{1}{c} - 1) > 0$, $\operatorname{Im}\phi_+^*(x)$ is decreasing and $\operatorname{Re}\phi_+^*(x)$ is increasing.

Coupling (3.7) and (3.9), we have

$$\tilde\phi(z) = \phi^*(z) \pm \pi i\left(1 - \frac{1}{c}\right),$$

from which it is easy to verify the image of the upper half-plane under the mapping $\tilde\phi$ shown in Fig. 2(c).

To verify the graph in Fig. 2(d), we first show that for $x \in (-\infty,0)$, $\operatorname{Re}\phi_+(x) < \operatorname{Re}\phi_+(0) < 0$ and $\operatorname{Im}\phi_+(x) = -\pi$. Then, we show that for $x \in (0,1)$, $\operatorname{Im}\phi_+(x) < 0$ and $\operatorname{Im}\phi_+(x)$ is an increasing function. To deal with $\operatorname{Re}\phi_+(x)$ in $(0,1)$, we consider three intervals $(0,a)$, $(a,b)$ and $(b,1)$, separately.

For $x \in (b,1)$, we show that $(\operatorname{Re}\phi_+(x))' = (\phi^*(x))' < 0$. Thus, $\operatorname{Re}\phi_+(x)$ is a decreasing function. For $x \in (a,b)$, we show that $\operatorname{Re}\phi_+(x) = 0$. For $x \in (0,a)$, we first show $\operatorname{Re}\phi_+(x) = \tilde{\phi}(x)$. By (3.11) and (3.19), $\operatorname{Re}\phi_+(x) < 0$ and $(\operatorname{Re}\phi_+(x))' = (\tilde{\phi}(x))' > 0$. Thus, $\operatorname{Re}\phi_+(x)$ is an increasing function. This ends the verification of Fig. 2(d), and completes the proof of the proposition. □

To construct our global parametrix, we first define the function

$$N(z) = \begin{pmatrix} \frac{\sqrt{z-a}+\sqrt{z-b}}{2(z-a)^{1/4}(z-b)^{1/4}} & -\mathrm{i}\frac{\sqrt{z-a}-\sqrt{z-b}}{2(z-a)^{1/4}(z-b)^{1/4}} \\ \mathrm{i}\frac{\sqrt{z-a}-\sqrt{z-b}}{2(z-a)^{1/4}(z-b)^{1/4}} & \frac{\sqrt{z-a}+\sqrt{z-b}}{2(z-a)^{1/4}(z-b)^{1/4}} \end{pmatrix}. \tag{3.24}$$

It is readily verified that $N(z)$ is analytic in $\mathbb{C}\setminus[a,b]$ and

$$N_+(x) = N_-(x)\begin{pmatrix} 0 & -1 \\ 1 & 0 \end{pmatrix}, \quad x \in (a,b). \tag{3.25}$$

Next, we introduce the Airy parametrix

$$A(z) := \begin{cases} \begin{pmatrix} \operatorname{Ai}(z) & \omega^2\operatorname{Ai}(\omega^2 z) \\ \mathrm{i}\operatorname{Ai}'(z) & \mathrm{i}\omega\operatorname{Ai}'(\omega^2 z) \end{pmatrix}, & z \in \mathbb{C}_+, \\ \begin{pmatrix} \operatorname{Ai}(z) & -\omega\operatorname{Ai}(\omega z) \\ \mathrm{i}\operatorname{Ai}'(z) & -\mathrm{i}\omega^2\operatorname{Ai}'(\omega z) \end{pmatrix}, & z \in \mathbb{C}_-, \end{cases} \tag{3.26}$$

where $\omega = \mathrm{e}^{2\pi\mathrm{i}/3}$. In view of the well-known identity [10], formula (9.2.12),

$$\operatorname{Ai}(z) + \omega\operatorname{Ai}(\omega z) + \omega^2\operatorname{Ai}(\omega^2 z) = 0, \tag{3.27}$$

it is clear that

$$A_+(z) = A_-(z)\begin{pmatrix} 1 & -1 \\ 0 & 1 \end{pmatrix}, \quad x \in \mathbb{R}. \tag{3.28}$$

By (3.27), we also have

$$A(z)\begin{pmatrix} 1 & 0 \\ \pm 1 & 1 \end{pmatrix} = \begin{cases} \begin{pmatrix} -\omega\operatorname{Ai}(\omega z) & \omega^2\operatorname{Ai}(\omega^2 z) \\ -\mathrm{i}\omega^2\operatorname{Ai}'(\omega z) & \mathrm{i}\omega\operatorname{Ai}'(\omega^2 z) \end{pmatrix}, & z \in \mathbb{C}_+, \\ \begin{pmatrix} -\omega^2\operatorname{Ai}(\omega^2 z) & -\omega\operatorname{Ai}(\omega z) \\ -\mathrm{i}\omega\operatorname{Ai}'(\omega^2 z) & -\mathrm{i}\omega^2\operatorname{Ai}'(\omega z) \end{pmatrix}, & z \in \mathbb{C}_-. \end{cases} \tag{3.29}$$

Recall the asymptotic expansions of the Airy function and its derivative [10], formulas (9.7.5) and (9.7.6),

$$\begin{aligned} \operatorname{Ai}(z) &\sim \frac{z^{-1/4}}{2\sqrt{\pi}}\mathrm{e}^{-(2/3)z^{3/2}}\sum_{s=0}^{\infty}\frac{(-1)^s u_s}{((2/3)z^{3/2})^s}, \\ \operatorname{Ai}'(z) &\sim -\frac{z^{1/4}}{2\sqrt{\pi}}\mathrm{e}^{-(2/3)z^{3/2}}\sum_{s=0}^{\infty}\frac{(-1)^s v_s}{((2/3)z^{3/2})^s} \end{aligned} \tag{3.30}$$

as $z \to \infty$ in $|\arg z| < \pi$, where $u_s, v_s$ are constants with $u_0 = v_0 = 1$. From (3.26) and (3.30), we obtain

$$A(z) = \frac{z^{-\sigma_3/4}}{2\sqrt{\pi}} \begin{pmatrix} 1 & -\mathrm{i} \\ -\mathrm{i} & 1 \end{pmatrix} (I + \mathrm{O}(|z|^{-3/2}))\mathrm{e}^{-(2/3)z^{3/2}\sigma_3} \tag{3.31}$$

as $z \to \infty$ in $|\arg z| < \pi$,

$$A(z) \begin{pmatrix} 1 & 0 \\ \pm 1 & 1 \end{pmatrix} = \frac{z^{-\sigma_3/4}}{2\sqrt{\pi}} \begin{pmatrix} 1 & -\mathrm{i} \\ -\mathrm{i} & 1 \end{pmatrix} (I + \mathrm{O}(|z|^{-(3/2)}))\mathrm{e}^{-(2/3)z^{3/2}\sigma_3} \tag{3.32}$$

as $z \to \infty$ with $\arg z \in (\pm\pi/3, \pm\pi]$. Finally, we introduce the $D$-functions:

$$D(z) := \frac{\mathrm{e}^{Nz}\Gamma(Nz + 1/2)}{\sqrt{2\pi}(Nz)^{Nz}} \tag{3.33}$$

and

$$\widetilde{D}(z) := \frac{\sqrt{2\pi}\mathrm{e}^{Nz}(-Nz)^{-Nz}}{\Gamma(-Nz + 1/2)}, \tag{3.34}$$

which were first used in [15] to construct global asymptotic formulas without any cut in the complex plane. For simplicity, we also introduce the notations

$$D^*(z) := D(1 - z), \qquad \widetilde{D}^*(z) := \widetilde{D}(1 - z). \tag{3.35}$$

By Euler's reflection formula, we have

$$\widetilde{D}(z) = 2\cos(N\pi z)\mathrm{e}^{\pm N\pi\mathrm{i}z} D(z) \tag{3.36}$$

for $z \in \mathbb{C}_\pm$. The reason why we consider $D(z)$ here is to make sure that the jump matrix $J_S(z)$ in Lemma 4.1 is asymptotically equal to the identity matrix. As $n \to \infty$, by applying Stirling's formula to (3.33) and (3.34), we obtain

$$D(z) = 1 + \mathrm{O}(1/n) \tag{3.37}$$

for $|\arg z| < \pi$, and

$$\widetilde{D}(z) = 1 + \mathrm{O}(1/n) \tag{3.38}$$

for $|\arg(-z)| < \pi$.

## 4. Construction of parametrix

First, we select an arbitrary fixed point $x_0 \in (a, b)$. Let $\Gamma$ be the line $\mathrm{Re}\, z = x_0$; see Fig. 3. To facilitate our discussion below, we divide the complex plane into four regions by using $\Gamma$ and the real line. With a similar technique as used in [3] and [15], we construct the parametrix of the RHP for $R$ in these four regions. Define

$$\widetilde{R}(z) := (-1)^n \sqrt{\pi} \left(W e^{nl}\right)^{\sigma_3/2} N(z) \begin{pmatrix} 1 & \mathrm{i} \\ \mathrm{i} & 1 \end{pmatrix} \tilde{f}(z)^{-\sigma_3/4} \sigma_1 A(\tilde{f}) \sigma_1 \times \left[ \frac{4\cos^2(N\pi z) D(z)^2}{W \prod_{j=0}^{N-1} (z - x_{N,j})^2} \right]^{\sigma_3/2} \tag{4.1}$$

for $z \in$ I and III,

$$\widetilde{R}(z) := (-1)^N \sqrt{\pi} \left(W e^{nl}\right)^{\sigma_3/2} N(z) \begin{pmatrix} 1 & -\mathrm{i} \\ -\mathrm{i} & 1 \end{pmatrix} f^*(z)^{-\sigma_3/4} \sigma_2 A(f^*) \sigma_2^{-1} \times \left[ \frac{4\cos^2(N\pi z) D^*(z)^2}{W \prod_{j=0}^{N-1} (z - x_{N,j})^2} \right]^{\sigma_3/2} \tag{4.2}$$

for $z \in$ II and IV, where $\tilde{f}(z)$ and $f^*(z)$ are given by

$$\tilde{f}(z) := \left(-\frac{3}{2} n \tilde{\phi}(z)\right)^{2/3}, \qquad f^*(z) := \left(-\frac{3}{2} n \phi^*(z)\right)^{2/3} \tag{4.3}$$

and

$$\sigma_1 := \begin{pmatrix} 0 & 1 \\ 1 & 0 \end{pmatrix}, \qquad \sigma_2 := \begin{pmatrix} 0 & 1 \\ -1 & 0 \end{pmatrix}.$$

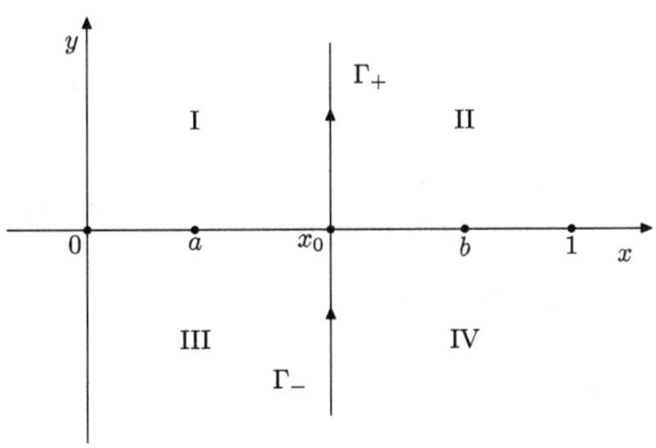


Fig. 3. The domains I, II, III and IV.

To show that $\widetilde{R}(z)$ has the same behavior as $R(z)$ as $z \to \infty$, we only consider the case when $z \in$ I. The other cases can be handled in a similar manner. A combination of (3.31) and (4.3) gives

$$\sqrt{\pi}\begin{pmatrix} 1 & \mathrm{i} \\ \mathrm{i} & 1 \end{pmatrix} \tilde{f}(z)^{-\sigma_3/4}\sigma_1 A(\tilde{f})\sigma_1 \sim (-1)^n \mathrm{e}^{-n(l/2-g)\sigma_3}\mathrm{e}^{N\pi\mathrm{i}z\sigma_3}. \tag{4.4}$$

Here, we have made use of (3.5) and (3.9). Thus, we obtain from (3.36) and (4.1)

$$\widetilde{R}(z) \sim \left(W\mathrm{e}^{nl}\right)^{\sigma_3/2} N(z)\mathrm{e}^{-n(l/2-g)\sigma_3}\widetilde{D}^{\sigma_3}W^{-\sigma_3/2}\prod_{j=0}^{N-1}(z-x_{N,j})^{-\sigma_3}. \tag{4.5}$$

Also note that $\widetilde{D}(z) \sim 1$ and $\prod_{j=0}^{N-1}(z-x_{N,j}) \sim z^N$ as $z \to \infty$. From (3.4) and the asymptotic behavior of $N(z)$, it follows that

$$\widetilde{R}(z) \sim \left(W\mathrm{e}^{nl}\right)^{\sigma_3/2} N(z)\left(W\mathrm{e}^{nl}\right)^{-\sigma_3/2}\mathrm{e}^{ng\sigma_3}z^{-N\sigma_3} = \left(I+\mathrm{O}(1/z)\right)z^{(n-N)\sigma_3} \tag{4.6}$$

as $z \to \infty$. Define the matrix

$$S(z) := \left(W\mathrm{e}^{nl}\right)^{-\sigma_3/2} R(z)\widetilde{R}(z)^{-1}\left(W\mathrm{e}^{nl}\right)^{\sigma_3/2}. \tag{4.7}$$

It is easy to see that

$$S(z) = I + \mathrm{O}(1/z) \tag{4.8}$$

as $z \to \infty$. The jump matrix of $S(z)$ is given by

$$J_S(z) := S_-(z)^{-1}S_+(z) = \left(W\mathrm{e}^{nl}\right)^{-\sigma_3/2}\widetilde{R}_-(z)J_R(z)\widetilde{R}_+(z)^{-1}\left(W\mathrm{e}^{nl}\right)^{\sigma_3/2}, \tag{4.9}$$

and the contour associated with the matrix $J_S(z)$ is $\Sigma_S = \Sigma + \Gamma + \mathbb{R}$; see Figs 1 and 3.

**Lemma 4.1.** *$J_S(z) = I + \mathrm{O}(\frac{1}{n})$ and $S(z) = I + \mathrm{O}(\frac{1}{n})$ as $n \to \infty$.*

**Proof.** For $z \in \Gamma$ and $\pm\operatorname{Im} z \in (0,\delta)$, we have $J_R(z) = I$. This together with (4.1), (4.2) and (4.9) gives

$$\begin{aligned} J_S(z) = (-1)^N N(z) &\left[\sqrt{\pi}\begin{pmatrix} 1 & -\mathrm{i} \\ -\mathrm{i} & 1 \end{pmatrix} f^*(z)^{-\sigma_3/4}\sigma_2 A(f^*)\sigma_2^{-1}\right]\left(D^*(z)/D(z)\right)^{\sigma_3} \\ &\times\left[\sqrt{\pi}\begin{pmatrix} 1 & \mathrm{i} \\ \mathrm{i} & 1 \end{pmatrix}\tilde{f}(z)^{-\sigma_3/4}\sigma_1 A(\tilde{f})\sigma_1\right]^{-1} N(z)^{-1}(-1)^n. \end{aligned} \tag{4.10}$$

On account of (4.3), we have from (3.32)

$$\sqrt{\pi}\begin{pmatrix} 1 & -\mathrm{i} \\ -\mathrm{i} & 1 \end{pmatrix} f^*(z)^{-\sigma_3/4}\sigma_2 A(f^*)\sigma_2^{-1} = \left(I+\mathrm{O}(1/n)\right)\mathrm{e}^{-n\phi^*\sigma_3}\begin{pmatrix} 1 & \pm 1 \\ 0 & 1 \end{pmatrix}$$

and

$$\left[\sqrt{\pi}\begin{pmatrix}1 & \mathrm{i}\\ \mathrm{i} & 1\end{pmatrix}\tilde{f}(z)^{-\sigma_3/4}\sigma_1 A(\tilde{f})\sigma_1\right]^{-1} = \begin{pmatrix}1 & \mp 1\\ 0 & 1\end{pmatrix}\mathrm{e}^{n\tilde{\phi}\sigma_3}\left(I+\mathrm{O}(1/n)\right)$$

as $n\to\infty$. Applying the above two equations to (4.10) gives

$$J_S(z) = N(z)\begin{pmatrix}\frac{D^*(z)}{D(z)} & (\mp\frac{D^*(z)}{D(z)} \pm \frac{D(z)}{D^*(z)})\mathrm{e}^{-2n\tilde{\phi}}\\ 0 & \frac{D(z)}{D^*(z)}\end{pmatrix}N(z)^{-1}\left(I+\mathrm{O}(1/n)\right) = I+\mathrm{O}(1/n).$$

Here, we have used the fact that $D(z)=1+\mathrm{O}(1/n)$, $D^*(z)=1+\mathrm{O}(1/n)$ and $\operatorname{Re}\tilde{\phi}>0$; see Fig. 2(c). (A corresponding drawing can be given for the image of the lower half-plane under the mapping $\tilde{\phi}$.)

For $z\in\Gamma$ and $\pm\operatorname{Im} z\notin(0,\delta)$, we have from (3.31) and (4.3)

$$\sqrt{\pi}\begin{pmatrix}1 & -\mathrm{i}\\ -\mathrm{i} & 1\end{pmatrix}f^*(z)^{-\sigma_3/4}\sigma_2 A(f^*)\sigma_2^{-1} = \mathrm{e}^{-n\phi^*\sigma_3}\left(I+\mathrm{O}(1/n)\right)$$

and

$$\left[\sqrt{\pi}\begin{pmatrix}1 & \mathrm{i}\\ \mathrm{i} & 1\end{pmatrix}\tilde{f}(z)^{-\sigma_3/4}\sigma_1 A(\tilde{f})\sigma_1\right]^{-1} = \mathrm{e}^{n\tilde{\phi}\sigma_3}\left(I+\mathrm{O}(1/n)\right).$$

Applying the last two equations to (4.10) yields

$$J_S(z) = (-1)^{N-n}N(z)\mathrm{e}^{-n\phi^*\sigma_3}\left(\frac{D^*(z)}{D(z)}\right)^{\sigma_3}\mathrm{e}^{n\tilde{\phi}\sigma_3}N(z)^{-1}\left(I+\mathrm{O}(1/n)\right) = I+\mathrm{O}(1/n).$$

Let us now consider $z$ in the left half-plane of $\Gamma$. For $z=x\in(-\infty,0)$, we have $J_R(x)=I$. Coupling (4.1) and (4.9) gives

$$\begin{aligned}J_S(z) = (-1)^n N(z)&\left[\begin{pmatrix}1 & \mathrm{i}\\ \mathrm{i} & 1\end{pmatrix}\tilde{f}_-(z)^{-\sigma_3/4}\sigma_1 A(\tilde{f}_-)\sigma_1\right]\left(D_-(z)/D_+(z)\right)^{\sigma_3}\\ &\times\left[\begin{pmatrix}1 & \mathrm{i}\\ \mathrm{i} & 1\end{pmatrix}\tilde{f}_+(z)^{-\sigma_3/4}\sigma_1 A(\tilde{f}_+)\sigma_1\right]^{-1}N(z)^{-1}(-1)^n.\end{aligned}\tag{4.11}$$

Note that $\arg\tilde{f}_\pm(x)\in(-\pi,\pi)$ in this case; see Fig. 2(c). Hence, we obtain from (3.31) and (4.3)

$$\begin{pmatrix}1 & \mathrm{i}\\ \mathrm{i} & 1\end{pmatrix}\tilde{f}_\pm(z)^{-\sigma_3/4}\sigma_1 A(\tilde{f}_\pm)\sigma_1 = \frac{1}{\sqrt{\pi}}\mathrm{e}^{-n\tilde{\phi}_\pm\sigma_3}\left(I+\mathrm{O}(1/n)\right).$$

Substituting the above equation in (4.11) yields

$$J_S(z) = N(z)\mathrm{e}^{-n\tilde{\phi}_-\sigma_3}\mathrm{e}^{2N\pi\mathrm{i}z\sigma_3}\mathrm{e}^{n\tilde{\phi}_+\sigma_3}N(z)^{-1}\left(I+\mathrm{O}(1/n)\right) = I+\mathrm{O}(1/n).$$

Here, we have used the facts $D_-/D_+=\mathrm{e}^{2N\pi\mathrm{i}z}$ and $\mathrm{e}^{n(\tilde{\phi}_+-\tilde{\phi}_-)}=\mathrm{e}^{-2N\pi\mathrm{i}z}$.

For $z = x \in [0, x_0]$, we have from (2.10) and (2.11)

$$J_R(x) = \begin{pmatrix} 1 & 0 \\ \frac{4\cos^2(N\pi x)}{W\prod_{j=0}^{N-1}(x-x_{N,j})^2} & 1 \end{pmatrix}.$$

Note that $\tilde{f}(z) \in \mathbb{C}_\mp$ when $z \in \mathbb{C}_\pm$. This together with (4.1) and (4.9) gives

$$\begin{aligned} J_S(x) = {} & \left[N(x)\begin{pmatrix} 1 & \mathrm{i} \\ \mathrm{i} & 1 \end{pmatrix} \tilde{f}(x)^{-\sigma_3/4}\sigma_1 A(\tilde{f}_-)\sigma_1\right] \begin{pmatrix} 1 & 0 \\ D^{-2} & 1 \end{pmatrix} \\ & \times \left[N(x)\begin{pmatrix} 1 & \mathrm{i} \\ \mathrm{i} & 1 \end{pmatrix} \tilde{f}(x)^{-\sigma_3/4}\sigma_1 A(\tilde{f}_+)\sigma_1\right]^{-1}. \end{aligned} \tag{4.12}$$

From (3.28), it follows that

$$\begin{aligned} J_S(x) = {} & \left[N(x)\begin{pmatrix} 1 & \mathrm{i} \\ \mathrm{i} & 1 \end{pmatrix} \tilde{f}(x)^{-\sigma_3/4}\sigma_1 A(\tilde{f}_+)\sigma_1\right] \left[\sigma_1 \begin{pmatrix} 1 & -1 \\ 0 & 1 \end{pmatrix} \sigma_1\right] \begin{pmatrix} 1 & 0 \\ D^{-2} & 1 \end{pmatrix} \\ & \times \left[N(x)\begin{pmatrix} 1 & \mathrm{i} \\ \mathrm{i} & 1 \end{pmatrix} \tilde{f}(x)^{-\sigma_3/4}\sigma_1 A(\tilde{f}_+)\sigma_1\right]^{-1} \\ = {} & \left[N(x)\begin{pmatrix} 1 & \mathrm{i} \\ \mathrm{i} & 1 \end{pmatrix} \tilde{f}(x)^{-\sigma_3/4}\sigma_1 A(\tilde{f}_+)\sigma_1\right] \begin{pmatrix} 1 & 0 \\ D^{-2}-1 & 1 \end{pmatrix} \\ & \times \left[N(x)\begin{pmatrix} 1 & \mathrm{i} \\ \mathrm{i} & 1 \end{pmatrix} \tilde{f}(x)^{-\sigma_3/4}\sigma_1 A(\tilde{f}_+)\sigma_1\right]^{-1}. \end{aligned} \tag{4.13}$$

Note that $N(z)$ and $\tilde{f}(z)^{-1/4}$ are both discontinues on the interval $(a, x_0]$. From Fig. 2(c) and (4.3), we observe that $\arg\tilde{\phi}_\pm(x) = \mp\pi/2$ and $\arg\tilde{f}_\pm(x) = \mp\pi$ for $x \in (a, x_0]$. Thus, we have $\tilde{f}_+(x)^{-\sigma_3/4} = \tilde{f}_-(x)^{-\sigma_3/4}\mathrm{e}^{\mathrm{i}\pi\sigma_3/2}$. By (3.25), it is readily seen that

$$N(z)\begin{pmatrix} 1 & \mathrm{i} \\ \mathrm{i} & 1 \end{pmatrix} \tilde{f}(z)^{-\sigma_3/4}$$

has no jump on the interval $(a, x_0]$. Applying (3.37) to (4.13) gives $J_S(x) = I + \mathrm{O}(1/n)$.

For $z \in \Sigma_\pm$ and $\operatorname{Re} z < x_0$, the jump matrix $J_R(z)$ is

$$J_R(z) = \begin{pmatrix} 1 & 0 \\ \frac{\mp 2\mathrm{e}^{\pm\mathrm{i}\pi Nz}\cos(N\pi z)}{W\prod_{j=0}^{N-1}(z-x_{N,j})^2} & 1 \end{pmatrix};$$

see (2.12). This together with (4.1) gives

$$\begin{aligned} J_S(z) = {} & N(z)\left[\begin{pmatrix} 1 & \mathrm{i} \\ \mathrm{i} & 1 \end{pmatrix} \tilde{f}(z)^{-\sigma_3/4}\sigma_1 A(\tilde{f})\sigma_1\right] \begin{pmatrix} 1 & 0 \\ \frac{\mp\mathrm{e}^{\pm\mathrm{i}\pi Nz}}{2\cos(N\pi z)D^2} & 1 \end{pmatrix} \\ & \times \left[\begin{pmatrix} 1 & \mathrm{i} \\ \mathrm{i} & 1 \end{pmatrix} \tilde{f}(z)^{-\sigma_3/4}\sigma_1 A(\tilde{f})\sigma_1\right]^{-1} N^{-1}(z). \end{aligned} \tag{4.14}$$

From (3.31) and (4.3), we then obtain

$$\begin{pmatrix} 1 & \mathrm{i} \\ \mathrm{i} & 1 \end{pmatrix} \tilde{f}(z)^{-\sigma_3/4}\sigma_1 A(\tilde{f})\sigma_1 = \frac{\mathrm{e}^{-n\tilde{\phi}\sigma_3}}{\sqrt{\pi}}\big(I + \mathrm{O}(1/n)\big).$$

Applying the above formula to (4.14) yields

$$\begin{aligned} J_S(z) &= N(z)\begin{pmatrix} 1 & 0 \\ \frac{\mp \mathrm{e}^{2n\tilde{\phi}}\mathrm{e}^{\pm 2\mathrm{i}\pi Nz}}{2\cos(N\pi z)\mathrm{e}^{\pm \mathrm{i}\pi Nz}D(z)^2} & 1 \end{pmatrix} N(z)^{-1}\big(I + \mathrm{O}(1/n)\big) \\ &= N(z)\begin{pmatrix} 1 & 0 \\ \frac{\mp \mathrm{e}^{2n\phi}}{\widetilde{D}(z)D(z)} & 1 \end{pmatrix} N(z)^{-1}\big(I + \mathrm{O}(1/n)\big). \end{aligned}$$

One can show that $\operatorname{Re}\phi < 0$ in this case; see Fig. 2(d). Thus, we have $J_S(z) = I + \mathrm{O}(1/n)$ as $n \to \infty$.

In a similar manner, we can prove that for $z$ in the right half-plane of $\Gamma$, the corresponding jump matrix $J_S(z)$ on $\Sigma$ and on the real line tends to the identity matrix. Hence, we have shown that $J_S(z) = I + \mathrm{O}(1/n)$ as $n \to \infty$ on the contour $\Sigma_S$. By the main result in [13], we conclude that $S(z) = I + \mathrm{O}(1/n)$ as $n \to \infty$. □

## 5. Main results

**Theorem 5.1.** *Let $l$, $D(z)$ and $\tilde{f}(z)$ be defined as in* (3.5), (3.33) *and* (4.3), *respectively. We have*

$$\begin{aligned} &\pi_{N,n}(Nz - 1/2) \\ &\quad = (-N)^n\sqrt{\pi}\mathrm{e}^{nl/2}\big\{\big[\sin(N\pi z)\operatorname{Ai}\big(\tilde{f}(z)\big) + \cos(N\pi z)D(z)\operatorname{Bi}\big(\tilde{f}(z)\big)\big]\widetilde{A}(z,n) \\ &\qquad + \big[\sin(N\pi z)\operatorname{Ai}'\big(\tilde{f}(z)\big) + \cos(N\pi z)D(z)\operatorname{Bi}'\big(\tilde{f}(z)\big)\big]\widetilde{B}(z,n)\big\} \end{aligned} \tag{5.1}$$

*for $z \in I$ and $III$, where*

$$\widetilde{A}(z,n) = \frac{(z-b)^{1/4}}{(z-a)^{1/4}}\tilde{f}(z)^{1/4}\big[1 + \mathrm{O}(1/n)\big], \qquad \widetilde{B}(z,n) = \frac{(z-a)^{1/4}}{(z-b)^{1/4}}\tilde{f}(z)^{-1/4}\big[1 + \mathrm{O}(1/n)\big].$$

*Similarly, with $D^*(z)$ and $f^*(z)$ defined in* (3.35) *and* (4.3),

$$\begin{aligned} &\pi_{N,n}(Nz - 1/2) \\ &\quad = (-1)^N N^n\sqrt{\pi}\mathrm{e}^{nl/2}\big\{\big[\cos(N\pi z)D^*(z)\operatorname{Bi}\big(f^*(z)\big) - \sin(N\pi z)\operatorname{Ai}\big(f^*(z)\big)\big]A^*(z,n) \\ &\qquad + \big[\cos(N\pi z)D^*(z)\operatorname{Bi}'\big(f^*(z)\big) - \sin(N\pi z)\operatorname{Ai}'\big(f^*(z)\big)\big]B^*(z,n)\big\} \end{aligned} \tag{5.2}$$

*for $z \in II$ and $IV$, where*

$$A^*(z,n) = \frac{(z-a)^{1/4}}{(z-b)^{1/4}}f^*(z)^{1/4}\big[1 + \mathrm{O}(1/n)\big], \qquad B^*(z,n) = \frac{(z-b)^{1/4}}{(z-a)^{1/4}}f^*(z)^{-1/4}\big[1 + \mathrm{O}(1/n)\big].$$

**Proof.** From the definition of $S(z)$ in (4.7), we have

$$R(z) = \left(W e^{nl}\right)^{\sigma_3/2} S(z) \left(W e^{nl}\right)^{-\sigma_3/2} \widetilde{R}(z).$$

For any matrix $X$, we denote its $(i, j)$ element by $X_{ij}$. The above formula then gives

$$R_{11}(z) = S_{11}(z)\widetilde{R}_{11}(z) + S_{12}(z)\widetilde{R}_{21}(z) W e^{nl}$$

and

$$R_{12}(z) = S_{11}(z)\widetilde{R}_{12}(z) + S_{12}(z)\widetilde{R}_{22}(z) W e^{nl}.$$

First, let us restrict $z$ to the region I indicated in Fig. 3. A combination of (3.24), (3.26) and (4.1) gives

$$R_{11}(z) = (-1)^n \sqrt{\pi} e^{nl/2} \left[-i\omega \operatorname{Ai}(\omega \tilde{f}) \tilde{a}(z, n) - i\omega^2 \operatorname{Ai}'(\omega \tilde{f}) \tilde{b}(z, n)\right] \times \frac{2\cos(N\pi z) D(z)}{\prod_{j=0}^{N-1} (z - x_{N,j})}$$

and

$$R_{12}(z) = (-1)^n \sqrt{\pi} e^{nl/2} \left[i \operatorname{Ai}(\tilde{f}) \tilde{a}(z, n) + i \operatorname{Ai}'(\tilde{f}) \tilde{b}(z, n)\right] \times \frac{W \prod_{j=0}^{N-1} (z - x_{N,j})}{2\cos(N\pi z) D(z)},$$

where

$$\tilde{a}(z, n) = \frac{(z - b)^{1/4}}{(z - a)^{1/4}} \tilde{f}(z)^{1/4} \left(S_{11}(z) - i S_{12}(z)\right)$$

and

$$\tilde{b}(z, n) = \frac{(z - a)^{1/4}}{(z - b)^{1/4}} \tilde{f}(z)^{-1/4} \left(S_{11}(z) + i S_{12}(z)\right).$$

From (2.5) and (2.6), we know that $H_{11}(z)$ has different expressions in different parts of region I. Let us first consider the regions $\mathrm{I} \cap \Omega_+$ and $\mathrm{III} \cap \Omega_-$. For $z \in \mathrm{I} \cap \Omega_+$, we have from (2.5)

$$\begin{aligned} H_{11}(z) &= R_{11}(z) - \frac{\mp i e^{\pm N\pi i z} \cos(N\pi z)}{N\pi \prod_{j=0}^{N-1} (z - x_{N,j})^2} R_{12}(z) \\ &= (-1)^n \sqrt{\pi} e^{nl/2} \prod_{j=0}^{N-1} (z - x_{N,j})^{-1} \\ &\qquad \times \left\{\cos(N\pi z) D(z) \left[-2i\omega \operatorname{Ai}(\omega \tilde{f}) \tilde{a}(z, n) - 2i\omega^2 \operatorname{Ai}'(\omega \tilde{f}) \tilde{b}(z, n)\right] \right. \\ &\qquad \left. - i e^{i\pi N z} D(z)^{-1} \left[\operatorname{Ai}(\tilde{f}) \tilde{a}(z, n) + \operatorname{Ai}'(\tilde{f}) \tilde{b}(z, n)\right]\right\}. \end{aligned} \tag{5.3}$$

The terms in curly brackets in (5.3) can be rewritten as

$$\begin{aligned}\cos(N\pi z)D(z)\big[&\big(-2\mathrm{i}\omega\,\mathrm{Ai}(\omega\tilde f)-\mathrm{i}\,\mathrm{Ai}(\tilde f)-\mathrm{i}\big(D(z)^{-2}-1\big)\,\mathrm{Ai}(\tilde f)\big)\tilde a(z,n)\\&+\big(-2\mathrm{i}\omega^2\,\mathrm{Ai}'(\omega\tilde f)-\mathrm{i}\,\mathrm{Ai}'(\tilde f)-\mathrm{i}\big(D(z)^{-2}-1\big)\,\mathrm{Ai}'(\tilde f)\big)\tilde b(z,n)\big]\\&+\sin(N\pi z)D(z)^{-1}\big[\mathrm{Ai}(\tilde f)\tilde a(z,n)+\mathrm{Ai}'(\tilde f)\tilde b(z,n)\big].\end{aligned}\tag{5.4}$$

Recall the well-known formula of the Airy functions [10], formula (9.2.11),

$$\mathrm{Bi}(z)=\pm\mathrm{i}\big[2\mathrm{e}^{\mp\pi\mathrm{i}/3}\,\mathrm{Ai}\big(\omega^{\pm1}z\big)-\mathrm{Ai}(z)\big].\tag{5.5}$$

Also, note that $D(z)\sim 1$ for $z\neq 0$ and $\mathrm{Ai}(\tilde f)$ and $\mathrm{Ai}'(\tilde f)$ are exponentially small as $n\to\infty$ when $z$ is close to the origin. Hence, we can always neglect the terms $(D(z)^{-2}-1)\,\mathrm{Ai}(\tilde f)$ and $(D(z)^{-2}-1)\,\mathrm{Ai}'(\tilde f)$ in (5.4). This together with Lemma 4.1 and (5.5) gives

$$\begin{aligned}H_{11}(z)=&(-1)^n\sqrt{\pi}\mathrm{e}^{nl/2}\\&\times\prod_{j=0}^{N-1}(z-x_{N,j})^{-1}\big\{\big[\sin(N\pi z)\,\mathrm{Ai}\big(\tilde f(z)\big)+\cos(N\pi z)D(z)\,\mathrm{Bi}\big(\tilde f(z)\big)\big]\widetilde A(z,n)\\&+\big[\sin(N\pi z)\,\mathrm{Ai}'\big(\tilde f(z)\big)+\cos(N\pi z)D(z)\,\mathrm{Bi}'\big(\tilde f(z)\big)\big]\widetilde B(z,n)\big\}.\end{aligned}\tag{5.6}$$

Similarly, for $z\in\mathrm{III}\cap\Omega_-$

$$H_{11}(z)=(-1)^n\sqrt{\pi}\mathrm{e}^{nl/2}\prod_{j=0}^{N-1}(z-x_{N,j})^{-1}\tag{5.7}$$

$$\begin{aligned}&\times\big\{\cos(N\pi z)D(z)\big[2\mathrm{i}\omega^2\,\mathrm{Ai}\big(\omega^2\tilde f\big)\tilde a(z,n)+2\mathrm{i}\omega\,\mathrm{Ai}'\big(\omega^2\tilde f\big)\tilde b(z,n)\big]\\&+\mathrm{i}\mathrm{e}^{-\mathrm{i}\pi Nz}D(z)^{-1}\big[\mathrm{Ai}(\tilde f)\tilde a(z,n)+\mathrm{Ai}'(\tilde f)\tilde b(z,n)\big]\big\}.\end{aligned}\tag{5.8}$$

Again, by (5.5), we get exactly the same formula given in (5.6). Using Lemma 4.1, we can obtain the desired result (5.1). Now, we show that the asymptotic formula of $H_{11}(z)$ in (5.6) holds not only for $z$ in $\Omega_\pm$, but also for $z$ in the whole I and III. From the relation between $R(z)$ and $H(z)$ in (2.6), we know that $H_{11}(z)=R_{11}(z)$ for $z\in\mathrm{I}\setminus\Omega_+$. In contrast to the expansion in (5.3), for $z\in\mathrm{I}\setminus\Omega_+$ the term

$$(-1)^n\sqrt{\pi}\mathrm{e}^{nl/2}\prod_{j=0}^{N-1}(z-x_{N,j})^{-1}\times\big\{-\mathrm{i}\mathrm{e}^{\mathrm{i}\pi Nz}D(z)^{-1}\big[\mathrm{Ai}(\tilde f)\,\tilde a(z,n)+\mathrm{Ai}'(\tilde f)\,\tilde b(z,n)\big]\big\}$$

does not appear, since the quantity $\mathrm{e}^{\mathrm{i}\pi Nz}\,\mathrm{Ai}(\tilde f)$ is exponentially small as $n$ goes to infinity, in comparison with the other term in (5.3). This suggests that the region of validity of the expansion given in (5.6) can be extended to $z\in\mathrm{I}\cup\mathrm{III}$. As a consequence, (5.1) holds for $z\in\mathrm{I}\cup\mathrm{III}$.

In a similar manner, we have

$$H_{11}(z) = (-1)^N \sqrt{\pi} e^{nl/2} \times \prod_{j=0}^{N-1} (z - x_{N,j})^{-1} \{ \cos(N\pi z) D^*(z) [2i\omega^2 \operatorname{Ai}(\omega^2 f^*) a^*(z,n) + 2i\omega \operatorname{Ai}'(\omega^2 f^*) b^*(z,n)] + i e^{i\pi N z} D^*(z)^{-1} [\operatorname{Ai}(f^*) a^*(z,n) + \operatorname{Ai}'(f^*) b^*(z,n)] \},$$

where

$$a^*(z,n) = \frac{(z-a)^{1/4}}{(z-b)^{1/4}} f^*(z)^{1/4} (S_{11} + iS_{12})$$

and

$$b^*(z,n) = \frac{(z-b)^{1/4}}{(z-a)^{1/4}} f^*(z)^{-1/4} (S_{11} - iS_{12})$$

for $z \in \mathrm{II} \cap \Omega_+$. Following the same argument as given above, one can obtain (5.2) for $z \in \mathrm{II}$ and IV. This completes the proof of Theorem 5.1. □

## 6. Comparison with earlier results

To compare our formulas in Theorem 5.1 with those given in [12], we first derive from (5.1) two simple asymptotic formulas for $\pi_{N,n}(Nz - 1/2)$ when $z$ is real and less than $a$; cf. (3.2).

**Theorem 6.1.** *Let $l$, $\tilde{\phi}(z)$, $D(z)$ and $\widetilde{D}(z)$ be defined as in* (3.5), (3.9), (3.33) *and* (3.34), *respectively. We have*

$$\pi_{N,n}(Nz - 1/2) = (-N)^n e^{nl/2} \left\{ e^{-n\tilde{\phi}(z)} D(z) \cos(N\pi z) \frac{(z-a)^{1/2} + (z-b)^{1/2}}{(z-a)^{1/4}(z-b)^{1/4}} [1 + O(1/n)] + O\left(e^{n \operatorname{Re} \tilde{\phi}(z)}\right) \right\} \tag{6.1}$$

*for* $0 < z \leqslant a - \delta < a$, *and*

$$\pi_{N,n}(Nz - 1/2) = N^n e^{nl/2} \left\{ e^{-n\phi(z)} \widetilde{D}(z) \frac{(z-a)^{1/2} + (z-b)^{1/2}}{2(z-a)^{1/4}(z-b)^{1/4}} [1 + O(1/n)] + O\left(e^{n \operatorname{Re} \tilde{\phi}(z)}\right) \right\} \tag{6.2}$$

*for* $z < 0$.

**Proof.** From the well-known asymptotic expansions of the Airy function [10], formulas (9.7.5)–(9.7.8), we have

$$\begin{aligned} \mathrm{Ai}(\tilde{f}(z)) &\sim \frac{\tilde{f}(z)^{-1/4}}{2\sqrt{\pi}} e^{n\tilde{\phi}(z)} \sum_{k=0}^{\infty} \frac{(-1)^k u_k}{(-n\tilde{\phi}(z))^k}, \\ \mathrm{Ai}'(\tilde{f}(z)) &\sim -\frac{\tilde{f}(z)^{1/4}}{2\sqrt{\pi}} e^{n\tilde{\phi}(z)} \sum_{k=0}^{\infty} \frac{(-1)^k v_k}{(-n\tilde{\phi}(z))^k} \end{aligned} \tag{6.3}$$

and

$$\begin{aligned} \mathrm{Bi}(\tilde{f}(z)) &\sim \frac{\tilde{f}(z)^{-1/4}}{\sqrt{\pi}} e^{-n\tilde{\phi}(z)} \sum_{k=0}^{\infty} \frac{u_k}{(-n\tilde{\phi}(z))^k}, \\ \mathrm{Bi}'(\tilde{f}(z)) &\sim \frac{\tilde{f}(z)^{1/4}}{\sqrt{\pi}} e^{-n\tilde{\phi}(z)} \sum_{k=0}^{\infty} \frac{v_k}{(-n\tilde{\phi}(z))^k}, \end{aligned} \tag{6.4}$$

where $\tilde{f}(z)$ is defined in (4.3). Note that $\tilde{\phi}(z)$ is analytic for real $z \in (0, a)$; see the paragraph containing (3.9). Applying (6.3)–(6.4) to (5.1) gives

$$\begin{aligned} &\pi_{N,n}(Nz - 1/2) \\ &\quad = (-N)^n e^{nl/2} \bigg\{ e^{n\tilde{\phi}(z)} \sin(N\pi z) \frac{(z-b)^{1/2} - (z-a)^{1/2}}{2(z-a)^{1/4}(z-b)^{1/4}} \big[1 + \mathrm{O}(1/n)\big] \\ &\qquad + e^{-n\tilde{\phi}(z)} D(z) \cos(N\pi z) \frac{(z-a)^{1/2} + (z-b)^{1/2}}{(z-a)^{1/4}(z-b)^{1/4}} \big[1 + \mathrm{O}(1/n)\big] \bigg\}. \end{aligned} \tag{6.5}$$

Since $\tilde{\phi}(z) < 0$ in this case on account of (3.11), $e^{n\tilde{\phi}(z)}$ is exponentially small as $n$ goes to infinity. Thus, (6.1) follows from (6.5).

Note that $\tilde{\phi}(z)$ has a jump across the negative real axis. For real $z < 0$, we obtain from (6.3)–(6.4) and (5.1)

$$\begin{aligned} &\pi^+_{N,n}(Nz - 1/2) \\ &\quad = (-N)^n e^{nl/2} \bigg\{ e^{n\tilde{\phi}_+(z)} \sin(N\pi z) \frac{(z-b)^{1/2} - (z-a)^{1/2}}{2(z-a)^{1/4}(z-b)^{1/4}} \big[1 + \mathrm{O}(1/n)\big] \\ &\qquad + e^{-n\tilde{\phi}_+(z)} D_+(z) \cos(N\pi z) \frac{(z-a)^{1/2} + (z-b)^{1/2}}{(z-a)^{1/4}(z-b)^{1/4}} \big[1 + \mathrm{O}(1/n)\big] \bigg\} \end{aligned} \tag{6.6}$$

as $n \to \infty$, where $\pi^+_{N,n}(Nz - 1/2)$ denotes the limiting value of $\pi_{N,n}(Nz - 1/2)$ as $z$ approaches the real line from above. Observe from Fig. 2(c) that $\mathrm{Re}\,\tilde{\phi}_+(z) < 0$. Thus, it follows from (6.6) that

$$\begin{aligned} \pi^+_{N,n}(Nz - 1/2) = (-N)^n e^{nl/2} \bigg\{ & e^{-n\tilde{\phi}_+(z)} D_+(z) \cos(N\pi z) \frac{(z-a)^{1/2} + (z-b)^{1/2}}{(z-a)^{1/4}(z-b)^{1/4}} \big[1 + \mathrm{O}(1/n)\big] \\ & + \mathrm{O}\big(e^{n\,\mathrm{Re}\,\tilde{\phi}(z)}\big) \bigg\}. \end{aligned} \tag{6.7}$$

Similarly, one can see that $\pi^-_{N,n}(Nz-1/2)$ is given by

$$(-N)^n e^{nl/2}\Big\{e^{-n\tilde{\phi}_-(z)}D_-(z)\cos(N\pi z)\frac{(z-a)^{1/2}+(z-b)^{1/2}}{(z-a)^{1/4}(z-b)^{1/4}}[1+O(1/n)]+O\big(e^{n\operatorname{Re}\tilde{\phi}(z)}\big)\Big\}. \tag{6.8}$$

From the definition of $\phi$ in (3.5), we note that $e^{n\phi(z)}$ can be analytically extended to $(-\infty,0)$. Thus, from (3.9) and (3.36) it follows that

$$e^{-n\tilde{\phi}_\pm(z)}D_\pm(z)\cos(N\pi z)=\frac{(-1)^n}{2}e^{-n\phi(z)}\widetilde{D}(z). \tag{6.9}$$

In view of (6.9), the two asymptotic formulas (6.7) and (6.8) are exactly the same. Hence,

$$\pi_{N,n}(Nz-1/2)=N^n e^{nl/2}\Big[e^{-n\phi(z)}\widetilde{D}(z)\frac{(z-a)^{1/2}+(z-b)^{1/2}}{2(z-a)^{1/4}(z-b)^{1/4}}[1+O(1/n)]+O\big(e^{n\operatorname{Re}\tilde{\phi}(z)}\big)\Big], \tag{6.10}$$

for real $z<0$. This gives (6.2). □

We will now show that our asymptotic formulas for the discrete Chebyshev polynomials are the same as those given by Pan and Wong [12]. First, we introduce the notation

$$x:=Nz-1/2. \tag{6.11}$$

Two different asymptotic approximations for $t_n(x,N+1)$ are given in [12], formulas (8.13) and (8.6), one for $x$ negative and the other for $x$ positive. Changing $N+1$ to $N$, they read

$$\begin{aligned}t_n(x,N)=&\frac{(-1)^{n+1}\Gamma(n+N+1)(N-1)^x n^{-2x-2}\Gamma(x+1)}{\Gamma(N)\pi}\\&\times\Big\{\sin(\pi x)\Big[1+O\Big(\frac{1}{N}\Big)\Big]+O(e^{N\eta})\Big\}\end{aligned} \tag{6.12}$$

for fixed $x>0$ and

$$t_n(x,N)=\frac{(-1)^n\Gamma(n+N+1)(N-1)^x n^{-2x-2}}{\Gamma(N)\Gamma(-x)}\Big[1+O\Big(\frac{1}{N}\Big)\Big] \tag{6.13}$$

for fixed $x<0$.

We next derive from (6.1) and (6.2) asymptotic formulas for $t_n(x,N)$ when $x$ is fixed (i.e., $z=O(1/N)$). For $x>0$, i.e., $z>0$, substituting (6.11) in (3.33) gives

$$D(z)\cos(N\pi z)=\frac{e^{Nz}\Gamma(Nz+1/2)}{\sqrt{2\pi}(Nz)^{Nz}}\cos(N\pi z)=-\frac{e^{x+1/2}\Gamma(x+1)}{\sqrt{2\pi}(x+1/2)^{x+1/2}}\sin(\pi x). \tag{6.14}$$

Moreover, it is readily verified that

$$\frac{(2n)!}{n!^2} \sim \frac{1}{\sqrt{\pi}} 2^{2n} n^{-1/2} \tag{6.15}$$

as $n \to \infty$ and

$$\begin{aligned}\frac{(z-a)^{1/2}+(z-b)^{1/2}}{(z-a)^{1/4}(z-b)^{1/4}} &\sim \frac{a^{1/2}+b^{1//2}}{(ab)^{1/4}} = \frac{(a+b+2a^{1/2}b^{1/2})^{1/2}}{(ab)^{1/4}} \\ &= \sqrt{2}(1+c)^{1/2}c^{-1/2}\end{aligned} \tag{6.16}$$

as $z \to 0$. Here, we have made use of the fact that $a+b=1$ and $ab=c^2/4$; see (3.2).

Note that $z \to 0$ as $n \to \infty$; a combination of (2.1), (6.14)–(6.16) and (6.1) yields

$$t_n(x,N) \sim (-1)^{n+1} N^n 2^{2n} n^{-1/2} \frac{\mathrm{e}^{x+1/2}\Gamma(x+1)}{\pi(x+1/2)^{x+1/2}} (1+c)^{1/2} c^{-1/2} \sin(\pi x) \mathrm{e}^{n(l/2-\tilde{\phi}(z))} \tag{6.17}$$

as $n \to \infty$. We now derive an explicit formula for $l/2 - \tilde{\phi}(z)$; see (6.26) below. From (3.5) and (3.9), one has

$$l/2 - \tilde{\phi}(z) = l/2 - \phi_+(z) - \pi\mathrm{i}\left(1 - \frac{1}{c}z\right) = \operatorname{Re} g_+(z) \tag{6.18}$$

for $0 < z \leqslant a - \delta$. This fact evokes us to calculate $g(z)$ first. From (3.3) and (3.4), it is easily seen that

$$g(z) = \frac{1}{c}\int_0^a \log(z-s)\,\mathrm{d}s + \frac{1}{c}\int_b^1 \log(z-s)\,\mathrm{d}s + \int_a^b \log(z-s)\mu(s)\,\mathrm{d}s. \tag{6.19}$$

By integration by parts twice, we obtain

$$\begin{aligned}&\int_a^b \log(z-s)\mu(s)\,\mathrm{d}s \\ &\quad = \mu(s)\log(z-s)s|_a^b - z\mu(s)\log(z-s)|_a^b - \int_a^b \mu(s)\,\mathrm{d}s \\ &\qquad + z\int_a^b \mu(s)'\log(z-s)\,\mathrm{d}s - \int_a^b s\log(z-s)\mu(s)'\,\mathrm{d}s.\end{aligned} \tag{6.20}$$

By the definition of $\mu(x)$, we have

$$\mu(x)' = -\frac{1}{2\pi}\frac{1-2x}{(x-x^2)\sqrt{(x-a)(b-x)}}. \tag{6.21}$$

Here, we have again used the relations $a+b=1$ and $ab=c^2/4$. Moreover, one can show that

$$\begin{aligned}\int_a^b \log(z-s)\mu(s)\,ds = &\left[\frac{1}{c}(z-a)\log(z-a)-\frac{1}{c}(z-b)\log(z-b)\right]-\left(1-\frac{2a}{c}\right)\\ &-\frac{z}{2\pi}\int_a^b \frac{\log(z-s)}{s\sqrt{(s-a)(b-s)}}\,ds\\ &+\frac{1}{2\pi}(z-1)\int_a^b \frac{\log(z-s)}{(1-s)\sqrt{(s-a)(b-s)}}\,ds\\ &+\frac{1}{\pi}\int_a^b \frac{\log(z-s)}{\sqrt{(s-a)(b-s)}}\,ds.\end{aligned} \tag{6.22}$$

Let the integrals on the right-hand side of the equality be denoted by $I_1$, $I_2$ and $I_3$, respectively. To evaluate $I_1$, we note that from [16] we have

$$\int_a^b \frac{\log s}{(s-z)\sqrt{(s-a)(b-s)}}\,ds = \frac{2\pi}{(z-a)^{1/2}(z-b)^{1/2}}\log\frac{z+\sqrt{ab}+(z-a)^{1/2}(z-b)^{1/2}}{(\sqrt{a}+\sqrt{b})z}. \tag{6.23}$$

Making the change of variable $s=z-x$, (6.23) gives

$$\begin{aligned}I_1 &= \int_{z-a}^{z-b} \frac{\log x}{(z-x)\sqrt{(z-x-a)(b-z+x)}}\,d(z-x)\\ &= -\frac{4\pi}{c}\log\frac{z+\sqrt{(z-a)(z-b)}+c/2}{(\sqrt{z-a}+\sqrt{z-b})z}.\end{aligned} \tag{6.24}$$

Similarly, we can evaluate $I_2$ and $I_3$. Thus, a straightforward calculation shows that

$$\begin{aligned}g(z) = &-1-2\log 2+\frac{1}{c}(z-1)\log(z-1)-\frac{1}{c}z\log z+\left(2-\frac{2}{c}\right)\log\left[(z-a)^{1/2}+(z-b)^{1/2}\right]\\ &+\frac{2}{c}z\log\left[z+(z-a)^{1/2}(z-b)^{1/2}+c/2\right]\\ &+\frac{2}{c}(1-z)\log\left[z-1+(z-a)^{1/2}(z-b)^{1/2}-c/2\right].\end{aligned} \tag{6.25}$$

By (6.18) and (6.25), $l/2-\tilde{\phi}(z)$ can be explicitly given by

$$\begin{aligned}&-1-2\log 2+\frac{1}{c}(z-1)\log(1-z)-\frac{1}{c}z\log z+\left(2-\frac{2}{c}\right)\log\left[(a-z)^{1/2}+(b-z)^{1/2}\right]\\ &+\frac{2}{c}z\log\left[z-(a-z)^{1/2}(b-z)^{1/2}+c/2\right]\\ &+\frac{2}{c}(1-z)\log\left[1-z+(a-z)^{1/2}(b-z)^{1/2}+c/2\right].\end{aligned} \tag{6.26}$$

Note that $n = cN$ and $z \to 0$ as $n \to \infty$. From (6.26), we have

$$l/2 - \tilde{\phi}(z) = \left[-1 - 2\log 2 + \left(1 + \frac{1}{c}\right)\log(1+c)\right] + z\left[\frac{1}{c}\log(z) - \frac{2}{c}\log c - \frac{1}{c}\right] + \mathrm{O}(z^2) \quad (6.27)$$

and so

$$\mathrm{e}^{n(l/2-\tilde{\phi}(z))} \sim \mathrm{e}^{-n}2^{-2n}(1+c)^{n+N}(x+1/2)^{x+1/2}\mathrm{e}^{-x-1/2}N^{x+1/2}n^{-2x-1}. \quad (6.28)$$

Substituting (6.28) in (6.17) yields

$$t_n(x,N) \sim \frac{(-1)^{n+1}\Gamma(x+1)n^{-2x-2}}{\pi}\sin(\pi x)N^{n+x+1}(1+c)^{n+N+1/2}\mathrm{e}^{-n} \quad (6.29)$$

which will agree with (6.12) to leading order if we can show that

$$\frac{\Gamma(n+N+1)(N-1)^x}{\Gamma(N)} \sim N^{n+x+1}(1+c)^{n+N+1/2}\mathrm{e}^{-n}. \quad (6.30)$$

The last result is a direct consequence of Stirling's formula.

Next, we consider the case when $x$ is negative, but for a moment let's restrict $-\infty < x < -\frac{1}{2}$ (i.e., $z < 0$). Inserting (6.11) in (3.34), we have

$$\widetilde{D}(z) = \frac{\sqrt{2\pi}\mathrm{e}^{x+1/2}}{(-x-1/2)^{x+1/2}\Gamma(-x)}. \quad (6.31)$$

A combination of (6.2), (6.15), (6.16) and (6.31) gives

$$t_n(x,N) \sim N^n 2^{2n} n^{-1/2}\frac{\mathrm{e}^{x+1/2}}{(-x-1/2)^{x+1/2}\Gamma(-x)}(1+c)^{1/2}c^{-1/2}\mathrm{e}^{n(l/2-\phi(z))}. \quad (6.32)$$

Since $\mathrm{e}^{n\phi(z)}$ is analytic in the interval $(-\infty, 0)$, we obtain from (3.5) that $\mathrm{e}^{n(l/2-\phi(z))} = \mathrm{e}^{n(l/2-\phi_+(z))} = \mathrm{e}^{ng_+(z)}$. On account of (6.25), it can be shown that

$$g_+(z) = \left[\pi\mathrm{i} - 1 - 2\log 2 + \left(1 + \frac{1}{c}\right)\log(1+c)\right] + z\left[\frac{1}{c}\log(-z) - \frac{2}{c}\log c - \frac{1}{c}\right] + \mathrm{O}(z^2) \quad (6.33)$$

as $z \to 0$. Thus, letting $n \to \infty$, we have from (6.33)

$$\mathrm{e}^{n(l/2-\phi(z))} \sim (-1)^n 2^{-2n}(1+c)^{n+N}\mathrm{e}^{-n}c^{-2x-1}N^{-x-1/2}(-x-1/2)^{x+1/2}\mathrm{e}^{-x-1/2}. \quad (6.34)$$

From (6.32) and (6.34), we obtain

$$t_n(x,N) \sim (-N)^n \frac{n^{-2x-2}}{\Gamma(-x)}(1+c)^{n+N+1/2}N^{x+1}\mathrm{e}^{-n}. \quad (6.35)$$

In the case $-\frac{1}{2} < x < 0$, formula (6.35) follows directly from (6.29) on an appeal to the reflection formula $\Gamma(1+x)\Gamma(-x) = -\pi/\sin(\pi x)$. On the other hand, coupling (6.13) and (6.30) gives

$$t_n(x, N) \sim (-N)^n \frac{n^{-2x-2}}{\Gamma(-x)}(1+c)^{n+N+1/2}N^{x+1}e^{-n} \tag{6.36}$$

as $n \to \infty$. Therefore, our results (6.29) and (6.35) agree with those of Pan and Wong [12] stated in (6.12) and (6.13).

## References


[1] J. Baik, T. Kriecherbauer, K.T.-R. McLaughlin and P.D. Miller, *Discrete Orthogonal Polynomials: Asymptotics and Applications*, Annals of Mathematics Studies, Vol. 164, Princeton Univ. Press, Princeton, NJ, 2007.

[2] R. Beals and R. Wong, *Special Functions: A Graduate Text*, Cambridge Univ. Press, Cambridge, 2010.

[3] D. Dai and R. Wong, Global asymptotics of Krawtchouk polynomials – a Riemann–Hilbert approach, *Chin. Ann. Math. Ser. B* **28** (2007), 1–34.

[4] P. Deift and X. Zhou, A steepest descent method for oscillatory Riemann–Hilbert problems, Asymptotics for the MKdV equation, *Ann. Math.* **137** (1993), 295–368.

[5] A.S. Fokas, A.R. Its and A.V. Kitaev, The isomonodromy approach to matrix models in 2D quantum gravity, *Comm. Math. Phys.* **147** (1992), 395–430.

[6] W. Gautschi, *Orthogonal Polynomials: Computation and Approximation*, Oxford Univ. Press, Oxford, 2004.

[7] F.B. Hildebrand, *Introduction to Numerical Analysis*, 2nd edn, McGraw-Hill, New York, 1974.

[8] S. Karlin and J.L. McGregor, The Hahn polynomials, formulas and an application, *Scripta Math.* **26** (1961), 33–46.

[9] A.B.J. Kuijlaars and W. Van Assche, The asymptotic zero distribution of orthogonal polynomials with varying recurrence coefficients, *J. Approx. Theory* **99** (1999), 167–197.

[10] F.W.J. Olver, D.W. Lozier, R.F. Boisvert and C.W. Clark, *NIST Handbook of Mathematical Functions*, Cambridge Univ. Press, Cambridge, 2010.

[11] C. Ou and R. Wong, The Riemann–Hilbert approach to global asymptotics of discrete orthogonal polynomials with infinite nodes, *Anal. Appl.* **8** (2010), 247–286.

[12] J.H. Pan and R. Wong, Uniform asymptotic expansions for the discrete Chebyshev polynomials, *Stud. Appl. Math.* **128** (2012), 337–384.

[13] W.-Y. Qiu and R. Wong, Asymptotic expansions for Riemann–Hilbert problems, *Anal. Appl.* **6** (2008), 269–298.

[14] G. Szegö, *Orthogonal Polynomials*, 4th edn, AMS Colloquium Publications, Vol. 23, Amer. Math. Soc., Providence, RI, 1975.

[15] X.S. Wang and R. Wong, Global asymptotics of the Meixner polynomials, *Asymptotic Analysis* **75** (2011), 211–231.

[16] Z. Wang and R. Wong, Uniform asymptotics of the Stieltjes–Wigert polynomials via the Riemann–Hilbert approach, *J. Math. Pures Appl.* **85** (2006), 698–718.

[17] R. Wong, *Asymptotic Approximations of Integrals*, Academic Press, Boston, MA, 1989; reprinted by SIAM, Philadelphia, PA, 2001.

# Global Asymptotics for Meixner-Pollaczek Polynomials with a Varying Parameter

*By Jun Wang, Weiyuan Qiu, and R. Wong*

In this paper, we study the uniform asymptotics of the Meixner-Pollaczek polynomials $P_n^{(\lambda_n)}(z;\phi)$ with varying parameter $\lambda_n = (n + \frac{1}{2})A$ as $n \to \infty$, where $A > 0$ is a constant. Two asymptotic expansions are obtained, which hold uniformly for $z$ in two overlapping regions which together cover the whole complex plane. One involves parabolic cylinder functions, and the other is in terms of elementary functions only. Our approach is based on the steepest descent method for oscillatory Riemann-Hilbert problems first introduced by Deift and Zhou [1].

## 1. Introduction

This paper is mainly concerned with the asymptotic behavior of the Meixner-Pollaczek polynomials, which were first discovered by Meixner [2] in 1934 and later studied by Pollaczek [3] in 1950. The major properties of these polynomials can be found by Chihara [4, p. 179–180] and Koekoek and Swarttouw [5]. The applications of the Meixner-Pollaczek polynomials arose in areas of mathematics and mathematical physics, such as Heisenberg algebra [6, 7], Lie algebra [8], unitary random-matrix ensemble [9], continued fractions of Ramanujan [10], etc.

Address for correspondence: R. Wong, Department of Mathematics, City University of Hong Kong, Hong Kong; e-mail: mawong@cityu.edu.hk

DOI: 10.1111/j.1467-9590.2012.00570.x
*STUDIES IN APPLIED MATHEMATICS* 130:345–392

The Meixner-Pollaczek polynomials $P_n^{(\lambda)}(x;\phi)$ with parameters $\lambda > 0$ and $\phi \in (0, \pi)$ can be defined by the hypergeometric function

$$P_n^{(\lambda)}(x;\phi) = \frac{(2\lambda)_n}{n!} e^{in\phi} {}_2F_1\left(\begin{matrix}-n, \lambda + ix \\ 2\lambda\end{matrix}; 1 - e^{-2i\phi}\right). \tag{1}$$

They are orthogonal on the real line with respect to the weight function

$$w(x;\lambda,\phi) = |\Gamma(\lambda + ix)|^2 \exp\{(2\phi - \pi)x\}, \tag{2}$$

and we have the orthogonality

$$\frac{1}{2\pi}\int_{-\infty}^{+\infty} P_m^{(\lambda)}(x;\phi)P_n^{(\lambda)}(x;\phi)w(x;\lambda,\phi)dx = \frac{\Gamma(n+2\lambda)}{(2\sin\phi)^{2\lambda}n!}\delta_{mn}. \tag{3}$$

Since the Gamma function satisfies $\Gamma(\overline{z}) = \overline{\Gamma(z)}$, we have

$$|\Gamma(\lambda + ix)|^2 = \Gamma(\lambda + ix)\Gamma(\lambda - ix).$$

Thus, $w(x;\lambda,\phi)$ can be analytically extended to the complex plane, except for the points $z = \pm(\lambda + k)i$ $(k = 0, 1, 2, \ldots)$, where it possesses a simple pole.

As far as we are aware, there are only a few results in the literature concerning the asymptotic behavior of the Meixner-Pollaczek polynomials. For asymptotics of the extreme zeros and distributions of zeros, see Chen and Ismail [11], Ismail and Li [12] and Krasovsky [13]. For uniform asymptotics of the polynomials and improved asymptotic approximations of zeros, see Li and Wong [14]. In the works mentioned above, the parameter $\lambda$ of the polynomials $P_n^{(\lambda)}(x;\phi)$ is fixed. In this paper, we are interested in the behavior of these polynomials when the parameter $\lambda$ depends on $n$ and tends to $+\infty$ as $n \to \infty$. We restrict our attention to the case

$$\lambda = \lambda_n = \left(n + \frac{1}{2}\right)A, \tag{4}$$

where $A > 0$ is a constant.

Uniform asymptotics of orthogonal polynomials with varying weights have been investigated by many authors; e.g., Deift et al. [15] for varying exponential weights; Bosbach and Gawronski [16], Kuijlaars and McLaughlin [17], and Aptekarev and Khabibullin [18] for Laguerre polynomials $L_n^{(\alpha_n)}(x)$; Bosback and Gawronski [19], Kuijlaars and Martinez-Finkelshtein [20], Martinez-Finkelshtein and Orive [21] and Wong and Zhang [22] for Jacobi polynomials $P_n^{(\alpha_n,\beta_n)}(x)$. Also, Baik and Suidan [23] have studied the Stieltjes–Wigert polynomials with varying parameters.

In this paper, we divide the complex plane into two regions, a bounded region containing all zeros of the polynomials, and an unbounded region disjoint from the interval containing the zeros. In the bounded region, we obtain a uniform asymptotic expansion of $P_n^{(\lambda_n)}(x;\phi)$ involving parabolic cylinder functions, and in the unbounded region we have a uniform asymptotic expansion of $P_n^{(\lambda_n)}(x;\phi)$ in terms of elementary functions only. The main results presented in Theorem 1 in Section 5 were announced in Ref. [24]. Our method is based on the Riemann-Hilbert approach first introduced by Deift and Zhou in Ref. [1], further developed in Refs. [15, 25], and with modifications made in Refs. [26], [22], and [27]. The starting point of the method is a theorem of Fokas, Its and Kitaev [28], which connects orthogonal polynomials to Riemann-Hilbert problems.

## 2. The Riemann-Hilbert problem

We consider the monic polynomials

$$\pi_n(z) = \pi_n(z, \lambda) = \frac{n!}{(2\sin\phi)^n} P_n^{(\lambda)}(z;\phi), \quad n = 0, 1, 2, \dots, \tag{5}$$

which are characterized by a Riemann-Hilbert problem(RHP) for $Y$:

Determine a $2 \times 2$ matrix-valued function $Y : \mathbb{C} \setminus \mathbb{R} \to \mathbb{C}^{2\times 2}$ such that

($\mathrm{Y}_a$) $Y(z)$ is analytic in $\mathbb{C} \setminus \mathbb{R}$;

($\mathrm{Y}_b$) $Y(z)$ possesses continuous boundary values $Y_+(x)$ and $Y_-(x)$ for $x \in \mathbb{R}$, and

$$Y_+(x) = Y_-(x) \begin{pmatrix} 1 & w(x;\lambda,\phi) \\ 0 & 1 \end{pmatrix},$$

where $Y_+(x)$ and $Y_-(x)$ denote the limiting values of $Y(z)$ as $z$ approaches the real line from the upper half plane and the lower half plane, respectively;

($\mathrm{Y}_c$) for $z \in \mathbb{C} \setminus \mathbb{R}$,

$$Y(z) = \left(I + O(1/z)\right) \begin{pmatrix} z^n & 0 \\ 0 & z^{-n} \end{pmatrix}$$

as $z \to \infty$.

By the well-known result of Fokas, Its, and Kitaev [28], we have

LEMMA 1. *The unique solution to the Riemann-Hilbert problem $Y$ is given by*

$$Y(z) = \begin{pmatrix} \pi_n(z) & C[\pi_n w](z) \\ c_n \pi_{n-1}(z) & c_n C[\pi_{n-1} w](z) \end{pmatrix},$$

*where $w$ is the weight function $w(x;\lambda,\phi)$, $C[f](z)$ is the Cauchy transform of $f$ on $\mathbb{R}$, and $c_n = -i(2\sin\phi)^{2(n+\lambda-1)}/[(n-1)!\,\Gamma(n+2\lambda-1)]$, which can be* *calculated from*

$$c_n^{-1} = -\frac{1}{2\pi i}\int_{-\infty}^{+\infty} \pi_{n-1}^2(x) w(x;\lambda,\phi)dx.$$

The following is an outline of our derivation of uniform asymptotic expansions of the Meixner-Pollaczek polynomials with a varying parameter by using the Riemann-Hilbert method. It consists of a sequence of transformations and a (heuristic) asymptotic approximation:

$$Y \to U \to T \to S \sim N.$$

The first transformation $Y \to U$ is a rescaling; the second transformation $U \to T$ normalizes the behavior of the solution at $z = \infty$, which can be done by using the logarithmic potential (also called the $g$-function); the third transformation $T \to S$ involves a factorization of the jump matrix and a deformation of the contour. The matrix $S$ has a limit $N$, and the RHP for $N$ is simple enough to be solved explicitly. The last step is the construction of a parametrix $\tilde{U}$ and a rigorous proof of $U \sim \tilde{U}$.

We end this section with the first transformation $Y \to U$. Set $\lambda = \lambda_n = \tau A$, where $\tau = n + \frac{1}{2}$ and $A > 0$. Define

$$U(z) := \begin{pmatrix} \tau^{-n} & 0 \\ 0 & \tau^n \end{pmatrix} Y(\tau z). \tag{6}$$

Then, $U(z)$ satisfies the following RHP:

(U$_a$) $U(z)$ is analytic in $\mathbb{C} \setminus \mathbb{R}$;

(U$_b$) $U(z)$ possesses continuous boundary values $U_+(x)$ and $U_-(x)$ for $x \in \mathbb{R}$, and

$$U_+(x) = U_-(x) \begin{pmatrix} 1 & w_n(x) \\ 0 & 1 \end{pmatrix},$$

where we let $w_n(x) = w(\tau x; \tau A, \phi)$;

(U$_c$) for $z \in \mathbb{C} \setminus \mathbb{R}$,

$$U(z) = (I + O(1/z)) \begin{pmatrix} z^n & 0 \\ 0 & z^{-n} \end{pmatrix} \quad \text{as} \quad z \to \infty.$$

From Lemma 1 and (6), it is easy to see

$$U_{11}(z) = \tau^{-n}\pi_n(\tau z), \tag{7}$$

where $U_{11}(z)$ denotes (1,1) entry of the matrix $U$.

## 3. Equilibrium measure and auxiliary functions

The distribution of zeros plays an important role in asymptotic analysis of orthogonal polynomials. Before normalizing the condition ($U_c$) in the RHP for $U$, we need to find the equilibrium measure and the logarithmic potential in advance. These are basic in the potential-theoretic method of obtaining weak asymptotics.

### 3.1. The auxiliary functions

Let $\mu_n(x)$ denote the equilibrium measure related to the weight function $w_n(x)$ which is supported on the interval $[\alpha_n, \beta_n]$. The constants $\alpha_n$ and $\beta_n$ are usually known as the Mhaskar-Rakhmanov-Saff (MRS) numbers. To determine the MRS numbers $\alpha_n, \beta_n$ and the measure $\mu_n(x)$, we introduce some auxiliary functions given below.

The first auxiliary function is the complex logarithmic potential of the corresponding measure $\mu_n$, that is

$$g_n(z) := \int_{\alpha_n}^{\beta_n} \log(z-s)\mu_n(s)ds, \tag{8}$$

where $\log(z-s)$ takes the principal branch on $\mathbb{C} \setminus (-\infty, s]$. The function $g_n(z)$ is analytic in the cut plane $\mathbb{C} \setminus (-\infty, \beta_n]$, and behaves like $\log z + O(z^{-1})$ as $z \to \infty$.

Define

$$G_n(z) := \frac{i}{\pi} g_n'(z) = \frac{1}{\pi i} \int_{\alpha_n}^{\beta_n} \frac{\mu_n(s)}{s-z} ds, \tag{9}$$

which is analytic in $\mathbb{C} \setminus [\alpha_n, \beta_n]$. Since $G_{n,\pm}(x) = \lim_{\varepsilon \to 0} G_n(x \pm i\varepsilon)$ for $x \in (\alpha_n, \beta_n)$, it follows that

$$G_{n,\pm}(x) = \lim_{\varepsilon \to 0} \frac{1}{\pi i} \int_{\alpha_n}^{\beta_n} \frac{\mu_n(s)(s-x \pm i\varepsilon)}{(s-x)^2+\varepsilon^2} ds = \pm \mu_n(x) + \frac{i}{\pi} \text{P.V.} \int_{\alpha_n}^{\beta_n} \frac{\mu_n(s)}{s-x} ds.$$

This means

$$\operatorname{Re} G_{n,\pm}(x) = \pm \mu_n(x), \quad \operatorname{Im} G_{n,+}(x) = \operatorname{Im} G_{n,-}(x), \quad \text{for} \quad x \in (\alpha_n, \beta_n). \tag{10}$$

As done in Ref. [29, p.703], we make the transformation

$$T(z) = e^{-\frac{\tau}{2}\ell_n \sigma_3} U(z)\, e^{[-\tau g_n(z) + \frac{1}{2}\log \frac{\varphi_n(z)}{2}]\sigma_3}\, e^{\frac{\tau}{2}\ell_n \sigma_3} \tag{11}$$

to normalize the condition ($U_c$) in the RHP for $U$, where $\sigma_3$ is the Pauli matrix $\binom{1\ \ 0}{0\ -1}$, $\ell_n$ is a constant to be determined,

$$\varphi_n(z) := z - \frac{\alpha_n+\beta_n}{2} + \zeta_n(z), \tag{12}$$

and

$$\zeta_n(z) := \sqrt{(z-\alpha_n)(z-\beta_n)}, \quad z \in \mathbb{C} \setminus [\alpha_n, \beta_n]. \tag{13}$$

The branch of $\zeta_n(z)$ in $\mathbb{C} \setminus [\alpha_n, \beta_n]$ is chosen so that $\zeta_n(z) \sim z$ as $z \to \infty$. Thus, $\varphi_n(z)$ is analytic in $\mathbb{C} \setminus [\alpha_n, \beta_n]$, and conformally maps $\mathbb{C} \setminus [\alpha_n, \beta_n]$ to $\mathbb{C} \setminus \{z : |z| \le \frac{\beta_n - \alpha_n}{2}\}$. (To see this, we note that the function $\varphi_n(z)$ in (12) can be written as $\varphi_n(z) = \frac{1}{2}(\sqrt{z-\alpha_n} + \sqrt{z-\beta_n})^2$, from which it is readily seen that $|\varphi_n(z)| > \frac{1}{2}(\beta_n - \alpha_n)$ for $z \in \mathbb{C} \setminus [\alpha_n, \beta_n]$.)

A straightforward calculation shows that $T$ is the unique solution of the following RHP:

($\mathrm{T}_a$) $T(z)$ is analytic in $\mathbb{C} \setminus \mathbb{R}$;
($\mathrm{T}_b$) $T_+(x) = T_-(x) J_T(x)$ for $x \in \mathbb{R}$, where the entries of $J_T(x)$ are given by

$$J_{11}(x) = e^{-\tau(g_{n,+}(x) - g_{n,-}(x)) + \frac{1}{2}\log(\varphi_{n,+}/\varphi_{n,-})},$$
$$J_{12}(x) = e^{\tau(g_{n,+}(x) + g_{n,-}(x)) + \log w_n(x) - \frac{1}{2}\log(\varphi_{n,+}\varphi_{n,-}/4) - \tau\ell_n},$$
$$J_{22}(x) = e^{\tau(g_{n,+}(x) - g_{n,-}(x)) - \frac{1}{2}\log(\varphi_{n,+}/\varphi_{n,-})},$$

and $J_{21}(x) = 0$;
($\mathrm{T}_c$) for $z \in \mathbb{C} \setminus \mathbb{R}$,

$$T(z) = I + O(1/z) \quad \text{as} \quad z \to \infty.$$

We choose the measure $\mu_n(x)$ so that the entry $J_{12}(x)$ in the jump matrix $J_T(x)$ becomes 1 for $x \in (\alpha_n, \beta_n)$, that is

$$\tau(g_{n,+}(x) + g_{n,-}(x)) + \log w_n(x) - \log\frac{\beta_n - \alpha_n}{4} - \tau\ell_n = 0 \quad \text{for } x \in (\alpha_n, \beta_n); \tag{14}$$

here, we have used the fact that $\varphi_{n,+}(x)\varphi_{n,-}(x) = (\beta_n - \alpha_n)^2/4$ for $x \in (\alpha_n, \beta_n)$. Letting $x \to \beta_n$ yields

$$\ell_n = 2g_n(\beta_n) + \frac{1}{\tau}\left(\log w_n(\beta_n) - \log\frac{\beta_n - \alpha_n}{4}\right). \tag{15}$$

We set

$$\begin{aligned} h_n(z) &= -\frac{1}{\tau}\frac{d}{dz}\log w_n(z) \\ &= i[\psi(\tau A - i\tau z) - \psi(\tau A + i\tau z)] + (\pi - 2\phi), \end{aligned} \tag{16}$$

where $\psi(z)$ is the logarithmic derivative of the Gamma function. Clearly, $h_n(z)$ is analytic in $\mathbb{C} \setminus \{\pm(\frac{k}{\tau} + A)i : k = 0, 1, 2, \ldots\}$, and $h_n(x)$ is a real-valued function for $x \in \mathbb{R}$ since $\psi(\bar{z}) = \overline{\psi(z)}$. Differentiating both sides of (14) gives

$$g'_{n,+}(x) + g'_{n,-}(x) = h_n(x), \quad x \in (\alpha_n, \beta_n). \tag{17}$$

Coupling (9) and (17), we obtain

$$G_{n,+}(x) + G_{n,-}(x) = \frac{i}{\pi} h_n(x) \quad \text{for} \quad x \in (\alpha_n, \beta_n). \tag{18}$$

From (10) and (18), it follows that

$$\operatorname{Im} G_{n,\pm}(x) = \frac{1}{2\pi} h_n(x) \quad \text{for} \quad x \in (\alpha_n, \beta_n). \tag{19}$$

Let

$$\nu_n(z) := \pi i G_n(z) + \frac{1}{2} h_n(z). \tag{20}$$

Then, $\nu_n(z)$ is analytic in $\mathbb{C} \setminus [\alpha_n, \beta_n] \cup \{\pm(\frac{k}{\tau} + A)i, k = 0, 1, 2, \ldots\}$. From (10), (19), and (20), it is readily seen that $\nu_{n,\pm}(x) = \pm \pi i \mu_n(x)$ for $x \in (\alpha_n, \beta_n)$. We define another auxiliary function

$$\phi_n(z) := \int_{\beta_n}^{z} \nu_n(s) ds, \tag{21}$$

where the path of integration lies entirely in $\mathbb{C} \setminus \{(-\infty, \beta_n] \cup (-i\infty, -iA] \cup [iA, i\infty)\}$, except for the initial point $\beta_n$. Integrating both sides of (20) from $\beta_n$ to $z$, we have

$$g_n(z) + \phi_n(z) = -\frac{1}{2\tau} \log w_n(z) + \frac{1}{2\tau} \log \frac{\beta_n - \alpha_n}{4} + \frac{\ell_n}{2}. \tag{22}$$

Note that $g_n(z) + \phi_n(z)$ is analytic in the interval $(-\infty, \beta_n]$.

### *3.2. The MRS numbers*

Equation (18) is actually a scalar RHP, and can be solved explicitly as

$$\begin{aligned} G_n(z) &= \frac{\sqrt{(z-\alpha_n)(z-\beta_n)}}{2\pi^2 i} \int_{\alpha_n}^{\beta_n} \frac{h_n(s)}{\sqrt{(\beta_n - s)(s-\alpha_n)}} \frac{1}{s-z} ds \\ &= \frac{\sqrt{(z-\alpha_n)(z-\beta_n)}}{2\pi^2 z i} \int_{\alpha_n}^{\beta_n} \frac{s h_n(s)}{\sqrt{(\beta_n - s)(s-\alpha_n)}} \left( \frac{1}{s-z} - \frac{1}{s} \right) ds. \end{aligned} \tag{23}$$

From (9), it is easy to see that $G_n(z) \to 0$ and $zG_n(z) \to i/\pi$ as $z \to \infty$. Thus, from (23) we obtain two equations

$$\int_{\alpha_n}^{\beta_n} \frac{i[\psi(\tau A - i\tau s) - \psi(\tau A + i\tau s)] + (\pi - 2\phi)}{\sqrt{(\beta_n - s)(s - \alpha_n)}} ds = 0, \tag{24}$$

$$\int_{\alpha_n}^{\beta_n} \frac{i[\psi(\tau A - i\tau s) - \psi(\tau A + i\tau s)] + (\pi - 2\phi)}{\sqrt{(\beta_n - s)(s - \alpha_n)}} s\, ds = 2\pi. \tag{25}$$

It is well known that

$$\begin{aligned} \int_{\alpha_n}^{\beta_n} \frac{1}{\sqrt{(\beta_n - s)(s - \alpha_n)}} ds &= \pi, \\ \int_{\alpha_n}^{\beta_n} \frac{s}{\sqrt{(\beta_n - s)(s - \alpha_n)}} ds &= \frac{\alpha_n + \beta_n}{2}\pi. \end{aligned} \tag{26}$$

To get the MRS numbers, we only need to estimate the following two integrals

$$\int_{\alpha_n}^{\beta_n} \frac{i[\psi(\tau A - i\tau s) - \psi(\tau A + i\tau s)]}{\sqrt{(\beta_n - s)(s - \alpha_n)}} ds, \tag{27}$$

$$\int_{\alpha_n}^{\beta_n} \frac{i[\psi(\tau A - i\tau s) - \psi(\tau A + i\tau s)]}{\sqrt{(\beta_n - s)(s - \alpha_n)}} s\, ds. \tag{28}$$

Recall the asymptotic expansion of $\psi(z)$ (refer to [30, p. 295])

$$\psi(z) = \log z - \frac{1}{2z} - \sum_{k=1}^{m-1} \frac{B_{2k}}{2k} z^{-2k} - U_m(z), \quad m \geq 1, \quad \text{for } |\arg z| < \pi, \tag{29}$$

where $\log z$ takes the principal branch,

$$U_m(z) = \int_0^\infty \frac{B_{2m} - B_{2m}(x - [x])}{(x + z)^{2m+1}} dx, \tag{30}$$

and $B_{2m}$, $B_{2m}(x)$ are the Bernoulli numbers and the Bernoulli polynomials, respectively. From [30, p. 294 (4.03)], one can get the estimate

$$|U_m(z)| \leq \frac{|B_{2m}| \sec^{2m+1}\left(\frac{1}{2}\arg z\right)}{2m|z|^{2m}}, \quad |\arg z| < \pi. \tag{31}$$

To estimate (27) and (28), we first calculate

$$I_1(A) = i \int_{\alpha_n}^{\beta_n} \frac{\log(A - is) - \log(A + is)}{\sqrt{(\beta_n - s)(s - \alpha_n)}} ds, \tag{32}$$

$$I_2(A) = i \int_{\alpha_n}^{\beta_n} \frac{\log(A - is) - \log(A + is)}{\sqrt{(\beta_n - s)(s - \alpha_n)}} s\, ds. \tag{33}$$

LEMMA 2. *Let $A$ be a positive real number. Then,*

$$I_1(A) = -\pi^2 + 2\pi \arg \varphi_n(iA), \tag{34}$$

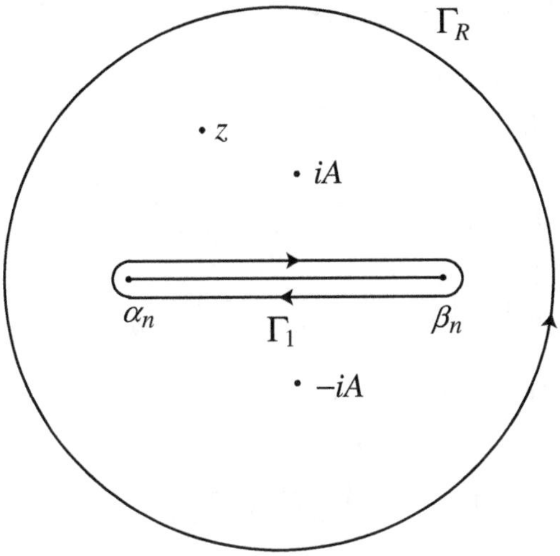


**Figure 1.** Contour $\Gamma$.

$$I_2(A) = -\frac{\alpha_n + \beta_n}{2}\pi^2 - 2\pi A - i\pi(\zeta_n(iA) - \zeta_n(-iA)) + (\alpha_n + \beta_n)\pi \arg \varphi_n(iA), \tag{35}$$

*where $\varphi_n(z)$ and $\zeta_n(z)$ are given, respectively, in (12) and (13), and* $0 < \arg \varphi_n(iA) < \pi$.

*Proof*: Differentiating both sides of (32) gives

$$\begin{aligned} I_1'(A) &= i\int_{\alpha_n}^{\beta_n}\left(\frac{1}{A-is} - \frac{1}{A+is}\right)\frac{ds}{\sqrt{(\beta_n - s)(s-\alpha_n)}} \\ &= -\int_{\alpha_n}^{\beta_n}\left(\frac{1}{s+iA} + \frac{1}{s-iA}\right)\frac{ds}{\sqrt{(\beta_n - s)(s-\alpha_n)}}. \end{aligned} \tag{36}$$

In order to find $I_1'(A)$, we choose a contour $\Gamma$ shown in Figure 1 which consists of a large circle $\Gamma_R = \{z : |z| = R\}$ and a closed curve $\Gamma_1$ surrounding the interval $[\alpha_n, \beta_n]$.

By the Cauchy residue theorem, we have

$$\frac{1}{2\pi i}\int_\Gamma \left(\frac{1}{s+iA} + \frac{1}{s-iA}\right)\frac{ds}{\sqrt{(s-\alpha_n)(s-\beta_n)}} = \frac{1}{\zeta_n(iA)} + \frac{1}{\zeta_n(-iA)}. \tag{37}$$

Since the integral on $\Gamma_R$ tends to zero as $R \to \infty$, we also have

$$\begin{aligned}
&\lim_{R\to\infty} \int_{\Gamma} \left( \frac{1}{s+iA} + \frac{1}{s-iA} \right) \frac{ds}{\sqrt{(s-\alpha_n)(s-\beta_n)}} \\
&= \int_{\Gamma_1} \left( \frac{1}{s+iA} + \frac{1}{s-iA} \right) \frac{ds}{\sqrt{(s-\alpha_n)(s-\beta_n)}} \\
&= \frac{2}{i} \int_{\alpha_n}^{\beta_n} \left( \frac{1}{s+iA} + \frac{1}{s-iA} \right) \frac{ds}{\sqrt{(\beta_n - s)(s-\alpha_n)}}. \qquad (38)
\end{aligned}$$

A combination of (36), (37), and (38) yields

$$I_1'(A) = \pi \left( \frac{1}{\zeta_n(iA)} + \frac{1}{\zeta_n(-iA)} \right). \qquad (39)$$

Integrating both sides of (39) from $A$ to a large positive number $R$ leads to

$$\begin{aligned}
I_1(A) &= I_1(R) - \pi \int_A^R \left[ \frac{1}{\sqrt{(si-\alpha_n)(si-\beta_n)}} + \frac{1}{\sqrt{(-si-\alpha_n)(-si-\beta_n)}} \right] ds \\
&= I_1(R) - i\pi \left( \int_{-iA}^{-iR} - \int_{iA}^{iR} \right) \frac{ds}{\sqrt{(s-\alpha_n)(s-\beta_n)}} \\
&= I_1(R) + i\pi \log \frac{\varphi_n(iR)}{\varphi_n(-iR)} - i\pi \log \frac{\varphi_n(iA)}{\varphi_n(-iA)}, \qquad (40)
\end{aligned}$$

where we have used the facts that

$$[\log \varphi_n(z)]' = \frac{1}{\sqrt{(z-\alpha_n)(z-\beta_n)}}, \qquad z \in \mathbb{C} \setminus (-\infty, \beta_n], \qquad (41)$$

and $\log \varphi_n(z)$ takes the branch on $\mathbb{C} \setminus (-\infty, \beta_n]$ such that $-\pi < \arg \varphi_n(z) < \pi$. Noting that $\varphi_n(\overline{z}) = \overline{\varphi_n(z)}$, we have

$$\log \frac{\varphi_n(iA)}{\varphi_n(-iA)} = 2i \arg \varphi_n(iA) \qquad \text{and} \qquad 0 < \arg \varphi_n(iA) < \pi. \qquad (42)$$

It is easy to verify that

$$\lim_{R\to\infty} I_1(R) = 0 \quad \text{and} \quad \lim_{R\to\infty} \log \frac{\varphi_n(iR)}{\varphi_n(-iR)} = \pi i. \qquad (43)$$

Letting $R \to \infty$ in (40) immediately yields (34).

Similarly, we have from (33)

$$\begin{aligned} I_2'(A) &= i\int_{\alpha_n}^{\beta_n}\left(\frac{1}{A-is}-\frac{1}{A+is}\right)\frac{s\,ds}{\sqrt{(\beta_n-s)(s-\alpha_n)}} \\ &= -2\int_{\alpha_n}^{\beta_n}\frac{s^2}{s^2+A^2}\frac{ds}{\sqrt{(\beta_n-s)(s-\alpha_n)}} \\ &= -2\pi - iA\int_{\alpha_n}^{\beta_n}\left(\frac{1}{s-iA}-\frac{1}{s+iA}\right)\frac{ds}{\sqrt{(\beta_n-s)(s-\alpha_n)}} \\ &= -2\pi + iA\pi\left(\frac{1}{\zeta_n(iA)}-\frac{1}{\zeta_n(-iA)}\right). \end{aligned} \tag{44}$$

Integrating both sides of (44) from $A$ to $R$ gives

$$\begin{aligned} I_2(A) &= I_2(R) + 2\pi R - 2\pi A - i\pi\int_A^R\left(\frac{1}{\zeta_n(is)}-\frac{1}{\zeta_n(-is)}\right)s\,ds \\ &= I_2(R) + 2\pi R - 2\pi A + i\pi\left(\int_{iA}^{iR}-\int_{-iA}^{-iR}\right)\frac{t}{\zeta_n(t)}dt \\ &= I_2(R) + 2\pi R - 2\pi A + i\pi(\zeta_n(iR)-\zeta_n(-iR)) \\ &\quad + i\pi\frac{\alpha_n+\beta_n}{2}\log\frac{\varphi_n(iR)}{\varphi_n(-iR)} \\ &\quad - i\pi(\zeta_n(iA)-\zeta_n(-iA)) - i\pi\frac{\alpha_n+\beta_n}{2}\log\frac{\varphi_n(iA)}{\varphi_n(-iA)}, \end{aligned} \tag{45}$$

where we have used the fact that

$$\left[\zeta_n(z)+\frac{\alpha_n+\beta_n}{2}\log\varphi_n(z)\right]' = \frac{z}{\sqrt{(z-\alpha_n)(z-\beta_n)}}, \quad z\in\mathbb{C}\setminus(-\infty,\beta_n].$$

Since

$$\begin{aligned} \zeta_n(iR)-\zeta_n(-iR) &= iR\left[\sqrt{\left(1-\frac{\alpha_n}{iR}\right)\left(1-\frac{\beta_n}{iR}\right)}+\sqrt{\left(1+\frac{\alpha_n}{iR}\right)\left(1+\frac{\beta_n}{iR}\right)}\right] \\ &= iR\left[1-\frac{\alpha_n+\beta_n}{2}\frac{1}{iR}+O\left(\frac{1}{R^2}\right)+1+\frac{\alpha_n+\beta_n}{2}\frac{1}{iR}\right. \\ &\qquad \left.+O\left(\frac{1}{R^2}\right)\right] \\ &= 2iR + O\left(\frac{1}{R}\right), \end{aligned}$$

we have

$$2\pi R + i\pi[\zeta_n(iR)-\zeta_n(-iR)] \to 0 \quad \text{as} \quad R\to\infty. \tag{46}$$

At the same time, we again have

$$\lim_{R\to\infty} I_2(R) = 0. \tag{47}$$

Letting $R \to \infty$ in (45), Equation (35) follows from (42), (43), (46), and (47). ■

LEMMA 3. *Let $A$ be a positive real number. Then,*

$$\int_{\alpha_n}^{\beta_n} \frac{i[\psi(\tau A - i\tau s) - \psi(\tau A + i\tau s)]}{\sqrt{(\beta_n - s)(s-\alpha_n)}} ds = I_1(A) - \frac{1}{2\tau} I_1'(A) + O\left(\frac{1}{\tau^2}\right), \tag{48}$$

$$\int_{\alpha_n}^{\beta_n} \frac{i[\psi(\tau A - i\tau s) - \psi(\tau A + i\tau s)]}{\sqrt{(\beta_n - s)(s-\alpha_n)}} s\, ds = I_2(A) - \frac{1}{2\tau} I_2'(A) + O\left(\frac{1}{\tau^2}\right). \tag{49}$$

*Proof*: Inserting (29) with $m = 1$ into (27) and (28) leads to

$$i\int_{\alpha_n}^{\beta_n} \frac{\psi(\tau A - i\tau s) - \psi(\tau A + i\tau s)}{\sqrt{(\beta_n - s)(s-\alpha_n)}} ds$$
$$= I_1(A) - \frac{1}{2\tau} I_1'(A) - \int_{\alpha_n}^{\beta_n} \frac{i[U_1(\tau A - i\tau s) - U_1(\tau A + i\tau s)]}{\sqrt{(\beta_n - s)(s-\alpha_n)}} ds, \tag{50}$$

$$i\int_{\alpha_n}^{\beta_n} \frac{\psi(\tau A - i\tau s) - \psi(\tau A + i\tau s)}{\sqrt{(\beta_n - s)(s-\alpha_n)}} s\, ds$$
$$= I_2(A) - \frac{1}{2\tau} I_2'(A) - \int_{\alpha_n}^{\beta_n} \frac{i[U_1(\tau A - i\tau s) - U_1(\tau A + i\tau s)]}{\sqrt{(\beta_n - s)(s-\alpha_n)}} s\, ds. \tag{51}$$

From (31), we have for $s \in \mathbb{R}$

$$|U_1(\tau A \pm i\tau s)| \le \frac{|B_2| \sec^3\left(\frac{1}{2}\arg(\tau A \pm i\tau s)\right)}{2\tau^2(A^2+s^2)} \le \frac{|B_2|\sec^3\left(\frac{\pi}{4}\right)}{2\tau^2 A^2}. \tag{52}$$

Inserting it into the integral on the the right-hand sides of (50) and (51) yields

$$\left|\int_{\alpha_n}^{\beta_n} \frac{i[U_1(\tau A - i\tau s) - U_1(\tau A + i\tau s)]}{\sqrt{(\beta_n - s)(s-\alpha_n)}} ds\right|,$$
$$\le \int_{\alpha_n}^{\beta_n} \frac{|U_1(\tau A - i\tau s)| + |U_1(\tau A + i\tau s)|}{\sqrt{(\beta_n - s)(s-\alpha_n)}} ds,$$
$$\le \frac{\pi |B_2| \sec^3\left(\frac{\pi}{4}\right)}{\tau^2 A^2}, \tag{53}$$

$$\begin{aligned}
&\left| \int_{\alpha_n}^{\beta_n} \frac{i[U_1(\tau A - i\tau s) - U_1(\tau A + i\tau s)]}{\sqrt{(\beta_n - s)(s - \alpha_n)}} s\, ds, \right| \\
&\quad \le \int_{\alpha_n}^{\beta_n} \frac{|U_1(\tau A - i\tau s)| + |U_1(\tau A + i\tau s)|}{\sqrt{(\beta_n - s)(s - \alpha_n)}} |s|\, ds, \\
&\quad \le \frac{|B_2| \sec^3\left(\frac{\pi}{4}\right)}{\tau^2} \int_{\alpha_n}^{\beta_n} \frac{|s|}{s^2 + A^2} \frac{ds}{\sqrt{(\beta_n - s)(s - \alpha_n)}}, \\
&\quad \le \frac{|B_2| \sec^3\left(\frac{\pi}{4}\right)}{2A\tau^2} \int_{\alpha_n}^{\beta_n} \frac{ds}{\sqrt{(\beta_n - s)(s - \alpha_n)}}, \\
&\quad = \frac{\pi |B_2| \sec^3\left(\frac{\pi}{4}\right)}{2\tau^2 A}.
\end{aligned} \tag{54}$$

Thus, (48) and (49) follow from a combination of (50) and (51) with (53) and (54). ■

Substituting (34) and (39) into (48) gives

$$\begin{aligned}
&\int_{\alpha_n}^{\beta_n} \frac{i[\psi(\tau A - i\tau s) - \psi(\tau A + i\tau s)]}{\sqrt{(\beta_n - s)(s - \alpha_n)}} ds \\
&\quad = -\pi^2 + 2\pi \arg \varphi_n(iA) - \frac{\pi}{2\tau} \left( \frac{1}{\zeta_n(iA)} + \frac{1}{\zeta_n(-iA)} \right) + O\left(\frac{1}{\tau^2}\right),
\end{aligned} \tag{55}$$

and inserting (35) and (44) in (49) leads to

$$\begin{aligned}
&\int_{\alpha_n}^{\beta_n} \frac{i[\psi(\tau A - i\tau s) - \psi(\tau A + i\tau s)]}{\sqrt{(\beta_n - s)(s - \alpha_n)}} s\, ds \\
&\quad = -\frac{\alpha_n + \beta_n}{2}\pi^2 - 2\pi A - i\pi(\zeta_n(iA) - \zeta_n(-iA)) \\
&\qquad + (\alpha_n + \beta_n)\pi \arg \varphi_n(iA) \\
&\qquad + \frac{\pi}{\tau} - \frac{\pi}{2\tau} iA \left( \frac{1}{\zeta_n(iA)} - \frac{1}{\zeta_n(-iA)} \right) + O\left(\frac{1}{\tau^2}\right).
\end{aligned} \tag{56}$$

Coupling (55) and (56) with (24) and (25), respectively, we obtain

$$\arg \varphi_n(iA) = \phi + \frac{1}{4\tau} \left( \frac{1}{\zeta_n(iA)} + \frac{1}{\zeta_n(-iA)} \right) + O\left(\frac{1}{\tau^2}\right), \tag{57}$$

$$\begin{aligned}
(\alpha_n + \beta_n) \arg \varphi_n(iA) &= (\alpha_n + \beta_n)\phi + 2(A + 1) + i(\zeta_n(iA) - \zeta_n(-iA)) \\
&\quad - \frac{1}{\tau} + \frac{iA}{2\tau} \left( \frac{1}{\zeta_n(iA)} - \frac{1}{\zeta_n(-iA)} \right) + O\left(\frac{1}{\tau^2}\right).
\end{aligned} \tag{58}$$

Next, we show that both $\alpha_n$ and $\beta_n$ are bounded as $n \to \infty$. If one of them tends to infinity as $n \to \infty$ while the other is bounded, then by the definition of $\varphi_n(z)$ in (12) it is easy to verify that

$$\arg \varphi_n(iA) \to \begin{cases} 0, & \alpha_n \to -\infty,\ \beta_n \text{ is bounded}, \\ \pi, & \beta_n \to \infty,\ \alpha_n \text{ is bounded}, \end{cases}$$

as $n \to \infty$, which contradicts the result $\arg \varphi_n(iA) \to \phi\,(0 < \phi < \pi)$ stated in (57).

If both $\alpha_n$ and $\beta_n$ tend to infinity as $n \to \infty$, then we divide the discussion into three cases: (i) $\alpha_n < \beta_n \le 0$, (ii) $0 \le \alpha_n < \beta_n$, (iii) $\alpha_n < 0 < \beta_n$.

In case (i), we have $\alpha_n, \beta_n \to -\infty$ as $n \to \infty$. Note that

$$\begin{aligned} \varphi_n(iA) &= iA - \frac{\alpha_n + \beta_n}{2} + \sqrt{(iA - \alpha_n)(iA - \beta_n)}, \\ &= iA + \frac{(-\alpha_n) + (-\beta_n)}{2} + \sqrt{(-\alpha_n)(-\beta_n)}\sqrt{\left(1 + \frac{iA}{-\alpha_n}\right)\left(1 + \frac{iA}{-\beta_n}\right)}, \\ &= \left\{\frac{1}{2}(\sqrt{-\alpha_n} + \sqrt{-\beta_n})^2 + iA\left[1 + \frac{1}{2}\left(\sqrt{\frac{-\alpha_n}{-\beta_n}} + \sqrt{\frac{-\beta_n}{-\alpha_n}}\right)\right]\right\}(1 + o(1)). \end{aligned}$$

Thus, $\operatorname{Re} \varphi_n(iA) > 0$ and $\operatorname{Im} \varphi_n(iA) > 0$ for large $n$, and

$$\frac{\operatorname{Im} \varphi_n(iA)}{\operatorname{Re} \varphi_n(iA)} \to 0 \quad \text{as} \quad n \to \infty.$$

This implies that $\arg \varphi_n(iA) \to 0$ as $n \to \infty$. Similarly, for case (ii),

$$\varphi_n(iA) = \left\{-\frac{1}{2}(\sqrt{\alpha_n} + \sqrt{\beta_n})^2 + iA\left[1 - \frac{1}{2}\left(\sqrt{\frac{\alpha_n}{\beta_n}} + \sqrt{\frac{\beta_n}{\alpha_n}}\right)\right]\right\}(1 + o(1)),$$

and we have $\arg \varphi_n(iA) \to \pi$ as $n \to \infty$. Both cases contradict the result $\arg \varphi_n(iA) \to \phi$ given in (57).

In case (iii), we have $\alpha_n \to -\infty$ and $\beta_n \to +\infty$ as $n \to \infty$. If $\phi = \pi/2$ then, by the symmetry of the weight function, we have $\beta_n = -\alpha_n$, in which case (58) becomes

$$2(A + 1) + i(\zeta_n(iA) - \zeta_n(-iA)) = O\left(\frac{1}{\tau}\right).$$

However, $\zeta_n(iA) - \zeta_n(-iA) \sim 2i\beta_n \to \infty$, which is a contradiction. Thus, we assume $\phi \neq \pi/2$. By the definition of $\varphi_n(z)$ given in (12) again, we have

$$\begin{aligned}\varphi_n(iA) &= iA + \frac{(-\alpha_n) - \beta_n}{2} + i\sqrt{(-\alpha_n)\beta_n}\sqrt{\left(1 + \frac{iA}{-\alpha_n}\right)\left(1 - \frac{iA}{\beta_n}\right)},\\ &= \left[\frac{(-\alpha_n) - \beta_n}{2} + i\sqrt{(-\alpha_n)\beta_n}\right](1 + o(1)) \quad \text{as} \quad n \to \infty,\end{aligned}$$

and

$$\begin{aligned}\arg\varphi_n(iA) &\sim \arctan\frac{2\sqrt{(-\alpha_n)\beta_n}}{(-\alpha_n) - \beta_n}\\ &\sim \arctan\frac{2\sqrt{\beta_n/(-\alpha_n)}}{1 - \beta_n/(-\alpha_n)} \quad \text{as} \quad n \to \infty. \end{aligned} \tag{59}$$

From (57) and (59), it follows that

$$\lim_{n\to\infty} \arctan\frac{2\sqrt{\beta_n/(-\alpha_n)}}{1 - \beta_n/(-\alpha_n)} = \lim_{n\to\infty} \arg\varphi_n(iA) = \phi,$$

which implies that $\lim_{n\to\infty}\sqrt{\beta_n/(-\alpha_n)} = \tan\frac{\phi}{2}$, or equivalently

$$\alpha_n \sim -\beta_n \cot^2\frac{\phi}{2} \quad \text{as} \quad n \to \infty. \tag{60}$$

Dividing both sides of (58) by $\alpha_n + \beta_n$, and using (60), we obtain

$$\frac{i(\zeta_n(iA) - \zeta_n(-iA))}{\left(1 - \cot^2\frac{\phi}{2}\right)\beta_n} = o(1) \quad \text{as} \quad n \to \infty,$$

which is impossible since $\zeta_n(iA) - \zeta_n(-iA) = 2i\sqrt{(-\alpha_n)\beta_n}(1 + o(1)) \sim 2i\beta_n \cot\frac{\phi}{2}$ as $n \to \infty$.

We have now proved that both $\alpha_n$ and $\beta_n$ are bounded. From (57), we have

$$\arg\varphi_n(iA) = \phi + O\left(\frac{1}{\tau}\right). \tag{61}$$

Inserting (61) in (58), and using the boundedness of $\alpha_n$ and $\beta_n$, it follows

$$2(A + 1) + i(\zeta_n(iA) - \zeta_n(-iA)) = O\left(\frac{1}{\tau}\right), \tag{62}$$

which in turn gives

$$\operatorname{Im}\zeta_n(iA) = A + 1 + O\left(\frac{1}{\tau}\right), \tag{63}$$

since $\zeta_n(iA)$ and $\zeta_n(-iA)$ are complex conjugates. By (12) and (61),

$$\arg\varphi_n(iA) = \arctan\frac{A + \operatorname{Im}\zeta_n(iA)}{-(\alpha_n+\beta_n)/2 + \operatorname{Re}\zeta_n(iA)} = \phi + O\left(\frac{1}{\tau}\right). \tag{64}$$

Inserting (63) in (64) yields

$$\operatorname{Re}\zeta_n(iA) = \frac{\alpha_n+\beta_n}{2} + (2A+1)\cot\phi + O\left(\frac{1}{\tau}\right). \tag{65}$$

Coupling (63) and (65), we obtain

$$\begin{aligned}\zeta_n(iA) &= \frac{\alpha_n+\beta_n}{2} + (2A+1)\cot\phi + i(A+1) + O\left(\frac{1}{\tau}\right),\\ \zeta_n(-iA) &= \frac{\alpha_n+\beta_n}{2} + (2A+1)\cot\phi - i(A+1) + O\left(\frac{1}{\tau}\right).\end{aligned} \tag{66}$$

By (13), we have

$$\zeta_n^2(iA) + \zeta_n^2(-iA) = -2A^2 + 2\alpha_n\beta_n, \quad \zeta_n^2(iA) - \zeta_n^2(-iA) = -2iA(\alpha_n+\beta_n).$$

A combination of the last four equations gives

$$\begin{aligned}-A^2 + \alpha_n\beta_n &= \left(\frac{\alpha_n+\beta_n}{2} + (2A+1)\cot\phi\right)^2 - (A+1)^2 + O\left(\frac{1}{\tau}\right),\\ -iA(\alpha_n+\beta_n) &= 2i(A+1)\left(\frac{\alpha_n+\beta_n}{2} + (2A+1)\cot\phi\right) + O\left(\frac{1}{\tau}\right),\end{aligned}$$

which imply

$$\begin{aligned}\alpha_n+\beta_n &= -2(A+1)\cot\phi + O\left(\frac{1}{\tau}\right),\\ \alpha_n\beta_n &= \frac{A^2}{\sin^2\phi} - (A+1)^2 + O\left(\frac{1}{\tau}\right).\end{aligned}$$

Solving for $\alpha_n$ and $\beta_n$ gives

$$\begin{aligned}\alpha_n &= \frac{-(A+1)\cos\phi - \sqrt{2A+1}}{\sin\phi} + O\left(\frac{1}{\tau}\right),\\ \beta_n &= \frac{-(A+1)\cos\phi + \sqrt{2A+1}}{\sin\phi} + O\left(\frac{1}{\tau}\right).\end{aligned} \tag{67}$$

We can also obtain full asymptotic expansions for $\alpha_n$ and $\beta_n$ by using the asymptotic expansion of $\psi(z)$ given in (29).

LEMMA 4. *The MRS numbers $\alpha_n$, $\beta_n$ have the asymptotic expansions*

$$\alpha_n \sim \sum_{j=0}^{\infty} \frac{a_j}{\tau^j} \quad \textit{and} \quad \beta_n \sim \sum_{j=0}^{\infty} \frac{b_j}{\tau^j}, \tag{68}$$

*where*

$$a_0 = \frac{-(A+1)\cos\phi - \sqrt{2A+1}}{\sin\phi}, \qquad b_0 = \frac{-(A+1)\cos\phi + \sqrt{2A+1}}{\sin\phi}, \tag{69}$$

*and the coefficients $a_k$, $b_k$ can be determined iteratively for $k = 1, 2, 3, \ldots$.*

*Proof*: We proceed by induction. Similar to the proof of Lemma 3, we substitute (29), (31) into (24) and (25), and obtain

$$\pi(\pi - 2\phi) + I_1(A) - \frac{1}{2\tau} I_1'(A) + \sum_{j=1}^{m} \frac{(-1)^{2j} B_{2j}}{(2j)!\tau^{2j}} I_1^{(2j)}(A) = O\left(\frac{1}{\tau^{2m+2}}\right), \tag{70}$$

$$\begin{aligned} -2\pi + \frac{\alpha_n + \beta_n}{2}\pi(\pi - 2\phi) + I_2(A) - \frac{1}{2\tau} I_2'(A) \\ + \sum_{j=1}^{m} \frac{(-1)^{2j} B_{2j}}{(2j)!\tau^{2j}} I_2^{(2j)}(A) = O\left(\frac{1}{\tau^{2m+2}}\right), \end{aligned} \tag{71}$$

since the integrals involving $U_{m+1}(\tau A \pm i\tau s)$ are $O(\tau^{-(2m+2)})$.

From (67), we already have

$$\alpha_n = a_0 + R_1^{\alpha}(\tau), \quad \beta_n = b_0 + R_1^{\beta}(\tau),$$

where $R_1^{\alpha}(\tau) = O(1/\tau)$ and $R_1^{\beta}(\tau) = O(1/\tau)$.

Suppose that

$$\alpha_n = \sum_{j=0}^{k-1} \frac{a_j}{\tau^j} + R_k^{\alpha}(\tau), \quad \beta_n = \sum_{j=0}^{k-1} \frac{b_j}{\tau^j} + R_k^{\beta}(\tau), \tag{72}$$

and

$$R_k^{\alpha}(\tau) = O\left(\frac{1}{\tau^k}\right), \quad R_k^{\beta}(\tau) = O\left(\frac{1}{\tau^k}\right). \tag{73}$$

Take $m$ in (70) and (71) such that $2m > k$, and then substitute (72) into (70) and (71). Expanding the left hand sides of the resulting equations in terms of $1/\tau^j$, we obtain

$$\sum_{j=0}^{k} \frac{\tilde{a}_j}{\tau^j} + r_1^{\alpha} R_k^{\alpha}(\tau) + r_1^{\beta} R_k^{\beta}(\tau) = O\left(\frac{1}{\tau^{k+1}}\right), \tag{74}$$

$$\sum_{j=0}^{k}\frac{\tilde{b}_j}{\tau^j}+r_2^{\alpha}R_k^{\alpha}(\tau)+r_2^{\beta}R_k^{\beta}(\tau)=O\left(\frac{1}{\tau^{k+1}}\right), \tag{75}$$

where we have also used (34) and (35), the coefficients $r_1^{\alpha}, r_2^{\alpha}, r_1^{\beta}, r_2^{\beta}$, and $\tilde{a}_j, \tilde{b}_j (j = 0, \dots, k)$ are determined by $a_j, b_j (j = 0, \dots, k-1)$. By induction assumption (73), and comparing the coefficients of $1/\tau^j$ on both sides of the equations for $0 \le j \le k-1$, we conclude that both $\tilde{a}_j$ and $\tilde{b}_j$ must be equal to 0 for $0 \le j \le k-1$. Thus, upon solving (74) and (75), we obtain

$$R_k^{\alpha}(\tau)=\frac{a_k}{\tau^k}+O\left(\frac{1}{\tau^{k+1}}\right), \quad R_k^{\beta}(\tau)=\frac{b_k}{\tau^k}+O\left(\frac{1}{\tau^{k+1}}\right).$$

The coefficients $a_k, b_k$ are readily seen to be dependent on $a_j, b_j$ for $j = 0, \dots, k-1$; hence, $a_k, b_k$ can be determined successively. ■

### 3.3. *The equilibrium measure*

The equilibrium measure $\mu_n(x)$ can be defined by $\mu_n(x) = \mathrm{Re}\, G_{n,+}(x)$ or $\mu_n(x) = \frac{1}{\pi i}\nu_{n,+}(x)$ for $x \in (\alpha_n, \beta_n)$, where $G_n(z)$ and $\nu_n(z)$ are given in (23) and (20), respectively. Using contour integration on $\Gamma$ shown in Figure 1 and the Cauchy residue theorem, we can obtain

$$\int_{\alpha_n}^{\beta_n}\frac{1}{(s-z)\sqrt{(s-\alpha_n)(\beta_n-s)}}ds=-\frac{\pi}{\zeta_n(z)}, \quad z\in\mathbb{C}\setminus[\alpha_n,\beta_n]. \tag{76}$$

Inserting (16) and (76) into (23) yields

$$G_n(z)=\frac{i(\pi-2\phi)}{2\pi}+\frac{\zeta_n(z)}{2\pi^2}\int_{\alpha_n}^{\beta_n}\frac{\psi(\tau A-i\tau s)-\psi(\tau A+i\tau s)}{(s-z)\sqrt{(s-\alpha_n)(\beta_n-s)}}ds, \tag{77}$$

for $z \in \mathbb{C}\setminus[\alpha_n, \beta_n]$. Substituting (77) and (16) in (20) gives

$$\begin{aligned}\nu_n(z)&=\frac{i\zeta_n(z)}{2\pi}\int_{\alpha_n}^{\beta_n}\frac{\psi(\tau A-i\tau s)-\psi(\tau A+i\tau s)}{(s-z)\sqrt{(s-\alpha_n)(\beta_n-s)}}ds\\&\quad+\frac{i}{2}[\psi(\tau A-i\tau z)-\psi(\tau A+i\tau z)], \qquad (78)\\&=\frac{i\zeta_n(z)}{2\pi}\int_{\alpha_n}^{\beta_n}\frac{[\psi(\tau A-i\tau s)-\psi(\tau A+i\tau s)]-[\psi(\tau A-i\tau z)-\psi(\tau A+i\tau z)]}{(s-z)\sqrt{(s-\alpha_n)(\beta_n-s)}}ds,\end{aligned}$$

for $z \in \mathbb{C}\setminus\{[\alpha_n, \beta_n] \cup (-i\infty, -iA] \cup [iA, i\infty)\}$. Here, we have also used (76) in the second equality in (78). We need to calculate the integral on the right-hand side of (78). To this end, we appeal to the asymptotic expansion of the Psi function given in (29), and first calculate the integral

$$I_3(z,A)=\int_{\alpha_n}^{\beta_n}\frac{\log(A-is)-\log(A+is)}{\sqrt{(s-\alpha_n)(\beta_n-s)}}\frac{1}{(s-z)}ds. \tag{79}$$

Let $\tilde{\varphi}_n(z) = \sqrt{\frac{z-\beta_n}{z-\alpha_n}}$ for $z \in \mathbb{C} \setminus [\alpha_n, \beta_n]$, where we take the branch of the square root such that $\operatorname{Re} \tilde{\varphi}_n(z) > 0$. For a given $z \in \mathbb{C} \setminus \{[\alpha_n, \beta_n] \cup (-i\infty, -iA] \cup [iA, i\infty)\}$, let

$$\log C_n(z, s) := \log \left( \frac{\tilde{\varphi}_n(s) + \tilde{\varphi}_n(z)}{\tilde{\varphi}_n(s) - \tilde{\varphi}_n(z)} \right), \quad s \in \mathbb{C} \setminus \{[\alpha_n, \beta_n] \cup \gamma\}, \tag{80}$$

where $\gamma$ is an arc connecting $\beta_n$ to $z$ which lies in $\mathbb{C} \setminus \{[\alpha_n, \beta_n] \cup (-i\infty, -iA] \cup [iA, i\infty)\}$, except for the end point $\beta_n$, and where $\log C_n(z, s)$ takes the branch such that $\log C_n(z, s) \to 0$ as $s \to \alpha_n$. Furthermore, we take the branch of $\log \frac{A-iz}{A+iz}$ in $\mathbb{C} \setminus \{(-i\infty, -iA] \cup [iA, i\infty)\}$ such that it assumes the value 0 when $z = 0$.

LEMMA 5. *For given* $z \in \mathbb{C} \setminus \{[\alpha_n, \beta_n] \cup (-i\infty, -iA] \cup [iA, i\infty)\}$, *we have*

$$I_3(z, A) = -\frac{\pi}{\zeta_n(z)} \log \frac{A - iz}{A + iz} + \frac{\pi}{\zeta_n(z)} \log \frac{C_n(z, iA)}{C_n(z, -iA)}. \tag{81}$$

*Proof*: Taking partial derivative with respect to $A$ on both sides of (79) yields

$$\begin{aligned}\frac{\partial I_3(z, A)}{\partial A} &= \int_{\alpha_n}^{\beta_n} \left( \frac{1}{A - is} - \frac{1}{A + is} \right) \frac{1}{(s - z)\sqrt{(s - \alpha_n)(\beta_n - s)}} ds, \\ &= i \int_{\alpha_n}^{\beta_n} \left( \frac{1}{s + iA} + \frac{1}{s - iA} \right) \frac{1}{(s - z)\sqrt{(s - \alpha_n)(\beta_n - s)}} ds. \end{aligned} \tag{82}$$

Again, using the contour integration on $\Gamma$ shown in Figure 1 and the Cauchy residue theorem, we get

$$\begin{aligned}\frac{\partial I_3(z, A)}{\partial A} &= -\frac{1}{2} \int_{\Gamma} \left( \frac{1}{s + iA} + \frac{1}{s - iA} \right) \frac{1}{(s - z)\zeta_n(s)} ds, \\ &= -i\pi \left[ \left( \frac{1}{z + iA} + \frac{1}{z - iA} \right) \frac{1}{\zeta_n(z)} + \frac{1}{(iA - z)\zeta_n(iA)} \right. \\ &\qquad \left. - \frac{1}{(iA + z)\zeta_n(-iA)} \right] \end{aligned} \tag{83}$$

for $z \in \mathbb{C} \setminus \{[\alpha_n, \beta_n] \cup (-i\infty, -iA] \cup [iA, i\infty)\}$. It is easy to verify that $I_3(z, \infty) = 0$. Integrating both sides of (83) from $A$ to $\infty$ gives

$$\begin{aligned} I_3(z, A) = i\pi \int_A^{\infty} &\left[ \left( \frac{1}{z + is} + \frac{1}{z - is} \right) \frac{1}{\zeta_n(z)} + \frac{1}{(is - z)\zeta_n(is)} \right. \\ &\left. - \frac{1}{(is + z)\zeta_n(-is)} \right] ds. \end{aligned} \tag{84}$$

It is readily seen that

$$\int_A^{\infty}\left(\frac{1}{z-is}+\frac{1}{z+is}\right)ds = i\int_A^{\infty}\left(\frac{1}{s+iz}-\frac{1}{s-iz}\right)ds$$
$$= i\log\frac{A-iz}{A+iz}. \tag{85}$$

Since

$$\frac{1}{\zeta_n(z)}\frac{\partial}{\partial t}\left[\log\left(\frac{\tilde{\varphi}_n(t)+\tilde{\varphi}_n(z)}{\tilde{\varphi}_n(t)-\tilde{\varphi}_n(z)}\right)\right] = -\frac{1}{(t-z)\zeta_n(t)},$$

we have

$$\int_A^{\infty}\frac{1}{(is-z)\zeta_n(is)}ds = -i\int_{iA}^{i\infty}\frac{1}{(t-z)\zeta_n(t)}dt$$
$$= \frac{i}{\zeta_n(z)}\left(\log\frac{\tilde{\varphi}_n(t)+\tilde{\varphi}_n(z)}{\tilde{\varphi}_n(t)-\tilde{\varphi}_n(z)}\right)\Bigg|_{iA}^{i\infty}. \tag{86}$$

Noting that $\lim_{t\to\infty}\tilde{\varphi}_n(t) = 1$ for $t \in \mathbb{C}\setminus[\alpha_n, \beta_n]$, we obtain

$$\int_A^{\infty}\frac{1}{(is-z)\zeta_n(is)}ds = \frac{i}{\zeta_n(z)}\left(\log\frac{1+\tilde{\varphi}_n(z)}{1-\tilde{\varphi}_n(z)} - \log\frac{\tilde{\varphi}_n(iA)+\tilde{\varphi}_n(z)}{\tilde{\varphi}_n(iA)-\tilde{\varphi}_n(z)}\right),$$
$$= \frac{i}{\zeta_n(z)}\left(\log\frac{1+\tilde{\varphi}_n(z)}{1-\tilde{\varphi}_n(z)} - \log C_n(z, iA)\right). \tag{87}$$

Similarly,

$$\int_A^{\infty}\frac{1}{(is+z)\zeta_n(-is)}ds = -i\int_{-iA}^{-i\infty}\frac{1}{(t-z)\zeta_n(t)}dt,$$
$$= \frac{i}{\zeta_n(z)}\left(\log\frac{\tilde{\varphi}_n(t)+\tilde{\varphi}_n(z)}{\tilde{\varphi}_n(t)-\tilde{\varphi}_n(z)}\right)\Bigg|_{-iA}^{-i\infty},$$
$$= \frac{i}{\zeta_n(z)}\left(\log\frac{1+\tilde{\varphi}_n(z)}{1-\tilde{\varphi}_n(z)} - \log C_n(z, -iA)\right). \tag{88}$$

Substituting (85), (87), and (88) into (84), the desired result in (81) immediately follows. ■

For given $\varepsilon \in (0, \frac{\pi}{2})$ and $c \in (0, 1)$, we define two sectors

$$S(\pm icA) = S(\pm icA, \varepsilon) = \left\{z \in \mathbb{C} : \left|\arg(z \mp icA) \mp \frac{\pi}{2}\right| \geq \varepsilon\right\}, \tag{89}$$

and use the notation

$$\Omega = \Omega(c, \varepsilon) = S(icA) \cap S(-icA). \tag{90}$$

LEMMA 6. *For any given $0 < \varepsilon < \frac{\pi}{2}$, we have*

$$\nu_n(z) = \frac{i}{2}\log\frac{C_n(z, iA)}{C_n(z, -iA)} + \frac{\zeta_n(z)}{4\tau}F_n(z), \tag{91}$$

*where $F_n(z)$ is analytic in $\mathbb{C} \setminus \{(-i\infty, -iA] \cup [iA, i\infty)\}$, and has the asymptotic expansion*

$$F_n(z) \sim \frac{1}{(z - iA)\zeta_n(iA)} + \frac{1}{(z + iA)\zeta_n(-iA)} + \sum_{k=1}^{\infty}\frac{(-1)^k B_{2k}}{k\tau^{2k-1}}\chi_k(z) \tag{92}$$

*as $n \to \infty$, uniformly for $z \in \Omega$. The coefficient functions $\chi_k(z)$ are defined by*

$$\chi_k(z) := \frac{i}{(2k-1)!}\frac{d^{2k-1}}{ds^{2k-1}}\left(\frac{1}{(s - z)\zeta_n(s)}\right)\Bigg|_{iA}^{-iA}, \quad k = 1, 2, \ldots. \tag{93}$$

*Proof*: Using the asymptotic expansion (29) of $\psi(z)$, we have

$$\begin{aligned}\psi(\tau A - i\tau s) - \psi(\tau A + i\tau s) &= \log\frac{A - is}{A + is} - \frac{1}{2\tau}\left(\frac{1}{A - is} - \frac{1}{A + is}\right)\\ &- \sum_{k=1}^{m-1}\frac{B_{2k}}{2k\tau^{2k}}\left(\frac{1}{(A - is)^{2k}} - \frac{1}{(A + is)^{2k}}\right)\\ &- [U_m(\tau A - i\tau s) - U_m(\tau A + i\tau s)],\end{aligned} \tag{94}$$

for $s \in \Omega = \Omega(c, \varepsilon)$. With the notation of $I_3(z, A)$ given in (79), we obtain

$$\begin{aligned}\int_{\alpha_n}^{\beta_n}\frac{\psi(\tau A - i\tau s) - \psi(\tau A + i\tau s)}{(s - z)\sqrt{(s - \alpha_n)(\beta_n - s)}}ds &= I_3(z, A) - \frac{1}{2\tau}\frac{\partial I_3(z, A)}{\partial A}\\ &+ \sum_{k=1}^{m-1}\frac{B_{2k}}{(2k)!\tau^{2k}}\frac{\partial^{2k}I_3(z, A)}{\partial A^{2k}} - \int_{\alpha_n}^{\beta_n}\frac{U_m(\tau A - i\tau s) - U_m(\tau A + i\tau s)}{(s - z)\sqrt{(s - \alpha_n)(\beta_n - s)}}ds.\end{aligned} \tag{95}$$

On account of (76) and (94) (in (94), we use variable $z$ instead of $s$), we also have

$$\begin{aligned}\int_{\alpha_n}^{\beta_n}\frac{\psi(\tau A - i\tau z) - \psi(\tau A + i\tau z)}{(s - z)\sqrt{(s - \alpha_n)(\beta_n - s)}}ds &= -\frac{\pi}{\zeta_n(z)}\log\frac{A - iz}{A + iz}\\ &+ \frac{\pi}{2\tau\zeta_n(z)}\left(\frac{1}{A - iz} - \frac{1}{A + iz}\right)\\ &+ \sum_{k=1}^{m-1}\frac{\pi B_{2k}}{2k\tau^{2k}\zeta_n(z)}\left(\frac{1}{(A - iz)^{2k}} - \frac{1}{(A + iz)^{2k}}\right)\\ &- \int_{\alpha_n}^{\beta_n}\frac{U_m(\tau A - i\tau z) - U_m(\tau A + i\tau z)}{(s - z)\sqrt{(s - \alpha_n)(\beta_n - s)}}ds.\end{aligned} \tag{96}$$

Inserting (95) and (96) into (78) gives

$$\begin{aligned}\nu_n(z) = \frac{i\zeta_n(z)}{2\pi}\left(I_3(z,A) - \frac{1}{2\tau}\frac{\partial I_3(z,A)}{\partial A} + \sum_{k=1}^{m-1}\frac{B_{2k}}{(2k)!\tau^{2k}}\frac{\partial^{2k} I_3(z,A)}{\partial A^{2k}}\right) \\ - \frac{i\zeta_n(z)}{2\pi}\left[-\frac{\pi}{\zeta_n(z)}\log\frac{A-iz}{A+iz} + \frac{\pi}{2\tau\zeta_n(z)}\left(\frac{1}{A-iz} - \frac{1}{A+iz}\right)\right. \\ \left. + \sum_{k=1}^{m-1}\frac{\pi B_{2k}}{2k\tau^{2k}\zeta_n(z)}\left(\frac{1}{(A-iz)^{2k}} - \frac{1}{(A+iz)^{2k}}\right)\right] \\ + \frac{\zeta_n(z)}{4\tau}(\kappa_m^+(z) - k_m^-(z))\end{aligned} \tag{97}$$

for $z \in \Omega$, where $\kappa_m^{\pm}(z)$ is defined by

$$\kappa_m^{\pm}(z) = \frac{2\tau i}{\pi}\int_{\alpha_n}^{\beta_n}\frac{U_m(\tau A \pm i\tau s) - U_m(\tau A \pm i\tau z)}{(s-z)\sqrt{(s-\alpha_n)(\beta_n - s)}}ds, \quad |\arg(A \mp iz)| < \pi. \tag{98}$$

From (83), it follows that

$$\begin{aligned}\frac{\partial^{2k} I_3(z,A)}{\partial A^{2k}} &= -i\pi\frac{\partial^{2k-1}}{\partial A^{2k-1}}\left(\frac{1}{(iA-z)\zeta_n(iA)} + \frac{1}{(-iA-z)\zeta_n(-iA)}\right) \\ &\quad - \frac{\pi}{\zeta_n(z)}\frac{\partial^{2k-1}}{\partial A^{2k-1}}\left(\frac{1}{A-iz} - \frac{1}{A+iz}\right), \\ &= \frac{\pi(2k-1)!}{\zeta_n(z)}\left(\frac{1}{(A-iz)^{2k}} - \frac{1}{(A+iz)^{2k}}\right) \\ &\quad + (-1)^k\pi\frac{d^{2k-1}}{ds^{2k-1}}\left(\frac{1}{(s-z)\zeta_n(s)}\right)\Big|_{iA}^{-iA}, \\ &= \frac{\pi(2k-1)!}{\zeta_n(z)}\left(\frac{1}{(A-iz)^{2k}} - \frac{1}{(A+iz)^{2k}}\right) \\ &\quad - i(-1)^k\pi(2k-1)!\chi_k(z),\end{aligned} \tag{99}$$

where $\chi_k(z)$ is given in (93). Substituting (81), (83), and (99) in (97) yields

$$\nu_n(z) = \frac{i}{2}\log\frac{C_n(z,iA)}{C_n(z,-iA)} + \frac{\zeta_n(z)}{4\tau}F_n(z), \quad z \in \Omega, \tag{100}$$

where $F_n(z)$ is an analytic function in $\mathbb{C}\setminus\{(-i\infty,-iA]\cup[iA,i\infty)\}$ given by

$$\begin{aligned}F_n(z) = \frac{1}{(z-iA)\zeta_n(iA)} + \frac{1}{(z+iA)\zeta_n(-iA)} + \sum_{k=1}^{m-1}\frac{(-1)^k B_{2k}}{k\tau^{2k-1}}\chi_k(z) \\ + \kappa_m^+(z) - \kappa_m^-(z).\end{aligned} \tag{101}$$

Now, we estimate $\kappa_m^{\pm}(z)$. Since $a^n - b^n = (a-b)\sum_{k=1}^{n} a^{n-k}b^{k-1}$, from (30) we immediately see

$$\begin{aligned} U_m(z_1) - U_m(z_2) &= \int_0^\infty [B_{2m} - B_{2m}(x-[x])] \\ &\quad \times \left(\frac{1}{(x+z_1)^{2m+1}} - \frac{1}{(x+z_2)^{2m+1}}\right) dx, \\ &= (z_2 - z_1)\sum_{k=1}^{2m+1}\int_0^\infty \frac{B_{2m} - B_{2m}(x-[x])}{(x+z_1)^k (x+z_2)^{2m+2-k}}\, dx, \end{aligned} \tag{102}$$

for $|\arg z_i| \le \pi - \varepsilon$ $(i = 1, 2)$. By Theorem 1.1(iii) in [30, p. 283 ], we have

$$|B_{2m}(x) - B_{2m}| \le (2 - 2^{1-2m})|B_{2m}| \le 2|B_{2m}|, \qquad x \in [0, 1]. \tag{103}$$

Note that for $z = re^{i\theta}$,

$$\begin{aligned} |x+z|^2 &= x^2 + 2xr\cos\theta + r^2 \\ &= (x+r)^2 - 4xr\sin^2\left(\frac{1}{2}\theta\right) \ge (x+r)^2\cos^2\left(\frac{1}{2}\theta\right). \end{aligned} \tag{104}$$

Substituting (103) and (104) in (102) gives

$$\begin{aligned} \left|\frac{U_m(z_1) - U_m(z_2)}{(z_2 - z_1)}\right| &\le \sum_{k=1}^{2m+1}\int_0^\infty \frac{|B_{2m} - B_{2m}(x-[x])|}{|x+z_1|^k |x+z_2|^{2m+2-k}}\, dx \\ &\le \sum_{k=1}^{2m+1}\int_0^\infty \frac{2|B_{2m}|\sec^k\left(\frac{1}{2}\arg z_1\right)\sec^{2m+2-k}\left(\frac{1}{2}\arg z_2\right)}{(x+|z_1|)^k (x+|z_2|)^{2m+2-k}}\, dx \\ &\le \int_0^\infty \frac{2(2m+1)|B_{2m}|\sec^{2m+2}\left(\frac{\pi-\varepsilon}{2}\right)}{(x+\min\{|z_1|, |z_2|\})^{2m+2}}\, dx \\ &\le 2|B_{2m}|\left(\sin\frac{\varepsilon}{2}\right)^{-(2m+2)} (\min\{|z_1|, |z_2|\})^{-(2m+1)} \end{aligned} \tag{105}$$

for $|\arg z_i| \le \pi - \varepsilon$ and $z_i \ne 0\,(i = 1, 2)$. From the definition of $\Omega$, it follows that

$$\min_{z\in\Omega} |A \mp iz| \ge (1-c)A\sin\varepsilon, \qquad |\arg(A \mp iz)| \le \pi - \varepsilon. \tag{106}$$

Taking $z_1 = \tau A \pm i\tau s$ and $z_2 = \tau A \pm i\tau z$ in (105), and applying (105) and (106) to (98), we obtain

$$\begin{aligned} |\kappa_m^{\pm}(z)| &\le \frac{4\tau^2|B_{2m}|\left(\sin\frac{\varepsilon}{2}\right)^{-(2m+2)}}{\pi(\min_{z\in\Omega}|\tau A \pm i\tau z|)^{2m+1}}\int_{\alpha_n}^{\beta_n}\frac{ds}{\sqrt{(s-\alpha_n)(\beta_n - s)}} \\ &\le \frac{4|B_{2m}|\left(\sin\frac{\varepsilon}{2}\right)^{-(2m+2)}}{\tau^{2m-1}(1-c)^{2m+1}A^{2m+1}\sin^{2m+1}\varepsilon}, \quad z \in \Omega. \end{aligned} \tag{107}$$

Therefore, (91) - (92) follows from a combination of (100), (101), and (107). ∎

Note that $(z-\alpha_n)\tilde{\varphi}_n(z)=\zeta_n(z)$ for $z\in\mathbb{C}\setminus[\alpha_n,\beta_n]$. If we put

$$D_n(z)=(z-\alpha_n)\sqrt{\beta_n^2+A^2}-(z-\beta_n)\sqrt{\alpha_n^2+A^2},\quad z\in\mathbb{C},\tag{108}$$

$$\rho_n=i\sqrt{\alpha_n^2+A^2}\,(\tilde{\varphi}_n(-iA)-\tilde{\varphi}_n(iA))=2\sqrt{\alpha_n^2+A^2}\,\operatorname{Im}\tilde{\varphi}_n(iA),\tag{109}$$

then from (80)

$$\begin{aligned}\frac{C_n(z,iA)}{C_n(z,-iA)}&=\left(\frac{\tilde{\varphi}_n(iA)+\tilde{\varphi}_n(z)}{\tilde{\varphi}_n(iA)-\tilde{\varphi}_n(z)}\right)\left(\frac{\tilde{\varphi}_n(-iA)-\tilde{\varphi}_n(z)}{\tilde{\varphi}_n(-iA)+\tilde{\varphi}_n(z)}\right)\\&=\frac{D_n(z)+(z-\alpha_n)\tilde{\varphi}_n(z)\sqrt{A^2+\alpha_n^2}(\tilde{\varphi}_n(-iA)-\tilde{\varphi}_n(iA))}{D_n(z)-(z-\alpha_n)\tilde{\varphi}_n(z)\sqrt{A^2+\alpha_n^2}(\tilde{\varphi}_n(-iA)-\tilde{\varphi}_n(iA))}\\&=\frac{D_n(z)-i\rho_n\zeta_n(z)}{D_n(z)+i\rho_n\zeta_n(z)},\quad z\in\mathbb{C}\setminus[\alpha_n,\beta_n].\end{aligned}\tag{110}$$

Thus, (91) can be written as

$$\nu_n(z)=\frac{i}{2}\log\frac{D_n(z)-i\rho_n\zeta_n(z)}{D_n(z)+i\rho_n\zeta_n(z)}+\frac{\zeta_n(z)}{4\tau}F_n(z),\tag{111}$$

for $z\in\mathbb{C}\setminus\{[\alpha_n,\beta_n]\cup(-i\infty,-iA]\cup[iA,i\infty)\}$. Noting that $\zeta_{n,\pm}(x)=\pm i\sqrt{(x-\alpha_n)(\beta_n-x)}$ for $x\in(\alpha_n,\beta_n)$, we conclude that $\mu_n(x)=\frac{1}{\pi i}\nu_{n,+}(x)$ has the asymptotic expansion given in the following lemma.

LEMMA 7. *The equilibrium measure $\mu_n(x)$ has the uniform asymptotic expansion*

$$\begin{aligned}\mu_n(x)&=\frac{1}{2\pi}\log\frac{D_n(x)+\rho_n\sqrt{(\beta_n-x)(x-\alpha_n)}}{D_n(x)-\rho_n\sqrt{(\beta_n-x)(x-\alpha_n)}}+\frac{\sqrt{(x-\alpha_n)(\beta_n-x)}}{4\tau\pi}F_n(x),\\F_n(x)&\sim\frac{1}{(x-iA)\zeta_n(iA)}+\frac{1}{(x+iA)\zeta_n(-iA)}+\sum_{k=1}^{\infty}\frac{(-1)^kB_{2k}}{k\tau^{2k-1}}\chi_k(x),\quad n\to\infty,\end{aligned}\tag{112}$$

*for $x\in[\alpha_n,\beta_n]$.*

For the symmetric case, that is when $\phi=\frac{\pi}{2}$ or equivalently $\alpha_n=-\beta_n$, the behavior of $\mu_n(x)$ has a simple form

$$\mu_n(x)\sim\frac{1}{2\pi}\log\frac{\sqrt{\beta_n^2+A^2}+\sqrt{\beta_n^2-x^2}}{\sqrt{\beta_n^2+A^2}-\sqrt{\beta_n^2-x^2}}\quad\text{as}\quad n\to\infty,\quad x\in[-\beta_n,\beta_n].$$

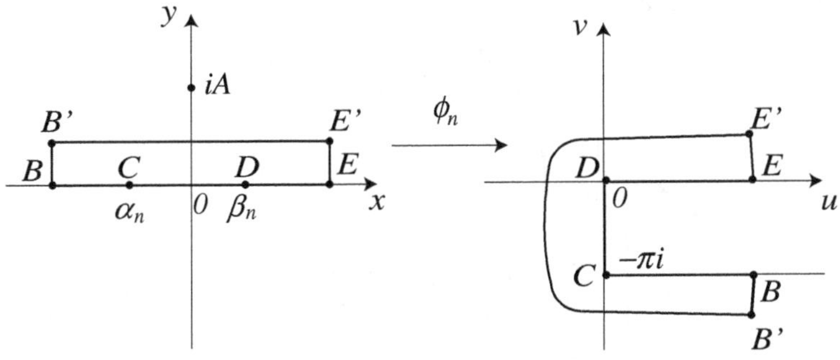


**Figure 2.** The image of $K_+$ under the conformal mapping $\phi_n(z)$.

### 3.4. *Properties of* $\phi_n(z)$

In this section, we investigate the mapping properties of $\phi_n(z)$ since it plays a significant role in our later discussion. We first restrict the variable of $\phi_n$ to the real axis $(-\infty, +\infty)$.

From the definition of $\phi_n$ given in (21) and the fact that $\nu_{n,\pm}(x) = \pm\pi i\mu_n(x)$ for $x \in (\alpha_n, \beta_n)$, it is easy to verify that

$$\begin{aligned} &\phi_n(\beta_n) = 0,\ \phi_{n,\pm}(\alpha_n) = \mp\pi i; \\ &\operatorname{Re}\phi_{n,\pm}(x) = 0 \text{ and } \arg\phi_{n,\pm}(x) = \pm\frac{3}{2}\pi \text{ for } x \in (\alpha_n, \beta_n); \end{aligned} \tag{113}$$

see Figure 2 below. Furthermore, since $\mu_n(x) > 0$, when $x$ moves from $\beta_n$ to $\alpha_n$, $\phi_{n,+}(x)$ moves from 0 to $-i\pi$ monotonically on the imaginary axis.

Since $\psi(\overline{z}) = \overline{\psi(z)}$, we immediately see that $i[\psi(\tau A + i\tau x) - \psi(\tau A - i\tau x)]$ is real when $x$ is real. From (78), it follows that $\nu_n(x)$ is real for $x < \alpha_n$ and $x > \beta_n$. From the definition of $\phi_n(x)$, the function is real for $x \in (\beta_n, \infty)$, and that $\phi_{n,+}(x) - \phi_{n,+}(\alpha_n) = \phi_{n,+}(x) + \pi i$ is also real for $x \in (-\infty, \alpha_n)$. We will show that $\phi_n(x)$ and $\phi_{n,+}(x) - \phi_{n,+}(\alpha_n)$ are monotonic in $(\beta_n, \infty)$ and $(-\infty, \alpha_n)$, respectively. To do this, we only need to show that $\nu_n(x)$ is positive on $(\beta_n, \infty)$ and negative on $(-\infty, \alpha_n)$.

Divide the interval $(\beta_n, \infty)$ into two parts: $x$ near $\beta_n$ and $x$ away from $\beta_n$, and recall the power series

$$\log\frac{1+z}{1-z} = 2z + \frac{2z^3}{3} + \frac{2z^5}{5} + \frac{2z^7}{7} + \cdots, \quad |z| < 1.$$

From (111), we can expand $\nu_n(z)$ as

$$\begin{aligned}\nu_n(z) &= \frac{i}{2}\log\frac{1-i\rho_n\zeta_n(z)/D_n(z)}{1+i\rho_n\zeta_n(z)/D_n(z)}+\frac{\zeta_n(z)}{4\tau}F_n(z)\\ &= \sum_{k=0}^{\infty}\frac{(-1)^k}{2k+1}\left(\frac{\rho_n\zeta_n(z)}{D_n(z)}\right)^{2k+1}+\frac{\zeta_n(z)}{4\tau}F_n(z)\\ &= \zeta_n(z)\left[\frac{\rho_n}{D_n(z)}\sum_{k=0}^{\infty}\frac{(-1)^k}{2k+1}\left(\frac{\rho_n\zeta_n(z)}{D_n(z)}\right)^{2k}+\frac{1}{4\tau}F_n(z)\right],\\ &\quad\text{if } \left|\frac{\rho_n\zeta_n(z)}{D_n(z)}\right|<1. \end{aligned}\tag{114}$$

We now estimate $\rho_n$, $D_n(x)$ and $\zeta_n(x)$ for $x\in(\beta_n,\infty)$ and close to $\beta_n$ and for large $n$. From (67),

$$\begin{aligned}\sqrt{\alpha_n^2+A^2}\tilde{\varphi}_n(iA) &= \sqrt{(-iA-\alpha_n)(iA-\beta_n)}\\ &= \sqrt{A^2+\alpha_n\beta_n-iA(\alpha_n-\beta_n)}\\ &= \frac{A}{\sin\phi}+i\sqrt{2A+1}+O\left(\frac{1}{\tau}\right)\quad\text{as}\quad n\to\infty,\end{aligned}$$

which implies that by (109)

$$\rho_n = 2\sqrt{\alpha_n^2+A^2}\,\mathrm{Im}\,\tilde{\varphi}_n(iA) = 2\sqrt{2A+1}+O(\tau^{-1}).$$

Thus, there exists $N_1>0$ such that for $n>N_1$,

$$\sqrt{2A+1}\le\rho_n\le 3\sqrt{2A+1}.\tag{115}$$

From the definition of $D_n(z)$ given in (108), we have

$$D_n(z) = (z-\beta_n)\big(\sqrt{\beta_n^2+A^2}-\sqrt{\alpha_n^2+A^2}\big)+(\beta_n-\alpha_n)\sqrt{\beta_n^2+A^2}.\tag{116}$$

Note that $-(\sqrt{\beta_n^2+A^2}+\sqrt{\alpha_n^2+A^2})\le\sqrt{\beta_n^2+A^2}-\sqrt{\alpha_n^2+A^2}\le\sqrt{\beta_n^2+A^2}+\sqrt{\alpha_n^2+A^2}$. Hence, Equation (116) gives

$$0\le|D_n(x)-(\beta_n-\alpha_n)\sqrt{\beta_n^2+A^2}|\le(x-\beta_n)\big(\sqrt{\beta_n^2+A^2}+\sqrt{\alpha_n^2+A^2}\big)\tag{117}$$

for $x\in[\beta_n,\infty)$. From (117), it follows that if

$$0\le x-\beta_n\le\delta_1(n):=\frac{(\beta_n-\alpha_n)\sqrt{\beta_n^2+A^2}}{2\big(\sqrt{\beta_n^2+A^2}+\sqrt{\alpha_n^2+A^2}\big)},\tag{118}$$

then

$$\frac{1}{2}(\beta_n-\alpha_n)\sqrt{\beta_n^2+A^2}\le D_n(x)\le\frac{3}{2}(\beta_n-\alpha_n)\sqrt{\beta_n^2+A^2}.\tag{119}$$

Again by (67), $\alpha_n \to a_0$ and $\beta_n \to b_0$ as $n \to \infty$, where $a_0, b_0$ are given by (69), and we have

$$m_1 = \lim_{n\to\infty} (\beta_n - \alpha_n)\sqrt{\beta_n^2 + A^2} = (b_0 - a_0)\sqrt{b_0^2 + A^2} > 0$$

and

$$\delta_1 = \lim_{n\to\infty} \delta_1(n) = \frac{m_1}{2\left(\sqrt{b_0^2 + A^2} + \sqrt{a_0^2 + A^2}\right)} > 0.$$

Thus, there exists $N_2 > 0$ such that for $n > N_2$, $\delta_2 := \frac{1}{2}\delta_1 < \delta_1(n)$ and

$$\frac{1}{2}m_1 \le (\beta_n - \alpha_n)\sqrt{\beta_n^2 + A^2} \le 2m_1. \tag{120}$$

By (118), (119) and (120), we obtain

$$\frac{1}{4}m_1 \le D_n(x) \le 3m_1 \tag{121}$$

for $x \in [\beta_n, \beta_n + \delta_2]$ and $n \ge N_2$. Coupling (115) and (121) gives

$$\frac{\rho_n}{D_n(x)} \ge m_2 := \frac{\sqrt{2A+1}}{3m_1} > 0 \tag{122}$$

for $x \in [\beta_n, \beta_n + \delta_2]$ and $n \ge N_3 = \max\{N_1, N_2\}$. On the other hand, since we have

$$\begin{aligned}\frac{\rho_n\zeta_n(x)}{D_n(x)} &\le \frac{3\sqrt{2A+1}}{\frac{1}{4}m_1}\sqrt{x-\beta_n}\sqrt{(\beta_n-\alpha_n)+(x-\beta_n)}\\ &\le \frac{12\sqrt{2A+1}}{m_1}\sqrt{x-\beta_n}\sqrt{(\beta_n-\alpha_n)+\delta_2}\end{aligned}$$

for $x \in [\beta_n, \beta_n + \delta_2]$ and $n \ge N_3$, there exists $0 < \delta_3 \le \delta_2$ such that for $x \in [\beta_n, \beta_n + \delta_3]$ and $n \ge N_3$,

$$0 < \frac{\rho_n\zeta_n(x)}{D_n(x)} \le \frac{1}{2}. \tag{123}$$

Substituting (122) and (123) in (114) yields

$$\begin{aligned}\nu_n(x) &\ge \zeta_n(x)\left[m_2\left(1 - \sum_{k=1}^{\infty}\frac{1}{4^k}\right) - \frac{1}{4\tau}|F_n(x)|\right]\\ &\ge \zeta_n(x)\left[\frac{2m_2}{3} - \frac{1}{4\tau}|F_n(x)|\right]\end{aligned} \tag{124}$$

for $x \in [\beta_n, \beta_n + \delta_3]$ and $n \geq N_3$. Since $F_n(z)$ is analytic in $\mathbb{C} \setminus \{(-i\infty, -iA] \cup [iA, i\infty)\}$ and has uniform asymptotic approximation (92)

$$F_n(z) \sim \frac{1}{(z - iA)\zeta_n(iA)} + \frac{1}{(z + iA)\zeta_n(-iA)}, \qquad n \to \infty,$$

for $z \in \Omega$, $|F_n(x)|$ has an upper bound on $[\beta_n, \beta_n + \delta_3]$ which is independent of $n$. Thus, from (124), there exists $N_4 \geq N_3 > 0$ such that for $n \geq N_4$ and $x \in (\beta_n, \beta_n + \delta_3]$,

$$\nu_n(x) \geq \zeta_n(x)\left[\frac{2m_2}{3} - \frac{1}{4\tau}|F_n(x)|\right] > 0. \tag{125}$$

Next, we estimate $\nu_n(x)$ when $x \in [\beta_n + \delta_3, \infty)$. We again use the asymptotic formula (111) of $\nu_n(z)$. When $x > \beta_n$, $\rho_n\zeta_n(x) > 0$; see (115). Hence, it follows that

$$\arg(D_n(x) + i\rho_n\zeta_n(x)) \in (0, \pi),$$

while

$$\arg(D_n(x) - i\rho_n\zeta_n(x)) = -\arg(D_n(x) + i\rho_n\zeta_n(x)) \in (-\pi, 0).$$

Since $|\frac{D_n(x) - i\rho_n\zeta_n(x)}{D_n(x) + i\rho_n\zeta_n(x)}| = 1$, we get

$$\begin{aligned} \nu_n^0(x) &:= \frac{i}{2}\log\frac{D_n(x) - i\rho_n\zeta_n(x)}{D_n(x) + i\rho_n\zeta_n(x)} \\ &= \frac{i}{2}(-2i)\arg(D_n(x) + i\rho_n\zeta_n(x)) \\ &= \arg(D_n(x) + i\rho_n\zeta_n(x)) > 0 \end{aligned} \tag{126}$$

for all finite $x$. Let $x \to \infty$. On account of (116), we have

$$\begin{aligned} \nu_n^0(x) &= \arg(D_n(x) + i\rho_n\zeta_n(x)) = \arg\frac{D_n(x) + i\rho_n\zeta_n(x)}{x} \\ &\to \arg(\sqrt{\beta_n^2 + A^2} - \sqrt{\alpha_n^2 + A^2} + i\rho_n) > 0 \quad \text{as } x \to \infty. \end{aligned}$$

Thus, when $x - \beta_n \geq \delta_3$ or, equivalently, $x \in [\beta_n + \delta_3, \infty)$, by the continuity of $\nu_n^0(x)$

$$\varepsilon_n := \inf\left\{\nu_n^0(x),\ x \in [\beta_n + \delta_3, \infty)\right\} > 0. \tag{127}$$

Take a fixed $\beta^*$ such that $\beta_n < \beta^* < \beta_n + \delta_3$ with sufficiently large $n$. For $x \in [\beta^*, +\infty)$, $\nu_n^0(x)$ is uniformly convergent to

$$\nu_\infty^0(x) = \frac{i}{2}\log\frac{D_\infty(x) - 2i\sqrt{2A+1}\zeta_\infty(x)}{D_\infty(x) + 2i\sqrt{2A+1}\zeta_\infty(x)} \qquad \text{as } n \to \infty,$$

where $D_\infty(x)$ and $\zeta_\infty(x)$ are defined in the same manner as $D_n(x)$ and $\zeta_n(x)$, except with $\alpha_n$, $\beta_n$ replaced by $a_0$, $b_0$, respectively. By the same argument as before, we can show that

$$\varepsilon_\infty := \inf\left\{\nu_\infty^0(x),\ x \in [b_0+\delta_3,\infty)\right\} > 0.$$

Hence, there must be a positive constant $\tilde{\varepsilon}$ independent of $n$ such that

$$\nu_n^0(x) \geq \tilde{\varepsilon} > 0 \quad \text{for } x \in [\beta_n+\delta_3,\infty). \tag{128}$$

From (101) with $m=1$, we have

$$F_n(z) = \frac{1}{(z-iA)\zeta_n(iA)} + \frac{1}{(z+iA)\zeta_n(-iA)} + \kappa_1^+(z) - \kappa_1^-(z), \tag{129}$$

where

$$\kappa_1^{\pm}(z) = \frac{2\tau i}{\pi}\int_{\alpha_n}^{\beta_n} \frac{U_1(\tau A \pm i\tau s) - U_1(\tau A \pm i\tau z)}{(s-z)\sqrt{(s-\alpha_n)(\beta_n - s)}}ds, \ |\arg(A \mp iz)| < \pi.$$

Using (31) and (26), we have for $x > \beta_n$

$$\begin{aligned}|\kappa_1^{\pm}(x)| &\leq \frac{2\tau}{\pi}\frac{2B_2\sec^3(\frac{\pi}{4})}{2\tau^2A^2}\frac{1}{x-\beta_n}\int_{\alpha_n}^{\beta_n}\frac{1}{\sqrt{(s-\alpha_n)(\beta_n-s)}}ds \\ &= \frac{2B_2\sec^3(\frac{\pi}{4})}{\tau A^2}\frac{1}{x-\beta_n}. \end{aligned} \tag{130}$$

Coupling (129) and (130) yields

$$|\zeta_n(x)F_n(x)| \leq m_3 \quad \text{for} \quad x \in [\beta_n+\delta_3,\infty), \tag{131}$$

where $m_3$ is a positive constant independent of $n$. A combination of (111), (128), and (131) gives

$$\begin{aligned}\nu_n(x) &= \nu_n^0(x) + \frac{1}{4\tau}\zeta_n(x)F_n(x) \\ &\geq \tilde{\varepsilon} - \frac{1}{4\tau}m_3 > 0 \end{aligned} \tag{132}$$

for $x \in [\beta_n+\delta_3,\infty)$ and sufficiently large $n$.

From the definition (21) of $\phi_n$, and on account of (125) and (132), $\phi_n(x) = \int_{\beta_n}^x \nu_n(s)ds$ increases from 0 to $\infty$ as $x$ moves from $\beta_n$ to $\infty$ for sufficiently large $n$.

By using the fact that

$$\zeta_n(x) = -\sqrt{(\alpha_n - x)(\beta_n - x)} \quad \text{and} \quad \nu_n^0(x) < 0$$

for $x \in (-\infty,\alpha_n]$, one can show in a similar way that $\nu_n(x) < 0$ for $x < \alpha_n$ and sufficiently large $n$. Hence, it follows that $\phi_{n,+}(x) - \phi_{n,+}(\alpha_n) = \int_{\alpha_n}^x \nu_n(s)ds$ is increasing from 0 to $\infty$ as $x$ moves from $\alpha_n$ to $-\infty$. Summing up the above results, we obtain the following lemma.

LEMMA 8. *If $x \in [\beta_n, \infty)$, then $\phi_n(x) \in [0, \infty)$ and, when $x$ moves from $\infty$ to $\beta_n$, $\phi_n(x)$ decreases from $\infty$ to 0. If $x \in [\alpha_n, \beta_n]$, then $\phi_{n,+}(x) \in [-i\pi, 0]$ and, when $x$ moves from $\beta_n$ to $\alpha_n$, $\phi_{n,+}(x)$ moves from 0 to $-i\pi$ monotonically. If $x \in (-\infty, \alpha_n]$, then $\phi_{n,+}(x) \in [-i\pi, \infty - i\pi)$ and, when $x$ moves from $\alpha_n$ to $-\infty$, $\phi_{n,+}(x)$ moves from $-i\pi$ to $\infty - i\pi$ also monotonically.*

From Lemma 8, we know that $\phi_n$ is conformal in a neighborhood of the real line on the upper half plane. Let

$$K = K(c, M) = \{z \in \mathbb{C} : |\operatorname{Re} z| \leq M, |\operatorname{Im} z| \leq cA\}, \tag{133}$$

where $0 < c < 1$ and $M > \max\{|\alpha_n|, |\beta_n|\}$, and let $K_\pm$ denote the upper and lower half of $K$ accordingly. Then, for any $M > \max\{|\alpha_n|, |\beta_n|\}$, there is a constant $c \in (0, 1)$ such that $\phi_n(z)$ conformally maps the upper-half rectangle $K_+ = K_+(c, M)$ into a region in $\mathbb{C} \setminus \{z : \operatorname{Re} z \geq 0, -\pi \leq \operatorname{Im} z \leq 0\}$, which is shown in Figure 2.

### 3.5. *Asymptotic expansion of $\ell_n$*

In this section, we investigate the asymptotic behavior of $\ell_n$. If we define

$$H_n(z) := \frac{g_n(z) - \log(z - \alpha_n)}{\sqrt{(z - \alpha_n)(z - \beta_n)}},$$

where $\log(z - \alpha_n)$ takes the principal branch on $\mathbb{C} \setminus (-\infty, \alpha_n]$, then

$$H_{n,+}(x) - H_{n,-}(x) = \frac{g_{n,+}(x) + g_{n,-}(x) - 2\log(x - \alpha_n)}{i\sqrt{(x - \alpha_n)(\beta_n - x)}}, \quad x \in (\alpha_n, \beta_n).$$

Solving this scalar RHP gives

$$\begin{aligned} g_n(z) - \log(z - \alpha_n) = & \frac{\sqrt{(z - \alpha_n)(z - \beta_n)}}{-2\pi} \\ & \times \int_{\alpha_n}^{\beta_n} \frac{g_{n,+}(s) + g_{n,-}(s) - 2\log(s - \alpha_n)}{(s - z)\sqrt{(s - \alpha_n)(\beta_n - s)}} ds \end{aligned}$$

for $z \in \mathbb{C} \setminus (-\infty, \alpha_n]$. Due to the fact that $g_n(z) - \log(z - \alpha_n) \to 0$ as $z \to \infty$, we have

$$\int_{\alpha_n}^{\beta_n} \frac{g_{n,+}(s) + g_{n,-}(s) - 2\log(s - \alpha_n)}{\sqrt{(s - \alpha_n)(\beta_n - s)}} ds = 0. \tag{134}$$

Combining (14), (26) and (134) yields

$$\begin{aligned} \ell_n = & \frac{1}{\tau\pi} \int_{\alpha_n}^{\beta_n} \frac{\log\Gamma(\tau A + i\tau s) + \log\Gamma(\tau A - i\tau s)}{\sqrt{(s - \alpha_n)(\beta_n - s)}} ds - \frac{\alpha_n + \beta_n}{2}(\pi - 2\phi) \\ & + \frac{2}{\pi} \int_{\alpha_n}^{\beta_n} \frac{\log(s - \alpha_n)}{\sqrt{(s - \alpha_n)(\beta_n - s)}} ds - \frac{1}{\tau} \log \frac{\beta_n - \alpha_n}{4}. \end{aligned} \tag{135}$$

Note that

$$\int_a^b \frac{\log s}{\sqrt{(s-a)(b-s)}}\,dx = 2\pi \log \frac{\sqrt{a}+\sqrt{b}}{2}, \qquad 0<a<b<\infty; \tag{136}$$

see [29, (3.13)]. Hence,

$$\int_{\alpha_n}^{\beta_n} \frac{\log(s-\alpha_n)}{\sqrt{(s-\alpha_n)(\beta_n-s)}}ds = 2\pi \log \frac{\sqrt{\beta_n-\alpha_n}}{2} = \pi \log \frac{\beta_n-\alpha_n}{4}. \tag{137}$$

Recall the asymptotic expansion

$$\log\Gamma(z) = \left(z-\frac{1}{2}\right)\log z - z + \frac{1}{2}\log(2\pi) + \sum_{k=1}^{m-1} \frac{B_{2k}}{2k(2k-1)z^{2k-1}} + R_m(z), \tag{138}$$

where $R_m(z)$ satisfies

$$|R_m(z)| \le \frac{|B_{2m}|\sec^{2m}\left(\frac{1}{2}\arg z\right)}{2m(2m-1)|z|^{2m-1}}, \qquad |\arg z| < \pi; \tag{139}$$

see [30, p. 294, (4.03), (4.06)]. We need to calculate the integral involving $\log\Gamma(\tau A \mp i\tau s)$ on the right-hand side of (135). To this end, we denote

$$I_4(A) = \int_{\alpha_n}^{\beta_n} \frac{\log(A^2+s^2)}{\sqrt{(s-\alpha_n)(\beta_n-s)}}ds. \tag{140}$$

Lemma 9. *Let $A$ be a positive real number. Then,*

$$\begin{aligned}\int_{\alpha_n}^{\beta_n} &\frac{\log\Gamma(\tau A+i\tau s)+\log\Gamma(\tau A-i\tau s)}{\sqrt{(s-\alpha_n)(\beta_n-s)}}ds \\ &\sim \left(\tau A-\frac{1}{2}\right)\pi\left[\log\varphi_n(iA)+\log\varphi_n(-iA)+2\log\frac{\tau}{2}\right] \\ &\quad + \tau\pi\left[\frac{\alpha_n+\beta_n}{2}\pi + i(\zeta_n(iA)-\zeta_n(-iA)) - (\alpha_n+\beta_n)\arg\varphi_n(iA)\right] \\ &\quad + \pi\log(2\pi) + i\pi\sum_{k=1}^{\infty}\frac{(-1)^{k-1}B_{2k}}{(2k)!\tau^{2k-1}}\vartheta_k(A), \qquad n\to\infty,\end{aligned} \tag{141}$$

*where $\varphi_n(z)$, $\zeta_n(z)$ are defined in (12) and (13), respectively, and*

$$\vartheta_k(A) = \frac{d^{2k-2}}{dz^{2k-2}}\left(\frac{1}{\zeta_n(z)}\right)\Bigg|_{-iA}^{iA}. \tag{142}$$

*Proof*: Regard $I_4(A)$ as a function of $A$. Using (138), we have

$$\begin{aligned}
&\int_{\alpha_n}^{\beta_n} \frac{\log\Gamma(\tau A + i\tau s) + \log\Gamma(\tau A - i\tau s)}{\sqrt{(s-\alpha_n)(\beta_n - s)}} ds \\
&= \left(\tau A - \frac{1}{2}\right)(I_4(A) + 2\pi \log\tau) - \tau I_2(A) - 2\tau A\pi + \pi\log(2\pi) \\
&\quad + \sum_{k=1}^{m-1} \frac{B_{2k}}{(2k)!\tau^{2k-1}} \frac{d^{2k-1} I_4(A)}{dA^{2k-1}} \\
&\quad + \int_{\alpha_n}^{\beta_n} \frac{R_m(\tau A + i\tau s) + R_m(\tau A - i\tau s)}{\sqrt{(s-\alpha_n)(\beta_n - s)}} ds. \qquad (143)
\end{aligned}$$

Differentiating both sides of (140) yields

$$\begin{aligned}
I_4'(A) &= \int_{\alpha_n}^{\beta_n} \frac{2A}{A^2 + s^2} \frac{ds}{\sqrt{(s-\alpha_n)(\beta_n - s)}}, \\
&= -i\int_{\alpha_n}^{\beta_n} \left(\frac{1}{s - iA} - \frac{1}{s + iA}\right) \frac{ds}{\sqrt{(s-\alpha_n)(\beta_n - s)}}, \\
&= i\pi\left(\frac{1}{\zeta_n(iA)} - \frac{1}{\zeta_n(-iA)}\right); \qquad (144)
\end{aligned}$$

see (44). Integrating both sides of (144) from $A$ to a large positive number $R$, and using (41), we get

$$\begin{aligned}
I_4(A) &= I_4(R) - i\pi\int_A^R \left[\frac{1}{\zeta_n(is)} - \frac{1}{\zeta_n(-is)}\right] ds, \\
&= I_4(R) - \pi\left[\int_{iA}^{iR} \frac{1}{\zeta_n(t)} dt + \int_{-iA}^{-iR} \frac{1}{\zeta_n(t)} dt\right], \\
&= I_4(R) - \pi\left[\log\frac{\varphi_n(iR)}{\varphi_n(iA)} + \log\frac{\varphi_n(-iR)}{\varphi_n(-iA)}\right]. \qquad (145)
\end{aligned}$$

Applying Maclaurin's expansion of $\log(1+z)$ for $|z| < 1$ to (140) gives

$$\begin{aligned}
I_4(R) &= \pi\log R^2 + \int_{\alpha_n}^{\beta_n} \frac{\log(1 + s^2/R^2)}{\sqrt{(s-\alpha_n)(\beta_n - s)}} ds, \\
&= \pi\log R^2 + \sum_{k=1}^{\infty} \frac{(-1)^{k-1}}{kR^{2k}} \int_{\alpha_n}^{\beta_n} \frac{s^{2k}}{\sqrt{(s-\alpha_n)(\beta_n - s)}} ds, \qquad (146)
\end{aligned}$$

where $R > \max\{|\alpha_n|, |\beta_n|\}$ and

$$\int_{\alpha_n}^{\beta_n} \frac{s^{2k}}{\sqrt{(s-\alpha_n)(\beta_n - s)}} ds = 2\int_0^{\frac{\pi}{2}} (\alpha_n\cos^2\theta + \beta_n\sin^2\theta)^{2k} d\theta$$

is a constant independent of $R$. At the same time, it is easy to verify from (12) and (13) that

$$\varphi_n(\pm iR) = \pm iR\left[1 \mp \frac{\alpha_n + \beta_n}{2iR} + \sqrt{\left(1 \mp \frac{\alpha_n}{iR}\right)\left(1 \mp \frac{\beta_n}{iR}\right)}\right],$$

which implies

$$\varphi_n(iR)\varphi_n(-iR) = 4R^2 + (\alpha_n + \beta_n)^2 + \frac{1}{2}(\alpha_n - \beta_n)^2 + O\left(\frac{1}{R}\right) \quad (147)$$

as $R \to \infty$. Coupling (146) and (147) yields

$$\lim_{R\to\infty} [I_4(R) - \pi \log(\varphi_n(iR)\varphi_n(-iR))] = -2\pi \log 2. \quad (148)$$

Letting $R \to \infty$ in (145), and applying (148), we immediately obtain

$$I_4(A) = \pi(\log \varphi_n(iA) + \log \varphi_n(-iA)) - 2\pi \log 2. \quad (149)$$

From (144), we have for any given $k \in \mathbb{N}$,

$$\frac{d^{2k-1} I_4(A)}{dA^{2k-1}} = i\pi \frac{d^{2k-2}}{dA^{2k-2}}\left(\frac{1}{\zeta_n(iA)} - \frac{1}{\zeta_n(-iA)}\right) = i\pi(-1)^{k-1}\vartheta_k(A). \quad (150)$$

From (140), it is readily seen that the quantity on the right-hand side of (150) is a real number. From (139), it follows that for real variable $s$,

$$|R_m(\tau A \pm i\tau s)| \le \frac{|B_{2m}| \sec^{2m}\left(\frac{\pi}{4}\right)}{2m(2m-1)|\tau A \pm i\tau s|^{2m-1}} \le \frac{2^m |B_{2m}|}{2m(2m-1)\tau^{2m-1}A^{2m-1}},$$

and so

$$\left| \int_{[alpha_n}^{\beta_n} \frac{R_m(\tau A + i\tau s) + R_m(\tau A - i\tau s)}{\sqrt{(s-\alpha_n)(\beta_n - s)}} ds \right| \le \frac{2^m \pi |B_{2m}|}{m(2m-1)\tau^{2m-1}A^{2m-1}}. \quad (151)$$

Therefore, substituting (35) and (149)–(151) in (143) gives (141). ■

Inserting (137) and (141) into (135) yields

$$\begin{aligned}\ell_n \sim {} & \left(A - \frac{1}{2\tau}\right)\left[\log \varphi_n(iA) + \log \varphi_n(-iA) + 2\log \frac{\tau}{2}\right] + i(\zeta_n(iA) - \zeta_n(-iA)) \\ & - (\alpha_n + \beta_n) \arg \varphi_n(iA) + \frac{1}{\tau}\log(2\pi) + (\alpha_n + \beta_n)\phi \\ & + \left(2 - \frac{1}{\tau}\right) \log \frac{\beta_n - \alpha_n}{4} + i\sum_{k=1}^{\infty} \frac{(-1)^{k-1} B_{2k}}{(2k)!\tau^{2k}} \vartheta_k(A). \end{aligned} \quad (152)$$

From (66), it is easy to verify that

$$\log \varphi_n(iA) + \log \varphi_n(-iA) = \log((2A+1)i + (2A+1)\cot\phi) + \log(-(2A+1)i + (2A+1)\cot\phi) + O\left(\frac{1}{\tau}\right)$$
$$= 2\log\left(\frac{2A+1}{\sin\phi}\right) + O\left(\frac{1}{\tau}\right). \tag{153}$$

Inserting (61), (62), (67), and (153) into (152), it leads to

$$\begin{aligned}
\ell_n &= \left(A - \frac{1}{2\tau}\right)\left[2\log\frac{2A+1}{\sin\phi} + 2\log\frac{\tau}{2}\right] \\
&\quad - 2(A+1) + 2\log\frac{\beta_n - \alpha_n}{4} + O(\tau^{-1}), \\
&= 2A\log\frac{2A+1}{\sin\phi} + \left(2A - \frac{1}{\tau}\right)\log\frac{\tau}{2} - 2(A+1) \\
&\quad + 2\log\frac{\sqrt{2A+1}}{2\sin\phi} + O(\tau^{-1}).
\end{aligned} \tag{154}$$

By an argument similar to that in obtaining the expansions of $\alpha_n, \beta_n$ given in (68), one can also show that $\ell_n$ has the asymptotic expansion

$$\begin{aligned}
\ell_n &\sim 2A\log\frac{2A+1}{\sin\phi} + \left(2A - \frac{1}{\tau}\right)\log\frac{\tau}{2} - 2(A+1) \\
&\quad + 2\log\frac{\sqrt{2A+1}}{2\sin\phi} + \sum_{j=1}^{\infty}\frac{c_j}{\tau^j},
\end{aligned} \tag{155}$$

where the coefficients $c_j$ are constants determined by $a_k, b_k (k = 0, 1, \ldots, j)$.

## 4. Construction of the parametrix

We define

$$\tilde{\phi}_n(z) = \int_{\alpha_n}^{z} \nu_n(s)ds, \quad z \in \mathbb{C} \setminus \{[\alpha_n, \infty) \cup (-i\infty, -iA] \cup [iA, i\infty)\}. \tag{156}$$

If we let $\mathbb{C}_\pm$ denote the upper and lower half plane accordingly, then for $z \in \mathbb{C}_\pm$, we have

$$\tilde{\phi}_n(z) = \int_{\alpha_n}^{\beta_n} \nu_n(s)ds + \int_{\beta_n}^{z} \nu_n(s)ds = \phi_n(z) \pm \pi i. \tag{157}$$

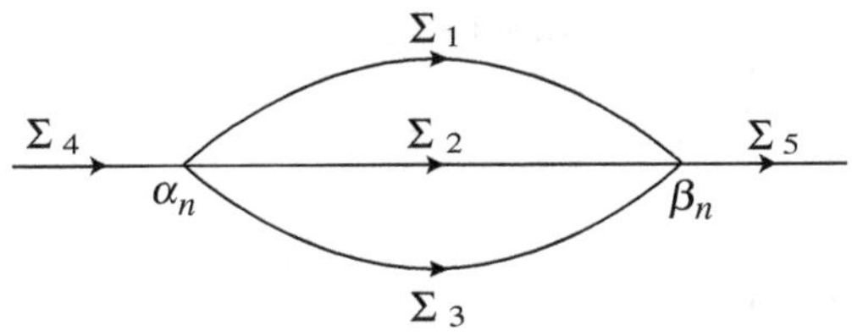


**Figure 3.** Contour $\Sigma_i (i = 1, 2, 3, 4, 5)$.

From (22) and the fact that $\phi_{n,+}(x) = -\phi_{n,-}(x)$ for $x \in (\alpha_n, \beta_n)$, the jump matrix $J_T$ in ($\mathrm{T}_b$) satisfies

$$J_T = \begin{pmatrix} e^{-2\tau\phi_{n,-}(x)+\frac{1}{2}\log(\varphi_{n,+}/\varphi_{n,-})} & 1 \\ 0 & e^{-2\tau\phi_{n,+}(x)-\frac{1}{2}\log(\varphi_{n,+}/\varphi_{n,-})} \end{pmatrix}$$

for $x \in (\alpha_n, \beta_n)$, and

$$J_T = \begin{pmatrix} 1 & e^{-2\tau\phi_n(x)+\log\frac{\beta_n-\alpha_n}{4}-\frac{1}{2}\log\frac{\varphi_n^2(x)}{4}} \\ 0 & 1 \end{pmatrix} \quad \text{for } x > \beta_n,$$

$$J_T = \begin{pmatrix} 1 & e^{-2\tau\tilde{\phi}_n(x)+\log\frac{\beta_n-\alpha_n}{4}-\frac{1}{2}\log\frac{\varphi_n^2(x)}{4}} \\ 0 & 1 \end{pmatrix} \quad \text{for } x < \alpha_n.$$

Let $\Sigma$ denote the contour $\Sigma_1 \cup \cdots \cup \Sigma_5$ shown in Figure 3. The two curves $\Sigma_1$ and $\Sigma_3$ enclose a lens-shaped region. Now, we are ready to define the transformation $T \to S$ as

$$S(z) := \begin{cases} T(z), & \text{for } z \text{ outside the lens-shaped region,} \\ T(z)\begin{pmatrix} 1 & 0 \\ -e^{2\tau\phi_n(z)+\log[2\varphi_n/(\beta_n-\alpha_n)]} & 1 \end{pmatrix} & \text{in the upper lens region,} \\ T(z)\begin{pmatrix} 1 & 0 \\ e^{2\tau\phi_n(z)+\log[2\varphi_n/(\beta_n-\alpha_n)]} & 1 \end{pmatrix} & \text{in the lower lens region;} \end{cases} \tag{158}$$

see [29]. Furthermore, define

$$J_S(z) = \begin{pmatrix} 1 & 0 \\ e^{2\tau\phi_n(z)+\log[2\varphi_n/(\beta_n-\alpha_n)]} & 1 \end{pmatrix}, \quad z \in \Sigma_1 \cup \Sigma_3; \tag{159}$$

$$J_S(x) = \begin{pmatrix} 0 & 1 \\ -1 & 0 \end{pmatrix}, \quad x \in \Sigma_2; \tag{160}$$

$$J_S(z) = \begin{pmatrix} 1 & e^{-2\tau\tilde{\phi}_n(z)+\log\frac{\beta_n-\alpha_n}{4}-\frac{1}{2}\log\frac{\varphi_n^2(x)}{4}} \\ 0 & 1 \end{pmatrix}, \quad z \in \Sigma_4; \tag{161}$$

$$J_S(z) = \begin{pmatrix} 1 & e^{-2\tau\phi_n(z)+\log\frac{\beta_n-\alpha_n}{4}-\frac{1}{2}\log\frac{\varphi_n^2(x)}{4}} \\ 0 & 1 \end{pmatrix}, \quad z \in \Sigma_5. \tag{162}$$

Then, it is readily verified that $S(z)$ satisfies the RHP:

($S_a$) $S(z)$ is analytic in $\mathbb{C} \setminus \Sigma$;
($S_b$) $S_+(z) = S_-(z)J_S(z)$ for $z \in \Sigma$, where $J_S(z)$ is defined in (159)–(162);
($S_c$) $S(z) = I + O(1/z)$ as $z \to \infty, z \in \mathbb{C} \setminus \Sigma$.

From the mapping properties of $\phi_n$, we know that $\operatorname{Re}\phi_n(z) < 0$ for $z \in \Sigma_1 \cup \Sigma_3, \phi_n(x) > 0$ for $x \in \Sigma_5$, and $\tilde{\phi}_n(x) > 0$ for $x \in \Sigma_4$. At the same time, for $x \in \Sigma_4 \cup \Sigma_5$, we have $\varphi_n^2(x) > \frac{(\beta_n-\alpha_n)^2}{4}$; that is, $\log\frac{\beta_n-\alpha_n}{4} - \frac{1}{2}\log\frac{\varphi_n^2(x)}{4} < 0$; see the statement following (13). Therefore, $J_S$ tends to the identity matrix as $n \to \infty$ for $z \in \Sigma \setminus \Sigma_2$. It is now natural to suggest that for large $n$, the solution of the RHP for $S$ may behave asymptotically like the solution of the following RHP for $N(z)$:

($N_a$) $N(z)$ is analytic in $\mathbb{C} \setminus [\alpha_n, \beta_n]$;
($N_b$) for $x \in (\alpha_n, \beta_n)$,

$$N_+(x) = N_-(x)\begin{pmatrix} 0 & 1 \\ -1 & 0 \end{pmatrix};$$

($N_c$) as $z \to \infty$,

$$N(z) = I + O(1/z).$$

The RHP for $N(z)$ can be solved explicitly, and its solution is

$$N(z) = \frac{1}{2}\begin{pmatrix} a(z) + a(z)^{-1} & -i(a(z) - a(z)^{-1}) \\ i(a(z) - a(z)^{-1}) & a(z) + a(z)^{-1} \end{pmatrix}, \tag{163}$$

where

$$a(z) = \left(\frac{z-\beta_n}{z-\alpha_n}\right)^{1/4}, \quad z \in \mathbb{C} \setminus [\alpha_n, \beta_n]. \tag{164}$$

Since $S(z) \sim N(z)$ as $n \to \infty$, on account of (158) we have $T(z) = S(z) \sim N(z)$ outside a neighborhood of $[\alpha_n, \beta_n]$, say, outside the rectangle $K$. In view of (11), the asymptotic behavior of $U(z)$ outside $K$ is given by

$$U(z) \sim e^{\frac{\tau}{2}\ell_n\sigma_3} N(z) e^{(\tau g_n(z) - \frac{1}{2}\log\frac{\varphi_n}{2} - \frac{\tau}{2}\ell_n)\sigma_3}. \tag{165}$$

By recalling the equation following (13), $N(z)$ in (163) can also be expressed as

$$N(z)=\frac{1}{[(z-\alpha_n)(z-\beta_n)]^{1/4}}\begin{pmatrix}\left(\frac{\varphi_n}{2}\right)^{1/2} & i\frac{\beta_n-\alpha_n}{4}\left(\frac{\varphi_n}{2}\right)^{-1/2}\\ -i\frac{\beta_n-\alpha_n}{4}\left(\frac{\varphi_n}{2}\right)^{-1/2} & \left(\frac{\varphi_n}{2}\right)^{1/2}\end{pmatrix}; \tag{166}$$

see [29, (3.56) and (3.57)]. Thus, using (22) and (166), we have

$$\begin{aligned}&N(z)e^{(\tau g_n-\frac{1}{2}\log\frac{\varphi_n(z)}{2}-\frac{\tau}{2}\ell_n)\sigma_3}w_n(z)^{\frac{1}{2}\sigma_3}\\&=N(z)e^{(-\tau\phi_n-\frac{1}{2}\log\frac{\varphi_n(z)}{2}+\frac{1}{2}\log\frac{\beta_n-\alpha_n}{4})\sigma_3}\\&=\frac{1}{2}\begin{pmatrix}1 & 0\\ \dfrac{-i(2z-\alpha_n-\beta_n)}{\beta_n-\alpha_n} & -2i\end{pmatrix}b_n(z)^{-\sigma_3}\begin{pmatrix}1 & i\\ -1 & i\end{pmatrix}e^{-\tau\phi_n(z)\sigma_3},\end{aligned} \tag{167}$$

where

$$b_n(z)=\frac{[(z-\alpha_n)(z-\beta_n)]^{1/4}}{\sqrt{\beta_n-\alpha_n}}\qquad z\in\mathbb{C}\setminus(-\infty,\beta_n]; \tag{168}$$

cf.[29, (3.58)]. Coupling (165) and (167) gives

$$U(z)\sim e^{\frac{1}{2}\tau\ell_n\sigma_3}\tilde{V}_{out}(z)w_n(z)^{-\frac{1}{2}\sigma_3},\quad z\in\mathbb{C}\setminus(K\cup\mathbb{R}),\quad n\to\infty, \tag{169}$$

where

$$\tilde{V}_{out}(z)=\frac{1}{2}\begin{pmatrix}1 & 0\\ \dfrac{-i(2z-\alpha_n-\beta_n)}{\beta_n-\alpha_n} & -2i\end{pmatrix}b_n(z)^{-\sigma_3}\begin{pmatrix}1 & i\\ -1 & i\end{pmatrix}e^{-\tau\phi_n(z)\sigma_3}. \tag{170}$$

To obtain the asymptotic behavior of $U(z)$ inside $K$, we need to construct a parametrix $\tilde{V}_{in}(z)$ in $K$. Note that using the jump condition in ($U_b$), the matrix

$$V(z)=e^{-\frac{\tau}{2}\ell_n\sigma_3}U(z)w_n(x)^{\frac{1}{2}\sigma_3} \tag{171}$$

satisfies the jump condition

$$V_+(x)=V_-(x)\begin{pmatrix}1 & 1\\ 0 & 1\end{pmatrix},\qquad x\in\mathbb{R}. \tag{172}$$

This requires us to construct the parametrix $\tilde{V}_{in}(z)$, which also satisfies the jump condition (172), and behaves like $\tilde{V}_{out}(z)$ on the boundary of $K$. The mapping properties of $\phi_n(z)$ invoke us to construct our approximate solution by using the parabolic cylinder function $U(a,z)$, since these properties are very similar to those of the function

$$f(\xi):=\xi\sqrt{\xi^2-1}-\log(\xi+\sqrt{\xi^2-1}),\quad \xi\in\mathbb{C}\setminus(-\infty,1],$$

which conformally maps the upper-half plane $\mathbb{C}_+$ onto the region $\mathbb{C} \setminus \{z : \operatorname{Re} z \geq 0, -\pi \leq \operatorname{Im} z \leq 0\}$. The function $f(\xi)$ plays an important role in describing the asymptotic behavior of the parabolic cylinder function $U(-\tau, 2\sqrt{\tau}\xi)$ as $\xi \to \infty$; see [31] and [32]. Comparing the mapping properties of $f(\xi)$ and $\phi_n(z)$, we can establish a one-to-one correspondence $\xi \leftrightarrow z$ defined by

$$f(\xi(z)) = \phi_n(z) \quad \text{or} \quad \xi(z) = (f^{-1} \circ \phi_n)(z) \quad \text{for} \quad z \in K. \tag{173}$$

With this, we can make use of the connection formula (see [31, (19.4.7)] or [32, p. 133])

$$\begin{aligned}\sqrt{2\pi}\, U(a, \pm x) = \Gamma\left(\frac{1}{2} - a\right) \{&e^{-i\pi(\frac{1}{2}a+\frac{1}{4})} U(-a, \pm ix) \\ &+ e^{i\pi(\frac{1}{2}a+\frac{1}{4})} U(-a, \mp ix)\},\end{aligned} \tag{174}$$

which provides the matrix equation

$$\begin{aligned}&\begin{pmatrix} U(-\tau, 2\sqrt{\tau}\,\xi) & \dfrac{\Gamma(n+1)}{\sqrt{2\pi}} e^{-i\pi n/2}\, U(\tau, -2i\sqrt{\tau}\,\xi) \\ \dfrac{1}{\sqrt{\tau}} U'(-\tau, 2\sqrt{\tau}\,\xi) & \dfrac{\Gamma(n+1)}{\sqrt{2\pi\tau}} e^{-i\pi(n+1)/2}\, U'(\tau, -2i\sqrt{\tau}\,\xi) \end{pmatrix} \\ &= \begin{pmatrix} U(-\tau, 2\sqrt{\tau}\,\xi) & -\dfrac{\Gamma(n+1)}{\sqrt{2\pi}} e^{i\pi n/2}\, U(\tau, 2i\sqrt{\tau}\,\xi) \\ \dfrac{1}{\sqrt{\tau}} U'(-\tau, 2\sqrt{\tau}\,\xi) & -\dfrac{\Gamma(n+1)}{\sqrt{2\pi\tau}} e^{i\pi(n+1)/2}\, U'(\tau, 2i\sqrt{\tau}\,\xi) \end{pmatrix} \begin{pmatrix} 1 & 1 \\ 0 & 1 \end{pmatrix}.\end{aligned} \tag{175}$$

To match $\tilde{V}_{out}(z)$ on the boundary of $K$, we use the asymptotic approximations (see [32, (4.3),(4.13),(6.2)])

$$U(-\tau, 2\sqrt{\tau}\,\xi) \sim 2^{-\frac{1}{2}} e^{-\frac{\tau}{2}} \tau^{\frac{\tau}{2}-\frac{1}{4}} \frac{e^{-\tau f(\xi)}}{(\xi^2-1)^{1/4}}, \tag{176a}$$

$$U'(-\tau, 2\sqrt{\tau}\,\xi) \sim -\sqrt{\tau}\, 2^{-\frac{1}{2}} e^{-\frac{\tau}{2}} \tau^{\frac{\tau}{2}-\frac{1}{4}} (\xi^2-1)^{1/4} e^{-\tau f(\xi)}, \tag{176b}$$

$$U(\tau, \mp 2i\sqrt{\tau}\,\xi) \sim 2^{-\frac{1}{2}} e^{\pm\pi i(\frac{\tau}{2}+\frac{1}{4})} e^{\frac{\tau}{2}} \tau^{-\frac{\tau}{2}-\frac{1}{4}} \frac{e^{\tau f(\xi)}}{(\xi^2-1)^{1/4}}, \tag{176c}$$

$$U'(\tau, \mp 2i\sqrt{\tau}\,\xi) \sim \pm i\sqrt{\tau} 2^{-\frac{1}{2}} e^{\pm\pi i(\frac{\tau}{2}+\frac{1}{4})} e^{\frac{\tau}{2}} \tau^{-\frac{\tau}{2}-\frac{1}{4}} (\xi^2-1)^{1/4} e^{\tau f(\xi)}, \tag{176d}$$

uniformly for $\xi \in \mathbb{C} \setminus (-\infty, 1]$, as $n \to \infty$. Moreover, applying Stirling's formula [32, (2.22)], we have

$$\Gamma(n+1) \sim (2\pi)^{\frac{1}{2}} e^{-\tau} \tau^{\tau} \sum_{s=0}^{\infty} \frac{\gamma_s}{\tau^s} \quad \text{as } n \to \infty, \tag{177}$$

where $\gamma_0 = 1, \gamma_1 = -\frac{1}{24}, \gamma_2 = \frac{1}{1152}, \gamma_3 = \frac{1003}{414720}, \ldots$. With the above results, straight calculation shows that

$$\begin{aligned} &\begin{pmatrix} U(-\tau, 2\sqrt{\tau}\,\xi) \pm \dfrac{\Gamma(n+1)}{\sqrt{2\pi}} e^{\mp i\pi n/2}\, U(\tau, \mp 2i\sqrt{\tau}\,\xi) \\ \dfrac{1}{\sqrt{\tau}} U'(-\tau, 2\sqrt{\tau}\,\xi) \pm \dfrac{\Gamma(n+1)}{\sqrt{2\pi\tau}} e^{\mp i\pi(n+1)/2}\, U'(\tau, \mp 2i\sqrt{\tau}\,\xi) \end{pmatrix} \\ &\sim \frac{1}{\sqrt{2}} e^{-\frac{\tau}{2}} \tau^{\frac{\tau}{2}-\frac{1}{4}} (\xi^2-1)^{-\frac{1}{4}\sigma_3} \begin{pmatrix} 1 & i \\ -1 & i \end{pmatrix} e^{-\tau f(\xi)\sigma_3} \\ &= \frac{1}{\sqrt{2}} e^{-\frac{\tau}{2}} \tau^{\frac{\tau}{2}-\frac{1}{4}} (\xi^2-1)^{-\frac{1}{4}\sigma_3} \begin{pmatrix} 1 & i \\ -1 & i \end{pmatrix} e^{-\tau\phi_n(z)\sigma_3} \quad \text{as } n \to \infty, \end{aligned} \tag{178}$$

uniformly for $\xi \in \mathbb{C}_{\pm}$, or equivalently $z \in K_{\pm}$. Define

$$\begin{aligned} \tilde{V}_{in}(z) := \frac{1}{\sqrt{2}} e^{\frac{\tau}{2}} \tau^{-\frac{\tau}{2}+\frac{1}{4}} &\begin{pmatrix} 1 & 0 \\ \dfrac{-i(2z-\alpha_n-\beta_n)}{\beta_n-\alpha_n} & -2i \end{pmatrix} \begin{pmatrix} \dfrac{(\xi^2-1)^{\frac{1}{4}}}{b_n(z)} \end{pmatrix}^{\sigma_3} \\ \times &\begin{pmatrix} U(-\tau, 2\sqrt{\tau}\,\xi) & \pm\dfrac{\Gamma(n+1)}{\sqrt{2\pi}} e^{\mp i\pi n/2}\, U(\tau, \mp 2i\sqrt{\tau}\,\xi) \\ \dfrac{1}{\sqrt{\tau}} U'(-\tau, 2\sqrt{\tau}\,\xi) & \pm\dfrac{\Gamma(n+1)}{\sqrt{2\pi\tau}} e^{\mp i\pi(n+1)/2}\, U'(\tau, \mp 2i\sqrt{\tau}\,\xi) \end{pmatrix} \end{aligned} \tag{179}$$

for $z \in K_{\pm}$. From (178) and (170), it is easy to see that $\tilde{V}_{in}(z) \sim \tilde{V}_{out}(z)$ as $n \to \infty$, uniformly for $z \in \partial K$. Since $(\xi^2-1)^{\frac{1}{4}}/b_n(z)$ is analytic in $\mathbb{C}$ even when $z \in [\alpha_n, \beta_n]$, it follows from (175) that

$$\tilde{V}_{in,+}(x) = \tilde{V}_{in,-}(x) \begin{pmatrix} 1 & 1 \\ 0 & 1 \end{pmatrix}, \quad x \in \mathbb{R} \cap K = (-M, M). \tag{180}$$

Thus, the reasonable asymptotic solution to the RHP for $U(z)$ is given by

$$\tilde{U}(z) = \begin{cases} e^{\frac{1}{2}\tau\ell_n\sigma_3} \tilde{V}_{out}(z) w_n(z)^{-\frac{1}{2}\sigma_3}, & z \in \mathbb{C} \setminus (K \cup \mathbb{R}), \\ e^{\frac{1}{2}\tau\ell_n\sigma_3} \tilde{V}_{in}(z) w_n(z)^{-\frac{1}{2}\sigma_3}, & z \in K \setminus (-M, M). \end{cases} \tag{181}$$

Using (180) and the fact that $w_n(z)$ is analytic for $z$ in $\mathbb{R}$, it is easily verified that

$$\tilde{U}_+(x) = \tilde{U}_-(x)\begin{pmatrix} 1 & w_n(x) \\ 0 & 1 \end{pmatrix}, \qquad x \in (-M, M).$$

However, there is an additional jump for $\tilde{U}$ on $\partial K$ and $(-\infty, -M] \cup [M, \infty)$. Since $\tilde{V}_{in}(z) \sim \tilde{V}_{out}(z)$ as $n \to \infty$, from (181) we know that $\tilde{U}_+(z) \sim \tilde{U}_-(z)$ uniformly for $z \in \partial K$ as $n \to \infty$, where $\tilde{U}_+(z)$ and $\tilde{U}_-(z)$ denote the limiting values of $\tilde{U}(z)$ as $z$ approaches $\partial K$ from outside and inside of $K$, respectively. The computation of the jump on $\partial K \cup (-\infty, -M] \cup [M, \infty)$ is given in the section below.

## 5. Uniform asymptotic expansions

In this section, we will show that the additional jump matrix has a uniform asymptotic expansion as $n \to \infty$, whose leading term is the identity matrix. Define

$$R(z) := e^{-\frac{\tau}{2}\ell_n\sigma_3} U(z)\tilde{U}(z)^{-1} e^{\frac{\tau}{2}\ell_n\sigma_3}. \tag{182}$$

Since $\tilde{U}(z)$ has the same jump as $U(z)$ in $(-M, M)$, $R(z)$ satisfies the relation

$$R_+(x) = R_-(x), \qquad x \in (-M, M).$$

In view of the definition of $\tilde{U}(z)$ in (181), it is easy to verify that $R(z)$ is a solution of the following RHP:

$(R_a)$ $R(z)$ is analytic in $z \in \mathbb{C} \setminus \Sigma^*$, where $\Sigma^* = \partial K \cup (-\infty, -M] \cup [M, \infty)$;
$(R_b)$ $R(z)$ satisfies the jump condition $R_+(z) = R_-(z)J_R(z)$, where

$$J_R(x) = e^{-\frac{\tau}{2}\ell_n\sigma_3}\tilde{U}_-(x)\begin{pmatrix} 1 & w_n(x) \\ 0 & 1 \end{pmatrix}\tilde{U}_+(x)^{-1} e^{\frac{\tau}{2}\ell_n\sigma_3} \tag{183}$$

for $x \in (-\infty, -M] \cup [M, \infty)$, and

$$J_R(z) = \tilde{V}_{in}(z)\tilde{V}_{out}(z)^{-1} \tag{184}$$

for $z \in \partial K$, with $\partial K$ oriented in the clockwise direction;
$(R_c)$ for $z \in \mathbb{C} \setminus \Sigma^*$,

$$R(z) = I + O(1/z) \qquad \text{as} \qquad z \to \infty.$$

The last equation follows from (169) and (181).

To solve this problem, we need to derive the asymptotic expansion of $J_R(z)$. First, for $z \in \partial K$, we insert (170) and (179) into (184) and obtain

$$J_R(z) = \frac{1}{\sqrt{2}} e^{\frac{\tau}{2}} \tau^{-\frac{\tau}{2}+\frac{1}{4}} \begin{pmatrix} 1 & 0 \\ \dfrac{-i(2z-\alpha_n-\beta_n)}{\beta_n-\alpha_n} & -2i \end{pmatrix} \left( \frac{(\xi^2-1)^{\frac{1}{4}}}{b_n(z)} \right)^{\sigma_3}$$
$$\times \begin{pmatrix} U(-\tau, 2\sqrt{\tau}\,\xi) & \pm\dfrac{\Gamma(n+1)}{\sqrt{2\pi}} e^{\mp i\pi n/2}\, U(\tau, \mp 2i\sqrt{\tau}\,\xi) \\ \frac{1}{\sqrt{\tau}} U'(-\tau, 2\sqrt{\tau}\,\xi) & \pm\dfrac{\Gamma(n+1)}{\sqrt{2\pi\tau}} e^{\mp i\pi(n+1)/2}\, U'(\tau, \mp 2i\sqrt{\tau}\,\xi) \end{pmatrix}$$
$$\times\, e^{\tau\phi_n(z)\sigma_3} \begin{pmatrix} \frac{1}{2} & \frac{i}{2} \\ -\frac{1}{2} & \frac{i}{2} \end{pmatrix}^{-1} b_n(z)^{\sigma_3} \begin{pmatrix} 1 & 0 \\ \dfrac{-i(2z-\alpha_n-\beta_n)}{\beta_n-\alpha_n} & -2i \end{pmatrix}^{-1} \tag{185}$$

for $z \in \partial K \cap \mathbb{C}_\pm$. Recall that the parabolic cylinder functions have asymptotic expansions [32, p. 140–143]

$$U(-\tau, 2\sqrt{\tau}\,\xi) \sim \rho(\sqrt{2\tau}) \frac{e^{-\tau f(\xi)}}{(\xi^2-1)^{1/4}} \sum_{s=0}^{\infty} \frac{A_s(\xi)}{(2\tau)^s}, \tag{186a}$$

$$U'(-\tau, 2\sqrt{\tau}\,\xi) \sim -\sqrt{\tau}\, \rho(\sqrt{2\tau}) (\xi^2-1)^{1/4} e^{-\tau f(\xi)} \sum_{s=0}^{\infty} \frac{B_s(\xi)}{(2\tau)^s}, \tag{186b}$$

$$U(\tau, \mp 2i\sqrt{\tau}\,\xi) \sim \rho(\mp i\sqrt{2\tau}) \frac{e^{\tau f(\xi)}}{(\xi^2-1)^{1/4}} \sum_{s=0}^{\infty} \frac{A_s(\xi)}{(-2\tau)^s}, \tag{186c}$$

$$U'(\tau, \mp 2i\sqrt{\tau}\,\xi) \sim \pm i\sqrt{\tau}\, \rho(\mp i\sqrt{2\tau}) (\xi^2-1)^{1/4} e^{\tau f(\xi)} \sum_{s=0}^{\infty} \frac{B_s(\xi)}{(-2\tau)^s} \tag{186d}$$

as $n \to \infty$, which hold uniformly for $\xi \in \mathbb{C} \setminus (-\infty, 1]$, where $\rho(\sqrt{2\tau})$, $\rho(\mp i\sqrt{2\tau})$ are independent of $\xi$ and satisfy

$$\rho(\sqrt{2\tau}) \sim 2^{-\frac{1}{2}} e^{-\frac{\tau}{2}} \tau^{\frac{\tau}{2}-\frac{1}{4}} \left( 1 + \sum_{s=0}^{\infty} \frac{\rho_{2s+1}}{(2\tau)^{2s+1}} \right)^{-1}, \tag{187a}$$

$$\rho(\mp i\sqrt{2\tau}) \sim 2^{-\frac{1}{2}} e^{\pm\pi i(\frac{\tau}{2}+\frac{1}{4})} e^{\frac{\tau}{2}} \tau^{-\frac{\tau}{2}-\frac{1}{4}} \left( 1 - \sum_{s=0}^{\infty} \frac{\rho_{2s+1}}{(2\tau)^{2s+1}} \right)^{-1} \tag{187b}$$

as $n \to \infty$. The coefficients in (186) are given by

$$A_s(\xi) = u_s(\xi)/(\xi^2-1)^{\frac{3}{2}s}, \qquad B_s(\xi) = v_s(\xi)/(\xi^2-1)^{\frac{3}{2}s}, \tag{188}$$

with $u_0(\xi) = v_0(\xi) = 1$, and both $u_s(\xi)$ and $v_s(\xi)$ being polynomials of degree $3s$ if $s$ is odd and $3s-2$ if $s$ is even. Using (186), (187) and (177), we have

$$\begin{pmatrix} U(-\tau, 2\sqrt{\tau}\,\xi) \pm \dfrac{\Gamma(n+1)}{\sqrt{2\pi}} e^{\mp i\pi n/2}\, U(\tau, \mp 2i\sqrt{\tau}\,\xi) \\ \dfrac{1}{\sqrt{\tau}} U'(-\tau, 2\sqrt{\tau}\,\xi) \pm \dfrac{\Gamma(n+1)}{\sqrt{2\pi\tau}} e^{\mp i\pi(n+1)/2}\, U'(\tau, \mp 2i\sqrt{\tau}\,\xi) \end{pmatrix} \tag{189}$$
$$\sim \frac{1}{\sqrt{2}} e^{-\frac{\tau}{2}} \tau^{\frac{\tau}{2}-\frac{1}{4}} (\xi^2-1)^{-\frac{1}{4}\sigma_3} \sum_{s=0}^{\infty} \frac{1}{\tau^s} \begin{pmatrix} \tilde{A}_{1,s}(\xi) & \tilde{A}_{2,s}(\xi) \\ \tilde{B}_{1,s}(\xi) & \tilde{B}_{2,s}(\xi) \end{pmatrix} e^{-\tau\phi_n\sigma_3}$$

as $n \to \infty$, where $\tilde{A}_{i,s}(\xi)(i = 1, 2)$ are determined by $A_j(\xi), \rho_{2j+1}, \gamma_j$ for $j = 0, \ldots, s$, $\tilde{B}_{i,s}(\xi)(i = 1, 2)$ are determined by $B_j(\xi), \rho_{2j+1}, \gamma_j$ for $j = 0, \ldots, s$, and

$$\begin{pmatrix} \tilde{A}_{1,0}(\xi) & \tilde{A}_{2,0}(\xi) \\ \tilde{B}_{1,0}(\xi) & \tilde{B}_{2,0}(\xi) \end{pmatrix} = \begin{pmatrix} 1 & i \\ -1 & i \end{pmatrix}. \tag{190}$$

Substituting (189) and (190) in (185) gives

$$J_R(z) \sim I + \sum_{s=1}^{\infty} \frac{1}{\tau^s} J_s(z), \tag{191}$$

where

$$J_s(z) = \begin{pmatrix} 1 & 0 \\ \dfrac{-i(2z-\alpha_n-\beta_n)}{\beta_n-\alpha_n} & -2i \end{pmatrix} b_n(z)^{-\sigma_3} \begin{pmatrix} \tilde{A}_{1,s}(\xi) & \tilde{A}_{2,s}(\xi) \\ \tilde{B}_{1,s}(\xi) & \tilde{B}_{2,s}(\xi) \end{pmatrix}$$
$$\times \begin{pmatrix} 1 & i \\ -1 & i \end{pmatrix}^{-1} b_n(z)^{\sigma_3} \begin{pmatrix} 1 & 0 \\ \dfrac{-i(2z-\alpha_n-\beta_n)}{\beta_n-\alpha_n} & -2i \end{pmatrix}^{-1}. \tag{192}$$

Next, for $x \in \mathbb{R} \setminus (-M, M)$, we combine (170), (181), and (183) to obtain

$$J_R(x) = \begin{pmatrix} 1 & 0 \\ \dfrac{-i(2x-\alpha_n-\beta_n)}{\beta_n-\alpha_n} & -2i \end{pmatrix} b_{n,-}(x)^{-\sigma_3} \begin{pmatrix} 1 & i \\ -1 & i \end{pmatrix} e^{-\tau\phi_{n,-}\sigma_3} \begin{pmatrix} 1 & 1 \\ 0 & 1 \end{pmatrix}$$
$$\times e^{\tau\phi_{n,+}\sigma_3} \begin{pmatrix} 1 & i \\ -1 & i \end{pmatrix}^{-1} b_{n,+}(x)^{\sigma_3} \begin{pmatrix} 1 & 0 \\ \dfrac{-i(2x-\alpha_n-\beta_n)}{\beta_n-\alpha_n} & -2i \end{pmatrix}^{-1}. \tag{193}$$

When $x \in [M,\infty) \subset (\beta_n,\infty)$, we have $\phi_{n,+}(x) = \phi_{n,-}(x)$ and $b_{n,+}(x) = b_{n,-}(x)$. Note that $\binom{1\ 1}{0\ 1} = I + \binom{0\ 1}{0\ 0}$. By straightforward computation, we can write (193) as

$$J_R(x) = I + \frac{1}{4}e^{-2\tau\phi_n(x)}L_n(x) \qquad \text{for } x \in [M,\infty), \tag{194}$$

where each entry of the $2\times 2$ matrix $L_n(x)$ is a linear combination of $b_n^{-2}$, $b_n^2$,

$$\frac{i(2x-\alpha_n-\beta_n)}{\beta_n-\alpha_n}, \quad \frac{i(2x-\alpha_n-\beta_n)}{\beta_n-\alpha_n}b_n^{-2}, \quad \frac{(2x-\alpha_n-\beta_n)^2}{(\beta_n-\alpha_n)^2}b_n^{-2}$$

and a constant which is independent of $n$. The coefficients in the linear combinations are also independent of $n$. We also need the behavior of $\phi_{n,+}(x)$ as $x \to \pm\infty$. Note that (22) gives the relationship between $\phi_n$, $g_n$ and $\log w_n$. From (138) with $m = 1$, we have

$$\begin{aligned}\log w_n(x) &= \log\Gamma(\tau A - i\tau x) + \log\Gamma(\tau A + i\tau x) - (\pi - 2\phi)\tau x,\\ &= \log(2\pi) - 2\tau A - (\pi-2\phi)\tau x + \left(\tau A - \frac{1}{2}\right)\log(\tau^2A^2+\tau^2x^2),\\ &\quad - i\tau x\log\left(\frac{A-ix}{A+ix}\right) + R_1(\tau A - i\tau x) + R_1(\tau A + i\tau x),\end{aligned} \tag{195}$$

where $R_1(\tau A - i\tau x) + R_1(\tau A + i\tau x)$ is real and satisfies

$$|R_1(\tau A - i\tau x) + R_1(\tau A + i\tau x)| \le \frac{|B_2|\sec^2\left(\frac{\pi}{4}\right)}{\tau\sqrt{A^2+x^2}} \tag{196}$$

by (139). Thus, inserting (195) into (22) leads to

$$\begin{aligned}\phi_n(x) &= \left(\frac{\pi}{2}-\phi\right)x - g_n(x) - \left(\frac{A}{2}-\frac{1}{4\tau}\right)[\log\tau^2 + \log(A^2+x^2)]\\ &\quad + A - \frac{1}{2\tau}\log(2\pi) + \frac{1}{2}ix\log\left(\frac{A-ix}{A+ix}\right) - \frac{1}{2\tau}(R_1(\tau A - i\tau x)\\ &\quad + R_1(\tau A + i\tau x)) + \frac{1}{2\tau}\log\frac{\beta_n-\alpha_n}{4} + \frac{\ell_n}{2}.\end{aligned} \tag{197}$$

Note that

$$\frac{1}{2}ix\log\left(\frac{A-ix}{A+ix}\right) = x\arg(A+ix) \sim \pm\frac{\pi}{2}x \quad \text{as } x \to \pm\infty, \tag{198}$$

and from (154)

$$\frac{\ell_n}{2} - \left(\frac{A}{2} - \frac{1}{4\tau}\right) \log \tau^2 = A \log \frac{2A+1}{\sin \phi} - A \log 2 - (A+1) + \log \frac{\sqrt{2A+1}}{2 \sin \phi} + O(\tau^{-1}) \tag{199}$$

as $n \to \infty$. It is also easy to see from (8) that $g_n(x)$ is real for $x > \beta_n$, $g_{n,+}(x) - i\pi$ takes real value for $x < \alpha_n$, and $g_n(z) \sim \log z$ for large $z$ (uniform in $n$ since both $\alpha_n, \beta_n$ and $\mu_n(x)$ are convergent uniformly as $n \to \infty$). Inserting (196), (198), and (199) into (197), and noting the properties of $g_n(z)$ given above, we have shown that there exist $X_1 > 0$ and $N_1 > 0$ such that when $n \geq N_1$, we have

$$\phi_n(x) \geq \frac{\pi - \phi}{2} x \quad \text{for } x \geq X_1; \quad \phi_{n,+}(x) + i\pi \geq \frac{-\phi}{2} x \quad \text{for } x \leq -X_1. \tag{200}$$

Since $b_n(z) = [(z-\alpha_n)(z-\beta_n)]^{\frac{1}{4}} / \sqrt{\beta_n - \alpha_n}$ and $\alpha_n, \beta_n$ satisfy (67), there exist $X_2 \geq X_1 > 0$ and $N_2 \geq N_1 > 0$ such that the matrix $L_n(x)$ in (194) satisfies

$$|L_n(x)| \leq C|x| \quad \text{for } n \geq N_2 \text{ and } |x| \geq X_2, \tag{201}$$

where $C$ is a positive constant independent of $n$. From (200), we immediately have $e^{-2\tau\phi_n(x)} \leq e^{-\tau(\pi-\phi)x}$ for $n \geq N_1$ and $x \geq X_1$. On account of this and (201), we can deduce that for any given $k \in \mathbb{N}$, there exists $X_3 \geq X_2 > 0$ such that for $n \geq N_2$,

$$\left|\tfrac{1}{4} e^{-2\tau\phi_n(x)} L_n(x)\right| \leq C e^{-\tau(\pi-\phi)x} x \leq C\tau^{-k} x^{-k}, \tag{202}$$

uniformly for $x \in [M, \infty)$ with $M \geq X_3$.

When $x \in (-\infty, -M] \subset (-\infty, \alpha_n]$, it is easy to see that $\phi_{n,+}(x) - \phi_{n,-}(x) = -2\pi i$ and $b_{n,-}(x) = e^{-\pi i} b_{n,+}(x)$. Again, from (193), it follows that

$$J_R(x) = I + \tfrac{1}{4} e^{-2\tau\phi_{n,+}(x)} L_n^*(x) \quad \text{for } x \in (-\infty, -M], \tag{203}$$

where $L_n^*(x)$ is defined in the same way as $L_n(x)$, except with $b_n$ replaced by $b_{n,+}$. Hence, $L_n^*(x)$ also satisfies an inequality similar to (201). By using the second inequality in (200), we have $|e^{-2\tau\phi_{n,+}(x)}| \leq e^{\tau\phi x}$ for $n \geq N_1$ and $x \leq -X_1$, where $-\phi x > 0$. Similarly, as in the case $x \in [M, \infty)$, we can deduce that for any given $k \in \mathbb{N}$, there exists $X_3 > 0$ such that for $n \geq N_2$,

$$\left|\tfrac{1}{4} e^{-2\tau\phi_{n,+}(x)} L_n^*(x)\right| \leq C e^{\tau\phi x} |x| \leq C\tau^{-k} |x|^{-k}, \tag{204}$$

uniformly for $x \in (-\infty, -M]$.

The two inequalities in (202) and (204) show that for sufficiently large $|x|$, the second term in (194) and (203) are small in comparison with identity matrix. Summing up the above analysis, we have derived the asymptotic expansion for the jump matrix $J_R(z)$:

$$J_R(z) \sim I + \sum_{k=1}^{\infty} \frac{J_k(z)}{\tau^k}, \qquad z \in \Sigma^*, \quad n \to \infty.$$

The coefficients $J_k(z)$ all satisfy $J_k(z) = O(1/z)$ as $z \to \infty$. By Theorem 3.8 in [33] concerning asymptotic solutions to Riemann-Hilbert problems, we have rigorously established that $R(z)$ has the uniform asymptotic expansion

$$R(z) \sim I + \sum_{k=1}^{\infty} \frac{R_k(z)}{\tau^k}, \qquad n \to \infty, \tag{205}$$

on $\mathbb{C} \setminus \Sigma^*$, where all $R_k(z)$ are analytic in $\mathbb{C} \setminus \Sigma^*$. Taking the (1, 1)-entry in the matrix $U(z)$, we obtain

THEOREM 1. *With the notations given above, we have*

$$\begin{aligned}\pi_n(\tau z) = \frac{1}{\sqrt{2}} e^{\frac{\tau}{2}(\ell_n+1)} \tau^{\frac{n}{2}} w_n(z)^{-\frac{1}{2}} [\, U(-\tau, 2\sqrt{\tau}\xi(z))\, A(z,n) \\ + U'(-\tau, 2\sqrt{\tau}\,\xi(z))\, B(z,n)\,]\end{aligned} \tag{206}$$

*for $z \in K$, where $A(z,n)$ and $B(z,n)$ are analytic functions of $z$, and as $n \to \infty$*

$$A(z,n) \sim \frac{(\xi^2-1)^{\frac{1}{4}}}{b_n(z)} \left[ 1 + \sum_{k=1}^{\infty} \frac{A_k(z)}{\tau^k} \right], \quad B(z,n) \sim \frac{b_n(z)}{(\xi^2-1)^{\frac{1}{4}}} \sum_{k=1}^{\infty} \frac{B_k(z)}{\tau^{k+\frac{1}{2}}} \tag{207}$$

*uniformly in $K$.* The coefficients $A_k(z)$ and $B_k(z)$ are all analytic functions in $K$.
*When $z \in \mathbb{C} \setminus K$, we also have the asymptotic expansion*

$$\pi_n(\tau z) \sim \frac{\tau^n}{\sqrt{\beta_n - \alpha_n}} e^{\tau g_n(z)} b_n(z)^{-1} \left[ 1 + \sum_{k=1}^{\infty} \frac{C_k(z)}{\tau^k} \right], \tag{208}$$

*which holds uniformly in $\mathbb{C} \setminus K$. The coefficients $C_k(z)$ are analytic functions in $\mathbb{C} \setminus K$. The behavior of $\pi_n(\tau z)$ on the boundary of $K$ can be obtained by taking the limit from either inside or outside of $K$.*

*Proof*: Let $U_{ij}(z)$ and $\tilde{U}_{ij}(z)$ denote $(i, j)$ entries in $U(z)$ and $\tilde{U}(z)$, respectively. By (182), $U(z) = e^{\frac{1}{2}\tau \ell_n \sigma_3} R(z) e^{-\frac{1}{2}\tau \ell_n \sigma_3} \tilde{U}(z)$. Noting (7) and taking the (1,1)-entry, we get

$$\pi_n(\tau z) = \tau^n U_{11}(z) = \tau^n (R_{11}(z)\tilde{U}_{11}(z) + R_{12}(z) e^{\tau \ell_n} \tilde{U}_{21}(z)). \tag{209}$$

By (205), we have

$$R_{11}(z) \sim 1 + \sum_{k=1}^{\infty} \frac{R_{11}^{(k)}(z)}{\tau^k}, \quad z \in \mathbb{C} \setminus \Sigma^*, \tag{210}$$

$$R_{12}(z) \sim \sum_{k=1}^{\infty} \frac{R_{12}^{(k)}(z)}{\tau^k}, \quad z \in \mathbb{C} \setminus \Sigma^*, \tag{211}$$

where $R_{ij}^{(k)}(z)$ are the $(i, j)$-entry of the matrix $R_k(z)$ in (205).

When $z \in K$, from the definition of $\tilde{U}(z)$ in (181), we have

$$\tilde{U}_{11}(z) = \frac{1}{\sqrt{2}} e^{\frac{\tau}{2}} \tau^{-\frac{n}{2}} e^{\frac{\tau}{2}\ell_n} w_n(z)^{-\frac{1}{2}} \cdot \frac{(\xi^2-1)^{\frac{1}{4}}}{b_n(z)} U(-\tau, 2\sqrt{\tau}\,\xi), \tag{212}$$

$$\tilde{U}_{21}(z) = \frac{1}{\sqrt{2}} e^{\frac{\tau}{2}} \tau^{-\frac{n}{2}} e^{-\frac{\tau}{2}\ell_n} w_n(z)^{-\frac{1}{2}} \Bigg( \frac{-i(2z-\alpha_n-\beta_n)}{(\beta_n-\alpha_n)} \cdot \frac{(\xi^2-1)^{\frac{1}{4}}}{b_n(z)} U(-\tau, 2\sqrt{\tau}\,\xi) + \frac{-2i}{\sqrt{\tau}} \cdot \frac{b_n(z)}{(\xi^2-1)^{\frac{1}{4}}} U'(-\tau, 2\sqrt{\tau}\,\xi) \Bigg). \tag{213}$$

Substituting (212) and (213) in (209), we easily obtain (206) with

$$A(z,n) = \frac{(\xi^2-1)^{\frac{1}{4}}}{b_n(z)} \left[ R_{11}(z) + \frac{-i(2z-\alpha_n-\beta_n)}{(\beta_n-\alpha_n)} R_{12}(z) \right], \tag{214}$$

$$B(z,n) = \frac{(-2i) b_n(z)}{(\xi^2-1)^{\frac{1}{4}} \sqrt{\tau}} R_{12}(z). \tag{215}$$

Here, $A(z, n)$ and $B(z, n)$ are analytic in $K$. Inserting (210) and (211) into (214) and (215), we get the asymptotic expansion given in (207).

When $z \notin K$, we have from (167), (170), and (181)

$$\tilde{U}(z) = e^{\frac{1}{2}\tau l_n \sigma_3} N(z) e^{(\tau g_n - \frac{1}{2}\log\frac{\varphi_n}{2} - \frac{\tau}{2} l_n)\sigma_3}.$$

From (166), it is easy to verify

$$\tilde{U}_{11}(z) = \frac{1}{\sqrt{\beta_n-\alpha_n}} b_n(z)^{-1} e^{\tau g_n(z)}, \tag{216}$$

$$\tilde{U}_{21}(z) = \left[ -i \frac{1}{\sqrt{\beta_n-\alpha_n}} b_n(z)^{-1} \left(\frac{\varphi_n}{2}\right)^{-1} \frac{\beta_n-\alpha_n}{4} \right] e^{\tau(g_n(z)-\ell_n)}. \tag{217}$$

Inserting (216) and (217) into (209) gives

$$\pi_n(\tau z) = \tau^n e^{\tau g_n(z)} [A(z,n) + B(z,n)] \tag{218}$$

with

$$A(z,n) = \frac{1}{\sqrt{\beta_n-\alpha_n}} b_n(z)^{-1} R_{11}(z),$$

$$B(z,n) = -i \frac{\sqrt{\beta_n-\alpha_n}}{4} b_n(z)^{-1} \left(\frac{\varphi_n}{2}\right)^{-1} R_{12}(z).$$

Again, by (210) and (211), we obtain the asymptotic expansion (208). ■

## Acknowledgment

The work of the first author was supported by National Natural Science Foundation of China (No. 11001057).

## References

1. P. Deift and X. Zhou, A steepest descent method for oscillatory Riemann-Hilbert problems, Asymptotics for the MKdV equation, *Ann. Math.* 137(2):295–368 (1993).
2. J. Meixner, Orthogonale Polynomsysteme mit einer besonderen Gestalt der erzeugenden Funktion, *J. London Math. Soc.* 9:6–13 (1934).
3. F. Polaczek, Sur une famille de polynomes orthogonaux qui continent les polynomes d'Hermite et de Laguerre comme cas limites, *C. R. Acad. Sci. Paris, Ser. I* 230:1563–1565 (1950).
4. T. S. Chihara, An Introduction to Orthogonal Polynomials, Gordon and Breach, Science Publisher, New York, 1978.
5. R. Koekoek and R. F. Swarttouw, The Askey-Scheme of Hypergeometric Orthogonal Polynomials and its q-Analogue, Delft University of Technology and Systems, Department of Technical Mathematics and Informatics, Report no. 98–17 (1998).
6. C. M. Bender, L. R. Mead, and S. S. Pinsky, Continuous Hahn polynomials and the Heisenberg algebra, *J. Math. Phys.* 28:509–513 (1987).
7. T. H. Kriecherbauer, Meixner-Pollaczek polynomials and the Heisenberg algebra, *J. Math. Phys.* 30:767–769 (1989).
8. H. T. Koelink and J. Van der Jeugt, Convolutions for orthogonal polynomials from Lie and quantum algbra representations, *SIAM J. Math. Anal.* 29:794–882 (1998).
9. V. Freilikher, E. Kanzieper, I. Yurkevich, Unitary random-matrix ensemble with governable level confinement, *Phys. Rev. E* 53:2200–2209 (1996).
10. D. R. Masson, Some continued fractions of Ramanujan and Meixner-Pollaczek polynomials, *Canad. Math. Bull.* 32:177–181 (1989).
11. Y. Chen and M. E. H. Ismail, Asymptotics of the extreme zeros of the Meixner-Pollaczek polynomials, *J. Comput. App. Math.* 82:59–78 (1997).
12. M. E. H. Ismail and X. Li, Bound on the extreme zeros of orthogonal polynomials, *Proc. Amer. Math. Soc.* 115:131–140 (1992).
13. I. V. Krasovsky, Asymptotic distribution of zeros of polynomials satisfying difference equations, *J. Comput. Appl. Math.* 150(1):57–70 (2003).
14. X. Li and R. Wong, On the asymptotics of the Mexiner-Pollaczek polynomials and their zeros, *Constr. Approx.* 17:59–90 (2001).
15. P. Deift, T. Kriecherbauer, K. T-R McLaughlin, S. Venakides, and X. Zhou, Uniform asymptotics for polynomials orthogonal with respect to varying exponential weights and applications to universality questions in random matrix theory, *Comm. Pure Appl. Math.* 52:1335–1425 (1999).
16. C. Bosbach and W. Gawronski, Strong asymptotics for Laguerre polynomials with varying weights, *J. Comput. Appl. Math.* 99:77–89 (1998).
17. A. B. J. Kuijlaars and K. McLaughlin, Riemann-Hilbert analysis for Laguerre polynomials with large negative parameters, *Comput. Meth. Funct. Theory* 1:205–233 (2001).
18. A. Aptekarev and R. Khabibullin, Asymptotic expansions for polynomials orthogonal with respect to a complex non-constant weight function, *Trans. Moscow Math. Soc.* 68:1–37 (2007).

19. C. Bosbach and W. Gawronski, Strong asymptotics for Jacobi polynomials with varying weights, *Methods Appl. Math.* 6:39–54 (1999).
20. A. B. J. Kuijlaars and A. Martinez-Finkelshtein, Strong asymptotics for Jacobi polynomials with varying nonstandard parameters, *J d'Analyse Math.* 94:195–234 (2004).
21. A. Martinez-Finkelshtein and R. Orive, Riemann-Hilbert analysis for Jacobi polynomials orthogonal on a single contour, *J. Approx. Theory* 134:137–170 (2005).
22. R. Wong and W. J. Zhang, Uniform asymptotics for Jacobi polynomials with varying large negative parameters – a Riemann-Hilbert approach, *Trans. Amer. Math. Soc.* 358:2663–2694 (2006).
23. J. Baik and T. M. Suidan, Random matrix central limit theorems for nonintersecting random walks, *Ann. Probab.* 35:1807–1834 (2007).
24. J. Wang, W. Y. Qiu and R. Wong, Uniform asymptotics for Meixner-Pollaczek polynomials with varying parameters, *C. R. Acad. Sci. Paris, Ser. I* 349:1031–1035 (2011).
25. P. Deift, T. Kriecherbauer, K. T-R McLaughlin, S. Venakides, and X. Zhou, Strong asymptotics of orthogonal polynomials with respect to exponential weights, *Comm. Pure Appl. Math.* 52:1491–1552 (1999).
26. W. Y. Qiu and R. Wong, Global asymptotic expansions of the Laguerre polynomials – a Riemann-Hilbert approach, *Numer. Algorithms* 49:331–372 (2008).
27. R. Wong and L. Zhang, Global asymptotics of Hermite polynomials via Riemann-Hilbert approach, *Discrete Contin. Dyn. Syst. B* 7(3):661–682 (2007).
28. A. S. Fokas, A. R. Its, and A. V. Kitaev, The isomonodromy approach to matrix models in 2D quantum gravity, *Comm. Math. Phys.* 147:395–430 (1992).
29. Z. Wang and R. Wong, Uniform asymptotics of the Stieltjes-Wigert polynomials via Riemann-Hilbert approach, *J. Math. Pures Appl.* 85:698–718 (2006).
30. F. W. J. Olver, *Asymptotics and Special Functions*, Academic Press, New York, 1974.
31. M. Abramowitz and I. A. Stegun, Handbook of Mathematical Functions with Formulas, Graphs, and Mathematical Tables, reprint of the 1972 edition, Dover Publications, New York, 1992.
32. F. W. J. Olver, Uniform asymptotic expansions for Weber parabolic cylinder functions of large orders, *J. Res. Nat. Bur. Standards Sect. B* 63B:131–169 (1959).
33. W. Y. Qiu and R. Wong, Asymptotic expansions for Riemann-Hilbert problems, *Anal. Appl.* 6:269–298 (2008).

Fudan University
Fudan University
City University of Hong Kong

(Received July 20, 2012)

# Asymptotics of the Discrete Chebyshev Polynomials

By J. H. Pan and R. Wong

The discrete Chebyshev polynomials $t_n(x, N)$ are orthogonal with respect to a distribution, which is a step function with jumps one unit at the points $x = 0, 1, \ldots, N-1$, $N$ being a fixed positive integer. By using a double integral representation, we have recently obtained asymptotic expansions for $t_n(aN, N+1)$ in the double scaling limit, namely, $N \to \infty$ and $n/N \to b$, where $b \in (0, 1)$ and $a \in (-\infty, \infty)$; see [8]. In this paper, we continue to investigate the behavior of these polynomials when the parameter $b$ approaches the endpoints of the interval $(0, 1)$. While the case $b \to 1$ is relatively simple (because it is very much like the case when $b$ is fixed), the case $b \to 0$ is quite complicated. The discussion of the latter case is divided into several subcases, depending on the quantities $n$, $x$, and $xN/n^2$, and different special functions have been used as approximants, including Airy, Bessel, and Kummer functions.

## 1. Introduction

The discrete Chebyshev polynomials $t_n(x, N)$ can be defined as a special case of the Hahn polynomials [1, p. 174]

$$
\begin{aligned}
Q_n(x; \alpha, \beta, N) &= {}_3F_2(-n, -x, n+\alpha+\beta+1; -N, \alpha+1; 1) \\
&= \sum_{k=0}^{n} \frac{(-n)_k(-x)_k(n+\alpha+\beta+1)_k}{(-N)_k(\alpha+1)_k k!}.
\end{aligned} \tag{1}
$$

Address for correspondence: R. Wong, Liu Bie Ju Centre for Mathematical Sciences, City University of Hong Kong, Tat Chee Avenue, Kowloon, Hong Kong; e-mail: mawong@cityu.edu.hk

DOI: 10.1111/sapm.12018
STUDIES IN APPLIED MATHEMATICS 132:13–49

With $\alpha = \beta = 0$ in (1), we have

$$\begin{aligned} t_n(x, N) &= (-1)^n (N-n)_n Q_n(x; 0, 0, N-1) \\ &= (-1)^n (N-n)_n \sum_{k=0}^{n} \frac{(-n)_k (-x)_k (n+1)_k}{(-N+1)_k k! k!}; \end{aligned} \tag{2}$$

see [1, p. 176].

In a recent paper [8], we have studied the asymptotic behavior of $t_n(x, N+1)$ as $n, N \to \infty$ in such a way that the ratios

$$a = x/N \quad \text{and} \quad b = n/N, \tag{3}$$

satisfy the inequalities

$$-\infty < a \le \frac{1}{2} \quad \text{and} \quad 0 < b < 1. \tag{4}$$

In view of the symmetry relation [8, eq. (8)]

$$t_n(x, N+1) = (-1)^n t_n(N-x, N+1), \tag{5}$$

our study actually covers the entire real-axis $-\infty < x < \infty$ or, equivalently, the entire parameter range $-\infty < a < \infty$. For a discussion of the asymptotic behavior of the Hahn polynomials, see [4].

In this paper, we shall investigate the behavior of the polynomials $t_n(x, N+1)$, when the parameter $b$ given in (3) either tends to 0 or tends to 1. The case $b \to 1$ turns out to be rather simple, because the ultimate expansions and their derivations are similar to those in the case when $b$ is fixed; see (27) for $0 \le a \le \frac{1}{2}$ and (34) for $a < 0$.

To derive asymptotic approximation for $t_n(x, N+1)$ when $b \to 0$, we divide our discussion into several cases, depending on the quantities $n$, $x$, and $xN/n^2$. A summary of our findings is given in Table 1 below.

Some explanation is needed for this table. By "small", "fixed", and "large", we mean, respectively, $x = o(1)$, $x \in [\delta, M]$ and $x \gg O(1)$ as $N \to \infty$, where $\delta$ is a small positive number and $M$ is a large positive number. By "Airy", we mean an asymptotic expansion whose associated approximants are the Airy

**Table 1**

Asymptotic Expansions in Different Cases when $\frac{n}{N} \to 0$ and $N \to \infty$

| $x$ \ $xN/n^2$ | Small | Fixed | Large |
|---|---|---|---|
| Large | Kummer | Airy | Bessel |
| Fixed | Kummer | Airy | Series(2) |
| Small | Kummer/series(2) | Airy | Series(2) |

functions Ai($z$) and Bi($z$). In the same manner, by "Bessel" and "Kummer", we mean asymptotic expansions whose associated approximants are, respectively, the Bessel function $J_\nu(z)$ and the Kummer function $M(a, c, z)$. The Kummer function used in this paper is defined by

$$\mathbf{M}(a, c, z) = \frac{1}{\Gamma(c)} M(a, c, z) = \frac{1}{\Gamma(c)} \sum_{s=0}^{\infty} \frac{(a)_s}{(c)_s} \frac{z^s}{s!}, \tag{6}$$

which was introduced in [6, p. 255]. The case involving Bessel functions covers an earlier result of Sharapudinov [10], where, instead of Bessel functions, Jacobi polynomials are used as an approximant. In the case when $xN/n^2$ is large and $x$ is either fixed or small, the series in (2) is itself an asymptotic expansion (in the generalized sense [11, p. 10]) as $N \to \infty$. Hence, there is no need to seek for another asymptotic representation.

The arrangement of the present paper is as follows. In Section 2, we recall the major results in [8]; to facilitate our presentation for later sections, we also include here brief derivations of the expansions given in [8]. In Section 3, we consider the case when $xN/n^2$ is small and show that the expansions are of Kummer-type. In Section 4, we consider the case when the quantity $xN/n^2$ is fixed; that is, when $xN/n^2$ is bounded away from zero and infinity. There are three subcases, depending on $x$ being large, fixed or small. In all three subcases, we will show that the expansions are of Airy-type; see Table 1. The case when both quantities $x$ and $xN/n^2$ are large is dealt with in Section 5, where we show that the expansion of $t_n(x, N+1)$ can be expressed in terms of Bessel functions. The final section is devoted to two remaining cases, namely, (i) when $xN/n^2$ is large and $x$ is either fixed or small, and (ii) $x < 0$.

## 2. Results in Ref. [8]

In [8], we have divided our discussion into two cases: (i) $0 \le a \le \frac{1}{2}$, and (ii) $a < 0$. As mentioned in Section 1, by virtue of the symmetry relation (5), these two cases cover the entire range $-\infty < a < \infty$.

In case (i), we started with the integral representation

$$t_n(x, N+1) = \frac{(-1)^n}{2\pi i} \frac{\Gamma(n+N+2)}{\Gamma(n+1)\Gamma(N-n+1)} \int_0^1 \int_{\gamma_1} \frac{1}{w-1} e^{Nf(t,w)} \mathrm{d}w \mathrm{d}t, \tag{7}$$

where

$$f(t, w) = b\ln(1-t) + (1-b)\ln t + a\ln w - a\ln(w-1) + b\ln[1-(1-t)w], \tag{8}$$

and the curve $\gamma_1$ starts at $w = 0$, runs along the lower edge of the positive real line toward $w = 1$, encircles the point $w = 1$ in the counterclockwise direction and returns to the origin along the upper edge of the positive real line. As a function of two variables, the partial derivatives $\partial f/\partial t$ and $\partial f/\partial w$ vanish at $(t, w) = (t_\pm, w_\pm)$, where

$$t_+ = \frac{2 - 2a - b^2 + b\sqrt{b^2 - 4a + 4a^2}}{2(1-a)(1+b)}, \qquad w_+ = \frac{b + \sqrt{b^2 - 4a + 4a^2}}{2b}, \tag{9}$$

and

$$t_- = \frac{2 - 2a - b^2 - b\sqrt{b^2 - 4a + 4a^2}}{2(1-a)(1+b)}, \qquad w_- = \frac{b - \sqrt{b^2 - 4a + 4a^2}}{2b}. \tag{10}$$

For each fixed $w \in \gamma_1$, we now find a steepest descent path of the phase function $f(t, w)$ in the variable $t$, which passes through a saddle point $t_0(w)$ depending on $w$. (Note that not only the saddle point $t_0(w)$, but all points $t$ on this steepest descent path depend on $w$.) The function $f(t_0(w), w)$ is a function of $w$ alone. To find the relevant saddle point $t_0(w)$, we solve the equation $\partial f(t, w)/\partial t = 0$, and obtain

$$t_0(w) = \frac{2w - 1 + \sqrt{1 + 4b^2(w-1)w}}{2(1+b)w}. \tag{11}$$

It can be shown that

$$t_0(w_+) = t_+, \quad t_0(w_-) = t_-. \tag{12}$$

Define the standard transformation $t \to \tau$ by

$$\tau^2 = f(t_0(w), w) - f(t, w); \tag{13}$$

see [11, p. 88]. Note that we have $\tau = -\infty$ when $t = 0$, and $\tau = +\infty$ when $t = 1$. Furthermore, this mapping takes $t = t_0(w)$ to $\tau = 0$. Coupling (7) and (13) gives

$$\begin{aligned} t_n(x, N+1) = &\frac{(-1)^n}{2\pi i} \frac{\Gamma(n+N+2)}{\Gamma(n+1)\Gamma(N-n+1)} \\ &\times \int_{-\infty}^{+\infty} \int_{\gamma_1} \frac{1}{w-1} e^{Nf(t_0(w),w)} e^{-N\tau^2} \frac{dt}{d\tau} dw d\tau, \end{aligned} \tag{14}$$

where $\gamma_1$ is the integration path of the variable $w$ in (7). Note that the first variable of $f$ in (14) is now $t_0(w)$, instead of $t$. Hence, the phase function is a function of $w$ alone. Setting $\partial f(t_0(w), w)/\partial w = 0$, we obtain the saddle points

$$w_\pm = \frac{b \pm \sqrt{b^2 - 4a + 4a^2}}{2b}; \tag{15}$$

cf. (9) and (10). Motivated by (8), define the new phase function

$$\psi(u) = a \ln u - a \ln(u-1) + \eta u. \tag{16}$$

The saddle points of $\psi$ are given by

$$u_{\pm} = \frac{\eta \pm \sqrt{\eta^2 + 4a\eta}}{2\eta}. \tag{17}$$

To reduce the double integral in (7) into a canonical form, we define the second mapping $w \to u$ by

$$\begin{aligned} f(t_0(w), w) &= \psi(u) + \gamma \\ &= a \ln u - a \ln(u-1) + \eta u + \gamma \end{aligned} \tag{18}$$

with

$$u(w_+) = u_-, \quad u(w_-) = u_+, \tag{19}$$

where $\eta$ and $\gamma$ are real numbers depending on the parameters $a$ and $b$ in (8). From (18), we have

$$\begin{aligned} \frac{\mathrm{d}w}{\mathrm{d}u} &= \frac{\mathrm{d}\psi}{\mathrm{d}u} \Big/ \frac{\mathrm{d}f(t_0(w), w)}{\mathrm{d}w} \\ &= \frac{\eta(u-u_+)(u-u_-)w(1-w)\left[(2a-1) - \sqrt{1+4b^2w^2-4b^2w}\,\right]}{-2b^2u(u-1)(w-w_+)(w-w_-)} \end{aligned} \tag{20}$$

for $u \neq u_{\pm}$, and by L'Hôpital's rule,

$$\frac{\mathrm{d}w}{\mathrm{d}u} = \left[\frac{\sqrt{\eta^2+4a\eta}(1-2a)(1-a)(-\eta)}{b^3\sqrt{b^2-4a+4a^2}}\right]^{1/2} \tag{21}$$

for $u = u_{\pm}$. With the change of variable $w \to u$ defined in (18), the representation in (14) becomes

$$\begin{aligned} t_n(x, N+1) = \frac{(-1)^n}{2\pi i} \frac{\Gamma(n+N+2)}{\Gamma(n+1)\Gamma(N-n+1)} e^{N\gamma} \\ \times \int_{-\infty}^{+\infty} \int_{\gamma_u} \frac{h(u,\tau)}{u-1} e^{N\psi(u) - N\tau^2} \mathrm{d}u \mathrm{d}\tau, \end{aligned} \tag{22}$$

where

$$h(u, \tau) = \frac{u-1}{w-1} \frac{\mathrm{d}w}{\mathrm{d}u} \frac{\mathrm{d}t}{\mathrm{d}\tau}, \tag{23}$$

and $\gamma_u$ is the steepest descent path of $\psi(u)$ in the $u$-plane. An asymptotic expansion, holding uniformly for $a \in [0, \frac{1}{2}]$, was then derived in [8] by an integration-by-parts technique. To state the result, we define recursively $h_0(u, \tau) = h(u, \tau)$,

$$h_l(u, \tau) = a_l(\tau) + b_l(\tau)u - (u-u_-)(u-u_+)g_l(u, \tau) \tag{24}$$

and

$$h_{l+1}(u,\tau)=(u-1)\left[g_l(u,\tau)+u\frac{\partial g_l(u,\tau)}{\partial u}\right],\quad l=0,1,2,\ldots. \tag{25}$$

Furthermore, we write

$$a_l(\tau)=\sum_{j=0}^{\infty}a_{l,j}\tau^j,\qquad b_l(\tau)=\sum_{j=0}^{\infty}b_{l,j}\tau^j. \tag{26}$$

The resulting expansion takes the form

$$\begin{aligned}t_n(x,N+1)\sim{}&\frac{(-1)^n\Gamma(n+N+2)e^{N\gamma}}{\Gamma(n+1)\Gamma(N-n+1)\sqrt{N}}\\&\times\left\{\mathbf{M}(aN+1,1,\eta N)\sum_{l=0}^{\infty}\frac{c_l}{(\eta N)^l}+\mathbf{M}'(aN+1,1,\eta N)\sum_{l=0}^{\infty}\frac{d_l}{(\eta N)^l}\right\},\end{aligned} \tag{27}$$

where $\mathbf{M}(a,c,z)$ is the Kummer function defined in (6), and the coefficients $c_l$ and $d_l$ are given explicitly by

$$c_l=\sum_{m=0}^{l}a_{l-m,2m}\Gamma\left(m+\frac{1}{2}\right)\eta^m,\quad d_l=\sum_{m=0}^{l}b_{l-m,2m}\Gamma\left(m+\frac{1}{2}\right)\eta^m. \tag{28}$$

Case (ii) was dealt with in a similar manner. We started with the double-integral representation

$$t_n(x,N+1)=\frac{(-1)^n}{2\pi i}\frac{\Gamma(n+N+2)}{\Gamma(n+1)\Gamma(N-n+1)}\int_0^1\int_{\gamma_2}\frac{1}{w-1}e^{N\tilde{f}(t,w)}\mathrm{d}w\mathrm{d}t, \tag{29}$$

where

$$\tilde{f}(t,w)=b\ln(1-t)+(1-b)\ln t+a\ln(-w)-a\ln(1-w)+b\ln[1-(1-t)w], \tag{30}$$

and the integration path $\gamma_2$ starts at $w=1$, traverses along the upper edge of the positive real line toward $w=0$, encircles the origin in the counterclockwise direction and returns to $w=1$ along the lower edge of the positive real line.

Now we follow the same argument as given in Case (i), and use the same mapping $t\to\tau$ defined in (13), except with $f(t,w)$ replaced by $\tilde{f}(t,w)$. The result is

$$\begin{aligned}t_n(x,N+1)={}&\frac{(-1)^n}{2\pi i}\frac{\Gamma(n+N+2)}{\Gamma(n+1)\Gamma(N-n+1)}\\&\times\int_{-\infty}^{+\infty}\int_{\gamma_2}\frac{1}{w-1}e^{N\tilde{f}(t_0(w),w)}e^{-N\tau^2}\frac{\mathrm{d}t}{\mathrm{d}\tau}\mathrm{d}w\mathrm{d}\tau.\end{aligned} \tag{31}$$

Because of the shape of the contour $\gamma_2$, it turns out that the phase function $\widetilde{f}(t_0(w), w)$ has only one relevant saddle point, namely $w_-$; see (15). As a consequence, we define the mapping $w \to u$ by

$$\widetilde{f}(t_0(w), w) = a\ln(-u) - u + \gamma \tag{32}$$

with

$$u(w_-) = a, \tag{33}$$

where $\gamma$ is a constant depending on the parameters $a$ and $b$ in (3). The final expansion is in terms of the gamma function, and we have

$$t_n(x, N+1) \sim \frac{(-1)^n\Gamma(n+N+2)N^{-aN}e^{N\gamma}}{\Gamma(n+1)\Gamma(N-n+1)\Gamma(-aN+1)}\sum_{l=0}^{\infty}\frac{c_l}{N^{l+\frac{1}{2}}}, \tag{34}$$

where $c_l$ are constants that can be given recursively.

## 3. Kummer-type expansion

As $b \to 0$, some of the steps in Section 2 are no longer valid. Let us first examine the mapping $t \to \tau$ given in (13). In this mapping, we have used the fact that for each fixed $w \in \gamma_1$, the saddle point $t = t_0(w)$ in (11) is bounded away from $t = 0$ and $t = 1$, and the steepest descent path from $t = 0$ to $t = 1$, passing through $t_0(w)$, is mapped onto $(-\infty, \infty)$ in the $\tau$-plane. However, from (11), we have, for any fixed $w$,

$$t_0(w) = 1 - b + O\left(b^2\right) \quad \text{as } b \to 0; \tag{35}$$

that is, as $b \to 0$, $t_0(w)$ approaches the branch point $t = 1$ in the $t$-plane. Thus, the mapping (13) is no longer suitable in this case. Because we are more interested in the neighborhood of $t = 1$, where the term $b\ln(1-t)$ in the phase function (8) becomes singular, we introduce the mapping

$$f(t, w) = b\ln\tau - \tau + A, \tag{36}$$

where the constant $A$ does not depend on $t$ or $\tau$ (but may depend on $w$); see (38) below. To make the mapping $t \to \tau$ defined in (36) one-to-one and analytic, we prescribe $t = t_0(w)$ to correspond to $\tau = b$, which is the saddle point of $b\ln\tau - \tau + A$; that is,

$$\tau(t_0(w)) = b. \tag{37}$$

This gives

$$A = f(t_0(w), w) - b\ln b + b. \tag{38}$$

Note that we have $\tau = 0$ when $t = 1$, and $\tau = +\infty$ when $t = 0$. Furthermore, from (36), we have

$$\frac{dt}{d\tau} = \frac{b/\tau - 1}{\partial f(t,w)/\partial t} = \frac{(\tau - b)t(1-t)[1-(1-t)w]}{\tau(1+b)w(t - t_0^+(w))(t - t_0^-(w))}, \qquad \tau \neq b,$$
$$\frac{dt}{d\tau} = \left\{ \frac{t_0^+(w)(1 - t_0^+(w))[1-(1-t_0^+(w))w]}{b\sqrt{1+4b^2(w-1)w}} \right\}^{1/2}, \qquad \tau = b, \tag{39}$$

where we have used L'Hôpital's rule for $\tau = b$ and

$$t_0^{\pm}(w) = \frac{2w - 1 \pm \sqrt{1+4b^2(w-1)w}}{2(1+b)w};$$

cf. (11). Coupling (7) and (36), we have, instead of (14),

$$t_n(x, N+1) = \frac{(-1)^{n+1}}{2\pi i} \frac{\Gamma(n+N+2)e^n b^{-n}}{\Gamma(n+1)\Gamma(N-n+1)} \times \int_0^{+\infty} \int_{\gamma_1} \frac{1}{w-1} e^{Nf(t_0(w),w)} e^{N(b \ln \tau - \tau)} \frac{dt}{d\tau} dw d\tau. \tag{40}$$

Next, let us examine the mapping (18). We note that this mapping still works in the present case; the only difference is that for fixed $n/N \in (0,1)$, the constant $\eta$ in the mapping (18) is bounded away from 0, whereas when $n/N \to 0$, $\eta$ approaches 0. More precisely, we have $\eta \sim -b^2$. This can roughly be seen from the saddle points (15) and (17), together with their correspondence relation (19) under the mapping $w \to u$ given in (18). To prove this rigorously, we give the following lemma.

LEMMA 1. *Let $\eta$ be the constant defined in the mapping* (18). *If $a/b^2 = o(1)$ (or, equivalently, $xN/n^2 = o(1)$), then we have*

$$\eta = -b^2 + o(b^2) \quad \text{as } b \to 0. \tag{41}$$

*Proof*: From (11), we have, for any fixed $w$,

$$t_0(w) = \frac{1}{1+b} - \frac{b^2(1-w)}{1+b} - \frac{b^4 w(w-1)^2}{1+b} + O(b^6 w^2 (w-1)^3) \quad \text{as } b \to 0, \tag{42}$$

where the term $O(b^4 w(w-1)^2)$ holds uniformly when either $w$ or $1-w$ is small. If $a/b^2 \to 0$, from (15) we have $w_- = 1 - w_+ \to 0^+$ and both $w_\pm$ bounded. Thus,

$$t_0(w_-) = 1 - b + b^2(1-b)w_- + O(b^4 w_-^2),$$
$$t_0(w_+) = \frac{1}{1+b} - \frac{b^2}{1+b} w_- + O(b^4 w_-^2). \tag{43}$$

To obtain $\eta$, we use (18) and (19). First, substituting (43) in (8) gives

$$f(t_0(w_+), w_+) - f(t_0(w_-), w_-) = 2a \ln \frac{1-w_-}{w_-} - b^2 + O(b^3). \tag{44}$$

Here, we have made use of the fact that if $w_\pm$ lie on the upper edge of the cut along the interval $0 < w < 1$, then we have $\ln(w_\pm - 1) = \ln(1 - w_\pm) + \pi i$. Similarly, if $w_\pm$ lie on the lower edge of the cut, then $\ln(w_\pm - 1) = \ln(1 - w_\pm) - \pi i$. Next, we have from (16) and (17)

$$\psi(u_-) - \psi(u_+) = 2a \ln \frac{1-u_+}{u_+} + \eta\sqrt{1+\frac{4a}{\eta}}. \tag{45}$$

Note that $u_+ + u_- = 1$ and $\eta < 0$. Formula (41) now follows from a combination of (18), (19), (44), and (45). ■

What Lemma 1 says is that if $a/b^2 \to 0$ and $x = O(1)$ or $x \gg O(1)$, then we have $|\eta|N \sim b^2 N = xb^2/a \to \infty$, that is, $\eta N$ is large.

Coupling (40) and (18), we have

$$\begin{aligned} t_n(x, N+1) = {} & \frac{(-1)^{n+1}\Gamma(n+N+2)e^n b^{-n} e^{N\gamma}}{2\pi i \Gamma(n+1)\Gamma(N-n+1)} \\ & \times \int_0^{+\infty} \int_{\gamma_1} \frac{h(u,\tau)}{u-1} e^{N\psi(u)} e^{N(b\ln\tau - \tau)} \mathrm{d}w \mathrm{d}\tau, \end{aligned} \tag{46}$$

where

$$h(u,\tau) = \frac{u-1}{w-1}\frac{\mathrm{d}w}{\mathrm{d}u}\frac{\mathrm{d}t}{\mathrm{d}\tau}, \tag{47}$$

$\mathrm{d}w/\mathrm{d}u$ is given by (20) and (21) and $\mathrm{d}t/\mathrm{d}\tau$ is given by (39). Following the same integration-by-parts procedure outlined in Section 2, we let $h_0(u,\tau) = h(u,\tau)$ in (47), and define $h_l(u,\tau)$ recursively by (24), (25), and (26). The final result is

$$\begin{aligned} t_n(x, N+1) = {} & \frac{(-1)^{n+1}\Gamma(n+N+2)e^n b^{-n} e^{N\gamma}}{\Gamma(N-n+1)N^{n+1}} \\ & \times \left[ \mathbf{M}(aN+1, 1, \eta N) \sum_{l=0}^{p-1} \frac{c_l}{(\eta N)^l} + \mathbf{M}'(aN+1, 1, \eta N) \sum_{l=0}^{p-1} \frac{d_l}{(\eta N)^l} + \varepsilon_p \right], \end{aligned} \tag{48}$$

where

$$c_l = \frac{N^{n+1}}{\Gamma(n+1)} \int_0^\infty a_l(\tau)\tau^n e^{-N\tau} \mathrm{d}\tau \sim \sum_{m=0}^\infty a_{l,m} \frac{\Gamma(n+m+1)}{\Gamma(n+1)N^m}, \tag{49}$$

$$d_l = \frac{N^{n+1}}{\Gamma(n+1)} \int_0^\infty b_l(\tau)\tau^n e^{-N\tau} \mathrm{d}\tau \sim \sum_{m=0}^\infty b_{l,m} \frac{\Gamma(n+m+1)}{\Gamma(n+1)N^m}, \tag{50}$$

and

$$\varepsilon_p = \frac{N^{n+1}}{2\pi i\Gamma(n+1)(\eta N)^p}\int_0^{+\infty}\int_{\gamma_1}\frac{h_p(u,\tau)}{u-1}e^{N\psi(u)}e^{N(b\ln\tau-\tau)}\mathrm{d}w\mathrm{d}\tau. \tag{51}$$

Note that because $b = n/N \to 0$, the series in (49) and (50) are asymptotic. Moreover, because $\eta N$ is large in this case, (48) is indeed a compound asymptotic expansion as $N \to \infty$.

Now, we consider the subcase in which $a/b^2 \to 0$ and $x = o(1)$. In this case, the expansion in (48) is still asymptotic as long as $|\eta|N \sim b^2 N \to \infty$ or, equivalently, $n \gg O(\sqrt{N})$. If $n = O(\sqrt{N})$, then the series in (2) is itself an asymptotic expansion as $N \to \infty$. This completes our discussion of all three cases listed under the condition "$xN/n^2$ small" in Table 1.

## 4. Airy-type expansion

For the case $a/b^2 \in [\gamma, M]$, where $\gamma$ is a small positive number and $M$ is a large positive number, the saddle points $w_\pm$ in (15) are bounded away from the singularities $w = 0$, $w = 1$, and $w = \infty$, and coalesce with each other when $a/b^2 \to \frac{1}{4(1-a)}$. Therefore, to derive an asymptotic expansion uniformly for $a/b^2 \in [\gamma, M]$, we need the cubic transformation (see (57) below) introduced by Chester, Friedman, and Ursell [2]. The resulting expansion is in terms of the Airy function $\mathrm{Ai}(\cdot)$.

Following the same argument as in Section 3, we again use the mapping from $t \to \tau$ in (36), and start with the integral representation (40)

$$\begin{aligned} t_n(x, N+1) = {} & \frac{(-1)^{n+1}}{2\pi i}\frac{\Gamma(n+N+2)e^n b^{-n}}{\Gamma(n+1)\Gamma(N-n+1)} \\ & \times \int_0^{+\infty}\int_{\gamma_1}\frac{1}{w-1}e^{Nf(t_0(w),w)}e^{N(b\ln\tau-\tau)}\frac{\mathrm{d}t}{\mathrm{d}\tau}\mathrm{d}w\mathrm{d}\tau, \end{aligned} \tag{52}$$

where the integration path is described in the line below (8). To proceed further, we divide $\gamma_1$ into two parts, and denote the part in the upper half of the plane by $\gamma^{(+)}$, and the other part in the lower half of the plane by $\gamma^{(-)}$. Recall from (8) that the phase function in (52) is given by

$$\begin{aligned} f(t_0(w), w) = {} & b\ln(1-t_0(w)) + (1-b)\ln t_0(w) + a\ln w \\ & - a\ln(w-1) + b\ln[1-(1-t_0(w))w], \end{aligned} \tag{53}$$

which has a cut $(-\infty, 1]$ in $w$-plane. Put

$$\begin{aligned} \overline{f}(t_0(w), w) = {} & b\ln(1-t_0(w)) + (1-b)\ln t_0(w) + a\ln w \\ & - a\ln(1-w) + b\ln[1-(1-t_0(w))w], \end{aligned} \tag{54}$$

which has cuts $(-\infty, 0]$ and $[1, \infty)$. From (53) and (54), we have

$$f(t_0(w), w) = \overline{f}(t_0(w), w) + a\pi i \tag{55}$$

for $w \in \gamma^{(+)}$, and

$$f(t_0(w), w) = \overline{f}(t_0(w), w) - a\pi i \tag{56}$$

for $w \in \gamma^{(-)}$. We now make the standard transformation

$$\overline{f}(t_0(w), w) = \frac{1}{3}u^3 - \zeta u + A, \tag{57}$$

with the correspondence between the critical points of the two sides prescribed by

$$w_+ \leftrightarrow \sqrt{\zeta}, \quad w_- \leftrightarrow -\sqrt{\zeta}. \tag{58}$$

If $\zeta > 0$, then $\pm\sqrt{\zeta}$ are both real; if $\zeta < 0$, then $\pm\sqrt{\zeta}$ are complex conjugates and purely imaginary. The values of $A$ and $\zeta$ can be obtained by using (57) and (58). We also have

$$\begin{aligned} \frac{dw}{du} &= \\ &\frac{(u-\sqrt{\zeta})(u+\sqrt{\zeta})w(1-w)\left[(2a-1)-\sqrt{1+4b^2w^2-4b^2w}\,\right]}{-2b^2(w-w_+)(w-w_-)}, \quad u \neq \pm\sqrt{\zeta}, \\ \frac{dw}{du} &= \left\{\frac{2\sqrt{\zeta}a(1-2a)(1-a)}{b^3\sqrt{b^2-4a+4a^2}}\right\}^{1/2}, \qquad u = \pm\sqrt{\zeta}, \end{aligned} \tag{59}$$

where we have again used L'Hôpital's rule for $u = \pm\sqrt{\zeta}$. Let us first consider the case $\zeta < 0$, and deform the image of the contour $\gamma^{(+)}$ under the mapping $w \to u$ defined in (57) to the steepest descent path of $\frac{1}{3}u^3 - \zeta u + A$ in the $u$-plane which passes through $\sqrt{\zeta}$. We denote the path by $C_2$. Similarly, we deform the image of the contour $\gamma^{(-)}$, and denote the steepest descent path passing through $-\sqrt{\zeta}$ by $C_3$; see Figure 1. Next, we consider the case $\zeta > 0$. Note that here the saddle points $\pm\sqrt{\zeta}$ are real, and that the contours $C_2$ and $C_3$ pass through both of them; see Figure 2. Clearly, in both cases, $C_3$ is the reflection of $C_2$ with respect to the real axis in the $u$-plane.

Let $C_1$ denote the dotted curve shown in Figures 1 and 2, and recall the identities $e^{iaN\pi} = \cos aN\pi + i\sin aN\pi$ and $e^{-iaN\pi} = \cos aN\pi - i\sin aN\pi$.

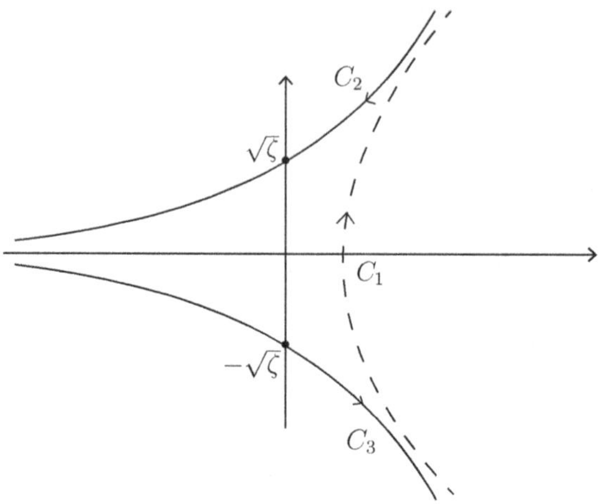


**Figure 1.** $C_2$ and $C_3$ ($\zeta < 0$).

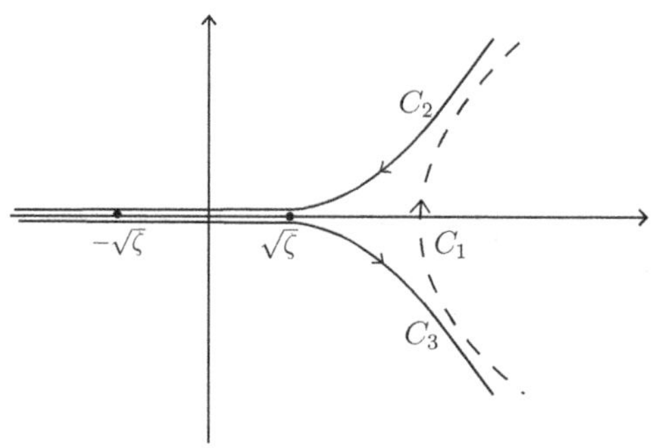


**Figure 2.** $C_2$ and $C_3$ ($\zeta > 0$).

A combination of (52), (55), (56), and (57) then gives

$$t_n(x, N+1) = \frac{(-1)^n \Gamma(n+N+2) e^n b^{-n} e^{NA}}{\Gamma(n+1)\Gamma(N-n+1)} \times \left\{ \frac{\cos aN\pi}{2\pi i} \int_0^{+\infty} \int_{C_1} h_0(u,\tau) e^{N\left(\frac{1}{3}u^3 - \zeta u\right)} e^{N(b\ln\tau - \tau)} \mathrm{d}u\mathrm{d}\tau + \frac{\sin aN\pi}{2\pi} \int_0^{+\infty} \int_{C_3 - C_2} h_0(u,\tau) e^{N\left(\frac{1}{3}u^3 - \zeta u\right)} e^{N(b\ln\tau - \tau)} \mathrm{d}u\mathrm{d}\tau \right\}, \tag{60}$$

where

$$h_0(u,\tau) = \frac{1}{w-1} \frac{\mathrm{d}t}{\mathrm{d}\tau} \frac{\mathrm{d}w}{\mathrm{d}u}, \tag{61}$$

$\mathrm{d}t/\mathrm{d}\tau$, $\mathrm{d}w/\mathrm{d}u$ are given, respectively, by (39) and (59).

### *4.1. When $\zeta$ is bounded*

Following the standard integration-by-parts procedure [11, p. 368], we define recursively

$$h_l(u,\tau) = a_l(\tau) + b_l(\tau)u - (u^2 - \zeta)g_l(u,\tau), \tag{62}$$

$$h_{l+1}(u,\tau) = \frac{\partial g_l(u,\tau)}{\partial u}. \tag{63}$$

Furthermore, expand

$$a_l(\tau) = \sum_{j=0}^{\infty} a_{l,j}\tau^j, \qquad b_l(\tau) = \sum_{j=0}^{\infty} b_{l,j}\tau^j. \tag{64}$$

From (60), (62), (63), and (64), it follows

$$\begin{aligned} t_n(x, N+1) = {} & \frac{(-1)^n\Gamma(n+N+2)e^n b^{-n} e^{NA}}{\Gamma(N-n+1)N^{n+1}} \\ & \times \left[\cos aN\pi \left(\mathrm{Ai}(N^{2/3}\zeta)\sum_{l=0}^{p-1}\frac{c_l}{N^{l+1/3}} - \mathrm{Ai}'(N^{2/3}\zeta)\sum_{l=0}^{p-1}\frac{d_l}{N^{l+2/3}}\right)\right. \\ & \left. + \sin aN\pi \left(\mathrm{Bi}(N^{2/3}\zeta)\sum_{l=0}^{p-1}\frac{c_l}{N^{l+1/3}} - \mathrm{Bi}'(N^{2/3}\zeta)\sum_{l=0}^{p-1}\frac{d_l}{N^{l+2/3}}\right) + \varepsilon_p\right], \end{aligned} \tag{65}$$

where

$$\begin{aligned} \varepsilon_p = {} & \frac{\cos(aN\pi)N^{n+1}}{2\pi i N^p\Gamma(n+1)}\int_0^{+\infty}\int_{C_1} h_p(u,\tau)e^{N\left(\frac{1}{3}u^3-\zeta u\right)}e^{N(b\ln\tau-\tau)}\mathrm{d}u\mathrm{d}\tau \\ & + \frac{\sin(aN\pi)N^{n+1}}{2\pi i N^p\Gamma(n+1)}\int_0^{+\infty}\int_{C_3-C_2} h_p(u,\tau)e^{N\left(\frac{1}{3}u^3-\zeta u\right)}e^{N(b\ln\tau-\tau)}\mathrm{d}u\mathrm{d}\tau, \end{aligned} \tag{66}$$

and the coefficients are given by

$$\begin{aligned} c_l &= \frac{N^{n+1}}{\Gamma(n+1)}\int_0^{\infty} a_l(\tau)\tau^n e^{-N\tau}\mathrm{d}\tau \sim \sum_{m=0}^{\infty} a_{l,m}\frac{\Gamma(n+m+1)}{\Gamma(n+1)N^m}, \\ d_l &= \frac{N^{n+1}}{\Gamma(n+1)}\int_0^{\infty} b_l(\tau)\tau^n e^{-N\tau}\mathrm{d}\tau \sim \sum_{m=0}^{\infty} b_{l,m}\frac{\Gamma(n+m+1)}{\Gamma(n+1)N^m}. \end{aligned} \tag{67}$$

In the present case, $a/b^2 \in [\delta, M]$ or, equivalently, $w_\pm$ are bounded. Thus, $\zeta$ is also bounded in view of (58). Therefore, it is easy to prove that (65) is asymptotic; for details, see [11, p. 371-372].

### 4.2. When $\zeta \to -\infty$

When $a/b^2 \to \infty$, by (15) the saddle points $w_\pm$ approach $\frac{1}{2} \pm i\infty$, respectively, along the line Re $w = \frac{1}{2}$. In view of (58), this is equivalent to saying that $\zeta \to -\infty$. To prove that (65) is also asymptotic in this case, we rewrite the expansion in (65) as

$$
\begin{aligned}
t_n(x, N+1) = {} & \frac{(-1)^n \Gamma(n+N+2) e^n b^{-n} e^{NA}}{\Gamma(N-n+1) N^{n+1}} \\
& \times \left[ \cos aN\pi \left( \frac{\mathrm{Ai}(N^{2/3}\zeta)}{N^{1/3}} \sum_{l=0}^{p-1} \frac{\widetilde{c}_l}{(aN)^l} - \frac{\mathrm{Ai}'(N^{2/3}\zeta)}{N^{2/3}} \sum_{l=0}^{p-1} \frac{\widetilde{d}_l}{(aN)^l} \right) \right. \\
& + \sin aN\pi \left( \frac{\mathrm{Bi}(N^{2/3}\zeta)}{N^{1/3}} \sum_{l=0}^{p-1} \frac{\widetilde{c}_l}{(aN)^l} - \frac{\mathrm{Bi}'(N^{2/3}\zeta)}{N^{2/3}} \right. \\
& \left. \left. \times \sum_{l=0}^{p-1} \frac{\widetilde{d}_l}{(aN)^l} \right) + \varepsilon_p \right],
\end{aligned}
\tag{68}
$$

where $\varepsilon_p$ is the same as in (66) and

$$
\widetilde{c}_l = a^l c_l \quad \text{and} \quad \widetilde{d}_l = a^l d_l. \tag{69}
$$

Thus, it is sufficient to first prove the boundedness of the coefficients $\widetilde{c}_l$ and $\widetilde{d}_l$, and then establish the asymptotic nature of the error term in (68), namely, to prove that there exist positive constants $A_p$, $A'_p$, $B_p$, and $B'_p$ such that

$$
\begin{aligned}
|\varepsilon_p^1| &\le \frac{A_p}{(aN)^p} \frac{\widetilde{\mathrm{Ai}}(N^{2/3}\zeta)}{N^{1/3}} + \frac{A'_p}{(aN)^p} \frac{\widetilde{\mathrm{Ai}}'(N^{2/3}\zeta)}{N^{2/3}}, \\
|\varepsilon_p^2| &\le \frac{B_p}{(aN)^p} \frac{\widetilde{\mathrm{Bi}}(N^{2/3}\zeta)}{N^{1/3}} + \frac{B'_p}{(aN)^p} \frac{\widetilde{\mathrm{Bi}}'(N^{2/3}\zeta)}{N^{2/3}},
\end{aligned}
\tag{70}
$$

where

$$
\begin{aligned}
\varepsilon_p^1 &= \frac{\cos(aN\pi) N^{n+1}}{2\pi i N^p \Gamma(n+1)} \int_0^{+\infty} \int_{C_1} h_p(u,\tau) e^{N\left(\frac{1}{3}u^3 - \zeta u\right)} e^{N(b \ln \tau - \tau)} \mathrm{d}u \mathrm{d}\tau, \\
\varepsilon_p^2 &= \frac{\sin(aN\pi) N^{n+1}}{2\pi i N^p \Gamma(n+1)} \int_0^{+\infty} \int_{C_3 - C_2} h_p(u,\tau) e^{N\left(\frac{1}{3}u^3 - \zeta u\right)} e^{N(b \ln \tau - \tau)} \mathrm{d}u \mathrm{d}\tau,
\end{aligned}
\tag{71}
$$

and $\widetilde{\mathrm{Ai}}(z) = \widetilde{\mathrm{Bi}}(z) = [\mathrm{Ai}^2(z) + \mathrm{Bi}^2(z)]^{1/2}$ for $z < 0$. Note that in the present case, $\zeta < 0$.

From (59), we recall that $\mathrm{d}w/\mathrm{d}u$ depends on $\zeta$. The value of $\zeta$ is obtained by solving the two equations gotten from (57) with $w$ and $u$ replaced, respectively, by $w_\pm$ and $\pm\sqrt{\zeta}$; see [11, p. 367]. Thus, $\zeta$ and hence $\mathrm{d}w/\mathrm{d}u$ both depend on the parameters $a$ and $b$ in (3). As $a$ and $b$ approach zero, $\zeta$ may tend to infinity. When the saddle points $u_\pm = \pm\sqrt{\zeta}$ are bounded, the mapping $w \to u$ defined by the cubic transformation (57) is analytic, and its derivative $dw/du$ is bounded. However, in the present case, the saddle points $u_\pm = \pm\sqrt{-\zeta}i$ go to infinity as $\zeta \to -\infty$. Hence, the coefficients $a_l(\tau)$, $b_l(\tau)$ and the function $h_l(u,\tau)$ given recursively in (62) and (63) may all blow up as $\zeta \to -\infty$, where $h_0(u,\tau)$ is defined by (61). Because the coefficients $c_l$ and $d_l$ in (65) are related to $a_l(\tau)$ and $b_l(\tau)$ via (67) and (64), to prove that the expansion in (68) is asymptotic when $\zeta \to -\infty$, we must first give estimates for the coefficient functions $a_l(\tau)$ and $b_l(\tau)$. To this end, we shall adopt a method introduced by Olde Daalhuis and Temme [5]. To begin with, we define

$$A_0(u,\zeta) = \frac{u}{u^2-\zeta}, \qquad B_0(u,\zeta) = \frac{1}{u^2-\zeta}. \tag{72}$$

Using (62) and the Cauchy residue theorem, it can be verified that

$$\begin{aligned} a_0(\tau) &= \frac{1}{2\pi i}\int_\Gamma h_0(u,\tau)A_0(u,\zeta)du, \\ b_0(\tau) &= \frac{1}{2\pi i}\int_\Gamma h_0(u,\tau)B_0(u,\zeta)du, \end{aligned} \tag{73}$$

where $\Gamma$ is the contour consisting of two circles, centering at $\pm\sqrt{-\zeta}i$, both with radius $R$, where $R$ could be as large as possible until the circles reach the singularities of $h_0(u,\tau)$ in the $u$-plane.

We note that $w = w_\pm$ are removable singularities of $dw/du$. However, $dw/du$ blows up at the points $w = w_\pm e^{2k\pi i}$, where $k \neq 0$ is any integer; see (59). To find their image points in the $u$-plane under the mapping (57), we take $w_k = w_+ e^{2k\pi i}$ as an example and denote its image point by $u_k$. The image point of $w_- e^{2k\pi i}$ can be treated in a similar manner. By (57), we have

$$\overline{f}(t_0(w_k), w_k) = \frac{1}{3}u_k^3 - \zeta u_k + A. \tag{74}$$

From (57) and (58), we also have

$$\overline{f}(t_0(w_+), w_+) = \frac{1}{3}u_+^3 - \zeta u_+ + A. \tag{75}$$

Subtracting (75) from (74) gives

$$2k\pi a i = i(-\zeta)^{3/2}\left(-\frac{1}{3}z_k^3 - z_k^2\right). \tag{76}$$

Because $0 < a \leq \frac{1}{2}$ and $-\zeta \gg O(1)$, from (76) it follows that $|z_k| \sim \sqrt{2k\pi}a^{1/2}(-\zeta)^{-3/4}$. Therefore, there exists a constant $0 < c_0 < 1$, independent of $a$, $b$, and $\zeta$, such that the interior of the two circles with centers at $u = \pm\sqrt{-\zeta}i$ and radius $R = c_0\sqrt{a}(-\zeta)^{-1/4}$ is free of the singularities of

$h_0(u,\tau)$ in the $u$-plane. Because $h_0(u,\tau)$ is now analytic inside the contour $\Gamma$, there exists a constant $c_h$ such that

$$|h_0(u,\tau)| \le c_h h(\tau) \tag{77}$$

for $u$ in the domain enclosed by the contour $\Gamma$, where $h(\tau)$ denotes the maximum of the two functions $|h_0(\pm\sqrt{-\zeta}i,\tau)|$. Note that as functions of $\tau$, $h_0(\pm\sqrt{-\zeta}i,\tau)$ are analytic in the neighborhood of steepest descent path in the $\tau$-plane.

We further introduce rational functions $A_k$ and $B_k$, $k=0,1,2...$, defined recursively by

$$\begin{aligned} A_{k+1}(u,\zeta) &= \frac{1}{u^2-\zeta}\frac{\mathrm{d}}{\mathrm{d}u}A_k(u,\zeta), \\ B_{k+1}(u,\zeta) &= \frac{1}{u^2-\zeta}\frac{\mathrm{d}}{\mathrm{d}u}B_k(u,\zeta), \end{aligned} \tag{78}$$

where $A_0$ and $B_0$ are given in (72). By induction, we can show that $A_k$ and $B_k$ are expressible as

$$\begin{aligned} A_k(u,\zeta) &= \sum_{i=0}^{[(k+1)/2]} \frac{c^1_{k,i}u^{k+1-2i}}{(u^2-\zeta)^{2k+1-i}}, \\ B_k(u,\zeta) &= \sum_{i=0}^{[k/2]} \frac{c^2_{k,i}u^{k-2i}}{(u^2-\zeta)^{2k-i}}, \end{aligned} \tag{79}$$

where $c^1_{k,i}$ and $c^2_{k,i}$ are constants independent of $u$ and $\zeta$. As in (73), by Cauchy's theorem, we have from Equations (62) and (63)

$$\begin{aligned} a_k(\tau) &= \frac{1}{2\pi i}\int_\Gamma h_k(u,\tau)A_0(u,\zeta)\mathrm{d}u \\ &= \frac{1}{2\pi i}\int_\Gamma h_{k-1}(u,\tau)A_1(u,\zeta)\mathrm{d}u - \frac{1}{2\pi i}\int_\Gamma (a_{k-1}(\tau)+b_{k-1}(\tau)u)\,A_1(u,\zeta)\mathrm{d}u \\ &= \frac{1}{2\pi i}\int_\Gamma h_{k-1}(u,\tau)A_1(u,\zeta)\mathrm{d}u \\ &\ \vdots \\ &= \frac{1}{2\pi i}\int_\Gamma h_0(u,\tau)A_k(u,\zeta)\mathrm{d}u, \end{aligned} \tag{80}$$

where we have used integration-by-parts to derive the second equality. The second term in the second equality vanishes because $(a_{k-1}(\tau)+b_{k-1}(\tau)u)\,A_1(u,\zeta)$ is $O\left(u^{-2}\right)$ as $|u|\to\infty$ and all poles of that function lie inside $\Gamma$; see (79).

Similarly, we also have

$$b_k(\tau) = \frac{1}{2\pi i}\int_\Gamma h_0(u,\tau)B_k(u,\zeta)\mathrm{d}u. \tag{81}$$

Using (79), it is easy to obtain the estimates

$$|A_k(u,\zeta)| \le C_k a^{-k-1/2}(-\zeta)^{1/4} \quad \text{and} \quad |B_k(u,\zeta)| \le C_k a^{-k} \tag{82}$$

for $u$ on and inside the contour $\Gamma$ and $-\zeta \to \infty$. Here and thereafter, $C_k$ is used as a generic symbol for constants independent of $u$, $\zeta$, $a$, and $b$. Substituting (82) into (80) and (81) gives

$$|a_k(\tau)| \le C_k h(\tau) a^{-k}, \tag{83}$$

and

$$|b_k(\tau)| \le C_k h(\tau) a^{-k+1/2}(-\zeta)^{-1/4}. \tag{84}$$

Therefore, $a^l a_l(\tau)$ and $a^l b_l(\tau)$ are both bounded for $\tau$ in the neighborhood of steepest descent path in the $\tau$-plane, and of course for $\tau$ in the neighborhood of $\tau = 0$. Thus, we have the boundedness of the coefficients $\widetilde{c}_l$ and $\widetilde{d}_l$.

To estimate $h_k(u,\tau)$, we use the rational functions $R_k(u,w,\zeta)$, $k = 0, 1, 2...$, defined recursively by

$$\begin{aligned} R_0(u,w,\zeta) &= \frac{1}{u-w}, \\ R_{k+1}(u,w,\zeta) &= \frac{1}{u^2-\zeta}\frac{\mathrm{d}}{\mathrm{d}u}R_k(u,w,\zeta). \end{aligned} \tag{85}$$

These functions were also introduced by Olde Daalhuis and Temme [5]. They showed by induction that $R_k(u,w,\zeta)$ can be written as

$$R_k(u,w,\zeta) = \sum_{i=0}^{k-1}\sum_{j=0}^{\min\{i,k-1-i\}} \frac{C_{ij}u^{i-j}}{(u-w)^{k+1-i-j}\left(u^2-\zeta\right)^{k+i}}, \qquad k = 1, 2..., \tag{86}$$

where $C_{ij}$ do not depend on $u$, $w$, and $\zeta$. Similar to (80), we have

$$h_k(w,\tau) = \frac{1}{2\pi i}\int_\Gamma h_0(u,\tau)R_k(u,w,\zeta)\mathrm{d}u, \tag{87}$$

where $\Gamma$ is the same contour used in (80) and $w$ lies inside two disks centered at $\pm\sqrt{-\zeta}i$ and with radius $\frac{1}{2}c_0\sqrt{a}(-\zeta)^{-1/4}$. It is easy to verify from (86) that

$$|R_k(u,w,\zeta)| \le C_k a^{-k-1/2}(-\zeta)^{1/4}, \tag{88}$$

and from (87) that

$$|h_k(w,\tau)| \le C_k a^{-k} h(\tau). \tag{89}$$

Substituting (89) into (71) gives (70); for details, see [5, pp. 311–312]. Note that to make the expansion (68) asymptotic, we require $x = aN$ to be large.

## 5. Bessel-type expansion

In the case of Hahn Polynomials $Q_n(x; \alpha, \beta, N)$ given in (1), Sharapudinov [10] has given an asymptotic formula involving Jacobi polynomials when the parameters satisfy $\alpha, \beta \geq \frac{1}{2}$ and $2 \leq n \leq c\sqrt{N}$, where $c$ is a positive constant. The value of the variable $x$ is also required to be large; more precisely, $x \geq \varepsilon N$ and $\varepsilon > 0$ is a small number. Although discrete Chebyshev polynomials $t_n(x, N)$ given in (2) is a special case of the Hahn polynomials, the values of the parameters are $\alpha = \beta = 0$; that is, Sharapudinov's result does not include our case. However, because the leading term in the uniform asymptotic expansion of the Jacobi polynomials is a Bessel function (see [7, p. 451]), the work of Sharapudinov did inspire us to look for an asymptotic expansion for $t_n(x, N+1)$ involving Bessel functions, when the parameters $a$ and $b$ in (3) satisfy $a/b^2 \to \infty$ and the variable $x$ is large. Our method differs completely from that of Sharapudinov.

Returning to (7), and making the change of variable $v = 1/w$, we have

$$t_n(x, N+1) = \frac{(-1)^n}{2\pi i} \frac{\Gamma(n+N+2)}{\Gamma(n+1)\Gamma(N-n+1)} \int_0^1 \int_{\gamma_3} \frac{1}{v(1-v)} e^{N\widehat{f}(t,v)} dv dt, \tag{90}$$

where

$$\widehat{f}(t, v) = b\ln(1-t) + (1-b)\ln t - a\ln(1-v) - b\ln v + b\ln[t+v-1]; \tag{91}$$

the curve $\gamma_3$ starts at $v = +\infty$, runs along the lower edge of the positive real line toward $v = 1$, encircles the point $v = 1$ in the clockwise direction and returns to $+\infty$ along the upper edge of the positive real line. Note that because $e^{Nb\ln v} = v^n$, there is no need to have a cut from the origin to infinity.

The saddle point of $\widehat{f}(t, v)$ in the $t$-plane is given by

$$t_0(v) = \frac{2 - v + \sqrt{4b^2 - 4b^2 v + v^2}}{2(1+b)}; \tag{92}$$

cf. (11), where the $v$-plane is cut along two line segments joining 0 to the two conjugate points $2b^2 \pm 2b\sqrt{1-b^2}i$, and the branch of the square root is chosen so that $\sqrt{4b^2 - 4b^2 v + v^2} \sim v$ as $v \to \infty$. From (92), we have

$$\begin{aligned} t_0(v) &= 1 + O(b) \qquad \text{as } |v| \sim cb, \\ t_0(v) &\sim 1 - b \qquad \text{as } v \gg b, \end{aligned} \tag{93}$$

where $c$ is a positive constant, and for $v \ll b$,

$$\begin{aligned} t_0(v) &\sim 1 - \frac{1}{2}v \qquad \text{as } v \to 0^+, \\ t_0(v) &\to 1 - 2b \qquad \text{as } v \to 0^-, \end{aligned} \tag{94}$$

where $0^+$ and $0^-$ mean limits approaching 0 from the right-hand side of the cut and the left-hand side of the cut, respectively. Moreover, easy calculation shows that for any $v$, $t_0(v)$ is not a real number on the cut $(1, \infty)$ in the $t$-plane. Following the same argument given prior to (36), we introduce the mapping

$$\widehat{f}(t, v) = b \ln \tau - \tau + A \tag{95}$$

with the correspondence between the saddle points $t = t_0(v)$ and $\tau = b$ given by

$$\tau(t_0(v)) = b. \tag{96}$$

Coupling (95) and (96) yields

$$A = \widehat{f}(t_0(v), v) - b \ln b + b. \tag{97}$$

The zeros of $\partial \widehat{f}(t, v)/\partial t = 0$ are given by

$$t_0^{\pm}(v) = \frac{2 - v \pm \sqrt{4b^2 - 4b^2 v + v^2}}{2(1+b)}, \tag{98}$$

where $t_0^+(v)$ is the relevant saddle point $t_0(v)$ given in (92). By straightforward calculation, we have

$$\begin{aligned} \frac{dt}{d\tau} &= \frac{b/\tau - 1}{\partial \widehat{f}(t, v)/\partial t} = \frac{(\tau - b)t(1-t)(t + v - 1)}{\tau(1+b)(t - t_0^+(v))(t - t_0^-(v))}, \qquad \tau \neq b, \\ \frac{dt}{d\tau} &= \left\{ \frac{t_0^+(v)(1 - t_0^+(v))(t_0^+(v) + v - 1)}{b\sqrt{4b^2 - 4b^2 v + v^2}} \right\}^{1/2}, \qquad \tau = b. \end{aligned} \tag{99}$$

Note that we have $\tau = 0$ when $t = 1$, and $\tau = +\infty$ when $t = 0$. For any fixed $v$, we can deform the original interval of integration $0 \leq t \leq 1$ into a steepest descent path $\Gamma_v$, passing through $t_0(v)$. Also note that $t_0^-(v)$ is not on $\Gamma_v$, unless $0 < v < 1$. Moreover, if $0 < v < 1$, then $0 < t_0^-(v) < t_0^+(v) < 1$ and $\Gamma_v$ is the real interval $[0, 1]$. In our case, there is only one point on the path $\gamma_3$ in the $v$-plane (see (90)), where it crosses the real line. Let us denote this point by $v_0$. When $v = v_0$, the mapping which we have introduced in (95) becomes singular at the point $t_0^-(v)$, because $dt/d\tau$ in (99) blows up. However, for this particular case, we only need to slightly modify the path by replacing part of the original path near this point by a small half circle as shown in Figure 3. The corresponding integration path in the $\tau$-plane following the mapping (95) also needs to be modified. However, this small modification on the integration path will not affect the following argument and calculation. With this in mind,

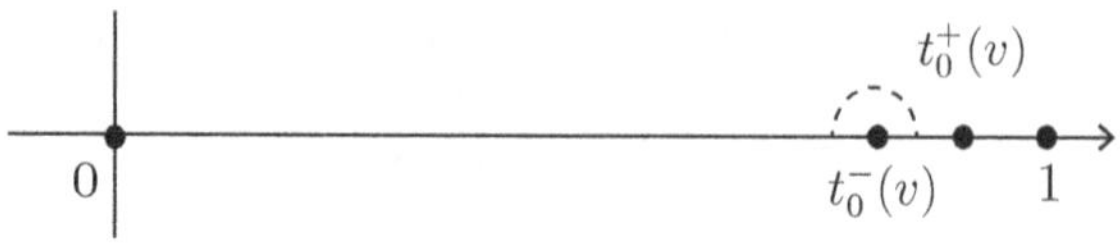


**Figure 3.** The indentation of the integration path in the $t$-plane for $0 < v < 1$.

we will simply ignore this particular case, and proceed with the assumption that the integration path in the $\tau$-plane is always $[0, +\infty)$ for all $w$, and that $dt/d\tau$ will not blow up in the neighborhood of the path.

Thus, from (90) and (95), we have

$$t_n(x, N+1) = \frac{(-1)^{n+1}}{2\pi i} \frac{\Gamma(n+N+2)e^n b^{-n}}{\Gamma(n+1)\Gamma(N-n+1)} \times \int_0^{+\infty} \int_{\gamma_3} \frac{1}{v(1-v)} e^{N\widehat{f}(t_0(v),v)} e^{N(b\ln\tau-\tau)} \frac{dt}{d\tau} dv d\tau. \tag{100}$$

Here, we rewrite the phase function $\widehat{f}(t_0(v), v)$ as

$$\widehat{f}(t_0(v), v) = (1-b)\ln t_0(v) + b\ln(1-t_0(v)) - a\ln(1-v) - b\ln\left(\frac{v}{t_0(v)+v-1}\right). \tag{101}$$

Recalling the statements following (91) and (92), we know that there are only two cuts in the $v$-plane: one along the infinite interval $[1, +\infty)$ and the other along the bent line joining the conjugate points $2b^2 \pm 2b\sqrt{1-b^2}i$ and passing through the origin. To find the saddle points of $\widehat{f}(t_0(v), v)$, we set

$$\frac{\partial \widehat{f}(t_0(v), v)}{\partial v} = 0,$$

and obtain

$$v_\pm = \frac{b^2 \mp bi\sqrt{4a - 4a^2 - b^2}}{2a(1-a)}. \tag{102}$$

Because (90) is obtained from (7) by making the change of variable $v = 1/w$, (102) can also be derived from (15). Note that in this case, $b^2/a \to 0$. Furthermore, because $a < \frac{1}{2}$, the quantity inside the square root is positive. Hence, $v_\pm$ are distinct, and approach $v = 0$.

Define

$$g(u) = m\left(u - \frac{1}{u}\right),$$

where $m > 0$ is some constant to be determined. The saddle points of $g(u)$ are

$$u_\pm = \pm i. \tag{103}$$

Make the transformation

$$\widehat{f}(t_0(v), v) = g(u) + \gamma = m\left(u - \frac{1}{u}\right) + \gamma \tag{104}$$

with

$$u(v_+) = u_-, \qquad u(v_-) = u_+. \tag{105}$$

Note the fact that $t_0(v_\pm) = t_\pm$, where $t_\pm$ are given in (9) and (10). This can be seen from (12) and the change of variable $v = 1/w$ that we have made. Thus, substituting (105) into (104) gives

$$\begin{aligned}\gamma &= \frac{1}{2}\ln\frac{1-b}{1+b} + \frac{1}{2}b\ln\frac{b^2}{1-b^2} \\ &= b\ln b - b + O(b^3)\end{aligned} \tag{106}$$

and

$$\begin{aligned} m = &-\frac{1}{2}\left\{(1-b)\arctan\frac{b\sqrt{4a-4a^2-b^2}}{2-2a-b^2} - a\arctan\frac{b\sqrt{4a-4a^2-b^2}}{2a-2a^2-b^2}\right. \\ &\left. -2b\arctan\frac{\sqrt{4a-4a^2-b^2}}{2+b-2a}\right\} \\ &\sim b\arctan\sqrt{\frac{a}{1-a}}, \end{aligned} \tag{107}$$

when $b^2/a \to 0$ and $b \to 0$. Moreover, from (104) we have

$$\begin{aligned} \frac{\mathrm{d}v}{\mathrm{d}u} &= \frac{m(u-i)(u+i)v(1-v)[(1-2a)v + \sqrt{4b^2 - 4b^2v + v^2}]}{2a(1-a)u^2(v-v_+)(v-v_-)}, \qquad u \neq u_\pm, \\ \left.\frac{\mathrm{d}v}{\mathrm{d}u}\right|_{u=u_\pm} &= \left\{\frac{-2m(v_\mp)^2(1-v_\mp)(1-2a)}{b\sqrt{4a-4a^2-b^2}}\right\}^{1/2}, \qquad u = u_\pm. \end{aligned} \tag{108}$$

Here, we have made use of the equality $\sqrt{4b^2 - 4b^2v_\mp + v_\mp^2} = (1-2a)v_\mp$.

Coupling (100) and (104), the integral representation of $t_n(x, N+1)$ becomes

$$\begin{aligned} t_n(x, N+1) = &\frac{(-1)^{n+1}}{2\pi i}\frac{\Gamma(n+N+2)e^n b^{-n} e^{N\gamma}}{\Gamma(n+1)\Gamma(N-n+1)} \\ &\times \int_0^{+\infty}\int_{\widehat{\gamma}} \frac{h(u,\tau)}{u} e^{N\left[m\left(u-\frac{1}{u}\right)\right]} e^{N(b\ln\tau - \tau)}\mathrm{d}u\mathrm{d}\tau, \end{aligned} \tag{109}$$

where

$$h(u,\tau) = \frac{u}{v(1-v)}\frac{\mathrm{d}t}{\mathrm{d}\tau}\frac{\mathrm{d}v}{\mathrm{d}u}, \tag{110}$$

and the contour $\widehat{\gamma}$ starts from $-\infty$, encircles the point $u=0$ in the counterclockwise direction and returns to $-\infty$. For $l=0,1,2,\ldots$, we define recursively

$$h_l(u,\tau)=a_l(\tau)+\frac{b_l(\tau)}{u}+\left(1+\frac{1}{u^2}\right)g_l(u,\tau), \tag{111}$$

$$h_{l+1}(u,\tau)=-\frac{u}{m}\frac{\mathrm{d}}{\mathrm{d}u}\left\{\frac{g_l(u,\tau)}{u}\right\}, \tag{112}$$

where $h_0(u,\tau)=h(u,\tau)$ given in (110). Furthermore, we expand $a_l(\tau)$ and $b_l(\tau)$ at $\tau=0$, and write

$$a_l(\tau)=\sum_{j=0}^{\infty}a_{l,j}\tau^j, \qquad b_l(\tau)=\sum_{j=0}^{\infty}b_{l,j}\tau^j. \tag{113}$$

It is easy to see that

$$\begin{aligned} a_l(\tau)&=\frac{1}{2}\left[h_l(i,\tau)+h_l(-i,\tau)\right],\\ b_l(\tau)&=\frac{i}{2}\left[h_l(i,\tau)-h_l(-i,\tau)\right]. \end{aligned} \tag{114}$$

From (109), (111), and (112), we have

$$\begin{aligned} t_n(x,N+1)&=\frac{(-1)^{n+1}}{2\pi i}\frac{\Gamma(n+N+2)e^n b^{-n}e^{N\gamma}}{\Gamma(n+1)\Gamma(N-n+1)}\\ &\quad\times\int_0^{+\infty}\int_{-\infty}^{(0^+)}\left[a_0(\tau)+\frac{b_0(\tau)}{u}+\left(1+\frac{1}{u^2}\right)g_0(u,\tau)\right]\\ &\quad\times\frac{e^{Nm\left(u-\frac{1}{u}\right)}}{u}\tau^n e^{-N\tau}\mathrm{d}u\mathrm{d}\tau\\ &=\frac{(-1)^{n+1}\Gamma(n+N+2)e^n b^{-n}e^{N\gamma}}{\Gamma(N-n+1)N^{n+1}}\\ &\quad\times\left[c_0J_0(2Nm)+d_0J_1(2Nm)+\varepsilon_1^+\right], \end{aligned} \tag{115}$$

where

$$\begin{aligned} c_0&=\frac{N^{n+1}}{\Gamma(n+1)}\int_0^{\infty}a_0(\tau)\tau^n e^{-N\tau}\mathrm{d}\tau\sim\sum_{m=0}^{\infty}a_{0,m}\frac{\Gamma(n+m+1)}{\Gamma(n+1)N^m},\\ d_0&=\frac{N^{n+1}}{\Gamma(n+1)}\int_0^{\infty}b_0(\tau)\tau^n e^{-N\tau}\mathrm{d}\tau\sim\sum_{m=0}^{\infty}b_{0,m}\frac{\Gamma(n+m+1)}{\Gamma(n+1)N^m}, \end{aligned} \tag{116}$$

and

$$\varepsilon_1^+=\frac{N^{n+1}}{2\Gamma(n+1)\pi i}\int_0^{+\infty}\int_{-\infty}^{(0^+)}\left(1+\frac{1}{u^2}\right)\frac{g_0(u,\tau)}{u}e^{Nm\left(u-\frac{1}{u}\right)}\tau^n e^{-N\tau}\mathrm{d}u\mathrm{d}\tau. \tag{117}$$

In (115), we have made use of the integral representations of Bessel function and Gamma function

$$J_v(z) = \frac{1}{2\pi i}\int_{-\infty}^{(0^+)} u^{-v-1}\exp\left\{\frac{z}{2}\left(u-\frac{1}{u}\right)\right\}du$$

and

$$\Gamma(n+1) = N^{n+1}\int_0^{+\infty} e^{-N\tau}\tau^n d\tau;$$

see [7], (10.9.19) and (5.9.1). Using (112) and integration by parts, we can rewrite $\varepsilon_1^+$ in (117) as

$$\varepsilon_1^+ = \frac{N^n}{2\Gamma(n+1)\pi i}\int_0^{+\infty}\int_{-\infty}^{(0^+)}\frac{h_1(u,\tau)}{u}e^{Nm(u-\frac{1}{u})}\tau^n e^{-N\tau}dud\tau. \tag{118}$$

Repeating the procedure above, we obtain

$$\begin{aligned} t_n(x, N+1) = {} & \frac{(-1)^{n+1}\Gamma(n+N+2)e^n b^{-n}e^{N\gamma}}{\Gamma(N-n+1)N^{n+1}} \\ & \times\left[J_0(2Nm)\textstyle\sum_{l=0}^{p-1}\frac{c_l}{N^l} + J_1(2Nm)\displaystyle\sum_{l=0}^{p-1}\frac{d_l}{N^l} + \varepsilon_p^+\right], \end{aligned} \tag{119}$$

where

$$c_l = \frac{N^{n+1}}{\Gamma(n+1)}\int_0^{\infty} a_l(\tau)\tau^n e^{-N\tau}d\tau \sim \sum_{m=0}^{\infty} a_{l,m}\frac{\Gamma(n+m+1)}{\Gamma(n+1)N^m}, \tag{120}$$

$$d_l = \frac{N^{n+1}}{\Gamma(n+1)}\int_0^{\infty} b_l(\tau)\tau^n e^{-N\tau}d\tau \sim \sum_{m=0}^{\infty} b_{l,m}\frac{\Gamma(n+m+1)}{\Gamma(n+1)N^m}, \tag{121}$$

$$\varepsilon_p^+ = \frac{N^{n+1-p}}{2\Gamma(n+1)\pi i}\int_0^{+\infty}\int_{-\infty}^{(0^+)}\frac{h_p(u,\tau)}{u}e^{Nm(u-\frac{1}{u})}\tau^n e^{-N\tau}dud\tau. \tag{122}$$

Because $h_0(u,\tau)$ involves $dv/du$ and $dv/du$ depends on the parameters $a$ and $b$ in (3), the coefficients $c_l$ and $d_l$ in (120) and (121) also depend on $a$ and $b$. For an estimate on these coefficients, see (146) below.

To facilitate the application of expansion (119), we recall that the constants $\gamma$ and $m$ are explicitly given in (106) and (107), and note that the leading coefficient can be (asymptotically) calculated by using (113) and (114). Indeed, we have

$$c_0 \sim a_{0,0} = \frac{1}{2}\left[h_0(i,0) + h(-i,0)\right]$$

and

$$d_0 \sim b_{0,0} = \frac{i}{2}\left[h_0(i,0) - h(-i,0)\right]$$

as $b \to 0$ and $N \to \infty$. Furthermore, Equation (128) gives

$$h_0(\pm i, 0) = \frac{-(1-a)v_{\mp}}{b}\sqrt{\frac{2(1-2a)m}{b\sqrt{4a-4a^2-b^2}(1-v_{\mp})}}e^{\pm \Delta i}, \tag{123}$$

where $\Delta = 2\arctan\sqrt{\frac{a}{1-a}} - \sqrt{\frac{a}{1-a}}$.

### *5.1. The mapping $v \to u$ in* (104)

In the case under discussion, $a/b^2 \to \infty$ and $x \to \infty$. Thus, because $0 < a < \frac{1}{2}$, we have $4a - 4a^2 - b^2 > 0$, and the saddle points in (102) are complex.

THEOREM 1. *When $a \in (0, \frac{1}{2}]$, $b \to 0^+$, and $b^2/a \to 0$, the mapping $v \to u$ defined in* (104) *is one-to-one and analytic for $v \in D_v$ in the $v$-plane and $u \in D_u$ in the $u$-plane, where the image of the boundary of $D_u$ in the $Z$-plane is given in* (125)*, and $D_v$ is the image of $D_u$ under the mapping* (104).

*Proof*: As in the cases of Charlier polynomials [9] and Meixner polynomials [3], we introduce an intermediate variable $Z$ defined by

$$b\ln(1-t_0(v)) + (1-b)\ln t_0(v) - a\ln(1-v) - b\ln\left(\frac{v}{t_0(v)+v-1}\right) - \gamma$$
$$= Z = m\left(u - \frac{1}{u}\right), \tag{124}$$

where $t_0(v)$ is the relevant saddle point given in (92).

We consider the upper half of the $v$-plane. Because the functions $\widehat{f}(t_0(v), v) - \gamma$ and $g(u)$ are both symmetric with respect to the real line, the case for the lower half of the $v$-plane can be handled in the same manner. To avoid multivaluedness around the saddle point, we divide the upper half of the $v$-plane into two parts by using the steepest descent path through $v_-$, that is, the point denoted by $D$; see Figure 4. Call the two parts Region I and Region III. Note that there are two branch cuts: one along the infinite interval $[1, \infty)$, and the other along the line segment joining $v = 0$ to $v = 2b^2 + i2b\sqrt{1-b^2}$, which was introduced in (92). We denoted the point $2b^2 + i2b\sqrt{1-b^2}$ by the letter $T$ in Figure 4.

As $v$ traverses along the boundary $ABCDEFGA$ of Region I, the image point $Z$ traverses along the corresponding boundary of a region in the $Z$-plane; see Figure 5. In Figure 6, we draw the boundary of the region, corresponding to Region I, in the $u$-plane. The image point $Z = \psi(u) + \gamma$ traverses along the boundary of the same region shown in Figure 5. Note that we have used an arc

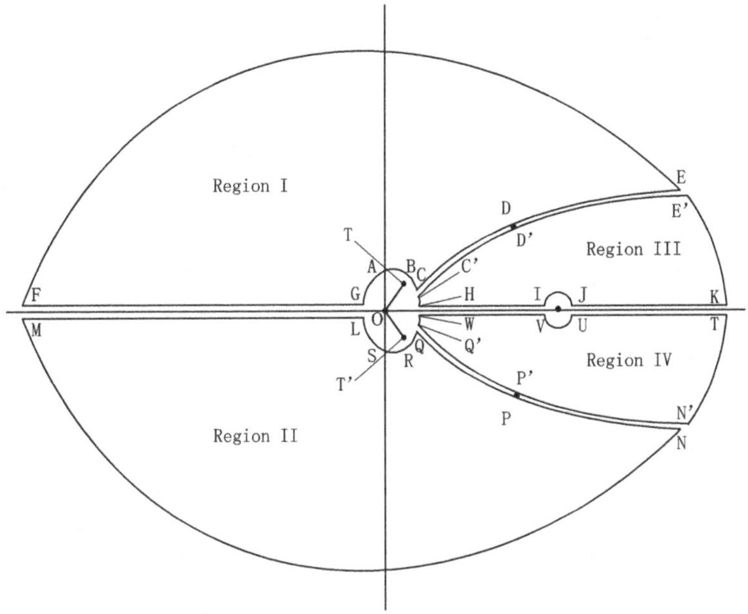


**Figure 4.** The upper half of the $v$-plane.

$GABC$ to avoid the cut $OT$. The boundary curves $EF$ and $GABC$ in the $v$-plane are rather arbitrary; for convenience, we choose them to be the ones whose images in the $Z$-plane are as indicated in Figure 5.

From (104), it is readily seen that the mapping $v \to u$ is the composite function of $g^{-1}$ and $\widehat{f}(t_0(v), v) - \gamma$. We have just verified that this mapping is one-to-one on the boundary of Region I. By the same argument, one can prove that this mapping is also one-to-one on the boundary of Region III. For the image of Region III in the intermediate $Z$-plane, see Figure 7. In the $u$-plane, we let $D_u$ denote the union of the Regions I–IV, with outer boundary $FEE'KJIVUTN'NM$, and inner boundary $GABCC'HWQ'QRSL$; see Figure 6. Under the mapping $Z = g(u)$, we express the images of parts of the outer and inner boundaries of $D_u$ in the $Z$-plane as follows:

$$\begin{aligned}
&GA : \text{Re } Z = bM && 0 \le \text{Im } Z \le \theta b\pi; \\
&AB : \ -bM \le \text{Re } Z \le bM && \text{Im } Z = \theta b\pi; \\
&BC \text{ and } C'H : \text{Re } Z = -bM && 0 \le \text{Im } Z \le \theta b\pi; \\
&IJ : \text{Re } Z = aM && 0 \le \text{Im } Z \le a\pi; \\
&JK : \ -aM \le \text{Re } Z \le aM && \text{Im } Z = a\pi; \\
&KF : \text{Re } Z = -aM && 0 \le \text{Im } Z \le a\pi,
\end{aligned} \tag{125}$$

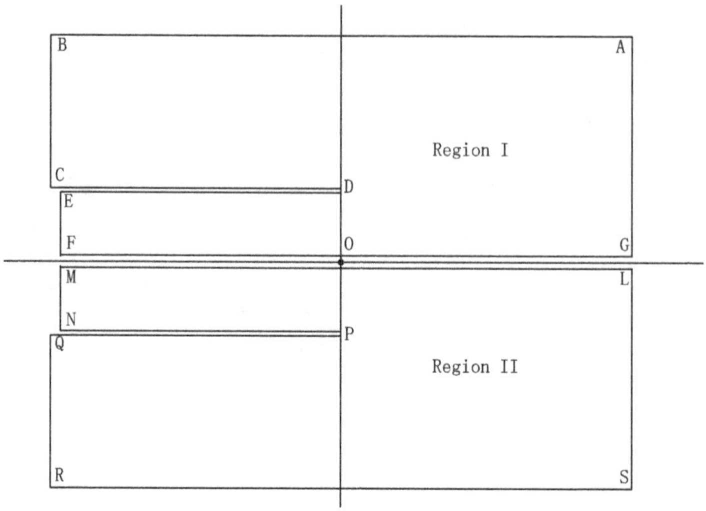


**Figure 5.** The image of Region I in the $Z$-plane.

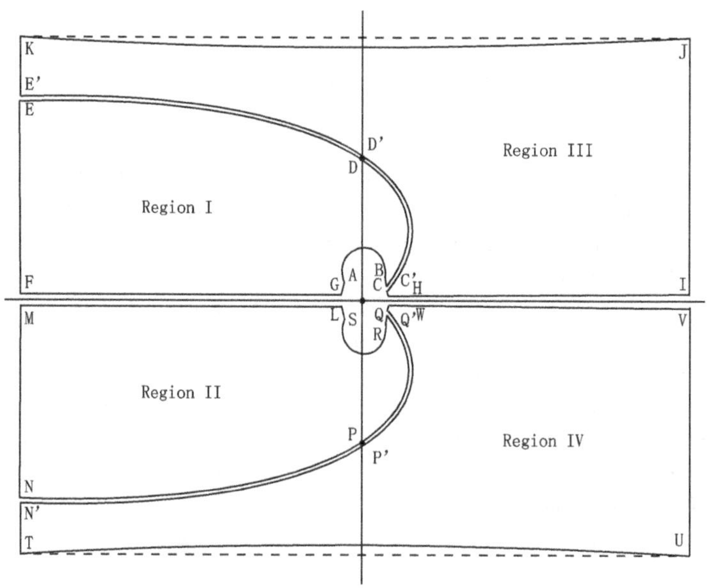


**Figure 6.** The upper half of the $u$-plane.

where $M > 0$ is any fixed constant, and $\theta$ is a generic symbol for a constant in (0, 1). Furthermore, we denote by $D_v$ the corresponding region of $D_u$ in the $v$-plane.

By Theorem 1.2.2 of [13, p. 12], the mapping $v \to u$ is one-to-one in the interior of both Regions I and III in the upper half of the $v$-plane. As explained

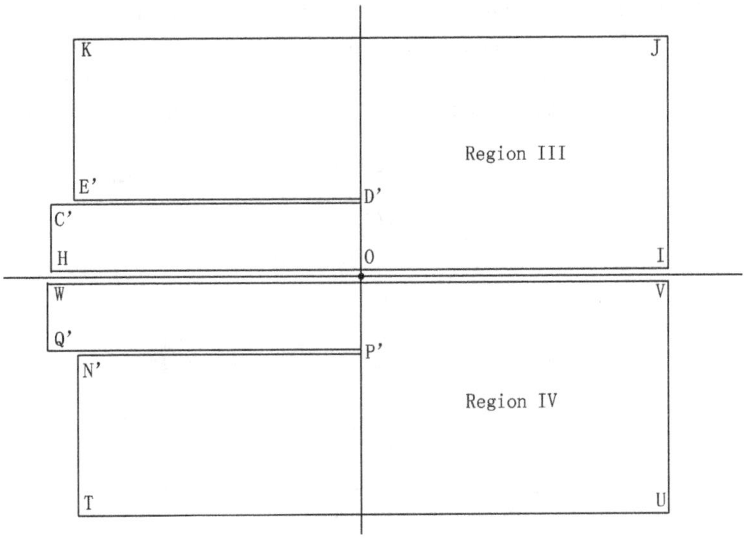


**Figure 7.** The image of Region II in the $Z$-plane.

earlier, the one-to-one property of this mapping in the lower half of the $v$-plane can be established by using the symmetry of the functions with respect to the real axis. Note that the only possible singular points of the mapping $v \to u$ in $D_v$ are at $v = v_{\mp}$. Because the images of these points in the $u$-plane (i.e., $u = \pm i$) are bounded, the mapping is in fact one-to-one and analytic in $D_v$. ■

### *5.2. Analyticity of $h_0(u, \tau)$*

We now investigate the function $h_0(u, \tau) = h(u, \tau)$ given in (110). By (99) and (108), we have

$$h_0(u,\tau) = \frac{m(\tau - b)t(1-t)(t+v-1)(u-i)(u+i)\left[(1-2a)v + \sqrt{4b^2 - 4b^2 v + v^2}\right]}{2a(1-a)\tau(1+b)(t - t_0^+(v))(t - t_0^-(v))u(v - v_+)(v - v_-)}, \tag{126}$$

when $\tau \neq b$ and $u \neq \pm i$, and

$$h_0(u,\tau) = \frac{m(u-i)(u+i)\left[(1-2a)v + \sqrt{4b^2 - 4b^2 v + v^2}\right]}{2a(1-a)u(v-v_+)(v-v_-)} \times \sqrt{\frac{t_0(v)(1-t_0(v))(t_0(v)+v-1)}{b\sqrt{4b^2 - 4b^2 v + v^2}}}, \tag{127}$$

when $\tau = b$ and $u \neq \pm i$, where $u \in D_u$ and $D_u$ is defined in the previous subsection. Note that $u = 0$ is excluded from $D_u$. It is easy to see that $(t-1)/\tau$ is bounded by a constant, independent of $b$ and $a$, as $\tau \to 0$ for $v \in D_v$ and $u \in D_u$. Therefore, $h_0(u, \tau)$ is analytic in $\tau$ for $\tau$ in the neighborhood of $[0, \infty)$. Moreover, $h_0(u, \tau)$ is also analytic in $u$ for $u \in D_u$, because the only singularities $u = \pm i$ in $D_u$ are both removable; indeed, we have

$$\begin{aligned} h_0(u,\tau) = &\frac{(\tau-b)t(1-t)(t+v_{\mp}-1)}{\tau(1+b)(t-t_0^+(v_{\mp}))(t-t_0^-(v_{\mp}))} \\ &\times\sqrt{\frac{2(1-2a)m}{(1-v_{\mp})b\sqrt{4a-4a^2-b^2}}}, \end{aligned} \tag{128}$$

when $\tau \neq b$ and $u = \pm i$, and by the equation following (108) we also have

$$h_0(u,\tau) = \sqrt{\frac{2mt_{\mp}(1-t_{\mp})(t_{\mp}+v_{\mp}-1)}{b^2 v_{\mp}(1-v_{\mp})\sqrt{4a-4a^2-b^2}}}, \tag{129}$$

when $\tau = b$ and $u = \pm i$. To reach (129), we have written $u_\pm$ as $u_\pm = \sqrt{u_\pm^2}$ and move $u_\pm^2$ inside the square root in $dv/du$ at $u = u_\pm$; see (108).

For the analysis to be used in the next subsection, we now give an estimate for $h_0(u, 0)$ when $u$ lies on the boundaries of $D_u$, that is, the outer boundary $FEE'KJIVUTN'NM$ and the inner boundary $GABCC'HWQ'QRSL$ shown in Figure 6. For convenience, let us denote the outer boundary by $C_O$ and the inner boundary by $C_I$.

First, we show that for $u$ on the inner boundary $C_I$,

$$C_1\frac{m}{b} < |u| < C_2\frac{m}{b}, \tag{130}$$

where $C_1 < C_2$ and $C$, $C_1$, $C_2$ are used, here and thereafter, as generic symbols for constants independent of $u$, $v$, $m$, $a$, and $b$. Recall from (107) that $m \sim b\arctan\sqrt{\frac{a}{1-a}}$ as $b \to 0$. To prove (130), we take the part $AB$ of the inner boundary $C_I$ as an illustration. Using (125), we obtain

$$\begin{aligned} &\left| m\left(1-\frac{1}{|u|^2}\right)\operatorname{Re} u\right| \le bM \quad \text{and} \\ &m\left(1+\frac{1}{|u|^2}\right)\operatorname{Im} u = \theta b\pi. \end{aligned} \tag{131}$$

Because $|\operatorname{Im} u| \le |u|$, from the equality in (131) we have $|u|^2 - \frac{\theta b\pi}{m}|u| + 1 \ge 0$. Therefore,

$$|u| \le \frac{\theta b\pi}{2m} - \sqrt{\left(\frac{\theta b\pi}{2m}\right)^2 - 1} \le \frac{Cm}{b\pi}. \tag{132}$$

Rewriting (131) gives

$$\frac{m}{|u|^2}\,|\mathrm{Re}\ u| \le bM + m\,|\mathrm{Re}\ u| \quad \text{and} \quad \frac{m}{|u|^2}\,|\mathrm{Im}\ u| \le \theta b\pi + m\,|\mathrm{Im}\ u|\,, \tag{133}$$

which lead to $\frac{m}{|u|} \le bM + \theta b\pi + m|u|$. Substituting (132) into the right-hand side of the last inequality, and combining the resulting inequality with (132), we obtain (130) immediately. Similarly, for $u$ on the outer boundary $C_{\mathrm{O}}$, we have the estimates

$$C_1\frac{a}{m} < |u| < C_2\frac{a}{m}. \tag{134}$$

Hence, it follows from (130), (134), and (126) that

$$|h(u,0)| \le C \qquad \text{for } u \in C_{\mathrm{I}} \cup C_{\mathrm{O}}. \tag{135}$$

*5.3. Error bounds for the remainder*

To prove the asymptotic nature of the expansion in (119), we need to give precise estimates for its coefficients $c_l$, $d_l$ in (120) and (121), and the error term $\varepsilon_p^+$ given in (122), because the derivative $\mathrm{d}v/\mathrm{d}u$ in (108) may blow up as $v_\pm$ approach 0, just like the case in Section 4. Because the coefficients $c_l$ and $d_l$ are related to $a_p(\tau)$ and $b_p(\tau)$ by (113), (120), and (121), let us first estimate $a_p(\tau)$ and $b_p(\tau)$. To this end, we define recursively

$$A_0(u) = \frac{u}{1+u^2}, \qquad B_0(u) = -\frac{1}{1+u^2},$$

and

$$A_{p+1}(u) = \frac{1}{m}\left(1+\frac{1}{u^2}\right)^{-1}\left\{\frac{1}{u}+\frac{\mathrm{d}}{\mathrm{d}u}\right\}A_p(u),$$

$$B_{p+1}(u) = \frac{1}{m}\left(1+\frac{1}{u^2}\right)^{-1}\left\{\frac{1}{u}+\frac{\mathrm{d}}{\mathrm{d}u}\right\}B_p(u)$$

for $p = 0, 1, 2, \ldots$ . By induction, it can be shown that

$$\begin{aligned} A_p(u) &= \left(\frac{1}{2m}\right)^p\left(\frac{1}{u}\right)^{p+1}\left(\frac{u^2}{1+u^2}\right)^{p+1}\sum_{l=0}^{p} c_{p,l}\left(\frac{u^2}{1+u^2}\right)^l, \\ B_p(u) &= \left(\frac{1}{2m}\right)^p\left(\frac{1}{u}\right)^{p+2}\left(\frac{u^2}{1+u^2}\right)^{p+1}\sum_{l=0}^{p} d_{p,l}\left(\frac{u^2}{1+u^2}\right)^l \end{aligned} \tag{136}$$

for $k = 0, 1, 2, \ldots$ , where $c_{k,l}$ and $d_{k,l}$ are constants independent of $m$ and $u$ (but dependent of $\tau$); see [12]. Furthermore, by (114), (111), and (112), it can be proved that

$$a_p(\tau) = \frac{1}{2\pi i} \int_{C_{\rm O} \bigcup C_{\rm I}} h_0(u, \tau) A_p(u) \mathrm{d}u \tag{137}$$

and

$$b_p(\tau) = \frac{1}{2\pi i} \int_{C_{\rm O} \bigcup C_{\rm I}} h_0(u, \tau) B_p(u) \mathrm{d}u; \tag{138}$$

see also [12]. Define $q := \min\{a, b\}$; because in this case both $n$ and $x \to \infty$, and $b^2/a \to 0$, we have $qN \to \infty$, $m/q = O(b/\sqrt{a}) \to 0$, if $a \le b$, and $m/q \sim \arctan \sqrt{a/(1-a)}$ if $a > b$.

Using (136), (130), and (134), it is easily verified that

$$\left|A_p(u)\right| \le \frac{Cm}{b^{p+1}}, \quad \left|B_p(u)\right| \le \frac{C}{b^p} \tag{139}$$

for $u$ on the inner boundary $C_{\rm I}$, and

$$\left|A_p(u)\right| \le \frac{Cm}{a^{p+1}}, \quad \left|B_p(u)\right| \le \frac{Cm^2}{a^{p+2}} \tag{140}$$

for $u$ on the outer boundary $C_{\rm O}$, where $C$ is used again as a generic symbol for constants independent of $u$, $a$, and $b$. Also, we have the estimates

$$\int_{C_{\rm I}} |\mathrm{d}s| \le \frac{Cm}{b} \quad \text{and} \quad \int_{C_{\rm O}} |\mathrm{d}s| \le \frac{Ca}{m}. \tag{141}$$

To see this, we take one part of $C_{\rm I}$, namely, $AB$ as an example; see (124) and (125), where $Z$ in (124) satisfies $\operatorname{Im} Z = \theta b\pi$. Using the second equality in (125), and in polar coordinates, $u = re^{i\beta}$, the last equation is equivalent to

$$\frac{(r^2+1)}{r} \sin\beta = \frac{\theta b\pi}{m}. \tag{142}$$

Thus,

$$\mathrm{d}r = \frac{(1+r^2)^2 m}{(1-r^2)\theta b\pi} \cos\beta \mathrm{d}\beta. \tag{143}$$

Because in this case $r < 1$, we have $|\mathrm{d}r| \le \frac{Cm}{b} |\mathrm{d}\beta|$. By (130), $|r| \le C_2 m/b$ and

$$|\mathrm{d}s| = \sqrt{(\mathrm{d}r)^2 + r^2(\mathrm{d}\beta)^2} \le \frac{Cm}{b} |\mathrm{d}\beta|, \tag{144}$$

from which the first inequality in (141) follows. The second inequality can be established in a similar manner.

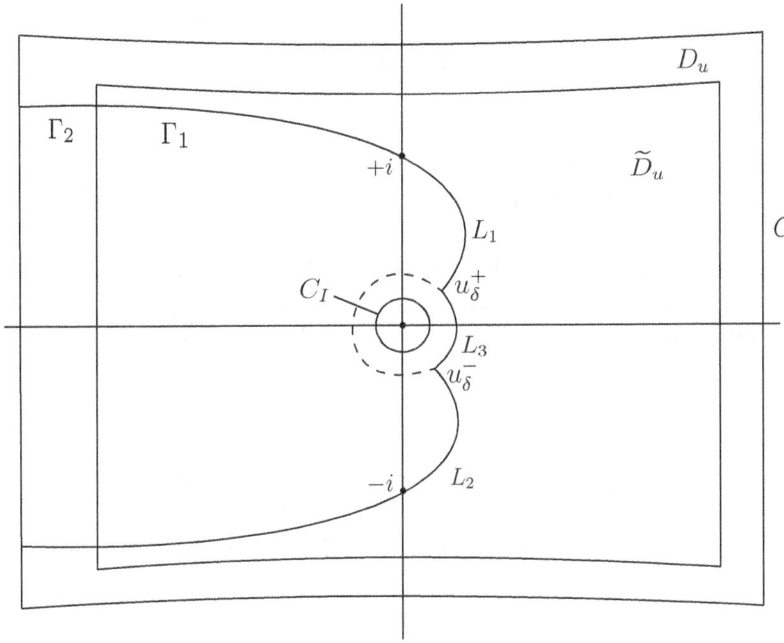


**Figure 8.** The contour in $u$-plane.

Because $h_0(u, 0) \neq 0$ and $h_0(u, \tau)$ is analytic in $\tau$, we have $|h_0(u, \tau)| \leq C|h_0(u, 0)|$ for some constant $C > 0$ and for $\tau$ in the neighborhood of the origin. By a combination of (137), (138), (139), (140), (141), and (135), we obtain

$$|a_p(\tau)| \leq \frac{C}{q^p} \quad \text{and} \quad |b_p(\tau)| \leq \frac{C}{q^p} \tag{145}$$

for $\tau$ near $\tau = 0$. Furthermore, by (113), (120), (121), and (145), we have

$$\left|c_p\right| \leq \frac{C}{q^p} \quad \text{and} \quad \left|d_p\right| \leq \frac{C}{q^p}. \tag{146}$$

To estimate the remainder in (122), we split the loop contour into two parts; see Figure 8. The bounded part of the contour, denoted by $\Gamma_1$, is contained in a subdomain $\widetilde{D}_u$ of $D_u$, which has a distance $c_1 a/m$ from the outer boundary of $D_u$ (i.e., $C_{\rm O}$), and has a distance $c_2 m/b$ from the inner boundary of $D_u$ (i.e., $C_{\rm I}$), $c_1$ and $c_2$ being two constants independent of $u$, $a$ and $b$. Note that $c_1 a/m$ may become large, whereas $c_2 m/b$ may approach zero. The unbounded part of the contour, denoted by $\Gamma_2$, is the rest of the loop outside the subdomain $\widetilde{D}_u$. Put

$$\varepsilon_{p,1}^{+} = \frac{N^{n+1-p}}{2\Gamma(n+1)\pi i} \int_0^{+\infty} \int_{\Gamma_1} \frac{h_p(u,\tau)}{u} e^{Nm(u-\frac{1}{u})} \tau^n e^{-N\tau} \mathrm{d}u\mathrm{d}\tau, \quad (147)$$

$$\varepsilon_{p,2}^{+} = \frac{N^{n+1-p}}{2\Gamma(n+1)\pi i} \int_0^{+\infty} \int_{\Gamma_2} \frac{h_p(u,\tau)}{u} e^{Nm(u-\frac{1}{u})} \tau^n e^{-N\tau} \mathrm{d}u\mathrm{d}\tau. \quad (148)$$

As in [5] and [12], it can be shown that $\varepsilon_{p,2}^{+}$ is exponentially small in comparison with $\varepsilon_{p,1}^{+}$. To estimate $\varepsilon_{p,1}^{+}$, we also follow [5] and [12] by first deriving an integral representation for $h_k(u,\tau)$. Define recursively

$$\begin{aligned} Q_0(u,w) &= \frac{1}{u-w}, \\ Q_{p+1}(u,w) &= \frac{1}{m}\left(1+\frac{1}{u^2}\right)^{-1}\left\{\frac{1}{u}+\frac{\mathrm{d}}{\mathrm{d}u}\right\} Q_p(u,w) \end{aligned} \quad (149)$$

for $p = 0, 1, 2, \ldots$ . Using induction, one can easily verify that

$$Q_p(u,w) = \frac{1}{(2m)^p} \sum_{i=0}^{p-1}\sum_{j=0}^{p-i} q_{p,i,j} \left(\frac{1}{1+u^2}\right)^{p+i} u^{2p+i-j} \left(\frac{1}{u-w}\right)^{p+1-i-j}, \quad (150)$$

where $q_{p,i,j}$ are some constants depending only on $i$, $j$, and $p$. Similar to (87), for $w \in \widetilde{D}_u$ we have from (111), (112), and (149) the integral representation

$$h_p(w,\tau) = \frac{1}{2\pi i} \int_{C_I \cup C_O} h_0(u,\tau) Q_p(u,w) \mathrm{d}u. \quad (151)$$

From (150), it is easy to see that

$$\begin{aligned} |Q_p(u,w)| &\le C\frac{m}{a^{p+1}} \quad \text{for } u \in C_O \text{ and } w \in \widetilde{D}_u, \\ |Q_p(u,w)| &\le C\frac{1}{mb^{p-1}} \quad \text{for } u \in C_I \text{ and } w \in \widetilde{D}_u. \end{aligned} \quad (152)$$

Coupling the above inequalities with (141) and (151), we obtain

$$|h_p(w,\tau)| \le \frac{C}{q^p} \quad (153)$$

for $w \in \widetilde{D}_u$ and $\tau$ in the neighborhood of $\tau = 0$.

If $Nm$ is bounded, one can easily show from (147) and (153) that

$$\left|\varepsilon_{p,1}^{+}\right| = O\left(\frac{1}{q^p N^p}\right). \quad (154)$$

If $Nm \to \infty$, we divide the integration path $\Gamma_1$ into three pieces: the steepest descent path $L_1$ through $u = u_+ = +i$ in the upper half of the

$u$-plane, the steepest descent path $L_2$ through $u = u_- = -i$ in the lower half of the $u$-plane, and the circular arc $|u| = 1 - \delta_0$, denoted by $L_3$, which joins the steepest descent paths $L_1$ and $L_2$ at $u = u_\delta^+$ and $u = u_\delta^-$, respectively; see Figure 8. Hence,

$$\varepsilon_{p,1}^+ = I_1 + I_2 + I_3, \tag{155}$$

where $I_i$, $i = 1, 2, 3$, denotes the integral over the subcontour $L_i$. Applying the steepest descent method [11], p.84], and using (153), it can be easily verified that

$$|I_{1,2}| = O\left(\frac{1}{q^p N^p}\sqrt{\frac{1}{Nm}}\right). \tag{156}$$

For $I_3$, because

$$\begin{aligned} \text{Re}\left(u - \frac{1}{u}\right) &= -\frac{(2-\delta)\delta}{(1-\delta)^2}\text{Re } u \le -\frac{(2-\delta)\delta}{(1-\delta)^2}\text{Re } u_\delta^\pm \\ &= \text{Re}\left(u - \frac{1}{u}\right)\bigg|_{u=u_\delta^\pm} < \text{Re}\left(u - \frac{1}{u}\right)\bigg|_{u=u_\pm}, \end{aligned}$$

$I_2$ is exponentially small in comparison with $I_{1,2}$. Therefore,

$$|\varepsilon_{p,1}^+| = O\left(\frac{1}{q^p N^p}\sqrt{\frac{1}{Nm}}\right). \tag{157}$$

It is well-known that the Bessel function $J_\alpha(t)$ is bounded when $t$ is bounded, and that as $t \to \infty$,

$$J_\alpha(t) \sim \sqrt{\frac{2}{\pi t}}\cos\left(t - \frac{1}{2}\alpha\pi - \frac{1}{4}\pi\right). \tag{158}$$

Coupling the estimates (154) and (157), and together with (158) and (146), we can find a constant $M$ independent of $a$ and $b$ such that

$$|\varepsilon_p^+| \le \frac{M}{qN}\left(|J_0(2Nm)|\frac{|c_{p-1}|}{N^{p-1}} + |J_1(2Nm)|\frac{|d_{p-1}|}{N^{p-1}}\right), \tag{159}$$

which establishes the asymptotic nature of the expansion in (119), given that $qN \to \infty$. In conclusion, in the case $n^2/xN \to 0$ and $x \to \infty$, (119) is an asymptotic expansion.

## 6. Remaining cases

We are now left with only two easy cases to consider.

### 6.1. $a > 0$, $a/b^2$ large and $x$ bounded

In this case, the series in (2) is itself an asymptotic expansion, because

$$\varphi_k(N) := \frac{(-n)_k(-x)_k(n+1)_k}{(-N+1)_k k!k!}, \qquad k = 0, 1, \ldots, n,$$

is an asymptotic sequence when $xN/n^2 \to \infty$; see Table 1 and also [11, p. 10].

### 6.2. $a < 0$

The case $a < 0$ is relatively easy, compared with the case $a > 0$. Let us start with the integral representation (29). For fixed $w \in \gamma_2$, the phase function $\widetilde{f}(t, w)$ in (29) has a saddle point

$$t_0(w) = \frac{2w - 1 + \sqrt{1 + 4b^2(w-1)w}}{2(1+b)w}; \tag{160}$$

cf. (11). Note that the only difference between $f(t, w)$ in (8) and $\widetilde{f}(t, w)$ in (30) is the choice of branch cuts in the $w$-plane due to the term $a \ln w - a \ln(w - 1)$ in $f(t, w)$ and the term $a \ln(-w) - a \ln(1 - w)$ in $\widetilde{f}(t, w)$. Similar to (35), for any fixed $w \in \gamma_2$, we have $t_0(w) = 1 - b + O(b^2)$ as $b \to 0$; in particular, $t_0(0) = 1 - b$. By the same reasoning given for (36), we introduce the mapping $t \to \tau$ defined by

$$b \ln \tau - \tau + A = \widetilde{f}(t, w) - \widetilde{f}(t_0(w), w) \tag{161}$$

with

$$\tau(t_0(w)) = b. \tag{162}$$

Coupling (162) and (161) yields $A = b - b \ln b$. From (161), we have

$$\begin{aligned} \frac{dt}{d\tau} &= \frac{b/\tau - 1}{\partial \widetilde{f}(t, w)/\partial t} = \frac{(\tau - b)t(1-t)[1 - (1-t)w]}{\tau(1+b)w(t - t_0^+(w))(t - t_0^-(w))}, && \tau \neq b, \\ \frac{dt}{d\tau} &= \left\{ \frac{t_0^+(w)(1 - t_0^+(w))[1 - (1 - t_0^+(w))w]}{b\sqrt{1 + 4b^2(w-1)w}} \right\}^{1/2}, && \tau = b, \end{aligned} \tag{163}$$

where we have used L'Hôspital's rule for $\tau = b$ and

$$t_0^{\pm}(w) = \frac{2w - 1 \pm \sqrt{1 + 4b^2(w-1)w}}{2(1+b)w}; \tag{164}$$

cf. (39). Furthermore, it follows from (161) and (29) that

$$\begin{aligned} t_n(x, N+1) = & \frac{(-1)^n}{2\pi i} \frac{\Gamma(n+N+2)e^n b^{-n}}{\Gamma(n+1)\Gamma(N-n+1)} \\ & \times \int_0^{+\infty} \int_{\gamma_2} \frac{1}{w-1} e^{N\widetilde{f}(t_0(w),w)} e^{N(b\ln\tau - \tau)} \frac{\mathrm{d}t}{\mathrm{d}\tau} \mathrm{d}w \mathrm{d}\tau; \end{aligned} \tag{165}$$

cf. (40). Solving the equation $\partial f(t_0(w), w)/\partial w = 0$, we obtain the saddle points

$$\begin{aligned} w_\pm &= \frac{b \pm \sqrt{b^2 - 4a + 4a^2}}{2b} \\ &= \frac{1}{2} \pm \frac{1}{2}\sqrt{1 - \frac{4a(1-a)}{b^2}}; \end{aligned} \tag{166}$$

cf. (15). The relevant saddle point on the integral path $\gamma_2$ is the negative saddle point $w_-$.

The following procedure is the same as that given in [8]. Recall the Hankel integral for the Gamma function

$$\frac{1}{\Gamma(z)} = \frac{1}{2\pi i} \int_{-\infty}^{(0+)} e^u u^{-z} \mathrm{d}u, \tag{167}$$

where the contour is a loop starting at $-\infty$, encircling the origin in the counterclockwise direction and returning to $-\infty$. With $z = -aN + 1$ and $u$ replaced by $-u$, we obtain

$$\frac{1}{\Gamma(-aN+1)} = \frac{N^{aN}}{2\pi i} \int_{\infty}^{(0+)} \frac{1}{u} e^{N\widetilde{\psi}(u)} \mathrm{d}u, \tag{168}$$

where $\widetilde{\psi}(u) = a\ln(-u) - u$. Make the transformation $w \to u$ defined by

$$\widetilde{f}(t_0(w), w) = a\ln(-u) - u + \gamma \tag{169}$$

with

$$u(w_-) = a, \tag{170}$$

where $\gamma$ is a constant to be determined. We have from (169)

$$\begin{aligned} \frac{\mathrm{d}w}{\mathrm{d}u} &= \frac{(a-u)w(w-1)\left[(1-2a) + \sqrt{4b^2w^2 - 4b^2w + 1}\right]}{2ub^2(w-w_+)(w-w_-)}, \qquad u \neq a, \\ \frac{\mathrm{d}w}{\mathrm{d}u} &= \left\{\frac{(1-a)(1-2a)}{b^3\sqrt{b^2 - 4a + 4a^2}}\right\}^{1/2}, \quad u = a, \end{aligned} \tag{171}$$

where we again have used L'Hôspital's rule for $u = a$.

By (165) and (169), we have

$$t_n(x, N+1) = \frac{(-1)^n}{2\pi i}\frac{\Gamma(n+N+2)e^n b^{-n} e^{N\gamma}}{\Gamma(n+1)\Gamma(N-n+1)} \times \int_0^{+\infty}\int_{\infty}^{(0^+)} \frac{h(u,\tau)}{u} e^{N(a\ln(-u)-u)} e^{N(b\ln\tau-\tau)} du d\tau, \tag{172}$$

where

$$h(u,\tau) = \frac{u}{w-1}\frac{dt}{d\tau}\frac{dw}{du}. \tag{173}$$

Define recursively

$$h_l(u,\tau) = a_l(\tau) + (u-a)g_l(u,\tau), \tag{174}$$

$$h_{l+1}(u,\tau) = u\frac{\partial g_l(u,\tau)}{\partial u}, \tag{175}$$

where $h_0(u,\tau) = h(u,\tau)$, and expand $a_l(\tau)$ into a Maclaurin series

$$a_l(\tau) = \sum_{j=0}^{\infty} a_{l,j}\tau^j. \tag{176}$$

By an integration-by-parts procedure, we obtain

$$t_n(x, N+1) = \frac{(-1)^n\Gamma(n+N+2)e^n b^{-n} e^{N\gamma}}{\Gamma(N-n+1)\Gamma(-aN+1)N^{aN+n+1}}\left[\sum_{l=0}^{p-1}\frac{c_l}{N^l} + \varepsilon_p^-\right], \tag{177}$$

where

$$c_l = \frac{N^{n+1}}{\Gamma(n+1)}\int_0^{\infty} a_l(\tau)\tau^n e^{-N\tau}d\tau \sim \sum_{m=0}^{\infty} a_{l,m}\frac{\Gamma(n+m+1)}{\Gamma(n+1)N^m} \tag{178}$$

and

$$\varepsilon_p^- = \frac{N^{aN+n+1-p}\Gamma(-aN+1)}{2\Gamma(n+1)\pi i}\int_0^{+\infty}\int_{-\infty}^{(0^+)} \frac{h_p(u,\tau)}{u} e^{N\widetilde{\psi}(u)}\tau^n e^{-N\tau} du d\tau. \tag{179}$$

Estimation of the error term given in (179) is exactly the same as what we have done in [8].

## References

1. R. Beals and R. Wong, *Special Functions, A Graduate Text*, Cambridge University Press, Cambridge, 2010.

2. C. Chester, B. Friedman, and F. Ursell, An extension of the method of steepest descents, *Proc. Cambridge Philos. Soc.* 53:599–611 (1957).
3. X.-S. Jin and R. Wong, Uniform asymptotic expansions for Meixner polynomials, *Constr. Approx.* 14:113–150 (1998).
4. Y. Lin and R. Wong, Global asymptotics of the Hahn polynomials, *Analysis and Applications*, 11(3), 1350018 (47 pages) (2013).
5. A. B. Olde Daalhuis and N. M. Temme, Uniform Airy-type expansions of integrals, *SIAM J. Math. Anal.* 25:304–321 (1994).
6. F. W. J. Olver, *Asymptotics and Special Functions*, Academic Press, New York, 1974. (Reprinted by A. K. Peters, Wellesley, MA, 1997.)
7. F. W. J. Olver, D. W. Lozier, C. W. Clark, and R. F. Boisvert, *NIST Handbook of Mathematical Functions*, Cambridge University Press, Cambridge, 2010.
8. J. H. Pan and R. Wong, Uniform asymptotic expansions for the discrete Chebyshev polynomials, *Stud. Appl. Math.* 128:337–384 (2012).
9. Rui Bo and R. Wong, Uniform asymptotic expansion of Charlier polynomials. *Meth. Applic. Anal.* 1:294–313 (1994).
10. I. I. Sharapudinov, Asymptotic properties of orthogonal Hahn polynomials in a discrete variable, *Sbornik: Mathematics*, 681:111–132 (1991).
11. R. Wong, *Asymptotic Approximations of Integrals*, Academic Press, Boston, 1989. (Reprinted by SIAM, Philadelphia, PA, 2001)
12. R. Wong and Y. Q. Zhao, Estimates for the error term in a uniform asymptotic expansion of the Jacobi polynomials. *Anal. Appl.* (Singap.) 1:213–241 (2003).
13. R. Wong, *Lecture Notes on Applied Analysis*, World Scientific, Singapore, 2010.

City University of Hong Kong

(Received April 24, 2013)

Journal of Computational and Applied Mathematics 265 (2014) 220–242

Contents lists available at ScienceDirect

Journal of Computational and Applied Mathematics

journal homepage: www.elsevier.com/locate/cam

# Asymptotic expansion of the Tricomi–Carlitz polynomials and their zeros

K.F. Lee, R. Wong *

*Department of Mathematics, City University of Hong Kong, 83 Tat Chee Avenue, Kowloon, Hong Kong*

ARTICLE INFO

*Article history:*
Received 27 September 2012
Received in revised form 23 April 2013

*MSC:*
41A60
33C45

*Keywords:*
Tricomi–Carlitz polynomials
Uniform asymptotic expansions
Airy function

ABSTRACT

The Tricomi–Carlitz polynomials $f_n^{(\alpha)}(x)$ are non-classical discrete orthogonal polynomials on the real line with respect to the step function whose jumps are

$$d\psi^{(\alpha)}(x) = \frac{(k+\alpha)^{k-1}e^{-k}}{k!} \quad \text{at } x = x_k = \pm(k+\alpha)^{-1/2},\ k = 0, 1, 2, \ldots.$$

In this paper, we derive an asymptotic expansion for $f_n^{(\alpha)}(t/\sqrt{\nu})$ as $n \to \infty$, valid uniformly for bounded real $t$, where $\nu = n + 2\alpha - 1/2$ and $\alpha$ is a positive parameter. The validity for bounded $t$ can be extended to unbounded $t$ by using a sequence of rational functions introduced by Olde Daalhuis and Temme. The expansion involves the Airy functions and their derivatives. Error bounds are given for one-term and two-term approximations. Asymptotic formulas are also presented for the zeros of $f_n^{(\alpha)}(t/\sqrt{\nu})$.



## 1. Introduction

The Tricomi–Carlitz polynomials [1] satisfy the three-term recurrence relation

$$(n+1)f_{n+1}^{(\alpha)}(x) - (n+\alpha)\,xf_n^{(\alpha)}(x) + f_{n-1}^{(\alpha)}(x) = 0, \quad n \geq 1, \tag{1.1}$$

with initial values $f_0^{(\alpha)}(x) = 1$ and $f_1^{(\alpha)}(x) = \alpha x$. In this paper, we restrict ourselves to the case $\alpha > 0$. These polynomials satisfy the orthogonal relation

$$\int_{-\infty}^{\infty} f_m^{(\alpha)}(x) f_n^{(\alpha)}(x)\, d\psi^{(\alpha)}(x) = \frac{2\,e^{\alpha}}{(n+\alpha)n!}\,\delta_{mn}, \tag{1.2}$$

where $\psi^{(\alpha)}(x)$ is the step function with jumps

$$d\psi^{(\alpha)}(x) = \frac{(k+\alpha)^{k-1}e^{-k}}{k!} \quad \text{at } x = x_k = \pm(k+\alpha)^{-1/2},\ k = 0, 1, 2, \ldots. \tag{1.3}$$

Their generating function

$$\exp\left\{\frac{w}{x} + \frac{1-\alpha x^2}{x^2}\log(1-wx)\right\} = \sum_{n=0}^{\infty} f_n^{(\alpha)}(x)\, w^n \tag{1.4}$$

* Corresponding author.
*E-mail addresses:* rscwong@cityu.edu.hk, mawong@cityu.edu.hk (R. Wong).

0377-0427/$ – see front matter © 2013 Elsevier B.V. All rights reserved.
http://dx.doi.org/10.1016/j.cam.2013.09.047

can be derived from the recurrence relation (1.1) when $x \neq 0$. The series in (1.4) is convergent for $|wx| < 1$. The values $x_k$ in (1.3) play a special role in the generating function (1.4) since for these $x$ values we have

$$e^{w/x_k}(1-x_k w)^k = \sum_{n=0}^{\infty} f_n^{(\alpha)}(x_k)\, w^n \tag{1.5}$$

and now the series converges for all values of $w$. If $x = 0$, then (1.4) reduces to

$$e^{-\frac{1}{2}w^2} = \sum_{n=0}^{\infty} f_n^{(\alpha)}(0)\, w^n. \tag{1.6}$$

By expanding $e^{-\frac{1}{2}w^2}$ at $w = 0$, one obtains

$$f_{2n}^{(\alpha)}(0) = \frac{(-1)^n}{2^n n!}, \qquad f_{2n+1}^{(\alpha)}(0) = 0, \quad n = 0, 1, 2, \ldots. \tag{1.7}$$

These polynomials can also be expressed in terms of the generalized hypergeometric series

$$f_n^{(\alpha)}(x) = \frac{1}{x^n n!}\, {}_2F_0(-n, \alpha - x^{-2}; -x^2). \tag{1.8}$$

Independently and simultaneously, Karlin and McGregor [2] studied a stochastic (Markovian) birth-and-death process, and discovered the random walk polynomials $r_n(x, a)$ derived from an infinitely many server process, which are related to the Tricomi–Carlitz polynomials via the identity $f_n^{(\alpha)}(x) = \alpha^{n/2} r_n(\sqrt{\alpha}x, \alpha)/n!$. Therefore, the Tricomi–Carlitz polynomials are also known as the Carlitz–Karlin–McGregor polynomials [3]. In 1984, Askey and Ismail [3] have given a generalization of the Tricomi–Carlitz polynomials, namely, $G_n(x; a, b) = r_n(x, a)a^n(b/a)_n/n!$ with $\alpha^{-n/2}G_n(x\sqrt{\alpha}; 0, \alpha) = f_n^{(\alpha)}(x)$. Other properties of $G_n(x; a, b)$ can be found in [3,4].

With regard to the asymptotics of $f_n^{(\alpha)}(x)$, we have from a direct application of Darboux's method [5, p. 116]

$$f_n^{(\alpha)}(x) = \frac{1}{\Gamma(\alpha - 1/x^2)} x^n e^{1/x^2} n^{\alpha-1-1/x^2} \left[1 + O(n^{-1})\right], \tag{1.9}$$

which is valid uniformly for all $x$ in any compact set $K \subseteq \mathbb{C} \setminus [-1/\sqrt{\alpha}, 1/\sqrt{\alpha}]$. Using elementary approaches, the same result has been given by Goh and Wimp [6]; see also [3]. In another paper [7], Goh and Wimp investigated the asymptotics of $f_n^{(\alpha)}(z/\sqrt{n})$ for $z$ in neighborhoods of $z = \pm i/2$. The behavior of $f_n^{(\alpha)}(x)$ as $\alpha \to \infty$ has been studied by López and Temme [8]. Their result is expressed in terms of Hermite polynomials.

Recently, Lee and Wong [1] derived an infinite asymptotic expansion for $f_n^{(\alpha)}(t/\sqrt{\nu})$, which holds uniformly for $t$ in $[0, \infty)$, where $\nu = n + 2\alpha - 1/2$. In view of the reflection formula $f_n^{(\alpha)}(-x) = (-1)^n f_n^{(\alpha)}(x)$, a corresponding result can be given for $t$ in $(-\infty, 0]$.

Note that the two regions of validity contain the critical values $x = \pm 2/\sqrt{\nu}$. Also, the result of Lee and Wong is not covered in the two papers of Goh and Wimp; in fact, it complements those in [6,7]. The method used by Lee and Wong is based on a turning-point theory for three-term recurrence relations developed by Wang and Wong [9,10].

The purpose of this paper is to confirm that the expansion of Lee and Wong is valid for unbounded values of $t$ and to show that numerically computable bounds can be constructed for the error terms associated with the expansion. The approach taken in this paper is based on the modified steepest-decent method of Chester, Friedman and Ursell [11], and the use of a sequence of rational functions introduced by Olde Daalhuis and Temme [12].

The arrangement of the present paper is as follows. In Section 2, we present an integral representation for Tricomi–Carlitz polynomials $f_n^{(\alpha)}(x)$, and transform it into a canonical form by using a cubic transformation. We prove the cubic transformation is one-to-one and analytic in a domain containing the path of integration. In Section 3, we use the Bleistein technique of integration by parts to derive an Airy-type expansion with an explicit remainder. In Section 4, we show that the asymptotic expansion derived in Section 3 in fact holds for $t$ tending to infinity. In Section 5, error bounds are given for one-term and two-term approximations. Asymptotic formulas for the zeros of $f_n^{(\alpha)}(t/\sqrt{\nu})$ are presented in the final section.

## 2. Reduction to a canonical integral

From (1.4), we have by the Cauchy integral formula

$$f_n^{(\alpha)}(x) = \frac{1}{2\pi i}\int_{\mathcal{C}} \exp\left\{\frac{w}{x} + \frac{1-\alpha x^2}{x^2}\log(1-wx)\right\}\frac{dw}{w^{n+1}}, \tag{2.1}$$

where $\mathcal{C}$ is a circle centered at the origin and oriented in the counter-clockwise direction with radius less than $1/|x|$, $x \neq 0$. Rescaling the variables

$$x = t/\sqrt{\nu}, \qquad w = s\sqrt{\nu} \quad \text{and} \quad \nu = n + 2\alpha - 1/2, \tag{2.2}$$

Eq. (2.1) becomes

$$f_n^{(\alpha)}(t/\sqrt{\nu}) = \frac{\nu^{-n/2}}{2\pi i}\int_{\mathcal{C}} e^{\nu\phi(s;t)}\frac{1}{s^{-2\alpha+1/2}s(1-ts)^{\alpha}}ds, \tag{2.3}$$

where

$$\phi(s;t) = \frac{s}{t} + \frac{1}{t^2}\log(1-ts) - \log s. \tag{2.4}$$

For definiteness, we choose for all logarithmic functions in (2.4) the principal branch

$$\log\zeta = \log|\zeta| + i\arg\zeta, \quad -\pi < \arg\zeta \le \pi. \tag{2.5}$$

The *phase function* $\phi(s;t)$ has branch points at $s = 0$ and $s = 1/t$, although $s = 0$ is not a branch point of the integrand in (2.3). This function is analytic and single-valued in the complex $s$-plane with cuts along the intervals $(-\infty, 0]$ and $[1/t, \infty)$. For later discussion, we also need to specify the values of the argument along the upper and lower edges of the cuts. To this end, we let $s^{\pm} = u + i0^{\pm}$, $u > 1/t$, denote points on the upper and lower edges of the cut along $[1/t, \infty)$. For instance, we have

$$\arg(1-ts^+) = -\pi \quad \text{and} \quad \arg(1-ts^-) = \pi. \tag{2.6}$$

From (1.4), it is easily seen that

$$\sum_{n=0}^{\infty} f_n^{(\alpha)}(-x)\,(-w)^n = \sum_{n=0}^{\infty} f_n^{(\alpha)}(x)\,w^n,$$

from which it follows that

$$f_n^{(\alpha)}(-x) = (-1)^n f_n^{(\alpha)}(x). \tag{2.7}$$

For this reason, we may restrict $x$ to the interval $[0, \infty)$.

The saddle points of $\phi(s;t)$, i.e., zeros of $\partial\phi/\partial s$, are at

$$s = s_{\pm} = \frac{1}{2}\left(t \pm \sqrt{t^2-4}\right).$$

It is convenient to consider two separate cases: (i) $0 < t < 2$ and (ii) $t \ge 2$. The last equation is equivalent to

$$s_{\pm} = \begin{cases} \frac{1}{2}\left(t \pm i\sqrt{4-t^2}\right) & \text{if } 0 < t < 2, \\ \frac{1}{2}\left(t \pm \sqrt{t^2-4}\right) & \text{if } t \ge 2. \end{cases} \tag{2.8}$$

Clearly, the saddle points coalesce at $s = 1$ when $t = 2$. Note that for $t \ge 2$, we have $1/t < s_- < s_+ < \infty$. The steepest descent (and ascent) paths are given by

$$\Im\phi(s;t) = \Im\phi(s_{\pm};t). \tag{2.9}$$

Case 1: $t > 2$. Since $s_{\pm}^{+}$ are on the upper edge of the cut, we have from (2.6)

$$\log(1-s_{\pm}^{+}t) = -i\pi + \log(s_{\pm}^{+}t - 1) = -i\pi + 2\log s_{\pm}^{+}, \tag{2.10}$$

where use has been made of the fact that $s_{\pm}$ are zeros of $\partial\phi/\partial s$, i.e., $s_{\pm}^2 - t\,s_{\pm} + 1 = 0$. Similarly, since $s_{\pm}^{-}$ are on the lower edge of the cut, we have

$$\log(1-s_{\pm}^{-}t) = +i\pi + \log(s_{\pm}^{-}t - 1) = +i\pi + 2\log s_{\pm}^{-}. \tag{2.11}$$

If we write $s = u + iv$, then Eq. (2.9) can be written as

$$\frac{v}{t} + \frac{1}{t^2}\arg(1-tu-itv) - \arg(u+iv) = \Im\phi(s_{\pm}^{+};t) = -\frac{\pi}{t^2} \tag{2.12}$$

and

$$\frac{v}{t} + \frac{1}{t^2}\arg(1-tu-itv) - \arg(u+iv) = \Im\phi(s_{\pm}^{-};t) = +\frac{\pi}{t^2}. \tag{2.13}$$

Using Maple, the steepest paths given by (2.12) and (2.13) are shown in Fig. 1, where arrows are used to indicate the direction of descent. (For a detailed analysis of directions of descent and ascent, see [13].)

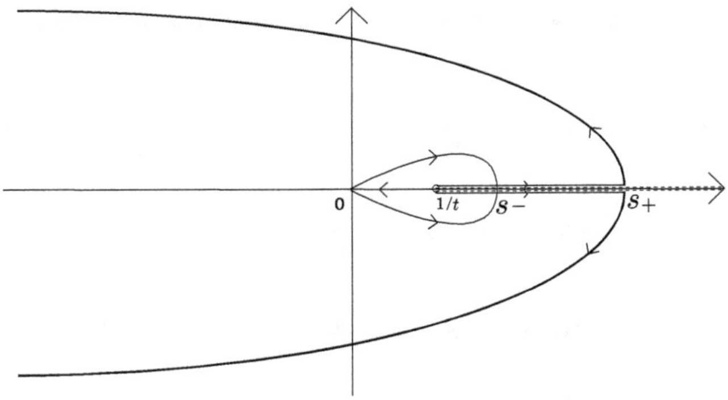


**Fig. 1.** Steepest paths when $t \geq 2$.

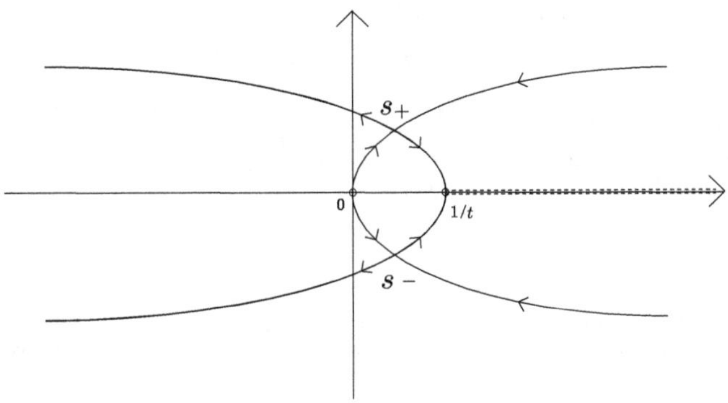


**Fig. 2.** Steepest path when $0 < t < 2$.

Case 2: $0 < t < 2$. For $u \in (-\infty, \infty)$, we take the branch of $\tan^{-1} u$ with values in $(-\frac{\pi}{2}, \frac{\pi}{2})$. If $z = x + iy$, it is easily seen that

$$\arg(x+iy) = \begin{cases} \tan^{-1}(y/x) & \text{if } x > 0 \text{ and } y > 0 \text{ or } < 0, \\ \pi + \tan^{-1}(y/x) & \text{if } x < 0 \text{ and } y > 0, \\ -\pi + \tan^{-1}(y/x) & \text{if } x < 0 \text{ and } y < 0. \end{cases} \tag{2.14}$$

By again using the fact that $s_\pm$ are zeros of $\partial\phi/\partial s$, i.e., $s_\pm^2 - t\,s_\pm + 1 = 0$, Eq. (2.9) can be written as

$$\frac{v}{t} + \frac{1}{t^2}\arg(1 - tu - itv) - \arg(u + iv) = \Im\phi(s_\pm; t) = \pm\theta_t, \tag{2.15}$$

where $\theta_t = \frac{1}{2t}\sqrt{4-t^2} - \frac{\pi}{t^2} + (\frac{2}{t^2} - 1)\tan^{-1}\frac{\sqrt{4-t^2}}{t}$. The graph of the curves given by (2.15) are shown in Fig. 2, with arrows again indicating directions of ascent and descent. (Detailed analysis can be found in [13].)

Since the two saddle points $s_\pm$ coalesce at $s = 1$, let us expand $\phi(s; t)$ into a Taylor series around this point. The result is

$$\begin{aligned}\phi(s;t) &= \frac{1}{t} + \frac{1}{t^2}\log(1-t) + \left[\frac{1}{t} + \frac{1}{t(t-1)} - 1\right](s-1) \\ &\quad + \frac{1}{2!}\left(1 - \frac{1}{(t-1)^2}\right)(s-1)^2 + \frac{1}{3!}\left[\frac{2t}{(t-1)^3} - 2\right](s-1)^3 + \cdots.\end{aligned}$$

When $t \to 2$, the linear and quadratic terms in the above expansion both vanish, and the first nonvanishing term is the cubic. This suggests that we make the cubic transformation

$$\phi(s;t) = \frac{u^3}{3} - \zeta(t)u + \eta(t) \tag{2.16}$$

introduced by Chester, Friedman and Ursell [11]; see also [14, p. 351]. Differentiating (2.16) with respect to $s$ gives

$$\frac{\partial s}{\partial u} = \frac{u^2 - \zeta(t)}{\partial\phi/\partial s} = (u^2 - \zeta(t))\frac{s(ts-1)}{(s-s_-)(s-s_+)}.$$

To ensure that the mapping $s \mapsto u$ is one-to-one and analytic, $\partial s/\partial u$ can neither vanish nor become infinite. Hence, we must assign the saddle points $s_\pm$ of $\phi(s;t)$ to correspond to the saddle points $\pm\sqrt{\zeta(t)}$ of the cubic polynomial on the right-hand side of (2.16). This gives

$$\begin{aligned} -\frac{2}{3}\zeta^{3/2}(t) + \eta(t) &= \frac{s_+}{t} + \frac{1}{t^2}\log(1 - ts_+) - \log s_+, \\ \frac{2}{3}\zeta^{3/2}(t) + \eta(t) &= \frac{s_-}{t} + \frac{1}{t^2}\log(1 - ts_-) - \log s_-. \end{aligned} \tag{2.17}$$

Note that, in addition to (2.10) and (2.11), we also have

$$s_+ + s_- = t, \qquad s_+s_- = 1, \qquad \frac{s_+}{s_-} = \frac{s_+^2}{s_-s_+} = ts_+ - 1,$$

$$s_+ - s_- = \begin{cases} \sqrt{t^2-4} & \text{if } t \geq 2, \\ i\sqrt{4-t^2} & \text{if } 0 < t < 2, \end{cases} \qquad \frac{s_+t-1}{s_-t-1} = (s_+t-1)^2 = s_+^4.$$

With the aid of these identities, we obtain upon solving the two equations in (2.17)

$$\zeta^{3/2}(t) = \begin{cases} \frac{3}{4}\left[\left(2 - \frac{4}{t^2}\right)\log\left(\frac{1}{2}\left(t + \sqrt{t^2-4}\right)\right) - \frac{1}{t}\sqrt{t^2-4}\right] & \text{if } t > 2, \\ -i\frac{3}{4}\left[\frac{1}{t}\sqrt{4-t^2} + \frac{2\pi}{t^2} - \frac{4}{t^2}\sin^{-1}\frac{t}{2} - 2\cos^{-1}\frac{t}{2}\right] & \text{if } 0 < t < 2 \end{cases} \tag{2.18}$$

and

$$\eta(t) = \frac{1}{2} \mp \frac{i\pi}{t^2}, \tag{2.19}$$

where the minus and plus signs in $\eta(t)$ correspond to the paths in the upper and the lower half of the plane, respectively. (For detailed calculations, we again refer to [13].) Note that as $t \to 2$, we have $\zeta(t) \to 0$. In fact, near $t = 2$, we have the Taylor expansion

$$\zeta(t) = (t-2)\left[1 - \frac{29}{60}(t-2) + \frac{799}{3150}(t-2)^2 + \cdots\right]. \tag{2.20}$$

The cubic equation (2.16) can be solved explicitly by using trigonometric functions; the solution is

$$u(s,t) = 2\zeta^{1/2}(t)\sin\frac{1}{3}\gamma, \tag{2.21}$$

where $\gamma$ satisfies

$$\sin\gamma = \frac{3}{2\zeta^{3/2}(t)}[\eta(t) - \phi(s;t)]; \tag{2.22}$$

see [5, p. 374]. The properties of the transformation $s \mapsto u$ are best seen by introducing an intermediate variable $Z$ defined by

$$u = 2\sqrt{\zeta(t)}\sin\frac{\gamma}{3}, \qquad \frac{2}{3}\zeta^{3/2}(t)\sin\gamma = Z, \quad Z = \eta(t) - \phi(s;t). \tag{2.23}$$

Note that here, for convenience, we simply write $u = u(s;t)$, $\gamma = \gamma(s;t)$ and $Z = Z(s;t)$. Since we require $s_+ \leftrightarrow u_+ = \zeta^{1/2}(t)$ and $s_- \leftrightarrow u_- = -\zeta^{1/2}(t)$, the transformation $s \mapsto Z$ take $s_+$ to $\frac{2}{3}\zeta^{3/2}(t)$ and $s_-$ to $-\frac{2}{3}\zeta^{3/2}(t)$.

The mapping $s \mapsto u$ defined in (2.16) is well studied in the literature; see, e.g. [11,5]. Here, we give only a brief sketch of the argument. (Details are given in [13].) First, we consider

Case 1: $t \geq 2$. To make the function $Z = Z(s,t)$ single-valued, we divide the upper $s$-plane into two regions by using part of the steepest-descent path passing through $s_+$; see Fig. 3.

Through the sequence of mappings $s \mapsto Z \mapsto \gamma \mapsto u$ in (2.23), the two regions bounded by ABB′CC′DD′EFGA and E′FGHIJE′ are mapped into two corresponding regions in the $u$-plane; see Fig. 4. Note that the function $\phi(s;t)$ has singularities at $s = 0$, $s = 1/t$ and saddle points $s = s_\pm$. One can show that the mapping $s \mapsto u$ is one-to-one and analytic on the boundaries of these two regions. Hence, it also has these properties in the interior of the regions; see [15, p. 12]. This

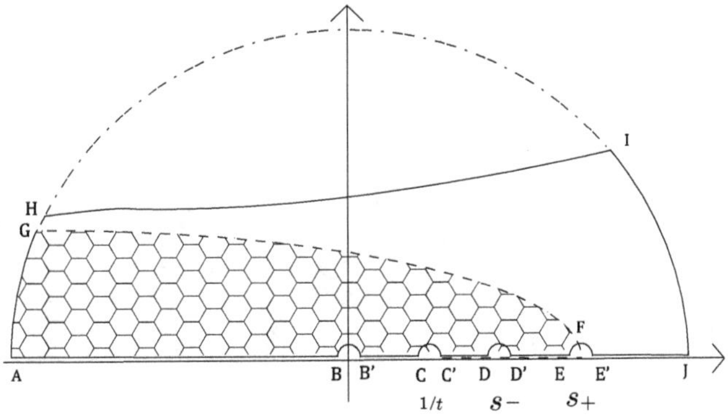


**Fig. 3.** $s$-plane ($t > 2$).

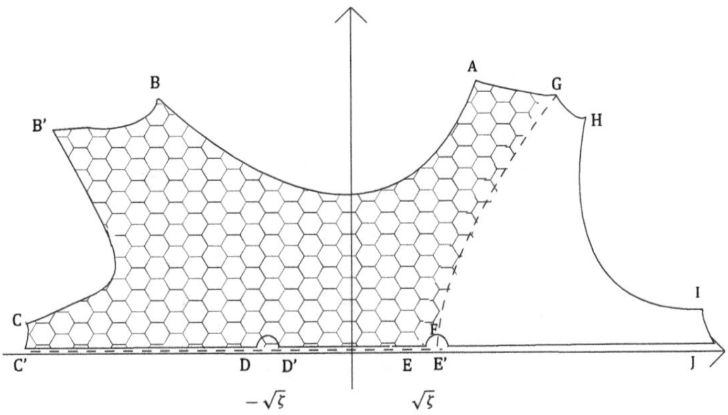


**Fig. 4.** $u$-plane ($t > 2$).

establishes the one-to-one and analytic nature of the mapping in the upper half-plane. The mapping properties of the lower half-plane can be obtained by using reflection with respect to the real-axis.

Case 2: $0 < t < 2$. This case is treated in a similar manner. We again divide the upper half of the $s$-plane into two pieces by the steepest-descent path passing through $s_+$; see Fig. 5.

The two regions with boundaries ABB′CDEA and $DC'$HGFED are mapped into two corresponding regions in the $u$-plane in a one-to-one analytic manner; see Fig. 6.

This property of the mapping $s \mapsto u$ in the lower half of the $s$-plane is again established by using reflection with respect to the real-axis.

We now return to the integral in (2.3)

$$f_n^{(\alpha)}(t/\sqrt{\nu}) = \frac{\nu^{-n/2}}{2\pi i}\int_{\mathcal{C}} e^{\nu\phi(s;t)}\frac{1}{s^{-2\alpha+1/2}s(1-ts)^{\alpha}}ds,$$

where $\mathcal{C}$ is a circle centered at the origin and with radius less than $1/t$. For $t \geq 2$, we see from Fig. 1 that $\mathcal{C}$ can be deformed into the steepest-descent paths shown in Fig. 7.

In the case when $0 < t < 2$, we deform $\mathcal{C}$ into the steepest-descent paths depicted in Fig. 2. In either case, the new contour will consist of two parts $\mathcal{C}_+$ and $\mathcal{C}_-$; $\mathcal{C}_+$ is the part in the upper half-plane and $\mathcal{C}_-$ is the part in the lower half-plane. Making the change of variable $s \mapsto u$ given in (2.16), we obtain

$$f_n^{(\alpha)}(t/\sqrt{\nu}) = \frac{\nu^{-n/2}}{2\pi i}e^{\nu\left(\frac{1}{2}-i\pi/t^2\right)}\int_{L_+} e^{\nu\left(\frac{1}{3}u^3-\zeta u\right)}h(u)\,du - \frac{\nu^{-n/2}}{2\pi i}e^{\nu\left(\frac{1}{2}+i\pi/t^2\right)}\int_{L_-} e^{\nu\left(\frac{1}{3}u^3-\zeta u\right)}h(u)\,du, \tag{2.24}$$

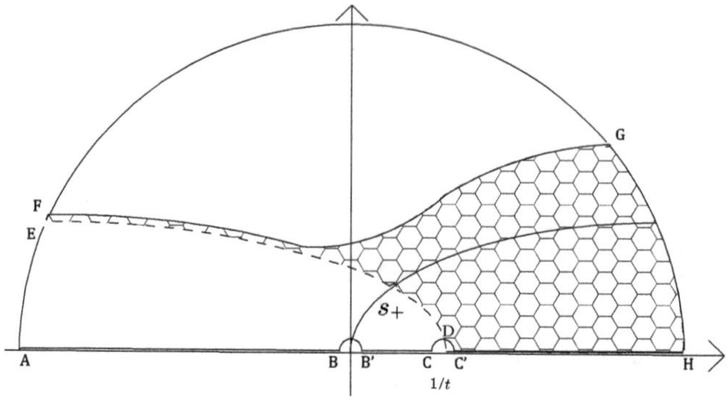


**Fig. 5.** $s$-plane $(0 < t < 2)$.

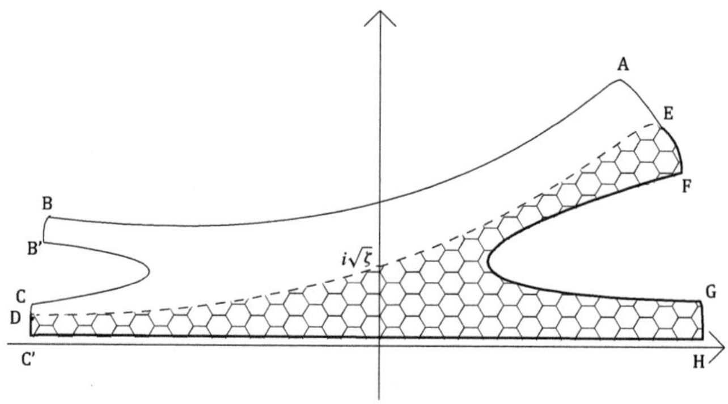


**Fig. 6.** $u$-plane $(0 < t < 2)$.

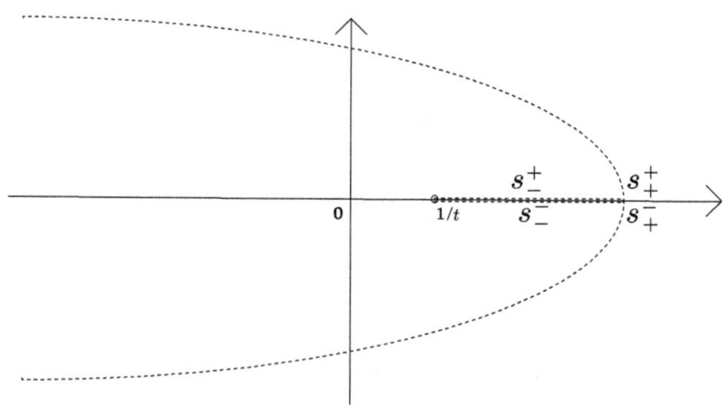


**Fig. 7.** Steepest-descent $(t \geq 2)$.

where $L_\pm$ are the images of $\mathcal{C}_\pm$ under the transformation and

$$h(u) = \frac{1}{s^{-2\alpha+1/2}s(1-ts)^\alpha}\frac{ds}{du}. \tag{2.25}$$

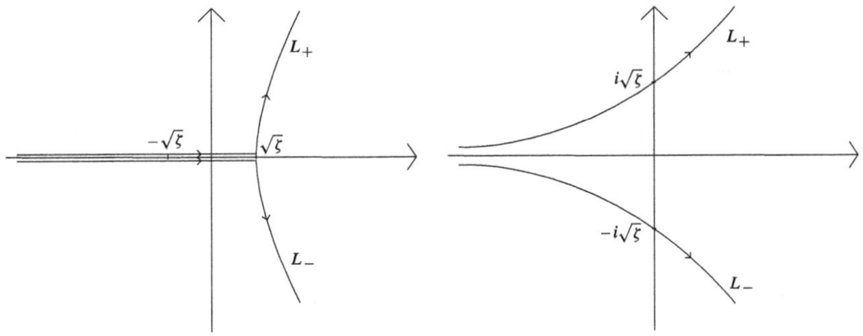


(a) $u$-plane ($t \geq 2$). (b) $u$-plane ($0 < t < 2$).

**Fig. 8.** Contours $L_+$ and $L_-$.

In (2.24), we have also made use of (2.19). Recall that a one-to-one analytic map takes steepest-descent paths to steepest-descent paths. Thus, the curves $L_+$ and $L_-$ are the steepest descent paths of the phase function $\frac{1}{3}u^3 - \zeta u$. Sketches of these curves are shown in Fig. 8.

Since $(1-ts)^\alpha = e^{-i\alpha\pi}(ts-1)^\alpha$ for $s \in \mathbb{C}_+$, (2.24) can be written as

$$f_n^{(\alpha)}(t/\sqrt{\nu}) = \frac{\nu^{-n/2}}{2\pi i} e^{\nu/2} e^{i(\alpha\pi - \pi\nu/t^2)} \int_{L_+} e^{\nu\left(\frac{1}{3}u^3 - \zeta u\right)} h(u)\, du - \frac{\nu^{-n/2}}{2\pi i} e^{\nu/2} e^{-i(\alpha\pi - \pi\nu/t^2)} \int_{L_-} e^{\nu\left(\frac{1}{3}u^3 - \zeta u\right)} h(u)\, du, \tag{2.26}$$

where

$$h(u) = \frac{1}{s^{-2\alpha+1/2} s(ts-1)^\alpha} \frac{ds}{du}. \tag{2.27}$$

## 3. Derivation of expansion

To derive the asymptotic expansion of $f_n^{(\alpha)}(t/\sqrt{\nu})$ from the integral representation (2.26), we apply the integration-by-parts technique developed by Bleistein [16]. First, we write

$$h_0(u) = h(u) = \alpha_0 + \beta_0 u + (u^2 - \zeta(t)) g_0(u). \tag{3.1}$$

Setting $u = \pm\sqrt{\zeta(t)}$ on both sides of the equation, we have

$$h_0\left(\sqrt{\zeta(t)}\right) = \alpha_0 + \beta_0\sqrt{\zeta(t)}, \qquad h_0\left(-\sqrt{\zeta(t)}\right) = \alpha_0 - \beta_0\sqrt{\zeta(t)}.$$

Hence,

$$\alpha_0 = \frac{1}{2}\left[h_0\left(\sqrt{\zeta(t)}\right) + h_0\left(-\sqrt{\zeta(t)}\right)\right], \qquad \beta_0 = \frac{1}{2\sqrt{\zeta(t)}}\left[h_0\left(\sqrt{\zeta(t)}\right) - h_0\left(-\sqrt{\zeta(t)}\right)\right]. \tag{3.2}$$

To obtain the values of $h_0(\pm\sqrt{\zeta(t)})$, we let $u \to -\sqrt{\zeta(t)}$ in the equation following (2.16). This gives

$$\left.\frac{ds}{du}\right|_{u=-\sqrt{\zeta(t)}} = \left[\frac{2\sqrt{\zeta(t)}\, s_-(1-ts_-)}{s_- - s_+}\right]^{1/2} = \left[\frac{2\sqrt{\zeta(t)}\, s_-(ts_- - 1)}{s_+ - s_-}\right]^{1/2}. \tag{3.3}$$

Similarly, we have

$$\left.\frac{ds}{du}\right|_{u=\sqrt{\zeta(t)}} = \left[\frac{2\sqrt{\zeta(t)}\, s_+(1-ts_+)}{s_- - s_+}\right]^{1/2} = \left[\frac{2\sqrt{\zeta(t)}\, s_+(ts_+ - 1)}{s_+ - s_-}\right]^{1/2}. \tag{3.4}$$

Note that the above two quantities become indeterminate (i.e. 0/0), when $t \to 2$. Thus, we must treat the case $t = 2$ with care. For $t > 2$, we have from (2.8) and (2.18)

$$\left.\frac{ds}{du}\right|_{u=\pm\sqrt{\zeta(t)}} = \frac{2^{\frac{1}{2}}}{(t^2-4)^{\frac{1}{4}}}\left[\frac{1}{2}\left(t \pm \sqrt{t^2-4}\right)\right]^{\frac{3}{2}}\left(\frac{3}{4}\right)^{\frac{1}{6}}\left[\left(2 - \frac{4}{t^2}\right)\log\left(\frac{1}{2}\left(t + \sqrt{t^2-4}\right)\right) - \frac{1}{t}\sqrt{t^2-4}\right]^{\frac{1}{6}}. \tag{3.5}$$

Similarly, for $0 < t < 2$, we have

$$\left.\frac{ds}{du}\right|_{u=\pm\sqrt{\zeta(t)}} = \frac{2^{\frac{1}{2}}}{(4-t^2)^{\frac{1}{4}}}\left[\frac{1}{2}\left(t \pm i\sqrt{4-t^2}\right)\right]^{\frac{3}{2}}\left(\frac{3}{4}\right)^{\frac{1}{6}}\left[\frac{1}{t}\sqrt{4-t^2}+\frac{2\pi}{t^2}-\frac{4}{t^2}\sin^{-1}\frac{t}{2}-2\cos^{-1}\frac{t}{2}\right]^{\frac{1}{6}}. \tag{3.6}$$

In the case when $t = 2$, we apply l'Hôpital's rule and obtain

$$\left.\frac{ds}{du}\right|_{u=\pm\sqrt{\zeta(2)}} = \lim_{t\to 2}\left.\frac{ds}{du}\right|_{u=\pm\sqrt{\zeta(t)}} = 1. \tag{3.7}$$

From (2.27), it follows that

$$h_0\left(\pm\sqrt{\zeta(t)}\right) = \frac{\sqrt{2}}{(t^2-4)^{1/4}}\left(\frac{3}{4}\right)^{\frac{1}{6}}\left[\left(2-\frac{4}{t^2}\right)\log\left(\frac{1}{2}\left(t+\sqrt{t^2-4}\right)\right)-\frac{1}{t}\sqrt{t^2-4}\right]^{\frac{1}{6}}$$

for $t > 2$,

$$h_0\left(\pm\sqrt{\zeta(t)}\right) = \frac{\sqrt{2}}{(4-t^2)^{1/4}}\left(\frac{3}{4}\right)^{\frac{1}{6}}\left[\frac{1}{t}\sqrt{4-t^2}+\frac{2\pi}{t^2}-\frac{4}{t^2}\sin^{-1}\frac{t}{2}-2\cos^{-1}\frac{t}{2}\right]^{\frac{1}{6}}$$

for $0 < t < 2$ and

$$h(\pm\sqrt{\zeta(t)}) = 1 \quad \text{for } t = 2.$$

Note that $u = \pm\sqrt{\zeta(t)}$ correspond to $s = s_\pm$. A combination of (3.2) and the last three equations gives

$$\alpha_0 = \frac{\sqrt{2}}{(t^2-4)^{1/4}}\left(\frac{3}{4}\right)^{\frac{1}{6}}\left[\left(2-\frac{4}{t^2}\right)\log\left(\frac{1}{2}\left(t+\sqrt{t^2-4}\right)\right)-\frac{1}{t}\sqrt{t^2-4}\right]^{\frac{1}{6}} \tag{3.8}$$

for $t > 2$,

$$\alpha_0 = \frac{\sqrt{2}}{(4-t^2)^{1/4}}\left(\frac{3}{4}\right)^{\frac{1}{6}}\left[\frac{1}{t}\sqrt{4-t^2}+\frac{2\pi}{t^2}-\frac{4}{t^2}\sin^{-1}\frac{t}{2}-2\cos^{-1}\frac{t}{2}\right]^{\frac{1}{6}} \tag{3.9}$$

for $0 < t < 2$ and

$$\alpha_0 = 1 \quad \text{for } t = 2. \tag{3.10}$$

Since $h(+\sqrt{\zeta(t)}) = h(-\sqrt{\zeta(t)})$ for all $t \in (0, \infty)$, we have

$$\beta_0 = 0 \quad \text{for } t \in (0, \infty). \tag{3.11}$$

Substituting (3.1) in (2.27), and integrating term-by-term, we encounter integrals which can be expressed in terms of the function

$$V(\lambda) = \int_\gamma e^{v^3/3-\lambda v}dv,$$

where $\gamma$ is one of the contours $\gamma_0$, $\gamma_1$ and $\gamma_2$ shown in Fig. 9.

In terms of the Airy function, we have

$$V(\lambda) = \int_{\gamma_j} e^{v^3/3-\lambda v}dv = -2\pi i\omega^j \mathrm{Ai}(\lambda\omega^j),$$

where $\omega$ is the cube root of unity (i.e. $\omega = e^{2\pi i/3}$). In view of the connection formulas

$$\mathrm{Ai}(z) + \omega\mathrm{Ai}(\omega z) + \omega^2\mathrm{Ai}(\omega^2 z) = 0, \tag{3.12}$$

$$\mathrm{Bi}(z) + i\omega\mathrm{Ai}(\omega z) - i\omega^2\mathrm{Ai}(\omega^2 z) = 0, \tag{3.13}$$

after rearranging the terms, Eq. (2.26) gives

$$\begin{aligned} e^{-\nu/2}\nu^{n/2}f_n^{(\alpha)}\left(\frac{t}{\sqrt{\nu}}\right) &= \cos\left(\pi\alpha-\frac{\pi\nu}{t^2}\right)\left[\mathrm{Ai}(\nu^{2/3}\zeta)\frac{\alpha_0}{\nu^{1/3}}-\mathrm{Ai}'(\nu^{2/3}\zeta)\frac{\beta_0}{\nu^{2/3}}\right] \\ &\quad + \sin\left(\pi\alpha-\frac{\pi\nu}{t^2}\right)\left[\mathrm{Bi}(\nu^{2/3}\zeta)\frac{\alpha_0}{\nu^{1/3}}-\mathrm{Bi}'(\nu^{2/3}\zeta)\frac{\beta_0}{\nu^{2/3}}\right]+\epsilon_1+\delta_1, \end{aligned} \tag{3.14}$$

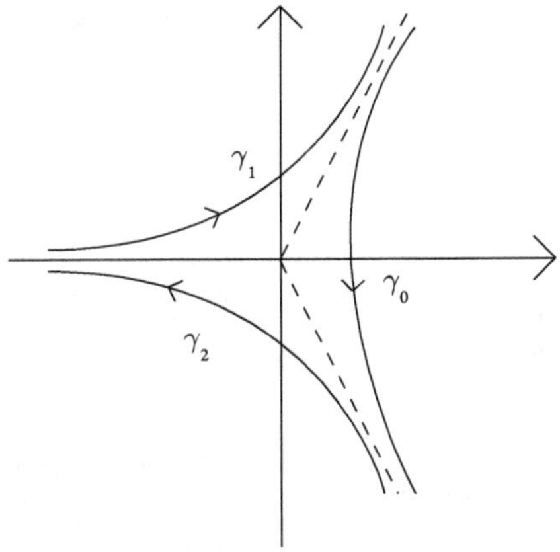


**Fig. 9.** Contours $\gamma_j$, $j = 0, 1, 2$.

where

$$\epsilon_1 = e^{i(\pi\alpha - \pi\nu/t^2)} \frac{1}{2\pi i} \int_{\gamma_1} e^{\nu(u^3/3 - \zeta u)} (u^2 - \zeta) g_0(u)\, du,$$
$$\delta_1 = e^{-i(\pi\alpha - \pi\nu/t^2)} \frac{1}{2\pi i} \int_{\gamma_2} e^{\nu(u^3/3 - \zeta u)} (u^2 - \zeta) g_0(u)\, du.$$

An integration by parts shows that

$$\begin{aligned} \epsilon_1 &= -e^{i(\pi\alpha - \pi\nu/t^2)} \frac{1}{2\pi i\nu} \int_{\gamma_1} e^{\nu(u^3/3 - \zeta u)} g_0'(u)\, du, \\ \delta_1 &= -e^{-i(\pi\alpha - \pi\nu/t^2)} \frac{1}{2\pi i\nu} \int_{\gamma_2} e^{\nu(u^3/3 - \zeta u)} g_0'(u)\, du. \end{aligned} \tag{3.15}$$

The error terms $\epsilon_1$ and $\delta_1$ now have an extra decaying factor $1/\nu$.

The above procedure can be repeated, and we define inductively

$$h_m(u) = \alpha_m + \beta_m u + (u^2 - \zeta) g_m(u), \tag{3.16}$$
$$h_{m+1}(u) = g_m'(u) \tag{3.17}$$

for $m = 0, 1, 2, \ldots$. As in (3.2), we have

$$\alpha_m = \frac{1}{2}\left[h_m\left(\sqrt{\zeta(t)}\right) + h_m\left(-\sqrt{\zeta(t)}\right)\right], \qquad \beta_m = \frac{1}{2\sqrt{\zeta(t)}}\left[h_m\left(\sqrt{\zeta(t)}\right) - h_m\left(-\sqrt{\zeta(t)}\right)\right]. \tag{3.18}$$

The final result is

$$\begin{aligned} e^{-\nu/2}\nu^{n/2} f_n^{(\alpha)}\left(\frac{t}{\sqrt{\nu}}\right) &= \cos\left(\pi\alpha - \frac{\pi\nu}{t^2}\right)\left[\mathrm{Ai}(\nu^{2/3}\zeta)\sum_{k=0}^{p-1}\frac{(-1)^k\alpha_k}{\nu^{k+1/3}} - \mathrm{Ai}'(\nu^{2/3}\zeta)\sum_{k=0}^{p-1}\frac{(-1)^k\beta_k}{\nu^{k+2/3}}\right] + \epsilon_p \\ &\quad + \sin\left(\pi\alpha - \frac{\pi\nu}{t^2}\right)\left[\mathrm{Bi}(\nu^{2/3}\zeta)\sum_{k=0}^{p-1}\frac{(-1)^k\alpha_k}{\nu^{k+1/3}} - \mathrm{Bi}'(\nu^{2/3}\zeta)\sum_{k=0}^{p-1}\frac{(-1)^k\beta_k}{\nu^{k+2/3}}\right] + \delta_p, \end{aligned} \tag{3.19}$$

where

$$\epsilon_p = e^{i(\pi\alpha - \pi\nu/t^2)} \frac{(-1)^p}{2\pi i\nu^p} \int_{\gamma_1} h_p(u) e^{\nu(u^3/3 - \zeta u)}\, du \tag{3.20}$$

and

$$\delta_p = e^{-i(\pi\alpha - \pi\nu/t^2)} \frac{(-1)^p}{2\pi i\nu^p} \int_{\gamma_2} h_p(u) e^{\nu(u^3/3 - \zeta u)}\, du \tag{3.21}$$

for $p = 1, 2, 3, \ldots$. To establish that (3.19) is indeed a uniform asymptotic expansion in $t$ (or $\zeta$), it still remains to show that there are constants $M_p$, $N_p$, $P_p$ and $Q_p$ such that

$$|\epsilon_p| \leq \frac{M_p}{\nu^{p+1/3}}|\mathrm{Ai}(\nu^{2/3}\zeta)| + \frac{N_p}{\nu^{p+2/3}}|\mathrm{Ai}'(\nu^{2/3}\zeta)| \tag{3.22}$$

and

$$|\delta_p| \leq \frac{P_p}{\nu^{p+1/3}}|\mathrm{Bi}(\nu^{2/3}\zeta)| + \frac{Q_p}{\nu^{p+2/3}}|\mathrm{Bi}'(\nu^{2/3}\zeta)|. \tag{3.23}$$

For bounded values of $\zeta$, the proofs of (3.22) and (3.23) are now routine, and we refer to [17, pp. 374–376] and [5, pp. 371–372] for details. The essential arguments here are: (i) to find the behaviors of the two integrals in (3.20) and (3.21), (ii) to compare them with the behaviors of the Airy functions Ai, Ai′, Bi and Bi′, and (iii) to observe that Ai and Ai′ do not have a common zero; neither do Bi and Bi′.

## 4. For unbounded values of $\zeta$

The advantage of the above derivation, based on Bleistein's method of integration by parts, is that it provides exact expressions for the error terms $\epsilon_p$ and $\delta_p$, and in addition that these expressions exhibit the decaying factor $1/\nu^p$ explicitly. Since the integrand $h_p(u) = h_p(u; t)$ in (3.20) and (3.21) is analytic in $t$, it is evident from these expressions that the expansion in (3.19) is asymptotic for bounded values of $t$. To show that it is an asymptotic expansion even for unbounded values of $t$, we use a more recent method of Olde Daalhuis and Temme [12] which was specifically designed for this kind of purpose. Thus, following [12], we introduce the sequence of rational functions

$$\begin{aligned} R_0(w, u, \sqrt{\zeta}) &= \frac{1}{w-u}, \\ R_{n+1}(w, u, \sqrt{\zeta}) &= \frac{-1}{w^2-\zeta}\frac{d}{dw}R_n(w, u, \sqrt{\zeta}), \quad n = 0, 1, \ldots, \end{aligned} \tag{4.1}$$

where $w, u, \sqrt{\zeta} \in \mathbb{C}$, $w \neq u$, and $w^2 \neq \zeta$. It can be shown that the function $h_n(u)$ defined by the recursive formula (4.1) has the representation

$$h_n(u) = \frac{1}{2\pi i}\int_\Gamma R_n(w, u, \sqrt{\zeta})\, h_0(w)dw, \tag{4.2}$$

where $\Gamma$ is any simple closed contour in the domain of analyticity of $h_0(u)$, enclosing the points $u$ and $\pm\sqrt{\zeta}$. Furthermore, it can be verified that there are constants $C_{ij}$, independent of $w$, $u$ and $\sqrt{\zeta}$ such that

$$R_n(w, u, \sqrt{\zeta}) = \sum_{i=0}^{n-1}\sum_{j=0}^{k_{n,i}} \frac{C_{ij}w^{i-j}}{(w-u)^{n+1-i-j}(w^2-\zeta)^{n+i}}, \quad n = 1, 2, \ldots, \tag{4.3}$$

where $k_{n,i} = \min(i, n-1-i)$. Note that the rational functions $R_n(w, u, \sqrt{\zeta})$ defined in (4.1) are independent of the function $h_0(u)$. Hence, (4.2) can be considered as an analogue of the Cauchy integral representation for the remainder in a Taylor expansion. To extend the validity of the asymptotic expansion in (3.19) from bounded $t$ to unbounded $t$, we first define

$$\rho_0(\sqrt{\zeta}) = \min\{|u \pm \sqrt{\zeta}| : u \text{ is a singularity of } h_0(u)\}. \tag{4.4}$$

For $t \geq 2$, the points $s_\pm = \frac{1}{2}(t \pm \sqrt{t^2-4})$ in the $s$-plane are mapped into $u = \pm\sqrt{\zeta}$, respectively, under the mapping $s \leftrightarrow u$ given by

$$\frac{s}{t} + \frac{1}{t^2}\log(1-ts) - \log s = \frac{1}{3}u^3 - \zeta(t)u + \eta(t), \tag{4.5}$$

where the logarithmic function is restricted to its principal value; see (2.4) and (2.16). However, the points $s_\pm$ on other sheets of the Riemann surface of logarithmic function are singular points of the mapping (4.5).

**Lemma 4.1.** *As $t \to \infty$ (and hence $\sqrt{\zeta} \to \infty$), we have*

$$\rho_0(\sqrt{\zeta}) \sim \frac{\sqrt{2\pi}}{\zeta^{1/4}}\left(1 - \frac{2}{t^2}\right)^{1/2}. \tag{4.6}$$

**Proof.** The cubic transformation (4.5) can be solved explicitly in terms of trigonometric functions. The solution that takes $s_\pm$ to $\pm\sqrt{\zeta}$, respectively, is given by (2.21) and (2.22); that is,

$$u(s, t) = 2\zeta^{1/2}(t)\sin\frac{1}{3}\gamma, \qquad \sin\gamma = \frac{3}{2\zeta^{3/2}(t)}[\eta(t) - \phi(s; t)].$$

Since $s = s_+$ is mapped into $u = \sqrt{\zeta}$, we obtain $\sqrt{\zeta} = 2\sqrt{\zeta}\sin\frac{1}{3}\gamma$ and

$$\sin\gamma = \frac{3}{2\zeta^{3/2}(t)}\left[\frac{1}{2} - \frac{s_+}{t} - \left(\frac{2}{t^2} - 1\right)\log s_+\right], \tag{4.7}$$

where we have made use of (2.10) and (2.19). Let $S_{\pm k}$ denote the singular points in the $u$-plane. When $u = S_{\pm k}$, we have $S_{\pm k} = 2\sqrt{\zeta}\sin\frac{1}{3}\gamma_{\pm k}$, and

$$\sin\gamma_{\pm k} = \frac{3}{2\zeta^{3/2}(t)}\left[\frac{1}{2} - \frac{s_+}{t} - \left(\frac{2}{t^2} - 1\right)(\log s_+ \pm 2k\pi i)\right]. \tag{4.8}$$

Note that the left-hand side of (4.7) is actually equal to 1. Hence, subtracting (4.7) from (4.8) gives

$$\sin\gamma_{\pm k} = 1 \pm \frac{3k\pi i}{\zeta^{3/2}(t)}\left(1 - \frac{2}{t^2}\right) \tag{4.9}$$

and

$$\gamma_{\pm k} = \arcsin\left[1 \pm \frac{3k\pi i}{\zeta^{3/2}(t)}\left(1 - \frac{2}{t^2}\right)\right] \sim \frac{\pi}{2} - \sqrt{\mp\frac{6k\pi i}{\zeta^{3/2}(t)}\left(1 - \frac{2}{t^2}\right)} \tag{4.10}$$

as $t \to \infty$ (or equivalently $\zeta(t) \to \infty$). The last asymptotic equality follows from an expansion for the arcsine function; see [18, p. 81, (4.4.41)]. Using the addition formula of the sine function and the Maclaurin expansions of sine and cosine, we have from the definition of $S_{\pm k}$ and asymptotic formula (4.10)

$$\begin{aligned} S_{\pm k} &= 2\sqrt{\zeta(t)}\sin\frac{1}{3}\gamma_{\pm k} \\ &\sim 2\sqrt{\zeta(t)}\left[\frac{1}{2}\cos\frac{1}{3}\sqrt{\mp\frac{6k\pi i}{\zeta^{3/2}(t)}\left(1 - \frac{2}{t^2}\right)} - \frac{\sqrt{3}}{2}\sin\frac{1}{3}\sqrt{\mp\frac{6k\pi i}{\zeta^{3/2}(t)}\left(1 - \frac{2}{t^2}\right)}\right] \\ &\sim \sqrt{\zeta(t)} - \sqrt{\mp\frac{2k\pi i}{\zeta^{1/2}(t)}\left(1 - \frac{2}{t^2}\right)}. \end{aligned} \tag{4.11}$$

Clearly, the two closest singularities to $\sqrt{\zeta(t)}$ are

$$S_1 \sim \sqrt{\zeta(t)} - \sqrt{-\frac{2\pi i}{\zeta^{1/2}(t)}\left(1 - \frac{2}{t^2}\right)} \quad \text{and} \quad S_{-1} \sim \sqrt{\zeta(t)} - \sqrt{\frac{2\pi i}{\zeta^{1/2}(t)}\left(1 - \frac{2}{t^2}\right)}. \tag{4.12}$$

Similarly, $s = s_-$ is mapped into $u = -\sqrt{\zeta(t)}$, and the two closest singularities to $-\sqrt{\zeta(t)}$ are

$$S_1' \sim -\sqrt{\zeta(t)} + \sqrt{\frac{2\pi i}{\zeta^{1/2}(t)}\left(1 - \frac{2}{t^2}\right)} \quad \text{and} \quad S_{-1}' \sim -\sqrt{\zeta(t)} + \sqrt{-\frac{2\pi i}{\zeta^{1/2}(t)}\left(1 - \frac{2}{t^2}\right)}; \tag{4.13}$$

thus, proving the lemma. □

To estimate the remainder $\epsilon_p$ in (3.20), we first deform the path of integration $\gamma_1$ into the contour $\gamma_{1a} + \gamma_{1b}$, where $\gamma_{1b}$ runs from $-\infty$ to $\sqrt{\zeta}$ along the real axis and $\gamma_{1a}$ continues on from $\sqrt{\zeta}$ to $\infty e^{i\pi/3}$ along the steepest descent path of the phase function $\frac{1}{3}u^3 - \zeta u$ in the first quadrant of the $u$-plane; see Fig. 10.

Next, we split $\gamma_{1a}$ into $\gamma_{1a}'$ and $\gamma_{1a}''$, where $\gamma_{1a}' = \{u \in \gamma_{1a} : |u - \zeta^{\frac{1}{2}}| \le c\,\zeta^{\frac{1}{2}\theta}\}$ and $\gamma_{1a}''$ is the complement of $\gamma_{1a}'$ on $\gamma_{1a}$ (i.e., $\gamma_{1a}'' = \gamma_{1a} \setminus \gamma_{1a}'$), $c$ and $\theta$ being constants satisfying $c > 0$ and $-1/2 < \theta \le 1$. Let $L_{1a}$ be a closed curve embracing the contour $\gamma_{1a}'$ such that as $\sqrt{\zeta} \to \infty$,

$$\text{length of } L_{1a} = O\left(\zeta^{\frac{1}{2}\theta}\right) \quad \text{and} \quad \text{distance } (L_{1a}, \gamma_{1a}') \sim c\,\zeta^{-\frac{1}{4}}. \tag{4.14}$$

Let $\Omega_{1a}$ denote the domain bounded by $L_{1a}$ and its closure by $\overline{\Omega}_{1a}$; see Fig. 11(a). Similarly, let $\gamma_{1b}' = \{u \in \gamma_{1b} : |u + \zeta^{\frac{1}{2}}| \le c\,\zeta^{\frac{1}{2}\theta}\}$ and $\gamma_{1b}'' = \gamma_{1b} \setminus \gamma_{1b}'$. Also, let $L_{1b}$ denote a closed curve embracing the contour $\gamma_{1b}'$ such that as $\sqrt{\zeta} \to \infty$,

$$\text{length of } L_{1b} = O\left(\zeta^{\frac{1}{2}\theta}\right) \quad \text{and} \quad \text{distance } (L_{1b}, \gamma_{1b}') \sim c\,\zeta^{-\frac{1}{4}}. \tag{4.15}$$

Let $\Omega_{1b}$ denote the domain bounded by $L_{1b}$ and its closure by $\overline{\Omega}_{1b}$; see Fig. 11(b). Put $\overline{\Omega}_{1,\theta} = \overline{\Omega}_{1a} \cup \overline{\Omega}_{1b}$. On account of Lemma 4.1, $h_0(u)$ is analytic in $\overline{\Omega}_{1,\theta}$, and hence bounded on $\overline{\Omega}_{1,\theta}$.

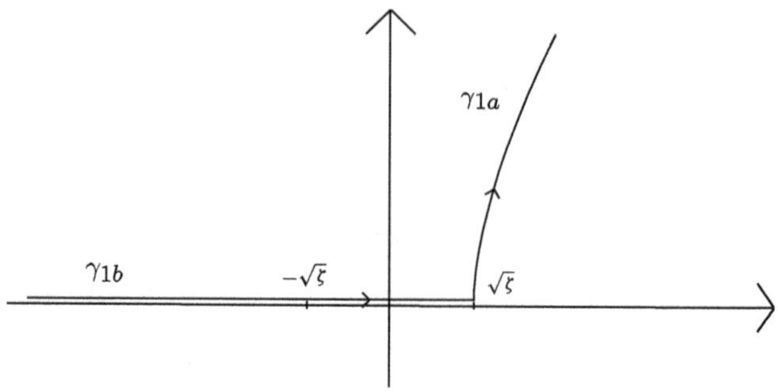


**Fig. 10.** Contour $\gamma_{1a}$ and $\gamma_{1b}$.

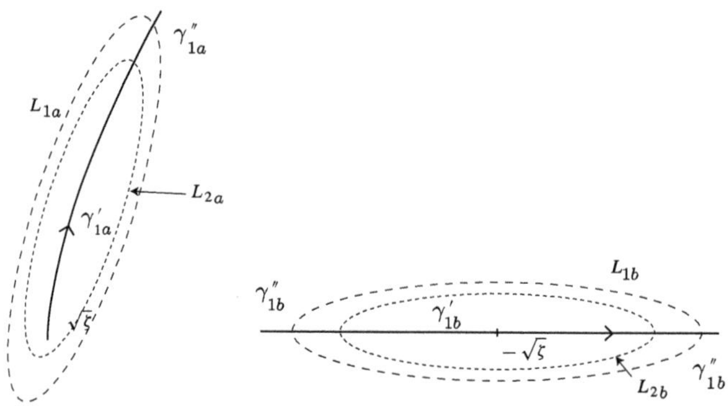


(a) Curves $L_{1a}$ and $L_{2a}$. (b) Curves $L_{1b}$ and $L_{2b}$.

**Fig. 11.** Curve $L$.

Similar to (4.14), we let $L_{2a}$ be a closed curve embracing the contour $\gamma'_{1a}$ such that as $\sqrt{\zeta} \to \infty$,

$$\text{length of } L_{2a} = O\left(\zeta^{\frac{1}{2}\theta}\right) \quad \text{and} \quad \text{distance } (L_{2a}, \gamma'_{1a}) \sim \frac{1}{2} c\, \zeta^{-\frac{1}{4}}. \tag{4.16}$$

Let $\Omega_{2a}$ denote the domain bounded by $L_{2a}$ and its closure by $\overline{\Omega}_{2a}$. Note that $L_{2a}$ lies inside $L_{1a}$. Furthermore, let $L_{2b}$ be a closed curve embracing the contour $\gamma'_{1b}$ such that as $\sqrt{\zeta} \to \infty$,

$$\text{length of } L_{2b} = O\left(\zeta^{\frac{1}{2}\theta}\right) \quad \text{and} \quad \text{distance } (L_{2b}, \gamma'_{1a}) \sim \frac{1}{2} c\, \zeta^{-\frac{1}{4}}. \tag{4.17}$$

Let $\Omega_{2b}$ denote the domain bounded by $L_{2b}$ and its closure by $\overline{\Omega}_{2b}$. Note that $L_{2b}$ lies inside $L_{1b}$. Put $\overline{\Omega}_{2,\theta} = \overline{\Omega}_{2a} \cup \overline{\Omega}_{2b}$. And $L_{1,\theta} = L_{1a} \cup L_{1b}$.

**Lemma 4.2.** *There exists a constant $c_0$ independent of $\sqrt{\zeta}$ such that as $\sqrt{\zeta} \to \infty$,*

$$\sup_{u \in \overline{\Omega}_{1,\theta}} |h_0(u)| \le c_0 |h_0(\sqrt{\zeta})| \left(\zeta^{\frac{1}{2}\theta + \frac{1}{4}}\right)^{2\alpha - 2}. \tag{4.18}$$

**Proof.** Note that from (2.18) and the equations following (3.7), we have

$$h_0(\pm\sqrt{\zeta}) = \frac{\sqrt{2}\zeta^{\frac{1}{4}}}{(t^2 - 4)^{\frac{1}{4}}}. \tag{4.19}$$

To bound $h_0(u)$, first we consider $u \in \overline{\Omega}_{1a}$ and estimate $s - s_+$. Replacing $s$ by $s_+$ in (4.5) and subtracting the resulting equation from (4.5), we obtain

$$\frac{s-s_+}{t} + \frac{1}{t^2}\log\left(1+\frac{t(s-s_+)}{s_+s_+}\right) - \log\left(\frac{s-s_+}{t}+1\right) = \frac{1}{3}(u+2\sqrt{\zeta})(u-\sqrt{\zeta})^2, \tag{4.20}$$

where use has been made of two identities following (2.17). Since there could be points $u \in \overline{\Omega}_{1a}$ such that $|u-\sqrt{\zeta}| \sim c\,\zeta^{\frac{1}{2}\theta}$ with $-\frac{1}{2} < \theta \leq 1$, for these $u$ the right-hand side of (4.20) behaves like $c^2\zeta^{2\theta+\frac{1}{2}}$. If $\frac{1}{t}(s-s_+) = O(1)$, then the left-hand side of (4.20) is bounded, which is a contradiction. Thus, (4.20) gives

$$\frac{1}{t}|s-s_+| \sim c^2\zeta^{2\theta+\frac{1}{2}} \quad \text{as } t \to \infty. \tag{4.21}$$

In view of the first equation following (2.17), it is easy to show that

$$\frac{1}{t}|s-s_-| \sim c^2\zeta^{2\theta+\frac{1}{2}} \quad \text{as } t \to \infty. \tag{4.22}$$

A combination of (2.27), (4.21), (4.22) and the equation following (2.16) gives

$$\left|\frac{h_0(u)}{h_0(\sqrt{\zeta})}\right| \sim \frac{1}{t^{\frac{1}{2}}c^{-2\alpha+1}\left(\zeta^{\frac{1}{2}\theta+\frac{1}{4}}\right)^{-2\alpha+1}}\,\frac{\sqrt{2}(t^2-4)^{\frac{1}{4}}}{c\,\zeta^{\frac{1}{2}\theta+\frac{1}{4}}} \sim \sqrt{2}c^{2\alpha-2}\left(\zeta^{\theta+\frac{1}{2}}\right)^{\alpha-1}, \tag{4.23}$$

which certainly implies (4.18). □

Returning to (3.20), we write

$$\epsilon_p = \epsilon_p'(1a) + \epsilon_p''(1a) + \epsilon_p'(1b) + \epsilon_p''(1b), \tag{4.24}$$

where

$$\epsilon_p'(1a) = e^{i(\pi\alpha-\pi\nu/t^2)}\frac{(-1)^p}{2\pi i\nu^p}\int_{\gamma_{1a}'} h_p(u)e^{\nu(u^3/3-\zeta u)}\,du, \tag{4.25}$$

$$\epsilon_p''(1a) = e^{i(\pi\alpha-\pi\nu/t^2)}\frac{(-1)^p}{2\pi i\nu^p}\int_{\gamma_{1a}''} h_p(u)e^{\nu(u^3/3-\zeta u)}\,du, \tag{4.26}$$

and $\epsilon_p'(1b)$, $\epsilon_p''(1b)$ are defined in a similar manner. By simple calculation, it can easily be shown that $\epsilon_p''(1a)$ and $\epsilon_p''(1b)$ are exponentially small in comparison with $\epsilon_p'(1a)$ and $\epsilon_p'(1b)$; see [12]. Furthermore, for $w \in L_{1a}$ and $u \in \overline{\Omega}_{2a}$, we have

$$|w-u| \geq \frac{1}{2}c\,\zeta^{-\frac{1}{4}}, \qquad |w-\sqrt{\zeta}| \geq \frac{1}{2}c\,\zeta^{-\frac{1}{4}} \quad \text{and} \quad |w+\sqrt{\zeta}| \geq \sqrt{\zeta}. \tag{4.27}$$

Thus, from (4.3) with $n$ replaced by $p$, we obtain

$$|R_p(w,u,\sqrt{\zeta})| \leq \sum_{i=0}^{p-1}\sum_{j=0}^{k_{p,i}} \frac{|C_{ij}|(2\sqrt{\zeta})^{i-j}}{|w-u|^{p+1-i-j}|w-\sqrt{\zeta}|^{p+i}(\sqrt{\zeta})^{p+i}}, \tag{4.28}$$

$p = 1, 2, \ldots$. Do the same for $w \in L_{1b}$ and $u \in \overline{\Omega}_{2b}$, and obtain an estimate corresponding to (4.28). Then, we have

$$\sup_{w\in L_{1,\theta},\, u\in\overline{\Omega}_{2,\theta}} |R_p(w,u,B)| \leq \sum_{i=0}^{p-1}\sum_{j=0}^{k_{p,i}} \widetilde{C}_{ij}\,\zeta^{-\frac{3i}{4}+\frac{1}{4}} \leq A_p\,\zeta^{\frac{1}{4}} \tag{4.29}$$

as $\sqrt{\zeta} \to \infty$, where the constants $\widetilde{C}_{ij}$ and $A_p$ are constants independent of $\sqrt{\zeta}$. A combination of (4.18), (4.29) and (4.14) gives

$$\left|\frac{1}{2\pi i}\int_{L_{1,\theta}} R_p(w,u,\sqrt{\zeta})\,h_0(w)\,dw\right| \leq A_p'\left(\zeta^{\theta+\frac{1}{2}}\right)^{\alpha-\frac{1}{2}}|h_0(\sqrt{\zeta})|, \tag{4.30}$$

where $A_p'$ is again a constant. By the Cauchy integral formula, and (3.17), we have upon integration by parts

$$h_p(u) = -\frac{1}{2\pi i}\int_{L_{1,\theta}} \frac{d}{dw}R_0(w,u,\sqrt{\zeta})\,g_{p-1}(w)\,dw. \tag{4.31}$$

Furthermore, from (4.1) and (3.16) it follows that

$$h_p(u) = \frac{1}{2\pi i}\int_{L_{1,\theta}} (w^2-\zeta)R_1(w,u,\sqrt{\zeta})\,g_{p-1}(w)\,dw$$
$$= \frac{1}{2\pi i}\int_{L_{1,\theta}} R_1(w,u,\sqrt{\zeta})\,h_{p-1}(w)\,dw - \frac{1}{2\pi i}\int_{L_{1,\theta}} R_1(w,u,\sqrt{\zeta})\,(\alpha_{p-1}+\beta_{p-1}w)\,dw. \tag{4.32}$$

The coefficients $\alpha_{p-1}$ and $\beta_{p-1}$ are given in (3.18). In [12, (4.6)], it has been shown that for $p = 1, 2, \ldots,$

$$\frac{1}{2\pi i}\int_{L_{1,\theta}} R_p(w,u,\sqrt{\zeta})\,h_{p-1}(w)\,dw = O(\zeta^{-3p/2}). \tag{4.33}$$

Since the expression for $\beta_{p-1}$ has a factor of $\zeta^{-1/2}$, in view of (3.18) and (4.33), the last integral in (4.32) is dominated by $\widetilde{h}_p\,O(\zeta^{-3/2})$, where we define

$$\widetilde{h}_p := \sup_{u\in\overline{\Omega}_{1,\theta}} |h_p(u)|. \tag{4.34}$$

The above process can be repeated, and we obtain

$$h_p(u) = \frac{1}{2\pi i}\int_{L_{1,\theta}} R_p(w,u,\sqrt{\zeta})\,h_0\,dw + \widetilde{h}_{p-1}O(\zeta^{-3/2}) + \cdots + \widetilde{h}_0 O(\zeta^{-3p/2}). \tag{4.35}$$

From (4.30) and (4.35), we have by induction

$$\widetilde{h}_p \le \widetilde{c}_p\left(\zeta^{\theta+\frac{1}{2}}\right)^{\alpha-\frac{1}{2}} |h_0(\sqrt{\zeta})|, \tag{4.36}$$

where $\widetilde{c}_p$ is a constant independent of $\sqrt{\zeta}$.

Returning to (4.24)–(4.26), we have

$$|\epsilon_p| \le \nu^{-p-\frac{1}{3}}(\widetilde{c}_p+1)\left(\zeta^{\theta+\frac{1}{2}}\right)^{\alpha-\frac{1}{2}} |h_0(\sqrt{\zeta})|\mathrm{Ai}(\nu^{2/3}\zeta), \tag{4.37}$$

since $\epsilon''_p(1a)$ and $\epsilon''_p(1b)$ are exponentially small in comparison with $\epsilon'_p(1a)$ and $\epsilon'_p(1b)$; see the statement following (4.25)–(4.26). A similar estimate, with $\mathrm{Ai}(\nu^{2/3}\zeta)$ in (4.37) replaced by $\mathrm{Bi}(\nu^{2/3}\zeta)$, holds for the remainder $\delta_p$ in (3.21). We have thus established the following result.

**Theorem 4.3.** *The expansion in* (3.19) *holds with the reminders $\epsilon_p$ and $\delta_p$ satisfying*

$$|\epsilon_p| + |\delta_p| \le \nu^{-p-\frac{1}{3}}(\widetilde{c}_p+1)\left(\zeta^{\theta+\frac{1}{2}}\right)^{\alpha-\frac{1}{2}} |h_0(\sqrt{\zeta})|[\mathrm{Ai}(\nu^{2/3}\zeta) + \mathrm{Bi}(\nu^{2/3}\zeta)], \tag{4.38}$$

*where $|h_0(\sqrt{\zeta})|$ is given by* (4.19) *and $\widetilde{c}_p$ is a constant.*

We have now extended the validity of the expansion (3.19) to all positive values of $t$; that is, (3.19) is indeed an asymptotic expansion which holds uniformly with respect to all $t \ge 0$.

## 5. Error bounds

To construct an error bound for the asymptotic expansion in (3.19) means to find a value for the constant $\widetilde{c}_p$ in (4.38), or to find values for the constants $M_p$, $N_p$, $P_p$ and $Q_p$ in (3.22) and (3.23). In general, this problem is very difficult if not impossible, and the desire to construct such bounds has been expressed in a survey article of Wong [19]. In the advent of powerful computers, the problem has now become more feasible. In this section, we will provide some basic steps in getting computable values for the constants mentioned above.

To simplify the matter, and to make the problem more tangible, we take $p = 1$ so that we need only deal with the two integrals

$$\epsilon_1 = -\frac{1}{\nu}e^{i(\pi\alpha-\pi\nu/t^2)}\frac{1}{2\pi i}\int_{\gamma_1} h_1(u)e^{\nu(u^3/3-\zeta u)}\,du \tag{5.1}$$

and

$$\delta_1 = -\frac{1}{\nu}e^{-i(\pi\alpha-\pi\nu/t^2)}\frac{1}{2\pi i}\int_{\gamma_2} h_1(u)e^{\nu(u^3/3-\zeta u)}\,du. \tag{5.2}$$

**Table 1**
Numerical values of $c_1$ with different values of $n$ and $\alpha = 1$.

| $t$ | $n = 100$ | $n = 500$ | $n = 1000$ |
|---|---|---|---|
| 0.5 | 0.3165 | N.A. | N.A. |
| 1.5 | 0.3705 | 0.2487 | 0.0937 |
| 2.0 | 0.000001543 | 0.000007818 | 0.00001915 |
| 2.5 | 0.6845 | 0.6846 | 0.6843 |
| 3.0 | 0.7679 | 0.7666 | 0.7663 |
| 3.5 | 0.7849 | 0.1913 | 0.6955 |
| 4.0 | 0.8866 | 0.8832 | 0.9580 |

As in Section 4, we deform the path of integration $\gamma_1$ in (5.1) into the contour $\gamma_{1a} + \gamma_{1b}$; see Fig. 10. If we write $u = \sigma + i\tau$, then the parametric equation for the steepest descent path $\gamma_{1a}$ passing through $\sqrt{\zeta}$ is given by $\tau^2 = 3\sigma^2 - 3\sqrt{\zeta}$. From (3.16) and (3.17), we have

$$h_1(u) = g_0'(u) = \frac{1}{(u^2-\zeta)^2}\left[h_0'(u)(u^2-\zeta) - 2u\,(h_0(u) - \alpha_0)\right]. \tag{5.3}$$

(Note that in our case $\beta_0 = 0$.) Since expansion (3.19) was purposely designed to be uniformly valid in a neighborhood of the critical value $t = 2$, let us restrict it to the interval $0 < t < 4$. Since $\alpha$ is a fixed number, it is relatively small in comparison to the large asymptotic variable $\sqrt{\nu}$ in the function $f_n^{(\alpha)}(t/\sqrt{\nu})$ under consideration. For simplicity, we also restrict it to $0 < \alpha < 4$. What we shall show in the following is that numerical evidence indicates that with these restrictions, there is a positive constant $c_1 < 2$ such that

$$\sup_{u\in\gamma_{1a}} |h_1(u)| \le c_1|h_1(\sqrt{\zeta})| \quad\text{and}\quad \sup_{u\in\gamma_{1b}} |h_1(u)| \le c_1|h_1(-\sqrt{\zeta})|. \tag{5.4}$$

Applying (5.4) to (5.1) and (5.2) will yield the error bound

$$|\epsilon_1 + \delta_1| \le |\epsilon_1| + |\delta_1| \le \frac{c_1}{\nu^{4/3}}\left[\left|h_1\left(\sqrt{\zeta}\right)\right|\,\left|\mathrm{Ai}\left(\nu^{2/3}\zeta\right)\right| + \left|h_1\left(-\sqrt{\zeta}\right)\right|\,\left|\mathrm{Bi}\left(\nu^{2/3}\zeta\right)\right|\right]; \tag{5.5}$$

cf. (3.22) and (3.23). Here,

$$\begin{aligned} h_1(\pm\sqrt{\zeta}) &= \pm\frac{h_0(\sqrt{\zeta})}{4\zeta}\left\{\frac{2\zeta}{s_\pm\sqrt{t^2-4}}\left[\left(2\alpha-\frac{3}{2}\right)\left(2\alpha-\frac{5}{2}\right)s_\pm^2 - 2\left(2\alpha-\frac{3}{2}\right)\alpha t s_\pm + \alpha(\alpha+1)t^2\right]\right.\\ &\quad\left.\mp\frac{\zeta}{t^2-4}\frac{1}{s_\pm^2}\left[4(t-s_\pm^3)\left(\left(2\alpha-\frac{3}{2}\right)s_\pm - \alpha t\right) + 3(s_\pm^4 - t^2)\right] + \frac{10\zeta\,(t-s_\pm^3)^2}{3(t^2-4)^{\frac{3}{2}}s_\pm^3} - \frac{5}{12\sqrt{\zeta}}\right\}. \end{aligned} \tag{5.6}$$

Now, we proceed to describe how to obtain a value for the constant $c_1$. We only consider the case $2 \le t < 4$, since the other case $0 < t \le 2$ can be handled in a similar manner. Using a computer, one can calculate the values of $h_1(u)$ along the contours $\gamma_{1a}$ and $\gamma_{1b}$, except possibly at the saddle points $u = \pm\sqrt{\zeta}$. Note that the denominator of $h_1(u)$ vanishes at $u = \pm\sqrt{\zeta}$; see (5.3). Numerical computation indicates that the supremum of $h_1(u)$ on $\gamma_{1a}$ occurs near $u = \sqrt{\zeta}$, and that the supremum of $h_1(u)$ on $\gamma_{1b}$ occurs near $u = -\sqrt{\zeta}$. To avoid calculation of indeterminate forms near $\pm\sqrt{\zeta}$, we draw a small circle with the center at each of these two points. Since $h_1(u)$ is analytic along the contour $\gamma_{1a}$ and $\gamma_{1b}$, by the maximum modulus principle the maximum of $h_1(u)$ inside the circles occur on the boundaries of these circles. Comparing the maximal values of $h_1(u)$ on the circles and those on the steepest descent paths $\gamma_{1a}$ and $\gamma_{1b}$ outside the circles, we find that the former are bigger than the latter. This led us to the claim stated in (5.4). To get an estimate on the constant $c_1$, we examine the ratio

$$\left|\frac{\text{exact value} - \text{leading term}}{\text{error term in(5.5)}}\right|, \tag{5.7}$$

where the exact value is $f_n^{(\alpha)}(t/\sqrt{\nu})$, the leading term is

$$e^{\nu/2}\nu^{-n/2-1/3}\alpha_0\left[\cos\left(\pi\alpha - \frac{\pi\nu}{t^2}\right)\mathrm{Ai}(\nu^{2/3}\zeta) + \sin\left(\pi\alpha - \frac{\pi\nu}{t^2}\right)\mathrm{Bi}(\nu^{2/3}\zeta)\right] \tag{5.8}$$

obtained from (3.14) with $\beta_0 = 0$, and the error estimate in (5.5) is

$$e^{\nu/2}\nu^{-n/2-4/3}\left(|h_1(\sqrt{\zeta})|\,|\mathrm{Ai}(\nu^{2/3}\zeta)| + |h_1(-\sqrt{\zeta})|\,|\mathrm{Bi}(\nu^{2/3}\zeta)|\right). \tag{5.9}$$

Numerical computation shows that for $0 < \alpha < 4$ and $0 < t < \infty$, the values of the ratio in (5.7) lie between 0 and 2; see Tables 1 and 2. This confirms the estimate $0 < c_1 < 2$ mentioned earlier.

**Table 2**
Numerical values of $c_1$ with different values of $n$ and $\alpha = 3$.

| $t$ | $n = 100$ | $n = 500$ | $n = 1000$ |
|---|---|---|---|
| 0.5 | 0.4685 | N.A. | N.A. |
| 1.5 | 0.7641 | 0.1298 | 0.6106 |
| 2.0 | 0.00004671 | 0.00002387 | 0.000009098 |
| 2.5 | 0.4176 | 0.3775 | 0.3727 |
| 3.0 | 0.9348 | 0.5197 | 0.7809 |
| 3.5 | 1.2532 | 0.7817 | 0.2606 |
| 4.0 | 1.4302 | 1.0293 | 0.4887 |

**Table 3**
Numerical values of $c_2$ when $n = 100$.

| $t$ | $\alpha = 1$ | $\alpha = 3$ |
|---|---|---|
| 0.5 | 0.2218 | 0.7571 |
| 1.5 | 0.5866 | 0.4855 |
| 2.0 | 0.0319 | 0.9556 |
| 2.5 | 0.6565 | 0.4162 |
| 3.0 | 0.7193 | 0.9272 |
| 3.5 | 0.7126 | 1.2348 |
| 4.0 | 0.7781 | 1.3975 |

In principle, one can construct error bounds for higher order asymptotic approximations, but the amount of work would be tremendous. With the aid of a computer, we also find the estimate

$$|\epsilon_2 + \delta_2| \le \frac{c_2}{\nu^{7/3}} \left[ |h_2(\sqrt{\zeta})|\,|\mathrm{Ai}(\nu^{2/3}\zeta)| + |h_2(-\sqrt{\zeta})|\,|\mathrm{Bi}(\nu^{2/3}\zeta)| \right]. \tag{5.10}$$

Table 3 provides the numerical evidence that the constant $c_2$ satisfies the estimate $0 < c_2 < 2$.

To conclude this section, we wish to point out that it is difficult to compute the exact values of $f_n^{(\alpha)}(t/\sqrt{\nu})$ for large values of $n$ (say, e.g., $n = 500, 1000$) and small positive values of $t$. This is mainly because of the factor $x^{-n}$ in the exact expression of the Tricomi–Carlitz polynomial

$$\begin{aligned} f_n^{(\alpha)}(x) &= \frac{1}{x^n} \sum_{k=0}^{n} (-1)^k \binom{x^{-2} - \alpha}{k} \frac{x^{2k}}{(n-k)!} \\ &= \frac{1}{n!x^n} \sum_{k=0}^{n} (-1)^k \binom{n}{k} (1 - \alpha x^2) \cdots (1 - (\alpha + k - 1)x^2), \end{aligned} \tag{5.11}$$

which can be easily derived from the generalized hypergeometric function in (1.8) or its relation with the Tricomi polynomials (given in [1, (1.1)]). Note that the second sum in (5.11) is a polynomial of degree $2n$, and that its coefficients of $x^k$, $k = 0, 1, \ldots, n-1$, are all zero, that is, the right-hand side of (5.11) is indeed a polynomial of degree $n$. In view of this difficulty, in the construction of Tables 1 and 2 we have used the three-term recurrence relation (1.1) to compute the exact value of $f_n^{(\alpha)}(t/\sqrt{\nu})$ for $t = 0.5$ and $n = 100$. (When $n = 500$ or $n = 1000$ and $t = 0.5$, the computation is just too big for our computer to produce a reasonable figure even by using (1.1).) The uniform asymptotic approximation (3.14) now becomes very useful, since its validity includes neighborhoods of $t = 0$ and $t = 2$, and there is no problem with calculating $\sin(\pi\alpha - \pi\nu/t^2)$ and $\cos(\pi\alpha - \pi\nu/t^2)$ even for small values of $t$.

## 6. Zeros

It is well known that the zeros of the Tricomi–Carlitz Polynomials $f_n^{(\alpha)}(x)$ all lie in the interval $-1/\sqrt{\alpha} \le x \le 1/\sqrt{\alpha}$; see [7]. After the rescaling $x = t/\sqrt{\nu}$ in (2.2), the zeros of $f_n^{(\alpha)}(t/\sqrt{\nu})$ all lie in the interval $-\sqrt{\nu/\alpha} \le t \le \sqrt{\nu/\alpha}$. In this section, we shall use the asymptotic approximation given in (3.14) to derive asymptotic formulas for the zeros of $f_n^{(\alpha)}(t/\sqrt{\nu})$ as $n \to \infty$. We divide our discussion into three cases: (i) large zeros in $2 < t < \sqrt{\nu/\alpha}$; (ii) small zeros in $0 < t < 2$; and (iii) zeros on both sides of $t = 2$. Several tables are provided to give a comparison of the asymptotic zeros and the true zeros.

Case (i). Let $t_{n,m}$ denote the $m$-th zero of $f_n^{(\alpha)}(t/\sqrt{\nu})$, arranged in ascending order

$$-\sqrt{\frac{\nu}{\alpha}} \le t_{n,1} < t_{n,2} < \cdots < t_{n,n-1} < t_{n,n} \le \sqrt{\frac{\nu}{\alpha}}. \tag{6.1}$$

For large zeros, we have the following result.

**Table 4**
True values and approximate values of the large zeros evaluated with Maple, when $n = 100$ and $\alpha = 1$.

| m-th | True | Approx. | Error % |
|---|---|---|---|
| $t_{100,91}$ | 3.1859 | 3.1859 | $5.3246 \times 10^{-9}$ |
| $t_{100,92}$ | 3.3582 | 3.3582 | $1.0000 \times 10^{-8}$ |
| $t_{100,93}$ | 3.5620 | 3.5620 | $-1.9025 \times 10^{-9}$ |
| $t_{100,94}$ | 3.8079 | 3.8079 | $2.4413 \times 10^{-9}$ |
| $t_{100,95}$ | 4.1130 | 4.1130 | $-8.8517 \times 10^{-9}$ |
| $t_{100,96}$ | 4.5056 | 4.5056 | $-9.4228 \times 10^{-12}$ |
| $t_{100,97}$ | 5.0374 | 5.0374 | 0.0 |
| $t_{100,98}$ | 5.8166 | 5.8167 | $-1.7967 \times 10^{-9}$ |
| $t_{100,99}$ | 7.1239 | 7.1239 | $-1.9025 \times 10^{-9}$ |
| $t_{100,100}$ | 10.0747 | 10.0747 | 0.0 |

**Theorem 6.1.** *The asymptotic behavior of the large zeros of $f_n^{(\alpha)}(t/\sqrt{\nu})$ can be approximated by*

$$t_{n,n-k} = \sqrt{\frac{\nu}{\alpha+k}}\left[1 + O(\nu^{-3/4}\log^{1/6}\nu)\right] \tag{6.2}$$

*as $n \to \infty$, for fixed $k = 0, 1, 2, \ldots$, where $\nu = n + 2\alpha - \frac{1}{2}$.*

**Proof.** Returning to (3.14), we note that the coefficient $\beta_0$ is zero; see (3.11). Hence, it follows from (3.22) and (3.23) that

$$e^{-\nu/2}\nu^{n/2}f_n^{(\alpha)}(t/\sqrt{\nu}) = \left[\cos\left(\pi\alpha - \frac{\pi\nu}{t^2}\right)\mathrm{Ai}(\nu^{2/3}\zeta) + \sin\left(\pi\alpha - \frac{\pi\nu}{t^2}\right)\mathrm{Bi}(\nu^{2/3}\zeta)\right]\frac{\alpha_0}{\nu^{1/3}} + \epsilon_1 + \delta_1, \tag{6.3}$$

where

$$|\epsilon_1| \le \frac{M_1}{\nu^{4/3}}|\mathrm{Ai}(\nu^{2/3}\zeta)| + \frac{N_1}{\nu^{5/3}}|\mathrm{Ai}'(\nu^{2/3}\zeta)| \tag{6.4}$$

and

$$|\delta_1| \le \frac{P_1}{\nu^{4/3}}|\mathrm{Bi}(\nu^{2/3}\zeta)| + \frac{Q_1}{\nu^{5/3}}|\mathrm{Bi}'(\nu^{2/3}\zeta)|. \tag{6.5}$$

Next, we recall the well-known asymptotic results

$$\mathrm{Ai}(\eta) \sim \frac{\eta^{-\frac{1}{4}}}{2\sqrt{\pi}}\exp\left(-\frac{2}{3}\eta^{\frac{3}{2}}\right), \qquad \mathrm{Ai}'(\eta) \sim -\frac{\eta^{\frac{1}{4}}}{2\sqrt{\pi}}\exp\left(-\frac{2}{3}\eta^{\frac{3}{2}}\right), \tag{6.6}$$

and

$$\mathrm{Bi}(\eta) \sim \frac{\eta^{-\frac{1}{4}}}{\sqrt{\pi}}\exp\left(\frac{2}{3}\eta^{\frac{3}{2}}\right), \qquad \mathrm{Bi}'(\eta) \sim \frac{\eta^{\frac{1}{4}}}{\sqrt{\pi}}\exp\left(\frac{2}{3}\eta^{\frac{3}{2}}\right), \tag{6.7}$$

as $\eta \to +\infty$; see [14, pp. 392–393]. Furthermore, since $\nu = n + 2\alpha - \frac{1}{2}$ and $\zeta > 0$ for $t > 2$, the terms involving $\mathrm{Ai}(\nu^{2/3}\zeta)$ and $\mathrm{Ai}'(\nu^{2/3}\zeta)$ are exponentially small, while the terms involving $\mathrm{Bi}(\nu^{2/3}\zeta)$ and $\mathrm{Bi}'(\nu^{2/3}\zeta)$ are exponentially large. Thus, (6.3) can be reduced to

$$e^{-\nu/2}\nu^{n/2}f_n^{(\alpha)}(t/\sqrt{\nu}) = \sin\left(\pi\alpha - \frac{\pi\nu}{t^2}\right)\mathrm{Bi}(\nu^{2/3}\zeta)\frac{\alpha_0}{\nu^{1/3}} + O\left\{\nu^{-\frac{3}{2}}\zeta^{\frac{1}{4}}\exp\left(\frac{2}{3}\nu\zeta^{3/2}\right)\right\}.$$

If $t$ is a zero of $f_n^{(\alpha)}(t/\sqrt{\nu})$, the left-hand side of the equation is equal to zero, and we have

$$\sin\left(\pi\alpha - \frac{\pi\nu}{t^2}\right) = O(\nu^{-3/4}\log^{1/6}\nu) \tag{6.8}$$

on account of (6.7). In view of (6.1), the solutions of (6.8) are given in (6.2). □

For $n = 100$ and $\alpha = 1$, the numerical values of the large zeros are given in Table 4.

Case (ii). To obtain the small zeros when $0 < t < 2$, we recall the asymptotic formulas

$$\mathrm{Ai}(-\eta) = \frac{\eta^{-\frac{1}{4}}}{\sqrt{\pi}}\left[\cos\left(\frac{2}{3}\eta^{\frac{3}{2}} - \frac{1}{4}\pi\right) + O\left(\eta^{-\frac{3}{2}}\right)\right], \tag{6.9}$$

$$\mathrm{Bi}(-\eta) = \frac{\eta^{-\frac{1}{4}}}{\sqrt{\pi}}\left[-\sin\left(\frac{2}{3}\eta^{\frac{3}{2}} - \frac{1}{4}\pi\right) + O\left(\eta^{-\frac{3}{2}}\right)\right], \tag{6.10}$$

and the corresponding formulas of $\mathrm{Ai}'(-\eta)$ and $\mathrm{Bi}'(-\eta)$; see [14, pp. 392–393]. From these formulas, it is evident that if $t$ is a zero of the Tricomi–Carlitz polynomials $f_n^{(\alpha)}(t/\sqrt{\nu})$, then Eq. (6.3) can be reduced to

$$\cos\left(\pi\alpha - \frac{\pi\nu}{t^2}\right)\mathrm{Ai}(\nu^{2/3}\zeta) + \sin\left(\pi\alpha - \frac{\pi\nu}{t^2}\right)\mathrm{Bi}(\nu^{2/3}\zeta) = O\left(\frac{1}{\nu}\right). \tag{6.11}$$

Let $\widetilde{t}_{n,m}$ denote the $m$-th zero of $f_n^{(\alpha)}(t/\sqrt{\nu})$, arranged in ascending order

$$0 \le \widetilde{t}_{n,1} < \widetilde{t}_{n,2} < \cdots < 2. \tag{6.12}$$

Since $\zeta(t) < 0$ when $0 < t < 2$, it is readily seen from (6.9) and (6.10) that the left-hand side of (6.11) can be written as

$$\begin{aligned}
&\cos\left(\pi\alpha - \frac{\pi\nu}{t^2}\right)\mathrm{Ai}(\nu^{2/3}\zeta) + \sin\left(\pi\alpha - \frac{\pi\nu}{t^2}\right)\mathrm{Bi}(\nu^{2/3}\zeta) \\
&= \frac{1}{\pi^{1/2}\nu^{1/6}|\zeta(t)|^{1/4}}\left[\cos\left(\pi\alpha - \frac{\pi\nu}{t^2}\right)\cos\left(\frac{2}{3}\nu|\zeta(t)|^{\frac{3}{2}} - \frac{1}{4}\pi\right)\right. \\
&\quad \left. - \sin\left(\pi\alpha - \frac{\pi\nu}{t^2}\right)\sin\left(\frac{2}{3}\nu|\zeta(t)|^{\frac{3}{2}} - \frac{1}{4}\pi\right) + O\left(\frac{1}{\nu}\right)\right] \\
&= \frac{1}{\pi^{1/2}\nu^{1/6}|\zeta(t)|^{1/4}}\left[\cos\left(\pi\alpha - \frac{\pi\nu}{t^2} + \frac{2}{3}\nu|\zeta(t)|^{\frac{3}{2}} - \frac{1}{4}\pi\right) + O\left(\frac{1}{\nu}\right)\right].
\end{aligned}$$

Therefore, from (6.11) we have

$$\cos\left(\pi\alpha - \frac{\pi\nu}{t^2} + \frac{2}{3}\nu|\zeta(t)|^{\frac{3}{2}} - \frac{1}{4}\pi\right) = O\left(\frac{1}{\nu}\right). \tag{6.13}$$

Note that near $t = 0$, $|\zeta(t)|^{\frac{3}{2}}$ has the expansion

$$|\zeta(t)|^{\frac{3}{2}} = \frac{3\pi}{2t^2} - \frac{3\pi}{4} + \frac{1}{2}t + \frac{1}{80}t^3 + \frac{9}{8960}t^5 + O(t^6). \tag{6.14}$$

Upon solving (6.13) and (6.14), we have

$$-\frac{n}{2} + \left(\frac{\nu}{3\pi}t + \frac{\nu}{120\pi}t^3 + \frac{3\nu}{4480\pi}t^5 + \cdots\right) = 2j \pm \frac{1}{2} + O\left(\frac{1}{\nu}\right). \tag{6.15}$$

As a first approximation, we take just the first two terms on the left-hand side of the equation and obtain

$$-\frac{n}{2} + \frac{\nu t}{3\pi} = 2j \pm \frac{1}{2} + O\left(\frac{1}{\nu}\right). \tag{6.16}$$

The solutions of Eq. (6.16) are presented in two separate cases. In the order arranged in (6.12), when $n$ is even, the small zeros are given by

$$\widetilde{t}_{n,2k+1} = \frac{3\pi}{2\nu}\left[4k + 1 + O\left(\frac{1}{\nu}\right)\right] \tag{6.17}$$

and

$$\widetilde{t}_{n,2k+2} = \frac{3\pi}{2\nu}\left[4k + 3 + O\left(\frac{1}{\nu}\right)\right] \tag{6.18}$$

for fixed $k = 0, 1, 2, \ldots$. Note that $\widetilde{t}_{n,1} \neq 0$. When $n$ is odd, the small zeros are given by $\widetilde{t}_{n,1} = 0$,

$$\widetilde{t}_{n,2k+2} = \frac{3\pi}{2\nu}\left[4k + 2 + O\left(\frac{1}{\nu}\right)\right] \tag{6.19}$$

and

$$\widetilde{t}_{n,2k+3} = \frac{3\pi}{2\nu}\left[4k + 4 + O\left(\frac{1}{\nu}\right)\right] \tag{6.20}$$

for fixed $k = 0, 1, 2, \ldots$.

**Theorem 6.2.** *The asymptotic behavior of the small zeros of* $f_n^{(\alpha)}(t/\sqrt{\nu})$, *arranged in ascending order in* (6.12), *are given by the formulas in* (6.17)–(6.20).

The above result can be improved if we pick three terms in (6.15), in which case we have

$$-\frac{n}{2}+\frac{\nu}{3\pi}t+\frac{\nu}{120\pi}t^3=2j\pm\frac{1}{2}+O\left(\frac{1}{\nu}\right). \tag{6.21}$$

This is a cubic equation, and we are looking for only the real root. Recall that the real root of a cubic equation of the form

$$t^3+pt+q=0 \tag{6.22}$$

is given explicitly by

$$t=\left(-\frac{q}{2}+\sqrt{\frac{q^2}{4}+\frac{p^3}{27}}\right)^{\frac{1}{3}}+\left(-\frac{q}{2}-\sqrt{\frac{q^2}{4}+\frac{p^3}{27}}\right)^{\frac{1}{3}}, \tag{6.23}$$

provided that

$$\frac{q^2}{4}+\frac{p^3}{27}>0. \tag{6.24}$$

The solutions of (6.21) are again presented in two separate cases. Let $K_m=4k+m+O(1/\nu)$, $m=1,2,3$ and 4. When $n$ is even, the small zeros are given by

$$\widetilde{t}_{n,2k+1}=\left[30K_1\frac{\pi}{\nu}+30\sqrt{\left(\frac{K_1\pi}{\nu}\right)^2+\frac{640}{243}}\right]^{\frac{1}{3}}+\left[30K_1\frac{\pi}{\nu}-30\sqrt{\left(\frac{K_1\pi}{\nu}\right)^2+\frac{640}{243}}\right]^{\frac{1}{3}} \tag{6.25}$$

and

$$\widetilde{t}_{n,2k+2}=\left[30K_3\frac{\pi}{\nu}+30\sqrt{\left(\frac{K_3\pi}{\nu}\right)^2+\frac{640}{243}}\right]^{\frac{1}{3}}+\left[30K_3\frac{\pi}{\nu}-30\sqrt{\left(\frac{K_3\pi}{\nu}\right)^2+\frac{640}{243}}\right]^{\frac{1}{3}} \tag{6.26}$$

for fixed $k=0,1,2,\ldots$. Note that, $\widetilde{t}_{n,1}\neq 0$. When $n$ is odd, the small zeros are given by $\widetilde{t}_{n,1}=0$,

$$\widetilde{t}_{n,2k+2}=\left[30K_2\frac{\pi}{\nu}+30\sqrt{\left(\frac{K_2\pi}{\nu}\right)^2+\frac{640}{243}}\right]^{\frac{1}{3}}+\left[30K_2\frac{\pi}{\nu}-30\sqrt{\left(\frac{K_2\pi}{\nu}\right)^2+\frac{640}{243}}\right]^{\frac{1}{3}} \tag{6.27}$$

and

$$\widetilde{t}_{n,2k+3}=\left[30K_4\frac{\pi}{\nu}+30\sqrt{\left(\frac{K_4\pi}{\nu}\right)^2+\frac{640}{243}}\right]^{\frac{1}{3}}+\left[30K_4\frac{\pi}{\nu}-30\sqrt{\left(\frac{K_4\pi}{\nu}\right)^2+\frac{640}{243}}\right]^{\frac{1}{3}} \tag{6.28}$$

for fixed $k=0,1,2,\ldots$.

It can be shown that as $\nu\to\infty$, the quantities on the right-hand sides of Eqs. (6.25)–(6.28) are asymptotically equal to the quantities on the right-hand sides of (6.17)–(6.20), respectively. More precisely, we have

$$\left[30K_m\frac{\pi}{\nu}+30\sqrt{\left(\frac{K_m\pi}{\nu}\right)^2+\frac{640}{243}}\right]^{\frac{1}{3}}+\left[30K_m\frac{\pi}{\nu}-30\sqrt{\left(\frac{K_m\pi}{\nu}\right)^2+\frac{640}{243}}\right]^{\frac{1}{3}}$$
$$\sim\frac{3\pi}{2\nu}\left(4k+m+O\left(\frac{1}{\nu}\right)\right),\quad m=1,2,3 \text{ and } 4. \tag{6.29}$$

Note that the $O$—term in (6.29) is still $O(1/\nu)$. To improve the order estimate from $O(1/\nu)$ to $O(1/\nu^2)$, one should use (3.19) with $p=2$, instead of (3.14).

Numerical computation shows that the solutions obtained in (6.25)–(6.28) are more accurate than those given in (6.17)–(6.20). Use of the solutions of Eq. (6.13) produces even better results; see the numerics in Table 5.

Case (iii). Now, we consider the zeros near the turning point $t=2$; that is, on both sides of $\sigma=0$, where $\sigma=\nu^{2/3}\zeta(t)$ is the variable inside the Airy functions in (6.3). If $t$ is a zero of $f_n^{(\alpha)}(t/\sqrt{\nu})$ and $\sigma$ is bounded, then Eq. (6.3) gives

$$\cos\left(\pi\alpha-\frac{\pi\nu}{t^2}\right)\mathrm{Ai}(\sigma)+\sin\left(\pi\alpha-\frac{\pi\nu}{t^2}\right)\mathrm{Bi}(\sigma)=O\left(\frac{1}{\nu}\right) \tag{6.30}$$

**Table 5**
True values and approximate values of small zeros evaluated with Maple, when $n = 100$ and $\alpha = 1$.

| $m$-th | True | (6.17)–(6.20) | (6.25)–(6.28) | Solution of (6.13) |
|---|---|---|---|---|
| $\tilde{t}_{100,1}$ | 0.0464 | 0.0464 | 0.0464 | 0.0464 |
| $\tilde{t}_{100,2}$ | 0.1392 | 0.1393 | 0.1393 | 0.1392 |
| $\tilde{t}_{100,3}$ | 0.2318 | 0.2321 | 0.2318 | 0.2318 |
| $\tilde{t}_{100,4}$ | 0.3241 | 0.3250 | 0.3241 | 0.3241 |
| $\tilde{t}_{100,5}$ | 0.4160 | 0.4178 | 0.4160 | 0.4160 |
| $\tilde{t}_{100,6}$ | 0.5074 | 0.5107 | 0.5074 | 0.5074 |
| $\tilde{t}_{100,7}$ | 0.5981 | 0.6036 | 0.5982 | 0.5980 |
| $\tilde{t}_{100,8}$ | 0.6880 | 0.6964 | 0.6883 | 0.6879 |
| $\tilde{t}_{100,9}$ | 0.7770 | 0.7893 | 0.7775 | 0.7769 |
| $\tilde{t}_{100,10}$ | 0.8649 | 0.8821 | 0.8659 | 0.8649 |
| $\tilde{t}_{100,11}$ | 0.9517 | 0.9750 | 0.9533 | 0.9517 |
| $\tilde{t}_{100,12}$ | 1.0372 | 1.0678 | 1.0397 | 1.0372 |

which can be written as

$$\sin\left[\pi\alpha - \frac{\pi\nu}{t^2} + \arctan\frac{\mathrm{Ai}(\sigma)}{\mathrm{Bi}(\sigma)}\right] = O\left(\frac{1}{\nu}\right). \tag{6.31}$$

Since

$$\arctan\frac{\mathrm{Ai}(\sigma)}{\mathrm{Bi}(\sigma)} = \frac{1}{6}\pi - \frac{3}{4}\frac{3^{\frac{1}{3}}(\Gamma(2/3))^2\sigma}{\pi} + \frac{9}{16}\frac{3^{\frac{1}{6}}(\Gamma(2/3))^4\sigma^2}{\pi^2} + O(\sigma^4) \tag{6.32}$$

for small $\sigma$, replacing arctan in (6.31) by its leading term given in (6.32) yields

$$\sin\left(\pi\alpha - \frac{\pi\nu}{t^2} + \frac{\pi}{6}\right) = O\left(\frac{1}{\nu}\right). \tag{6.33}$$

Recall that here we are only concerned with the zeros of the polynomials near $t = 2$ (i.e. $\sigma$ close to 0). From (6.33), we obtain

$$\alpha - \frac{\nu}{t^2} + \frac{1}{6} = m + O\left(\frac{1}{\nu}\right), \tag{6.34}$$

where $m$ is an integer. Recall that $\sigma = \nu^{2/3}\zeta(t)$. Hence, by Lagrange's inversion formula, we have from (2.20)

$$t = 2 + \sum_{j=1}^{\infty} A_j\left(\sigma\nu^{-\frac{2}{3}}\right)^j, \tag{6.35}$$

where $A_1 = 1, A_2 = \frac{29}{60}, A_3 = \frac{299}{1400}, \ldots$. Substituting (6.35) in (6.34) gives

$$\alpha - \frac{\nu}{4}\left(1 - \sigma\nu^{-\frac{2}{3}} + \cdots\right) + \frac{1}{6} = m + O\left(\frac{1}{\nu}\right), \tag{6.36}$$

which in turn gives

$$\frac{1}{4}\sigma\nu^{\frac{1}{3}} = m - \left(\alpha + \frac{1}{6} - \frac{\nu}{4}\right) + O\left(\frac{1}{\nu}\right). \tag{6.37}$$

Now, we choose $m$ to be a large negative integer so that $m - (\alpha + \frac{1}{6} - \frac{\nu}{4})$ is a bounded quantity. Let us write (6.37) in the form

$$\sigma = \frac{t_1}{\nu^{1/3}} + \cdots \tag{6.38}$$

with

$$\frac{1}{4}t_1 = m - \left(\alpha + \frac{1}{6} - \frac{\nu}{4}\right). \tag{6.39}$$

If we denote by $\nu_\alpha$ the fractional part of $\frac{1}{4}\nu - \frac{1}{6} - \alpha$ (i.e., $\nu_\alpha = \frac{1}{4}\nu - \frac{1}{6} - \alpha - [\frac{1}{4}\nu - \frac{1}{6} - \alpha]$), then (6.39) can be expressed as

$$t_1 = 4(k + \nu_\alpha), \tag{6.40}$$

**Table 6**
True values and approximate values of zeros near the turning point $t = 2$ evaluated with Maple, when $n = 100$ and $\alpha = 1$.

| | True | Approx. | Error % |
|---|---|---|---|
| $t^{(l)}_{100,3}$ | 1.8506 | 1.8900 | −2.1289 |
| $t^{(l)}_{100,2}$ | 1.9079 | 1.9294 | −1.1283 |
| $t^{(l)}_{100,1}$ | 1.9610 | 1.9688 | −0.3970 |
| | 2.0 | 2.0 | 0.0 |
| $t^{(r)}_{100,1}$ | 2.0098 | 2.0082 | 0.0812 |
| $t^{(r)}_{100,2}$ | 2.0554 | 2.0476 | 0.3797 |
| $t^{(r)}_{100,3}$ | 2.1006 | 2.0870 | 0.6463 |

where $k$ is an integer. Let $t^{(l)}_{n,k+1}$ and $t^{(r)}_{n,k+1}$, $k = 0, 1, 2, \ldots$, denote, respectively, the $(k+1)$th zero on the left and on the right side of the turning point $t = 2$, arranged in the order

$$\cdots < t^{(l)}_{n,2} < t^{(l)}_{n,1} < 2 < t^{(r)}_{n,1} < t^{(r)}_{n,2} < \cdots. \tag{6.41}$$

We can now write down the asymptotic formulas as $\nu \to \infty$

$$t^{(r)}_{n,k+1} = 2 + \frac{4}{\nu}(k + \nu_\alpha) + O\left(\frac{1}{\nu^2}\right) \tag{6.42}$$

and

$$t^{(l)}_{n,k+1} = 2 + \frac{4}{\nu}(-k - 1 + \nu_\alpha) + O\left(\frac{1}{\nu^2}\right) \tag{6.43}$$

for fixed $k = 0, 1, 2, \ldots$, where $\nu_\alpha$ denotes the fractional part of $\frac{1}{4}\nu - \frac{1}{6} - \alpha$ (i.e., $\nu_\alpha = \frac{1}{4}\nu - \frac{1}{6} - \alpha - [\frac{1}{4}\nu - \frac{1}{6} - \alpha]$).

**Remark 6.3.** Using (6.38), the right-hand side of (6.32) is equal to $\frac{1}{6}\pi + O(\nu^{-1/3})$. With this modification, the order term in (6.33) can be improved to $O(\nu^{-1/3})$. But, this does not change the order estimate in (6.42) and (6.43).

**Theorem 6.4.** *The asymptotic behavior of the zeros of $f_n^{(\alpha)}(t/\sqrt{\nu})$ on two sides of the turning point $t = 2$, as arranged in (6.41), are given in (6.42) and (6.43).*

For $n = 100$ and $\alpha = 1$, the numerical values of the first three zeros on the two sides of $t = 2$ are given in Table 6. Approximation of the zeros of $f_n^{(\alpha)}(x)$ as $\alpha \to \infty$ has been studied by López and Temme [8].

Asymptotic zero distribution of the Tricomi–Carlitz polynomials has been investigated by Goh and Wimp [6,7] and Kuijlaars and Van Assche [20]. Goh and Wimp investigated the zero distribution of $f_n^{(\alpha)}(z/\sqrt{n})$ as $n \to \infty$, by using a probability approach; while Kuijlaars and Van Assche used the three-term recurrence relation (1.1) to establish the zero distribution of $f_n^{(\alpha)}(t/\sqrt{N})$ with $n/N \to c$ as $n \to \infty$. The zero distribution obtained by Kuijlaars and Van Assche [20] is

$$\mu_c(t) = \begin{cases} \dfrac{4}{\pi c|t|^3}\left[\arcsin\left(\dfrac{|t|\sqrt{c}}{2}\right) - \dfrac{|t|}{4}\sqrt{c(4 - t^2 c)}\right], & |t| < \dfrac{2}{\sqrt{c}}, \\ \dfrac{2}{c|t|^3}, & |t| \geq \dfrac{2}{\sqrt{c}}, \end{cases} \tag{6.44}$$

which agrees with the result

$$\mu(t) = \begin{cases} \dfrac{4}{\pi|t|^3}\left[\arcsin\left(\dfrac{|t|}{2}\right) - \dfrac{|t|}{4}\sqrt{4 - t^2}\right], & |t| < 2, \\ \dfrac{2}{|t|^3}, & |t| \geq 2, \end{cases} \tag{6.45}$$

of Goh and Wimp [6,7] if we take $c = 1$ (i.e., $n/N \to 1$ as $n \to \infty$). The graph of the distribution function (6.45) is shown in Fig. 12. Note that in our case $N = \nu$. The area under the curve of $\mu(t)$ in (6.45) in the interval $[a, b]$ represents the ratio

$$\frac{\text{the number of zeros in } [a, b]}{\text{total number of zeros}}. \tag{6.46}$$

As an illustration, we take $n = 100$ and $\alpha = 1$. We find that the polynomial $f_n^{(\alpha)}(t/\sqrt{\nu})$ has 25 zeros in the interval $0 < t < 2$, and that the value of the integral of $\mu(t)$ on the interval $[0, 2]$ is 0.25.

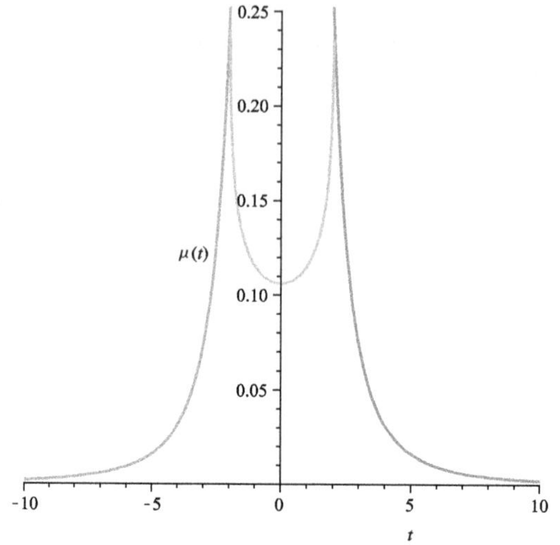


**Fig. 12.** The zero distribution of $f_n^{\alpha}(t/\sqrt{\nu})$.


## Acknowledgment

We would like to thank one of the referees for a very careful reading of the manuscript. In the original manuscript, he pointed out that there was an error in the order estimate in (6.2).


## References


[1] K.F. Lee, R. Wong, Uniform asymptotic expansions of the Tricomi–Carlitz polynomials, Proc. Amer. Math. Soc. 138 (2010) 2513–2519. MR2607881 (2011d:41050).
[2] S. Karlin, J. McGregor, Many server queueing processes with Poisson input and exponential service times, Pacific J. Math. 8 (1958) 87–118. MR0097132 (20 #3611).
[3] R.A. Askey, M.E.H. Ismail, Recurrence Relations, Continued Fractions and Orthogonal Polynomials, in: Memoirs Amer. Math. Soc., vol. 300, Providence, RI, 1984. MR0743545 (85g:33008).
[4] M.E.H. Ismail, Classical and Quantum Orthogonal Polynomials in One Variable, Cambridge University Press, Cambridge, 2005. MR2191786 (2007f:33001).
[5] R. Wong, Asymptotic Approximations of Integrals, Academic Press, Boston, 1989. MR1016818 (90j:41061).
[6] W.M.Y. Goh, J. Wimp, On the asymptotics of the Tricomi–Carlitz polynomials and their zero distribution, I, SIAM J. Math. Anal. 25 (1994) 420–428. MR1266567 (95b:42025).
[7] W.M.Y. Goh, J. Wimp, The zero distribution of the Tricomi–Carlitz polynomials, Comput. Math. Appl. 33 (1997) 119–127. MR1442066 (99e:33006).
[8] J.L. López, N.M. Temme, Approximation of orthogonal polynomials in terms of hermite polynomials. Dedicated to Richard A. Askey on the occasion of his 65th birthday, part II, Methods Appl. Anal. 6 (1999) 131–146. MR1803886 (2001m:33013).
[9] Z. Wang, R. Wong, Uniform asymptotic expansion of $J_\nu(\nu a)$ via a difference equation, Numer. Math. 91 (2002) 147–193. MR1896091 (2003g:33008).
[10] Z. Wang, R. Wong, Asymptotic expansions for second-order linear difference equations with a turning point, Numer. Math. 94 (2003) 147–194. MR1971216 (2004c:39012).
[11] C. Chester, B. Friedman, F. Ursell, An extension of the method of steepest descents, Proc. Cambridge Philos. Soc. 53 (1957) 599–611. MR0090690 (19,853a).
[12] A.B. Olde Daalhuis, N.M. Temme, Uniform Airy-type expansions of integrals, SIAM J. Math. Anal. 25 (2) (1994) 304–321. MR1266561 (95h:41056).
[13] K.F. Lee, Uniform asymptotic expansions of the Tricomi–Carlitz polynomials, Ph.D. Thesis, City University of Hong Kong, 2012.
[14] F.W.J. Olver, Asymptotics and Special Functions, Academic Press, New York, 1974. Reprinted by A. K. Peters, Ltd., Wellesley, 1997. MR1429619 (97i:41001).
[15] R. Wong, Lecture Notes on Applied Analysis, World Scientific, Singapore, 2010. MR2743248.
[16] N. Bleistein, Uniform asymptotic expansions of integrals with stationary point near algebraic singularity, Comm. Pure Appl. Math. 19 (1966) 353–370. MR0204943 (34 #4778).
[17] N. Bleistein, R.A. Handelsman, Asymptotic Expansions of Integrals, Dover Publications, Inc., New York, 1986. MR0863284 (89d:41049).
[18] Milton Abramowitz, Irene A. Stegun, Handbook of Mathematical Functions: with Formulas, Graphs, and Mathematical Tables, in: National Bureau of Standards Applied Mathematics Series, vol. 55, Washington, DC, 1964. MR0167642 (29#4914).
[19] R. Wong, Error bounds for asymptotic expansions of integrals, SIAM Rev. 22 (1980) 401–435. MR593856 (82a:41030).
[20] A.B.J. Kuijlaars, W. Van Assche, The asymptotic zero distribution of orthogonal polynomials with varying recurrence coefficients, J. Approx. Theory 99 (1999) 167–197. MR1696553 (2000h:42015).

Asymptotic Analysis 88 (2014) 111–123 111
DOI 10.3233/ASY-131212
IOS Press

# A novel generalized solution expansion for the Lagerstrom model

Chunhua Ou [a] and R. Wong [b,*]
[a] *Department of Mathematics and Statistics, Memorial University, St. John's, Canada*
*E-mail: ou@mun.ca*
[b] *Liu Bie Ju Centre for Mathematical Sciences, City University of Hong Kong, Hong Kong, China*
*E-mail: mawong@cityu.edu.hk*

**Abstract.** In this paper we re-visit the Lagerstrom problem

$$y'' + \frac{n-1}{r}y' + \varepsilon yy' = 0,$$
$$y(1) = 0, \qquad y(\infty) = 1,$$

where $\varepsilon$ is a small positive real number and $n$ is a positive integer (or any real number greater than 2). Using rigorous analysis, a generalized asymptotic expansion, as $\varepsilon \to 0$, is derived for the solution of this problem. A trans-series expansion of the solution for large values of $r$ is also presented; the leading term coefficient is determined by a connection formula between the values of the solution at the two points $r = 1$ and $r = \infty$. An extension and a discussion of the problem for $n \in [1, 2)$ is also given.

Keywords: boundary value problem, multiple scale, singular perturbation

## 1. Introduction

Lagerstrom's boundary value problem

$$y'' + \frac{n-1}{r}y' + \varepsilon yy' = 0, \tag{1.1}$$

subject to

$$y(1) = 0, \qquad y(\infty) = 1, \tag{1.2}$$

has drawn considerable attention in the past decades [1,3–9,12–15]. Here, $\varepsilon$ is a small real number, and the first two terms in (1.1) represent the radial Laplacian operator with $n$ being the spatial dimension. The variable $y$ denotes the temperature and the term $\varepsilon yy'$ stands for some fictitious heat loss; see [8, p. 92] for details.

*Corresponding author: R. Wong, Liu Bie Ju Centre for Mathematical Sciences, City University of Hong Kong, Tat Chee Avenue, Kowloon, Hong Kong, China. E-mail: mawong@cityu.edu.hk.

At a first glance, this problem may be viewed as a regular perturbation problem, and a naive expansion

$$y = y_0 + \varepsilon y_1 + \varepsilon^2 y_2 + \cdots$$

with boundary conditions

$$y_0(1) = 0, \qquad y_0(\infty) = 1; \qquad y_i(1) = y_i(\infty) = 0, \quad i \geqslant 1,$$

can be given to *approximate* the exact solution. We take $n = 3$ as an example, and substitute this expansion into (1.1)–(1.2) to obtain

$$y_0(r) = 1 - \frac{1}{r}$$

and

$$y_1(r) = -\left(1 + \frac{1}{r}\right) \log r + C\left(1 - \frac{1}{r}\right)$$

for some constant $C$. The solution $y_1$ becomes unbounded for any constant $C$ as $r \to \infty$, and the asymptotic expansion for the true solution fails. The cause of this failure is due to the fact that the interval under consideration is unbounded and the leading term $y_0$ does not have the exponential decay required by the original system. This failure had forced Kaplun and Lagerstrom [7] to use a "switchback" technique; i.e., by adding a term involving $\varepsilon \log \frac{1}{\varepsilon}$ to the expansion and matching the expansion with the outer solution near $r = 1/\sqrt{\varepsilon}$.

As mentioned at the beginning of this section, there has been a series of further study of this problem; see [1,3–9,12–15]. In particular, Hunter et al. [6] have constructed an iterative scheme (Oseen iteration) that defines a contraction mapping, and they showed that the problem has a unique solution for all $n > 1$ and $\varepsilon \geqslant 0$. Recently, Hastings and Mcleod [3,4] have provided a new proof of the existence and uniqueness of solution to (1.1)–(1.2) by using a shooting method. As to the construction of an asymptotic expansion, they transformed, after a scaling, the problem into an integral equation

$$u(\rho) - 1 = C\varepsilon^{n-2} \int_\infty^\rho \frac{1}{\tau^{n-1}} \mathrm{e}^{-\tau} \mathrm{e}^{-\int_\infty^\tau (u(\sigma)-1)\,\mathrm{d}\sigma}\,\mathrm{d}\tau,$$

where $C$ is given by

$$-1 = C\varepsilon^{n-2} \int_\infty^\varepsilon \frac{1}{\tau^{n-1}} \mathrm{e}^{-\tau} \mathrm{e}^{-\int_\infty^\tau (u(\sigma)-1)\,\mathrm{d}\sigma}\,\mathrm{d}\tau.$$

Note that $C$ is not an independent constant; it depends on $\varepsilon$, $n$ and the solution $u$. Because of this dependence, it is not known whether the function

$$\Phi(\varepsilon) = C\varepsilon^{n-2} \int_\varepsilon^\infty E_{n-1}(\tau)\,\mathrm{d}\tau \triangleq C\varepsilon^{n-2} \int_\varepsilon^\infty \int_s^\infty \frac{1}{\tau^{n-1}} \mathrm{e}^{-\tau}\,\mathrm{d}\tau\,\mathrm{d}s$$

tends to zero as $\varepsilon \to 0$ for $n \geqslant 2$, but the argument in [3] for the asymptotic expansion of the solution needs this prior assumption. (This deficiency is irrelevant in our analysis in the following sections.)

The purpose of this paper is threefold: (i) further the study of Hasting and Mcleod, by providing an even more rigorous and simpler derivation of an expansion for the solution; (ii) present a trans-series expansion [2] for the solution near infinity, the coefficient of leading term being determined by a connection formula between the values of the solution at the two points $r = 1$ and $r = \infty$; (iii) discuss possible extensions of the problem; for instance, we can extend our method to treat the problem

$$y'' + \frac{n-1}{r}y' + \varepsilon yy' + (y')^2 = 0, \quad y(1) = 0,\ y(\infty) = 1,$$

which is related to compressible flow.

## 2. Convergent asymptotic expansion

In this section, after converting (1.1)–(1.2) into an integral equation, we shall give the existence of a solution to the problem (1.1)–(1.2) for all real number $n \geqslant 2$. In particular, a convergent (generalized) asymptotic expansion is constructed, in terms of a successive iteration scheme.

Since the boundary condition at infinity requires $y(\infty) = 1$, we may let

$$y = 1 + u,$$

so that we obtain the alternative system

$$\begin{cases} u'' + \left(\dfrac{n-1}{r} + \varepsilon\right)u' + \varepsilon uu' = 0, \\ u(1) = -1, \qquad u(+\infty) = 0. \end{cases} \tag{2.1}$$

(Here, $u$ is dependent on $\varepsilon$, but for short we simply write it as $u(r)$.)

Since $u(+\infty) = 0$, after a simple asymptotic analysis of (2.1) it is not difficult to see that near infinity, the solution (if it exists) would decay exponentially to zero with

$$u'(r) = \mathrm{O}\left(\mathrm{e}^{-\varepsilon r}\frac{1}{r^{n-1}}\right) \quad \text{as } r \to \infty;$$

see (4.3).

To establish the existence of a solution to Eq. (2.1), we follow the usual approach of converting (2.1) into an integral equation so that the method of successive approximation (or fixed-point theorem of an operator) can be applied. To this end, we multiply both side of the first equation by $\mathrm{e}^{\varepsilon r}r^{n-1}$ and integrate it over the interval $(r, \infty)$ to obtain

$$u(r) = C\int_r^\infty \frac{\mathrm{e}^{-\varepsilon s}}{s^{n-1}}\,\mathrm{d}s - \varepsilon\int_\infty^r \frac{\mathrm{e}^{-\varepsilon s}}{s^{n-1}}\int_\infty^s \mathrm{e}^{\varepsilon\tau}\tau^{n-1}u(\tau)u'(\tau)\,\mathrm{d}\tau\,\mathrm{d}s, \tag{2.2}$$

where $C$ is a constant to be determined by the boundary condition $u(1) = -1$. Notice that when $r = 1$, the first integral on the right-hand side of (2.2) becomes

$$\rho_n(\varepsilon) = \int_1^\infty \frac{\mathrm{e}^{-\varepsilon s}}{s^{n-1}}\,\mathrm{d}s = \varepsilon^{n-2}\int_\varepsilon^\infty \frac{\mathrm{e}^{-s}}{s^{n-1}}\,\mathrm{d}s, \tag{2.3}$$

and has the expansion

$$\rho_n(\varepsilon) = \frac{(-\varepsilon)^{n-2}}{(n-2)!}\{-\ln\varepsilon + \psi(n-1)\} + {\sum_{s=0}^{\infty}}' \frac{(-\varepsilon)^s}{s!(n-s-2)}, \tag{2.4}$$

where the prime on the sum signifies that the term $s = n-2$ is omitted; see [11, Formula (8.19.8)]. Here, $\psi$ is the derivative of the logarithm Gamma function; see e.g [11, Formula (5.2.2)]. Also, we have

$$\rho_n(\varepsilon) = \varepsilon^{n-2}\Gamma(2-n) - \sum_{s=0}^{\infty} \frac{(-\varepsilon)^s}{s!(2-n+s)}, \quad n \in \mathbb{C}, \tag{2.5}$$

with the right-hand side being replaced by its limiting form (2.4) when $n = 2, 3, 4, \ldots$. In particular, it follows that

$$\rho_n(\varepsilon) \sim \begin{cases} \dfrac{1}{n-2}, & n > 2, \\ -\ln\varepsilon, & n = 2, \\ \varepsilon^{n-2}\Gamma(2-n), & n < 2 \end{cases} \tag{2.6}$$

as $\varepsilon \to 0$.

For later use, we also introduce the function

$$\delta(\varepsilon) = \frac{\varepsilon\rho_{n-1}(\varepsilon)}{\rho_n(\varepsilon)}. \tag{2.7}$$

From (2.6), one obtains

$$\delta(\varepsilon) \sim \begin{cases} \dfrac{n-2}{n-3}\varepsilon, & n > 3, \\ -\varepsilon\ln\varepsilon, & n = 3, \\ \varepsilon^{n-2}(n-2)\Gamma(3-n), & 2 < n < 3, \\ -\dfrac{1}{\ln}\varepsilon, & n = 2, \\ \dfrac{\Gamma(3-n)}{\Gamma(2-n)}, & n < 2 \end{cases} \tag{2.8}$$

as $\varepsilon \to 0$.

On account of the boundary condition $u(1) = -1$, the constant $C$ in (2.2) can be expressed as

$$C = \frac{1}{\rho_n(\varepsilon)}\left\{-1 + \varepsilon\int_\infty^1 \frac{\mathrm{e}^{-\varepsilon s}}{s^{n-1}}\int_\infty^s \mathrm{e}^{\varepsilon\tau}\tau^{n-1}u(\tau)u'(\tau)\,\mathrm{d}\tau\,\mathrm{d}s\right\}.$$

Hence, we can rewrite (2.2) as

$$u(r) = \frac{-1}{\rho_n(\varepsilon)} \int_r^\infty \frac{e^{-\varepsilon s}}{s^{n-1}}\, ds \left\{ 1 - \varepsilon \int_\infty^1 \frac{e^{-\varepsilon s}}{s^{n-1}} \int_\infty^s e^{\varepsilon\tau} \tau^{n-1} u(\tau) u'(\tau)\, d\tau\, ds \right\}$$
$$- \varepsilon \int_\infty^r \frac{e^{-\varepsilon s}}{s^{n-1}} \int_\infty^s e^{\varepsilon\tau} \tau^{n-1} u(\tau) u'(\tau)\, d\tau\, ds. \tag{2.9}$$

This integral equation differs from Eq. (11) in [3], since it does not involve a constant $C$ which depends on the solution; see our comment in the second last paragraph in Section 1.

To show that the integral Eq. (2.9) has a solution, we define the sequence $\{u_m(r)\}_{m=0}^\infty$ by

$$u_0(r) = \frac{-1}{\rho_n(\varepsilon)} \int_r^\infty \frac{e^{-\varepsilon s}}{s^{n-1}}\, ds \tag{2.10}$$

and

$$u_m(r) = u_0(r) - \varepsilon u_0(r) \int_\infty^1 \frac{e^{-\varepsilon s}}{s^{n-1}} \int_\infty^s e^{\varepsilon\tau} \tau^{n-1} u_{m-1}(\tau) u'_{m-1}(\tau)\, d\tau\, ds$$
$$- \varepsilon \int_\infty^r \frac{e^{-\varepsilon s}}{s^{n-1}} \int_\infty^s e^{\varepsilon\tau} \tau^{n-1} u_{m-1}(\tau) u'_{m-1}(\tau)\, d\tau\, ds \tag{2.11}$$

for $m \geqslant 1$.

We shall prove that the series $\sum_{m=1}^\infty [u_m(r) - u_{m-1}(r)]$ converges uniformly on $[1, \infty)$ so that the solution $u(r)$ is given by

$$u(r) = \lim_{m\to\infty} u_m(r) = u_0(r) + \sum_{m=1}^\infty \left[u_m(r) - u_{m-1}(r)\right]$$

with the asymptotic relation

$$u_{m+1}(r) - u_m(r) = O\big(\delta(\varepsilon)^m\big), \quad m \geqslant 1, \tag{2.12}$$

as $\varepsilon \to 0$. Recall that $\delta(\varepsilon)$ depends on the space dimension $n$; see (2.8). Our main result is the following theorem.

**Theorem.** *Assume that* $n \geqslant 2$. *There exists a constant* $\varepsilon_0 > 0$ *such that for* $0 < \varepsilon \leqslant \varepsilon_0$, *problem* (1.1)–(1.2) *has a solution* $y(r) = 1 + u(r)$, *where* $u(r)$ *has the asymptotic expansion*

$$u(r) = u_0(r) + \sum_{m=1}^\infty \left[u_m(r) - u_{m-1}(r)\right] \tag{2.13}$$

*with* $u_m(r)$, $m \geqslant 0$, *being defined by* (2.10)–(2.11) *and satisfying*

$$|u_m(r) - u_{m-1}(r)| \leqslant \big(8\delta(\varepsilon)\big)^m |u_0(r)|. \tag{2.14}$$

*Here,*

$$\delta(\varepsilon) = \frac{\varepsilon \rho_{n-1}(\varepsilon)}{\rho_n(\varepsilon)}; \qquad \rho_k(\varepsilon) = \int_1^\infty \frac{\mathrm{e}^{-\varepsilon s}}{s^{k-1}}\,\mathrm{d}s, \quad k = n-1, n. \tag{2.15}$$

Since $|\delta(\varepsilon) \leqslant 1$, in terms of *generalized asymptotic expansion* [11, p. 43], the result in (2.13)–(2.14) can be expressed as

$$u(r) \sim u_0(r) + \sum_{m=1}^{\infty} \left[u_m(r) - u_{m-1}(r)\right]; \quad \left\{\left(\delta(\varepsilon)\right)^m\right\},$$

as $\varepsilon \to 0^+$.

When $n < 2$, $\delta(\varepsilon)$ defined in (2.15) is not small as $\varepsilon \to 0$ (see (2.8)), so (2.14) cannot guarantee the convergence of the series in (2.13), and the series is not an asymptotic expansion; i.e., our result fails in this case.

When $n = 1$, (1.1)–(1.2) has the explicit solution

$$y(r) = 1 - \frac{2}{1 + \mathrm{e}^{\varepsilon(r-1)}}, \tag{2.16}$$

which cannot be obtained from our theorem. The case when $1 < n < 2$ will be discussed in our last section.

## 3. The proof

In this section, we prove our main result. First, we use induction to establish that there is a constant $\varepsilon_0$ such that for $0 < \varepsilon < \varepsilon_0$, we have

$$\left|u_m(r) - u_0(r)\right| \leqslant 4\delta(\varepsilon)\left|u_0(r)\right|, \qquad \left|u_m'(r) - u_0'(r)\right| \leqslant 4\delta(\varepsilon)\left|u_0'(r)\right| \tag{3.1}$$

and

$$\left|u_m(r) - u_{m-1}(r)\right| \leqslant \left(8\delta(\varepsilon)\right)^m\left|u_0(r)\right|, \qquad \left|u_m'(r) - u_{m-1}'(r)\right| \leqslant \left(8\delta(\varepsilon)\right)^m\left|u_0'(r)\right| \tag{3.2}$$

for $m \geqslant 0$.

Recall that $u_{-1}(r) \equiv 0$. Obviously, (3.1) and (3.2) are true for $m = 0$.

Assume that (3.1) and (3.2) hold for $m = k$; we want to show that they also hold for $m = k+1$. From (2.11), by using the induction hypothesis we have

$$\begin{aligned}
\left|u_{k+1}(r) - u_0(r)\right| &\leqslant \varepsilon\left|u_0(r)\right| \cdot \left|\int_\infty^1 \frac{\mathrm{e}^{-\varepsilon s}}{s^{n-1}} \int_\infty^s \mathrm{e}^{\varepsilon\tau}\tau^{n-1} u_k(\tau)u_k'(\tau)\,\mathrm{d}\tau\,\mathrm{d}s\right| \\
&\quad + \varepsilon\left|\int_\infty^r \frac{\mathrm{e}^{-\varepsilon s}}{s^{n-1}} \int_\infty^s \mathrm{e}^{\varepsilon\tau}\tau^{n-1} u_k(\tau)u_k'(\tau)\,\mathrm{d}\tau\,\mathrm{d}s\right| \\
&\leqslant 2\left(1 + 4\delta(\varepsilon)\right)^2 \frac{\rho_{n-1}(\varepsilon)}{\rho_n(\varepsilon)}\varepsilon\left|u_0(r)\right|,
\end{aligned} \tag{3.3}$$

where we have also made use of (2.10) couple times. Therefore, there exists a constant $\varepsilon_0$ such that for $0 < \varepsilon \leqslant \varepsilon_0$,

$$2\big(1 + 4\delta(\varepsilon)\big)^2 \frac{\rho_{n-1}(\varepsilon)}{\rho_n(\varepsilon)}\varepsilon < 4\delta(\varepsilon),$$

so that (3.3) implies

$$\big|u_{k+1}(r) - u_0(r)\big| < 4\delta(\varepsilon)\big|u_0(r)\big|.$$

Similarly we can prove

$$\big|u'_{k+1}(r) - u'_0(r)\big| < 4\delta(\varepsilon)\big|u'_0(r)\big|.$$

It now remains to show that (3.2) is true when $m = k + 1$. Indeed, we have

$$\begin{aligned}
\big|u_{k+1}(r) - u_k(r)\big| &\leqslant \varepsilon\big|u_0(r)\big| \cdot \left|\int_\infty^1 \frac{\mathrm{e}^{-\varepsilon s}}{s^{n-1}} \int_\infty^s \mathrm{e}^{\varepsilon\tau}\tau^{n-1}\big|u_k(\tau)u'_k(\tau) - u_{k-1}(\tau)u'_{k-1}(\tau)\big|\,\mathrm{d}\tau\,\mathrm{d}s\right| \\
&\quad + \varepsilon\left|\int_\infty^r \frac{\mathrm{e}^{-\varepsilon s}}{s^{n-1}} \int_\infty^s \mathrm{e}^{\varepsilon\tau}\tau^{n-1}\big|u_k(\tau)u'_k(\tau) - u_{k-1}(\tau)u'_{k-1}(\tau)\big|\,\mathrm{d}\tau\,\mathrm{d}s\right| \\
&\leqslant 4\frac{\rho_{n-1}(\varepsilon)}{\rho_n(\varepsilon)}\big(1 + 4\delta(\varepsilon)\big)\varepsilon \cdot \big(8\delta(\varepsilon)\big)^k\big|u_0(r)\big| \\
&\leqslant \big(8\delta(\varepsilon)\big)^{k+1}\big|u_0(r)\big|.
\end{aligned}$$

Here we have made use of the fact

$$\begin{aligned}
&\big|u_k(\tau)u'_k(\tau) - u_{k-1}(\tau)u'_{k-1}(\tau)\big| \\
&\quad = \big|u_k(\tau)u'_k(\tau) - u_k(\tau)u'_{k-1}(\tau) + u_k(\tau)u'_{k-1}(\tau) - u_{k-1}(\tau)u'_{k-1}(\tau)\big| \\
&\quad \leqslant \big|u_k(\tau)\big| \cdot \big|u'_k(\tau) - u'_{k-1}(\tau)\big| + \big|u'_{k-1}(\tau)\big| \cdot \big|u_k(\tau) - u_{k-1}(\tau)\big| \\
&\quad \leqslant 2\big(1 + 4\delta(\varepsilon)\big)\big(8\delta(\varepsilon)\big)^k\big|u_0(\tau)u'_0(\tau)\big|
\end{aligned}$$

via the induction hypothesis. The second inequality in (3.2) can be proved in a similar manner, and we omit the detail.

Since $|u_0(r)| \leqslant 1$, we can drop $|u_0(r)|$ on the right-hand side of the inequalities in (3.2).

With the Weierstrass M-test for uniform convergence, we obtain

$$\sum_{m=1}^{\infty}\big|u_m(r) - u_{m-1}(r)\big| \leqslant \frac{8\delta(\varepsilon)}{1 - 8\delta(\varepsilon)}\big|u_0(r)\big|,$$

and the solution $u(r)$ satisfies

$$u(r) = u_0(r) + \big|u_0(r)\big|\mathrm{O}\big(\delta(\varepsilon)\big)$$

uniformly for $r \in [1, \infty)$. This completes the proof of the theorem.

The leading term $u_0(r)$ has the formula

$$u_0(r) = \frac{-1}{\rho_n(\varepsilon)} \int_r^\infty \frac{\mathrm{e}^{-\varepsilon s}}{s^{n-1}} \, \mathrm{d}s;$$

see (2.10). When $n = 3$, $\rho_n(\varepsilon)$ has the following asymptotic expansion

$$\rho_3(\varepsilon) = 1 + \varepsilon \log \varepsilon + (\gamma - 1)\varepsilon - \frac{1}{2}\varepsilon^2 + \mathrm{o}(\varepsilon^2).$$

This explains the existence of $\varepsilon \log \varepsilon$ in the expansions in Refs [7–9] of the so-called switch-back method. Also, recall that when $n = 3$,

$$\delta(\varepsilon) = \frac{\varepsilon \rho_{n-1}(\varepsilon)}{\rho_n(\varepsilon)} \sim -\varepsilon \ln \varepsilon;$$

when $n = 2$,

$$\delta(\varepsilon) \sim -\frac{1}{\ln \varepsilon}.$$

## 4. Behavior near infinity

In this section, we wish to investigate the behavior of the solution as $r$ tends to infinity for a fixed small value of $\varepsilon$. In fact, we consider the slightly more general problem

$$\begin{cases} y'' + \dfrac{n-1}{r} y' + \varepsilon y y' = 0, \\ y(0) = a, \qquad y(\infty) = 1, \end{cases} \tag{4.1}$$

where $a$ is a real number. As before, a constant shift of the solution by $y = 1 + u$ gives a new problem

$$\begin{aligned} & u'' + \left(\frac{n-1}{r} + \varepsilon\right) u' + \varepsilon u u' = 0, \\ & u(1) = a - 1, \qquad u(+\infty) = 0. \end{aligned} \tag{4.2}$$

### 4.1. Connection formula

A direct linear analysis of (4.2) near infinity yields

$$u(r) \sim d \frac{\mathrm{e}^{-\varepsilon r}}{r^{n-1}}, \tag{4.3}$$

where $d$ is an arbitrary constant. Finding a value for the constant $d$ in terms of the initial value $a$ will establish a connection formula between $a$ and $d$.

Since we have changed the initial value to $u(1) = a - 1$, we need to make a minor modification to the iteration scheme (2.10)–(2.11). Define

$$u_0(r) = \frac{a-1}{\rho_n(\varepsilon)} \int_r^{\infty} \frac{\mathrm{e}^{-\varepsilon s}}{s^{n-1}}\, \mathrm{d}s$$

and

$$\begin{aligned} u_m(r) = u_0(r) &+ \varepsilon \frac{u_0(r)}{a-1} \int_{\infty}^{1} \frac{\mathrm{e}^{-\varepsilon s}}{s^{n-1}} \int_{\infty}^{s} \mathrm{e}^{\varepsilon\tau} \tau^{n-1} u_{m-1}(\tau) u'_{m-1}(\tau)\, \mathrm{d}\tau\, \mathrm{d}s \\ &- \varepsilon \int_{\infty}^{r} \frac{\mathrm{e}^{-\varepsilon s}}{s^{n-1}} \int_{\infty}^{s} \mathrm{e}^{\varepsilon\tau} \tau^{n-1} u_{m-1}(\tau) u'_{m-1}(\tau)\, \mathrm{d}\tau\, \mathrm{d}s \end{aligned} \tag{4.4}$$

for $m \geqslant 1$. By the same method as in the previous section, for $n \geqslant 2$ we can obtain the uniformly convergent series

$$u(r) = u_0(r) + \sum_{m=1}^{\infty} \left[ u_m(r) - u_{m-1}(r) \right]$$

with $u(r)$ being the solution to (4.2). For fixed $\varepsilon$, we have as $r \to \infty$

$$u_0(r) \sim \frac{a-1}{\rho_n(\varepsilon)\varepsilon} \frac{\mathrm{e}^{-\varepsilon r}}{r^{n-1}}$$

and

$$\begin{aligned} u_1(r) = u_0(r) &+ \varepsilon \frac{u_0(r)}{a-1} \int_{\infty}^{1} \frac{\mathrm{e}^{-\varepsilon s}}{s^{n-1}} \int_{\infty}^{s} \mathrm{e}^{\varepsilon\tau} \tau^{n-1} u_0(\tau) u'_0(\tau)\, \mathrm{d}\tau\, \mathrm{d}s \\ &- \varepsilon \int_{\infty}^{r} \frac{\mathrm{e}^{-\varepsilon s}}{s^{n-1}} \int_{\infty}^{s} \mathrm{e}^{\varepsilon\tau} \tau^{n-1} u_0(\tau) u'_0(\tau)\, \mathrm{d}\tau\, \mathrm{d}s \\ &\sim \frac{a-1}{\rho_n(\varepsilon)\varepsilon} \frac{\mathrm{e}^{-\varepsilon r}}{r^{n-1}} \left\{ 1 - \frac{\varepsilon(a-1)}{\rho_n^2(\varepsilon)} \alpha(\varepsilon) \right\} + \frac{(a-1)^2}{2\rho_n^2(\varepsilon)\varepsilon^2} \frac{\mathrm{e}^{-2\varepsilon r}}{r^{2n-2}} + \cdots, \end{aligned} \tag{4.5}$$

where

$$\alpha(\varepsilon) = \int_1^{\infty} \frac{\mathrm{e}^{-\varepsilon s}}{s^{n-1}} \int_s^{\infty} \frac{\mathrm{e}^{-\varepsilon v}}{v^{n-1}} (v - s)\, \mathrm{d}v\, \mathrm{d}s.$$

Thus, $d$ is given by

$$d = d(\varepsilon) \sim \frac{a-1}{\rho_n(\varepsilon)\varepsilon} \left\{ 1 - \frac{\varepsilon(a-1)}{\rho_n^2(\varepsilon)} \alpha(\varepsilon) \right\}. \tag{4.6}$$

### 4.2. Trans-series near infinity

The asymptotic expansion in (4.5) suggests that for $n \geqslant 2$, the solution $u(r)$ may have a trans-series expansion [2] (as $r \to \infty$) of the form

$$u(r) = \sum_{i=1}^{\infty} \mathrm{e}^{-\mathrm{i}\varepsilon r} v_i(\varepsilon, r). \tag{4.7}$$

Substituting (4.7) into (4.2) and regrouping terms containing the factor $\mathrm{e}^{-\mathrm{i}\varepsilon r}$, $i = 1, 2, \ldots$, we obtain

$$v_1'' + \left(\frac{n-1}{r} - \varepsilon\right) v_1' - \varepsilon \frac{n-1}{r} v_1 = 0, \tag{4.8}$$

$$v_2'' + \left(\frac{n-1}{r} - 3\varepsilon\right) v_2' + \left(2\varepsilon^2 - \frac{2(n-1)}{r}\right) v_2 = \varepsilon^2 v_1^2 - \varepsilon v_1 v_1', \tag{4.9}$$

and similar equations for $v_i$, $i = 3, 4, \ldots$. It can be shown that near infinity, the $v_i$'s have power-series expansions of the form

$$v_i(\varepsilon, r) = r^{\beta} \sum_{j=0}^{\infty} d_j^i(\varepsilon) r^{-j}, \quad i \geqslant 1,$$

where $\beta$ is a real number. In particular, when $i = 1$, we have the leading term

$$v_1(\varepsilon, r) \sim \frac{d(\varepsilon)}{r^{n-1}}$$

with $d(\varepsilon)$ given by (4.6).

## 5. Extension and discussion

### 5.1. Extension

The method presented in the previous sections can also be used to study the equation

$$y'' + \frac{n-1}{r} y' + \varepsilon y y' + (y')^2 = 0 \tag{5.1}$$

subject to

$$y(1) = 0, \qquad y(\infty) = 1, \tag{5.2}$$

where the additional term $(y')^2$ signifies that the flow is compressible [3].

By setting

$$s = \varepsilon r, \qquad \tau = \varepsilon t,$$

the integral equation corresponding to (2.2) for the re-scaled solution $u(s) = y(r)$ is given by

$$e^{u(s)} - e = C\varepsilon^{n-2} \int_{\infty}^{s} \frac{1}{\tau^{n-1}} e^{-\tau} e^{-\int_{\infty}^{\tau}(u-1)\,d\sigma}\, d\tau, \quad \varepsilon \leqslant s < \infty. \tag{5.3}$$

For more details, see formula (35) in [3]. The constant $C$ is determined by the scaled boundary condition $u(\varepsilon) = 0$, i.e.,

$$C = \frac{1 - e}{\varepsilon^{n-2} \int_{\infty}^{\varepsilon} 1/\tau^{n-1} e^{-\tau} e^{-\int_{\infty}^{\tau}(u-1)\,d\sigma}\, d\tau}. \tag{5.4}$$

Thus, (5.3) becomes

$$e^{u(s)} - e = \frac{(1 - e) \int_{\infty}^{s} 1/\tau^{n-1} e^{-\tau} e^{-\int_{\infty}^{\tau}(u-1)\,d\sigma}\, d\tau}{\int_{\infty}^{\varepsilon} 1/\tau^{n-1} e^{-\tau} e^{-\int_{\infty}^{\tau}(u-1)\,d\sigma}\, d\tau}, \quad \varepsilon \leqslant s < \infty. \tag{5.5}$$

Now, define the sequence $\{u_i(s)\}_{i=0}^{\infty}$ of successive approximations

$$e^{u_0(s)} - e = \frac{(1 - e) \int_{\infty}^{s} 1/\tau^{n-1} e^{-\tau}\, d\tau}{\int_{\infty}^{\varepsilon} 1/\tau^{n-1} e^{-\tau}\, d\tau} = (1 - e)\frac{\rho_n(s)}{\rho_n(\varepsilon)}$$

and

$$e^{u_{i+1}(s)} - e = \frac{(1 - e) \int_{\infty}^{s} 1/\tau^{n-1} e^{-\tau} e^{-\int_{\infty}^{\tau}(u_i-1)\,d\sigma}\, d\tau}{\int_{\infty}^{\varepsilon} 1/\tau^{n-1} e^{-\tau} e^{-\int_{\infty}^{\tau}(u_i-1)\,d\sigma}\, d\tau}, \quad i \geqslant 0.$$

When $n \geqslant 2$, it can be shown in a similar manner as in Section 3 that the solution $u(s)$ is given by

$$u(s) = \lim_{m\to\infty} u_m(s) = u_0(s) + \sum_{m=1}^{\infty} \left[u_m(s) - u_{m-1}(s)\right]$$

with the series being uniformly convergent in $[\varepsilon, \infty)$. An asymptotic relation such as the one in (2.12) can also be established.

An alternative method of rigorously deriving an asymptotic expansion for (5.1)–(5.2) is to use the transformation $y = \log v$; see [15]. Upon this transformation, (5.1)–(5.2) becomes

$$v'' + \frac{n-1}{r} v' + \varepsilon v' \log v = 0, \quad v(1) = 1, \ v(\infty) = e.$$

Using the same method as in Sections 2 and 3, one can rigorously obtain an asymptotic solution, but we shall omit the details.

### 5.2. The case when $1 \leqslant n < 2$

When $n \geqslant 2$, as was seen in Sections 2 and 3, the leading term (2.10) of the asymptotic expansion of the solution $u(r)$ is determined by the two linear terms in Eq. (2.1). This result fails when $1 \leqslant n < 2$, a case for which the nonlinear term also contributes. In particular, when $n = 1$, the middle term in the original Eq. (1.1) disappears and the exact solution can be found explicitly to depend strongly on the nonlinear term; see (2.16).

In the case $1 \leqslant n < 2$, the solution at $r = 0$ is not singular any more, and it can be continuously extended through $r = 0$ by solving the problem near $r = 0$ with the method of power series. In view of this, we expect that some information may be found via the following idea.

Set $s = \varepsilon r$ and $P(s) = y(r)$. After this scaling, the original problem is transformed into

$$\begin{cases} P'' + \dfrac{n-1}{s}P' + PP' = 0, \\ P(\varepsilon) = 0, \qquad P(\infty) = 1. \end{cases}$$

Since $P(s)$ may not blow up at $s = 0$, we conjecture that the solution satisfies

$$u(r) = P(\varepsilon r) = P_0(\varepsilon r) + o(1),$$

where $P_0$ is a solution of the unperturbed problem

$$\begin{cases} P_0'' + \dfrac{n-1}{s}P_0' + P_0P_0' = 0, \\ P_0(0) = 0, \qquad P_0(\infty) = 1. \end{cases} \tag{5.6}$$

Unfortunately, an explicit formula for the solution of (5.6) is not available, except in the case when $n = 1$.

## Acknowledgements

The authors would like to thank the referee for valuable comments on an earlier version of the paper.

## References

[1] D.S. Cohen, A. Fokas and P.A. Lagerstrom, Proof of some asymptotic results for a model equation for low Reynolds number flow, *SIAM J. Math. Anal.* **35**(1) (1978), 187–207.

[2] O. Costin, Exponential asymptotics, transseries, and generalized Borel summation for analytic, nonlinear, rank-one systems of ordinary differential equations, *Internat. Math. Res. Notices* **1995**(8) (1995), 377–417.

[3] S.P. Hastings and J.B. Mcleod, An elementary approach to a model problem of Lagerstrom, *SIAM J. Math. Anal.* **40**(6) (2009), 2421–2436.

[4] S.P. Hastings and J.B. Mcleod, *Classical Methods in Ordinary Differential Equations*, Amer. Math. Soc., Providence, RI, 2012.

[5] G.C. Hsiao, Singular perturbations for a nonlinear differential equation with a small parameter, *SIAM J. Math. Anal.* **4** (1973), 283–301.

[6] C. Hunter, M. Tajdari and S.D. Boyer, On Lagerstrom's model of slow incompressible viscous flow, *SIAM J. Appl. Math.* **50**(1) (1990), 48–63.

[7] S. Kaplun and P.A. Lagerstrom, Asymptotic expansions of Navier–Stokes solutions for small Reynolds numbers, *J. Math. Mech.* **6** (1957), 585–593.
[8] P.A. Lagerstrom and R.G. Casten, Basic concepts underlying singular perturbation techniques, *SIAM Rev.* **14** (1972), 63–120.
[9] P.A. Lagerstrom and D.A. Reinelt, Note on logarithmic switchback terms in regular and singular perturbation expansions, *SIAM J. Appl. Math.* **44**(3) (1984), 451–462.
[10] A.D. MacGillivray, On a model equation of Lagerstrom, *SIAM J. Appl. Math.* **34**(4) (1978), 804–812.
[11] F.W.J. Olver, D.W. Lozier, R.F. Boisvert and C.W. Clark, *NIST Handbook of Mathematical Functions*, Cambridge Univ. Press, Cambridge, 2010.
[12] N. Popovic and P. Szmolyan, Rigorous asymptotic expansions for Lagerstrom's model equation – a geometric approach, *Nonlinear Anal.* **59**(4) (2004), 531–565.
[13] N. Popovic and P. Szmolyan, A geometric analysis of the Lagerstrom model problem, *J. Differential Equations* **199**(2) (2004), 290–325.
[14] S. Rosenblat and J. Shepherd, On the asymptotic solution of the Lagerstrom model equation, *SIAM J. Appl. Math.* **29** (1975), 110–120.
[15] L.A. Skinner, Note on the Lagerstrom singular perturbation models, *SIAM J. Appl. Math.* **41**(2) (1981), 362–364.

PROCEEDINGS OF THE
AMERICAN MATHEMATICAL SOCIETY
Volume 142, Number 11, November 2014, Pages 3953–3964
S 0002-9939(2014)12134-4
Article electronically published on July 29, 2014

# ASYMPTOTIC EXPANSION OF THE MODIFIED LOMMEL POLYNOMIALS $h_{n,\nu}(x)$ AND THEIR ZEROS

K. F. LEE AND R. WONG

(Communicated by Walter Van Assche)

*Dedicated to the Lord*

ABSTRACT. The modified Lommel polynomials satisfy the second-order linear difference equation

$$h_{n+1,\nu}(x) - 2(n+\nu)\,x\,h_{n,\nu}(x) + h_{n-1,\nu}(x) = 0, \qquad n \geq 0,$$

with initial values $h_{-1,\nu}(x) = 0$ and $h_{0,\nu}(x) = 1$, where $x$ is a real variable and $\nu$ is a fixed positive parameter. An asymptotic expansion, as $n \to \infty$, is derived for these polynomials by using a turning-point theory for three-term recurrence relations developed by Wang and Wong [Numer. Math. **91** (2002) and **94** (2003)]. The result holds uniformly in the infinite interval $0 \leq x < \infty$, containing the critical value $x = 1/N$, where $N = n + \nu$. Behavior of the zeros of these polynomials is also studied.

## 1. Introduction

The Lommel polynomials $R_{n,\nu}(x)$ arise in the theory of Bessel functions. The modified Lommel polynomials $h_{n,\nu}(x)$ are defined by

$$h_{n,\nu}(x) = R_{n,\nu}(1/x); \tag{1.1}$$

see [8, §§9.6-9.73]. The modified Lommel polynomials satisfy the three-term recurrence relation

$$h_{n+1,\nu}(x) - 2(n+\nu)\,x\,h_{n,\nu}(x) + h_{n-1,\nu}(x) = 0, \qquad n \geq 0, \tag{1.2}$$

with initial values $h_{-1,\nu}(x) = 0$, $h_{0,\nu}(x) = 1$. The polynomials have the hypergeometric representation (for $n > 1$)

$$h_{n,\nu}(x) = (\nu)_n (2x)^n \, {}_2F_3(-n/2, (-n+1)/2; \nu, -n, 1-\nu-n; -1/x^2), \tag{1.3}$$

where $(\nu)_n = \Gamma(\nu+n)/\Gamma(\nu)$. Equation (1.3) implies the reflection formula of modified Lommel polynomials

$$h_{n,\nu}(-x) = (-1)^n \, h_{n,\nu}(x). \tag{1.4}$$

Furthermore, the correct orthogonality relation is

$$\sum_{k=1}^{\infty} h_{m,\nu}(\pm x_k) h_{n,\nu}(\pm x_k) x_k^2 = \frac{1}{2(n+\nu)} \delta_{mn}, \tag{1.5}$$

Received by the editors July 6, 2012 and, in revised form, December 31, 2012.
2010 *Mathematics Subject Classification.* Primary 41A60, 39A10; Secondary 33C45.
*Key words and phrases.* Modified Lommel polynomials, second-order linear difference equations, uniform asymptotic expansions, Airy function.

where $x_k = j_{\nu-1,k}^{-1}$ and $\{j_{\nu-1,k}\}$ are the zeros of the Bessel function $J_{\nu-1}(x)$, for $k = 1, 2, \cdots$; see [3, p. 197, (6.5.17)]. (Here we wish to point out that the orthogonality relations given in Dickinson [2] and Chihara [1] are incorrect.) From some formulas involving Lommel polynomials (cf., Watson [8, §9.6]), we may define the modified Lommel polynomials $h_{n,\nu}(x)$ in terms of Bessel functions by

$$J_{\nu+n}(1/x) = J_\nu(1/x)h_{n,\nu}(x) - J_{\nu-1}(1/x)h_{n-1,\nu+1}(x), \tag{1.6}$$

or, more explicitly,

$$\pi^{-1}(2x \sin \pi\nu)h_{n,\nu}(x) = J_{\nu+n}(1/x)J_{-\nu+1}(1/x) + (-1)^n J_{-\nu-n}(1/x)J_{\nu-1}(1/x), \tag{1.7}$$

where $n$ is an integer. The behavior of $h_{n,\nu}(x)$ has been investigated by Hurwitz (see Watson [8, §9.65]), and he used an elementary approach to show that

$$\lim_{n\to\infty} \frac{(2x)^{1-\nu-n}h_{n,\nu}(x)}{\Gamma(n+\nu)} = J_{\nu-1}(1/x) \tag{1.8}$$

uniformly for all $x$ in any closed and bounded annular region centered at the origin.

The purpose of this paper is to present an asymptotic expansion for $h_{n,\nu}(t/N)$, which holds uniformly for $t$ in $[0, \infty)$, where $N = n + \nu$. Our approach is to use a turning-point theory recently introduced by Wang and Wong [6, 7] for three-term recurrence relations. Asymptotic formulas are also obtained for the zeros of the modified Lommel polynomials. It should be mentioned that our result for $h_{n,\nu}(x)$ can also be obtained from the uniform asymptotic expansions of Bessel functions. But, the derivations of these expansions are themselves very sophisticated. Our approach is more direct from the viewpoint of difference equations.

## 2. Difference equation

Equation (1.2) is already in the canonical form considered in [7],

$$h_{n+1,\nu}(x) - (A_n x + B_n)h_{n,\nu}(x) + h_{n-1,\nu}(x) = 0, \tag{2.1}$$

with

$$A_n = 2(n+\nu) \quad \text{and} \quad B_n = 0. \tag{2.2}$$

In terms of the notation

$$A_n \sim n^{-\theta}\sum_{s=0}^{\infty}\frac{\alpha_s}{n^s} \quad \text{and} \quad B_n \sim \sum_{s=0}^{\infty}\frac{\beta_s}{n^s} \quad (\text{as } n\to\infty) \tag{2.3}$$

used in [7], we have

$$\theta = -1, \quad \alpha_0 = 2, \quad \alpha_1 = 2\nu, \quad \alpha_n = 0 \quad \text{for} \quad n = 2, 3, \cdots, \tag{2.4}$$

and $\beta_0 = \beta_1 = \beta_2 = \cdots = 0$. If these expansions are recast in the form

$$A_n \sim N^{-\theta}\sum_{s=0}^{\infty}\frac{\alpha'_s}{N^s} \quad \text{and} \quad B_n \sim \sum_{s=0}^{\infty}\frac{\beta'_s}{N^s}, \tag{2.5}$$

where $N = n + \tau_0$ and $\tau_0$ is some fixed real number to be determined, it is easily found that

$$\alpha'_0 = 2, \quad \alpha'_1 = 2\nu - 2\tau_0, \cdots, \tag{2.6}$$

and $\beta_0' = \beta_1' = \beta_2' = \cdots = 0$. To apply the result in [7], we first choose $\tau_0$ so that $\alpha_1' = 0$. From (2.6), it is obvious that the choice is

$$\tau_0 = \nu. \tag{2.7}$$

According to the equation in [7, (2.4)], the characteristic equation is

$$\lambda^2 - 2t\lambda + 1 = 0, \tag{2.8}$$

where $t$ is the rescaled variable $t = Nx$. The two roots of this equation are

$$\lambda_\pm = t \pm \sqrt{t^2 - 1}. \tag{2.9}$$

The points $t_\pm = \pm 1$, where the two roots coincide, are called the *turning points* of equation (2.1).

We now define the function $\zeta(t)$ introduced in [7, (4.10)]. With $t_+ = 1$, $\theta = -1$, $\alpha_0' = 2$ and $\beta_0' = 0$, this function is given by

$$\frac{2}{3}[\zeta(t)]^{3/2} := \log\left(t + \sqrt{t^2 - 1}\right) - \frac{1}{t}\int_1^t \frac{s}{\sqrt{s^2 - 1}}\, ds, \qquad t \geq 1,$$

and

$$\frac{2}{3}[-\zeta(t)]^{3/2} := t^{-1}\int_t^1 \frac{s}{\sqrt{1 - s^2}}\, ds - \cos^{-1} t, \qquad 0 < t < 1.$$

By direct calculation, one obtains

$$\frac{2}{3}[\zeta(t)]^{3/2} = \log\left(t + \sqrt{t^2 - 1}\right) - \frac{1}{t}(t^2 - 1)^{\frac{1}{2}} \tag{2.10}$$

for $t \geq 1$ and

$$\frac{2}{3}[-\zeta(t)]^{3/2} = \frac{1}{t}(1 - t^2)^{\frac{1}{2}} - \cos^{-1} t \tag{2.11}$$

for $0 < t < 1$. Note that as $t \to 1$, we have $\zeta(t) \to 0$. In fact, near $t = 1$, we have the Taylor expansion

$$\zeta(t) = 2^{\frac{1}{3}}(t - 1)\left[1 - \frac{7}{10}(t - 1) + \frac{102}{175}(t - 1)^2 + \cdots\right]. \tag{2.12}$$

We also define the functions $H_0(\zeta)$ and $\Phi(\zeta)$ introduced in [7, (4.19) and (4.28)]. In the present situation, these functions are given by

$$H_0(\zeta) := -\sqrt{\frac{t^2 - 1}{\zeta}} \qquad \text{and} \qquad \Phi(\zeta) = 0, \tag{2.13}$$

where $\zeta$ is the function defined in (2.10) and (2.11). Note that in our special case, $\alpha_1' = \beta_1' = 0$. Hence, according to the definition of $\Phi(\zeta)$ given in [7, (4.28)], the second equation in (2.13) holds for $0 < t < \infty$ instead of $t \geq \delta$, $0 < \delta < 1$.

With this preliminary work done, we can now apply the main result in [7] to conclude that there are constants $C_1(x)$ and $C_2(x)$ such that the polynomials $h_{n,\nu}(x)$ in (2.1) can be expressed as

$$h_{n,\nu}(x) = C_1(x)P_n(x) + C_2(x)Q_n(x), \tag{2.14}$$

where, with $x = N^{-1}t$, $P_n(x)$ and $Q_n(x)$ have the asymptotic expansions
(2.15)

$$P_n(x) = \left(\frac{\zeta}{t^2 - 1}\right)^{\frac{1}{4}}\left[\mathrm{Ai}\left(N^{\frac{2}{3}}\zeta\right)\sum_{s=0}^{p-1}\frac{\widetilde{A}_s(\zeta)}{N^{s-\frac{1}{6}}} + \mathrm{Ai}'\left(N^{\frac{2}{3}}\zeta\right)\sum_{s=0}^{p-1}\frac{\widetilde{B}_s(\zeta)}{N^{s+\frac{1}{6}}} + \epsilon_p(N,t)\right]$$

and

(2.16)

$$Q_n(x) = \left(\frac{\zeta}{t^2-1}\right)^{\frac{1}{4}} \left[\mathrm{Bi}\left(N^{\frac{2}{3}}\zeta\right)\sum_{s=0}^{p-1}\frac{\widetilde{A}_s(\zeta)}{N^{s-\frac{1}{6}}} + \mathrm{Bi}'\left(N^{\frac{2}{3}}\zeta\right)\sum_{s=0}^{p-1}\frac{\widetilde{B}_s(\zeta)}{N^{s+\frac{1}{6}}} + \delta_p(N,t)\right]$$

for large values of $n$ (or, equivalently, $N$). In (2.15) and (2.16), $\mathrm{Ai}(\cdot)$ and $\mathrm{Bi}(\cdot)$ are the Airy functions, the leading coefficients are given by

$$\widetilde{A}_0(\zeta) = 1 \quad \text{and} \quad \widetilde{B}_0(\zeta) = 0, \tag{2.17}$$

and there exists a positive constant $M_p$ such that

$$|\epsilon_p(N,t)| \le \frac{M_p}{N^{p-\frac{1}{6}}}|\widetilde{\mathrm{Ai}}(N^{\frac{2}{3}}\zeta)| \qquad \text{and} \qquad |\delta_p(N,t)| \le \frac{M_p}{N^{p-\frac{1}{6}}}|\widetilde{\mathrm{Bi}}(N^{\frac{2}{3}}\zeta)|, \tag{2.18}$$

where $\widetilde{\mathrm{Ai}}$ and $\widetilde{\mathrm{Bi}}$ are the modulus functions given in [7, (7.10) and (7.11)].

## 3. Determination of $C_1(x)$ and $C_2(x)$

First, we examine the behavior of $h_{n,\nu}(x)$ as $n \to \infty$. Here we do not use the result of (1.8) since it will provide only the value of $C_2(x)$ and not that of $C_1(x)$. We first note that when $N$ is large, we have from the well-known asymptotic formulas of $J_\nu(1/x)$ and $Y_\nu(1/x)$:

$$J_{\nu+n}\left(\frac{1}{x}\right) = J_N\left(\frac{1}{x}\right) \sim \frac{1}{\sqrt{2\pi N}}\left(\frac{e}{2Nx}\right)^N, \tag{3.1}$$

$$Y_{\nu+n}\left(\frac{1}{x}\right) = Y_N\left(\frac{1}{x}\right) \sim -\sqrt{\frac{2}{\pi N}}\left(\frac{e}{2Nx}\right)^{-N}; \tag{3.2}$$

see [5, p. 436]. For a negative order of the Bessel function, we know from [5, p. 243] that

$$J_{-\nu-n}\left(\frac{1}{x}\right) = J_{-N}\left(\frac{1}{x}\right) = \cos(N\pi)J_N(1/x) - \sin(N\pi)Y_N(1/x) \tag{3.3}$$

and

$$Y_{\nu-1}(z) = \frac{1}{\sin(\nu\pi)}\left(J_{\nu-1}(z)\cos(\nu\pi) + J_{-\nu+1}(z)\right). \tag{3.4}$$

Coupling (1.7), (3.1), (3.2), (3.3) and (3.4), we have

$$h_{n,\nu}(x) \sim \frac{e^N(2xN)^{-N}}{\sqrt{2\pi N}}\frac{\pi}{2x}Y_{\nu-1}(1/x) + (2x)^{N-1}\sqrt{2\pi}N^{N-\frac{1}{2}}e^{-N}J_{\nu-1}(1/x). \tag{3.5}$$

Next, we recall the well-known asymptotic formulas

$$\mathrm{Ai}(\eta) \sim \frac{\eta^{-\frac{1}{4}}}{2\sqrt{\pi}}\exp\left(-\frac{2}{3}\eta^{\frac{3}{2}}\right), \qquad \mathrm{Ai}'(\eta) \sim -\frac{\eta^{\frac{1}{4}}}{2\sqrt{\pi}}\exp\left(-\frac{2}{3}\eta^{\frac{3}{2}}\right), \tag{3.6}$$

and

$$\mathrm{Bi}(\eta) \sim \frac{\eta^{-\frac{1}{4}}}{\sqrt{\pi}}\exp\left(\frac{2}{3}\eta^{\frac{3}{2}}\right), \qquad \mathrm{Bi}'(\eta) \sim \frac{\eta^{\frac{1}{4}}}{\sqrt{\pi}}\exp\left(\frac{2}{3}\eta^{\frac{3}{2}}\right), \tag{3.7}$$

as $\eta \to +\infty$; see [5, pp. 392-393]. Furthermore, from (2.10) we have

$$\frac{2}{3}[\zeta(t)]^{3/2} \sim \log(2t) - 1 = \log(2xN) - 1. \tag{3.8}$$

Since $N = n + \nu$, it is readily seen from (3.6), (3.7) and (3.8) that

$$P_n(x) \sim \frac{1}{2\sqrt{\pi\, x\, N}} \frac{1}{(2\, x\, N)^N} e^N, \qquad Q_n(x) \sim \frac{2^N}{\sqrt{\pi}} (xN)^{N-\frac{1}{2}} e^{-N}.$$

Comparing the two sides of (2.14) gives

$$C_1(x) = \frac{\pi}{\sqrt{2\, x}} Y_{\nu-1}(1/x), \qquad C_2(x) = \frac{\pi}{\sqrt{2\, x}} J_{\nu-1}(1/x). \tag{3.9}$$

In summary, we have from (2.14), (2.15) and (2.16) that

$$\begin{aligned} h_{n,\nu}(x) = {} & \frac{\pi}{\sqrt{2\, x}} Y_{\nu-1}(1/x) \left( \frac{\zeta}{t^2 - 1} \right)^{\frac{1}{4}} N^{\frac{1}{6}} \left[ \mathrm{Ai}\left( N^{\frac{2}{3}} \zeta \right) + O\left( \frac{1}{N} \right) \right] \\ & + \frac{\pi}{\sqrt{2\, x}} J_{\nu-1}(1/x) \left( \frac{\zeta}{t^2 - 1} \right)^{\frac{1}{4}} N^{\frac{1}{6}} \left[ \mathrm{Bi}\left( N^{\frac{2}{3}} \zeta \right) + O\left( \frac{1}{N} \right) \right], \end{aligned} \tag{3.10}$$

where $x = N^{-1} t$ and $N = n + \nu$. This result holds uniformly for $0 \le t < \infty$.

## 4. Zeros

It is well known that the zeros of the modified Lommel polynomials $h_{n,\nu}(x)$ all lie in the interval $-j_{\nu-1,1}^{-1} \le x \le j_{\nu-1,1}^{-1}$. This can be seen from (1.5). After the rescaling $x = N^{-1} t$, the zeros of $h_{n,\nu}(t/N)$ all lie in the interval $-N/j_{\nu-1,1}^{-1} \le t \le N/j_{\nu-1,1}^{-1}$. In this section, we shall use the asymptotic approximation given in (3.10) to derive asymptotic formulas for the zeros of $h_{n,\nu}(t/N)$ as $n \to \infty$. We divide our discussion into three cases: (i) large zeros in $1 < t < N/j_{\nu-1,1}^{-1}$; (ii) small zeros in $0 < t < 1$; and (iii) zeros near the turning point $t = 1$ and on two sides of the point. Several tables are provided to give a comparison of the numerical values of the asymptotic zeros and the true zeros.

*Case* (i). Let $t_{n,m}$ denote the $m$-th zero of $h_{n,\nu}(t/N)$, arranged in ascending order:

$$-\frac{N}{j_{\nu-1,1}} \le t_{n,1} < t_{n,2} < \cdots < t_{n,n-1} < t_{n,n} \le \frac{N}{j_{\nu-1,1}}. \tag{4.1}$$

For large zeros, we have the following result.

**Theorem 4.1.** *The asymptotic behavior of the large zeros $t_{n,m}$ is given by*

$$t_{n,n-k} = \frac{N}{j_{\nu-1,k+1}} \left[ 1 + O\left( \frac{1}{N} \right) \right], \qquad n \to \infty, \tag{4.2}$$

*for fixed $k = 0, 1, 2, \cdots$, where $N = n + \nu$.*

*Proof.* Returning to (3.10), we have

$$\begin{aligned} h_{n,\nu}\left( \frac{t}{N} \right) = \frac{\pi}{\sqrt{2\, x}} \left( \frac{\zeta}{t^2 - 1} \right)^{\frac{1}{4}} \Bigg[ & N^{\frac{1}{6}} \left( Y_{\nu-1}\left( \frac{1}{x} \right) \mathrm{Ai}(N^{\frac{2}{3}} \zeta) \right. \\ & \left. + J_{\nu-1}\left( \frac{1}{x} \right) \mathrm{Bi}(N^{\frac{2}{3}} \zeta) \right) + O\left( \frac{1}{N} \right) \Bigg]. \end{aligned} \tag{4.3}$$

Since $N = n+\nu$ and $\zeta > 0$ for $t > 1$, the terms involving $\mathrm{Ai}(N^{2/3}\zeta)$ and $\mathrm{Ai}'(N^{2/3}\zeta)$ are exponentially small, while the terms involving $\mathrm{Bi}(N^{2/3}\zeta)$ and $\mathrm{Bi}'(N^{2/3}\zeta)$ are exponentially large; see (3.6) and (3.7). Thus, (4.3) reduces to

$$h_{n,\nu}\left(\frac{t}{N}\right)=\frac{\pi}{\sqrt{2x}}\left(\frac{\zeta}{t^2-1}\right)^{\frac{1}{4}}\left[N^{\frac{1}{6}}J_{\nu-1}\left(\frac{1}{x}\right)\mathrm{Bi}(N^{\frac{2}{3}}\zeta)+O\left\{N^{-1}\exp\left(\frac{2}{3}N\zeta^{\frac{3}{2}}\right)\right\}\right].$$

If $t$ is a zero of $h_{n,\nu}(t/N)$, the left-hand side of the equation is equal to zero, and we have

$$J_{\nu-1}\left(\frac{1}{x}\right)=O\left(\frac{1}{N}\right) \tag{4.4}$$

since $\mathrm{Bi}(N^{2/3}\zeta)$ has no zero for $\zeta>0$. In view of (4.1), the solutions of (4.4) are given in (4.2). □

For $n=100$ and $\nu=1$, the numerical values of the large zeros are given in Table 1. Note that in this case, the polynomial $h_{n,\nu}(t/N)$ has 100 zeros; half of them lie on the positive real axis, and the other half lie on the negative real axis.

TABLE 1. True values and approximate values of large zeros when $n=100$ and $\nu=1$ evaluated with Maple.

| $m$-th | true | (4.2) | error % |
|---|---|---|---|
| $t_{100,1,91}$ | 3.2969 | 3.2969 | $-1.3785\times10^{-8}$ |
| $t_{100,1,92}$ | 3.6736 | 3.6736 | $1.6330\times10^{-8}$ |
| $t_{100,1,93}$ | 4.1474 | 4.1474 | $-6.1147\times10^{-9}$ |
| $t_{100,1,94}$ | 4.7615 | 4.7615 | $6.8524\times10^{-9}$ |
| $t_{100,1,95}$ | 5.5890 | 5.5890 | $-5.8648\times10^{-9}$ |
| $t_{100,1,96}$ | 6.7645 | 6.7645 | $-1.3906\times10^{-8}$ |
| $t_{100,1,97}$ | 8.5655 | 8.5655 | $-1.4199\times10^{-8}$ |
| $t_{100,1,98}$ | 11.6713 | 11.6713 | $3.5866\times10^{-8}$ |
| $t_{100,1,99}$ | 18.2968 | 18.2968 | $3.0196\times10^{-8}$ |
| $t_{100,1,100}$ | 41.9989 | 41.9989 | $-3.7573\times10^{-9}$ |

*Case* (ii). For the small zeros in $0<t<1$, we recall the asymptotic formulas

$$\mathrm{Ai}(-\eta)=\frac{\eta^{-\frac{1}{4}}}{\sqrt{\pi}}\left[\cos(\frac{2}{3}\eta^{\frac{3}{2}}-\frac{1}{4}\pi)+O(\eta^{-\frac{3}{2}})\right], \tag{4.5}$$

$$\mathrm{Bi}(-\eta)=\frac{\eta^{-\frac{1}{4}}}{\sqrt{\pi}}\left[-\sin(\frac{2}{3}\eta^{\frac{3}{2}}-\frac{1}{4}\pi)+O(\eta^{-\frac{3}{2}})\right], \tag{4.6}$$

$$J_\nu(\eta)=\sqrt{\frac{2}{\pi\eta}}\left[\cos(\eta-\frac{\nu\pi}{2}-\frac{\pi}{4})+O(\eta^{-1})\right], \tag{4.7}$$

$$Y_\nu(\eta)=\sqrt{\frac{2}{\pi\eta}}\left[\sin(\eta-\frac{\nu\pi}{2}-\frac{\pi}{4})+O(\eta^{-1})\right], \tag{4.8}$$

and the corresponding formulas of $\mathrm{Ai}'(-\eta)$ and $\mathrm{Bi}'(-\eta)$; see [5, pp. 392-393]. From these formulas, it is evident that if $t$ is a zero of the modified Lommel polynomials $h_{n,\nu}(t/N)$, then equation (4.3) can be reduced to

$$\frac{\pi}{\sqrt{2x}}N^{\frac{1}{6}}\left[Y_{\nu-1}\left(\frac{1}{x}\right)\mathrm{Ai}(N^{\frac{2}{3}}\zeta)+J_{\nu-1}\left(\frac{1}{x}\right)\mathrm{Bi}(N^{\frac{2}{3}}\zeta)\right]=O\left(\frac{1}{N}\right). \tag{4.9}$$

Let $\widetilde{t}_{n,m}$ denote the $m$-th zero of $h_{n,\nu}(t/N)$, arranged in ascending order:

$$0 \le \widetilde{t}_{n,1} < \widetilde{t}_{n,2} < \cdots < 1. \tag{4.10}$$

Since $\zeta(t) < 0$ for $0 < t < 1$, it is readily seen from (4.5) and (4.6) that the left-hand side of (4.9) can be written as

$$\begin{aligned}
&\frac{\pi}{\sqrt{2x}} N^{\frac{1}{6}} Y_{\nu-1}\left(\frac{1}{x}\right) \mathrm{Ai}(N^{\frac{2}{3}}\zeta) + \frac{\pi}{\sqrt{2x}} N^{\frac{1}{6}} J_{\nu-1}\left(\frac{1}{x}\right) \mathrm{Bi}(N^{\frac{2}{3}}\zeta) \\
=&\frac{1}{|\zeta(t)|^{1/4}}\left[\sin\left(\frac{N}{t} - \frac{\nu\pi}{2} + \frac{\pi}{4}\right)\cos\left(\frac{2}{3}N|\zeta(t)|^{\frac{3}{2}} - \frac{\pi}{4}\right)\right. \\
&\qquad\left. - \cos\left(\frac{N}{t} - \frac{\nu\pi}{2} + \frac{\pi}{4}\right)\sin\left(\frac{2}{3}N|\zeta(t)|^{\frac{3}{2}} - \frac{\pi}{4}\right) + O\left(\frac{1}{N}\right)\right] \\
=&\frac{1}{|\zeta(t)|^{1/4}}\left[\sin\left(\frac{N}{t} - \frac{\nu\pi}{2} + \frac{\pi}{4} - \frac{2}{3}N|\zeta(t)|^{\frac{3}{2}} + \frac{\pi}{4}\right) + O\left(\frac{1}{N}\right)\right].
\end{aligned}$$

Therefore, from (4.9) we have

$$\sin\left(\frac{N}{t} - \frac{\nu\pi}{2} + \frac{\pi}{4} - \frac{2}{3}N|\zeta(t)|^{\frac{3}{2}} + \frac{\pi}{4}\right) = O\left(\frac{1}{N}\right). \tag{4.11}$$

Note that near $t = 0$, $|\zeta(t)|^{\frac{3}{2}}$ has the expansion

$$\frac{2}{3}|\zeta(t)|^{\frac{3}{2}} = \frac{1}{t} - \frac{\pi}{2} + \frac{1}{2}t + \frac{1}{24}t^3 + \frac{1}{80}t^5 + O\left(t^6\right). \tag{4.12}$$

Upon solving (4.11) and (4.12), we have

$$\frac{n+1}{2}\pi - \frac{1}{2}Nt - \frac{1}{24}Nt^3 - \frac{1}{80}Nt^5 + \cdots = k\pi + O\left(\frac{1}{N}\right). \tag{4.13}$$

As a first approximation, we take just the first two terms on the left-hand side of the equation and obtain

$$\frac{n+1}{2}\pi - \frac{1}{2}Nt = k\pi + O\left(\frac{1}{N}\right). \tag{4.14}$$

The solutions of equation (4.14) are presented in two separate cases. In the order arranged in (4.10), when $n$ is even, the small zeros are given by

$$\widetilde{t}_{n,m+1} = \frac{2}{N}\left[\frac{\pi}{2} + m\pi + O\left(\frac{1}{N}\right)\right] \tag{4.15}$$

for fixed $m = 0, 1, 2, \cdots$. Note that $\widetilde{t}_{n,1} \neq 0$. When $n$ is odd, the small zeros are given by $\widetilde{t}_{n,1} = 0$ and

$$\widetilde{t}_{n,m+1} = \frac{2}{N}\left[m\pi + O\left(\frac{1}{N}\right)\right] \tag{4.16}$$

for fixed $m = 0, 1, 2, \cdots$.

**Theorem 4.2.** *The asymptotic behavior of the small zeros of $h_{n,\nu}(t/N)$, arranged in ascending order in* (4.10)*, are given by the formulas in* (4.15)–(4.16).

The above result can be improved if we pick three terms in (4.13), in which case we have

$$\frac{n+1}{2}\pi - \frac{1}{2}Nt - \frac{1}{24}Nt^3 = k\pi + O\left(\frac{1}{N}\right). \tag{4.17}$$

This is a cubic equation, and we are looking for only the real root. Recall that the real root of the cubic equation

$$t^3 + pt + q = 0 \tag{4.18}$$

is given explicitly by

$$t = \left(-\frac{q}{2} + \sqrt{\frac{q^2}{4} + \frac{p^3}{27}}\right)^{\frac{1}{3}} + \left(-\frac{q}{2} - \sqrt{\frac{q^2}{4} + \frac{p^3}{27}}\right)^{\frac{1}{3}}, \tag{4.19}$$

provided that

$$\frac{q^2}{4} + \frac{p^3}{27} > 0. \tag{4.20}$$

The solutions of (4.17) are again presented in two separate cases. Let $K_1 = \frac{\pi}{2} + m\pi + O(\frac{1}{N})$ and $K_2 = m\pi + O(\frac{1}{N})$. When $n$ is even, the small zeros are given by

$$\widetilde{t}_{n,m+1} = \left[\frac{12}{N}K_1 + 12\sqrt{\left(\frac{K_1}{N}\right)^2 + \frac{4}{9}}\right]^{\frac{1}{3}} + \left[\frac{12}{N}K_1 - 12\sqrt{\left(\frac{K_1}{N}\right)^2 + \frac{4}{9}}\right]^{\frac{1}{3}} \tag{4.21}$$

for fixed $m = 0, 1, 2, \cdots$. Note that $\widetilde{t}_{n,1} \neq 0$. When $n$ is odd, the small zeros are given by $\widetilde{t}_{n,1} = 0$ and

$$\widetilde{t}_{n,m+1} = \left[\frac{12}{N}K_2 + 12\sqrt{\left(\frac{K_2}{N}\right)^2 + \frac{4}{9}}\right]^{\frac{1}{3}} + \left[\frac{12}{N}K_2 - 12\sqrt{\left(\frac{K_2}{N}\right)^2 + \frac{4}{9}}\right]^{\frac{1}{3}} \tag{4.22}$$

for fixed $m = 0, 1, 2 \cdots$.

It can be shown that as $n \to \infty$, the quantities on the right-hand sides of equations (4.21) – (4.22) are asymptotically equal to the quantities on the right-hand sides of (4.15) – (4.16), respectively. More precisely, we have

$$\begin{aligned}&\left[\frac{12}{N}K_1 + 12\sqrt{\left(\frac{K_1}{N}\right)^2 + \frac{4}{9}}\right]^{\frac{1}{3}} + \left[\frac{12}{N}K_1 - 12\sqrt{\left(\frac{K_1}{N}\right)^2 + \frac{4}{9}}\right]^{\frac{1}{3}}\\ &\sim \frac{2}{N}\left[\frac{\pi}{2} + m\pi + O\left(\frac{1}{N}\right)\right]\end{aligned} \tag{4.23}$$

and

$$\begin{aligned}&\left[\frac{12}{N}K_2 + 12\sqrt{\left(\frac{K_2}{N}\right)^2 + \frac{4}{9}}\right]^{\frac{1}{3}} + \left[\frac{12}{N}K_2 - 12\sqrt{\left(\frac{K_2}{N}\right)^2 + \frac{4}{9}}\right]^{\frac{1}{3}}\\ &\sim \frac{2}{N}\left[m\pi + O\left(\frac{1}{N}\right)\right],\end{aligned} \tag{4.24}$$

respectively. Note that the $O$ – terms in (4.23) and (4.24) are still $O(1/N)$. To improve the order estimate from $O(1/N)$ to $O(1/N^2)$, one should use (2.14) and (3.9) with $p = 2$ in (2.15) and (2.16), instead of (3.10).

Numerical computation shows that the solutions given in (4.21) and (4.22) are more accurate than those presented in (4.15) and (4.16); see the numerics in Table 2. Use of numerical solutions of equation (4.11) would produce even better results.

TABLE 2. True values and approximate values of zeros evaluated with Maple, when $n = 100$ and $\nu = 1$.

| $m$-th | true | (4.15), (4.16) | error % | (4.21), (4.22) | error % |
|---|---|---|---|---|---|
| $\tilde{t}_{100,1,1}$ | 0.0311 | 0.0311 | 0.0081 | 0.0311 | 0.0000 |
| $\tilde{t}_{100,1,2}$ | 0.0932 | 0.0933 | 0.0726 | 0.0932 | 0.0001 |
| $\tilde{t}_{100,1,3}$ | 0.1552 | 0.1555 | 0.2021 | 0.1552 | 0.0013 |
| $\tilde{t}_{100,1,4}$ | 0.2169 | 0.2177 | 0.3973 | 0.2169 | 0.0053 |
| $\tilde{t}_{100,1,5}$ | 0.2781 | 0.2799 | 0.6596 | 0.2782 | 0.0148 |

*Case* (iii). Now we consider the zeros near the turning point $t = 1$, that is, on both sides of $\sigma = 0$, where $\sigma = N^{2/3}\zeta(t)$ is the variable inside the Airy functions in (4.3). If $t$ is a zero of $h_{n,\nu}(t/N)$ and $\sigma$ is bounded, then equation (4.3) gives

$$\frac{\pi}{\sqrt{2x}} Y_{\nu-1}\left(\frac{1}{x}\right) \mathrm{Ai}(\sigma) + \frac{\pi}{\sqrt{2x}} J_{\nu-1}\left(\frac{1}{x}\right) \mathrm{Bi}(\sigma) = O\left(\frac{1}{N}\right), \tag{4.25}$$

which can be written as

$$\sin\left[\frac{N}{t} - \frac{\nu\pi}{2} + \frac{\pi}{4} + \arctan\frac{\mathrm{Bi}(\sigma)}{\mathrm{Ai}(\sigma)}\right] = O\left(\frac{1}{N}\right). \tag{4.26}$$

Since

$$\arctan\frac{\mathrm{Bi}(\sigma)}{\mathrm{Ai}(\sigma)} = \frac{1}{3}\pi + \frac{3}{4}\frac{3^{\frac{1}{3}}}{\pi}\left(\Gamma\left(\frac{2}{3}\right)\right)^3 \sigma - \frac{9}{16}\frac{3^{\frac{1}{6}}}{\pi^2}\left(\Gamma\left(\frac{2}{3}\right)\right)^4 \sigma^2 + O\left(\sigma^4\right) \tag{4.27}$$

for small $\sigma$, replacing arctan in (4.26) by its leading term given in (4.27) yields

$$\sin\left(\frac{N}{t} - \frac{\nu\pi}{2} + \frac{\pi}{4} + \frac{\pi}{3}\right) = O\left(\frac{1}{N}\right). \tag{4.28}$$

Recall that here we are only concerned with the zeros of the polynomials near $t = 1$ (i.e., $\sigma$ close to 0). From (4.28), we obtain

$$\frac{N}{t\pi} - \frac{\nu}{2} + \frac{1}{4} + \frac{1}{3} = k + O\left(\frac{1}{N}\right), \tag{4.29}$$

where $k$ is an integer. Recall that $\sigma = N^{2/3}\zeta(t)$. Hence, by Lagrange's inversion formula, we have from (2.12) that

$$t = 1 + \sum_{j=1}^{\infty} A_j (2^{-\frac{1}{3}}\sigma N^{-\frac{2}{3}})^j, \tag{4.30}$$

where $A_1 = 1$, $A_2 = \frac{7}{10}$, $A_3 = \frac{139}{350}$, $\cdots$. Substituting (4.30) in (4.29) gives

$$\frac{N}{\pi}(1 - 2^{-\frac{1}{3}}\sigma N^{-\frac{2}{3}} + \cdots) = k - \frac{7}{12} + \frac{\nu}{2} + O\left(\frac{1}{N}\right), \tag{4.31}$$

which in turn gives

$$2^{-\frac{1}{3}}\frac{\sigma}{\pi}N^{\frac{1}{3}} = \frac{N}{\pi} - k + \frac{7}{12} - \frac{\nu}{2} + O\left(\frac{1}{N}\right). \tag{4.32}$$

Now, we choose $k$ to be a large integer so that $\frac{N}{\pi} - k + \frac{7}{12} - \frac{\nu}{2}$ is a bounded quantity. Let us write (4.32) in the form

$$\sigma = \frac{t_1}{N^{1/3}} + \cdots \tag{4.33}$$

with

$$\frac{1}{2^{\frac{1}{3}}\pi} t_1 = \frac{N}{\pi} - k + \frac{7}{12} - \frac{\nu}{2}. \tag{4.34}$$

If we denote by $N_\nu$ the fractional part of $\frac{N}{\pi} + \frac{7}{12} - \frac{\nu}{2}$ (i.e., $N_\nu = \frac{N}{\pi} + \frac{7}{12} - \frac{\nu}{2} - [\frac{N}{\pi} + \frac{7}{12} - \frac{\nu}{2}]$), (4.34) can be expressed as

$$t_1 = 2^{\frac{1}{3}}\pi(m + N_\nu), \tag{4.35}$$

where $m$ is an integer. Let $t^{(l)}_{n,m+1}$ and $t^{(r)}_{n,m+1}$, $m = 0, 1, 2, \cdots$, denote, respectively, the $(m+1)$th zero on the left-hand and on the right-hand side of the turning point $t = 1$, arranged in the order

$$\cdots < t^{(l)}_{n,2} < t^{(l)}_{n,1} < 1 < t^{(r)}_{n,1} < t^{(r)}_{n,2} < \cdots . \tag{4.36}$$

We can now write the asymptotic formulas as $n \to \infty$

$$t^{(r)}_{n,m+1} = 1 + \frac{2^{\frac{1}{3}}\pi}{N}(m + N_\nu) \tag{4.37}$$

and

$$t^{(l)}_{n,m+1} = 1 + \frac{2^{\frac{1}{3}}\pi}{N}(-m - 1 + N_\nu) \tag{4.38}$$

for fixed $m = 0, 1, 2 \cdots$, where $N_\nu$ denotes the fractional part of $\frac{N}{\pi} + \frac{7}{12} - \frac{\nu}{2}$.

**Theorem 4.3.** *The asymptotic behavior of the zeros of $h_{n,\nu}(t/N)$ on two sides of the turning point $t = 1$, as arranged in* (4.36), *are given in* (4.37) *and* (4.38).

For $n = 100$ and $\nu = 1$, the numerical values of the first two zeros on the two sides of $t = 1$ are given in Table 3.

TABLE 3. True values and approximate values of zeros near the turning point $t = 1$ evaluated with Maple, when $n = 100$ and $\nu = 1$.

| $m$-th | true | approx. | error % |
|---|---|---|---|
| $t^{(l)}_{100,1,1}$ | 0.9703 | 0.9761 | 0.5993 |
| | 1.0 | 1.0 | |
| $t^{(r)}_{100,1,1}$ | 1.0086 | 1.0072 | -0.1391 |

Asymptotic zero distribution of the modified Lommel polynomials has been investigated by Kuijlaars and Van Assche [4]. They used the three-term recurrence relation (1.2) to establish the zero distribution of $h_{n,\nu}(t/N)$ with $n/N \to c$ as $n \to \infty$. The zero distribution obtained by Kuijlaars and Van Assche [4] is

$$\mu_c(t) = \begin{cases} \frac{1}{\pi c t^2}\left(1 - \sqrt{1 - t^2c^2}\right), & |t| < \frac{1}{c}, \\ \frac{1}{\pi c t^2}, & |t| \geq \frac{1}{c}. \end{cases} \tag{4.39}$$

(There is a minor error in the statement of their result; the functions $\alpha(t)$ and $\beta(t)$ should be $\alpha(t) = -1/t$ and $\beta(t) = 1/t$ in [4, p. 195]; therefore the zero distribution of the modified Lommel polynomials $h_{n,\nu}(t/N)$ should be the function given in (4.39), instead of the one in [4, (4.18)].) Note that in our case $N = n + \nu$, the ratio $n/N \to 1$ when $n \to \infty$, and the corresponding asymptotic zero distribution is

$$\mu(t) = \begin{cases} \frac{1}{\pi t^2}\left(1 - \sqrt{1-t^2}\right), & |t| < 1, \\ \frac{1}{\pi t^2}, & |t| \geq 1. \end{cases} \tag{4.40}$$

The graph of the distribution function (4.40) is shown in Figure 1. The area under the curve of $\mu(t)$ in (4.40) in an interval $[a, b]$ represents the ratio

$$\frac{\text{the number of zeros in } [a, b]}{\text{total number of zeros}}. \tag{4.41}$$

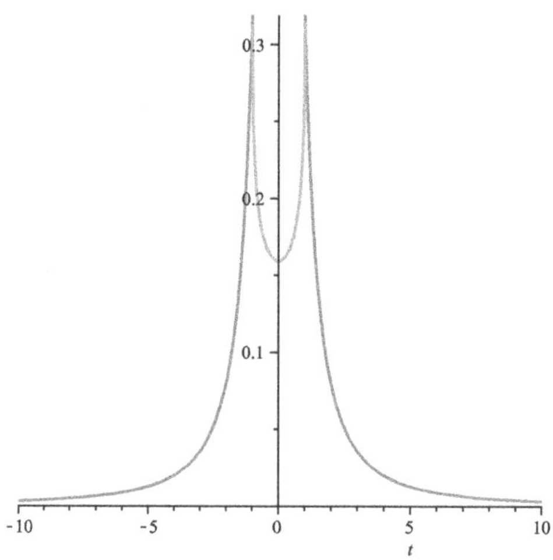


FIGURE 1. The zero distribution of $h_{n,\nu}(t/N)$.

As an illustration, we take $n = 100$ and $\nu = 1$. We find that the polynomial $h_{n,\nu}(t/N)$ has 18 zeros in the interval $0 < t < 1$ and that the value of the integral of $\mu(t)$ on the interval $[0, 1]$ is 0.1817.

## REFERENCES

[1] T. S. Chihara, *An introduction to orthogonal polynomials*, Gordon and Breach Science Publishers, New York, 1978. Mathematics and its Applications, Vol. 13. MR0481884 (58 #1979)

[2] David Dickinson, *On Lommel and Bessel polynomials*, Proc. Amer. Math. Soc. **5** (1954), 946–956. MR0086897 (19,263c)

[3] Mourad E. H. Ismail, *Classical and quantum orthogonal polynomials in one variable.* With two chapters by Walter Van Assche and a foreword by Richard A. Askey, Encyclopedia of Mathematics and its Applications, vol. 98, Cambridge University Press, Cambridge, 2005. MR2191786 (2007f:33001)

[4] A. B. J. Kuijlaars and W. Van Assche, *The asymptotic zero distribution of orthogonal polynomials with varying recurrence coefficients*, J. Approx. Theory **99** (1999), no. 1, 167–197, DOI 10.1006/jath.1999.3316. MR1696553 (2000h:42015)

[5] F. W. J. Olver, *Asymptotics and special functions.* Computer Science and Applied Mathematics, Academic Press [A subsidiary of Harcourt Brace Jovanovich, Publishers], New York-London, 1974. MR0435697 (55 #8655)

[6] Z. Wang and R. Wong, *Uniform asymptotic expansion of $J_\nu(\nu a)$ via a difference equation*, Numer. Math. **91** (2002), no. 1, 147–193, DOI 10.1007/s002110100316. MR1896091 (2003g:33008)

[7] Z. Wang and R. Wong, *Asymptotic expansions for second-order linear difference equations with a turning point*, Numer. Math. **94** (2003), no. 1, 147–194, DOI 10.1007/s00211-002-0416-y. MR1971216 (2004c:39012)

[8] G. N. Watson, *A Treatise on the Theory of Bessel Functions*, Cambridge University Press, Cambridge, England, 1944. MR0010746 (6,64a)

Department of Mathematics, City University of Hong Kong, 83 Tat Chee Avenue, Kowloon, Hong Kong

*E-mail address*: `charleslkf8571@gmail.com`

Department of Mathematics, City University of Hong Kong, 83 Tat Chee Avenue, Kowloon, Hong Kong

*E-mail address*: `rscwong@cityu.edu.hk`

Analysis and Applications, Vol. 12, No. 6 (2014) 727–746

DOI: 10.1142/S0219530514500547

# Global asymptotics of the Szegő–Askey polynomials

Y. Lin*,†,‡ and R. Wong†,§

*Department of Mathematics
South China University of Technology
Guangzhou, P. R. China

†Department of Mathematics
City University of Hong Kong
Kowloon, Hong Kong
‡scyulin@scut.edu.cn
§mawong@cityu.edu.hk

Received 14 March 2014
Accepted 14 September 2014
Published 14 October 2014

The Szegő–Askey polynomials are orthogonal polynomials on the unit circle. In this paper, we study their asymptotic behavior by knowing only their weight function. Using the Riemann–Hilbert method, we obtain global asymptotic formulas in terms of Bessel functions and elementary functions for $z$ in two overlapping regions, which together cover the whole complex plane. Our results agree with those obtained earlier by Temme [Uniform asymptotic expansion for a class of polynomials biorthogonal on the unit circle, *Constr. Approx.* **2** (1986) 369–376]. Temme's approach started from an explicit expression of the Szegő–Askey polynomials in terms of an ${}_2F_1$-function, and followed by integral methods.

*Keywords*: Global asymptotics; Szegő–Askey polynomials; Riemann–Hilbert problems; Bessel function.

Mathematics Subject Classification 2010: 41A60, 33C45

## 1. Introduction

Given a weight function on a curve $\Gamma$ in the complex plane $\mathbb{C}$, the definition of polynomials orthogonal on a real interval can easily be extended to the more general curve $\Gamma$. Since the unit circle is the simplest closed curve in $\mathbb{C}$, orthogonal polynomials on the unit circle naturally attract most of the attentions.

Let $w(z)$ be a weight function with support on a subset of the unit circle, and assume that $w(z)$ is normalized by

$$\frac{1}{2\pi i}\int_{|z|=1} w(z)\frac{dz}{z} = 1. \tag{1.1}$$

For each weight function $w(z)$, there exists a unique sequence of polynomials $\phi_n(z) = \kappa_n z^n + \sum_{\ell=1}^{n} \kappa_{n,n-l} z^{n-l}$, $\kappa_n > 0$, which is orthonormal with respect to

$w(z)$; that is,

$$\frac{1}{2\pi i}\int_{|z|=1}\phi_m(z)\overline{\phi_n(z)}w(z)\frac{dz}{z}=\delta_{mn}, \tag{1.2}$$

where the bar signifies complex conjugate. We define the *reversed polynomials* $\phi_n^*(z)$ by

$$\phi_n^*(z):=z^n\overline{\phi_n(1/\bar{z})} \tag{1.3}$$

or, equivalently,

$$\phi_n^*(z)=\kappa_n+\sum_{\ell=1}^{n}\bar{\kappa}_{n,n-l}z^{\ell}. \tag{1.4}$$

(Note: There is a typo in [9, (18.33.3), p. 475].) Moreover, they satisfy the following recurrence relations:

$$\kappa_n z\phi_n(z)=\kappa_{n+1}\phi_{n+1}(z)-\phi_{n+1}(0)\phi_{n+1}^*(z) \tag{1.5}$$

and

$$\kappa_n\phi_{n+1}(z)=\kappa_{n+1}z\phi_n(z)+\phi_{n+1}(0)\phi_n^*(z); \tag{1.6}$$

see [9, p. 475]. If $n+1$ is replaced by $n$ in (1.5), it follows from (1.6) that

$$\kappa_n\phi_n(0)\phi_{n+1}(z)+\kappa_{n-1}\phi_{n+1}(0)z\phi_{n-1}(z)=(\kappa_n\phi_{n+1}(0)+\kappa_{n+1}\phi_n(0)z)\phi_n(z). \tag{1.7}$$

For general information on orthogonal polynomials on the unit circle, we refer to [13, 12].

In this paper, we are concerned with an important special case of orthogonal polynomials on the unit circle, which was first introduced by Szegő [13], and further discussed in [1]. They are associated with the weight function

$$w(z)=\left(1-\frac{1}{2}(z+z^{-1})\right)^{\lambda},\quad \lambda>-\frac{1}{2}, \tag{1.8}$$

and are explicitly given by

$$\phi_n(z)=\sum_{\ell=0}^{n}\frac{(\lambda+1)_\ell(\lambda)_{n-\ell}}{\ell!(n-\ell)!}z^\ell=\frac{(\lambda)_n}{n!}{}_2F_1\left({-n,\lambda+1 \atop -\lambda-n+1};z\right); \tag{1.9}$$

see also NIST Handbook [9].

Comparing with orthogonal polynomials on the real line, the difficulty in obtaining asymptotic approximations for orthogonal polynomials on the unit circle lies in the fact that they do not satisfy a second-order linear differential equation, or a three-term recurrence relation. Nevertheless, in the case of the Szegő–Askey polynomials, studies can be made by using integral approaches. For instance, in [6–8] Olde Daalhuis has derived asymptotic expansions for the Gauss hypergeometric function

$${}_2F_1\left({a+e_1n,b+e_2n \atop c+e_3n};z\right)\quad \text{as } |n|\to\infty,\ \text{where } e_j=0 \text{ or } \pm1,\ j=1,2,3.$$

The expansions hold for fixed values of $a$, $b$, and $c$, and is uniformly valid for $z$ in the domain $|\arg(-z)| < \pi$. However, these results cannot be extended to the whole complex plane. Also, an even more direct derivation, based on the explicit formula

$$\phi_n(z) = \frac{(2\lambda+1)_n}{n!} {}_2F_1(-n, \lambda+1; 2\lambda+1; 1-z), \tag{1.10}$$

has been given by Temme [14]. (Note that the two equations in (1.9) and (1.10) are equivalent; see [9, (15.8.7), p. 390])

The objective of this paper is to derive uniform asymptotic expansions of the Szegő–Askey polynomials $\phi_n(z)$, based only on the knowledge of the weight function $w(z)$ in (1.8), and not making use of the explicit formula in (1.9) or (1.10). Our approach is to use a modified version of the nonlinear steepest descent method for Riemann–Hilbert problems, introduced by Deift and Zhou [4]. This version has been used successfully to obtain globally uniform asymptotic expansions of several orthogonal polynomials; see [3, 10, 16].

The presentation of this paper is arranged as follows. In Sec. 2, we start with the fundamental Riemann–Hilbert problem $Y$ associated with the Szegő–Askey polynomials. The first transformation $Y \to T$ is introduced to normalize the behavior at infinity. Then, we make the second transformation $T \to V$, which involves the factorization of the jump matrix of $T$ and deformation of contours. In Sec. 3, we construct the parametrix $\tilde{V}(z)$ by using Bessel functions and modified Bessel functions in a neighborhood of $z = 1$, and prove that this parametrix is asymptotically equal to the solution of the Riemann–Hilbert problem for $V(z)$ given in Sec. 2. Our main results are presented in Sec. 4. In Sec. 5, we show that our results agree with those obtained by Temme. In the final section, we briefly discuss the locations of the zeros of $\phi_n(z)$.

## 2. Riemann–Hilbert Problem

We begin with the Riemann–Hilbert problem (RHP) for the Szegő–Askey polynomials $\phi_n(z)$. Observe that from the orthogonality condition (1.2), it follows that

$$\frac{1}{2\pi i}\int_{|z|=1} \phi_n(z) z^{n-k}\frac{w(z)}{z}dz = 0, \quad k = 0, 1, \ldots, n-1.$$

Moreover, the reversed polynomials $\phi_n^*(z)$ also satisfy the following orthogonality condition:

$$\frac{1}{2\pi i}\int_{|z|=1} \phi_{n-1}^*(z) z^{k}\frac{w(z)}{z}dz = \begin{cases} 0, & k = 0, 1, \ldots, n-2, \\ 1/\kappa_{n-1}, & k = n-1. \end{cases}$$

Denote by $\pi_n(z) = \phi_n(z)/\kappa_n$ the corresponding monic orthogonal polynomials. We consider the following RHP for a $2 \times 2$ matrix-valued function $Y(z)$:

$(Y_a)$ $Y(z)$ is holomorphic in $\mathbb{C}\backslash\{z : |z| = 1\}$;

($Y_b$) $Y(z)$ satisfies

$$Y_+(z) = Y_-(z) \begin{pmatrix} 1 & w(z)/z^n \\ 0 & 1 \end{pmatrix} \quad \text{for } |z| = 1;$$

($Y_c$) as $z \to \infty$,

$$Y(z) = \left(I + O\left(\frac{1}{z}\right)\right) \begin{pmatrix} z^n & 0 \\ 0 & z^{-n} \end{pmatrix};$$

($Y_d$) for $\lambda \geq 0$, $Y(z)$ has the behavior

$$Y(z) = O\begin{pmatrix} 1 & 1 \\ 1 & 1 \end{pmatrix} \quad \text{as } z \to 1,$$

and for $\lambda < 0$, $Y(z)$ has the behavior

$$Y(z) = O\begin{pmatrix} 1 & |z-1|^{2\lambda} \\ 1 & |z-1|^{2\lambda} \end{pmatrix} \quad \text{as } z \to 1.$$

By a well-known theorem of Fokas, Its and Kitaev [5] (see also [2, p. 1135]), we have the following theorem.

**Theorem 2.1.** *The unique solution to the above RHP is given by*

$$Y(z) = \begin{pmatrix} \pi_n(z) & \dfrac{1}{2\pi i}\displaystyle\int_{|s|=1} \dfrac{\pi_n(s)w(s)ds}{s^n(s-z)} \\ -\kappa_{n-1}\phi^*_{n-1}(z) & -\dfrac{\kappa_{n-1}}{2\pi i}\displaystyle\int_{|s|=1} \dfrac{\phi^*_{n-1}(s)w(s)ds}{s^n(s-z)} \end{pmatrix}. \tag{2.1}$$

To normalize the solution at infinity, we need to introduce some notations. Let $\Sigma_1$ be the unit circle $\{z \in \mathbb{C} : |z| = 1\}$ and define

$$H(z) := \begin{cases} \begin{pmatrix} z^{-n} & 0 \\ 0 & z^n \end{pmatrix} & \text{if } |z| > 1, \\ I & \text{if } |z| < 1. \end{cases} \tag{2.2}$$

Consider the transformation $Y \to T$ defined by

$$T(z) := Y(z)H(z). \tag{2.3}$$

Note that $T(z)$ is bounded at infinity and its jump matrix has oscillatory diagonal entries. Also, it is readily seen that $T$ is the unique solution of the RHP:

($T_a$) $T(z)$ is analytic for $z$ in $\mathbb{C}\backslash\Sigma_1$;
($T_b$) $T(z)$ satisfies the jump condition

$$T_+(z) = T_-(z) \begin{pmatrix} z^n & w(z) \\ 0 & z^{-n} \end{pmatrix} \quad \text{for } z \in \Sigma_1;$$

($T_c$) as $z \to \infty$,

$$T(z) = I + O(1/z);$$

$(T_d)$ for $\lambda \geq 0$, $T(z)$ has the behavior

$$T(z) = O\begin{pmatrix} 1 & 1 \\ 1 & 1 \end{pmatrix} \quad \text{as } z \to 1,$$

and for $\lambda < 0$, $T(z)$ has the behavior

$$T(z) = O\begin{pmatrix} 1 & |z-1|^{2\lambda} \\ 1 & |z-1|^{2\lambda} \end{pmatrix} \quad \text{as } z \to 1.$$

Note that, for the jump matrix on the unit circle, we have the factorization

$$\begin{pmatrix} z^n & w(z) \\ 0 & z^{-n} \end{pmatrix} = \begin{pmatrix} 1 & 0 \\ 1/(z^n w(z)) & 1 \end{pmatrix} \begin{pmatrix} 0 & w(z) \\ -1/w(z) & 0 \end{pmatrix} \begin{pmatrix} 1 & 0 \\ z^n/w(z) & 1 \end{pmatrix}. \tag{2.4}$$

The first and third matrices on the right-hand side of (2.4) have, respectively, the analytic continuations $\left(\begin{smallmatrix} 1 & 0 \\ z^{-n}w^{-1} & 1 \end{smallmatrix}\right)$ and $\left(\begin{smallmatrix} 1 & 0 \\ z^{n}w^{-1} & 1 \end{smallmatrix}\right)$ for $z$ in $\{z \in \mathbb{C} : |z| < 1\}\backslash(0,1)$ and $\{z \in \mathbb{C} : |z| > 1\}\backslash(1,+\infty)$. Let $\Sigma = \Sigma_1 \cup \Sigma_2 \cup \Sigma_3$ denote the contours shown in Fig. 1.

Define

$$V(z) := T(z), \qquad z \in \Omega_1 \cup \Omega_4, \tag{2.5}$$

$$V(z) := T(z)\begin{pmatrix} 1 & 0 \\ z^n/w(z) & 1 \end{pmatrix}^{-1}, \quad z \in \Omega_2, \tag{2.6}$$

$$V(z) := T(z)\begin{pmatrix} 1 & 0 \\ 1/(z^n w(z)) & 1 \end{pmatrix}, \quad z \in \Omega_3, \tag{2.7}$$

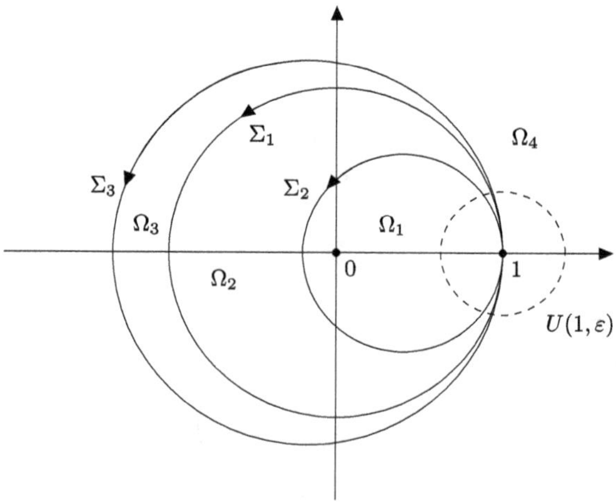


Fig. 1. The contours $\Sigma_i$ and the regions $\Omega_i$.

where regions $\Omega = \bigcup_{i=1}^{4} \Omega_i$ are also depicted in Fig. 1. It can be verified that the jump matrices of $V(z)$ are given by

$$J_V(z) = \begin{pmatrix} 0 & w(z) \\ -1/w(z) & 0 \end{pmatrix}, \quad z \in \Sigma_1, \tag{2.8}$$

$$J_V(z) := \begin{pmatrix} 1 & 0 \\ z^n/w(z) & 1 \end{pmatrix}, \quad z \in \Sigma_2, \tag{2.9}$$

$$J_V(z) := \begin{pmatrix} 1 & 0 \\ 1/(z^n w(z)) & 1 \end{pmatrix}, \quad z \in \Sigma_3. \tag{2.10}$$

Thus, $V(z)$ satisfies the following RHP:

($V_a$) $V(z)$ is analytic for $z \in \mathbb{C}\backslash\Sigma$;
($V_b$) $V(z)$ satisfies the jump condition,

$$V_+(z) = V_-(z) J_V(z), \quad z \in \Sigma;$$

($V_c$) as $z \to \infty$,

$$V(z) = I + O(1/z);$$

($V_d$) for $\lambda \geq 0$, $V(z)$ has the behavior

$$V(z) = \begin{cases} O\begin{pmatrix} 1 & 1 \\ 1 & 1 \end{pmatrix}, & z \in \Omega_1 \cup \Omega_4, \\ O\begin{pmatrix} |z-1|^{-2\lambda} & 1 \\ |z-1|^{-2\lambda} & 1 \end{pmatrix}, & z \in \Omega_2 \cup \Omega_3, \end{cases}$$

as $z \to 1$, and for $\lambda < 0$, $V(z)$ has the behavior

$$V(z) = O\begin{pmatrix} 1 & |z-1|^{2\lambda} \\ 1 & |z-1|^{2\lambda} \end{pmatrix} \quad \text{as } z \to 1.$$

Note from ($T_b$) and (2.9)–(2.10) that the oscillatory terms in the diagonal entries of the jump matrix of $T$ have been transformed into exponentially decaying terms in the off-diagonal entries of the jump matrix of $V$ on the contour $\Sigma_2$ and $\Sigma_3$. Thus, the jump matrix of $V$ tends exponentially fast to the identity matrix on $\Sigma_2 \cup \Sigma_3$. It is now natural to suggest that for large $n$, the solution of the RHP for $V$ may behave asymptotically like the solution of the following RHP for $N$:

($N_a$) $N(z)$ is analytic for $z$ in $\mathbb{C}\backslash\Sigma_1$;
($N_b$) $N(z)$ satisfies the jump condition,

$$N_+(z) = N_-(z) \begin{pmatrix} 0 & w(z) \\ -1/w(z) & 0 \end{pmatrix} \quad \text{for } z \in \Sigma_1;$$

($N_c$) as $z \to \infty$,

$$N(z) = I + O(1/z).$$

To solve the RHP for $N$, we first consider the functions

$$w_1(z) = (1-z)^\lambda, \quad z \in \mathbb{C}\backslash(1,+\infty), \tag{2.11}$$

and

$$w_2(z) = \left(\frac{z}{z-1}\right)^\lambda, \quad z \in \mathbb{C}\backslash(0,1). \tag{2.12}$$

Note that $w_2(z)$ behaves like 1 as $z \to \infty$. Moreover, it is easily shown that

$$w_1(z)w_2(z)^{-1} = 2^\lambda \left(1 - \frac{1}{2}\left(z + \frac{1}{z}\right)\right)^\lambda = 2^\lambda w(z). \tag{2.13}$$

Hence, with $w_1(z)$ and $w_2(z)$ defined in (2.11)–(2.12), we set

$$\widetilde{N}(z) = N(z)w_2(z)^{-\sigma_3} \quad \text{for } |z| > 1, \tag{2.14}$$

and

$$\widetilde{N}(z) = N(z)w_1(z)^{\sigma_3} \begin{pmatrix} 0 & -1 \\ 1 & 0 \end{pmatrix} 2^{\lambda\sigma_3} \quad \text{for } |z| < 1. \tag{2.15}$$

It follows from $(N_b)$ and (2.13) that $\widetilde{N}(z)$ is a solution of the following RHP:

$(\widetilde{N}_a)$ $\widetilde{N}(z)$ is analytic for $z$ in $\mathbb{C}\backslash\Sigma_1$;
$(\widetilde{N}_b)$ for $z \in \Sigma_1$,

$$\widetilde{N}_+(z) = \widetilde{N}_-(z);$$

$(\widetilde{N}_c)$ as $z \to \infty$,

$$\widetilde{N}(z) = I + O(1/z).$$

A straightforward calculation shows that $\widetilde{N}(z) = I$. Hence,

$$N(z) = \begin{cases} w_2(z)^{\sigma_3} & \text{if } |z| > 1, \\ 2^{-\lambda\sigma_3}\begin{pmatrix} 0 & 1 \\ -1 & 0 \end{pmatrix} w_1(z)^{-\sigma_3} & \text{if } |z| < 1. \end{cases} \tag{2.16}$$

However, since $z^n w^{-1} \not\to 0$ and $z^{-n}w^{-1} \not\to 0$ for $z$ near 1, $J_V(z) \not\to I$. That is, the jump matrices for $V$ and $N$ are not uniformly close to each other near the critical point $z = 1$, and special attention must be paid to the neighborhood of this point.

Let $U(1,\varepsilon) := \{z \in \mathbb{C} : |z-1| < \varepsilon\}$, where $\varepsilon$ denotes a small positive number. In view of (2.8)–(2.10), we wish to find a matrix-valued function $V_1(z)$ that satisfies the following RHP:

$(V_{1,a})$ $V_1(z)$ is analytic for $z \in U(1,\varepsilon)\backslash\Sigma$;
$(V_{1,b})$ $V_1(z)$ satisfies the following jump condition on $\Sigma \cap U(1,\varepsilon)$,

$$(V_1)_+(z) = (V_1)_-(z)J_V(z);$$

$(V_{1,c})$ for $z \in \partial U(1,\varepsilon)$,

$$V_1(z) \sim N(z)$$

as $n \to \infty$.

## 3. Construction of the Parametrix

In this section, we first introduce the function $\Psi$ associated with the Bessel functions, and then formulate a model RHP to be used in the formation of the parametrix in a region containing $z = 1$. Finally, we construct an approximation $\widetilde{V}(z)$ to the solution $V(z)$ of the RHP presented in Sec. 2 for large $n$.

Let $\Gamma = \bigcup_{i=1}^{8} \Gamma_i$ denote the contours shown in Fig. 2, which divide the complex plane into eight regions labeled I–VIII. We consider as in [15] the following RHP for $\Psi$ in the $\zeta$-plane:

($\Psi_a$) $\Psi(\zeta)$ is analytic for $\zeta \in \mathbb{C}\backslash\Gamma$;

($\Psi_b$) $\Psi(\zeta)$ satisfies the following jump relations on $\Gamma$:

$$\Psi_+(\zeta) = \Psi_-(\zeta)\begin{pmatrix} 0 & 1 \\ -1 & 0 \end{pmatrix} \qquad \text{for } \zeta \in \Gamma_1 \cup \Gamma_5,$$

$$\Psi_+(\zeta) = \Psi_-(\zeta)\begin{pmatrix} 1 & 0 \\ e^{-2\lambda\pi i} & 1 \end{pmatrix} \qquad \text{for } \zeta \in \Gamma_2 \cup \Gamma_6,$$

$$\Psi_+(\zeta) = \Psi_-(\zeta)e^{\lambda\pi i\sigma_3} \qquad \text{for } \zeta \in \Gamma_3 \cup \Gamma_7,$$

$$\Psi_+(\zeta) = \Psi_-(\zeta)\begin{pmatrix} 1 & 0 \\ e^{2\lambda\pi i} & 1 \end{pmatrix} \qquad \text{for } \zeta \in \Gamma_4 \cup \Gamma_8;$$

($\Psi_c$) for $\lambda < 0$, $\Psi(\zeta)$ has the behavior

$$\Psi(\zeta) = O\begin{pmatrix} |\zeta|^\lambda & |\zeta|^\lambda \\ |\zeta|^\lambda & |\zeta|^\lambda \end{pmatrix} \quad \text{as } \zeta \to 0, \tag{3.1}$$

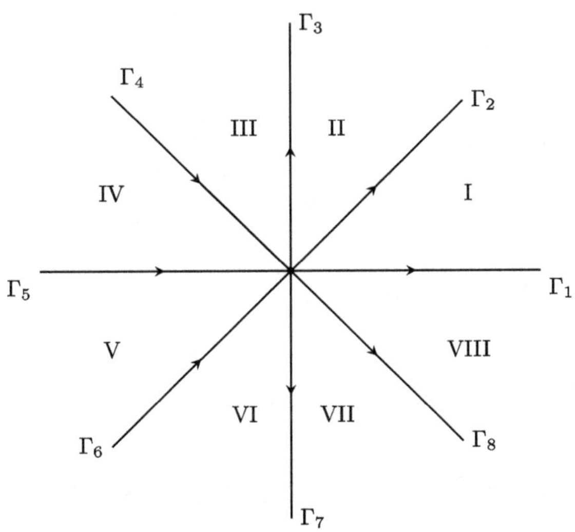


Fig. 2. The contours $\Gamma_i$ and the regions I–VIII.

and for $\lambda \geq 0$, $\Psi(\zeta)$ has the behavior

$$\Psi(\zeta) = \begin{cases} O\begin{pmatrix} |\zeta|^{\lambda} & |\zeta|^{-\lambda} \\ |\zeta|^{\lambda} & |\zeta|^{-\lambda} \end{pmatrix} & \text{as } \zeta \to 0,\ \zeta \in \text{II, III, VI, VII}, \\ O\begin{pmatrix} |\zeta|^{-\lambda} & |\zeta|^{-\lambda} \\ |\zeta|^{-\lambda} & |\zeta|^{-\lambda} \end{pmatrix} & \text{as } \zeta \to 0,\ \zeta \in \text{I, IV, V, VIII}. \end{cases} \tag{3.2}$$

In [15], Vanlessen has shown that there exists a unique solution to the RHP for $\Psi$; see [15, Theorem 4.2]. More precisely, he showed that for $\zeta \in \text{I}$,

$$\Psi(\zeta) := \frac{1}{2}\sqrt{\pi}\zeta^{1/2}\begin{pmatrix} H^{(2)}_{\lambda+\frac{1}{2}}(\zeta) & -iH^{(1)}_{\lambda+\frac{1}{2}}(\zeta) \\ H^{(2)}_{\lambda-\frac{1}{2}}(\zeta) & -iH^{(1)}_{\lambda-\frac{1}{2}}(\zeta) \end{pmatrix} e^{-(\lambda+\frac{1}{4})\pi i\sigma_3}; \tag{3.3}$$

for $\zeta \in \text{II}$,

$$\Psi(\zeta) := \begin{pmatrix} \sqrt{\pi}\zeta^{1/2}I_{\lambda+\frac{1}{2}}(\zeta e^{-\frac{\pi i}{2}}) & -\frac{1}{\sqrt{\pi}}\zeta^{1/2}K_{\lambda+\frac{1}{2}}(\zeta e^{-\frac{\pi i}{2}}) \\ -i\sqrt{\pi}\zeta^{1/2}I_{\lambda-\frac{1}{2}}(\zeta e^{-\frac{\pi i}{2}}) & -\frac{i}{\sqrt{\pi}}\zeta^{1/2}K_{\lambda-\frac{1}{2}}(\zeta e^{-\frac{\pi i}{2}}) \end{pmatrix} e^{-\frac{1}{2}\lambda\pi i\sigma_3}; \tag{3.4}$$

for $\zeta \in \text{III}$,

$$\Psi(\zeta) := \begin{pmatrix} \sqrt{\pi}\zeta^{1/2}I_{\lambda+\frac{1}{2}}(\zeta e^{-\frac{\pi i}{2}}) & -\frac{1}{\sqrt{\pi}}\zeta^{1/2}K_{\lambda+\frac{1}{2}}(\zeta e^{-\frac{\pi i}{2}}) \\ -i\sqrt{\pi}\zeta^{1/2}I_{\lambda-\frac{1}{2}}(\zeta e^{-\frac{\pi i}{2}}) & -\frac{i}{\sqrt{\pi}}\zeta^{1/2}K_{\lambda-\frac{1}{2}}(\zeta e^{-\frac{\pi i}{2}}) \end{pmatrix} e^{\frac{1}{2}\lambda\pi i\sigma_3}; \tag{3.5}$$

for $\zeta \in \text{IV}$,

$$\Psi(\zeta) := \frac{1}{2}\sqrt{\pi}(-\zeta)^{1/2}\begin{pmatrix} iH^{(1)}_{\lambda+\frac{1}{2}}(-\zeta) & -H^{(2)}_{\lambda+\frac{1}{2}}(-\zeta) \\ -iH^{(1)}_{\lambda-\frac{1}{2}}(-\zeta) & H^{(2)}_{\lambda-\frac{1}{2}}(-\zeta) \end{pmatrix} e^{(\lambda+\frac{1}{4})\pi i\sigma_3}; \tag{3.6}$$

for $\zeta \in \text{V}$,

$$\Psi(\zeta) := \frac{1}{2}\sqrt{\pi}(-\zeta)^{1/2}\begin{pmatrix} -H^{(2)}_{\lambda+\frac{1}{2}}(-\zeta) & -iH^{(1)}_{\lambda+\frac{1}{2}}(-\zeta) \\ H^{(2)}_{\lambda-\frac{1}{2}}(-\zeta) & iH^{(1)}_{\lambda-\frac{1}{2}}(-\zeta) \end{pmatrix} e^{-(\lambda+\frac{1}{4})\pi i\sigma_3}; \tag{3.7}$$

for $\zeta \in \text{VI}$,

$$\Psi(\zeta) := \begin{pmatrix} -i\sqrt{\pi}\zeta^{1/2}I_{\lambda+\frac{1}{2}}(\zeta e^{\frac{\pi i}{2}}) & -\frac{i}{\sqrt{\pi}}\zeta^{1/2}K_{\lambda+\frac{1}{2}}(\zeta e^{\frac{\pi i}{2}}) \\ \sqrt{\pi}\zeta^{1/2}I_{\lambda-\frac{1}{2}}(\zeta e^{\frac{\pi i}{2}}) & -\frac{1}{\sqrt{\pi}}\zeta^{1/2}K_{\lambda-\frac{1}{2}}(\zeta e^{\frac{\pi i}{2}}) \end{pmatrix} e^{-\frac{1}{2}\lambda\pi i\sigma_3}; \tag{3.8}$$

for $\zeta \in$ VII,

$$\Psi(\zeta) := \begin{pmatrix} -i\sqrt{\pi}\zeta^{1/2}I_{\lambda+\frac{1}{2}}(\zeta e^{\frac{\pi i}{2}}) & -\frac{i}{\sqrt{\pi}}\zeta^{1/2}K_{\lambda+\frac{1}{2}}(\zeta e^{\frac{\pi i}{2}}) \\ \sqrt{\pi}\zeta^{1/2}I_{\lambda-\frac{1}{2}}(\zeta e^{\frac{\pi i}{2}}) & -\frac{1}{\sqrt{\pi}}\zeta^{1/2}K_{\lambda-\frac{1}{2}}(\zeta e^{\frac{\pi i}{2}}) \end{pmatrix} e^{\frac{1}{2}\lambda\pi i\sigma_3}; \tag{3.9}$$

for $\zeta \in$ VIII,

$$\Psi(\zeta) := \frac{1}{2}\sqrt{\pi}\zeta^{1/2}\begin{pmatrix} -iH^{(1)}_{\lambda+\frac{1}{2}}(\zeta) & -H^{(2)}_{\lambda+\frac{1}{2}}(\zeta) \\ -iH^{(1)}_{\lambda-\frac{1}{2}}(\zeta) & -H^{(2)}_{\lambda-\frac{1}{2}}(\zeta) \end{pmatrix} e^{(\lambda+\frac{1}{4})\pi i\sigma_3}. \tag{3.10}$$

Furthermore, it was shown in [15] that as $\zeta \to \infty$, we have

$$\Psi(\zeta) = \frac{1}{\sqrt{2}}\begin{pmatrix} 1 & -i \\ -i & 1 \end{pmatrix}\left[I + O\left(\frac{1}{\zeta}\right)\right] e^{\frac{\pi i}{4}\sigma_3}e^{-i\zeta\sigma_3}e^{-\frac{1}{2}\lambda\pi i\sigma_3} \tag{3.11}$$

uniformly for $\zeta \in$ I or II,

$$\Psi(\zeta) = \frac{1}{\sqrt{2}}\begin{pmatrix} 1 & -i \\ -i & 1 \end{pmatrix}\left[I + O\left(\frac{1}{\zeta}\right)\right] e^{\frac{\pi i}{4}\sigma_3}e^{-i\zeta\sigma_3}e^{\frac{1}{2}\lambda\pi i\sigma_3} \tag{3.12}$$

uniformly for $\zeta \in$ III or IV,

$$\Psi(\zeta) = \frac{1}{\sqrt{2}}\begin{pmatrix} 1 & -i \\ -i & 1 \end{pmatrix}\left[I + O\left(\frac{1}{\zeta}\right)\right] e^{\frac{\pi i}{4}\sigma_3}e^{-i\zeta\sigma_3}e^{\frac{1}{2}\lambda\pi i\sigma_3}\begin{pmatrix} 0 & -1 \\ 1 & 0 \end{pmatrix} \tag{3.13}$$

uniformly for $\zeta \in$ V or VI, and

$$\Psi(\zeta) = \frac{1}{\sqrt{2}}\begin{pmatrix} 1 & -i \\ -i & 1 \end{pmatrix}\left[I + O\left(\frac{1}{\zeta}\right)\right] e^{\frac{\pi i}{4}\sigma_3}e^{-i\zeta\sigma_3}e^{-\frac{1}{2}\lambda\pi i\sigma_3}\begin{pmatrix} 0 & -1 \\ 1 & 0 \end{pmatrix} \tag{3.14}$$

uniformly for $\zeta \in$ VII or VIII.

With the above preparation, we are now ready to formulate a model RHP for our later use. First, define $\widetilde{\Omega}_i = \Omega_i \cap U(1,\varepsilon)$ and $\widetilde{\Sigma}_i = \Sigma_i \cap U(1,\varepsilon), i = 1,\dots,4$. Moreover, we set $\Gamma'_1 = \widetilde{\Sigma}_1 \cap \mathbb{C}_+$, $\Gamma'_2 = \widetilde{\Sigma}_2 \cap \mathbb{C}_+$, $\Gamma'_3 = (1-\varepsilon,1)$, $\Gamma'_4 = \widetilde{\Sigma}_2 \cap \mathbb{C}_-$, $\Gamma'_5 = \widetilde{\Sigma}_1 \cap \mathbb{C}_-$, $\Gamma'_6 = \widetilde{\Sigma}_3 \cap \mathbb{C}_-$, $\Gamma'_7 = (1, 1+\varepsilon)$, and $\Gamma'_8 = \widetilde{\Sigma}_3 \cap \mathbb{C}_+$, the exact shape of which will be determined later; see Fig. 3.

Now, put

$$\varphi(z) := \begin{cases} w^{1/2}(z)z^{-n/2}e^{-\lambda\pi i}, & z \in \mathrm{I}' \cup \mathrm{II}', \\ w^{1/2}(z)z^{-n/2}e^{\lambda\pi i}, & z \in \mathrm{III}' \cup \mathrm{IV}', \\ w^{1/2}(z)z^{n/2}e^{-\lambda\pi i}, & z \in \mathrm{V}' \cup \mathrm{VI}', \\ w^{1/2}(z)z^{n/2}e^{\lambda\pi i}, & z \in \mathrm{VII}' \cup \mathrm{VIII}'; \end{cases} \tag{3.15}$$

we seek the solution $V_1(z)$ to the last RHP in Sec. 2 in the form

$$V_1(z) = E(z)P(z)\varphi(z)^{-\sigma_3}, \quad z \in U(1,\varepsilon)\backslash\Gamma', \tag{3.16}$$

where $\Gamma' = \Gamma'_1 \cup \cdots \cup \Gamma'_8$, the matrix-valued function $E(z)$ is analytic in $U(1,\varepsilon)$, and $P(z)$ and $E(z)$ will be determined later.

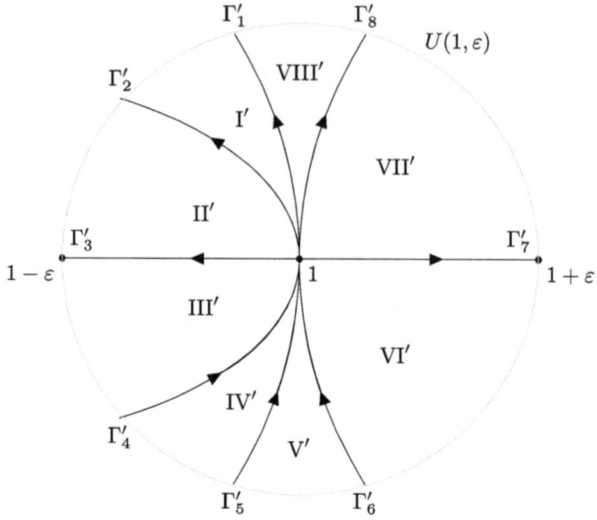


Fig. 3. The contours $\Gamma_i'$ and the regions I′–VIII′.

Note that $w(z)$ has a jump on $(1-\varepsilon, 1+\varepsilon)$. More precisely, we have

$$w_+(x)w_-(x)^{-1} = e^{-2\lambda\pi i}, \quad x \in \Gamma_7', \tag{3.17}$$

and

$$w_+(x)w_-(x)^{-1} = e^{-2\lambda\pi i}, \quad x \in \Gamma_3'. \tag{3.18}$$

Coupling these with (3.15), it follows that $P(z)$ satisfies the following RHP:

($P_a$) $P(z)$ is analytic for $z \in U(1,\varepsilon)\backslash\Gamma'$;
($P_b$) $P(z)$ satisfies the jump conditions

$$\begin{aligned} P_+(z) &= P_-(z)\begin{pmatrix} 0 & 1 \\ -1 & 0 \end{pmatrix} && \text{for } z \in \Gamma_1' \cup \Gamma_5', \\ P_+(z) &= P_-(z)\begin{pmatrix} 1 & 0 \\ e^{-2\lambda\pi i} & 1 \end{pmatrix} && \text{for } z \in \Gamma_2' \cup \Gamma_6', \\ P_+(z) &= P_-(z)e^{\lambda\pi i\sigma_3} && \text{for } z \in \Gamma_3' \cup \Gamma_7', \\ P_+(z) &= P_-(z)\begin{pmatrix} 1 & 0 \\ e^{2\lambda\pi i} & 1 \end{pmatrix} && \text{for } z \in \Gamma_4' \cup \Gamma_8'; \end{aligned}$$

($P_c$) for $\lambda < 0$, $P(z)$ has the behavior

$$P(z) = O\begin{pmatrix} |z-1|^\lambda & |z-1|^\lambda \\ |z-1|^\lambda & |z-1|^\lambda \end{pmatrix} \quad \text{as } z \to 1,$$

and for $\lambda \geq 0$, $P(z)$ has the behavior

$$P(z) = \begin{cases} O\begin{pmatrix} |z-1|^{\lambda} & |z-1|^{-\lambda} \\ |z-1|^{\lambda} & |z-1|^{-\lambda} \end{pmatrix} & \text{as } z \to 1,\ z \in \mathrm{II}', \mathrm{III}', \mathrm{VI}', \mathrm{VII}', \\ O\begin{pmatrix} |z-1|^{-\lambda} & |z-1|^{-\lambda} \\ |z-1|^{-\lambda} & |z-1|^{-\lambda} \end{pmatrix} & \text{as } z \to 1,\ z \in \mathrm{I}', \mathrm{IV}', \mathrm{V}', \mathrm{VIII}'. \end{cases} \tag{3.19}$$

To construct our parametrix, we need an auxiliary function

$$f(z) = -i \log z, \tag{3.20}$$

where $\log z$ denotes the principal branch of the logarithm. It is readily seen that $f$ is a one-to-one mapping of $U(1,\varepsilon)$ onto a neighborhood of the origin. Define

$$\zeta = \frac{n}{2} f(z). \tag{3.21}$$

Comparing the RHP for $\Psi$ with the RHP for $P$ invokes us to introduce

$$P(z) := \Psi(\zeta) = \Psi\left(\frac{n}{2} f(z)\right). \tag{3.22}$$

From (3.11)–(3.14), we have

$$\Psi(\zeta) \sim \frac{1}{\sqrt{2}} \begin{pmatrix} 1 & -i \\ -i & 1 \end{pmatrix} E_1(z) \quad \text{as } \zeta \to \infty,$$

where

$$E_1(z) := \begin{cases} e^{\frac{\pi i}{4}\sigma_3} e^{-i\zeta\sigma_3} e^{-\frac{1}{2}\lambda\pi i\sigma_3} & \text{if } z \in \mathrm{I}' \text{ or } \mathrm{II}', \\ e^{\frac{\pi i}{4}\sigma_3} e^{-i\zeta\sigma_3} e^{\frac{1}{2}\lambda\pi i\sigma_3} & \text{if } z \in \mathrm{III}' \text{ or } \mathrm{IV}', \\ e^{\frac{\pi i}{4}\sigma_3} e^{-i\zeta\sigma_3} e^{\frac{1}{2}\lambda\pi i\sigma_3} \begin{pmatrix} 0 & -1 \\ 1 & 0 \end{pmatrix} & \text{if } z \in \mathrm{V}' \text{ or } \mathrm{VI}', \\ e^{\frac{\pi i}{4}\sigma_3} e^{-i\zeta\sigma_3} e^{-\frac{1}{2}\lambda\pi i\sigma_3} \begin{pmatrix} 0 & -1 \\ 1 & 0 \end{pmatrix} & \text{if } z \in \mathrm{VII}' \text{ or } \mathrm{VIII}'. \end{cases}$$

Equations (3.16) and (3.22) lead us to set

$$E(z) := N(z)\varphi(z)^{\sigma_3} E_1(z)^{-1} \frac{1}{\sqrt{2}} \begin{pmatrix} 1 & i \\ i & 1 \end{pmatrix}. \tag{3.23}$$

The above heuristic argument now suggests that $V(z)$ is asymptotically approximated by

$$\widetilde{V}(z) := \begin{cases} N(z), & z \in \mathbb{C} \backslash U(1,\varepsilon), \\ E(z)\Psi\left(\frac{n}{2} f(z)\right) \varphi(z)^{-\sigma_3}, & z \in U(1,\varepsilon). \end{cases} \tag{3.24}$$

Finally, we come to the discussion of the contours $\Gamma_i'$, $i = 1, \ldots, 8$. On account of the mapping properties of $f(z)$, the curve $\Gamma_i'$ is chosen so that its image under the transformation $f(z)$ is $\Gamma_i$. Hence, the complex plane is divided into eight regions I′–VIII′; see Fig. 3.

From (3.16) and (3.23), one can see that $V_1$ and $\widetilde{V}$ are the same in $U(1,\varepsilon)$. By $(V_{1,b})$ in Sec. 2, $V_1$ and $V$ have the same jump matrix. Hence, $V$ and $\widetilde{V}$ satisfy the same jump conditions for $z \in U(1,\varepsilon)$, and the matrix

$$S(z) := V(z)\widetilde{V}(z)^{-1} \tag{3.25}$$

satisfies the relation

$$S_+(z) = S_-(z), \quad z \in \Gamma_i'. \tag{3.26}$$

Furthermore, it can be readily verified that $S(z)$ is a solution of the RHP:

$(S_a)$ $S(z)$ is analytic for $z \in \mathbb{C}\backslash\Sigma_S$; where $\Sigma_S = [(\Sigma_2 \cup \Sigma_3)\backslash U(1,\varepsilon)] \cup \partial U$;

$(S_b)$ for $z \in \Sigma_S$

$$S_+(z) = S_-(z)J_S(z),$$

where

$$J_s(z) = \begin{pmatrix} 1 & -\dfrac{z^n w_1^2}{2^{2\lambda} w} \\ 0 & 1 \end{pmatrix} \quad \text{for } z \in \Sigma_2\backslash U(1,\varepsilon).$$

Note that for $z \in \Sigma_2\backslash U(1,\varepsilon)$, $J_S(z) \sim I$ as $n \to \infty$. Similarly, for $z$ in other parts of the contour $\Sigma_S$, the jump matrix $J_S(z)$ also tends to the identity matrix.

$(S_c)$ as $z \to \infty$,

$$S(z) = I + O(1/z).$$

Since $J_S(z) = I + O(1/n)$ as $n \to \infty$ on the whole contour $\Sigma_S$, by [11, Theorem 3.8] the solution of the RHP for $S$ has the asymptotic behavior

$$S(z) = I + O(1/n) \tag{3.27}$$

as $n \to \infty$ uniformly for $z \in \mathbb{C}\backslash\Sigma_S$.

## 4. Main Results

We finally arrive at the following theorem.

**Theorem 4.1.** *With $w_1(z)$ and $w_2(z)$ defined in (2.11) and (2.12), we have the asymptotic formula for $z \in \Omega_1\backslash U(1,\varepsilon)$,*

$$\pi_n(z) = \frac{1}{n}\frac{\lambda}{(1-z)^{\lambda+1}} + O(1/n^2); \tag{4.1}$$

*for $z \in (\Omega_2 \cup \Omega_3)\backslash U(1,\varepsilon)$,*

$$\pi_n(z) = z^n w_2(z)(1 + O(1/n)) + O(1/n); \tag{4.2}$$

*and for $z \in \Omega_4\backslash U(1,\varepsilon)$,*

$$\pi_n(z) = z^{n+\lambda}(z-1)^{-\lambda}(1 + O(1/n)). \tag{4.3}$$

*Let $\zeta$ be as defined in* (3.21). *We have*

$$\pi_n(z) = z^{n/2}w_2(z)e^{\pi i/4}\sqrt{\frac{\pi}{2}}\zeta^{1/2}[I_{\lambda+\frac{1}{2}}(n\log z/2) + I_{\lambda-\frac{1}{2}}(n\log z/2)](1+O(1/n))$$
$$+ z^{n/2}2^{\lambda}w_1^{-1}(z)e^{\frac{3}{4}\pi i}\sqrt{\frac{\pi}{2}}\zeta^{1/2}e^{-(\lambda+1/2)\pi i}$$
$$\times [I_{\lambda+\frac{1}{2}}(n\log z/2) - I_{\lambda-\frac{1}{2}}(n\log z/2)]O(1/n) \tag{4.4}$$

*for $z \in U(1,\varepsilon)$.*

**Proof.** Since $V(z) = S(z)\widetilde{V}(z)$, we have

$$V_{11}(z) = S_{11}(z)\widetilde{V}_{11}(z) + S_{12}(z)\widetilde{V}_{21}(z) \tag{4.5}$$

and

$$V_{12}(z) = S_{11}(z)\widetilde{V}_{12}(z) + S_{12}(z)\widetilde{V}_{22}(z). \tag{4.6}$$

From (3.27), we also obtain

$$S_{11}(z) = 1 + O(1/n), \quad S_{12}(z) = O(1/n). \tag{4.7}$$

First, let us consider $z \in \Omega_4 \backslash U(1,\varepsilon)$. The definition of $\widetilde{V}(z)$ in (3.24) gives us

$$\widetilde{V}_{11}(z) = N_{11}(z) \quad \text{and} \quad \widetilde{V}_{12}(z) = N_{12}(z). \tag{4.8}$$

Note that in this case

$$\pi_n(z) = Y_{11}(z) = T_{11}(z)z^n = V_{11}(z)z^n. \tag{4.9}$$

Hence, for $z \in \Omega_4$

$$\pi_n(z) = z^n N_{11}(z)S_{11}(z) = z^{n+\lambda}(z-1)^{-\lambda}(1+O(1/n)). \tag{4.10}$$

Next, we consider $z \in U(1,\varepsilon)$. Let us first restrict $z \in \mathrm{II}'$; see Fig. 3. From (3.4) and (3.15), we have by a combination of (3.23)–(3.24), (2.2) and (2.5)

$$\pi_n(z) = S_{11}(z)N_{12}(z)\left[\varphi(z)^{\sigma_3}E_1(z)^{-1}\frac{1}{\sqrt{2}}\begin{pmatrix}1 & i\\ i & 1\end{pmatrix}\Psi\left(\frac{n}{2}f(z)\right)\varphi(z)^{-\sigma_3}\right]_{21}$$
$$+ S_{12}(z)\widetilde{V}_{21}(z)$$
$$= z^{n/2}w_2(z)e^{\pi i/4}e^{\lambda\pi i}\sqrt{\frac{\pi}{2}}\zeta^{1/2}i[I_{\lambda+\frac{1}{2}}(\zeta e^{-\pi i/2}) - I_{\lambda-\frac{1}{2}}(\zeta e^{-\pi i/2})](1+O(1/n))$$
$$+ z^{n/2}2^{\lambda}w_1^{-1}(z)e^{\frac{3}{4}\pi i}\sqrt{\frac{\pi}{2}}\zeta^{1/2}[I_{\lambda+\frac{1}{2}}(\zeta e^{-\pi i/2}) + I_{\lambda-\frac{1}{2}}(\zeta e^{-\pi i/2})]O(1/n). \tag{4.11}$$

Using $I_\nu(ze^{m\pi i}) = e^{m\nu\pi i}I_\nu(z)$ (see [9, (10.34.1)]) and (3.21), we obtain the asymptotic formula (4.4). In a similar manner, one can prove the exactly same formula (4.11) for $z \in \mathrm{I}', \mathrm{III}'$–$\mathrm{VIII}'$, thus proving (4.4).

For $z \in \Omega_1 \backslash U(1,\varepsilon)$, we have from (4.5) and (3.24)

$$\begin{aligned}\pi_n(z) &= N_{11}(z)S_{11}(z) + S_{12}(z)N_{21}(z) = N_{21}(z)O(1/n)\\ &= -\frac{2^\lambda}{n}w_1^{-1}(z)C(z) + O(1/n^2),\end{aligned} \tag{4.12}$$

where $C(z) = O(1)$ as $n \to \infty$. The second equality follows from the fact that $N_{11}(z) = 0$.

To determine the function $C(z)$, we consider the asymptotic formula

$$\begin{aligned}&I_{\lambda+\frac{1}{2}}(\zeta e^{-\pi i/2}) - I_{\lambda-\frac{1}{2}}(\zeta e^{-\pi i/2})\\ &= \frac{z^{-\frac{n}{2}}}{\sqrt{2\pi}\zeta^{1/2}}e^{\pi i/4}\left[\frac{a_1\left(\lambda+\frac{1}{2}\right)}{n/2\log z} - \frac{a_1\left(\lambda-\frac{1}{2}\right)}{n/2\log z} + O(1/n^2)\right]\\ &= \sqrt{\frac{2}{\pi}}\frac{z^{-\frac{n}{2}}}{\zeta^{1/2}}e^{\pi i/4}\left[\frac{\lambda}{n\log z} + O(1/n^2)\right],\end{aligned} \tag{4.13}$$

where $a_1(\nu) = \frac{4\nu^2-1}{8}$; see [9, (10.40.1) and (10.17.1)]. Coupling (4.13) with (4.11), we have for $z \in \mathrm{II}'$

$$\pi_n(z) \sim -\frac{z^\lambda}{(z-1)^\lambda}\frac{\lambda}{n\log z}e^{\lambda\pi i}. \tag{4.14}$$

Since $z$ is in the neighborhood of 1, we also have $z^\lambda \sim 1$ and $\log z = \log(z-1+1) \sim z-1$. It follows that

$$\pi_n(z) \sim \frac{\lambda}{(1-z)^{\lambda+1}}\frac{1}{n}, \tag{4.15}$$

thus giving $C(z) = \frac{\lambda 2^{-\lambda}}{z-1}$ and (4.1).

Note that for $z \in \Omega_2 \backslash U(1,\varepsilon)$,

$$Y(z) = S(z)N(z)\begin{pmatrix} 1 & 0\\ z^n/w(z) & 1\end{pmatrix}. \tag{4.16}$$

Hence, for $z \in \Omega_2 \backslash U(1,\varepsilon)$,

$$\begin{aligned}\pi_n(z) &= N_{12}(z)z^n/w(z)(1+O(1/n)) + O(1/n)\\ &= z^n 2^{-\lambda}\frac{w_1(z)}{w(z)}(1+O(1/n)) + O(1/n)\\ &= z^n w_2(z)(1+O(1/n)) + O(1/n).\end{aligned} \tag{4.17}$$

Here, we have made use of (2.13). Similarly, we can prove (4.17) for $z \in \Omega_3 \backslash U(1,\varepsilon)$.

The $O$-terms are uniform with respect to $z$, since the contours $\Sigma_2$, $\Sigma_3$ and the radius $\varepsilon$ of the disk $U(1,\varepsilon)$ are rather arbitrary; this completes the proof of the theorem. □

**Remark 4.2.** Since the size of contour $\Sigma_2$ in Fig. 1 is rather arbitrary, region $\Omega_1$ can be taken nearly as big as region $\Omega_2$. Furthermore, since $z^n$ is exponentially small when $|z| < 1$, the second $O(1/n)$ in (4.2) is actually the term given in (4.1). That is, for $z \in (\Omega_2 \cup \Omega_3)\backslash U(1, \varepsilon)$, we have

$$\pi_n(z) = z^n w_2(z)[1 + O(1/n)] + \frac{1}{n}\frac{\lambda}{(1-z)^{\lambda+1}} + O(1/n^2). \tag{4.18}$$

**Remark 4.3.** In view of (2.12), the two asymptotic formulas in (4.2) and (4.3) are essentially the same. By Remark 4.2, formula (4.18) in facts holds for $z \in (\Omega_2 \cup \Omega_3 \cup \Omega_4)\backslash U(1, \varepsilon)$.

If we take the contour $\Sigma_2$ sufficiently small so that it lies completely inside the disk $U(1, \varepsilon)$, then the result in Remark 4.3 will hold for all $z \in \mathbb{C}\backslash U(1, \varepsilon)$. Thus, we need only two formulas to describe the asymptotic behavior of the Szegő–Askey polynomials. That is, we have the following corollary.

**Corollary 4.4.** *For $z \in U(1, \varepsilon)$, the asymptotic formula of $\pi_n(z)$ is given by (4.4); for $z \in \mathbb{C}\backslash U(1, \varepsilon)$, the asymptotic formula of $\pi_n(z)$ is given by (4.18).*

## 5. Comparison

Now, we will show that our asymptotic formulas for the Szegő–Askey polynomials are the same as those given by Temme [14]. First, note that by (1.8) and [14, (1.1)]

$$P_n(z; \lambda, 0) = \frac{n!}{(2\lambda+1)_n}\phi_n(z). \tag{5.1}$$

Two different asymptotic approximations for $P_n(z; \alpha, \beta)$ are given in [14, (2.6) and (2.7)], one for $|z| < 1$ and the other for $|z| > 1$. Changing $n+1$ to $n$, they read

$$P_n(z; \lambda, 0) \sim \frac{\Gamma(2\lambda+1)}{\Gamma(\lambda)}[n(1-z)]^{-\lambda-1} \tag{5.2}$$

and

$$P_n(z; \lambda, 0) \sim \frac{\Gamma(2\lambda+1)}{\Gamma(\lambda+1)}z^{n+\lambda}(z-1)^{-\lambda}n^{-\lambda}. \tag{5.3}$$

By using (4.1) and (5.1), we have

$$\begin{aligned} P_n(z; \lambda, 0) &= \frac{n!}{(2\lambda+1)_n}\kappa_n\pi_n(z) \\ &\sim \frac{n!}{(2\lambda+1)_n}\frac{(\lambda+1)_n}{n!}\frac{1}{n}\frac{\lambda}{(1-z)^{\lambda+1}} \\ &\sim \frac{\Gamma(2\lambda+1)}{\Gamma(\lambda)}[n(1-z)]^{-\lambda-1}, \end{aligned} \tag{5.4}$$

which agrees with (5.2). The last result is a direct consequence of Stirling's formula.

In the case $|z| > 1$, it follows that

$$P_n(z;\lambda,0) \sim \frac{n!}{(2\lambda+1)_n}\frac{(\lambda+1)_n}{n!}z^{n+\lambda}(z-1)^{-\lambda}$$

$$\sim \frac{\Gamma(2\lambda+1)}{\Gamma(\lambda+1)}z^{n+\lambda}(z-1)^{-\lambda}n^{-\lambda}. \tag{5.5}$$

Therefore, our results (4.1) and (4.3) agree with those of Temme [14] stated in (5.2) and (5.3).

For $z \in U(1,\varepsilon)$, we use the identity

$$M\left(\nu+\frac{1}{2}, 2\nu+1, 2z\right) = \Gamma(\nu+1)e^z\left(\frac{z}{2}\right)^{-\nu} I_\nu(z), \tag{5.6}$$

where $M(a,b,z)$ is the confluent hypergeometric function ${}_1F_1(a;b;z)$ and $I_\nu(z)$ is a modified Bessel function; [9, (13.6.9), p. 328]. Also, it can be verified that ${}_1F_1(a;b;z)$ satisfies the equation

$${}_1F_1(\lambda+1;2\lambda+1;z) = {}_1F_1(\lambda;2\lambda;z) + \frac{z}{4\left(\lambda+\frac{1}{2}\right)}{}_1F_1(\lambda+1;2\lambda+2;z). \tag{5.7}$$

Using the two results (5.6) and (5.7) together, one can show that

$${}_1F_1(\lambda+1,2\lambda+1,z) = 2^{2\lambda-1}\Gamma(\lambda+1/2)\frac{e^{z/2}}{z^{\lambda-1/2}}[I_{\lambda-1/2}(z/2)+I_{\lambda+1/2}(z/2)]. \tag{5.8}$$

Coupling this with (3.20) again shows that (4.4) agrees with Temme's result (2.14) in [14].

## 6. Zeros

To describe the location of the zeros of the Szegő–Askey polynomials, we first recall the Verblunsky coefficients $\alpha_n$ defined in [12, p. 2]. By comparing Eq. (1.1.8) in [12] with our Eq. (1.6), it is easily seen that

$$\alpha_n = -\frac{1}{\kappa_{n+1}}\overline{\phi_{n+1}(0)},$$

where $\kappa_n$ is the leading coefficient of $\phi_n(z)$. From (1.9), we have

$$\kappa_{n+1} = \frac{(\lambda+1)_{n+1}}{(n+1)!} \quad \text{and} \quad \phi_{n+1}(0) = \frac{(\lambda)_{n+1}}{(n+1)!}.$$

Thus, in our case, the Verblunsky coefficient is explicitly given by

$$\alpha_n = -\frac{\lambda}{\lambda+n+1}.$$

A theorem of Mhaskar–Saff [12, p. 392] states that if

$$A = \lim_{n\to\infty}\sup|\alpha_n|^{1/n} = 1$$

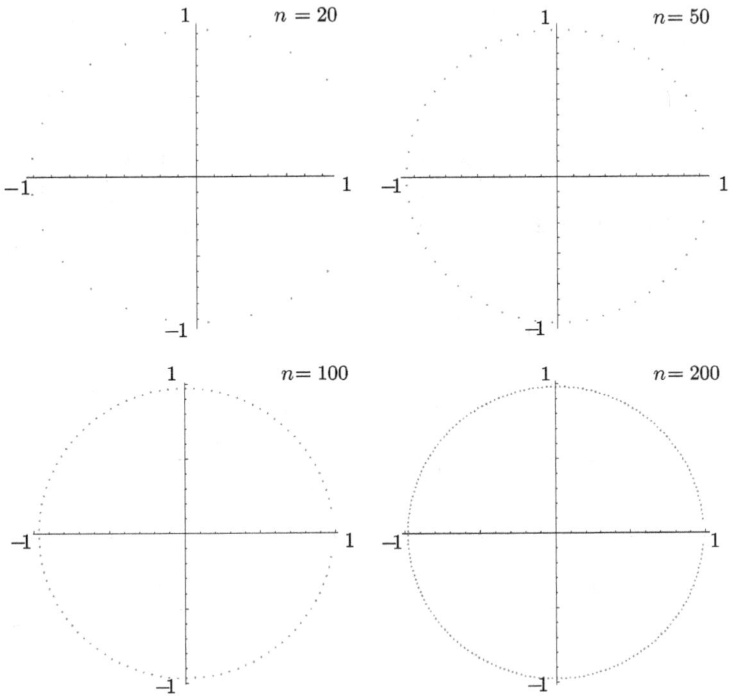


Fig. 4. Distribution of zeros, when $\lambda = 5$.

and

$$\lim_{n\to\infty} \frac{1}{n} \sum_{j=0}^{n-1} |\alpha_j| = 0,$$

then the counting measure of zeros converges to uniform measure on the unit circle (i.e. the zeros of $\phi_n(z)$ are uniformly distributed on the circle $\{z : |z| = 1\}$). For illustration, we display the zeros in Fig. 4, when $\lambda = 5$ and $n = 20, 50, 100, 200$.

From Fig. 4, we notice that there is a gap on the circle near the point $z = 1$, and that the gap gets narrower as $n \to \infty$. We now investigate how fast the two end-zeros tend to $z = 1$. To this end, we first use (5.8) to rewrite (4.4) as

$$\frac{1}{\sqrt{\pi}} 2^{2\lambda} \Gamma\left(\lambda + \frac{1}{2}\right) (n \log z)^{-\lambda} \left(\frac{z-1}{z}\right)^{\lambda} \pi_n(z)$$
$$= {}_1F_1(\lambda + 1, 2\lambda + 1, n \log z)[1 + O(1/n)] + O(1/n). \tag{6.1}$$

Note that the function in front of $\pi_n(z)$ on the left-hand side of (6.1) has no zero. Hence, the polynomial $\pi_n(z)$ vanishes only when $z$ is near a zero of ${}_1F_1(\lambda+1, 2\lambda + 1, n \log z)$. Now we recall the facts that for $a$ and $b - a \neq 0, -1, -2, \ldots$ the confluent hypergeometric function ${}_1F_1(a, b, z)$ has infinitely many $z$-zeros in $\mathbb{C}$, and that for $a \geq 0$ and $b \geq 0$ there are no real zeros. Furthermore, for fixed $a, b \in \mathbb{C}$ the large zeros of ${}_1F_1(a, b, z)$ satisfy

$$z = \pm(2n + a)\pi i + \ln\left(-\frac{\Gamma(a)}{\Gamma(b-a)}(\pm 2n\pi i)^{b-2a}\right) + O(n^{-1}\ln n), \tag{6.2}$$

where $n$ is a large positive integer; see [9, (13.9.9), p. 331]. In our case, $a = \lambda + 1$ and $b = 2\lambda + 1$ in (6.2). Let $n$ in (6.2) be the degree of the polynomial $\pi_n(z)$ in (6.1), and replace $z$ in (6.2) by the variable $n \log z$ in the function ${}_1F_1$ in (6.1). This gives

$$\log z = \pm 2\pi i \pm \frac{(\lambda+1)\pi}{n} i + \frac{1}{n}\left[\ln(-\lambda) - \ln(2n\pi) \mp i\frac{\pi}{2}\right] + O\left(\frac{\ln n}{n^2}\right). \tag{6.3}$$

If one denotes the end-zeros of $\pi_n(z)$ in the upper and lower complex plane by $z_{n,+}$ and $z_{n,-}$, respectively, then we are led to the result

$$\operatorname{Re} z_{n,+} = \operatorname{Re} z_{n,-} \sim 1 - \frac{\ln n}{n} + \frac{1}{n}\ln\left(\frac{|\lambda|}{2\pi}\right) + O\left(\frac{\ln n}{n^2}\right)$$

and

$$\operatorname{Im} z_{n,+} = \frac{\left(\lambda - \frac{1}{2}\right)\pi}{n} + O\left(\frac{\ln n}{n^2}\right), \quad \operatorname{Im} z_{n,-} = -\frac{\left(\lambda - \frac{1}{2}\right)\pi}{n} + O\left(\frac{\ln n}{n^2}\right),$$

that is,

$$z_{n,\pm} = 1 + O\left(\frac{\ln n}{n}\right). \tag{6.4}$$

If we write $z = e^{i\theta}$, then (6.3) gives

$$n\theta = \pm 2n\pi \pm (\lambda+1)\pi + \ln\left(-\frac{\lambda}{2n\pi} \mp i\frac{\pi}{2}\right) + O\left(\frac{\ln n}{n}\right).$$

Thus, the result in (6.4) only gives the rate in which the end-zeros of $\pi_n(z)$ approach the point $z = 1$, when $n\theta \to \pm\infty$.

In the case when $n\theta$ is bounded, one can use, instead of (6.1), the asymptotic formula in (4.4). In view of the behavior of the modified Bessel function near the origin, there seems to be no zero of $\pi_n(z)$ in a shrinking neighborhood of $z = 1$ with radius of $o(1/n)$.

## References


[1] R. Askey (ed.), *Gabor Szegő: Collected Papers*, Vol. 1 (Birkhäuser, Boston, 1982).

[2] J. Baik, P. Deift and K. Johansson, On the distribution of the length of longest increasing subsequence of random permutations, *J. Amer. Math. Soc.* **12** (1999) 1119–1178.

[3] D. Dai and R. Wong, Global asymptotics of Krawtchouk polynomials — A Riemann–Hilbert approach, *Chinese Ann. Math. Ser. B* **28** (2007) 1–34.

[4] P. Deift and X. Zhou, A steepest descent method for oscillatory Riemann–Hilbert problems, Asymptotics for the MKdV equation, *Ann. Math.* **137** (1993) 295–368.

[5] A. S. Fokas, A. R. Its and A. V. Kitaev, The isomonodromy approach to matrix models in 2D quantum gravity, *Comm. Math. Phys.* **147** (1992) 395–430.
[6] A. B. Olde Daalhuis, Uniform asymptotic expansions for hypergeometric functions with large parameters I, *Anal. Appl.* **1** (2003) 111–120.
[7] A. B. Olde Daalhuis, Uniform asymptotic expansions for hypergeometric functions with large parameters II, *Anal. Appl.* **1** (2003) 121–128.
[8] A. B. Olde Daalhuis, Uniform asymptotic expansions for hypergeometric functions with large parameters III, *Anal. Appl.* **8** (2010) 199–210.
[9] F. W. J. Olver, D. W. Lozier, R. F. Boisvert and C. W. Clark, *NIST Handbook of Mathematical Functions* (Cambridge University Press, Cambridge, 2010).
[10] C. Ou and R. Wong, The Riemann–Hilbert approach to global asymptotics of discrete orthogonal polynomials with infinite nodes, *Anal. Appl.* **8** (2010) 247–286.
[11] W.-Y. Qiu and R. Wong, Asymptotic expansions for Riemann–Hilbert problems, *Anal. Appl.* **6** (2008) 269–298.
[12] B. Simon, *Orthogonal Polynomials on the Unit Circle*, AMS Colloquium Publications, Vol. 54 (American Mathematical Society, Providence, RI, 2005).
[13] G. Szegő, *Orthogonal Polynomials*, 4th edn., AMS Colloquium Publications, Vol. 23 (American Mathematical Society, Providence, RI, 1975).
[14] N. M. Temme, Uniform asymptotic expansion for a class of polynomials biorthogonal on the unit circle, *Constr. Approx.* **2** (1986) 369–376.
[15] M. Vanlessen, Asymptotics of the recurrence coefficients of orthogonal polynomials associated to the generalized Jacobi weight, *J. Approx. Theory.* **125** (2003) 198–237.
[16] X. S. Wang and R. Wong, Global asymptotics of the Meixner polynomials, *Asymptot. Anal.* **75** (2011) 211–231.

J. Math. Anal. Appl. 436 (2016) 1149–1164

Contents lists available at ScienceDirect

Journal of Mathematical Analysis and Applications

www.elsevier.com/locate/jmaa

# Asymptotics of Racah polynomials with varying parameters

X.-S. Wang [a,*], R. Wong [b]

[a] Department of Mathematics, Southeast Missouri State University, Cape Girardeau, MO 63701, USA
[b] Department of Mathematics, City University of Hong Kong, Tat Chee Avenue, Kowloon, Hong Kong

ARTICLE INFO

*Article history:*
Received 6 September 2015
Available online 29 December 2015
Submitted by K. Driver

*Keywords:*
Asymptotics
Racah polynomials
Recurrence relation
Askey scheme

ABSTRACT

Within the Askey scheme of hypergeometric orthogonal polynomials, Racah polynomials stay on the top of the hierarchy and they generalize all of the discrete hypergeometric orthogonal polynomials. In this paper, we investigate asymptotic behaviors of Racah polynomials with varying parameters when the polynomial degree tends to infinity. Using the difference equation technique developed in our earlier papers, we obtain an asymptotic formula in the outer region via ratio asymptotics and then derive asymptotic formulas in the oscillatory region via a matching method. Our asymptotic formulas are explicitly given in terms of the polynomial degree, variable and parameters, using elementary functions such as logarithmic, exponential and rational functions. By taking limits, our results also yield asymptotic formulas for orthogonal polynomials in the lower hierarchy of the Askey scheme such as Hahn polynomials and Krawtchouk polynomials.


## 1. Introduction

The Racah polynomials are named after Racah, because the orthogonal relation is equivalent to that of Racah coefficients or 6-$j$ symbols; see [1]. In [11], Wilson defined the Racah polynomials in terms of a ${}_4F_3$ hypergeometric function. Let $\lambda(x) := x(x+\gamma+\delta+1)$ and $N$ be a nonnegative integer. Define

$$R_n(\lambda(x);\alpha,\beta,\gamma,\delta) := {}_4F_3\left(\begin{matrix} -n, & n+\alpha+\beta+1, & -x, & x+\gamma+\delta+1 \\ & \alpha+1, & \beta+\delta+1, & \gamma+1 \end{matrix}\middle| 1\right), \tag{1.1}$$

where $n = 0, \cdots, N$ and one of the three equalities is satisfied: $\alpha+1 = -N$ or $\beta+\delta+1 = -N$ or $\gamma+1 = -N$. Set

* Corresponding author.
*E-mail address:* xswang@semo.edu (X.-S. Wang).

http://dx.doi.org/10.1016/j.jmaa.2015.12.035
0022-247X/

$$A_n := -\frac{(n+\alpha+1)(n+\alpha+\beta+1)(n+\beta+\delta+1)(n+\gamma+1)}{(2n+\alpha+\beta+1)(2n+\alpha+\beta+2)},$$
$$C_n := -\frac{n(n+\alpha+\beta-\gamma)(n+\alpha-\delta)(n+\beta)}{(2n+\alpha+\beta)(2n+\alpha+\beta+1)}. \tag{1.2}$$

The Racah polynomials (1.1) satisfy the recurrence relation [7, (9.2.3)]

$$\lambda(x)R_n(\lambda(x)) = -A_nR_{n+1}(\lambda(x)) + (A_n + C_n)R_n(\lambda(x)) - C_nR_{n-1}(\lambda(x)).$$

This recurrence relation can be normalized as

$$\pi_{n+1}(z) = (z - A_n - C_n)\pi_n(z) - A_{n-1}C_n\pi_{n-1}(z), \quad \pi_0(z) = 1, \ \pi_1(z) = z - A_0, \tag{1.3}$$

where $z = \lambda(x)$ and

$$\pi_n(\lambda(x)) := \frac{(\alpha+1)_n(\beta+\delta+1)_n(\gamma+1)_n}{(n+\alpha+\beta+1)_n}R_n(\lambda(x)). \tag{1.4}$$

In addition to the variable $x$ and the degree $n$, the polynomials in (1.1) involve four free parameters. This inevitably makes the problem of deriving their asymptotic formulas much more complicated. The only result that we can find on this topic is that given by Chen, Ismail and Simeonov [2]. Their method starts with the hypergeometric representation in (1.1). By approximating the ratio of two shifted factorials, they obtain several asymptotic formulas in terms of hypergeometric function ${}_3F_2$ or ${}_2F_1$, when the parameters are fixed.

In the present paper, we are interested in the large-$n$ behavior of $\pi_n(z)$ with varying parameters $\alpha, \beta, \gamma, \delta$. More precisely, we set

$$\alpha + 1 = Na, \ \beta = Nb, \ \gamma + 1 = Nc, \ \delta + 1 = Nd, \tag{1.5}$$

where either $a = -1$ or $b + d = -1$ or $c = -1$. For simplicity, we assume $A_n > 0$ and $C_n > 0$. By Favard's theorem, these conditions guarantee that the zeros of $\pi_n(z)$ are all real and simple; see [3, Sections 1.4 and 1.5]. Thus, we require some additional conditions:

1. when $a = -1$, we assume $b, c, d > 0$ and $b > c + 1$;
2. when $b + d = -1$, we assume $a, b, c > 0$ and $a + b + 1 < c$;
3. when $c = -1$, we assume $a, b, d > 0$ and $a + 1 < d$.

Let $n/N = p$ be a fixed number in $(0, 1)$. We shall derive asymptotic formulas for $\pi_n(N^2y)$ as $N \to \infty$.

## 2. Main results

Define the ratio $w_k(z) := \pi_k(z)/\pi_{k-1}(z)$ for $k = 1, \cdots, n$. From (1.3), it follows that

$$w_{k+1}(z) = z - (A_k + C_k) - \frac{A_{k-1}C_k}{w_k(z)}. \tag{2.1}$$

Recall $p = n/N$. We obtain from (1.2) and (1.5) that

$$\lim_{N\to\infty}\frac{A_n}{N^2} = A(p); \quad \lim_{N\to\infty}\frac{C_n}{N^2} = C(p),$$

where

$$A(t) := -\frac{(t+a)(t+a+b)(t+b+d)(t+c)}{(2t+a+b)^2}, \qquad 0 < t < p,$$
$$C(t) := -\frac{t(t+a+b-c)(t+a-d)(t+b)}{(2t+a+b)^2}, \qquad 0 < t < p. \tag{2.2}$$

Now, we shall introduce the transition points (or turning points) when the characteristic roots of (1.3) coincide; see [9]. For $t \in [0,p]$, we define

$$y_\pm(t) := A(t) + C(t) \pm 2\sqrt{A(t)C(t)}; \tag{2.3}$$

these are the transition points when $t = p$. From (2.2), it is readily seen that $C(0) = 0$. Thus, $y_\pm(0) = A(0)$. For simplicity, we only consider the case when $y_+(t)$ is increasing and $y_-(t)$ is decreasing for $t \in [0,p]$. The other cases will be studied in forthcoming papers. If $z = N^2y$ with $y \in \mathbf{C} \setminus [y_-(p), y_+(p)]$, we apply a standard successive approximation technique to (2.1) and obtain

$$w_k(z) = w_k^0[1 + w_k^1 + O(1/n^2)] \text{ as } N \to \infty,$$

uniformly for $k = 1, \cdots, n$; see [10]. Here, the leading term is

$$w_k^0 = \frac{z - A_k - C_k + \sqrt{(z - A_k - C_k)^2 - 4A_{k-1}C_k}}{2}, \tag{2.4}$$

and the first-order term is

$$w_k^1 = \frac{1}{2} + \frac{A_{k+1} - A_k + C_{k+1} - C_k - \sqrt{(z - A_{k+1} - C_{k+1})^2 - 4A_kC_{k+1}}}{2\sqrt{(z - A_k - C_k)^2 - 4A_{k-1}C_k}}. \tag{2.5}$$

We can rewrite (2.5) as

$$w_k^1 = \frac{A_{k+1} - A_k + C_{k+1} - C_k + \sqrt{(z - A_k - C_k)^2 - 4A_{k-1}C_k} - \sqrt{(z - A_{k+1} - C_{k+1})^2 - 4A_kC_{k+1}}}{2\sqrt{(z - A_k - C_k)^2 - 4A_{k-1}C_k}}.$$

Note that $z = N^2y$ with $y \in \mathbf{C} \setminus [y_-(p), y_+(p)]$. From (1.2), it is readily seen that for any $k = 1, \cdots, n$, the denominator in the expression of $w_k^1$ is of order $O(n^2)$, while the numerator is of order $O(n)$. Thus, $w_k^1 = O(1/n)$. Since $\pi_n = w_1 \cdots w_n$, we have

$$\ln \pi_n = \sum_{k=1}^n \ln w_k = \sum_{k=1}^n \ln w_k^0 + \sum_{k=1}^n \ln(1 + w_k^1) + O(1/n) = \sum_{k=1}^n \ln w_k^0 + \sum_{k=1}^n w_k^1 + O(1/n). \tag{2.6}$$

To find an asymptotic formula for $\pi_n$, we only need to approximate $\sum_{k=1}^n \ln w_k^0$ and $\sum_{k=1}^n w_k^1$. First, we note that

$$\sqrt{[z - (A_k + C_k)]^2 - 4A_{k-1}C_k} \sim N^2\sqrt{[y - (A(k/N) + C(k/N))]^2 - 4A(k/N)C(k/N)}, \tag{2.7}$$

where $A(k/N)$ and $C(k/N)$ are given in (2.2) with $t = k/N$. Secondly, we have

$$\frac{A_{k+1} - A_k}{N} \sim A^*(k/N) := \frac{4\mathcal{A}_1\mathcal{A}_2\mathcal{A}_3\mathcal{A}_4 - (\mathcal{A}_1\mathcal{A}_2\mathcal{A}_3 + \mathcal{A}_1\mathcal{A}_2\mathcal{A}_4 + \mathcal{A}_1\mathcal{A}_3\mathcal{A}_4 + \mathcal{A}_2\mathcal{A}_3\mathcal{A}_4)\mathcal{D}}{\mathcal{D}^3},$$
$$\frac{C_{k+1} - C_k}{N} \sim C^*(k/N) := \frac{4\mathcal{C}_1\mathcal{C}_2\mathcal{C}_3\mathcal{C}_4 - (\mathcal{C}_1\mathcal{C}_2\mathcal{C}_3 + \mathcal{C}_1\mathcal{C}_2\mathcal{C}_4 + \mathcal{C}_1\mathcal{C}_3\mathcal{C}_4 + \mathcal{C}_2\mathcal{C}_3\mathcal{C}_4)\mathcal{D}}{\mathcal{D}^3}, \tag{2.8}$$

where

$$\mathcal{A}_1(t) := t + a,\ \mathcal{A}_2(t) := t + a + b,\ \mathcal{A}_3(t) := t + b + d, \tag{2.9a}$$

$$\mathcal{A}_4(t) := t + c,\ \mathcal{C}_1(t) := t,\ \mathcal{C}_2(t) := t + a + b - c, \tag{2.9b}$$

and

$$\mathcal{C}_3(t) := t + a - d,\ \mathcal{C}_4(t) := t + b,\ \mathcal{D}(t) := 2t + a + b. \tag{2.9c}$$

Using the above notations, we can rewrite

$$A(t) = -\mathcal{A}_1(t)\mathcal{A}_2(t)\mathcal{A}_3(t)\mathcal{A}_4(t)/\mathcal{D}(t)^2,$$

and

$$C(t) = -\mathcal{C}_1(t)\mathcal{C}_2(t)\mathcal{C}_3(t)\mathcal{C}_4(t)/\mathcal{D}(t)^2.$$

Finally, we approximate

$$\sqrt{[z - (A_k + C_k)]^2 - 4A_{k-1}C_k} - \sqrt{[z - (A_{k+1} + C_{k+1})]^2 - 4A_kC_{k+1}}$$

by multiplying and dividing by

$$\sqrt{[z - (A_k + C_k)]^2 - 4A_{k-1}C_k} + \sqrt{[z - (A_{k+1} + C_{k+1})]^2 - 4A_kC_{k+1}}.$$

In the resulting expression, we replace the denominator by using the approximant in (2.7), and replace $A_{k+1} - A_k$ and $C_{k+1} - C_k$ by using $A^*(k/N)$ and $C^*(k/N)$ given in (2.8). The final result is

$$\frac{N\{[y - A(k/N) - C(k/N)][A^*(k/N) + C^*(k/N)] + 2[A(k/N)C^*(k/N) + A^*(k/N)C(k/N)]\}}{\sqrt{[y - (A(k/N) + C(k/N))]^2 - 4A(k/N)C(k/N)}}. \tag{2.10}$$

For convenience, we introduce the notations

$$S(y;t) := \sqrt{(y - A(t) - C(t))^2 - 4A(t)C(t)} = \sqrt{[y - y_-(t)][y - y_+(t)]}, \tag{2.11}$$

$$T(y;t) := y - A(t) - C(t) + S(y;t) = [\sqrt{y - y_-(t)} + \sqrt{y - y_+(t)}]^2/2, \tag{2.12}$$

and

$$G(y;t) := T(y;t)[A^*(t) + C^*(t)] + 2[A(t)C^*(t) + A^*(t)C(t)]. \tag{2.13}$$

Note that $S(y;t)$ is just the denominator in (2.10) with $t = k/N$.

Substituting the estimates (2.7), (2.8) and (2.10) into (2.5), and using the notations (2.11)–(2.13), we obtain

$$w_k^1 = \frac{G(y;k/N)}{2NS(y;k/N)^2} + O(1/N^2).$$

It then follows from the trapezoidal rule (cf. [10]) that

$$\sum_{k=1}^{n} w_k^1 = \int_0^p \frac{G(y;t)}{2S(y;t)^2}dt + O(1/n). \tag{2.14}$$

Next, we approximate $\sum_{k=1}^{n} \ln w_k^0$. Note that

$$\frac{A_k}{N^2} \sim A(k/N)\left(1 - \frac{1}{N\mathcal{D}}\right), \qquad \frac{C_k}{N^2} \sim C(k/N)\left(1 + \frac{1}{N\mathcal{D}}\right),$$

and

$$\frac{A_{k-1}}{N^2} \sim A(k/N)\left(1 + \frac{3}{N\mathcal{D}} - \frac{1}{N\mathcal{A}_1} - \frac{1}{N\mathcal{A}_2} - \frac{1}{N\mathcal{A}_3} - \frac{1}{N\mathcal{A}_4}\right),$$

where $\mathcal{A}_1, \cdots, \mathcal{A}_4$ and $\mathcal{D}$ are given in (2.9).

Furthermore, we obtain from (2.4) that

$$\begin{aligned} \frac{2w_k^0}{N^2} &= \frac{z - A_k - C_k + \sqrt{(z - A_k - C_k)^2 - 4A_kC_k}}{N^2} \\ &+ \frac{\sqrt{(z - A_k - C_k)^2 - 4A_{k-1}C_k} - \sqrt{(z - A_k - C_k)^2 - 4A_kC_k}}{N^2}. \end{aligned} \tag{2.15}$$

The first fraction on the right-hand side of (2.15) can be approximated as

$$\begin{aligned} & y - A(k/N) + \frac{A(k/N)}{N\mathcal{D}} - C(k/N) - \frac{C(k/N)}{N\mathcal{D}} \\ &+ \sqrt{\left[y - A(k/N) - C(k/N) + \frac{A(k/N) - C(k/N)}{N\mathcal{D}}\right]^2 - A(k/N)C(k/N)} \\ \sim\, & y - A(k/N) - C(k/N) + \sqrt{[y - A(k/N) - C(k/N)]^2 - A(k/N)C(k/N)} + \frac{A(k/N) - C(k/N)}{N\mathcal{D}} \\ &+ \frac{[A(k/N) - C(k/N)][y - A(k/N) - C(k/N)]}{N\mathcal{D}\sqrt{[y - A(k/N) - C(k/N)]^2 - A(k/N)C(k/N)}} \\ \sim\, & T(y; k/N) + \frac{[A(k/N) - C(k/N)]T(y; k/N)}{N\mathcal{D}(k/N)S(y; k/N)}. \end{aligned}$$

The second fraction on the right-hand side of (2.15) can be approximated as

$$\begin{aligned} & \frac{-4(A_{k-1} - A_k)C_k}{N^2[\sqrt{(z - A_k - C_k)^2 - 4A_{k-1}C_k} + \sqrt{(z - A_k - C_k)^2 - 4A_kC_k}]} \\ \sim\, & \frac{-2A(k/N)C(k/N)}{S(y; k/N)}\left(\frac{4}{N\mathcal{D}} - \frac{1}{N\mathcal{A}_1} - \frac{1}{N\mathcal{A}_2} - \frac{1}{N\mathcal{A}_3} - \frac{1}{N\mathcal{A}_4}\right). \end{aligned}$$

Applying the above two estimates to (2.15) yields

$$\frac{2w_k^0}{N^2} \sim T(y; k/N) + \frac{[A(k/N) - C(k/N)]T(y; k/N)}{N\mathcal{D}(k/N)S(y; k/N)} - \frac{2A(k/N)C(k/N)\mathcal{E}(k/N)}{NS(y; k/N)},$$

where

$$\mathcal{E}(t) := \frac{4}{\mathcal{D}(t)} - \frac{1}{\mathcal{A}_1(t)} - \frac{1}{\mathcal{A}_2(t)} - \frac{1}{\mathcal{A}_3(t)} - \frac{1}{\mathcal{A}_4(t)}. \tag{2.16}$$

Thus, we obtain from the trapezoidal rule that

$$\sum_{k=1}^{n} \ln w_k^0 = n\ln(N^2/2) + N\int_0^p \ln T(y;t)dt + \frac{1}{2}\ln\frac{T(y;p)}{T(y;0)}$$
$$+ \int_0^p \left[\frac{A(t)-C(t)}{\mathcal{D}(t)S(y;t)} - \frac{2A(t)C(t)\mathcal{E}(t)}{S(y;t)T(y;t)}\right] dt + O(1/n). \tag{2.17}$$

Now, we are ready to state our first main result.

**Theorem 2.1.** *Let $\pi_n(z)$ be the monic Racah polynomials satisfying the recurrence relation* (1.3). *Assume $n/N = p$ is fixed in $(0,1)$. Let $A(t)$, $C(t)$ and $y_\pm(t)$ be defined as in* (2.2) *and* (2.3). *Also, let $\Omega_1(y)$ and $\Omega_0(y)$ denote, respectively, the integrals in* (2.14) *and* (2.17). *Assume that $y_+(t)$ is increasing and $y_-(t)$ is decreasing for $t \in [0,p]$. Then, for $y \in \mathbf{C} \setminus [y_-(p), y_+(p)]$, we have*

$$\pi_n(N^2y) = \left(\frac{N^2}{2}\right)^n e^{Ng(y)+r(y)}\left[1 + O\left(\frac{1}{n}\right)\right], \tag{2.18}$$

*where the main term $g(y)$ is given by*

$$g(y) := \int_0^p \ln T(y;t)dt, \tag{2.19}$$

*and the correction term $r(y)$ is given by*

$$r(y) := \frac{1}{2}\ln\frac{T(y;p)}{T(y;0)} + \Omega_0(y) + \Omega_1(y). \tag{2.20}$$

**Proof.** Use a combination of (2.6), (2.14) and (2.17). □

Next, we investigate $\pi_n(N^2y)$ for $y$ in a small neighborhood of the oscillatory interval $(y_-(p), y_+(p))$. For $y \in (y_-(p), y_+(p))$, we use the notation

$$F^\pm(y;t) := \lim_{\varepsilon\to 0^+} F(y \pm i\varepsilon; t). \tag{2.21}$$

Thus,

$$S^\pm(y) := \lim_{\varepsilon\to 0^+} S(y \pm i\varepsilon), \quad g^\pm(y) := \lim_{\varepsilon\to 0^+} g(y \pm i\varepsilon), \cdots, \text{etc.} \tag{2.22}$$

Recall that $y_\pm(0) = A(0)$. We consider two cases: $y \in (A(0), y_+(p))$ and $y \in (y_-(p), A(0))$, separately.

If $A(0) = y_+(0) < y < y_+(p)$, we choose $t_y \in (0,p)$ such that

$$y = y_+(t_y) = A(t_y) + C(t_y) + 2\sqrt{A(t_y)C(t_y)};$$

cf. (2.3). In this case, we have

$$S^\pm(y;t) = \begin{cases} \sqrt{[y - y_-(t)][y - y_+(t)]}, & 0 \le t \le t_y, \\ \pm i\sqrt{[y - y_-(t)][y_+(t) - y]}, & t_y \le t \le p. \end{cases} \tag{2.23a}$$

If $y_-(p) < y < y_-(0) = A(0)$, we choose $t_y \in (0,p)$ such that

$$y = y_-(t_y) = A(t_y) + C(t_y) - 2\sqrt{A(t_y)C(t_y)},$$

in which case we have

$$S^\pm(y;t) = \begin{cases} -\sqrt{[y_-(t) - y][y_+(t) - y]}, & 0 \le t \le t_y, \\ \pm i\sqrt{[y - y_-(t)][y_+(t) - y]}, & t_y \le t \le p. \end{cases} \tag{2.23b}$$

Using the definition of $F^\pm(y;t)$ given in (2.21), it follows from (2.19) and (2.20) that

$$g^\pm(y) = \int_0^p \ln T^\pm(y;t)dt, \tag{2.24}$$

where

$$T^\pm(y;t) = y - A(t) - C(t) + S^\pm(y;t); \tag{2.25}$$

and

$$r^\pm(y) = \frac{1}{2}\ln\frac{T^\pm(y;p)}{T^\pm(y;0)} + \Omega_0^\pm(y) + \Omega_1^\pm(y), \tag{2.26}$$

where

$$\Omega_0^\pm(y) = \int_0^p \left[\frac{A(t) - C(t)}{\mathcal{D}(t)S^\pm(y;t)} - \frac{2A(t)C(t)\mathcal{E}(t)}{S^\pm(y;t)T^\pm(y;t)}\right] dt, \tag{2.27}$$

and

$$\Omega_1^\pm(y) = P.V.\int_0^p \frac{G(y;t)}{2S(y;t)^2}dt \mp i\pi\frac{G(y;t_y)}{2y_-'(t_y)[y - y_+(t_y)]} \tag{2.28a}$$

if $y_-(p) < y < y_-(0) = A(0)$, and

$$\Omega_1^\pm(y) = P.V.\int_0^p \frac{G(y;t)}{2S(y;t)^2}dt \mp i\pi\frac{G(y;t_y)}{2[y - y_-(t_y)]y_+'(t_y)} \tag{2.28b}$$

if $A(0) = y_+(0) < y < y_+(p)$. In (2.28), "P.V." denotes the Cauchy principal value.

An explanation is needed for the results in (2.28). In view of the second equality in (2.11), the integral $\Omega_1(y)$ in (2.14) (see the statement of Theorem 2.1) has a singularity at $t = t_y$. For this reason, we digress briefly to discuss the Stieltjes transform

$$F(z) = \int_a^b \frac{f(s)}{s - z}ds, \quad z \in \mathbf{C} \setminus [a,b].$$

By using Cauchy's theorem, one can show that

$$F^\pm(x) = P.V.\int_a^b \frac{f(s)}{s - x}dx \pm i\pi f(x), \; a < x < b. \tag{2.29}$$

We first consider the case $A(0) = y_+(0) < y < y_+(p)$. The integral $\Omega_1(y)$ can be written as

$$\Omega_1(y) = \int_0^p \frac{(t - t_y)G(y;t)}{2S(y;t)^2(t - t_y)} dt.$$

From (2.29), it follows that

$$\begin{aligned}\Omega_1^{\pm}(y) &= P.V. \int_0^p \frac{G(y;t)}{2S(y;t)^2} dt \pm i\pi \lim_{t \to t_y} \frac{t - t_y}{2S(y;t)^2} G(y;t) \\ &= P.V. \int_0^p \frac{G(y;t)}{2S(y;t)^2} dt \mp i\pi \frac{G(y;t_y)}{2[y - y_-(t_y)]y_+'(t_y)}.\end{aligned}$$

In the case $y_-(p) < y < y_-(0) = A(0)$, the result is

$$\Omega_1^{\pm}(y) = P.V. \int_0^p \frac{G(y;t)}{2S(y;t)^2} dt \mp i\pi \frac{G(y;t_y)}{2y_-'(t_y)[y - y_+(t_y)]}.$$

Our second main theorem is stated below.

**Theorem 2.2.** *Let $\pi_n(z)$ be the monic Racah polynomials satisfying the recurrence relation* (1.3)*. Assume $n/N = p$ is fixed in $(0, 1)$. Let $A(t)$, $C(t)$ and $y_\pm(t)$ be defined as in* (2.2) *and* (2.3)*. We further assume that $y_+(t)$ is increasing and $y_-(t)$ is decreasing for $t \in [0, p]$. Then, we have*

$$\pi_n(N^2 y) = \left(\frac{N^2}{2}\right)^n \left\{ e^{Ng^+(y) + r^+(y)} \left[1 + O\left(\frac{1}{n}\right)\right] + e^{Ng^-(y) + r^-(y)} \left[1 + O\left(\frac{1}{n}\right)\right]\right\} \tag{2.30}$$

*for $y \in (y_-(p), A(0)) \cup (A(0), y_+(p))$.*

**Proof.** First, we note that the functions $g^\pm(y)$ and $r^\pm(y)$ can be analytically continued to a small neighborhood of any compact subset of $(y_-(p), A(0)) \cup (A(0), y_+(p))$ in the complex plane. In such a neighborhood, by (2.19) and (2.12),

$$g_+(y) - g_-(y) = \int_{t_y}^p \ln \frac{T^+(y;t)}{T^-(y;t)} dt$$

has positive real part if $y$ is in the upper-half plane and negative real part if $y$ is in the lower-half plane. Thus, the function $e^{N[g^+(y) - g^-(y)]}$ is exponentially large on the upper-half plane and exponentially small on the lower-half plane, which together with (2.18) implies that the asymptotic formula (2.30) is valid on both upper- and lower-half planes. By analytical continuation, this formula is also valid for $y \in (y_-(p), A(0)) \cup (A(0), y_+(p))$; see [8,10]. □

**Remark 2.3.** A standard application of the uniform technique developed by Wang and Wong [9] allows us to provide uniform asymptotic formulas of Racah polynomials near the turning points $y_-(p)$ and $y_+(p)$. For the sake of simplicity, we omit the details in this paper.

## 3. Limiting cases: Hahn polynomials

The Hahn polynomials are defined by the ${}_3F_2$ hypergeometric series [7, (9.5.1)]

$$Q_n(x;\alpha,\beta,N) = {}_3F_2\left(\begin{matrix} -n, & n+\alpha+\beta+1, & -x \\ & \alpha+1, & -N \end{matrix}\middle| 1\right), \quad n=0,\cdots,N. \tag{3.1}$$

They are the limiting cases of Racah polynomials in the following sense [7, (9.2.15)]

$$\lim_{\delta\to\infty} R_n(\lambda(x),\alpha,\beta,-N-1,\delta) = Q_n(x;\alpha,\beta,N). \tag{3.2}$$

Here $\lambda(x) = x(x+\gamma+\delta+1)$ and $\gamma = -N-1$. Since asymptotic formulas of Racah polynomials hold uniformly with respect to the parameter $\delta$, asymptotic formulas of the Hahn polynomials can be deduced from the formulas in (2.18) and (2.30). Let

$$\bar{\pi}_n(x) := \frac{(\alpha+1)_n(-N)_n}{(n+\alpha+\beta+1)_n} Q_n(x;\alpha,\beta,N) \tag{3.3}$$

be the monic Hahn polynomials. From (1.4), (3.1), (3.2) and (3.3), it follows that

$$\bar{\pi}_n(x) = \lim_{\delta\to\infty} \frac{\pi_n(\lambda(x))}{(\beta+\delta+1)_n}. \tag{3.4}$$

Recall that

$$z = N^2 y = \lambda(x), \ \alpha+1 = Na, \ \beta = Nb, \ \gamma+1 = Nc, \ \delta+1 = Nd.$$

With $\gamma = -N-1$ in (3.2), we must take $c=-1$. Since $\delta\to\infty$, we also have $d\to\infty$; cf. the equalities in the preceding line. Set $x = N\bar{y}$ and take logarithms on both sides of (3.4). Since $(\beta+\delta+1)_n = (Nb+Nd)_n \sim (Nd)^n$ as $d\to\infty$, we obtain

$$\ln\bar{\pi}_n(N\bar{y}) = \lim_{d\to\infty}[\ln\pi_n(N^2y) - n\ln N - n\ln d]. \tag{3.5}$$

Furthermore, since $z = N^2y = \lambda(N\bar{y}) = N\bar{y}(N\bar{y}+Nc+Nd-1)$, we have $y/d\to\bar{y}$ as $d\to\infty$. From (2.2), it follows that for $t\in[0,p]$,

$$\lim_{d\to\infty}\frac{A(t)}{d} = \frac{(t+a)(t+a+b)(1-t)}{(2t+a+b)^2} =: \bar{A}(t),$$
$$\lim_{d\to\infty}\frac{C(t)}{d} = \frac{t(t+a+b+1)(t+b)}{(2t+a+b)^2} =: \bar{C}(t). \tag{3.6}$$

Note from (2.9) that $\mathcal{A}_3(t)/d\to 1$ and $\mathcal{C}_3(t)/d\to -1$ as $d\to\infty$. Thus, we have from (2.8) and (2.9)

$$\bar{A}^*(t) := \lim_{d\to\infty}\frac{A^*(t)}{d} = \frac{4\mathcal{A}_1(t)\mathcal{A}_2(t)\mathcal{A}_4(t) - (\mathcal{A}_1(t)\mathcal{A}_2(t)+\mathcal{A}_1(t)\mathcal{A}_4(t)+\mathcal{A}_2(t)\mathcal{A}_4(t))\mathcal{D}(t)}{\mathcal{D}^3(t)},$$
$$\bar{C}^*(t) := \lim_{d\to\infty}\frac{C^*(t)}{d} = \frac{(\mathcal{C}_1(t)\mathcal{C}_2(t)+\mathcal{C}_1(t)\mathcal{C}_4(t)+\mathcal{C}_2(t)\mathcal{C}_4(t))\mathcal{D}(t) - 4\mathcal{C}_1(t)\mathcal{C}_2(t)\mathcal{C}_4(t)}{\mathcal{D}^3(t)}, \tag{3.7}$$

where $\mathcal{A}_i(t)$, $\mathcal{C}_i(t)$ (with $i=1,2,4$) and $\mathcal{D}(t)$ are defined as in (2.9) with $c$ replaced by $-1$. Finally, by (2.11) and (2.12), we obtain

$$\bar{S}(\bar{y};t) := \lim_{d\to\infty} \frac{S(y;t)}{d} = \sqrt{[\bar{y} - \bar{A}(t) - \bar{C}(t)]^2 - 4\bar{A}(t)\bar{C}(t)} \tag{3.8}$$

and

$$\bar{T}(\bar{y};t) := \lim_{d\to\infty} \frac{T(y;t)}{d} = \bar{y} - \bar{A}(t) - \bar{C}(t) + \bar{S}(\bar{y};t). \tag{3.9}$$

On account of (2.13) and (2.16), we also have

$$\bar{G}(\bar{y};t) := \lim_{d\to\infty} \frac{G(y;t)}{d^2} = \bar{T}(\bar{y};t)[\bar{A}^*(t) + \bar{C}^*(t)] + 2[\bar{A}(t)\bar{C}^*(t) + \bar{A}^*(t)\bar{C}(t)] \tag{3.10}$$

and

$$\bar{\mathcal{E}}(t) := \lim_{d\to\infty} \mathcal{E}(t) = \frac{4}{\mathcal{D}(t)} - \frac{1}{\mathcal{A}_1(t)} - \frac{1}{\mathcal{A}_2(t)} - \frac{1}{\mathcal{A}_4(t)}. \tag{3.11}$$

Combining (2.3) and (3.6) gives

$$\bar{y}_\pm(t) := \lim_{d\to\infty} \frac{y_\pm(t)}{d} = \bar{A}(t) + \bar{C}(t) \pm 2\sqrt{\bar{A}(t)\bar{C}(t)}. \tag{3.12}$$

From (2.18) and (3.5), we have the following result.

**Corollary 3.1.** *Let $\alpha = Na$ and $\beta = Nb$ with $a, b > 0$. Assume $n/N = p$ is a fixed number in $(0,1)$. Let $\bar{y}_\pm(t)$ be defined as in (3.12). Assume $\bar{y}_+(t)$ is increasing and $\bar{y}_-(t)$ is decreasing for $t \in [0,p]$. As $n \to \infty$, we have the following asymptotic formula for the monic Hahn polynomials $\bar{\pi}_n$ in (3.3):*

$$\bar{\pi}_n(N\bar{y}) = \left(\frac{N}{2}\right)^n e^{N\bar{g}(\bar{y}) + \bar{r}(\bar{y})} \left[1 + O\left(\frac{1}{n}\right)\right] \tag{3.13}$$

*for $\bar{y} \in \mathbf{C} \setminus [\bar{y}_-(p), \bar{y}_+(p)]$. Here, the main term $\bar{g}(\bar{y})$ is given by*

$$\bar{g}(\bar{y}) := \lim_{d\to\infty} [g(y) - p\ln d] = \int_0^p \ln \bar{T}(\bar{y};t)dt, \tag{3.14}$$

*and the correction term $\bar{r}(\bar{y})$ is given by*

$$\bar{r}(\bar{y}) := \lim_{d\to\infty} r(y) = \frac{1}{2}\ln\frac{\bar{T}(\bar{y};p)}{\bar{T}(\bar{y};0)} + \bar{\Omega}_0(\bar{y}) + \bar{\Omega}_1(\bar{y}), \tag{3.15}$$

*where*

$$\bar{\Omega}_0(\bar{y}) := \lim_{d\to\infty} \Omega_0(y) = \int_0^p \left[\frac{\bar{A}(t) - \bar{C}(t)}{\mathcal{D}(t)\bar{S}(\bar{y};t)} - \frac{2\bar{A}(t)\bar{C}(t)\bar{\mathcal{E}}(t)}{\bar{S}(\bar{y};t)\bar{T}(\bar{y};t)}\right] dt, \tag{3.16}$$

*and*

$$\bar{\Omega}_1(\bar{y}) := \lim_{d\to\infty} \Omega_1(y) = \int_0^p \frac{\bar{G}(\bar{y};t)}{2\bar{S}(\bar{y};t)^2} dt. \tag{3.17}$$

Equation (3.14) follows from (2.19) and (3.9), and equation (3.15) is obtained from (2.20), (2.17) and (2.14); see also (3.8), (3.10) and (3.11).

The asymptotic formula of monic Hahn polynomials $\bar{\pi}_n(N\bar{y})$ in the oscillatory interval can be obtained by using either the matching method or by taking a limit in the corresponding formula of Racah polynomials.

**Corollary 3.2.** *Let $\alpha = Na$ and $\beta = Nb$ with $a, b > 0$. Assume $n/N = p$ is a fixed number in $(0,1)$. Let $\bar{y}_\pm(t)$ be defined as in (3.12). Assume $\bar{y}_+(t)$ is increasing and $\bar{y}_-(t)$ is decreasing for $t \in [0,p]$. Let $\bar{S}^\pm(\bar{y};t)$ and $\bar{T}^\pm(\bar{y};t)$ be defined similarly as in (2.23) and (2.25). As $n \to \infty$, we have*

$$\bar{\pi}_n(N\bar{y}) = \left(\frac{N}{2}\right)^n \left\{e^{N\bar{g}^+(\bar{y})+\bar{r}^+(\bar{y})}\left[1+O\left(\frac{1}{n}\right)\right] + e^{N\bar{g}^-(\bar{y})+\bar{r}^-(\bar{y})}\left[1+O\left(\frac{1}{n}\right)\right]\right\} \tag{3.18}$$

*for $\bar{y} \in (\bar{y}_-(p), \bar{A}(0)) \cup (\bar{A}(0), \bar{y}_+(p))$. Here,*

$$\bar{g}^\pm(\bar{y}) = \int_0^p \ln \bar{T}^\pm(\bar{y};t)dt, \tag{3.19}$$

*and*

$$\bar{r}^\pm(\bar{y}) = \frac{1}{2}\ln\frac{\bar{T}^\pm(\bar{y};p)}{\bar{T}^\pm(\bar{y};0)} + \bar{\Omega}_0^\pm(\bar{y}) + \bar{\Omega}_1^\pm(\bar{y}), \tag{3.20}$$

*where*

$$\bar{\Omega}_0^\pm(\bar{y}) = \int_0^p \left[\frac{\bar{A}(t)-\bar{C}(t)}{\mathcal{D}(t)\bar{S}^\pm(\bar{y};t)} - \frac{2\bar{A}(t)\bar{C}(t)\bar{\mathcal{E}}(t)}{\bar{S}^\pm(\bar{y};t)\bar{T}^\pm(\bar{y};t)}\right]dt, \tag{3.21}$$

*and*

$$\bar{\Omega}_1^\pm(\bar{y}) = P.V.\int_0^p \frac{\bar{G}(\bar{y};t)}{2\bar{S}(\bar{y};t)^2}dt \mp i\pi\frac{\bar{G}(\bar{y};t_{\bar{y}})}{2\bar{y}'_-(t_{\bar{y}})[\bar{y}-\bar{y}_+(t_{\bar{y}})]} \tag{3.22a}$$

*if $\bar{y}_-(p) < \bar{y} < \bar{y}_-(0) = \bar{A}(0)$ with $\bar{y} = \bar{y}_-(t_{\bar{y}})$ for some $t_{\bar{y}} \in (0,p)$, and*

$$\bar{\Omega}_1^\pm(\bar{y}) = P.V.\int_0^p \frac{\bar{G}(\bar{y};t)}{2\bar{S}(\bar{y};t)^2}dt \mp i\pi\frac{\bar{G}(\bar{y};t_{\bar{y}})}{2[\bar{y}-\bar{y}_-(t_{\bar{y}})]\bar{y}'_+(t_{\bar{y}})} \tag{3.22b}$$

*if $\bar{A}(0) = \bar{y}_+(0) < \bar{y} < \bar{y}_+(p)$ with $\bar{y} = \bar{y}_+(t_{\bar{y}})$ for some $t_{\bar{y}} \in (0,p)$.*

## 4. Limiting cases: Krawtchouk polynomials

The Krawtchouk polynomials are defined by the ${}_2F_1$ hypergeometric series [7, (9.11.1)]

$$K_n(x;q,N) = {}_2F_1\left(\begin{matrix}-n, & -x \\ & -N\end{matrix}\middle|\frac{1}{q}\right), \quad n = 0,\cdots,N. \tag{4.1}$$

These polynomials are limiting cases of Hahn polynomials in the following sense. Fix $q \in (0,1)$, and let $\alpha/\beta = q/(1-q)$ and $\beta \to \infty$. We have [7, (9.5.16)]

$$\lim_{\beta\to\infty} Q_n(x;\beta q/(1-q),\beta,N) = K_n(x;q,N). \tag{4.2}$$

The monic Krawtchouk polynomials are given by

$$\tilde{\pi}_n(x) := (-N)_n q^n K_n(x;q,N). \tag{4.3}$$

Recall that $\alpha+1 = Na$ and $\beta = Nb$. We choose $a = mq$ and $b = m(1-q)$, and then let $m\to\infty$. In view of (3.3), (4.1), (4.2) and (4.3), we have the following limiting relation between monic Hahn polynomials and monic Krawtchouk polynomials.

$$\lim_{m\to\infty} \bar{\pi}_n(x) = \tilde{\pi}_n(x). \tag{4.4}$$

Furthermore, from (3.6), (3.7) and (2.9) with $a = mq$ and $b = m(1-q)$, we obtain for $t \in [0,p]$,

$$\begin{aligned}\lim_{m\to\infty} \bar{A}(t) &= q(1-t) =: \tilde{A}(t),\\ \lim_{m\to\infty} \bar{C}(t) &= t(1-q) =: \tilde{C}(t),\end{aligned} \tag{4.5}$$

and

$$\begin{aligned}\lim_{m\to\infty} \bar{A}^*(t) &= -q,\\ \lim_{m\to\infty} \bar{C}^*(t) &= 1-q.\end{aligned} \tag{4.6}$$

Recall that in the previous section, we have chosen $c = -1$ to obtain Hahn polynomials from Racah polynomials. In view of (2.9), we have as $m\to\infty$

$$1/\mathcal{D}(t)\to 0,\ 1/\mathcal{A}_1(t)\to 0,\ 1/\mathcal{A}_2(t)\to 0,\ 1/\mathcal{A}_4(t) = 1/(t-1). \tag{4.7}$$

Let $\tilde{y} = \bar{y}$. From (3.8), (3.9) and (4.5), it follows that

$$\tilde{S}(\tilde{y};t) := \lim_{m\to\infty} \bar{S}(\bar{y};t) = \sqrt{[\tilde{y}-\tilde{A}(t)-\tilde{C}(t)]^2 - 4\tilde{A}(t)\tilde{C}(t)}, \tag{4.8}$$

and

$$\tilde{T}(\tilde{y};t) := \lim_{m\to\infty} \bar{T}(\bar{y};t) = \tilde{y} - \tilde{A}(t) - \tilde{C}(t) + \tilde{S}(\tilde{y};t). \tag{4.9}$$

Similarly, from (3.10), (4.6) and (4.5), we obtain

$$\tilde{G}(\tilde{y};t) := \lim_{m\to\infty} \bar{G}(\bar{y};t) = \tilde{T}(\tilde{y};t)(1-2q) + 2q(1-q)(1-2t). \tag{4.10}$$

Furthermore, in view of (3.11) and (4.7), we also have

$$\tilde{\mathcal{E}}(t) := \lim_{m\to\infty} \bar{\mathcal{E}}(t) = -\frac{1}{t-1}. \tag{4.11}$$

Coupling (3.12) and (4.5) gives

$$\tilde{y}_\pm(t) := \lim_{m\to\infty} \bar{y}_\pm(t) = \tilde{A}(t) + \tilde{C}(t) \pm 2\sqrt{\tilde{A}(t)\tilde{C}(t)}, \tag{4.12}$$

and from (3.14) and (4.9) it follows

$$\tilde{g}(\tilde{y}) := \lim_{m\to\infty} \bar{g}(\bar{y}) = \int_0^p \ln \tilde{T}(\tilde{y};t)dt. \tag{4.13}$$

A combination of (3.15)–(3.17), (4.5), (4.7), (4.9) and (4.11) yields

$$\begin{aligned}\tilde{r}(\tilde{y}) &:= \lim_{m\to\infty} \bar{r}(\bar{y}) \\ &= \frac{1}{2}\ln\frac{\tilde{y}-\tilde{A}(p)-\tilde{C}(p)+\tilde{S}(\tilde{y};p)}{\tilde{y}-\tilde{A}(0)-\tilde{C}(0)+\tilde{S}(\tilde{y};0)} + \int_0^p \frac{\tilde{G}(\tilde{y};t)}{2\tilde{S}(\tilde{y};t)^2}dt - \int_0^p \frac{2tq(1-q)}{\tilde{S}(\tilde{y};t)\tilde{T}(\tilde{y};t)}dt.\end{aligned} \tag{4.14}$$

The last two integrals can be evaluated explicitly. Indeed, on account of (4.5), (4.8) and (4.9), they are equal to

$$\begin{aligned}&\int_0^p \frac{\tilde{y}-2\tilde{y}q+q-t}{2[t^2-2(\tilde{y}-2\tilde{y}q+q)t+(\tilde{y}-q)^2]}dt + \int_0^p \frac{1-2q}{2\sqrt{[t-(\tilde{y}-2\tilde{y}q+q)]^2+4q(1-q)\tilde{y}(\tilde{y}-1)}}dt \\ &= -\frac{1}{4}\ln\frac{p^2-2(\tilde{y}-2\tilde{y}q+q)p+(\tilde{y}-q)^2}{(\tilde{y}-q)^2} \\ &\quad + \frac{1-2q}{2}\ln\frac{p-(\tilde{y}-2\tilde{y}q+q)+\sqrt{[p-(\tilde{y}-2\tilde{y}q+q)]^2+4q(1-q)\tilde{y}(\tilde{y}-1)}}{2q(\tilde{y}-1)}\end{aligned}$$

and

$$\begin{aligned}&\int_0^p \frac{\tilde{y}-q-t(1-2q)-\sqrt{[t-(\tilde{y}-2\tilde{y}q+q)]^2+4q(1-q)\tilde{y}(\tilde{y}-1)}}{2(1-t)\sqrt{[t-(\tilde{y}-2\tilde{y}q+q)]^2+4q(1-q)\tilde{y}(\tilde{y}-1)}}dt \\ &= \frac{1}{2}\ln\frac{\tilde{y}^2+q^2-\tilde{y}-q-(\tilde{y}-2\tilde{y}q+q-1)p+(\tilde{y}+q-1)\sqrt{[p-(\tilde{y}-2\tilde{y}q+q)]^2+4q(1-q)\tilde{y}(\tilde{y}-1)}}{2\tilde{y}(\tilde{y}-1)} \\ &\quad + \frac{1-2q}{2}\ln\frac{p-(\tilde{y}-2\tilde{y}q+q)+\sqrt{[p-(\tilde{y}-2\tilde{y}q+q)]^2+4q(1-q)\tilde{y}(\tilde{y}-1)}}{2q(\tilde{y}-1)}\end{aligned}$$

respectively, where we have also made use of the following integral formulas [5]

$$\int \frac{1}{\sqrt{u^2+a}}du = \ln(u+\sqrt{u^2+a}),$$

$$\int \frac{1}{(b-u)\sqrt{u^2+a}}du = \frac{1}{\sqrt{a+b^2}}\ln\frac{2[a+bu+\sqrt{(a+b^2)(a+u^2)}]}{b-u}.$$

Finally, we substitute the above explicit formulas for the two integrals into (4.14). Upon simplification, we obtain

$$\tilde{r}(\tilde{y}) = \frac{1}{2}\ln\frac{\tilde{y}-q-(2\tilde{y}-1)p+\sqrt{[p-(\tilde{y}-2\tilde{y}q+q)]^2+4q(1-q)\tilde{y}(\tilde{y}-1)}}{2(1-p)\sqrt{[p-(\tilde{y}-2\tilde{y}q+q)]^2+4q(1-q)\tilde{y}(\tilde{y}-1)}}. \tag{4.15}$$

In view of (3.13) and (4.4), we have the following results.

**Corollary 4.1.** *Let $\tilde{\pi}_n(x)$ be the monic Krawtchouk polynomials as defined in (4.3). Assume that $n/N = p$ is a fixed number in $(0,1)$. Let $\tilde{y}_\pm$ be defined as in (4.12). Assume $\tilde{y}_+(t)$ is increasing and $\tilde{y}_-(t)$ is decreasing for $t \in [0,p]$. As $n \to \infty$, we have*

$$\tilde{\pi}_n(N\tilde{y}) = \left(\frac{N}{2}\right)^n e^{N\tilde{g}(\tilde{y})+\tilde{r}(\tilde{y})}\left[1+O\left(\frac{1}{n}\right)\right] \tag{4.16}$$

*for $\tilde{y} \in \mathbf{C} \setminus [\tilde{y}_-(p), \tilde{y}_+(p)]$. Here, $\tilde{g}(\tilde{y})$ and $\tilde{r}(\tilde{y})$ are given by (4.13) and (4.15), respectively.*

The asymptotic formula of monic Krawtchouk polynomials $\tilde{\pi}_n(N\tilde{y})$ in the oscillatory interval can again be obtained by either the matching method or taking limit in the corresponding formula of Hahn polynomials.

**Corollary 4.2.** *Assume that $n/N = p$ is a fixed number in $(0,1)$. Let $\tilde{y}_\pm(t)$ be defined as in (4.12). Assume $\tilde{y}_+(t)$ is increasing and $\tilde{y}_-(t)$ is decreasing for $t \in [0,p]$. Let $\tilde{S}^\pm(\tilde{y};t)$ and $\tilde{T}^\pm(\tilde{y};t)$ be defined similarly as in (2.23) and (2.25). As $n \to \infty$, we have*

$$\tilde{\pi}_n(N\tilde{y}) = \left(\frac{N}{2}\right)^n \left\{e^{N\tilde{g}^+(\tilde{y})+\tilde{r}^+(\tilde{y})}\left[1+O\left(\frac{1}{n}\right)\right] + e^{N\tilde{g}^-(\tilde{y})+\tilde{r}^-(\tilde{y})}\left[1+O\left(\frac{1}{n}\right)\right]\right\} \tag{4.17}$$

*for $\tilde{y} \in (\tilde{y}_-(p), \tilde{A}(0)) \cup (\tilde{A}(0), \tilde{y}_+(p))$. Here,*

$$\tilde{g}^\pm(\tilde{y}) = \int_0^p \ln \tilde{T}^\pm(\tilde{y};t)dt, \tag{4.18}$$

*and*

$$\tilde{r}^\pm(\tilde{y}) = \frac{1}{2}\ln\frac{\tilde{y} - q - (2\tilde{y}-1)p + \tilde{S}^\pm(\tilde{y};p)}{2(1-p)\tilde{S}^\pm(\tilde{y};p)}. \tag{4.19}$$

**Remark 4.3.** Our assumption that $\tilde{y}_+(t)$ is increasing and $\tilde{y}_-(t)$ is decreasing for $t \in [0,p]$ corresponds to the void-band-void (VBV) case considered in [4]. We have verified that the results in this section agree with those in [4] and [6], although the work is very involved.

## 5. Racah polynomials with fixed variable

In this section, we shall provide a simple formula for the Racah polynomials with fixed variable $z$. For simplicity, we assume $A(t) > C(t)$ for all $t \in [0,p]$. From (2.3), we have

$$\sqrt{y_\pm(t)} = \sqrt{A(t)} \pm \sqrt{C(t)}.$$

Since $0 < y_-(t) < y_+(t)$, we obtain from (2.11) and the above equation

$$S(0;t) = -\sqrt{y_-(t)}\sqrt{y_+(t)} = -[A(t) - C(t)] < 0.$$

Here, we remark that the square roots $\sqrt{y - y_-(t)}$ and $\sqrt{y - y_+(t)}$ in (2.11) are taking their principal values; thus $S(y;t) < 0$ for all real $y < y_-(t) < y_+(t)$. Consequently, we have

$$S(z/N^2;t) = S(0;t) + O(1/N^2) = C(t) - A(t) + O(1/N^2). \tag{5.1}$$

Coupling the above equation with (2.12) gives

$$T(z/N^2;t) = -2A(t) + O(1/N^2). \tag{5.2}$$

In view of $n = Np$, we substitute (5.2) into (2.19) to obtain

$$\exp[Ng(z/N^2)] = (-2)^n \exp\left[N\int_0^p \ln A(t)dt\right][1+O(1/N)]. \tag{5.3}$$

Next, we intend to estimate the correction function $r(z/N^2)$. Recall from (2.20) that

$$r(z/N^2) = \frac{1}{2}\ln\frac{T(z/N^2;p)}{T(z/N^2;0)} + \Omega_0(z/N^2) + \Omega_1(z/N^2), \tag{5.4}$$

where $\Omega_0$ and $\Omega_1$ are the two integrals defined in (2.17) and (2.14), respectively; namely,

$$\Omega_0(z/N^2) = \int_0^p \left[\frac{A(t)-C(t)}{\mathcal{D}(t)S(z/N^2;t)} - \frac{2A(t)C(t)\mathcal{E}(t)}{S(z/N^2;t)T(z/N^2;t)}\right]dt, \tag{5.5}$$

$$\Omega_1(z/N^2) = \int_0^p \frac{G(z/N^2;t)}{2S(z/N^2;t)^2}dt. \tag{5.6}$$

Substituting (5.1) and (5.2) into (5.5) yields

$$\Omega_0(z/N^2) = \int_0^p \left[\frac{-1}{\mathcal{D}(t)} + \frac{C(t)\mathcal{E}(t)}{C(t)-A(t)}\right]dt + O(1/N^2), \tag{5.7}$$

and substituting (5.2) into (2.13) gives

$$\begin{aligned} G(z/N^2;t) &= -2A(t)[A^*(t)+C^*(t)] + 2[A(t)C^*(t)+A^*(t)C(t)] + O(1/N^2) \\ &= 2A^*(t)[C(t)-A(t)] + O(1/N^2). \end{aligned} \tag{5.8}$$

From (5.1), (5.6) and (5.8), it then follows that

$$\Omega_1(z/N^2) = \int_0^p \frac{A^*(t)}{C(t)-A(t)}dt + O(1/N^2) = \int_0^p \frac{-A(t)\mathcal{E}(t)}{C(t)-A(t)}dt + O(1/N^2), \tag{5.9}$$

where we have used the identity $A^*(t) = -A(t)\mathcal{E}(t)$, which can be obtained via (2.2), (2.8), (2.9) and (2.16). Coupling (5.7) and (5.9), we obtain

$$\begin{aligned} &\Omega_0(z/N^2) + \Omega_1(z/N^2) \\ &= \int_0^p \left[\frac{-1}{\mathcal{D}(t)} + \mathcal{E}(t)\right]dt + O(1/N^2) \\ &= \int_0^p \left[\frac{3}{2t+a+b} - \frac{1}{t+a} - \frac{1}{t+a+b} - \frac{1}{t+b+d} - \frac{1}{t+c}\right]dt + O(1/N^2) \\ &= \frac{3}{2}\ln\frac{2p+a+b}{a+b} + \ln\frac{ac(a+b)(b+d)}{(p+a)(p+c)(p+a+b)(p+b+d)} + O(1/N^2). \end{aligned} \tag{5.10}$$

Here, we have made use of the definitions of $\mathcal{D}(t)$ and $\mathcal{E}(t)$ given in (2.9) and (2.16), respectively. On the other hand, we have from (2.2) and (5.2)

$$\ln\frac{T(z/N^2;p)}{T(z/N^2;0)} = \ln\frac{(p+a)(p+c)(p+a+b)(p+b+d)(a+b)}{ac(b+d)(2p+a+b)^2} + O(1/N^2). \tag{5.11}$$

Applying (5.10) and (5.11) to (5.4) yields

$$r(z/N^2) = \frac{1}{2}\ln\frac{ac(b+d)(2p+a+b)}{(p+a)(p+c)(p+b+d)(p+a+b)} + O(1/N^2). \tag{5.12}$$

By substituting (5.3) and (5.12) into (2.18), we obtain the following result.

**Corollary 5.1.** *Let $\pi_n(z)$ be the monic Racah polynomials satisfying the recurrence relation* (1.3). *Assume $n/N = p$ is fixed in* $(0,1)$. *Let $A(t)$, $C(t)$ and $y_\pm(t)$ be defined as in* (2.2) *and* (2.3). *We further assume that $A(t) > C(t)$, $y_+(t)$ is increasing and $y_-(t)$ is decreasing for $t \in [0,p]$. Then as $n \to \infty$, we have for any fixed $z$,*

$$\pi_n(z) = (-N^2)^n \exp\left[N\int_0^p \ln A(t)dt\right]\sqrt{\frac{ac(b+d)(2p+a+b)}{(p+a)(p+c)(p+b+d)(p+a+b)}} \times [1+O(1/N)]. \tag{5.13}$$

**Acknowledgments**

We would like to thank an anonymous referee for checking the derivations of our main theorems and for providing valuable suggestions which have helped to improve the presentation of this paper. XSW is partially supported by the GRFC (grant number 103417) and Summer Research Grant from Southeast Missouri State University.

**References**

[1] R. Askey, J. Wilson, A set of orthogonal polynomials that generalize the Racah coefficients or 6-j symbols, SIAM J. Math. Anal. 10 (1979) 1008–1016.

[2] L.-C. Chen, M.E.H. Ismail, P. Simeonov, Asymptotics of Racah coefficients and polynomials, J. Phys. A 32 (3) (1999) 537–553.

[3] T.S. Chihara, An Introduction to Orthogonal Polynomials, Math. Appl., vol. 13, Gordon and Breach Science Publishers, New York, London, Paris, 1978.

[4] D. Dai, R. Wong, Global asymptotics of Krawtchouk polynomials – a Riemann–Hilbert approach, Chin. Ann. Math. Ser. B 28 (1) (2007) 1–34.

[5] I.S. Gradshteyn, I.M. Ryzhik, Table of Integrals, Series, and Products, sixth edition, Academic Press, Inc., San Diego, CA, 2000.

[6] M.E.H. Ismail, P. Simeonov, Strong asymptotics for Krawtchouk polynomials, J. Comput. Appl. Math. 100 (1998) 121–144.

[7] R. Koekoek, P.A. Lesky, R.F. Swarttouw, Hypergeometric Orthogonal Polynomials and Their q-Analogues, Springer Monogr. Math., Springer-Verlag, Berlin, 2010.

[8] X.-S. Wang, Plancherel–Rotach asymptotics of second-order difference equations with linear coefficients, J. Approx. Theory 188 (2014) 1–18.

[9] Z. Wang, R. Wong, Asymptotic expansions for second-order linear difference equations with a turning point, Numer. Math. 94 (1) (2003) 147–194.

[10] X.-S. Wang, R. Wong, Asymptotics of orthogonal polynomials via recurrence relations, Anal. Appl. (Singap.) 10 (2012) 215–235.

[11] J.A. Wilson, Hypergeometric series recurrence relations and some new orthogonal functions, PhD thesis, Univ. Wisconsin, Madison, 1978.

PROCEEDINGS OF THE
AMERICAN MATHEMATICAL SOCIETY
Volume 144, Number 10, October 2016, Pages 4367–4380
http://dx.doi.org/10.1090/proc/13191
Article electronically published on June 3, 2016

# SPECIAL FUNCTIONS, INTEGRAL EQUATIONS AND A RIEMANN-HILBERT PROBLEM

R. WONG AND YU-QIU ZHAO

(Communicated by Walter Van Assche)

ABSTRACT. We consider a pair of special functions, $u_\beta$ and $v_\beta$, defined respectively as the solutions to the integral equations

$$u(x) = 1 + \int_0^\infty \frac{K(t)u(t)dt}{t+x} \text{ and } v(x) = 1 - \int_0^\infty \frac{K(t)v(t)dt}{t+x}, \ x \in [0, \infty),$$

where $K(t) = \frac{1}{\pi}\exp\left(-t^\beta \sin\frac{\pi\beta}{2}\right)\sin\left(t^\beta \cos\frac{\pi\beta}{2}\right)$ for $\beta \in (0,1)$. In this note, we establish the existence and uniqueness of $u_\beta$ and $v_\beta$ which are bounded and continuous in $[0, +\infty)$. Also, we show that a solution to a model Riemann-Hilbert problem in Kriecherbauer and McLaughlin [*Int. Math. Res. Not.*, 1999] can be constructed explicitly in terms of these functions. A preliminary asymptotic study is carried out on the Stokes phenomena of these functions by making use of their connection formulas.

Several open questions are also proposed for a thorough investigation of the analytic and asymptotic properties of the functions $u_\beta$ and $v_\beta$, and a related new special function $G_\beta$.

## 1. INTRODUCTION

Kriecherbauer and McLaughlin [8] used the Riemann-Hilbert approach to study the strong asymptotics of the polynomials orthogonal with respect to the Freud weight

$$w_\beta(x)dx = e^{-\kappa_\beta |x|^\beta}dx, \ x \in \mathbb{R}, \ \kappa_\beta = \frac{\Gamma(\frac{\beta}{2})\Gamma(\frac{1}{2})}{\Gamma(\frac{\beta+1}{2})}, \ \beta > 0. \tag{1}$$

When $0 < \beta < 1$, several representative results are obtained in terms of a model Riemann-Hilbert problem (RHP) for a certain $2\times 2$ matrix-valued function $L$:

$$L : \mathbb{C}\backslash\mathbb{R} \to \mathbb{C}^{2\times 2} \text{ is analytic,}$$

$$L_+(s) = L_-(s)v_L(s) \text{ for } s \in \mathbb{R}, \tag{2}$$

$$L(s) = O(\ln|s|) \text{ as } s \to 0, \tag{3}$$

$$L(s) \to I \text{ for } s \to \infty, \tag{4}$$

Received by the editors September 5, 2015 and, in revised form, December 17, 2015.
2010 *Mathematics Subject Classification.* Primary 33E30, 41A60, 45A05.
*Key words and phrases.* Special function, Freud weight, integral equation, Riemann-Hilbert problem, asymptotics, Stokes phenomenon.

where the jump matrix is given by

$$v_L(s) = \begin{pmatrix} 1 & -\eta_L(s) \\ \eta_L(-s) & 1 \end{pmatrix}, \quad s \in \mathbb{R},$$

with

$$\eta_L(s) = 2i(-1)^{n+1} e^{-|s|^\beta \sin(\pi\beta/2)} \sin\left(|s|^\beta \cos\frac{\pi\beta}{2}\right) \mathbf{1}_{[0,\infty)}(s), \quad s \in \mathbb{R},$$

and $n$ being the polynomial degree; see [8, (6.41)-(6.43)].

Later in this note, we will give a more natural description of the behavior of $L(s)$ at the origin. Instead of (3), we henceforth assume

$$L(s) = O\left(\epsilon_\beta(s)\right) \text{ as } s \to 0, \ \epsilon_\beta(s) = \begin{cases} 1, & 1/2 < \beta < 1, \\ \ln|s|, & \beta = 1/2, \\ |s|^{\beta-1/2}, & 0 < \beta < 1/2. \end{cases} \tag{5}$$

*Remark* 1. The uniqueness of the solution to the RHP for $L$, subject to either (3) or (5), can be justified by using Liouville's theorem. Now let $L_3(s)$ be the solution to the RHP with local behavior (3), and $L_5(s)$ be that with local behavior (5); then $L_3(s)L_5^{-1}(s)$ has no jump on the real line $\mathbb{R}$, and has only an isolated singularity at the origin. Since the singularity at the origin is not a pole or an essential singularity, it must be removable. In addition, since $L_3(s)L_5^{-1}(s)$ approaches $I$ as $s \to \infty$, by Liouville's theorem this matrix-valued entire function is the identity matrix. Thus $L_3(s) \equiv L_5(s)$ if both solutions exist. In view of Lemma 2 below, the local behavior in (5) is in a sense more natural. Later, in Section 4, we will see from Theorem 3 that we actually have $L(s) = O(1)$ as $s \to 0$ for the parametrix constructed in (13). Accordingly, both (3) and (5) are fulfilled.

In [8], the existence and uniqueness of $L$ were established in the Wiener class. For $0 < \beta < 1$, the leading order behavior of the solution to the model problem at the origin has not been determined explicitly; see the comment in Deift et al. [5, p. 62, Remark 3.6]. For a brief historical account of the asymptotics of the orthogonal polynomials associated with the Freud weight, the reader is referred to [8] and the references therein.

Recently, it was found there are other orthogonal polynomials with weights having similar singularities as the Freud weight. For example, Chen and Ismail [2] considered the Freud-like orthogonal polynomials arising from a recurrence relation. The polynomials are related to the indeterminate moment problems. The weight function, supported on $\mathbb{R}$, has a singularity at 0 which is similar to the Freud weight (1); see also [3, (1.6)] for the weight, and for a difference equation approach to obtain the Plancherel-Rotach asymptotics.

In [4], Deaño, Kuijlaars and Román consider the zero distribution of polynomials orthogonal with respect to the Bessel function $J_\nu(x)$. They encountered a varying exponential weight of Freud-type $e^{n\pi|x|}$ with a potential function $\pi|x|$. To justify the existence of a certain local parametrix, a quite unnatural restriction on $\nu$ has to be brought in. Again, the leading order behavior of the parametrix has not been explicitly given.

To solve those parametrices and model problems, one may either represent them in terms of known special functions, or alternatively use new special functions to serve the same purpose.

The objective of the present note is to introduce a pair of special functions, $u_\beta$ and $v_\beta$, determined by two linear integral equations. Based on these functions, the solution to the model problem of Kriecherbauer and McLaughlin [8] can readily be constructed.

In the rest of this note, we will address the unique solvability of the integral equations in $C[0,\infty)$, derive the behavior of the specific solutions $u_\beta$ and $v_\beta$ at infinity and at the origin, and carry out a brief discussion of the Stokes phenomena of these functions, taking the integral equations as resurgent relations. We will also propose a thorough investigation of the analytic and asymptotic properties of the new special functions $u_\beta$ and $v_\beta$, and a relevant new special function $G_\beta$.

## 2. Integral equations and the RH problem

First, let us derive the integral equations from the model problem for $L$ formulated in Section 1. From the jump condition (2) and the behavior (4)-(5), it is readily seen that the (1,1)-entry $L_{11}(s)$ is analytic in $\mathbb{C}\setminus(-\infty,0]$, such that

$$\begin{aligned}&(L_{11})_+(s)-(L_{11})_-(s)=\eta_L(-s)L_{12}(s),\ s\in(-\infty,0),\\&L_{11}(s)=1+o(1),\ s\to\infty,\\&L_{11}(s)=O(\epsilon_\beta(s)),\ s\to 0.\end{aligned}$$

Hence, in view of the behavior of $L_{12}$; cf. (4)-(5), we have

$$L_{11}(s)=1+\frac{1}{2\pi i}\int_{-\infty}^{0}\frac{\eta_L(-\tau)L_{12}(\tau)d\tau}{\tau-s}=1+\frac{1}{2\pi i}\int_{0}^{\infty}\frac{\eta_L(\tau)L_{12}(-\tau)d\tau}{-\tau-s}\tag{6}$$

for $s\in\mathbb{C}\setminus(-\infty,0]$, especially for $s\in(0,\infty)$. Similarly, it is also seen from (2) that $L_{12}(s)$ is analytic in $\mathbb{C}\setminus[0,\infty)$, and solves the scalar RH problem

$$\begin{aligned}&(L_{12})_+(s)-(L_{12})_-(s)=-\eta_L(s)L_{11}(s),\ s\in(0,\infty),\\&L_{12}(s)=o(1),\ s\to\infty,\\&L_{12}(s)=O(\epsilon_\beta(s)),\ s\to 0.\end{aligned}$$

Hence, we have

$$L_{12}(s)=\frac{1}{2\pi i}\int_{0}^{\infty}\frac{-\eta_L(\tau)L_{11}(\tau)d\tau}{\tau-s}\quad\text{for}\quad s\in\mathbb{C}\setminus[0,\infty),\tag{7}$$

especially for $s\in(-\infty,0)$. For $x\in(0,+\infty)$, now we define

$$u_\beta(x)=L_{11}(x)+(-1)^nL_{12}(-x),\ v_\beta(x)=L_{11}(x)-(-1)^nL_{12}(-x).\tag{8}$$

From (6)-(7) and (8), it is readily verified that $u_\beta$ and $v_\beta$ solve, respectively, the following integral equations:

$$u(x)=1+\int_0^\infty\frac{K(t)u(t)dt}{t+x},\ x\in(0,\infty),\tag{9}$$

$$v(x)=1-\int_0^\infty\frac{K(t)v(t)dt}{t+x},\ x\in(0,\infty),\tag{10}$$

where

$$K(t) = \frac{1}{2\pi i}\left[\exp\left(e^{(\frac{\pi}{2}+\frac{\beta\pi}{2})i}t^{\beta}\right) - \exp\left(e^{-(\frac{\pi}{2}+\frac{\beta\pi}{2})i}t^{\beta}\right)\right], \tag{11}$$

$$= \frac{1}{\pi}\exp\left(-t^{\beta}\sin\frac{\pi\beta}{2}\right)\sin\left(t^{\beta}\cos\frac{\pi\beta}{2}\right), \tag{12}$$

and $\eta_L(t) = 2\pi i(-1)^{n+1}K(t)$ for $t \in [0, +\infty)$.

It is worth noting that similar coupled scalar integral equations have been derived in [4], though the equations in that paper were used only to construct a contraction mapping. Also, it is mentioned in Fokas et al. [6, p. 161] that a function $u(x)$ can be parametrized via the solution of a linear integral equation. This is exactly what we are doing: We are defining a pair of new special functions, using the above integral equations, to construct a parametrix.

Conversely, if $u_\beta(x)$ and $v_\beta(x)$ do solve the integral equations, with $u_\beta - 1$, $v_\beta - 1 \in L_2[0, \infty)$, then we can deduce that $u_\beta(z)$, $v_\beta(z) = 1 + O(1/z)$ as $z \to \infty$ for $|\arg z| < 3\pi/2$ and are of the order $O(\epsilon_\beta(z))$ as $z \to 0$ for $|\arg z| \le \pi$; see the discussion in the next section, and especially Lemma 2. Thus, we have

**Theorem 1.** *The piece-wise analytic function*

$$L(s) = \begin{cases} \begin{pmatrix} L_\beta(s) & (-1)^n U_\beta(se^{-\pi i}) \\ (-1)^n U_\beta(s) & L_\beta(se^{-\pi i}) \end{pmatrix}, & 0 < \arg s < \pi; \\ \\ \begin{pmatrix} L_\beta(s) & (-1)^n U_\beta(se^{\pi i}) \\ (-1)^n U_\beta(s) & L_\beta(se^{\pi i}) \end{pmatrix}, & -\pi < \arg s < 0 \end{cases} \tag{13}$$

*solves the RH problem* (2), (4) *and* (5), *where*

(14)
$$L_\beta(z) = \frac{1}{2}\left(u_\beta(z) + v_\beta(z)\right), \; U_\beta(z) = \frac{1}{2}\left(u_\beta(z) - v_\beta(z)\right), \; \arg z \in (-\infty, \infty), \; z \neq 0.$$

## 3. Unique solvability of the integral equations

In [8], the unique solvability of the model problem for $L$ is justified in the Wiener class; i.e., the Fourier transform of each entry is in $L_1(\mathbb{R})$.

We note that in view of (14) the unique solvability of the integral equations (9)-(10) can accordingly be deduced from that of $L$.

Now we take an alternative approach, using the integral equations alone. We first show that both $u_\beta(x) - 1$ and $v_\beta(x) - 1$ are in $L_2[0, +\infty)$. We begin with the following lemma:

**Lemma 1.** *The operator $T$ is a compact operator on $L_2[0, +\infty)$, where*

$$(Tu)(x) = \int_0^\infty \frac{K(t)u(t)dt}{t+x}. \tag{15}$$

*Proof.* The lemma is easily justified by noting that $\frac{K(t)}{t+x} \in L_2([0, +\infty) \times [0, +\infty))$. Thus, $T$ is a Hilbert-Schmidt operator, and hence is compact; cf. e.g. [16, p. 277]. □

The next step is to show that the operator $I-T$ has trivial null space. To this aim, we need to know more about the analytic structure of the solutions to (9) and (10). Indeed, if $u(x)$ solves (9), then we see from equation (9) that $u$ can be extended analytically to $\mathbb{C}\setminus(-\infty,0]$, and that $u(z)=1+O(1/z)$ in subsectors. From (11), we see that the kernel $K(t)$ can be extended analytically to complex $t$ with $\arg t\in(-\infty,\infty)$, and is exponentially small for $|\arg t|<\pi/2$. Using Cauchy's theorem, it can be shown that

$$u(z)=1+\int_0^\infty \frac{K(te^{i\theta})u(te^{i\theta})dt}{t+e^{-i\theta}z},\quad |\theta|<\frac{\pi}{2}, \tag{16}$$

which enables us to extend $u(z)$ to the larger sector $|\arg z|<3\pi/2$, and

$$u(z)=1+O(1/z),\ \text{as } z\to\infty,\ |\arg z|<3\pi/2. \tag{17}$$

Similar results can be obtained for $v(z)$. This procedure can be continued, since by using the Plemelj formula we have from (9)-(10)

$$u(ze^{-\pi i})-u(ze^{\pi i})=2\pi iK(z)u(z),\ v(ze^{-\pi i})-v(ze^{\pi i})=-2\pi iK(z)v(z), \tag{18}$$

initially for real and positive $z$, and then for complex $z$. Appealing to the formulas in (18), we see that the above process of analytic continuation can keep on going until we have the results valid for $\arg z\in\mathbb{R}$. For later use, we need the following behavior of $(Tu)(z)$ at the origin:

**Lemma 2.** *For $u\in L_2[0,+\infty)$, we have $(Tu)(z)=O(\epsilon_\beta(z))$ for $|\arg z|\le\pi$ as $z\to 0$; cf. (5), where* $\epsilon_\beta(z)=\begin{cases}1, & 1/2<\beta<1,\\ \ln|z|, & \beta=1/2,\\ |z|^{\beta-1/2}, & 0<\beta<1/2.\end{cases}$

*Proof.* Recall the function $K(t)$ in (12). For real and positive $z$, we have by using Hölder's inequality and splitting the interval of integration

$$|(Tu)(z)|^2\le\|u\|_2^2\int_0^\infty\frac{K(t)^2dt}{(t+z)^2}=O\left(\int_0^1\frac{t^{2\beta}dt}{(t+z)^2}\right)+O(1)=O(\epsilon_\beta(z))$$

as $z\to 0$. The last equality can be obtained by calculating a residue. With slight modifications, and taking into account the rotation formula (16), we can extend the above inequality to $|\arg z|\le\pi$. □

We show that the operator $I-T$ has trivial null space, by using a vanishing lemma technique. A similar argument can be found in, e.g., [8].

**Lemma 3.** *Assume that $u\in L_2[0,+\infty)$ and $u-Tu=0$ in $L_2[0,+\infty)$. Then it holds $u\equiv 0$ for $x\in[0,\infty)$ in $L_2[0,+\infty)$.*

*Proof.* Indeed, from $u=Tu$ we see that the above analytic continuation procedure works also for the present $u(z)$, the connection formula (18) still holds, and instead of (17) we have $u(z)=O(1/z)$ as $z\to\infty$ for $|\arg z|\le\pi$. We introduce an auxiliary vector function

$$A(z)=\begin{cases}\left(\ u(z),\ \ u(ze^{-\pi i})\ \right), & \arg z\in(0,\pi),\\ \left(\ u(z),\ \ u(ze^{\pi i})\ \right), & \arg z\in(-\pi,0).\end{cases}$$

It is readily seen that $A(z)$ is analytic in both the upper and lower half plane, behaves uniformly as $O(1/z)$ at infinity, and has jumps along the real axis as

$$A_+(x) = A_-(x)J_A(x), \; J_A(x) = \begin{cases} \begin{pmatrix} 1 & 2\pi i K(x) \\ 0 & 1 \end{pmatrix} & \text{for } x \in (0, +\infty), \\ \begin{pmatrix} 1 & 0 \\ -2\pi i K(|x|) & 1 \end{pmatrix} & \text{for } x \in (-\infty, 0); \end{cases}$$

cf. (18).

Denote by $A^*$ the conjugate transpose of $A$. Since $A_+A_-^*$ has an analytic continuation to the upper half plane, and $A_-A_+^*$ to the lower half, in view of the behavior $A(z) = O(1/z)$ at infinity, by the Cauchy integral theorem we have

$$\int_{\mathbb{R}} A_-(x)J_A(x)A_-^*(x)dx = \int_{\mathbb{R}} A_+(x)A_-^*(x)dx = 0,$$

$$\int_{\mathbb{R}} A_-(x)J_A^*(x)A_-^*(x)dx = \int_{\mathbb{R}} A_-(x)A_+^*(x)dx = 0.$$

Here, use has also been made of the fact that $u(z) = O(\epsilon_\beta(z))$ for small $z$ with $|\arg z| \le \pi$; see Lemma 2. Summing up the integrals gives

$$\int_{\mathbb{R}} A_-(x)\left(J_A(x) + J_A^*(x)\right) A_-^*(x)dx = 0. \tag{19}$$

It can be seen from (12) that $|\pi i K(|x|)| < 1$ for $x \in \mathbb{R}$. Hence,

$$J_A(x) + J_A^*(x) = \begin{pmatrix} 2 & 2\pi i K(|x|) \\ -2\pi i K(|x|) & 2 \end{pmatrix}$$

is positive definite for $x \in \mathbb{R}$. From (19) one deduces that $A_-(x) \equiv 0$, and thus $u(x) \equiv 0$ for $x \in (0, +\infty)$. □

We note that $\int_0^\infty \frac{K(t)dt}{t+x} \in L_2[0, +\infty)$ since it is smooth on $(0, \infty)$, bounded at $0^+$, and of order $O(1/x)$ at $+\infty$. Hence, a combination of Lemmas 1 and 3 gives the unique solvability of the following integral equation:

$$u_1(x) - Tu_1(x) = \int_0^\infty \frac{K(t)dt}{t+x}$$

in the space $L_2[0, +\infty)$. Similar analysis can be carried out for the other equation (10). To conclude, we have the following:

**Theorem 2.** *There exist unique solutions $u(x)$ and $v(x)$ to the integral equations (9) and (10), respectively, such that $u(x) - 1 \in L_2[0, +\infty)$ and $v(x) - 1 \in L_2[0, +\infty)$.*

## 4. Boundedness of $u_\beta(x)$ and $v_\beta(x)$ on $[0, \infty)$

The main result in the last section is that $u_\beta(x) - 1 \in L_2[0, +\infty)$. By using successive approximation, one can actually show that $u_\beta(x)$ is continuous and bounded for $\beta > \beta_c$ :=0.4158853544. Indeed, it suffices to show that the operator $T$ defined in (15) is a contraction on the space $L_\infty$ for those $\beta$. Straightforward estimation gives

$$|(Tu)(x)| \le M_\beta \max_{[0,+\infty)} |u(t)| \text{ for } x \in [0, +\infty), \text{ with } M_\beta = \int_0^\infty \frac{|K(t)|dt}{t} \le \frac{\cot(\frac{\pi\beta}{2})}{\pi\beta}.$$

The bound $\cot(\frac{\pi\beta}{2})/\pi\beta$ is monotonically decreasing for $\beta \in (0,1)$, and is less than 1 for $\beta > \beta_c$.

In this section, we shall show that $u_\beta(x)$ and $v_\beta(x)$ are in fact bounded for all $\beta \in (0,1)$.

**Theorem 3.** *Let $u_\beta(x)$ and $v_\beta(x)$ be solutions of the integral equations in* (9) *and* (10), *respectively. Then, $u_\beta(x)$, $v_\beta(x) \in L_\infty[0,\infty)$ and $u_\beta(x)$, $v_\beta(x) \in C[0,\infty)$.*

*Proof.* Let $\tilde{u}(x) = u_\beta(x) - 1$. By Theorem 2, $\tilde{u}(x) \in L_2[0,\infty)$. Rewrite equation (9) as

$$\tilde{u}(x) = \int_0^\infty \frac{K(t)\tilde{u}(t)dt}{t+x} + u_0(x),\ x \in [0,\infty),$$

where

$$u_0(x) = \int_0^\infty \frac{K(t)}{t+x}dt.$$

Clearly, $u_0(x) \in L_2[0,\infty) \cap L_\infty[0,\infty)$. The function $\tilde{u}(x)$ is smooth in $(0,\infty)$, since it is an analytic function in the cut plane $\mathbb{C} \setminus (-\infty,0]$. To prove that $\tilde{u}(x)$ is also bounded there, we divide our discussion into several cases. In general, for all $\beta \in (0,1)$, we have

$$\begin{aligned}|\tilde{u}(x)| \le |u_0(x)| + \int_0^\infty \frac{|K(t)|\,|\tilde{u}(t)|dt}{t+x} &\le \frac{1}{x}\int_0^\infty |K(t)|(1+|\tilde{u}(t)|)dt \\ &\le \frac{1}{x}\left(\|K\|_1 + \|K\|_2\|\tilde{u}\|_2\right)\end{aligned}$$

for all positive $x$, and especially for $x \ge 1$. We may focus on $0 < x \le 1$.

*Case* (i). $\frac{1}{2} < \beta < 1$. For $0 < x \le 1$, we have

$$|\tilde{u}(x)| \le a + \int_0^\infty \frac{|K(t)|\,|\tilde{u}(t)|}{t+x}dt \le a + \int_0^\infty \frac{|K(t)|}{t}|\tilde{u}(t)|dt \le a + \left\|\frac{K(t)}{t}\right\|_2 \|\tilde{u}\|_2,$$

where $a = \|u_0\|_\infty$. Thus, $\tilde{u}(x)$ is bounded. Here, we have used the fact that $\frac{K(t)}{t} \in L_2[0,\infty)$.

*Case* (ii). $\frac{1}{4} < \beta \le \frac{1}{2}$. In this case, we write $\beta = \frac{1}{4} + \varepsilon$ with $\varepsilon \in (0, \frac{1}{4}]$. We first recall the inequality [10, p. 30]

$$x^{\frac{1}{p}} y^{\frac{1}{q}} \le \frac{x}{p} + \frac{y}{q} \le x + y, \tag{20}$$

where $x \ge 0$, $y \ge 0$ and $1/p + 1/q = 1$ with $p > 1$. From this inequality, we have

$$t + x \ge t^{\beta+\frac{1}{2}-\frac{\varepsilon}{2}} x^{\frac{1}{2}-\beta+\frac{\varepsilon}{2}}. \tag{21}$$

Note that $\beta + \frac{1}{2} - \frac{\varepsilon}{2} = \frac{3}{4} + \frac{\varepsilon}{2} > 0$ and $\frac{1}{2} - \beta + \frac{\varepsilon}{2} \ge \frac{\varepsilon}{2} > 0$. Consequently, we have

$$\begin{aligned}|\tilde{u}(x)| \le a + \int_0^\infty \frac{|K(t)|\,|\tilde{u}(t)|}{t+x}dt &\le a + \left\|\frac{K(t)}{t^{\beta+\frac{1}{2}-\frac{\varepsilon}{2}}}\right\|_2 \|\tilde{u}\|_2 \frac{1}{x^{\frac{1}{2}-\beta+\frac{\varepsilon}{2}}} \\ &\le \frac{c}{x^{\frac{1}{2}-\beta+\frac{\varepsilon}{2}}},\ x \in (0,1],\end{aligned} \tag{22}$$

where $c$ is a generic positive constant. The fact that $K(t)/t^{\beta+\frac{1}{2}-\frac{\varepsilon}{2}} \in L_2[0,\infty)$ is readily justified by its behavior $O(t^{\frac{\varepsilon}{2}-\frac{1}{2}})$ at $t = 0^+$, and its exponential decay at $t = +\infty$. This estimate can be used to self-improve and achieve

$$|\tilde{u}(x)| \le a + \int_0^\infty \frac{|K(t)|\,|\tilde{u}(t)|}{t}dt \le a + \left\|\frac{K(t)}{t^{1-\beta+\varepsilon}}\right\|_2 \|t^{-\beta+\varepsilon}\tilde{u}(t)\|_2$$

for $x \in (0,\infty)$. To see that the two quantities in the last term of the above inequality are finite, we just check their behavior as $t \to 0^+$ and as $t \to +\infty$. For instance, as $t \to 0^+$, we have $K(t)/t^{1-\beta+\varepsilon} = O(t^{2\beta-1-\varepsilon}) = O(t^{\varepsilon-\frac{1}{2}})$ since $\beta = \frac{1}{4}+\varepsilon$ and $t^{-\beta+\varepsilon}\tilde{u}(t) = O(t^{\frac{\varepsilon}{2}-\frac{1}{2}})$ by the estimate in (22).

*Case* (iii). $\frac{1}{2(k+1)} < \beta \le \frac{1}{2k}$, $k = 2,3,\cdots$. The analysis is very much the same as that given for the previous case, except that we need to repeat the self-improvement argument several times. First, let us write $\beta = \frac{1}{2(k+1)} + \varepsilon$ with $\varepsilon \in (0, \frac{1}{2k(k+1)}]$. Straightforward calculation shows

$$k\beta + \frac{1}{2} - \frac{(2k-1)\varepsilon}{2} > \cdots > 2\beta + \frac{1}{2} - \frac{3\varepsilon}{2} > \beta + \frac{1}{2} - \frac{\varepsilon}{2} = \frac{1}{2(k+1)} + \frac{1}{2} + \frac{\varepsilon}{2} > 0,$$

and

$$\frac{1}{2} - \beta + \frac{\varepsilon}{2} > \frac{1}{2} - 2\beta + \frac{3\varepsilon}{2} > \cdots > \frac{1}{2} - k\beta + \frac{(2k-1)\varepsilon}{2} = \frac{1}{2(k+1)} - \frac{\varepsilon}{2} \ge \frac{2k-1}{4k(k+1)} > 0.$$

Hence, for $x \in (0,1]$, repeated application of the inequality in (20) and the self-improvement steps in (21) and (22) lead to

$$\begin{cases} t + x \ge t^{\beta+\frac{1}{2}-\frac{\varepsilon}{2}} x^{\frac{1}{2}-\beta+\frac{\varepsilon}{2}} \text{ and} \\ |\tilde{u}(x)| \le a + \left\| \frac{K(t)}{t^{\beta+\frac{1}{2}-\frac{\varepsilon}{2}}} \right\|_2 \|\tilde{u}(t)\|_2 \frac{1}{x^{\frac{1}{2}-\beta+\frac{\varepsilon}{2}}} \le \frac{c}{x^{\frac{1}{2}-\beta+\frac{\varepsilon}{2}}}, \end{cases}$$

$$\begin{cases} t + x \ge t^{2\beta+\frac{1}{2}-\frac{3\varepsilon}{2}} x^{\frac{1}{2}-2\beta+\frac{3\varepsilon}{2}} \text{ and} \\ |\tilde{u}(x)| \le a + \left\| \frac{K(t)}{t^{\beta+\frac{1}{2}-\frac{\varepsilon}{2}}} \right\|_2 \left\| t^{-\beta+\varepsilon}\tilde{u}(t) \right\|_2 \frac{1}{x^{\frac{1}{2}-2\beta+\frac{3\varepsilon}{2}}} \le \frac{c}{x^{\frac{1}{2}-2\beta+\frac{3\varepsilon}{2}}}, \end{cases}$$

$$\vdots$$

$$\begin{cases} t + x \ge t^{k\beta+\frac{1}{2}-\frac{(2k-1)\varepsilon}{2}} x^{\frac{1}{2}-k\beta+\frac{(2k-1)\varepsilon}{2}} \text{ and} \\ |\tilde{u}(x)| \le a + \left\| \frac{K(t)}{t^{\beta+\frac{1}{2}-\frac{\varepsilon}{2}}} \right\|_2 \left\| t^{-(k-1)(\beta-\varepsilon)}\tilde{u}(t) \right\|_2 \frac{1}{x^{\frac{1}{2}-k\beta+\frac{(2k-1)\varepsilon}{2}}} \le \frac{c}{x^{\frac{1}{2}-k\beta+\frac{(2k-1)\varepsilon}{2}}}. \end{cases}$$

The last estimate ensures the finiteness of the two $L^2$-norms in the following inequalities:

$$|\tilde{u}(x)| \le a + \int_0^\infty \frac{|K(t)|\,|\tilde{u}(t)|}{t} dt \le a + \left\| \frac{K(t)}{t^{1-k\beta+k\varepsilon}} \right\|_2 \|t^{-k\beta+k\varepsilon}\tilde{u}(t)\|_2.$$

This establishes the boundedness of $u_\beta(x)$ for $x \in (0,1]$, and hence also for $x \in (0,\infty)$. Since $u_\beta(x)$ is bounded on $(0,\infty)$, the continuity of $u_\beta(x)$ at $x=0$ from the right now follows from equation (9) and the Lebesgue dominated convergence theorem.

An analogous argument gives $v_\beta(x) \in L_\infty[0,\infty)$ and $v_\beta(x) \in C[0,\infty)$. □

## 5. Asymptotic behavior and Stokes phenomenon

From the integral equations, and in view of (16), we readily obtain the asymptotic approximations

$$(23) \quad u_\beta(z) \sim 1 + \sum_{k=1}^\infty \frac{c_k}{z^k} \text{ and } v_\beta(z) \sim 1 + \sum_{k=1}^\infty \frac{d_k}{z^k} \text{ as } |z| \to +\infty,\ |\arg z| < \frac{3\pi}{2},$$

where

$$c_k = (-1)^{k-1} \int_0^\infty K(t)t^{k-1}u_\beta(t)dt \text{ and } d_k = (-1)^k \int_0^\infty K(t)t^{k-1}v_\beta(t)dt$$

for $k = 1, 2, \cdots$; cf. Boyd [1] for asymptotic approximations of Stieltjes transforms. See also Wong [12], and Wong and Zhao [14].

To satisfy certain jump conditions and matching conditions, the Stokes phenomenon plays a crucial role in the construction of local parametrices for RHPs. Now we turn to a brief discussion of the Stokes phenomena of $u_\beta$ and $v_\beta$, based on the connection formulas in (18), and regarding equations (9)-(10) as exact resurgence relations. A similar analysis was carried out in Boyd [1] for the modified Bessel function; see also Xu and Zhao [15] for an attempt to link together the RH approach with resurgence properties.

Assume that $u(x) = u_\beta(x)$ solves the integral equation (9) with $u-1 \in L_2[0,\infty)$, and $u(z)$ denotes its analytic continuation. In general, we see that $u(z) = 1+O(1/z)$ for large $z$ as $|\arg z| < \frac{3\pi}{2}$; cf. (17). Coupling (17) and (18), one can verify that

$$u(z) = 1 + O\left(\frac{1}{z}\right) + \exp\left(e^{-i\pi(1+\beta)/2}(ze^{-\pi i})^\beta\right)$$

for $\frac{\pi}{2} < \arg z \le \frac{3\pi}{2}$; see also (11). The term on the extreme right is exponentially large, when $\arg z$ goes beyond $\frac{3\pi}{2}$. According to Berry's Stokes smoothing, the switch-on of the exponential occurs when $1 + O(1/z)$ is the most dominant as compared with the exponential term, namely, when $-\left[\frac{\pi}{2} + \left(\frac{3\pi}{2} - \arg z\right)\beta\right] = -\pi$, i.e., $\arg z = \frac{3\pi}{2} - \frac{\pi}{2\beta}$. We give some details in what follows.

In view of (11), we may write

(24)

$$u(z) = 1 + \frac{1}{2\pi i}\int_0^\infty \frac{e^{ie^{\beta\pi i/2}t^\beta}u(t)dt}{t+z} - \frac{1}{2\pi i}\int_0^\infty \frac{e^{-ie^{-\beta\pi i/2}t^\beta}u(t)dt}{t+z} := 1 + I_P - I_N,$$

initially for real positive $z$, and then extended elsewhere. Now we assume that $\beta \in [1/3, 1)$, and take $I_N(z)$ as an example. Rotating the integration path clockwise to $\arg t = \phi \in [-\pi, 0)$, and making a change of variables $t = e^{-\pi i}z\tau^\alpha$ with $\alpha = 1/\beta$, we have

$$I_N(z) = \frac{\alpha}{2\pi i}\int_0^\infty \frac{e^{-\rho e^{i\theta}\tau}u(e^{-\pi i}z\tau^\alpha)\tau^{\alpha-1}d\tau}{\tau^\alpha - 1}, \tag{25}$$

where $\theta = \beta\left(\arg z - \frac{\pi}{2}(3-\alpha)\right)$, the large parameter is $\rho = |z|^\beta$, and the path of integration is indented to pass above the pole at $\tau = 1$. Before reaching the equality in (25), we have already rotated the path $\arg\tau = \phi + \pi$ to the positive real line.

Substitute the identity $\frac{1}{\tau^\alpha - 1} = -\sum_{k=1}^{N-1}\tau^{k\alpha} + \frac{\tau^{N\alpha}}{\tau^\alpha - 1}$ into the integral in (25). The remainder in the resulting expansion is an integral with a saddle point as well as a pole. Optimal truncation of the expansion occurs when the saddle point coalesces with the pole (see [1,13]); that is, when $\rho \approx N\alpha$ or, more precisely, $\alpha N = \rho + r$, with $r$ being bounded. The result we have now is an exponentially improved asymptotic expansion, whose error term is given by

$$\frac{1}{2\pi i}\int_0^\infty \left\{\frac{\tau - 1}{\tau^\alpha - 1}u(e^{-\pi i}z\tau^\alpha)\alpha\tau^{\alpha-1+r}\right\}\frac{e^{-\rho(e^{i\theta}\tau - \ln\tau)}d\tau}{\tau - 1},$$

which exhibits a Stokes smoothing of error-function type as $\theta \to 0^-$ when the saddle point $\tau = e^{-i\theta}$ coalesces with the simple pole $\tau = 1$. This indicates that in our case a Stokes line occurs when we have $\theta = 0$, i.e., $\arg z = \frac{\pi}{2}(3-\alpha)$. The analysis for the other integral $I_P(z)$ gives another Stokes line $\arg z = -\frac{\pi}{2}(3-\alpha)$.

A slight modification is needed for the case $\beta \in [1/5, 1/3)$, since $\phi_0 = -\frac{\pi}{2}(\alpha-1) \in [-2\pi, -\pi)$ in this case. Rotating the integration path clockwise and picking up the residue $r_N(z) = -\exp\left(e^{-(\frac{\pi}{2}+\frac{\beta\pi}{2})i}(ze^{-\pi i})^\beta\right)u(ze^{-\pi i})$ at $t = ze^{-\pi i}$, one has

$$
\begin{aligned}
I_N(z) &= r_N + \frac{1}{2\pi i}\int_0^{\infty e^{i\phi}} \frac{e^{-ie^{-\frac{\beta\pi i}{2}}t^\beta}u(t)dt}{t+z} \\
&= J_N(z) + \frac{1}{2\pi i}\int_0^{\infty e^{i\phi}} \frac{e^{-ie^{-\frac{\beta\pi i}{2}}t^\beta}u(te^{2\pi i})dt}{t+z},
\end{aligned}
$$

where $\phi < -\pi$, and $J_N(z)$ is a sum of exponentials and integrals. The second equality is obtained by substituting in $u(t) = u(te^{2\pi i}) + K(te^{\pi i})u(te^{\pi i})$; cf. (18). Now we focus on the last integral. Making a change of variables $t = e^{-\pi i}z\tau^\alpha$, and using the same optimal truncation $\alpha N = \rho + r$ with $\rho = |z|^\beta$, we obtain a slightly different error term

$$
\frac{1}{2\pi i}\int_0^\infty \left\{\frac{\tau-1}{\tau^\alpha-1}u(e^{\pi i}z\tau^\alpha)\alpha\tau^{\alpha-1+r}\right\}\frac{e^{-\rho\left(e^{i\theta}\tau-\ln\tau\right)}d\tau}{\tau-1},
$$

where again $\theta = \beta\left(\arg z - \frac{\pi}{2}(3-\alpha)\right)$, the large parameter $\rho = |z|^\beta$, but the path of integration is indented to pass below the pole at $\tau = 1$. As before, a Stokes line in this case is identified as $\arg z = -\frac{\pi}{2}(\alpha-3) \in (-\pi, 0)$. Another symmetric Stokes line $\arg z = \frac{\pi}{2}(\alpha-3)$ can be found by analyzing $I_P(z)$; see (24).

In general, for $\beta \in [\frac{1}{4l+3}, \frac{1}{4l-1})$, $l = 1, 2, \cdots$, repeated use of the connection formula (18) gives

$$
I_N(z) = J_N(z) + \frac{1}{2\pi i}\int_0^{\infty e^{i\phi}} \frac{e^{-ie^{-\frac{\beta\pi i}{2}}t^\beta}u(te^{2l\pi i})dt}{t+z},
$$

where $\phi < -(2l-1)\pi$. Here again $J_N(z)$ is a combination of exponentials and integrals. Similar analysis can be carried out for the last integral to locate the Stokes line. The same treatment also works for $v_\beta$. The reader is referred to [13, p. 360] for full details.

To summarize, we have

**Theorem 4.** *Assume that the functions $u_\beta(z)$ and $v_\beta(z)$, solving respectively* (9) *and* (10), *such that $u_\beta - 1$, $v_\beta - 1 \in L_2[0,\infty)$. Then the Stokes lines for both functions are*

$$
\arg z = \pm\left\{\frac{\pi}{2}(3-\alpha) + 2(l-1)\pi\right\},
$$

*where $\alpha = \frac{1}{\beta} \in (4l-3, 4l+1]$, $l = 1, 2, \cdots$.*

## 6. Discussion

In the previous sections, we have determined a pair of new special functions $u_\beta$ and $v_\beta$ via integral equations. Preliminary analysis reveals some of the analytic structures of the functions, such as their analytic continuations, connection formulas, and asymptotic results involving Stokes phenomenon. Still we have the following open problems.

**Problem 1.** Here we propose a thorough investigation of the analytic and asymptotic properties of the functions $u_\beta$ and $v_\beta$, such as their zeros, modulus function

[11, §10.18], kernel [9], and differential or difference equations, etc. For example, one may begin with

**(i):** Determine the coefficients $c_k$ and $d_k$ of the expansion at infinity given in (23), which are now expressed in terms of $u_\beta$ and $v_\beta$. It would be interesting and challenging to decode the coefficients from (9) and (10) in an explicit manner.

**(ii):** On account of (29), one may reasonably expect that the behaviors of $u_\beta$ and $v_\beta$ at the origin are of the form $\sum_{k=0} \tilde{c}_k z^k + \sum_{k=1} \tilde{d}_k z^{\beta k}$; see also Theorem 3 and Remark 1 for the boundedness of these functions. A natural problem is to determine explicitly the coefficients $\tilde{c}_k$ and $\tilde{d}_k$, at least for $k = 0$ and 1.

The Stieltjes transform of the function $K(t)$ given in (11) suggests that it would also be of interest to study the related integral

$$G_\beta(z) := \int_0^\infty \frac{e^{-t^\beta}}{t+z} dt \tag{26}$$

for $\beta > 0$, defined initially for $\arg z \in (-\pi, \pi)$ and then analytically continued to $\arg z \in \mathbb{R}$. The special case when $\beta = 2$ is termed the Goodwin-Staton integral [11, (7.2.12)]; see also Jones [7], where, by making use of the incomplete Gamma function, asymptotic formulas are obtained for a slightly more general integral.

The Mittag-Leffler function can be represented in terms of this function. For $\alpha = 1/\beta > 0$, the Mittag-Leffler function is given by

$$E_\alpha(z) := \sum_{n=0}^\infty \frac{z^n}{\Gamma(\alpha n+1)} = \beta \sum e^{Z_k} + \frac{1}{2\pi i} \int_C \frac{t^{\alpha-1} e^t dt}{t^\alpha - z} := \Sigma(z) + I(z), \tag{27}$$

where $Z_k = z^\beta e^{2\pi i k \beta}$ for integer $k$ and the summation is taken over all $k$ satisfying $|\arg Z_k| < \pi$. The path $C$ is a Hankel contour, and dented properly if $|\arg Z_k| = \pi$ for a certain integer $k$; cf. [13].

As in Wong and Zhao [13], it is appropriate to divide our discussion into several cases: (i) $\alpha$ is a positive integer, (ii) $\alpha \in (2l-1, 2l)$, $l = 1, 2, 3, \cdots$, and (iii) $\alpha \in (2l, 2l+1)$, $l = 0, 1, 2, \cdots$.

In case (i), i.e., when $\alpha$ is a positive integer, the contour integral in (27) vanishes; thus the function $E_\alpha(z)$ is an elementary function. In case (ii), we have

$$I(z) = \frac{\beta}{2\pi i} G_\beta(e^{(2l-1-\alpha)\pi i} z) - \frac{\beta}{2\pi i} G_\beta(e^{(\alpha-2l+1)\pi i} z)$$

initially for $\arg z = 0$, and then extend the region of validity by analytic continuation. Similarly, in case (iii),

$$I(z) = \frac{\beta}{2\pi i} G_\beta(e^{(2l+1-\alpha)\pi i} z) - \frac{\beta}{2\pi i} G_\beta(e^{(\alpha-2l-1)\pi i} z).$$

Asymptotic expansions of $G_\beta$ can be derived as follows. For example, at infinity, we have

$$G_\beta(z) \sim \sum_{n=1}^\infty \frac{(-1)^{n-1} \Gamma(n/\beta)}{\beta z^n} \tag{28}$$

initially for $|\arg z| < \pi$. The region of validity can be extended to $|\arg z| < \pi + \frac{\pi}{2\beta}$. Also, at the origin, we write $G_\beta(z) = \left\{ \int_0^1 + \int_1^\infty \right\} \frac{e^{-t^\beta} dt}{t+z}$. Expanding the exponential in the first integral in the convergent series, and using the Cauchy residue theorem, we have

$$(29)\qquad G_\beta(z) \sim -\ln z + \sum_{n=1}^{\infty} \frac{(-1)^{n+1}\pi}{n!\sin(\beta n\pi)} z^{\beta n} + \sum_{k=0}^{\infty} c_{\beta,k}(-1)^k z^k,\ z \to 0,$$

as long as $\beta$ is not a rational number, where

$$(30)\qquad c_{\beta,0} = \sum_{n=1}^{\infty} \frac{(-1)^n}{n!n\beta} + \frac{\Gamma(0,1)}{\beta} = \frac{1}{\beta}\left[\int_0^1 \frac{e^{-t}-1}{t}dt + \int_1^\infty \frac{e^{-t}}{t}dt\right] = -\frac{\gamma}{\beta}$$

and

$$c_{\beta,k} = \sum_{n=0}^{\infty} \frac{(-1)^n}{n!(n\beta - k)} + \frac{\Gamma(-k/\beta,1)}{\beta},\ k = 1, 2, \cdots,$$

where $\Gamma(a,z) = \int_z^\infty t^{a-1}e^{-t}dt$ is the incomplete Gamma function. For the last equality in (30), see [11, (6.2.1), (6.2.3)-(6.2.4)], $\gamma$ being Euler's constant. It can be shown that the expansion in (29) is valid in the sector $|\arg z| < \pi + \frac{\pi}{2\beta}$, as long as $\beta$ is a positive irrational number.

If $\beta > 0$ is a rational number, then we write $\beta = p/q$ with $p$ and $q$ being relatively prime. The change of variable $t^{1/q} = s$ gives

$$(31)\qquad G_\beta(z) = \int_0^\infty \frac{qs^{q-1}e^{-s^p}}{s^q + \zeta^q} d\zeta,\ \zeta^q = z.$$

Let $\omega_l = e^{(2l+1)\pi i/q}$, $l = 0, 1, \cdots, q-1$. Since

$$s^q + \zeta^q = (s - \omega_0\zeta)(s - \omega_1\zeta)\cdots(s - \omega_{q-1}\zeta),$$

taking logarithmic derivative yields

$$\frac{qs^{q-1}}{s^q + \zeta^q} = \sum_{l=0}^{q-1} \frac{1}{s - \omega_l\zeta}.$$

Thus, $G_\beta(z)$ in (31) can be written as

$$G_\beta(z) = \sum_{l=0}^{q-1} G_p\left(e^{(2l+1-q)\pi i/q}\zeta\right).$$

The problem is now reduced to consider the function $G_p(z)$, where $p$ is a positive integer. In this case, the integral for $G_p(z)$ can again be broken into two parts, one over the interval $(0,1)$ and the other on the infinite interval $(1,\infty)$. The integral on $(1,\infty)$ can be asymptotically evaluated as before. For the integral on $(0,1)$, we expand the exponential function $e^{-t^p}$ into a power series. This leads to infinitely many integrals of the form

$$I_n(z) = \int_0^1 \frac{t^{np}}{t+z} dt,\ n = 0, 1, 2, \cdots.$$

For each $n \geq 1$, we have

$$\frac{1}{t+z} = \frac{1}{t} - \frac{z}{t^2} + \cdots + (-1)^{np-1}\frac{z^{np-1}}{t^{np}} + (-1)^{np}\frac{z^{np}}{t^{np}(t+z)},$$

and hence

$$I_n(z) = \sum_{k=0}^{np-1} \frac{(-1)^k z^k}{np - k} + (-1)^{np} z^{np} \left[\ln(1+z) - \ln z\right].$$

Asymptotic expansion of $G_p(z)$ can thus be obtained for small values of $z$.

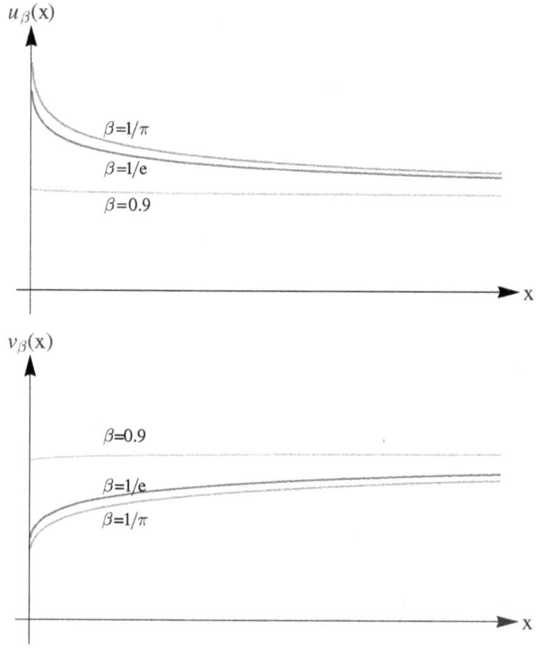


FIGURE 1. The solution profiles of $u_\beta(x)$ (left) and $v_\beta(x)$ (right).

**Problem 2.** We have used numerical simulation to draw the graphs of $u_\beta(x)$ for various values $\beta \in (0,1)$; see Figure 1. From the graphs, it is reasonable to conjecture that the function $u_\beta(x)$ is monotonically decreasing in $x \in [0,\infty)$, and also that the initial value $u_\beta(0)$ is decreasing in $\beta \in (0,1)$. What we need now are proofs of these statements. From Figure 1, corresponding conjectures can be made for the function $v_\beta(x)$.

## ACKNOWLEDGEMENTS

We would like to thank Dr. Xiang-Sheng Wang for drawing the graphs and helping us with the numerical simulation. The research of the second author was supported in part by the National Natural Science Foundation of China under grant numbers 10871212 and 11571375.

## REFERENCES

[1] W. G. C. Boyd, *Stieltjes transforms and the Stokes phenomenon*, Proc. Roy. Soc. London Ser. A **429** (1990), no. 1876, 227–246. MR1052129

[2] Yang Chen and M. E. H. Ismail, *Some indeterminate moment problems and Freud-like weights*, Constr. Approx. **14** (1998), no. 3, 439–458, DOI 10.1007/s003659900083. MR1626714

[3] Dan Dai, Mourad E. H. Ismail, and Xiang-Sheng Wang, *Plancherel-Rotach asymptotic expansion for some polynomials from indeterminate moment problems*, Constr. Approx. **40** (2014), no. 1, 61–104, DOI 10.1007/s00365-013-9215-1. MR3229995

[4] Alfredo Deaño, Arno B. J. Kuijlaars, and Pablo Román, *Asymptotic Behavior and Zero Distribution of Polynomials Orthogonal with Respect to Bessel Functions*, Constr. Approx. **43** (2016), no. 1, 153–196, DOI 10.1007/s00365-015-9300-8. MR3439237

[5] P. Deift, T. Kriecherbauer, K. T.-R. McLaughlin, S. Venakides, and X. Zhou, *A Riemann-Hilbert approach to asymptotic questions for orthogonal polynomials*, Proceedings of the Fifth International Symposium on Orthogonal Polynomials, Special Functions and their Applications (Patras, 1999), J. Comput. Appl. Math. **133** (2001), no. 1-2, 47–63, DOI 10.1016/S0377-0427(00)00634-8. MR1858269

[6] Athanassios S. Fokas, Alexander R. Its, Andrei A. Kapaev, and Victor Yu. Novokshenov, *Painlevé transcendents*, Mathematical Surveys and Monographs, vol. 128, American Mathematical Society, Providence, RI, 2006. The Riemann-Hilbert approach. MR2264522

[7] D. S. Jones, *The generalized Goodwin-Staton integral*, Proc. Edinb. Math. Soc. (2) **48** (2005), no. 3, 635–650, DOI 10.1017/S0013091504001087. MR2171190

[8] T. Kriecherbauer and K. T.-R. McLaughlin, *Strong asymptotics of polynomials orthogonal with respect to Freud weights*, Internat. Math. Res. Notices **6** (1999), 299–333, DOI 10.1155/S1073792899000161. MR1680380

[9] A. B. J. Kuijlaars and M. Vanlessen, *Universality for eigenvalue correlations from the modified Jacobi unitary ensemble*, Int. Math. Res. Not. **30** (2002), 1575–1600, DOI 10.1155/S1073792802203116. MR1912278

[10] D. S. Mitrinović, *Analytic inequalities*, Springer-Verlag, New York-Berlin, 1970. In cooperation with P. M. Vasić. Die Grundlehren der mathematischen Wissenschaften, Band 165. MR0274686

[11] *NIST handbook of mathematical functions*, U.S. Department of Commerce, National Institute of Standards and Technology, Washington, DC; Cambridge University Press, Cambridge, 2010. Edited by Frank W. J. Olver, Daniel W. Lozier, Ronald F. Boisvert and Charles W. Clark; With 1 CD-ROM (Windows, Macintosh and UNIX). MR2723248

[12] R. Wong, *Asymptotic approximations of integrals*, Classics in Applied Mathematics, vol. 34, Society for Industrial and Applied Mathematics (SIAM), Philadelphia, PA, 2001. Corrected reprint of the 1989 original. MR1851050

[13] R. Wong and Yu-Qiu Zhao, *Exponential asymptotics of the Mittag-Leffler function*, Constr. Approx. **18** (2002), no. 3, 355–385, DOI 10.1007/s00365-001-0019-3. MR1906764

[14] R. Wong and Yu-Qiu Zhao, *Gevrey asymptotics and Stieltjes transforms of algebraically decaying functions*, R. Soc. Lond. Proc. Ser. A Math. Phys. Eng. Sci. **458** (2002), no. 2019, 625–644, DOI 10.1098/rspa.2001.0883. MR1898087

[15] ShuaiXia Xu and YuQiu Zhao, *Resurgence relation and global asymptotic analysis of orthogonal polynomials via the Riemann-Hilbert approach*, Sci. China Math. **54** (2011), no. 4, 661–679, DOI 10.1007/s11425-010-4151-z. MR2786707

[16] Kôsaku Yosida, *Functional analysis*, 6th ed., Grundlehren der Mathematischen Wissenschaften [Fundamental Principles of Mathematical Sciences], vol. 123, Springer-Verlag, Berlin-New York, 1980. MR617913

Department of Mathematics, City University of Hong Kong, Kowloon, Hong Kong

Department of Mathematics, Sun Yat-sen University, GuangZhou 510275, People's Republic of China

*E-mail address*: stszyq@mail.sysu.edu.cn

# Asymptotics of Pseudo-Jacobi Polynomials with Varying Parameters

*By Z. Song and R. Wong*

In this paper, we study the asymptotic behavior of the Pseudo-Jacobi polynomials $P_n(z; a, b)$ as $n \to \infty$ for $z$ in the whole complex plane. These polynomials are also known as the Romanovski–Routh polynomials. They occur in quantum mechanics, quark physics, and random matrix theory. When the parameter $a$ is fixed or $a > -n$, there is no real-line orthogonality. Here, we consider the case when the parameters $a$ and $b$ depend on $n$; more precisely, we assume $a = -(An + A_0)$, $A > 1$ and $b = Bn + B_0$, where $A, B, A_0, B_0$ are real constants. Our main tool is the asymptotic method developed for differential equations with a large parameter.

## 1. Introduction

It is well-known that some second-order linear ordinary differential equations have polynomial solutions; for instance, those satisfied by the three classical orthogonal polynomials. The Pseudo-Jacobi (P-J) polynomials can be introduced as the eigenfunctions of the symmetric differential operator defined by

$$L = p\frac{d^2}{dx^2} + \frac{(pw)'}{w}\frac{d}{dx}, \tag{1}$$

where $p(x)$ is a polynomial of degree 2 and $w(x)$ is a continuous function called the weight function associated with $L$; see [1, p. 56]. This operator

Address for correspondence: R. Wong, Department of Mathematics, City University of Hong Kong, Hong Kong; e-mail: lbj@cityu.edu.hk

*In memory of Professor David J. Benney.*

DOI: 10.1111/sapm.12177



*STUDIES IN APPLIED MATHEMATICS* 139:179–217

can admit orthogonal polynomials as its eigenfunctions by proper choices of $p$ and $w$, and the three classical ones (namely, Hermite, Laguerre, Jacobi) correspond to $p = 1, x, 1 - x^2$, respectively. Another case is when $p(x)$ has distinct complex roots; after normalization, we can take $p = x^2 + 1$; see [1, p. 58]. A necessary condition that the operator $L$ takes the space of real quadratic polynomials into itself is that $(pw)'/w$ or $pw'/w$ is a polynomial of degree at most 1. Thus, one may assume

$$\frac{w'}{w} = \frac{2(ax + b)}{x^2 + 1}, \tag{2}$$

where $a$ and $b$ are two real parameters. This immediately gives the weight

$$w = (x^2 + 1)^a e^{2b \arctan x}, \qquad x \in (-\infty, \infty). \tag{3}$$

The associated operator is then given by

$$L = (x^2 + 1)\frac{d^2}{dx^2} + 2[(1 + a)x + b]\frac{d}{dx}. \tag{4}$$

If polynomials $P_n(x; a, b)$ of degree $n$, $n = 0, 1, 2, \ldots$, are its eigenfunctions, then one obtains

$$LP_n(x; a, b) - n(n + 2a + 1)P_n(x; a, b) = 0. \tag{5}$$

For the orthogonality of $P_n(x; a, b)$ to be well-defined on the real line, we require that

$$\int_{-\infty}^{+\infty} x^{2n} w(x) dx < \infty. \tag{6}$$

However, this is true only when $n < -a - 1/2$ with negative $a$. For example, if $a = -N - 1$ with $N$ being a positive integer, then the orthogonal polynomials $P_n(x; a, b)$ are well-defined with respect to $w(x)$ only for $n = 0, \ldots, N$. Therefore, for fixed $a < -1$, we have only a finite number of orthogonal polynomials, in contrast to the three classical ones.

The above polynomials $P_n(x; a, b)$ were first discovered by Romanovski [2] in 1929 (in French) within the context of probability distribution functions in statistics. In fact, they belong to a general family of Routh polynomials [3] introduced in 1884; so from a historical point of view, some authors prefer to name them Romanovski–Routh or just Romanovski polynomials. For some history, see [4, 5]. Because they have not been discussed in standard books on orthogonal polynomials and special functions (e.g., [6, 7]) and were mentioned only briefly by Askey [8, 9] in 1980s, they are not known to most of the researchers in the field. Another name for these polynomials is twisted Jacobi polynomials [10, 11]. In view of its close connection with the Jacobi polynomial of purely imaginary argument and complex parameters (see (8) below), in 1996, Lesky [12] (in German) coined the term "Pseudo-Jacobi polynomials." Later these polynomials are also included in the monograph [13]. It was probably the review article [14] that brought attention to the community of people working in orthogonal polynomials. This article discusses various applications of this polynomial ranging from quantum mechanics and quark physics to random matrix theory (see also [15] and references in [14]). For instance, they appear in the solutions of the Schrödinger equation with certain noncentral electric potential obtained by separation of variables in spherical coordinates; their role is similar to that of the associated Legendre functions; see [16] and [17].

The definition of monic P-J polynomials in terms of hypergeometric functions is given in [13], with parameters $N$ and $\nu$ related to the parameters $a$ and $b$ in (3); more precisely, with $a = -N - 1$ and $b = \nu$, the monic P-J polynomials are defined by

$$P_n(x; a, b) = \frac{(-2i)^n(1 + a + bi)_n}{(n + 2a + 1)_n} \, {}_2F_1\left( \begin{matrix} -n, n + 2a + 1 \\ 1 + a + bi \end{matrix} ; \frac{1 - ix}{2} \right). \tag{7}$$

By comparing the definitions of the P-J polynomials and Jacobi polynomials $P_n^{(\alpha,\beta)}$, we immediately get the relation

$$P_n(x; a, b) = (-i)^n P_n^{(a+bi, a-bi)}(ix). \tag{8}$$

This relation has also been used as the definition of P-J polynomials in [18, p. 508] and [19]. In view of the symmetry property of Jacobi polynomials, the P-J polynomials satisfy

$$P_n(z; a, -b) = (-1)^n P_n(-z; a, b); \tag{9}$$

see [19]. Because the coefficients of P-J polynomials are real, it also follows that

$$P_n(\bar{z}; a, b) = \overline{P_n(z; a, b)}, \tag{10}$$

where an overbar means the complex conjugate. For other properties such as orthogonality, recurrence relations and Rodrigues formula, we refer to [13, pp. 231–233].

Probably due to their close relation with Jacobi polynomials, recently P-J polynomials have attracted much attention [19, 20]. Orthogonality of the P-J polynomials and properties of their zeros have been studied in [19] for real parameter $a$ in different ranges such as $a < -n$ and $-n < a < -\frac{n}{2} - \frac{1}{2}$. As a special case, the properties of the zeros of the pseudo-ultraspherical polynomials have been independently investigated in [21] almost at the same time. In this work, we intend to study the asymptotic behavior of the P-J polynomials as the degree $n$ tends to infinity. Note that, for fixed $a$ or $a > -n$, there is no real-line orthogonality anymore; see [19, 22] for their orthogonality. For fixed parameters $a$ and $b$, in view of (7) the asymptotics of these polynomials can be obtained as a special case of that of the hypergeometric functions investigated in [23, 24]. In the present paper, we consider the case when the parameters depend on the degree $n$,

$$a_n = -(An + A_0), \qquad (A > 1); \qquad b_n = Bn + B_0, \tag{11}$$

where $A, A_0, B, B_0$ are real constants. Our aim is to study the asymptotic behavior of $P_n(z; a_n, b_n)$ (with varying parameters) for $z$ in the whole complex plane, as $n \to \infty$. Although asymptotic zero distribution, bounds for the extreme zeros as well as the monotonicity and convexity of the zeros have been investigated in [19], the asymptotic behavior of these polynomials for complex variable $z$ has not been studied. As a byproduct of the asymptotic behavior, approximation of the zeros can be easily obtained (see Corollary 2 in Section 4). The main tool used in this paper is the well-developed method for differential equations with a large parameter as presented in Olver [25]. Another aim of this paper is to use this example as an illustration to show that this method can be applied to many applied disciplines (at least formally), and deserves a larger audience. Although this method has been applied to orthogonal polynomials with exponential weights [26,27], the focus there was restricted to the case of real variable.

This paper is divided into two parts. The first part deals with the asymptotics when the variable $z$ lies outside the interval where the zeros are located. In the second part, we seek a uniform asymptotic approximation that is valid in a region which includes most of the interval of oscillation and an endpoint. From this approximation, one can deduce an asymptotic formula for the zeros of the P-J polynomials. With the aid of the symmetries in (9) and (10), the results in the two parts together cover the whole complex plane. The asymptotic problem of P-J polynomials can also be tackled by using the Riemann–Hilbert method; see, for instance, the papers [28] and [29] on Jacobi polynomials with varying parameters. However, that method is more involved in its formulation, while the present one is simpler to understand and sufficient to provide the information needed for practical applications.

## 2. Zero distribution

Recall that the monic P-J polynomials $P_n(z; a_n, b_n)$ with parameters $a_n$ and $b_n$ given in (11) satisfy the differential equation (see Eq. 5)

$$(z^2 + 1)P_n'' + 2[(1 + a_n)z + b_n]P_n' - n(n + 2a_n + 1)P_n = 0, \tag{12}$$

and that its associated weight function is

$$w(z) = (z^2 + 1)^{a_n} e^{2b_n \arctan z} = (1 - iz)^{a_n + ib_n}(1 + iz)^{a_n - ib_n}, \tag{13}$$

where the branch cuts are taken from $x = \pm i$ to infinity so that the function is well-defined on the real line. Let $\{x_{n,j}\}|_{j=1}^n$ be the zeros of $P_n(x; a_n, b_n)$. To study the zero distribution, we define a sequence of discrete measures $\lambda_n$ by

$$\lambda_n(f) \equiv \frac{1}{n}\sum_{j=1}^{n} f(x_{n,j}) = \int f(x)d\lambda_n(x), \tag{14}$$

where $f$ is any continuous function. We have the following proposition.

PROPOSITION 1. *Let the sequence of Pseudo-Jacobi polynomials $\{P_n(z; a_n, b_n)\}$ satisfy* (12) *and* (11). *Then there is a unique measure $\lambda$, which is the weak-$*$ limit of $\lambda_n$ as $n \to \infty$. The measure $\lambda$ is supported on $[\gamma_-, \gamma_+]$ and given by*

$$d\lambda = \frac{(A-1)\sqrt{(x-\gamma_-)(\gamma_+ - x)}}{\pi(x^2+1)}dx, \tag{15}$$

*where*

$$\gamma_\pm = \frac{AB \pm \sqrt{(2A-1)[(A-1)^2 + B^2]}}{(A-1)^2}. \tag{16}$$

This result has been proved in [19]. Because we need the information given in (16), to make the paper self-contained we give a brief derivation of the above proposition by following the analysis in [1]. All results in the subsequent sections are appearing for the first time.

Because the zeros of $P_n(z; a_n, b_n)$ are simple, we let

$$F_n(z) \equiv \frac{P_n'(z; a_n, b_n)}{nP_n(z; a_n, b_n)} = \frac{1}{n}\sum_{j=1}^{n}\frac{1}{z - x_{n,j}} = \int \frac{d\lambda_n(x)}{z - x}; \tag{17}$$

so $F_n$ is the Stieltjes transform of $\lambda_n$. For any polynomial $f$, it follows from the residue theorem that

$$\lambda_n(f) = \frac{1}{2\pi i}\int_C F_n(z)f(z)dz, \tag{18}$$

where $C$ is some counter-clockwise contour enclosing all the points $x_{n,j}$ ($j = 1, .., n$). In view of the Weierstrass approximation theorem, it suffices to demonstrate the results of the proposition for polynomials $f$. The asymptotic behavior of $F_n$ can be obtained from the differential equation (12) satisfied by $P_n$, and the limiting measure can be calculated from (18).

Indeed, by (12) and

$$F_n' = \frac{P_n''}{nP_n} - nF_n^2, \tag{19}$$

one can easily see that

$$(z^2+1)\left[F_n^2 + \frac{1}{n}F_n'\right] = (z^2+1)\frac{P_n''}{n^2P_n} = \frac{(n+2a_n+1)}{n} - \frac{2[(1+a_n)z+b_n]}{n}F_n. \tag{20}$$

Suppose that $F_n \to F$ as $n \to \infty$. Taking the limit $n \to \infty$ in the above equation together with the assumptions in (11), we obtain

$$(z^2+1)F^2 = (1-2A) + 2(Az-B)F. \tag{21}$$

Because $F_n$ is $\sim 1/z$ as $z \to \infty$, we choose the root

$$F(z) = \frac{Az - B - \sqrt{Q_0(z)}}{z^2+1}, \tag{22}$$

where

$$\begin{aligned} Q_0(z) &= (A-1)^2z^2 - 2ABz + B^2 + 1 - 2A \\ &= (A-1)^2(z-\gamma_-)(z-\gamma_+), \end{aligned} \tag{23}$$

and the constants $\gamma_\pm$ are given in (16). For the branch cuts, we take $\sqrt{Q_0(z)}$ to be equal to $(A-1)\sqrt{z-\gamma_-}\sqrt{z-\gamma_+}$. Also, by using the first equality in (23) one can verify that $F(z)$ has removable singularities at $z = \pm i$. From (18), we obtain

$$\lambda(f) = \frac{1}{2\pi i}\lim_{n\to\infty}\int_C F_n(z)f(z)dz = \int_{\gamma_-}^{\gamma_+} \frac{(A-1)\sqrt{(x-\gamma_-)(\gamma_+-x)}}{\pi(x^2+1)}f(x)dx. \tag{24}$$

## 3. Outer region

By outer region, we mean the complement of the interval where the zeros lie; i.e., $\mathbb{C}\setminus[\gamma_-, \gamma_+]$. In this section, we study the asymptotics of P-J polynomials in this region by using the Liouville–Green (L-G) approximation (also known as the WKB method). To fix ideas, we first analyze the general solution on a finite interval of the real line, and then extend the result to a complex domain by studying some associated holomorphic functions. The asymptotics of P-J polynomial in the outer region will be obtained by matching the approximate solution with the behavior of the polynomial at infinity.

### 3.1. General solution on a finite real interval

We start with the differential equation in (12). To remove the first-order derivative term, we make the transformation

$$\begin{aligned} y_n(z;a_n,b_n) &\equiv P_n(z;a_n,b_n)e^{b_n \arctan z}(1+z^2)^{(a_n+1)/2} \\ &= P_n(z;a_n,b_n)\sqrt{(1+z^2)w(z)}. \end{aligned} \tag{25}$$

The new function satisfies

$$y_n''(z) = n^2\left[f_0(z) + \frac{1}{n}f_1(z) + \frac{1}{n^2}f_2(z)\right]y_n(z), \qquad f_i(z) = \frac{Q_i(z)}{(1+z^2)^2}, \tag{26}$$

where

$$\begin{aligned} Q_0(z) &= (A-1)^2z^2 - 2ABz + B^2 + 1 - 2A = (A-1)^2(z-\gamma_-)(z-\gamma_+), \\ Q_1(z) &= (A-1)(2A_0-1)z^2 - 2(A_0B + AB_0)z + 1 - A - 2A_0 + 2BB_0, \\ Q_2(z) &= A_0(A_0-1)z^2 - 2A_0B_0z + B_0^2 - A_0 + 1. \end{aligned} \tag{27}$$

Now we restrict ourselves to the bounded interval $z \in I = [\gamma_+ + \delta, M]$, where $M > \gamma_+$ is a fixed finite number. Following the derivations in [25, p. 383], we make the transformation

$$\xi(z) = \int^z \sqrt{f_0(t)}dt, \qquad y_n(z) = f_0^{-1/4}(z)W_n(\xi). \tag{28}$$

Equation (26) then becomes

$$W_n''(\xi) = \left[n^2 + n\phi(\xi) + \psi(\xi)\right]W_n(\xi), \tag{29}$$

where

$$\phi(\xi) = \frac{f_1(z)}{f_0(z)}, \quad \psi(\xi) = \frac{f_2(z)}{f_0(z)} - \frac{1}{f_0^{3/4}(z)}\frac{d^2}{dz^2}\left[\frac{1}{f_0^{1/4}(z)}\right]. \tag{30}$$

To (29) we seek a formal series solution of the type

$$W_n^+(\xi) = e^{n\xi + X(\xi)}\sum_{s=0}^{\infty}\frac{w_s^+(\xi)}{n^s}, \qquad w_0^+ = 1. \tag{31}$$

Substituting it into (29), we find

$$X(\xi) = \frac{1}{2}\int^{\xi}\phi(v)dv, \tag{32}$$

and

$$w_{s+1}^{+}(\xi) = -\frac{1}{2}\phi w_s^{+}(\xi) - \frac{1}{2}[w_s^{+}]'(\xi) + \frac{1}{2}\int^{\xi}\left[\psi + \frac{1}{2}\phi' - \frac{1}{4}\phi^2\right] w_s^{+} dv,$$
$$s = 0, 1, \ldots \tag{33}$$

The other solution can be obtained by replacing $n$ and $\phi$ by their negatives, and we denote it by

$$W_n^{-}(\xi) = e^{-n\xi - X(\xi)} \sum_{s=0}^{\infty} \frac{(-1)^s w_s^{-}(\xi)}{n^s}, \quad w_0^{-} = 1, \tag{34}$$

with the same $\xi$ and $X(\xi)$ defined above. The coefficients $w_s^{-}$ can be calculated by the same recursive formula (33) with $\phi$ replaced by $-\phi$. If we truncate the series solutions in (31) or (34) at $s = p - 1$, the remainder $\varepsilon_p^{\pm}$ will satisfy the differential equation

$$[\varepsilon_p^{\pm}(\xi)]'' - \left[n^2 + n\phi(\xi) + \psi(\xi)\right] \varepsilon_p^{\pm}(\xi) = e^{\pm[n\xi + X(\xi)]} \frac{2[w_p^{\pm}(\xi)]'}{(\pm n)^{p-1}}. \tag{35}$$

A standard procedure of estimating $\varepsilon_p^{\pm}$ is the successive approximation method. Applying Theorem 9.1 in [25, p. 385], we obtain the following proposition.

Proposition 2. *For the compact interval $z \in I = [\gamma_+ + \delta, M]$ with finite $M$, the differential equation* (29) *has, for each value of $n$ and positive integer $p$, solutions $W_{n,p}^{\pm}$, which are holomorphic in the interval and given by*

$$W_{n,p}^{\pm}(\xi) = e^{\pm[n\xi + X(\xi)]} \sum_{s=0}^{p-1} \frac{(\pm 1)^s w_s^{\pm}(\xi)}{n^s} + \varepsilon_p^{\pm}(\xi), \quad w_0^{\pm} = 1. \tag{36}$$

*Here, the coefficients $w_s^{\pm}(\xi)$ are defined recursively by* (33) *with $\phi$ replaced by $\pm\phi$. The error terms $\varepsilon_p^{\pm}(\xi)$ are bounded by*

$$\frac{\left|e^{\pm[n\xi + X(\xi)]}\right|}{N_\phi} \frac{2\mathcal{V}_{\alpha_\pm,\xi}(w_p^{\pm})}{n^{p-1}} \exp\left\{\frac{\mathcal{V}_{\alpha_\pm,\xi}(H)}{N_\phi}\right\}, \tag{37}$$

*where $\alpha_\pm$ are some reference points, $\mathcal{V}_{\alpha_\pm,\xi}(\cdot)$ denotes the total variation from $\alpha_\pm$ to $\xi$, and*

$$N_\phi = \inf_{v \in \xi(I)} |n + \frac{1}{2}\phi(v)|,$$

$$H(\xi; n) = \int^{\xi}\left[\psi(v) - \frac{\phi^2(v)}{4} + \frac{\phi''(v)}{4n + 2\phi(v)} - \frac{3\phi'(v)^2}{[4n + 2\phi(v)]^2}\right] dv. \tag{38}$$

*Remark* 1. The bounds for derivatives of $\varepsilon_p^{\pm}(\xi)$ can also be given, but we have omitted these expressions. For large $n$ in the proposition, we have

$$N_\phi = O(n), \quad \mathcal{V}_{\alpha_\pm,\xi}\left(w_p^{\pm}\right) = O(1), \quad \mathcal{V}_{\alpha_\pm,\xi}(H) = O(1),$$
$$\varepsilon_p^{\pm} = |e^{\pm[n\xi+X(\xi)]}|O(n^{-p}); \tag{39}$$

the last estimate shows the asymptotic nature of the two solutions in (36). Recall that for a continuously differentiable function $f(v)$ in $[a, b]$, the total variation is defined by

$$\mathcal{V}_{a,b}(f) = \int_a^b |f'(v)|dv. \tag{40}$$

*Remark* 2. The result on the bounded interval $I$ can be extended to an unbounded complex domain, say $D$, if (i) the functions $\phi$ and $\psi$ are holomorphic in $\xi$(D), (ii) one can choose a progressive path in the $\xi$-plane from $\alpha_\pm$ to $\xi$, and (iii) one can show that the two total variations in (37) are bounded. Essentially, a progressive path consists of finite chains of smooth arcs with nonvanishing first-order derivative $\xi'(z)$, and along such a curve from $\alpha_+$ (or $\alpha_-$) to $\xi$, the quantity $\mathrm{Re}(n\xi + X)$ is nondecreasing (or nonincreasing); see [25, p. 222].

### 3.2. *Explicit expressions and the mapping $z \to \xi$*

The transformation $\xi(z)$ in (28) is given by

$$\xi(z) = \int^z \sqrt{f_0(t)}dt = \int^z \frac{\sqrt{Q_0(t)}}{1+t^2}dt = (A-1)\int^z \frac{\sqrt{(t-\gamma_-)(t-\gamma_+)}}{1+t^2}dt, \tag{41}$$

which can be evaluated to yield

$$\begin{aligned}\xi(z) &= 2(A-1)\log\left(\sqrt{z-\gamma_-}+\sqrt{z-\gamma_+}\right) - C_1\\ &\quad -\frac{1}{2}(A+iB)\Big[2\log\left(\sqrt{1+i\gamma_-}\sqrt{z-\gamma_+}+\sqrt{1+i\gamma_+}\sqrt{z-\gamma_-}\right)\\ &\quad -\log(1+iz)\Big] - \frac{1}{2}(A-iB)\Big[2\log\Big(\sqrt{1-i\gamma_-}\sqrt{z-\gamma_+}\\ &\quad +\sqrt{1-i\gamma_+}\sqrt{z-\gamma_-}\Big) - \log(1-iz)\Big],\end{aligned} \tag{42}$$

where the constant $C_1$ is uniquely determined by the condition

$$\xi(z) \sim (A-1)\log z + o(1), \quad z \to +\infty, \tag{43}$$

and is explicitly given by

$$C_1 = 2(A-1)\log 2 - \frac{1}{2}B\pi - (A+iB)\log\left(\sqrt{1+i\gamma_-}+\sqrt{1+i\gamma_+}\right) \\ - (A-iB)\log\left(\sqrt{1-i\gamma_-}+\sqrt{1-i\gamma_+}\right). \tag{44}$$

Similarly, from (26), (30), and (32) we have, as a function of $z$,

$$X(z) = \frac{1}{2}\int^{\xi}\phi(v)dv = \frac{1}{2}\int^{z}\frac{Q_1(t)}{\sqrt{Q_0(t)}(1+t^2)}dt, \tag{45}$$

which yields

$$\begin{aligned} X(z) &= (2A_0-1)\log\left(\sqrt{z-\gamma_-}+\sqrt{z-\gamma_+}\right) - C_2 \\ &\quad -\frac{1}{2}(A_0+iB_0)\left[2\log\left(\sqrt{1+i\gamma_-}\sqrt{z-\gamma_+}+\sqrt{1+i\gamma_+}\sqrt{z-\gamma_-}\right)\right. \\ &\quad \left.-\log(1+iz)\right] - \frac{1}{2}(A_0-iB_0)\left[2\log\left(\sqrt{1-i\gamma_-}\sqrt{z-\gamma_+}\right.\right. \\ &\quad \left.\left.+\sqrt{1-i\gamma_+}\sqrt{z-\gamma_-}\right) - \log(1-iz)\right], \end{aligned} \tag{46}$$

where the constant $C_2$ is determined by the condition

$$X(z) \sim \left(A_0 - \frac{1}{2}\right)\log z + o(1), \quad z \to +\infty, \tag{47}$$

and is explicitly given by

$$C_2 = (2A_0-1)\log 2 - \frac{1}{2}B_0\pi - (A_0+iB_0)\log\left(\sqrt{1+i\gamma_-}+\sqrt{1+i\gamma_+}\right) \\ - (A_0-iB_0)\log\left(\sqrt{1-i\gamma_-}+\sqrt{1-i\gamma_+}\right). \tag{48}$$

Note that in (45), we have used $X(z)$ to denote $X(\xi(z))$; for simplicity, we continue to use this notation.

Now, we analyze the mapping $\xi(z)$ for $z$ in (approximately) the first quadrant of the complex $z$-plane. First, we consider the case $\gamma_- \le 0 \le \gamma_+$, and observe:

(i) As $z$ tends to infinity in the sector $0 \le \arg z \le \pi/2$, the asymptotic formula in (43) continues to hold.

(ii) As $z \to i$ in the first quadrant, we have from (42)

$$\xi(z) \sim \frac{1}{2}(A+Bi)\log(1+iz) \quad \text{and} \quad \operatorname{Re}\xi \to -\infty. \tag{49}$$

(iii) For $z \in [\gamma_+, +\infty)$, $\xi$ is an increasing function of $z$ and $\xi \in [\xi_0, +\infty)$, where $\xi_0 \equiv \xi(\gamma_+)$. To see this, we note that when $z$ is real and

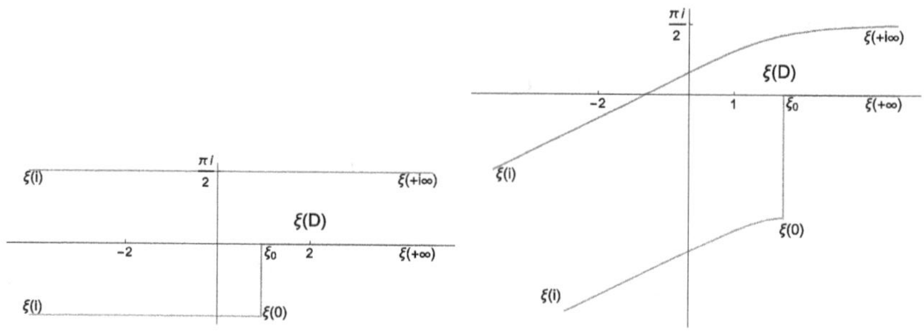


(a) $\xi$-plane, with $A = 2, B = 0$ (b) $\xi$-plane, with $A = 2, B = 1$

**Figure 1.** The mapping $\xi(z)$ for $z \in D$ in (50) with $\gamma_- \le 0 \le \gamma_+$.

large, the first term in (42) is real and the third term in (42) is the complex conjugate of the fourth term. Because the constant $C_1$ is also real by (44), the function $\xi(z)$ in (42) is real for large real $z$ and tends to $+\infty$ by condition (43). Furthermore, it follows from (41) that $\xi'(z) = \sqrt{f_0(z)}$ is real and positive in the interval $(\gamma_+, +\infty)$ and hence the function $\xi(z)$ is always real for $z \in [\gamma_+, +\infty)$.

(iv) As $z$ decreases from $\gamma_+$ to 0 (along the upper edge of the line segment $[0, \gamma_+]$), $\operatorname{Re} \xi = \xi_0$ whereas $\operatorname{Im} \xi$ decreases (see Figure 1); this is because $\xi'(z) = \sqrt{f_0(z)}$ is purely imaginary with a positive imaginary part.

One can similarly analyze the intervals $(0, i)$ and $(i, i\infty)$, but we will skip that here. Figure 1 gives two examples with $A = 2, B = 0$ and $A = 2, B = 1$, where the complex domain $D$ is defined as

$$D = \{z \mid \operatorname{Re} z \ge 0, \quad \operatorname{Im} z \ge 0\} \setminus [0, \gamma_+], \qquad \text{if} \quad \gamma_- \le 0 \le \gamma_+, \tag{50}$$

where exclusion of the interval $[0, \gamma_+]$ is understood to mean the exclusion of a small neighborhood in the first quadrant containing the interval $[0, \gamma_+]$. Note that the curves in the $\xi$-plane in Figure 1 (and in the following figures) are drawn directly from the explicit expression (42). In other words, they are the exact curves with specified parameters $A$ and $B$. These curves also indicate that all branch cuts are correctly chosen in (42).

Next, we consider the case $0 < \gamma_- < \gamma_+$. Here, we have the same properties as in the previous case, except property (iv). In this case (see Figure 2b), as $z$ decreases from $\gamma_+$ to $\gamma_-$ (along the upper edge of the line segment $[\gamma_-, \gamma_+]$) we again have $\operatorname{Re} \xi = \xi_0 \equiv \xi(\gamma_+)$ while $\operatorname{Im} \xi$ decreases. However, as $z$ decreases from $\gamma_-$ to 0, $\operatorname{Re} \xi$ increases while $\operatorname{Im} \xi$ stays as a constant. As $z$ increases from 0 to $i$ along the imaginary axis, both $\operatorname{Re} \xi$ and $\operatorname{Im} \xi$ will decrease to $-\infty$. Based on these observations, we can define

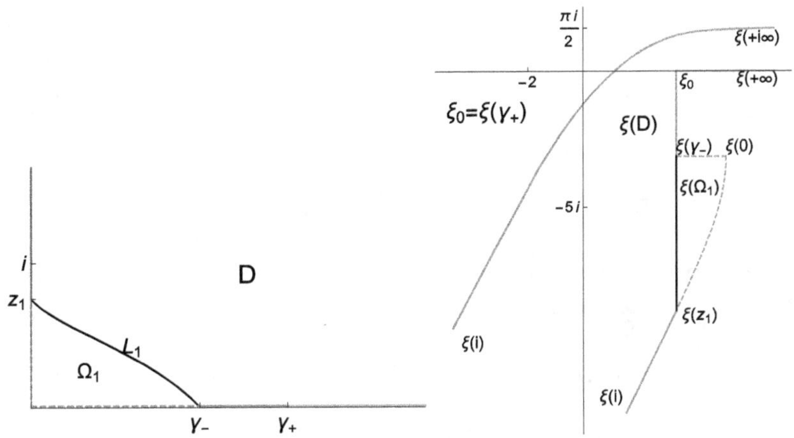


(a) z-plane, domain D (b) $\xi$-plane, with $A = 2, B = 4$

**Figure 2.** The mapping $\xi(z)$ for $z \in D$ in (53) with $0 < \gamma_- < \gamma_+$.

a curve $L_1$ in the first quadrant of the $z$-plane by

$$\operatorname{Re} \xi(z) = \xi_0, \quad \text{or equivalently} \quad \operatorname{Re}\left(\int_{\gamma_-}^{z} \frac{\sqrt{(t-\gamma_-)(t-\gamma_+)}}{1+t^2} dt\right) = 0, \tag{51}$$

which connects $\gamma_-$ and a point $z_1 = iy_1$ $(0 < y_1 < 1)$ on the positive imaginary axis, and denote the domain bounded by the axes and $L_1$ by $\Omega_1$; i.e.,

$$\begin{aligned}\Omega_1 = \Big\{ z \mid \operatorname{Re} z \geq 0, \quad \operatorname{Im} z \geq 0, \quad \xi_0 \leq \operatorname{Re} \xi(z) \leq \operatorname{Re} \xi(0), \\ \operatorname{Im} \xi(z_1) \leq \operatorname{Im} \xi(z) \leq \operatorname{Im} \xi(0) \Big\}.\end{aligned} \tag{52}$$

Our choice of the domain $D$ is defined by

$$D = \{z \mid \operatorname{Re} z \geq 0, \quad \operatorname{Im} z \geq 0\} \setminus ([\gamma_-, \gamma_+] \cup \Omega_1), \quad \text{if} \quad 0 < \gamma_- < \gamma_+, \tag{53}$$

where small neighborhoods of the interval $[\gamma_-, \gamma_+]$ and the curve $L_1$ are again excluded.

The final case $\gamma_- < \gamma_+ < 0$ can be dealt with in a similar manner. We define a curve $L_2$ in the second quadrant of $z$-plane (see Figure 3) by

$$\operatorname{Re} \xi(z) = \xi_0, \quad \text{or equivalently} \quad \operatorname{Re}\left(\int_{\gamma_+}^{z} \frac{\sqrt{(t-\gamma_-)(t-\gamma_+)}}{1+t^2} dt\right) = 0, \tag{54}$$

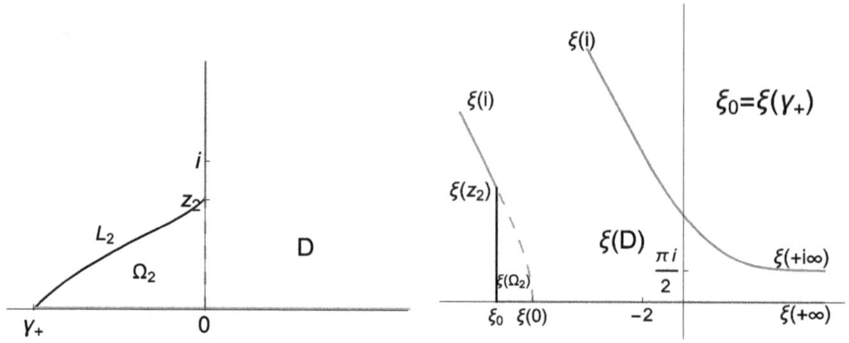


(a) z-plane, domain D ($\Omega_2 \subset D$) (b) $\xi$-plane, with $A = 2, B = -4$

**Figure 3.** The mapping $\xi(z)$ for $z \in D$ in (56) with $\gamma_- < \gamma_+ < 0$.

which connects $\gamma_+$ and a point $z_2 = iy_2$ $(0 < y_2 < 1)$ on the positive imaginary axis, and denote the domain bounded by the axes and $L_2$ by $\Omega_2$; i.e.,

$$\Omega_2 = \Big\{ z \,|\, \mathrm{Re} z \le 0, \quad \mathrm{Im} z \ge 0, \quad \xi_0 \le \mathrm{Re}(\xi(z)) \le \xi(0), \\ 0 \le \mathrm{Im}(\xi(z)) \le \mathrm{Im}(\xi(z_2)) \Big\}. \tag{55}$$

The relevant domain $D$ is defined by

$$D = \{z \,|\, \mathrm{Re} z \ge 0, \quad \mathrm{Im} z \ge 0\} \cup \Omega_2 \setminus [\gamma_+, \gamma_+ + \delta], \quad \text{if} \quad \gamma_- < \gamma_+ < 0, \tag{56}$$

where a small neighborhood of $L_2$ can be included but a small neighborhood of the point $\gamma_+$ is excluded $(\delta > 0)$.

Proposition 3. *Let the complex domain $D$ be defined as in* (50), (53), *and* (56) *for each of the three cases considered above. In view of the symmetry properties* (9) *and* (10) *of the P-J polynomials, by taking conjugation and negation this domain $D$ can be used to cover the whole complex plane except for a neighborhood of the critical interval* $[\gamma_-, \gamma_+]$.

*Proof*: First, by (10) it is sufficient to analyze the upper half plane $C^+$. Coupling (9) and (10), we have for $z \in C^+ \setminus D$ (approximately the second quadrant)

$$P_n(z; a_n, b_n) = (-1)^n \overline{P_n(\tilde{z}; a_n, -b_n)}, \qquad \tilde{z} = -\bar{z}. \tag{57}$$

For the polynomial $P_n(\tilde{z}; a_n, -b_n)$, the critical interval is $[\tilde{\gamma}_-, \tilde{\gamma}_+]$, where by Proposition 1 the end-points are related to the previous ones via

$$\tilde{\gamma}_- = -\gamma_+, \quad \tilde{\gamma}_+ = -\gamma_-. \tag{58}$$

For the case $\gamma_- \leq 0 \leq \gamma_+$, the region $z \in C^+ \setminus (D \cup [\gamma_-, \gamma_+])$ corresponds to the region $\tilde{z} \in \tilde{D}$ for $P_n(\tilde{z}; a_n, -b_n)$, where $\tilde{D}$ is naturally defined by (50) with quantities $z$ and $\gamma_+$ replaced by $\tilde{z}$ and $\tilde{\gamma}_+$, respectively. Thus, the region $C^+ \setminus [\gamma_-, \gamma_+]$ is fully addressed.

For the case $0 < \gamma_- < \gamma_+$, we only need to worry about the excluded region $\Omega_1$ or the curve $L_1$ defined by (51), because by conjugation and negation one can readily see that $\Omega_1$ is just the same region $\Omega_2$ to be considered in the final case $\gamma_- < \gamma_+ < 0$. Let

$$g(z) \equiv \frac{\sqrt{z - \gamma_-}\sqrt{z - \gamma_+}}{1 + z^2}. \tag{59}$$

Then, (51) becomes $\mathrm{Re}(\int_{\gamma_-}^{z} g(t)dt) = 0$. Upon differentiation, we have $\mathrm{Re}\,[g(z)dz] = 0$. If the points on $L_1$ are denoted by $z = x + iy$, then we obtain the differential equation

$$\frac{dx}{dy} = \frac{\mathrm{Im}(g(z))}{\mathrm{Re}(g(z))}, \quad x(0) = \gamma_-. \tag{60}$$

For the polynomial $P_n(\tilde{z}; a_n, -b_n)$ on right-hand side of (57), in view of (58) we should define the domain $\tilde{D}$ for $\tilde{z}$ by (56). The associated curve $\tilde{L}_2$ is defined by (54) with $z$ and $\gamma_+$ replaced by $\tilde{z}$ and $\tilde{\gamma}_+$, respectively. Similarly let

$$\tilde{g}(\tilde{z}) \equiv \frac{\sqrt{\tilde{z} - \tilde{\gamma}_-}\sqrt{\tilde{z} - \tilde{\gamma}_+}}{1 + \tilde{z}^2}. \tag{61}$$

If the points on $\tilde{L}_2$ are denoted by $\tilde{z} = \tilde{x} + i\tilde{y}$, then we obtain the differential equation

$$\frac{d\tilde{x}}{d\tilde{y}} = \frac{\mathrm{Im}(\tilde{g}(\tilde{z}))}{\mathrm{Re}(\tilde{g}(\tilde{z}))}, \quad \tilde{x}(0) = \tilde{\gamma}_+ = -\gamma_-. \tag{62}$$

With $\tilde{z} = -\bar{z}$, it is easy to verify that

$$\tilde{g}(\tilde{z}) = -\overline{g(z)}, \quad \text{and} \quad \frac{\mathrm{Im}(\tilde{g}(\tilde{z}))}{\mathrm{Re}(\tilde{g}(\tilde{z}))} = -\frac{\mathrm{Im}(g(z))}{\mathrm{Re}(g(z))}. \tag{63}$$

A combination of (60), (62), and (63) shows that the two curves $L_1$ and $\tilde{L}_2$ are symmetric with respect to the imaginary axis. Therefore, the excluded region $\Omega_1$ in (53) for $z$ corresponds to exactly the region $\tilde{\Omega}_2$ for $\tilde{z}$. This means that the above defined domains $D$ and $\tilde{D}$ are sufficient to cover $C^+ \setminus [\gamma_-, \gamma_+]$. Because the argument for the case $\gamma_- < \gamma_+ < 0$ is entirely similar, to save space we skip it all together. ∎

### 3.3. *Estimates in complex domain D*

In this subsection, we aim to extend the region of validity of Proposition 2 from the interval $I$ on the real line to the domain $D$ in the complex plane.

In this domain, there are two regular singular points $z = \infty$ and $z = i$, originating from the differential equation (26). We first present estimates for the coefficients $w_s^{\pm}(\xi)$ in $W_{n,p}^{\pm}$ near these two points. To this end, it is convenient to treat them as functions of the original variable $z$, and for simplicity, we continue to adopt the notations $w_s^{\pm}(z)$ for $w_s^{\pm}(\xi(z))$; see the sentence following (48). Now, we rewrite the recursive formula (33) as follows

$$w_{s+1}^{\pm}(z) = \mp\frac{1}{2}\phi(z)w_s^{\pm}(z) - \frac{1}{2}[w_s^{\pm}(z)]' f_0^{-1/2}(z) + \frac{1}{2}\int^z G_{\pm}(t)w_s^{\pm}(t)dt,$$
$$s = 0, 1, \ldots, \tag{64}$$

where $w_0^{\pm} = 1$ and

$$\begin{aligned} G_{\pm}(z) &\equiv \left(\psi - \frac{1}{4}\phi^2\right) f_0^{1/2} \mp \frac{1}{2}\frac{d\phi}{dz} \\ &= \frac{f_2(z)}{f_0^{1/2}(z)} - \frac{1}{f_0^{1/4}(z)}\left[\frac{1}{f_0^{1/4}(z)}\right]'' - \frac{f_1^2(z)}{4f_0^{3/2}(z)} \mp \frac{1}{2}\left(\frac{f_1(z)}{f_0(z)}\right)'. \end{aligned} \tag{65}$$

In the above, all functions $w_s^{\pm}, \phi, \psi$ are treated as functions of the original variable $z$, and prime denotes the derivative with respect to $z$. In deriving (64) and (65), we have made use of (28) and (30).

Lemma 1. *The coefficients $w_s^{\pm}(z)$ ($s \geq 1$) obtained recursively from* (64) *are bounded as $z \to \infty$, and can be uniquely determined by choosing specific integration constants such that*

$$w_s^{\pm}(z) = O(z^{-1}), \quad as \quad z \to \infty. \tag{66}$$

*Proof*: We consider only the case $w_s^{-}(z)$, as the proof for $w_s^{+}(z)$ is essentially the same. First, we examine the case $s = 0$. Because $f_i(z)$ ($i = 0, 1, 2$) are rational functions by (26) and (27), as $z \to \infty$ we have

$$f_0(z) \sim \frac{1}{z^2}\sum_{s=0}^{\infty} f_{0,s}z^{-s}, \quad f_1(z) \sim \frac{1}{z^2}\sum_{s=0}^{\infty} f_{1,s}z^{-s}, \quad f_2(z) \sim \frac{1}{z^2}\sum_{s=0}^{\infty} f_{2,s}z^{-s}, \tag{67}$$

where

$$f_{0,0} = (A-1)^2, \quad f_{1,0} = (A-1)(2A_0-1), \quad f_{2,0} = A_0(A_0-1). \tag{68}$$

It is clear that $\phi(z) = f_1(z)/f_0(z)$ is bounded for large $z$; in fact, we have

$$\phi(z) = \frac{f_{1,0}}{f_{0,0}} + O(z^{-1}), \quad \text{as} \quad z \to \infty. \tag{69}$$

Thus, it remains to consider the integral in (64). From the expansions in (67), it follows that as $z \to \infty$

$$\frac{f_2(z)}{f_0^{1/2}(z)} = f_{0,0}^{-1/2} f_{2,0} \frac{1}{z} + O(z^{-2}),$$

$$\frac{-1}{f_0^{1/4}(z)} \left[ \frac{1}{f_0^{1/4}(z)} \right]'' = \frac{1}{4} f_{0,0}^{-1/2} \frac{1}{z} + O(z^{-2}),$$

$$\frac{-f_1^2(z)}{4 f_0^{3/2}(z)} = -\frac{1}{4} f_{0,0}^{-3/2} f_{1,0}^2 \frac{1}{z} + O(z^{-2}), \quad \left( \frac{f_1(z)}{f_0(z)} \right)' = O(z^{-2}); \tag{70}$$

together with (65) and (68), we obtain

$$G_-(z) = f_{0,0}^{-1/2} \left[ f_{2,0} + \frac{1}{4} - \frac{1}{4} f_{0,0}^{-1} f_{1,0}^2 \right] \frac{1}{z} + O(z^{-2}) = O(z^{-2}). \tag{71}$$

Integrating this gives an $O(z^{-1})$ term, except for an integration constant. Therefore, from (64), $w_1^-$ is bounded and

$$w_1^-(z) = C + O(z^{-1}), \quad \text{as} \quad z \to +\infty. \tag{72}$$

One can choose a suitable integration constant to make $C = 0$, leading to $w_1^-(z) = O(z^{-1})$.

Now we proceed by mathematical induction. Suppose $w_s^-(z)$ satisfies (66) for some $s \geq 1$, and we need to show that $w_{s+1}^-(z)$ also satisfies (66). With the estimates (69), (67), and (71), it is clear that as $z \to \infty$,

$$\phi(z) w_s^-(z) = O(z^{-1}), \quad [w_s^-(z)]' f_0^{-1/2}(z) = O(z^{-1}), \tag{73}$$

$$\text{and} \int^z G_-(t) w_s^-(t) dt = C + O(z^{-2}),$$

respectively. Again, one can make $C = 0$ by choosing a suitable integration constant, which is actually achieved by setting the lower limit of the integral to be $+\infty$. Therefore it follows from (64) that $w_{s+1}^-$ satisfies (66). ∎

LEMMA 2. *The coefficients $w_s^{\pm}(z)$ $(s \geq 1)$ recursively determined by* (64) *satisfy*

$$w_s^{\pm}(z) = O(1), \qquad as \quad z \to i. \tag{74}$$

*Proof*: The argument for this lemma follows similar lines as that in Lemma 1; we simply change the singularity at $\infty$ to the point at $i$. Thus, we consider only the case $w_s^-(z)$ as an example, and prove it by mathematical induction. Suppose that $w_s^-(z)$ $(s \geq 0)$ satisfies the stronger condition

$$w_s^-(z) = \text{constant} + O(z - i), \quad \text{as} \quad z \to i, \tag{75}$$

which is obviously true for $s = 0$. By using the recursive formula (64), we aim to show that $w_{s+1}^-(z)$ also satisfies (75). By the formulas of $f_k(z)$ ($k = 0, 1, 2$) in (26) and (27), we have as $z \to i$

$$f_0(z) \sim \frac{1}{(z-i)^2} \sum_{s=0}^{\infty} \bar{f}_{0,s}(z-i)^s, \quad f_1(z) \sim \frac{1}{(z-i)^2} \sum_{s=0}^{\infty} \bar{f}_{1,s}(z-i)^s,$$

$$f_2(z) \sim \frac{1}{(z-i)^2} \sum_{s=0}^{\infty} \bar{f}_{2,s}(z-i)^s, \tag{76}$$

where

$$\bar{f}_{0,0} = \frac{1}{4}(A + iB)^2, \quad \bar{f}_{1,0} = \frac{1}{2}(A + iB)(A_0 + iB_0),$$

$$\bar{f}_{2,0} = \frac{1}{4}\left[(A_0 + iB_0)^2 - 1\right]. \tag{77}$$

From (75) and (76) and the definition $\phi = f_1/f_0$, it follows that

$$\phi(z)w_s^-(z) = \text{constant} + O(z-i), \quad [w_s^-(z)]' f_0^{-1/2}(z) = O(z-i),$$

$$\text{as} \quad z \to i. \tag{78}$$

Thus, the first two terms in (64) satisfy the estimate in (75). Similar to (71), the function $G_-(z)$ has the estimate

$$G_-(z) = \bar{f}_{0,0}^{-1/2}\left[\bar{f}_{2,0} + \frac{1}{4} - \frac{1}{4}\bar{f}_{0,0}^{-1}\bar{f}_{1,0}^2\right]\frac{1}{z-i} + O(1) = O(1), \tag{79}$$

where (77) has been used in the last equality. This together with (75) implies

$$\int^z G_-(t)w_s^-(t)dt = \text{constant} + O(z-i), \quad \text{as} \quad z \to i. \tag{80}$$

Coupling (78) and (80) completes the proof. ■

Now we summarize the result in the following proposition.

Proposition 4. *For the domain $D$ defined by* (50), (53), *and* (56) *for each of the three cases considered in Proposition 3, and for any positive integer $p$, the differential equation* (29) *has solutions $W_{n,p}^{\pm}$, which are holomorphic in $D$ and satisfy*

$$W_{n,p}^{\pm}(z) = e^{\pm[n\xi(z)+X(z)]} \sum_{s=0}^{p-1} \frac{(\pm 1)^s w_s^{\pm}(z)}{n^s} + \varepsilon_p^{\pm}(z), \quad w_0^{\pm} = 1, \tag{81}$$

*where $\xi(z)$ and $X(z)$ are given in* (42) *and* (46), *and $w_s^{\pm}(z)$ ($s \geq 1$) are recursively determined by* (64) *and Lemma 1. For $n \geq n_0$ ($n_0$ finite), the* *remainders* $\varepsilon_p^{\pm}(z)$ *are bounded by*

$$\left|e^{\pm[n\xi+X(\xi)]}\right| \frac{\kappa_1 \mathcal{V}_{\alpha_\pm,\xi}(w_p^{\pm})}{n^p} \exp\left\{\frac{\kappa_0 \mathcal{V}_{\alpha_\pm,\xi}(H)}{n}\right\}, \tag{82}$$

*where* $\alpha_+ = \xi(i), \alpha_- = +\infty$, *$H$ is defined in* (38), *and* $\kappa_0, \kappa_1$ *are positive constants.*

This proposition is a consequence of Theorem 9.1 in [25, p. 385]. Here, we only make a few remarks, following Remark 2.

*Remark* 3. First, it is easily verified that the two functions $\phi(\xi(z))$ and $\psi(\xi(z))$ in (30) (as a function of $z$) are holomorphic in $D$. Second, it is readily seen from (28) and the construction of $D$ that $\xi'(z) \neq 0$ for all $z \in D$ and one can choose a path such that from $\alpha_+$ and $\alpha_-$ (see the definitions following (82)) to $\xi$, the quantity Re $\xi$ is nondecreasing and nonincreasing, respectively. So, for $n \geq n_0$ (some finite $n_0$), the monotonicity requirement for $\mathrm{Re}(n\xi + X)$ in Remark 2 is satisfied in $D$ except for some finite region away from singularities at $\infty$ and $i$. Because for any finite region $\mathrm{Re}(n\xi)$ satisfies the above property, the function $X$ in this finite region contributes only some constants $\kappa_0, \kappa_1$ in (82) when estimating the kernel in the integral equation of $\varepsilon_p^{\pm}(z)$ (see [25, p. 384, (9.14)]). Another way to use Olver's Theorem 9.1 mentioned above is to apply his result directly without creating the constants $\kappa_1$ and $\kappa_0$ in (82). However, this will require a modification of the domain $D$. We will also have to choose a smaller domain $D_n$ for each $n$ in Proposition 3 to guarantee the monotonicity of Re $(n\xi + X)$ and to require $D_n$ to tend to $D$ for large $n$ so that this will not affect the result in Theorem 1 below.

*Remark* 4. The results in Lemmas 1 and 2 can be used to show that the total variations $\mathcal{V}_{\alpha_\pm,\xi}(w_p^{\pm})$ are finite for all $z \in D$ including $\infty$. The estimation of the term $\mathcal{V}_{\alpha_\pm,\xi}(H)$ is very similar to that for the coefficients $w_s^{\pm}$ in the two lemmas. If we suppress the function $\xi(z)$ in $H(\xi(z); n)$ and simply write $H(z)$ for $H(\xi(z); n)$, then the function $H(z)$ is given by

$$H(z) = \int^z \left\{\left[\psi(t) - \frac{1}{4}\phi^2(t)\right] f_0^{1/2}(t) + \frac{[\phi'(t) f_0^{-1/2}(t)]'}{4n + 2\phi(t)} - \frac{3\phi'^2(t) f_0^{-1/2}(t)}{[4n + 2\phi(t)]^2}\right\} dt, \tag{83}$$

where we have also written $\psi(t), \phi(t)$, and $f_0(t)$ for $\psi(\xi(t)), \phi(\xi(t))$, and $f_0(\xi(t))$, respectively. The term $[\psi - \frac{1}{4}\phi^2] f_0^{1/2}$ in (83) has already appeared in $G_\pm$ in (65), and other terms can be similarly estimated as

$$\left[\phi' f_0^{-1/2}\right]' = O(z^{-2}), \quad \phi'^2 f_0^{-1/2} = O(z^{-3}), \quad \text{as} \quad z \to \infty,$$

$$\left[\phi' f_0^{-1/2}\right]' = O(1), \quad \phi'^2 f_0^{-1/2} = O(z - i), \quad \text{as} \quad z \to i. \tag{84}$$

Also, because $\phi$ is finite, $1/(4n + 2\phi)$ is positive and finite for $n \geq n_0$ (some finite $n_0$). All these together show that the two total variations $\mathcal{V}_{\alpha_\pm,\xi}(w_p^\mp)$ and $\mathcal{V}_{\alpha_\pm,\xi}(H)$ are finite for $\xi$ in the image of $D$ under the mapping $\xi = \xi(z)$.

### 3.4. Matching

To express the P-J polynomial $P_n(z; a_n, b_n)$ in terms of the two solutions $W_{n,p}^\pm(z)$ in (81), now we analyze their behavior as $z \to +\infty$ (or $\xi \to +\infty$) and for fixed $n$. From (81), (43), (47) and Lemma 1, it follows that for any $p \geq 1$ and any $n \geq n_0$ (see Proposition 4), we have

$$W_{n,p}^-(z) \sim e^{-n\xi - X} \sim z^{(1-A)n+\frac{1}{2}-A_0}, \quad W_{n,p}^+(z) = O(z^{(A-1)n+A_0-\frac{1}{2}}),$$
$$\text{as} \quad z \to +\infty; \tag{85}$$

for the first formula we have also made use of the fact that $\varepsilon_p^-(z)$ tends to 0, because $\alpha_- = +\infty$ in the bound given in (82).

By the asymptotic behavior of $P_n(z; a_n, b_n)$ ($\sim z^n$) and the transformations (25) and (28), we must have

$$y_n(z) \sim e^{b_n\pi/2} z^{(1-A)n+1-A_0}, \quad W_n(\xi) \sim \sqrt{A-1}\, e^{b_n\pi/2} z^{(1-A)n+\frac{1}{2}-A_0},$$
$$z \to +\infty. \tag{86}$$

Because $W_n$ is a linear combination of the two solutions $W_{n,p}^\pm$, by matching the behavior we get

$$W_n(\xi) = \sqrt{A-1}\, e^{b_n\pi/2} W_{n,p}^-(z); \tag{87}$$

tracing backward, we obtain from (25) and (28)

$$\begin{aligned} P_n(z; a_n, b_n) &= \sqrt{A-1}\, e^{b_n\pi/2} [w(z)]^{-1/2} [Q_0(z)]^{-1/4} W_{n,p}^-(z) \\ &= e^{b_n\pi/2} [w(z)]^{-1/2} [(z-\gamma_-)(z-\gamma_+)]^{-1/4} W_{n,p}^-(z), \end{aligned} \tag{88}$$

where $w(z)$ is given by (13).

Note that although the functions $\xi(z)$, $X(z)$ in (42) and (46) have branch points at $z = \pm i$, in view of (13) the singular terms $\log(1 \pm iz)$ in the leading-order term of $W_{n,p}^-(z)$ will cancel out with the term $[w(z)]^{-1/2}$ in (88). Furthermore, by Lemma 2 the coefficients $w_s^-(z)$ are smooth at $z = i$. Therefore, the right-hand side of (88) is homomorphic in $D$ as expected. Because the above result is valid for any $p$, we have established the asymptotic nature of the infinite series. Now, we summarize our first main result in the following theorem.

THEOREM 1. *For $z \in D$ given by (50), (53), or (56), the P-J polynomials $P_n(z; a_n, b_n)$ with parameters in (11) have the uniform asymptotic expansion*

$$P_n(z; a_n, b_n) \sim e^{b_n\pi/2}[(z-\gamma_-)(z-\gamma_+)]^{-1/4} e^{-n\xi^*(z)-X^*(z)} \sum_{s=0}^{\infty} \frac{(-1)^s w_s^-(z)}{n^s}, \tag{89}$$

*as $n \to \infty$, where $\xi^*(z)$ and $X^*(z)$ are obtained by removing the* $\log(1 \pm iz)$ *terms from $\xi(z)$ and $X(z)$ in (42) and (46), and the coefficients $w_s^-(z)$ ($s \geq 1$) can be recursively determined by (64) and Lemma 1 with $w_0^-(z) = 1$.*

*Remark* 5. Because both singularities at $z = \infty$ and $z = i$ of eq. (26) are regular, it is known that they can be included in the outer region where the asymptotic formula involves only elementary functions. If $z = i$ were a simple pole rather than a double pole in (26), then the asymptotic formula in the neighborhood of this point would involve Bessel functions; see [25, Chapter 12]. Numerical evidence shows that the above result is valid in a much larger domain, i.e., the whole complex plane except for a neighborhood of the critical interval.

## 4. Uniform asymptotics at turning point

In this section, we intend to derive an asymptotic expansion for $P_n(z; a_n, b_n)$, which holds uniformly in a region that contains a turning point and the critical interval. By symmetry, we can concentrate on the turning point $z = \gamma_+$. In the previous section, the function $\xi(z)$ is not well-defined near $\gamma_+$, which is a branch point. So we need to modify the transformation $z \to \xi$ in (28) and make it holomorphic near $\gamma_+$. The method of deriving uniform expansions is clearly illustrated in Olver [25], and has been used in [26, 27] to deal with orthogonal polynomials with some exponential weights.

### *4.1. Formal series solutions*

As in previous section, we start with the canonical differential equation (26). Following the derivation in [25, p. 426], we make the transformation

$$\zeta(z) = \left[\frac{3}{2}\int_{\gamma_+}^{z} \sqrt{f_0(t)}dt\right]^{2/3}, \quad y_n(z) = \tilde{f}_0^{-1/4}\widetilde{W}_n(\zeta), \tag{90}$$

where $\tilde{f}_0 = f_0/\zeta$. The new function $\widetilde{W}_n(\zeta)$ satisfies

$$\widetilde{W}_n''(\zeta) = [n^2\zeta + n\tilde{\phi}(\zeta) + \tilde{\psi}(\zeta)]\widetilde{W}_n(\zeta), \tag{91}$$

where

$$\tilde{\phi}(\zeta) = \frac{f_1(z)}{\tilde{f}_0(z)}, \quad \tilde{\psi}(\zeta) = \frac{f_2(z)}{\tilde{f}_0(z)} - \frac{1}{\tilde{f}_0^{3/4}(z)} \frac{d^2}{dz^2}\left[\frac{1}{\tilde{f}_0^{1/4}(z)}\right]. \tag{92}$$

If $\tilde{\phi} = 0$, then the leading approximation would be a linear combination of $\mathrm{Ai}(n^{2/3}\zeta)$ and $\mathrm{Bi}(n^{2/3}\zeta)$. For the current situation, we need to define an additional function

$$\Phi(\zeta) = \frac{1}{2\zeta^{1/2}} \int_0^{\zeta} \frac{\tilde{\phi}(v)}{v^{1/2}} dv. \tag{93}$$

Now, we try a series solution of the form

$$\begin{aligned} \widetilde{W}_n^-(\zeta) &= \mathrm{Ai}\left(n^{2/3}\zeta + n^{-1/3}\Phi\right) \sum_{s=0}^{\infty} \frac{X_s(\zeta)}{n^s} \\ &\quad + n^{-4/3}\mathrm{Ai}'\left(n^{2/3}\zeta + n^{-1/3}\Phi\right) \sum_{s=0}^{\infty} \frac{Y_s(\zeta)}{n^s}, \end{aligned} \tag{94}$$

with $X_0 = 1$. The other solution $\widetilde{W}_n^+(\zeta)$ is obtained by replacing Ai by Bi in (94). The coefficients are determined recursively [25, p. 428] by

$$2\zeta Y_s' + Y_s = R_{1,s}, \quad 2X_{s+1}' = R_{2,s}, \quad s \geq 0, \tag{95}$$

where

$$\begin{aligned} R_{1,s} &= \left[\tilde{\psi} - (2\Phi\Phi' + \zeta\Phi'^2)\right] X_s - X_s'' - \Phi\Phi'^2 X_{s-1} - (2\Phi' + \zeta\Phi'')Y_{s-1} \\ &\quad -2(\zeta\Phi' + \Phi)Y_{s-1}' - (\Phi\Phi'' + \Phi'^2)Y_{s-2} - 2\Phi\Phi' Y_{s-2}', \\ R_{2,s} &= -\Phi'' X_s - 2\Phi' X_s' + \left[\tilde{\psi} - (2\Phi\Phi' + \zeta\Phi'^2)\right] Y_{s-1} - Y_{s-1}'' - \Phi\Phi'^2 Y_{s-2}. \end{aligned} \tag{96}$$

By integration of (95), we may choose

$$Y_s(\zeta) = \frac{1}{2\zeta^{1/2}} \int_0^{\zeta} R_{1,s}(v) \frac{dv}{v^{1/2}}, \quad X_{s+1}(\zeta) = \frac{1}{2} \int^{\zeta} R_{2,s}(v) dv, \tag{97}$$

as solutions of the equations in (95).

Lemma 3. *All the functions $\zeta(z), \tilde{\phi}(\zeta), \tilde{\psi}(\zeta), \Phi(\zeta), X_s(\zeta), Y_s(\zeta)$ $(s \geq 0)$ defined in* (90), (92), (93), *and* (97) *are analytic near $z = \gamma_+$, where functions of $\zeta$ are being considered as functions of $z$ through $\zeta(z)$.*

*Proof*: This can in fact be considered as a direct consequence of Lemma 3.1 and Lemma 7.1 in [25, p. 399 & p. 410]. Here, we present only a rough sketch of the proof. First, we examine $\zeta(z)$ defined in (90). By (26) and (27), the integrand $\sqrt{f_0}$ in (90) can be expanded in a neighborhood of $z = \gamma_+$ as

$$\sqrt{f_0(z)} = (z-\gamma_+)^{1/2} \sum_{s=0}^{\infty} c_s (z-\gamma_+)^s, \quad c_0 = \frac{(A-1)\sqrt{\gamma_+ - \gamma_-}}{1+\gamma_+^2} > 0. \quad (98)$$

Thus, the integral in (90) will have the form $\frac{2}{3}(z-\gamma_+)^{3/2}(c_0 + \cdots)$. Hence, it follows that $\zeta(z)$ is analytic near $z = \gamma_+$ and has the behavior

$$\zeta(z) \sim c_0^{2/3}(z-\gamma_+), \quad \text{as} \quad z \to \gamma_+. \quad (99)$$

From (98), (99) and the definition of $\tilde{f}_0$, we have

$$\tilde{f}_0(z) = f_0(z)/\zeta = c_0^{4/3} + O(z-\gamma_+), \quad \text{as} \quad z \to \gamma_+. \quad (100)$$

Because $f_1(z)$ is analytic at $z = \gamma_+$ on account of (26) and (27), it follows from (92) that $\tilde{\phi}$ is analytic near $z = \gamma_+$. Near $z = \gamma_+$, (99) gives

$$\tilde{\phi}(\zeta) = \sum_{s=0}^{\infty} d_s (z-\gamma_+)^s = \sum_{s=0}^{\infty} \tilde{d}_s \zeta^s. \quad (101)$$

Further, by definition (93), termwise integration shows that $\Phi(\zeta)$ is also analytic near $z = \gamma_+$ and

$$\Phi(0) = d_0 = f_1(\gamma_+) c_0^{-4/3}. \quad (102)$$

Similarly one can show that $X_s(\zeta)$ and $Y_s(\zeta)$ ($s \geq 0$) are also analytic near $z = \gamma_+$, because no singularity is involved in the functions $R_{1,s}$ and $R_{2,s}$ in (96) in a neighborhood of $z = \gamma_+$. ■

The two functions $\zeta(z)$ and $\Phi(\zeta)$ are related to the previously defined functions $\xi(z)$ and $X(z)$ in (42) and (46). More precisely, we have from definitions (90) and (93)

$$\zeta(z) = \left[\frac{3}{2}(\xi(z) - \xi_0)\right]^{2/3}, \quad \xi_0 \equiv \xi(\gamma_+), \quad (103)$$

and

$$\begin{aligned} \Phi(\zeta(z)) &= \frac{1}{2\zeta^{1/2}(z)} \int_{\gamma_+}^{z} \frac{f_1(t)}{\sqrt{f_0(t)}} dt = \frac{1}{2\zeta^{1/2}(z)} \int_{\gamma_+}^{z} \frac{Q_1(t)}{\sqrt{Q_0(t)}(1+t^2)} dt \\ &= \zeta^{-1/2}(z)\left[X(z) - X_0\right], \qquad X_0 \equiv X(\gamma_+), \end{aligned} \quad (104)$$

where we have used the identity $\zeta^{1/2}\zeta' = \xi' = \sqrt{f_0}$ in the first equality.

It is clear that based on Lemma 3, the solutions $\widetilde{W}_n^{\pm}$ are well-defined near $z = \gamma_+$. By following the discussions in [25, pp. 428–429], one can easily establish the asymptotic nature of the formal solutions $\widetilde{W}_n^{\pm}(\zeta(z))$ in the compact interval $[\gamma_- + \delta, M]$, where $\delta > 0$ and $\gamma_+ < M < \infty$. However, a more careful examination shows that these two solutions are not uniformly valid all the way to infinity; for instance, they are not valid on the interval $[\gamma_- + \delta, +\infty)$. In particular, the term $Y_0(\zeta)$ is not bounded as $\zeta$ or $z$ tends to infinity; see (96), (97), (115), and (119). So, before we analyze the error term in the next subsection, we will first modify the two solutions to make them valid uniformly at infinity. From a practical point of view, the compact interval $[\gamma_- + \delta, M]$ is sufficient to cover the critical interval. However, here we seek a solution, which is valid in a much more global region, because theoretically speaking, it is always good to have approximations valid in regions as global as possible and the method to be used may also provide some guidance for other similar applications in the future.

### 4.2. *Uniform solutions valid at infinity*

There are a few remedies that can be used to make an asymptotic approximation valid uniformly at infinity and the basic idea is to modify the transformation $\zeta(z)$ in (90) (or $\xi(z)$ in (28)) appearing as the argument in the Airy functions in (94) (or the exponential function in (31)). The simplest way is to change the large parameter $n$ by a small perturbation; e.g.,

$$n = N + \alpha, \tag{105}$$

with $\alpha$ being a constant to be determined later; see Lemma 4 below. Equation (91) then becomes

$$\widetilde{W}_n''(\zeta) = [N^2\zeta + N\hat{\phi}(\zeta) + \hat{\psi}(\zeta)]\widetilde{W}_n(\zeta), \tag{106}$$

where

$$\hat{\phi}(\zeta) = \tilde{\phi}(\zeta) + 2\alpha\zeta, \quad \hat{\psi}(\zeta) = \tilde{\psi}(\zeta) + \alpha\tilde{\phi}(\zeta) + \alpha^2\zeta. \tag{107}$$

We can similarly define an auxiliary function $\hat{\Phi}$ by

$$\hat{\Phi}(\zeta) = \frac{1}{2\zeta^{1/2}} \int_0^\zeta \frac{\hat{\phi}(v)}{v^{1/2}} dv = \Phi(\zeta) + \frac{2}{3}\alpha\zeta. \tag{108}$$

Accordingly, we try solutions of the form

$$\widehat{W}_n^-(\zeta) = \mathrm{Ai}\,(\zeta_N) \sum_{s=0}^{\infty} \frac{\hat{X}_s(\zeta)}{N^s} + N^{-\frac{4}{3}}\mathrm{Ai}'\,(\zeta_N) \sum_{s=0}^{\infty} \frac{\hat{Y}_s(\zeta)}{N^s}, \tag{109}$$

where

$$\zeta_N = N^{2/3}\zeta + N^{-1/3}\hat{\Phi}(\zeta), \tag{110}$$

and $\hat{X}_0 = 1$. The recursive formulas for $\hat{X}_{s+1}$ and $\hat{Y}_s$ $(s \geq 0)$ are similar to those given in (97) and we have

$$\hat{Y}_s(\zeta) = \frac{1}{2\zeta^{1/2}} \int_0^\zeta \hat{R}_{1,s}(v) \frac{dv}{v^{1/2}}, \quad \hat{X}_{s+1}(\zeta) = \frac{1}{2} \int^\zeta \hat{R}_{2,s}(v) dv, \tag{111}$$

where $\hat{R}_{1,s}$ and $\hat{R}_{2,s}$ are defined in (96) with $\Phi, \tilde{\psi}, X_k, Y_k$ replaced by $\hat{\Phi}, \hat{\psi}, \hat{X}_k, \hat{Y}_k$. In particular, for $s = 0$ we have

$$\hat{Y}_0(\zeta) = \frac{1}{2\zeta^{1/2}} \int_0^\zeta \left[\hat{\psi}(v) - 2\hat{\Phi}(v)\hat{\Phi}'(v) - v\hat{\Phi}'^2(v)\right] \frac{dv}{v^{1/2}}, \tag{112}$$

$$\hat{X}_1(\zeta) = \frac{1}{2}\int^\zeta -\hat{\Phi}''(v)dv = -\frac{1}{2}\hat{\Phi}'(\zeta) + \text{constant}.$$

The other solution $\widehat{W}_n^+(\zeta)$ is obtained by replacing Ai by Bi in (109). Lemma 3 implies that all functions $\hat{\phi}, \hat{\psi}, \hat{\Phi}, \hat{X}_s, \hat{Y}_s$ $(s \geq 0)$ are analytic near $z = \gamma_+$. As $z$ or $\zeta$ tends to infinity, we have the following lemma.

LEMMA 4. *With $\alpha = -(2A_0 - 1)/2(A - 1)$ in (105), the coefficients $\hat{X}_{s+1}(\zeta)$ and $\hat{Y}_s(\zeta)$ $(s \geq 0)$ in (111) are bounded as $\zeta \to +\infty$. By choosing specific integration constants in $\hat{X}_{s+1}$, they can be uniquely determined so that*

$$\hat{X}_{s+1}(\zeta) = O(\zeta^{-3/2}), \quad \hat{Y}_s(\zeta) = \hat{Y}_{s,0}\zeta^{-1/2} + O(\zeta^{-2}), \quad as \quad \zeta \to +\infty, \tag{113}$$

*where $\hat{Y}_{s,0}$ is some constant defined by* (128).

*Proof*: First, we estimate the two functions $\hat{\psi}(\zeta)$ and $\hat{\Phi}(\zeta)$ in formula (112) as $\zeta$ tends to infinity. It follows from definition (103) and condition (43) that

$$\zeta \sim \left[\frac{3}{2}(A-1)\log z - \frac{3}{2}\xi_0\right]^{2/3}, \quad \text{as} \quad z \to +\infty, \tag{114}$$

which implies that terms of order $O(z^{-1})$ are exponentially small in comparison with the function $\zeta(z)$. Based on the expansions in (67) for $f_i(z)$ $(i = 0, 1, 2)$ and definitions (92) and (93), we have

$$\tilde{\phi}(\zeta) = \frac{f_{1,0}}{f_{0,0}}\zeta + O(\zeta z^{-1}), \quad \Phi(\zeta) = \frac{f_{1,0}}{3f_{0,0}}\zeta + O(\zeta^{-1/2}), \quad \text{as} \quad \zeta \to +\infty, \tag{115}$$

where the second equation can also be derived from (104). Because

$$\zeta' = \sqrt{f_0}\zeta^{-1/2} \quad \text{and} \quad f_0^{-1/4} = f_{0,0}^{-1/4}\sqrt{z}[1 + O(z^{-1})] \tag{116}$$

by (90) and (67), it follows that as $\zeta \to +\infty$,

$$\begin{aligned}(\tilde{f}_0^{-1/4})' &= (\zeta^{1/4}f_0^{-1/4})' = \zeta^{1/4}(f_0^{-1/4})' + (\zeta^{1/4})'f_0^{-1/4} \\ &= \frac{1}{2}f_{0,0}^{-1/4}\zeta^{1/4}z^{-1/2}[1 + O(z^{-1})] + \frac{1}{4}f_{0,0}^{1/4}\zeta^{-5/4}z^{-1/2}[1 + O(z^{-1})].\end{aligned} \tag{117}$$

Differentiation again gives

$$(\tilde{f}_0^{-1/4})'' = -\frac{1}{4}f_{0,0}^{-1/4}\zeta^{1/4}z^{-3/2}\left[1 + O\left(\zeta^{-3}\right)\right], \quad \text{as} \quad \zeta \to +\infty, \tag{118}$$

where terms of order $O(\zeta^{-5/4}z^{-3/2})$ have canceled out in the calculation. Thus, we get from (92) and (67) that

$$\tilde{\psi}(\zeta) = \zeta\frac{f_{2,0}}{f_{0,0}} + O(\zeta z^{-1}) + \frac{1}{4}f_{0,0}^{-1}\zeta + O(\zeta^{-2}) = \frac{4f_{2,0}+1}{4f_{0,0}}\zeta + O(\zeta^{-2}),$$
$$\text{as } \zeta \to +\infty. \tag{119}$$

For large $\zeta$, by using (115) and (119), one can show that the functions $\hat{\Phi}$ and $\hat{\psi}$ in (107) and (108) have asymptotic formulas

$$\begin{aligned}\hat{\Phi}(\zeta) &= \left(\frac{f_{1,0}}{3f_{0,0}} + \frac{2}{3}\alpha\right)\zeta + O(\zeta^{-1/2}),\\ \hat{\psi}(\zeta) &= \left(\frac{4f_{2,0}+1}{4f_{0,0}} + \alpha\frac{f_{1,0}}{f_{0,0}} + \alpha^2\right)\zeta + O(\zeta^{-2})\\ &= \left(\frac{f_{1,0}}{2f_{0,0}} + \alpha\right)^2\zeta + O(\zeta^{-2}),\end{aligned} \tag{120}$$

where the last equality is obtained from (68). Hence, we can choose

$$\alpha = -\frac{f_{1,0}}{2f_{0,0}} = -\frac{2A_0 - 1}{2(A-1)}, \tag{121}$$

so that

$$\hat{\Phi}(\zeta) = O(\zeta^{-1/2}), \quad \hat{\psi}(\zeta) = O(\zeta^{-2}), \quad \text{as} \quad \zeta \to +\infty. \tag{122}$$

Based on above estimates, the integrand in $\hat{Y}_0$ in (112) is found to satisfy

$$\left[\hat{\psi}(v) - 2\Phi(v)\Phi'(v) - v\Phi'(v)^2\right]v^{-1/2} = O(v^{-5/2}), \quad \text{as} \quad v \to +\infty. \tag{123}$$

So the associated integral is convergent as $\zeta \to \infty$ and we get

$$\hat{Y}_0(\zeta) = \hat{Y}_{0,0}\zeta^{-1/2} + O(\zeta^{-2}), \quad \text{as} \quad \zeta \to +\infty, \tag{124}$$

where $\hat{Y}_{0,0}$ is some constant defined by a convergent integral (cf. (128)). It is easily seen from (112) and (122) that

$$\hat{X}_1(\zeta) = \hat{C}_1 + O(\zeta^{-3/2}), \quad \text{as} \quad \zeta \to +\infty, \tag{125}$$

and $\hat{X}_1(\zeta)$ can be uniquely determined by choosing an integration constant such that $\hat{C}_1 = 0$.

Finally, we establish the estimates in (113) for $\hat{X}_{s+1}$, $\hat{Y}_s$ $(s \geq 1)$ by using mathematical induction. Suppose that as $\zeta \to +\infty$,

$$\hat{X}_{k+1}(\zeta) = O(\zeta^{-3/2}), \quad \hat{Y}_k(\zeta) = \hat{Y}_{k,0}\zeta^{-1/2} + O(\zeta^{-2}), \quad 0 \leq k \leq s-1. \tag{126}$$

Together with (120), we can show that the functions $\hat{R}_{1,s}$ and $\hat{R}_{2,s}$ (defined by (96) with $\Phi$, $\tilde{\psi}$, $X_k$, $Y_k$ replaced by $\hat{\Phi}$, $\hat{\psi}$, $\hat{X}_k$, $\hat{Y}_k$) satisfy

$$\hat{R}_{1,s}(\zeta) = O(\zeta^{-3/2}), \quad \hat{R}_{2,s}(\zeta) = O(\zeta^{-5/2}), \quad \text{as} \quad \zeta \to +\infty. \tag{127}$$

By the first estimate, the associated integral in formula (111) is convergent and hence $\hat{Y}_s$ satisfies (113), where $\hat{Y}_{s,0}$ is given by

$$\hat{Y}_{s,0} = \frac{1}{2}\int_0^\infty \hat{R}_{1,s}(v)\frac{dv}{v^{1/2}}. \tag{128}$$

Coupling the second estimate with the recursive formula (111) gives

$$\hat{X}_{s+1}(\zeta) = \text{constant} + O(\zeta^{-3/2}), \quad \text{as} \quad \zeta \to +\infty. \tag{129}$$

By choosing a suitable integration constant, this establishes (113) in the lemma. ■

Before presenting bounds for the error terms of the truncated general solutions, we recall the modulus function $M(x)$ and the weight function $E(x)$ associated with Airy functions [25, p. 395]. Let $x = c$ denote the negative root of the equation $\mathrm{Ai}(x) = \mathrm{Bi}(x)$ with the smallest absolute value, and define

$$E(x) = [\mathrm{Bi}(x)/\mathrm{Ai}(x)]^{1/2} \quad (c \le x < +\infty), \qquad E(x) = 1 \quad (-\infty < x \le c), \tag{130}$$

and

$$M(x) = \left[E^2(x)\mathrm{Ai}^2(x) + E^{-2}(x)\mathrm{Bi}^2(x)\right]^{1/2}, \tag{131}$$

which have the asymptotic behaviors

$$\begin{aligned} E(x) \sim \sqrt{2}e^{\frac{2}{3}x^{3/2}}, \quad M(x) \sim \pi^{-1/2}x^{-1/4}, \quad &\text{as} \quad x \to +\infty, \\ M(x) \sim \pi^{-1/2}(-x)^{-1/4}, \quad &\text{as} \quad x \to -\infty. \end{aligned} \tag{132}$$

The Airy functions can be written as

$$\mathrm{Ai}(x) = E^{-1}(x)M(x)\sin\theta(x), \quad \mathrm{Bi}(x) = E(x)M(x)\cos\theta(x), \tag{133}$$

where $\theta(x)$ is the phase function. In a complex neighborhood of the real line, we can still define the above functions with $\mathrm{Ai}(x)$ and $\mathrm{Bi}(x)$ replaced by $|\mathrm{Ai}(z)|$ and $|\mathrm{Bi}(z)|$, and we have

$$|\mathrm{Ai}(z)| = E^{-1}(z)M(z)\sin\theta(z), \quad |\mathrm{Bi}(z)| = E(z)M(z)\cos\theta(z); \tag{134}$$

see [25, p. 415]. Let $D^\star$ denote the complex neighborhood of a real interval defined by

$$D^\star = \{z \mid \gamma_- + \delta \le \mathrm{Re}\, z < +\infty, \quad |\mathrm{Im}\, z| \le \delta_1\}, \tag{135}$$

where $\delta, \delta_1$ are some small positive numbers. Figure 4 presents an example of the domain $D^\star$ under the mapping $\zeta(z)$. The curves in the figure are

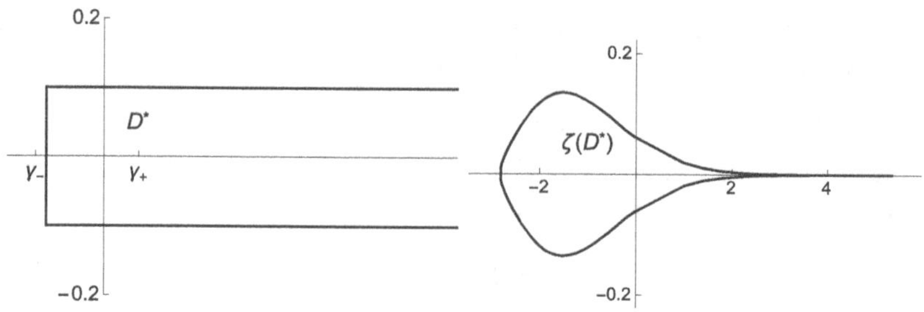


(a) $z$-plane, domain $D^\star$.

(b) $\zeta$-plane, with $A = 2, B = 0$.

**Figure 4.** The mapping $\zeta(z)$ for $z \in D^\star$.

the exact curves based on the definition (103), and one can see that when $z$ tends to infinity in $D^\star$ the function $\zeta(z)$ is almost on the real line.

We are now ready to state our result concerning the asymptotic solutions of differential equation (106). Note that this result is not proved in [25], so here we will give a brief sketch of its proof.

PROPOSITION 5. *For the domain $D^\star$ in* (135) *and for any positive integer $p$, the differential equation* (106) *has solutions $\widehat{W}^{\pm}_{n,p}(\zeta)$, which are holomorphic in $\zeta(D^\star)$ (the image of $D^\star$ under $\zeta = \zeta(z)$) and given by*

$$\begin{aligned} \widehat{W}^{-}_{n,p}(\zeta) &= \mathrm{Ai}\,(\zeta_N)\sum_{s=0}^{p-1}\frac{\hat{X}_s(\zeta)}{N^s} + N^{-\frac{4}{3}}\mathrm{Ai}'\,(\zeta_N)\sum_{s=0}^{p-2}\frac{\hat{Y}_s(\zeta)}{N^s} + \hat{\varepsilon}^{-}_p(\zeta), \\ \widehat{W}^{+}_{n,p}(\zeta) &= \mathrm{Bi}\,(\zeta_N)\sum_{s=0}^{p-1}\frac{\hat{X}_s(\zeta)}{N^s} + N^{-\frac{4}{3}}\mathrm{Bi}'\,(\zeta_N)\sum_{s=0}^{p-2}\frac{\hat{Y}_s(\zeta)}{N^s} + \hat{\varepsilon}^{+}_p(\zeta), \end{aligned} \tag{136}$$

*where $\hat{X}_0 = 1$, $\zeta_N$ is given by* (110) *and $N$ is related to $n$ by* (105) *and Lemma 4. The coefficients $\hat{X}_{s+1}$ and $\hat{Y}_s$ $(s \geq 0)$ are determined by* (111) *and Lemma 4, and for $N \geq N_0$ ($N_0$ finite) the remainders $\hat{\varepsilon}^{\pm}_p$ are bounded by*

$$\begin{aligned} |\hat{\varepsilon}^{\pm}_p| \leq E^{\pm 1}(\zeta_N)M(\zeta_N)\exp\left\{\frac{\hat{\kappa}_0\mathcal{V}_{\alpha_\pm,\zeta}(\hat{H})}{N}\right\}&\left[\frac{2\hat{\kappa}_1\mathcal{V}_{\alpha_\pm,\zeta}(\zeta^{1/2}\hat{Y}_{p-1}) + 2\hat{\kappa}_2\mathcal{V}_{\alpha_\pm,\zeta}(\hat{X}_p)}{N^p}\right. \\ &+\frac{\hat{\kappa}_1\mathcal{V}_{\alpha_\pm,\zeta}\left(\int^{\zeta} v^{-1/2}\hat{\Phi}(\hat{\Phi}')^2\hat{X}_{p-1} + 2v^{-1/2}(\hat{\Phi}\hat{\Phi}')^{1/2}((\hat{\Phi}\hat{\Phi}')^{1/2}\hat{Y}_{p-2})'dv\right)}{N^{p+1}} \\ &\left.+\ \frac{\hat{\kappa}_2\mathcal{V}_{\alpha_\pm,\zeta}\left(\int^{\zeta}\hat{\Phi}(\hat{\Phi}')^2\hat{Y}_{p-2}(v)dv\right)}{N^{p+1}}\right], \end{aligned} \tag{137}$$

*where* $\alpha_+ = \zeta(\gamma_- + \delta), \alpha_- = +\infty$, $\hat{\kappa}_i$ $(i = 0, 1, 2)$ *are some constants and* $\hat{H}$ *is given by* (150) *and* (142).

*Proof*: We only give a proof for $\hat{\varepsilon}_p^+$, because the other one is similar. The remainder satisfies the differential equation

$$(\hat{\varepsilon}_p^+)'' - \left(N^2\zeta + N\hat{\phi}(\zeta) + \hat{\psi}(\zeta)\right)\hat{\varepsilon}_p^+ = R(\zeta), \tag{138}$$

where

$$R(\zeta) = \mathrm{Bi}(\zeta_N)\left[\frac{2\zeta\hat{Y}'_{p-1} + \hat{Y}_{p-1}}{N^{p-1}} - \frac{\hat{\Phi}(\hat{\Phi}')^2\hat{X}_{p-1} + 2(\hat{\Phi}\hat{\Phi}')^{1/2}\left((\hat{\Phi}\hat{\Phi}')^{1/2}Y_{p-2}\right)'}{N^p}\right]$$

$$+ N^{-4/3}\mathrm{Bi}'(\zeta_N)\left[\frac{2X'_p}{N^{p-2}} - \frac{\hat{\Phi}(\hat{\Phi}')^2\hat{Y}_{p-2}}{N^{p-1}}\right], \tag{139}$$

and primes denote differentiations with respect to $\zeta$, except those in Ai and Bi. The two functions

$$Z_1(\zeta) = (N + \hat{\Phi}')^{-1/2}\mathrm{Ai}(\zeta_N), \quad Z_2(\zeta) = (N + \hat{\Phi}')^{-1/2}\mathrm{Bi}(\zeta_N), \tag{140}$$

satisfy a similar differential equation

$$Z_j''(\zeta) = [N^2\zeta + N\hat{\phi}(\zeta) + \bar{\psi}(\zeta)]Z_j(\zeta), \quad j = 1, 2, \tag{141}$$

where

$$\bar{\psi}(\zeta) = \zeta\hat{\Phi}'^2 + 2\hat{\Phi}\hat{\Phi}' + \frac{\hat{\Phi}\hat{\Phi}'}{N} + \frac{3\hat{\Phi}''^2 - 2\hat{\Phi}'\hat{\Phi}''' - 2N\hat{\Phi}'''}{4(N + \hat{\Phi}')^2}; \tag{142}$$

see [25, p. 429]. The Wronskian of $Z_1$ and $Z_2$ is

$$\mathcal{W}(Z_1, Z_2) = N^{-1/3}[\mathrm{Ai}(\zeta_N)\mathrm{Bi}'(\zeta_N) - \mathrm{Bi}(\zeta_N)\mathrm{Ai}'(\zeta_N)] = \frac{1}{\pi}N^{-1/3}. \tag{143}$$

Using $Z_1$ and $Z_2$ as complementary solutions and solving (138) by variation of parameters, we obtain

$$\hat{\varepsilon}_p^+(\zeta) = \pi N^{-2/3}\int_{\alpha_+}^{\zeta} K(\zeta, v)\left[[\hat{\psi}(v) - \bar{\psi}(v)]\hat{\varepsilon}_p^+(v) + R(v)\right]dv, \tag{144}$$

where

$$K(\zeta, v) = \frac{\mathrm{Ai}(v_N)\mathrm{Bi}(\zeta_N) - \mathrm{Ai}(\zeta_N)\mathrm{Bi}(v_N)}{[1 + \hat{\Phi}'(v)/N]^{1/2}[1 + \hat{\Phi}'(\zeta)/N]^{1/2}}, \quad v_N = N^{2/3}v + N^{-1/3}\hat{\Phi}(v); \tag{145}$$

see [25, p. 384, (9.14)]. For the real interval $z \in [\gamma_- + \delta, +\infty)$, it follows from the definition of the mapping $\zeta(z)$ (cf. (103)) that $\alpha_+ \le v \le \zeta$ and hence $v_N \le \zeta_N$ for $N \ge N_0$ (with some finite $N_0 > 0$) along the path of integration in (144). It is easy to show (cf. [25, p. 400]) that

$$|K(\zeta, v)| \le K_0 \frac{E(\zeta_N)}{E(v_N)} M(\zeta_N) M(v_N), \quad K_0 = \sup_{\zeta \in \zeta(D^\star)} |1 + \hat{\Phi}'(\zeta)/N|^{-1}, \tag{146}$$

where $K_0$ is a constant (well defined for $N \ge N_0$ with some $N_0$ because $\hat{\Phi}'$ is bounded). For $z \in D^\star$, a path from $\alpha_+$ to $\zeta(z)$ can be chosen such that $E(v_N) \le E(\zeta_N)$ along this path, so the above inequality still holds. To apply Theorem 10.1 in [25, pp. 218–219], we first identify the functions in this paper with the quantities in that theorem (indicated by a star)

$$\begin{aligned}
K^*(\zeta, v) &= \pi N^{1/3} v^{1/2} K(\zeta, v), \quad \psi_0^* = \frac{\hat{\psi}(v) - \bar{\psi}(v)}{N v^{1/2}}, \\
|K^*(\zeta, v)| &\le P_0^*(\zeta) Q^*(v), \quad P_0^*(\zeta) = E(\zeta_N) M(\zeta_N), \\
Q^*(v) &= K_0 \pi N^{1/3} |v|^{1/2} M(v_N)/E(v_N), \quad J_1^*(v) = \mathrm{Bi}(v_N), \\
\phi_1^*(v) &= \frac{2v\hat{Y}'_{p-1}(v) + \hat{Y}_{p-1}(v)}{N^p v^{1/2}}, \; J_2^*(v) = N^{-1/3} v^{-1/2} \mathrm{Bi}'(v_N), \\
\phi_2^*(v) &= \frac{2X'_p(v)}{N^p}, \; \phi_3^*(v) = \frac{\hat{\Phi}(\hat{\Phi}')^2 \hat{X}_{p-1}(v) + 2(\hat{\Phi}\hat{\Phi}')^{1/2}\left((\hat{\Phi}\hat{\Phi}')^{1/2}\hat{Y}_{p-2}\right)'}{v^{1/2} N^{p+1}}, \\
\phi_4^*(v) &= \frac{\hat{\Phi}(\hat{\Phi}')^2 \hat{Y}_{p-2}(v)}{N^{p+1}}.
\end{aligned} \tag{147}$$

By that theorem, we obtain (equivalent to (137) for $\hat{\varepsilon}_p^+$)

$$\begin{aligned}
|\hat{\varepsilon}_p^+| \le P_0^*(\zeta) &\exp\left\{\hat{\kappa}_0 \mathcal{V}_{\alpha_+,\zeta}(\hat{H})/N\right\} \\
&\times \left[\hat{\kappa}_1 \mathcal{V}_{\alpha_+,\zeta}\left(\int^\zeta \phi_1^*(v)dv\right) + \hat{\kappa}_2 \mathcal{V}_{\alpha_+,\zeta}\left(\int^\zeta \phi_2^*(v)dv\right)\right. \\
&\left. + \hat{\kappa}_1 \mathcal{V}_{\alpha_+,\zeta}\left(\int^\zeta \phi_3^*(v)dv\right) + \hat{\kappa}_2 \mathcal{V}_{\alpha_+,\zeta}\left(\int^\zeta \phi_4^*(v)dv\right)\right],
\end{aligned} \tag{148}$$

where

$$\begin{aligned}
\hat{\kappa}_0 &= \sup_{\zeta \in \zeta(D^\star)} \{P_0^*(\zeta) Q^*(\zeta)\}, \quad \hat{\kappa}_1 = \sup_{\zeta \in \zeta(D^\star)} \{|J_1^*(\zeta)| Q^*(\zeta)\}, \\
\hat{\kappa}_2 &= \sup_{\zeta \in \zeta(D^\star)} \{|J_2^*(\zeta)| Q^*(\zeta)\},
\end{aligned} \tag{149}$$

and

$$\hat{H}(\zeta) = \int^\zeta [\hat{\psi}(v) - \bar{\psi}(v)] v^{-1/2} dv. \tag{150}$$

One can show that the constants $\hat{\kappa}_i$ $(i = 0, 1, 2)$ are well-defined by the asymptotic behaviors of the pertinent functions in (149), (147), and (132). For example, we consider the case of $\hat{\kappa}_0$ . As $z \in D^\star$ tends to $\infty$, $\zeta(z)$ tends to $\infty$ almost along the positive real line and it follows from the definition (110) of $\zeta_N$ and (122) that

$$N^{1/3}\zeta^{1/2} = \zeta_N^{1/2} + O(\zeta^{-1/2}), \quad \text{as} \quad \zeta \to +\infty, \tag{151}$$

uniformly in $N$. Thus, as $\zeta \to +\infty$

$$P_0^*(\zeta)Q^*(\zeta) \sim K_0\pi\zeta_N^{1/2}M^2(\zeta_N), \tag{152}$$

which is bounded by (132). Also, one can show, in a very similar manner as in Lemma 4, that the five total variations $\mathcal{V}_{\alpha_+,\zeta}$ in (137) are finite as $\zeta \to +\infty$. Therefore, the error bound (137) holds uniformly for $\zeta \in \zeta(D^\star)$. ■

### *4.3. Matching*

To express the polynomial $P_n(z; a_n, b_n)$ in terms of $\widehat{W}^{\pm}_{n,p}(\zeta)$ in (136), one could achieve this by comparing their behavior as $z \to +\infty$ directly. However, to ease calculation, here we will use the result in (88) with the behavior of $W^-_{n,p}$ given in (85)

$$W^-_{n,p}(z) \sim e^{-n\xi - X}, \quad \text{as} \quad z \to +\infty. \tag{153}$$

Recall that, for large $z$, the Airy function has the asymptotic behavior

$$\begin{aligned} \mathrm{Ai}(z) &= \frac{1}{2\sqrt{\pi}} z^{-1/4} e^{-\frac{2}{3}z^{3/2}} \left[1 + O(z^{-3/2})\right], \\ \mathrm{Ai}'(z) &= -\frac{1}{2\sqrt{\pi}} z^{1/4} e^{-\frac{2}{3}z^{3/2}} \left[1 + O(z^{-3/2})\right]. \end{aligned} \tag{154}$$

With $\zeta_N = N^{2/3}\zeta + N^{-1/3}\hat{\Phi}$ in (110), we can use the results in (154), (103), (104), (105), (108), and (122) to arrive at

$$\begin{aligned} \mathrm{Ai}\,(\zeta_N) &= \frac{1}{2\sqrt{\pi}}\zeta_N^{-1/4} \exp\left\{-\frac{2}{3}N(\zeta + \hat{\Phi}/N)^{3/2}\right\} \left[1 + O(\zeta^{-3/2})\right] \\ &= \frac{1}{2\sqrt{\pi}} N^{-1/6}\zeta^{-1/4} \exp\left\{-\frac{2}{3}N\zeta^{3/2}\left[1 + \frac{3\hat{\Phi}}{2N\zeta}\right]\right\} \left[1 + O(\zeta^{-3/2})\right] \\ &= \frac{1}{2\sqrt{\pi}} N^{-1/6}\zeta^{-1/4} \exp\left\{-(N+\alpha)(\xi - \xi_0) - (X - X_0)\right\} \left[1 + O(\zeta^{-3/2})\right] \\ &= \frac{1}{2\sqrt{\pi}} N^{-1/6} e^{n\xi_0 + X_0} \zeta^{-1/4} e^{-n\xi - X} \left[1 + O(\zeta^{-3/2})\right], \quad \text{as} \quad \zeta \to +\infty. \end{aligned} \tag{155}$$

Similarly, we also have

$$\mathrm{Ai}'(\zeta_N) = -\frac{1}{2\sqrt{\pi}} N^{1/6} e^{n\xi_0+X_0} \zeta^{1/4} e^{-n\xi-X}\left[1+O(\zeta^{-3/2})\right], \quad \text{as} \quad \zeta \to +\infty. \tag{156}$$

Therefore, by Proposition 5 and Lemma 4, as $\zeta \to +\infty$ and for any $p \geq 1$ and $N \geq N_0$, the solution $\widehat{W}^{-}_{n,p}(\zeta)$ in (136) has the behavior

$$\widehat{W}^{-}_{n,p} \sim C_{n,p}\zeta^{-1/4}e^{-n\xi-X}, \quad C_{n,p} \equiv \frac{1}{2\sqrt{\pi}} e^{n\xi_0+X_0} N^{-1/6}\left[1 - N^{-1}\sum_{s=0}^{p-2}\frac{\hat{Y}_{s,0}}{N^s}\right], \tag{157}$$

where $\hat{Y}_{s,0}$ is given by (128). Note that $\hat{\varepsilon}^{-}_{p}$ in (136) has no contribution to the above leading-order behavior as $\zeta \to +\infty$, because the factor $E^{-1}(\zeta_N)M(\zeta_N)$ in (137) has the same order as $\mathrm{Ai}(\zeta_N)$ and all the total variations $\mathcal{V}_{\alpha_-,\zeta}(\cdot)$ in (137) tend to 0 by the definition $\alpha_- = +\infty$.

By definitions (90) and (28), we have

$$W_n(\xi) = \zeta^{1/4}\widetilde{W}_n(\zeta). \tag{158}$$

Because $W_n$ is approximated by $W^{\pm}_{n,p}$ and $\widetilde{W}_n$ is approximated by $\widehat{W}^{\pm}_{n,p}$ in Propositions 4 and 5, respectively, the solution $W^{-}_{n,p}$ should be a multiple of $\zeta^{1/4}\widehat{W}^{-}_{n,p}$ (both decay exponentially with respect to $\xi$ as $\xi \to +\infty$). Comparing (153) and (157) gives

$$W^{-}_{n,p}(z) = C^{-1}_{n,p}\zeta^{1/4}\widehat{W}^{-}_{n,p}(\zeta). \tag{159}$$

Replacing $W^{-}_{n,p}(z)$ in (88) by the above identity leads to

$$P_n(z; a_n, b_n) = C^{-1}_{n,p}e^{b_n\pi/2}[w(z)]^{-1/2}\left[\frac{\zeta}{(z-\gamma_-)(z-\gamma_+)}\right]^{1/4}\widehat{W}^{-}_{n,p}(\zeta). \tag{160}$$

Because this above formula holds for any positive integer $p$ and $n \geq n^*_0$ (some finite $n^*_0$), we can summarize the above result in the following theorem.

THEOREM 2. *For $z \in D^\star$ defined in (135), the P-J polynomials $P_n(z; a_n, b_n)$ with parameters in (11) have the uniform asymptotic expansion*

$$P_n(z; a_n, b_n) \sim C^{-1}_{n,\infty}e^{b_n\pi/2}[w(z)]^{-1/2}\left[\frac{\zeta}{(z-\gamma_-)(z-\gamma_+)}\right]^{1/4}\widehat{W}^{-}_{n}(\zeta), \tag{161}$$

*as $n \to \infty$, where $C_{n,\infty}$ is given in (157) and (128), $w(z)$ is given in (13), $\gamma_\pm$ are given in (16), $\zeta(z)$ is given by (103), and $\widehat{W}^{-}_{n}(\zeta)$ is given by (109).*

Corollary 1. *For $z \in D^\star$ defined in* (135), *the P-J polynomials $P_n(z; a_n, b_n)$ with parameters in* (11) *have the leading asymptotic behavior*

$$P_n(z; a_n, b_n) \sim 2\sqrt{\pi} N^{1/6} e^{-(n\xi_0+X_0)+b_n\pi/2}[w(z)]^{-1/2} \times \left[\frac{\zeta}{(z-\gamma_-)(z-\gamma_+)}\right]^{1/4} \mathrm{Ai}(\zeta_N), \tag{162}$$

*as $n \to \infty$, where $w(z)$ is given in* (13), *$\gamma_\pm$ are given in* (16), *$\zeta_N = N^{2/3}\zeta + N^{-1/3}\hat{\Phi}$ and $\zeta$, $\xi_0$, $\hat{\Phi}$, $X_0$, $N$ are given by* (103), (108), (104) *and* (105) *with $\alpha$ defined in Lemma 4.*

Based on the above result, we now analyze the zeros of $P_n(z; a_n, b_n)$ by following the argument in [25, p. 405]. Let the zeros be arranged in decreasing order:

$$\gamma_- < x_{n,n} < x_{n,n-1} < \cdots < x_{n,2} < x_{n,1} < \gamma_+. \tag{163}$$

Because the factor before the Airy function in Corollary 1 can never vanish in the interval $z \in [\gamma_-, \gamma_+]$, the zeros can be determined approximately by $\mathrm{Ai}(\zeta_N) = 0$. Taking into account of the remainder, the zeros are determined by

$$\mathrm{Ai}(\zeta_N) + \hat{\varepsilon}_1^-(\zeta) = 0. \tag{164}$$

Let $\tilde{a}_k$ denote the $k$th negative zero of $\mathrm{Ai}(x)$, and let $\zeta_{n,k}$ denote the $k$th root of Eq. (164). Because the term $\hat{\varepsilon}_1^-$ in (164) is of higher (asymptotic) order, it is reasonable to expect that

$$\zeta_{n,k} \approx \tilde{\zeta}_{n,k}, \tag{165}$$

where $\tilde{\zeta}_{n,k}$ satisfies

$$N^{2/3}\tilde{\zeta}_{n,k} + N^{-1/3}\hat{\Phi}(\tilde{\zeta}_{n,k}) = \tilde{a}_k. \tag{166}$$

Because $\zeta = \zeta(z)$ is finite on the interval $\gamma_- + \delta \le z \le \gamma_+$, the function $\hat{\Phi}(\zeta)$ is bounded for $\zeta(\gamma_- + \delta) \le \zeta \le \zeta(\gamma_+)$. Thus, because $\tilde{a}_k$ is negative, it follows that $\tilde{\zeta}_{n,k}$ is negative when $N$ or $n$ is sufficiently large.

In the following analysis, it is supposed that $N$ is sufficiently large so that the following inequality holds

$$\exp\left\{\frac{\hat{\kappa}_0 \mathcal{V}_{\alpha_-,\alpha_+}(\hat{H})}{N}\right\}\left[\frac{2\hat{\kappa}_1\mathcal{V}_{\alpha_-,\alpha_+}(\zeta^{1/2}\hat{Y}_0) + 2\hat{\kappa}_2\mathcal{V}_{\alpha_-,\alpha_+}(\hat{X}_1)}{N} + \frac{\hat{\kappa}_1\mathcal{V}_{\alpha_-,\alpha_+}\left(\int^\zeta v^{-1/2}\hat{\Phi}(\hat{\Phi}')^2 dv\right)}{N^2}\right] < \frac{1}{2}. \tag{167}$$

We write

$$\rho_1(N,\zeta) = \hat{\varepsilon}_1^-(\zeta)E(\zeta_N)/M(\zeta_N),$$

$$\sigma_1(N,\zeta) = \exp\left\{\frac{\hat{\kappa}_0 \mathcal{V}_{\alpha_-,\zeta}(\hat{H})}{N}\right\}\left[\frac{2\hat{\kappa}_1 \mathcal{V}_{\alpha_-,\zeta}(\zeta^{1/2}\hat{Y}_0) + 2\hat{\kappa}_2 \mathcal{V}_{\alpha_-,\zeta}(\hat{X}_1)}{N}\right.$$
$$\left. + \frac{\hat{\kappa}_1 \mathcal{V}_{\alpha_-,\zeta}\left(\int^\zeta v^{-1/2}\hat{\Phi}(\hat{\Phi}')^2 dv\right)}{N^2}\right]. \tag{168}$$

Then, by (137) and (167) we have

$$|\rho_1(N,\zeta)| \le \sigma_1(N,\zeta) < \frac{1}{2}. \tag{169}$$

In view of (133) and (168), Eq. (164) can be written as

$$\sin\theta(\zeta_N) = -\rho_1(N,\zeta). \tag{170}$$

When $\zeta_N \ge c$ with $c$ defined in (130), we have $\theta(\zeta_N) = \pi/4$ by the definition (131) and Eq. (133); see also [25, p. 395]. Thus, together with (170), there can be no zeros in this range.

When $\zeta_N < c$, we get from (170)

$$\theta(\zeta_N) - k\pi - (-1)^{k-1}\arcsin[\rho_1(N,\zeta)] = 0. \tag{171}$$

Let $\tilde{b}_k$ denote the $k$th negative zero of $\mathrm{Bi}(x)$. From the bound

$$|\arcsin[\rho_1(N,\zeta)]| < \arcsin\frac{1}{2} = \frac{1}{6}\pi, \tag{172}$$

and the result [25, p. 404]

$$\theta(\tilde{b}_k) = \left(k - \frac{1}{2}\right)\pi, \tag{173}$$

it is easily seen that the left-hand side of (171) is negative when $\zeta_N = \tilde{b}_k$ and positive when $\zeta_N = \tilde{b}_{k+1}$. Accordingly, there is a root in the interval

$$\tilde{b}_{k+1} < \zeta_N < \tilde{b}_k, \tag{174}$$

which also corresponds to an interval of $\zeta$ because $\zeta_N$ is monotonically increasing in $\zeta$ for sufficiently large $N$. In fact, from the monotonicity of the left-hand side of (171), there is only one root in this range; see [26] and [25, p. 406]. Because $\theta(x)$ is a decreasing function and $\theta(\tilde{b}_1) = \pi/2$, $\theta(c) = \pi/4$, it follows that there is no root for (170) in the interval $\tilde{b}_1 < \zeta_N < c$. Therefore, the root of (171) corresponds to $\zeta_{n,k}$.

By the mean-value theorem, we have for $\tilde{b}_{k+1} < \zeta_N < \tilde{b}_k$

$$\theta(\zeta_N) = \theta(\tilde{a}_k) + (\zeta_N - \tilde{a}_k)\theta'(\tau), \quad \tau \in (\tilde{b}_{k+1}, \tilde{b}_k). \tag{175}$$

Recall that $\theta(\tilde{a}_k) = k\pi$ (see [25, p. 404]) and $\zeta_N = N^{2/3}\zeta + N^{-1/3}\hat{\Phi}(\zeta)$ in (110). From (171) and (175), it follows that

$$N^{2/3}\zeta_{n,k} + N^{-1/3}\hat{\Phi}(\zeta_{n,k}) - \tilde{a}_k = (-1)^{k-1}\arcsin[\rho_1(N, \zeta_{n,k})]/\theta'(\tau). \quad (176)$$

Hence, with the aid of (169), we obtain

$$\begin{aligned}|N^{2/3}\zeta_{n,k} + N^{-1/3}\hat{\Phi}(\zeta_{n,k}) - \tilde{a}_k| &\le |\arcsin[\sigma_1(N, \zeta_{n,k})]/\theta'(\tau)| \\ &\le \frac{\pi}{3}\sigma_1(N, \zeta_{n,k})/|\theta'(\tau)|;\end{aligned} \quad (177)$$

cf. [26, p. 162]. In view of the identity [25, p. 404]

$$\theta'(x) = -1/[\pi M^2(x)], \quad (178)$$

the right-hand side of (177) becomes

$$\frac{\pi}{3}\sigma_1(N, \zeta_{n,k})/|\theta'(\tau)| = \frac{\pi^2}{3}M^2(\tau)\sigma_1(N, \zeta_{n,k}) = M^2(\tau)O(N^{-1}) = O(N^{-1}), \quad (179)$$

where the last two equalities are due to (168) and (132). Coupling (179) and (177) yields

$$N^{2/3}\zeta_{n,k} + N^{-1/3}\hat{\Phi}(\zeta_{n,k}) - \tilde{a}_k = O(N^{-1}). \quad (180)$$

By the mean-value theorem, we have

$$\zeta_{n,k} = \tilde{\zeta}_{n,k} + O(N^{-5/3}) = N^{-2/3}\tilde{a}_k + O(N^{-1}). \quad (181)$$

The last equality follows from (166). Let $z = \Omega(\zeta)$ denote the inverse of the function $\zeta = \zeta(z)$, and we obtain

$$x_{n,k} \equiv \Omega(\zeta_{n,k}) = \Omega(\tilde{\zeta}_{n,k}) + O(N^{-5/3}) = \Omega(N^{-2/3}\tilde{a}_k) + O(N^{-1}), \quad (182)$$

which holds uniformly for $x_{n,k} \in [\gamma_- + \delta, \gamma_+]$.

For fixed $k$, we can proceed further to get a more explicit formula. Because $\zeta_{n,k} \approx N^{-2/3}\tilde{a}_k$ and $\hat{\Phi}(0) = \Phi(0)$, we have from (180)

$$\zeta_{n,k} = N^{-2/3}\tilde{a}_k - \Phi(0)N^{-1} + O(N^{-5/3}), \quad (183)$$

where $\Phi(0)$ is given in (102). Next, we expand $\Omega(\zeta)$ as

$$\Omega(\zeta) = \Omega(0) + \Omega'(0)\zeta + \frac{1}{2}\Omega''(0)\zeta^2 + \cdots . \quad (184)$$

From (182), (183), and (184), we obtain

$$x_{n,k} = \Omega(0) + \Omega'(0)\tilde{a}_k N^{-2/3} - \Omega'(0)\Phi(0)N^{-1} + \frac{1}{2}\Omega''(0)\tilde{a}_k^2 N^{-4/3} + O(N^{-5/3}). \quad (185)$$

Finally we calculate the constants $\Omega(0)$, $\Omega'(0)$, $\Omega''(0)$. From (103), we have $\zeta(\gamma_+) = 0$; hence $\Omega(0) = \gamma_+$. Differentiation of (90) with respect to $\zeta$

gives

$$\frac{dz}{d\zeta} = \frac{\zeta^{1/2}}{\sqrt{f_0(z)}}, \tag{186}$$

and

$$\Omega'(0) = \left.\frac{dz}{d\zeta}\right|_{\zeta=0} = c_0^{-2/3}, \tag{187}$$

where $c_0$ is given in (98). In a similar manner, we get

$$\Omega''(0) = -\frac{1}{5}\frac{f_0''(\gamma_+)}{[f_0'(\gamma_+)]^{5/3}} = -\frac{f_0''(\gamma_+)}{5c_0^{10/3}}. \tag{188}$$

To summarize, we have the following corollary.

COROLLARY 2. *Let the parameters $a_n$ and $b_n$ be given as in* (11), *and let the zeros of $P_n(z; a_n, b_n)$ be arranged as in* (163). *For each fixed k, we have*

$$x_{n,k} = \gamma_+ + c_0^{-2/3}\tilde{a}_k N^{-2/3} - f_1(\gamma_+)c_0^{-2}N^{-1} - \frac{f_0''(\gamma_+)}{10c_0^{10/3}}\tilde{a}_k^2 N^{-4/3} + O(N^{-5/3}), \tag{189}$$

*as $n \to \infty$, where N is given by* (105) *and Lemma 4, $\tilde{a}_k$ is the kth negative zero of Airy function* Ai$(x)$ *and the O-symbol depends on k. The constants $\gamma_+$ and $c_0$ are given by* (16) *and* (98), *and the functions $f_i(z)$ $(i = 0, 1)$ are given by* (26).

*Remark* 6. The above formula is valid for zeros near $\gamma_+$. For zeros near the other turning point $\gamma_-$, one can use the symmetry property of the polynomial in (9).

## 5. Numerical examples

In this section, we give some numerical examples to verify the preceding asymptotic results, and to make some comments on how they can be used. As an illustrative example, we take

$$A = 2, \quad B = 0, \quad A_0 = 0, \quad B_0 = 0, \quad n = 10. \tag{190}$$

Then the critical interval is $[-\sqrt{3}, \sqrt{3}]$, and the exact P-J polynomial is given by

$$P_n(z) \equiv P_n(z; a_n, b_n) = z^{10} - \frac{15}{7}z^8 + \frac{30}{23}z^6 - \frac{6}{23}z^4 + \frac{1}{69}z^2 - \frac{1}{10005}. \tag{191}$$

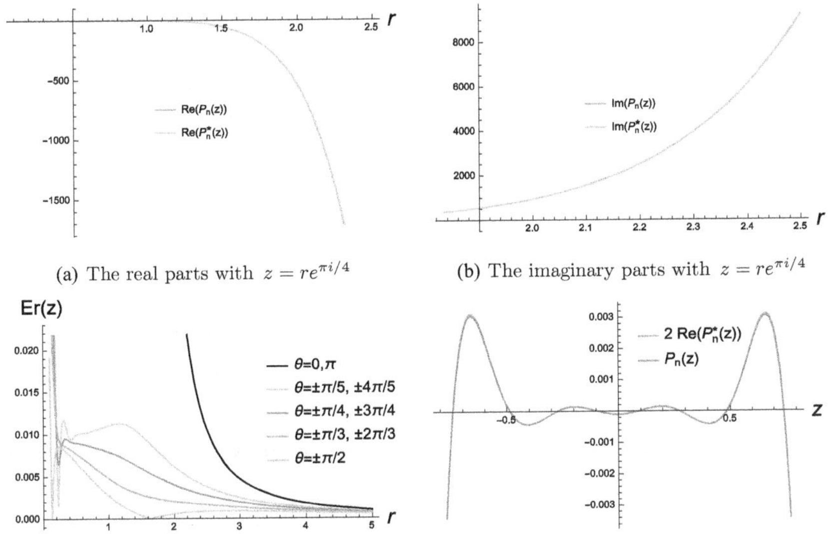


(a) The real parts with $z = re^{\pi i/4}$

(b) The imaginary parts with $z = re^{\pi i/4}$

(c) The relative error of outside region.

(d) The critical interval.

**Figure 5.** Comparison of $P_n(z; a_n, b_n)$ and $P_n^\star(z; a_n, b_n)$ in (192).

First, we examine the leading-order asymptotic approximation in Theorem 1, which is given by

$$P_n^\star(z) \equiv P_n^\star(z; a_n, b_n) = e^{b_n\pi/2}[(z-\gamma_-)(z-\gamma_+)]^{-1/4}e^{-n\xi^*(z)-X^*(z)}. \quad (192)$$

Figures 5(a) and 5(b) depict the real and imaginary parts of $P_n(z)$ and $P_n^\star(z)$ on a ray $z = re^{\pi i/4}$. Note that these figures show only the green graph and not the red graph, indicating that the two curves are so close that they coincide. The relative error is defined as

$$\mathrm{Er}(z) = \frac{|P_n^\star(z) - P_n(z)|}{|P_n(z)|}. \quad (193)$$

Figure 5(c) shows the relative errors for $z = re^{i\theta}$ with various values of $\theta$ and $r$. It indicates that the approximation $P_n^\star$ is valid in the whole complex plane except for a neighborhood of the critical interval $[-\sqrt{3}, \sqrt{3}]$. Only when $z$ is close to this critical interval, the error becomes large. For many orthogonal polynomials, it is believed that once the asymptotic formula for the outer region is obtained, the asymptotic formula for the critical interval is simply the sum (not average) of the upper and lower limits of the former (see, e.g., [30]). This is also the case for the P-J polynomials. As the lower limit is the complex conjugate of the upper limit, the sum of the upper and lower limits is just $2\mathrm{Re}(P_n^\star(z))$; cf. also [31]. Figure 5(d) shows the

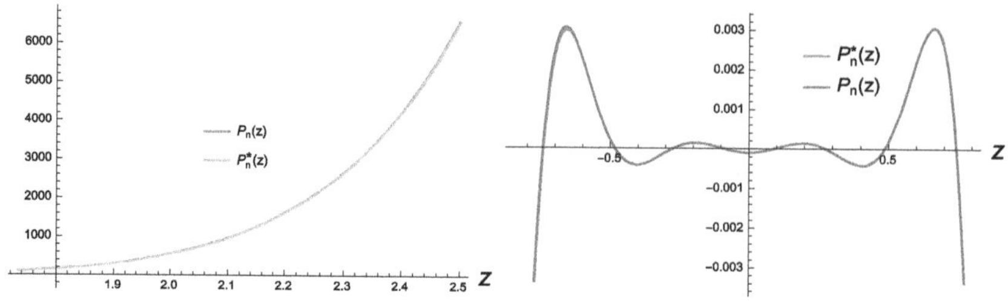


(a) For $z > \gamma_+$. (b) For $z$ inside the critical interval.

**Figure 6.** Comparison of $P_n(z)$ and $P_n^\star(z)$ in (194).

comparison between $2\text{Re}(P_n^\star(z))$ and $P_n(z)$ for a middle part of the critical interval.

Next, to verify the asymptotic formula in Corollary 1, we redefine

$$P_n^\star(z) \equiv P_n^\star(z; a_n, b_n) = 2\sqrt{\pi}N^{1/6}e^{-(n\xi_0+X_0)+b_n\pi/2}[w(z)]^{-1/2} \times \left[\frac{\zeta}{(z-\gamma_-)(z-\gamma_+)}\right]^{1/4} \text{Ai}(\zeta_N). \tag{194}$$

Figure 6 gives the comparison between $P_n(z)$ and the above $P_n^\star(z)$ for $z$ inside and outside the critical interval on the real line. As with Figures 5(a) and 5(b), the two curves in Figures 6(a) and 6(b) almost coincide, showing that the approximation is indeed very good. Note that one should be very cautious about the branch of the function $\zeta(z)$ in (103). For $z > \gamma_+$, the function $\zeta(z)$ is real and well-defined. Inside the critical interval $z \in [\gamma_- + \delta, \gamma_+]$, the function $\xi(z) - \xi_0$ is pure imaginary with argument $3\pi/2$ from upper limit and hence the function $\zeta(z)$ is real and negative. In practice, the computer will identify the argument of $\xi(z) - \xi_0$ as $-\pi/2$. Thus, one should multiply a factor $e^{4\pi i/3} = (e^{2\pi i})^{2/3}$ when computing the function $\zeta(z)$ in this interval by softwares. More numerical calculations indicate that the approximation $P_n^\star(z)$ in (194) holds in a much larger region of the complex plane, say $D^\star \cup D$, which can cover the whole complex plane with the help of symmetry conditions (9) and (10).

The zeros calculated from the term $\text{Ai}(\zeta_N)$ or Eq. (182) are very close to the exact ones, within an error of 2% for all the ten zeros in this case. However, the approximation by (189) is not very good for this example, because $n$ is not large enough. When $n$ is larger, say $n = 50$, formula (189) can indeed produce accurate results for the first few zeros.

## References


1. R. Beals and R. Wong, *Special Functions and Orthogonal Polynomials*, Cambridge University Press, Cambridge, 2016.
2. V. Romanovski, Sur quelques classes nouvelles de polynomes orthogonaux, *C. R. Acad. Sci. Paris*, 188:1023–1025 (1929).
3. E. J. Routh, On some properties of certain solutions of a differential equation of the second order, *Proc. Lond. Math. Soc.* 1:245–262 (1884).
4. G. Natanson, Exact quantization of the Milson potential via Romanovski-Routh polynomials, *arXiv:1310.0796*, 2013.
5. M. Masjed-Jamei, F. Marcellán, and E. J. Huertas, A finite class of orthogonal functions generated by Routh–Romanovski polynomials, *Complex Variab. Elliptic Equat.* 59:162–171 (2014).
6. G. Szego, *Orthogonal Polynomials*, 4th ed., American Mathematical Soc., Providence, RI, 1939.
7. A. F. Nikiforov and V. B. Uvarov, *Special Functions of Mathematical Physics*, Birkhäuser, Basel, 1988.
8. R. A. Askey, An integral of Ramanujan and orthogonal polynomials, *J. Indian Math. Soc.(NS)*, 51:27–36 (1987).
9. R. A. Askey, Beta integrals and the associated orthogonal polynomials, in *Number theory, Madras 1987, Lecture Notes in Math.*, Vol. 1395 (K. Alladi, Ed.), pp. 84–121, Springer, Berlin, 1989.
10. C. W. Cryer, Rodrigues' formula and the classical orthogonal polynomials, Technical report, DTIC Document, 1968.
11. K. H. Kwon and L. L. Littlejohn, Classification of classical orthogonal polynomials, *J. Korean Math. Soc*, 34:973–1008 (1997).
12. P. A. Lesky, Endliche und unendliche systeme von kontinuierlichen klassischen orthogonalpolynomen, *ZAMM-J. Appl. Math. Mech.* 76:181–184 (1996).
13. R. Koekoek, P. A. Lesky, and R. F. Swarttouw, *Hypergeometric Orthogonal Polynomials and Their q-Analogues*, Springer Science & Business Media, Heidelberg, Berlin, 2010.
14. A. Raposo, H. Weber, D. E. Alvarez-Castillo, and M. Kirchbach, Romanovski polynomials in selected physics problems, *Open Phys.* 5:253–284 (2007).
15. P. J. Forrester, *Log-gases and Random Matrices (LMS-34)*, Princeton University Press, Princeton, 2010.
16. D. E. Alvarez-Castillo, Exactly solvable potentials and Romanovski polynomials in quantum mechanics, Master's Thesis, San Luis Potosi U., 2007; arXiv:0808.1642.
17. R. Dutt, A. Gangopadhyaya, and U. P. Sukhatme, Noncentral potentials and spherical harmonics using supersymmetry and shape invariance, *Am. J. Phys.* 65:400–403 (1997).
18. M. E. H. Ismail, *Classical and Quantum Orthogonal Polynomials in One Variable*, Cambridge University Press, Cambridge, 2005.
19. K. Jordaan and F. Toókos, Orthogonality and asymptotics of Pseudo-Jacobi polynomials for non-classical parameters, *J. Approx. Theory*, 178:1–12 (2014).
20. M. A. Khan, A. H. Khan, and S. M. Abbas, A note on pseudo Jacobi polynomials, *Ain Shams Eng. J.* 4:127–131 (2013).
21. K. Driver and M. E. Muldoon, Zeros of pseudo-ultraspherical polynomials, *Anal. Appl.* 12:563–581 (2014).

22. A. B. J. Kuijlaars, A. Martinez-Finkelshtein, and R. Orive, Orthogonality of Jacobi polynomials with general parameters, *Electron. Trans. Numer. Anal.* 19:1–17 (2005).
23. D. S. Jones, Asymptotics of the hypergeometric function, *Math. Methods Appl. Sci.* 24:369–389 (2001).
24. S. F. Khwaja and A. B. Olde Daalhuis, Uniform asymptotic expansions for hypergeometric functions with large parameters IV, *Anal. Appl.* 12:667–710 (2014).
25. F. W. J. Olver, *Asymptotics and Special Functions*, Academic Press, New York, 1974. [Reprinted by A. K. Peters, Wellesley, 1997.]
26. R. Bo and R. Wong, A uniform asymptotic formula for orthogonal polynomials associated with $exp(-x^4)$, *J. Approx. Theory* 98:146–166 (1999).
27. W. Y. Qiu and R. Wong, Uniform asymptotic formula for orthogonal polynomials with exponential weight, *SIAM J. Math. Anal.* 31:992–1029 (2000).
28. A. Martínez-Finkelshtein and R. Orive, Riemann–Hilbert analysis for Jacobi polynomials orthogonal on a single contour, *J. Approx. Theory*, 134:137–170 (2005).
29. R. Wong and W. Zhang, Uniform asymptotics for Jacobi polynomials with varying large negative parameters—Riemann-Hilbert approach, *Trans. Am. Math. Soc.* 358:2663–2694 (2006).
30. X.-S. Wang and R. Wong, Asymptotics of orthogonal polynomials via recurrence relations, *Anal. Appl.* 10:215–235 (2012).
31. R. Wong and Y.-Q. Zhao, Uniform asymptotic expansion of the Jacobi polynomials in a complex domain, *Proc. Roy. Soc. Lond., Ser. A*, 460:2569-2586 (2004).


York University
City University of Hong Kong

(Received October 19, 2016)

PROCEEDINGS OF THE
AMERICAN MATHEMATICAL SOCIETY
Volume 146, Number 3, March 2018, Pages 1083–1096
http://dx.doi.org/10.1090/proc/13771
Article electronically published on September 28, 2017

# ASYMPTOTICS OF RACAH POLYNOMIALS WITH FIXED PARAMETERS

X.-S. WANG AND R. WONG

(Communicated by Mourad Ismail)

ABSTRACT. In this paper, we investigate asymptotic behaviors of Racah polynomials with fixed parameters and scaled variable as the polynomial degree tends to infinity. We start from the difference equation satisfied by the polynomials and derive an asymptotic formula in the outer region via ratio asymptotics. Next, we find the asymptotic formulas in the oscillatory region via a simple matching principle. Unlike the varying parameter case considered in a previous paper, the zeros of Racah polynomials with fixed parameters may not always be real. For this unusual case, we also provide a standard method to determine the oscillatory curve which attracts the zeros of Racah polynomials when the degree becomes large.

## 1. INTRODUCTION

The Racah polynomials, which lie on the top level of Askey hierarchy for hypergeometric orthogonal polynomials, have the following ${}_4F_3$ hypergeometric function representation [6, 8]:

$$
\begin{aligned}
(1.1) \qquad R_n(z) &= R_n(z;\alpha,\beta,\gamma,\delta)\\
&:= {}_4F_3\left(\begin{matrix} -n, & n+\alpha+\beta+1, & -x, & x+\gamma+\delta+1\\ & \alpha+1, & \beta+\delta+1, & \gamma+1\end{matrix}\,\middle|\, 1\right),
\end{aligned}
$$

where $n$ and $N$ are two non-negative integers with $0 \le n \le N$, $z = x(x+\gamma+\delta+1)$, and one of the following three equalities is satisfied: $\alpha+1=-N$ or $\gamma+1=-N$ or $\beta+\delta+1=-N$. The orthogonality relation for the Racah polynomials is [6, 8]

$$
\begin{aligned}
&\sum_{x=0}^{N} w(x) R_m(z) R_n(z)\\
&\qquad = M\frac{(n+\alpha+\beta+1)_n(\alpha+\beta-\gamma+1)_n(\alpha-\delta+1)_n(\beta+1)_n n!}{(\alpha+\beta+2)_{2n}(\alpha+1)_n(\beta+\delta+1)_n(\gamma+1)_n}\delta_{mn},
\end{aligned}
$$

where $z = x(x+\gamma+\delta+1)$,

$$
w(x) = \frac{(\alpha+1)_x(\beta+\delta+1)_x(\gamma+1)_x(\gamma+\delta+1)_x((\gamma+\delta+3)/2)_x}{(-\alpha+\gamma+\delta+1)_x(-\beta+\gamma+1)_x((\gamma+\delta+1)/2)_x(\delta+1)_x x!},
$$

Received by the editors January 17, 2017 and, in revised form, April 11, 2017.
2010 *Mathematics Subject Classification.* Primary 41A60; Secondary 33C45.
*Key words and phrases.* Asymptotics, Racah polynomials, difference equations, matching principle.

and

$$M=\begin{cases}\dfrac{(-\beta)_N(\gamma+\delta+2)_N}{(-\beta+\gamma+1)_N(\delta+1)_N}, & \text{if } \alpha+1=-N,\\ \dfrac{(-\alpha+\delta)_N(\gamma+\delta+2)_N}{(-\alpha+\gamma+\delta+1)_N(\delta+1)_N}, & \text{if } \beta+\delta+1=-N,\\ \dfrac{(\alpha+\beta+2)_N(-\delta)_N}{(\alpha-\delta+1)_N(\beta+1)_N}, & \text{if } \gamma+1=-N.\end{cases}$$

Here, we have made use of the Pochhammer symbol

$$(\alpha)_n:=\prod_{k=0}^{n-1}(\alpha+k).$$

The Racah polynomials can also be viewed as birth and death process polynomials in the sense that they satisfy the difference equation [4, 6]:

$$-zR_n(z)=\lambda_nR_{n+1}(z)+\mu_nR_{n-1}(z)-(\lambda_n+\mu_n)R_n(z), \tag{1.2}$$

together with initial conditions $R_0(z)=1$ and $R_1(z)=(\lambda_0+\mu_0-z)/\lambda_0$, where the birth and death rates are given by

$$\begin{aligned}\lambda_n&:=-\frac{(n+\alpha+1)(n+\alpha+\beta+1)(n+\beta+\delta+1)(n+\gamma+1)}{(2n+\alpha+\beta+1)(2n+\alpha+\beta+2)},\\ \mu_n&:=-\frac{n(n+\alpha+\beta-\gamma)(n+\alpha-\delta)(n+\beta)}{(2n+\alpha+\beta)(2n+\alpha+\beta+1)},\end{aligned} \tag{1.3}$$

respectively. By normalizing the leading coefficients, we have the monic Racah polynomials

$$\pi_n(z):=\frac{(\alpha+1)_n(\beta+\delta+1)_n(\gamma+1)_n}{(n+\alpha+\beta+1)_n}R_n(z),$$

which satisfy the difference equation

$$\pi_{n+1}(z)=(z-a_n)\pi_n(z)-b_n\pi_{n-1}(z),\ \pi_0(z)=1,\ \pi_1(z)=z-a_0, \tag{1.4}$$

where

$$a_n=\lambda_n+\mu_n,\ b_n=\lambda_{n-1}\mu_n. \tag{1.5}$$

In addition to the applications in birth and death processes, the Racah polynomials are also related to the so-called 6-$j$ symbols in the quantum mechanical theory of angular momentum [1, 8]. Asymptotic behaviors of Racah polynomials for large degree $n$ were studied by Chen, Ismail and Simeonov [2], where both the variable and parameters were fixed. Their formulas were given in terms of ${}_3F_2$ or ${}_2F_1$ hypergeometric functions. Ismail and Simeonov [5] studied inequalities and asymptotics for a terminating ${}_4F_3$ series. Their results can be applied to a special case of Racah polynomials with $\alpha=\beta=0$ and $\delta=-\gamma=N+1$. Note that for this special case, two parameters are fixed, while the other two are large and varying. In a recent work [7], we derived explicit asymptotic formulas, in terms of elementary functions such as logarithmic, exponential and rational functions, for Racah polynomials with scaled variable and large varying parameters. A simple formula for the fixed variable was also obtained. It remains to consider the case with scaled variable and fixed parameters, which is the main focus of this paper. To be specific, we will provide asymptotic analysis of the monic Racah polynomials $\pi_n(N^2y)$ as

$N \to \infty$, where $n/N = p$ is a fixed positive number. The following three cases will be investigated respectively: $\alpha + 1 = -N$, $\gamma + 1 = -N$, and $\beta + \delta + 1 = -N$.

## 2. Case I: $\alpha + 1 = -N$

In this section, we consider the case $\alpha + 1 = -N$ and $\beta, \delta, \gamma$ are fixed. First, we let $z = N^2 y$ and introduce the ratios

$$w_k(z) := \frac{\pi_k(z)}{\pi_{k-1}(z)} \tag{2.1}$$

for $k = 1, \cdots, n$. It is readily seen from (1.4) that $w_1(z) = z - a_0$ and

$$w_{k+1}(z) = z - a_k - \frac{b_k}{w_k(z)}, \tag{2.2}$$

where $a_k$ and $b_k$ are obtained by substituting $\alpha + 1 = -N$ in (1.3) and using (1.5):

$$\begin{aligned} a_k = &-\frac{(k-N)(k-N+\beta)(k+\beta+\delta+1)(k+\gamma+1)}{(2k-N+\beta)(2k-N+\beta+1)} \\ &-\frac{k(k-N+\beta-\gamma-1)(k-N-\delta-1)(k+\beta)}{(2k-N+\beta-1)(2k-N+\beta)}; \\ b_k = &\frac{k(k-N+\beta-\gamma-1)(k-N-\delta-1)(k+\beta)}{(2k-N+\beta-1)(2k-N+\beta)} \\ &\times \frac{(k-N-1)(k-N+\beta-1)(k+\beta+\delta)(k+\gamma)}{(2k-N+\beta-2)(2k-N+\beta-1)}. \end{aligned}$$

If $\beta, \delta, \gamma$ are positive, then for sufficiently large $N$ the coefficient $b_k$ with $1 \le k \le n$ is always positive as long as $n/N = p < 1/2$. By Favard's theorem [3, §§1.4 & 1.5], the zeros of $\pi_k(z)$ are all real and simple. Put $t := k/N$. It follows from the above expressions for $a_k$ and $b_k$ that

(2.3)
$$\frac{a_k}{N^2} = -2f(t) - \frac{f(t)h_1(t)}{N} + O\left(\frac{1}{N^2}\right); \quad \frac{b_k}{N^4} = f(t)^2 + \frac{f(t)^2 h_2(t)}{N} + O\left(\frac{1}{N^2}\right),$$

and

$$\frac{a_{k+1} - a_k}{N} = -2f(t)h_0(t) + O\left(\frac{1}{N}\right); \quad \frac{b_{k+1} - b_k}{N^3} = 2f(t)^2 h_0(t) + O\left(\frac{1}{N}\right), \tag{2.4}$$

where

$$f(t) := \frac{t^2(t-1)^2}{(2t-1)^2}; \tag{2.5}$$

$$h_1(t) := \frac{2\beta - \gamma - \delta - 2}{t-1} + \frac{2\beta + \gamma + \delta + 2}{t} - \frac{4\beta}{2t-1}; \tag{2.6}$$

$$h_2(t) := \frac{2\beta - \gamma - \delta - 4}{t-1} + \frac{2\beta + \gamma + \delta}{t} - \frac{4\beta - 4}{2t-1}; \tag{2.7}$$

$$h_0(t) := \frac{f'(t)}{f(t)} = \frac{2}{t-1} + \frac{2}{t} - \frac{4}{2t-1} = h_1(t) - h_2(t). \tag{2.8}$$

Formula (2.4) can be obtained from (2.3) by simply taking derivative with respect to $t$. When $n/N = p < 1/2$, the asymptotic formulas (2.3)-(2.4) are uniform in $k = 0, 1, \cdots, n$. Throughout this section, we shall always assume that $p \in (0, 1/2)$.

Note that if $p \geq 1/2$, then the asymptotic formulas (2.3)-(2.4) are no longer valid. It would be a challenge to derive the asymptotic formula of $\pi_n(N^2y)$ in this case.

**Lemma 2.1.** *Let $n/N = p$ be a fixed number in $(0, 1/2)$ and $z = N^2y$. For $y \in \mathbf{C} \setminus [y_p, 0]$ with $y_p = -4p^2(p-1)^2/(2p-1)^2$, we have*

$$w_k = w_k^0[1 + w_k^1 + O(1/N^2)] \tag{2.9}$$

*as $N \to \infty$, where the leading term is*

$$w_k^0 = \frac{z - a_k + \sqrt{(z-a_k)^2 - 4b_k}}{2}, \tag{2.10}$$

*and the first-order term is*

$$w_k^1 = \frac{a_{k+1} - a_k}{2\sqrt{(z-a_k)^2 - 4b_k}} + \frac{2(b_{k+1} - b_k) + (z - a_k)(a_{k+1} - a_k)}{2[(z-a_k)^2 - 4b_k]}. \tag{2.11}$$

*The error estimate is uniform for all $1 \leq k \leq n$.*

*Proof.* Formally, we can obtain the leading term by substituting $w_k^0$ for both $w_{k+1}$ and $w_k$ in (2.2) and then solving the resulting quadratic equation. To give a rigorous argument, we need to estimate the error bound; namely, we will show that if $w_k^0$ is defined as in (2.10), then the first-order error term $\varepsilon_k^1 := w_k/w_k^0 - 1$ is of order $O(1/N)$. Furthermore, if $w_k^1$ is defined as in (2.11), then we shall prove that the second-order error term $\varepsilon_k^2 := w_k/w_k^0 - 1 - w_k^1$ is of order $O(1/N^2)$.

We substitute $w_k = w_k^0(1+\varepsilon_k^1)$ in (2.2) and use (2.10) to find a difference equation for the first-order error term:

$$\varepsilon_{k+1}^1 = \frac{z - a_k - w_{k+1}^0 - (z - a_k - w_k^0)(1+\varepsilon_k^1)^{-1}}{w_{k+1}^0}.$$

For $z = N^2y$ with $y \in \mathbf{C} \setminus [y_p, 0]$, we shall prove by induction that $\varepsilon_k^1 = O(1/N)$ uniformly for all $k = 1, \cdots, n$. First, we rewrite the above equation as

$$\varepsilon_{k+1}^1 = \frac{w_k^0 - w_{k+1}^0}{w_{k+1}^0} + \frac{N^2y - a_k - w_k^0}{w_{k+1}^0}[1 - (1+\varepsilon_k^1)^{-1}]. \tag{2.12}$$

As $N \to \infty$, we have from (2.3), (2.4), (2.10)

$$\varepsilon_1^1 = \frac{w_1 - w_1^0}{w_1^0} = O\left(\frac{1}{N}\right),$$
$$\frac{w_k^0 - w_{k+1}^0}{w_{k+1}^0} = O\left(\frac{1}{N}\right),$$

and

$$\frac{N^2y - a_k - w_k^0}{w_{k+1}^0} = \frac{y + 2f(t) - \sqrt{y^2 + 4yf(t)}}{y + 2f(t) + \sqrt{y^2 + 4yf(t)}} + O\left(\frac{1}{N}\right).$$

By the definition of $f(t)$ in (2.5), we observe that $0 \leq f(t) \leq f(p)$ for all $0 \leq t \leq p < 1/2$. Moreover, $y_p = -4p^2(p-1)^2/(2p-1)^2 = -4f(p)$. Note that $|u + \sqrt{u^2-1}| > 1$ for all $u$ bounded away from $[-1, 1]$. If we take $u = [y + 2f(t)]/[2f(t)]$, then we have for any $y \in \mathbf{C} \setminus [-4f(p), 0]$,

$$\sup_{t\in[0,p]} \left| \frac{y + 2f(t) - \sqrt{y^2 + 4yf(t)}}{y + 2f(t) + \sqrt{y^2 + 4yf(t)}} \right| < 1.$$

Observe that the quantity inside the absolute-value signs above is $(u-\sqrt{u^2-1})/(u+\sqrt{u^2-1})$. Hence, there exist $\delta \in (0,1)$, $L>0$ and $N_1>L$ such that $|\varepsilon_1^1| \le L/N$ and

$$\left|\frac{N^2 y - a_k - w_k^0}{w_{k+1}^0}\right| \le 1-\delta,$$
$$\left|\frac{w_k^0 - w_{k+1}^0}{w_{k+1}^0}\right| \le \frac{\delta^2 L}{N},$$
$$\sup_{|s|\le L/N} \frac{|1-(1+s)^{-1}|}{|s|} \le 1+\delta,$$

for all $k=1,\cdots,n$ and $N>N_1$. It is readily seen from (2.12) that if $|\varepsilon_k^1| \le L/N$ for some $k \ge 1$, then

$$|\varepsilon_{k+1}^1| \le \frac{\delta^2 L}{N} + \frac{(1-\delta)(1+\delta)L}{N} = \frac{L}{N}.$$

By induction, we have $|\varepsilon_k^1| \le L/N$ for all $k=1,\cdots,n$ and $N>N_1$.

To derive the explicit formula (2.11) for the first-order term $w_k^1$, we substitute $w_k = w_k^0(1+w_k^1+\varepsilon_k^2)$ into (2.2) to obtain a difference equation for the second-order error term:

$$w_{k+1}^0(1+w_{k+1}^1+\varepsilon_{k+1}^2) = z - a_k - \frac{b_k}{w_k^0}(1+w_k^1+\varepsilon_k^2)^{-1}. \tag{2.13}$$

By (2.10), we can rewrite $b_k/w_k^0$ as $z-a_k-w_k^0$. Formally, let us for the moment treat $\varepsilon_k^2, \varepsilon_{k+1}^2, (w_k^1)^2$ and $w_{k+1}^1 - w_k^1$ as being of order $O(1/N^2)$. Equation (2.13) then becomes

$$w_{k+1}^0\bigl[1+w_k^1+O(1/N^2)\bigr] = z-a_k-(z-a_k-w_k^0)\bigl[1-w_k^1+O(1/N^2)\bigr].$$

Upon rearranging the terms, we obtain

$$\bigl[w_{k+1}^0-(z-a_k-w_k^0)\bigr]w_k^1 = w_k^0 - w_{k+1}^0 + O(1), \tag{2.14}$$

where we have made use of the facts that

$$z-a_k-w_k^0 = \frac{z-a_k-\sqrt{(z-a_k)^2-4b_k}}{2} = O(N^2)$$

and

$$w_{k+1}^0 = \frac{z-a_{k+1}+\sqrt{(z-a_{k+1})^2-4b_{k+1}}}{2} = O(N^2).$$

The coefficient of $w_k^1$ on the left-hand side of (2.14) can be expressed as

$$w_{k+1}^0-(z-a_k-w_k^0) = \frac{a_k-a_{k+1}+\sqrt{(z-a_{k+1})^2-4b_{k+1}}+\sqrt{(z-a_k)^2-4b_k}}{2}.$$

It is easily seen that the first square root on the right-hand side of the last equation is equal to

$$\sqrt{(z-a_k)^2-4b_k-4(b_{k+1}-b_k)-2(z-a_k)(a_{k+1}-a_k)+(a_{k+1}-a_k)^2}.$$

Using binomial expansion, we can also show that the above quantity is equal to

$$\sqrt{(z-a_k)^2-4b_k} - \frac{2(b_{k+1}-b_k)+(z-a_k)(a_{k+1}-a_k)}{\sqrt{(z-a_k)^2-4b_k}} + O(1).$$

Thus

$$\sqrt{(z-a_{k+1})^2-4b_{k+1}} = \sqrt{(z-a_k)^2-4b_k} - \frac{2(b_{k+1}-b_k)+(z-a_k)(a_{k+1}-a_k)}{\sqrt{(z-a_k)^2-4b_k}} + O(1).$$

Coupling the last two equations gives

$$\begin{aligned}\frac{w^0_{k+1}-(z-a_k-w^0_k)}{\sqrt{(z-a_k)^2-4b_k}} &= 1+\frac{a_{k+1}-a_k}{2\sqrt{(z-a_k)^2-4b_k}} \\ &\quad + \frac{\sqrt{(z-a_{k+1})^2-4b_{k+1}}-\sqrt{(z-a_k)^2-4b_k}}{2\sqrt{(z-a_k)^2-4b_k}} \\ &= 1+O\left(\frac{1}{N}\right)\end{aligned} \tag{2.15}$$

and

$$\begin{aligned}w^0_k - w^0_{k+1} &= \frac{z-a_k+\sqrt{(z-a_k)^2-4b_k}}{2} - \frac{z-a_{k+1}+\sqrt{(z-a_{k+1})^2-4b_{k+1}}}{2} \\ &= \frac{a_{k+1}-a_k}{2} + \frac{2(b_{k+1}-b_k)+(z-a_k)}{2\sqrt{(z-a_k)^2-4b_k}} + O(1).\end{aligned} \tag{2.16}$$

Substituting (2.15) and (2.16) in (2.14) yields

$$\begin{aligned}\left[1+O(1/N)\right]w^1_k &= \frac{a_{k+1}-a_k}{2\sqrt{(z-a_k)^2-4b_k}} \\ &\quad + \frac{2(b_{k+1}-b_k)+(z-a_k)(a_{k+1}-a_k)}{2\left[(z-a_k)^2-4b_k\right]} \\ &\quad + \frac{O(1)}{\sqrt{(z-a_k)^2-4b_k}}.\end{aligned}$$

By ignoring the $O(1/N)$ term in the last equation, we have at least formally derived the explicit formula in (2.11). This formula would have been proved rigorously, if we can show that the error term $\varepsilon^2_k$ is $O(1/N^2)$, which is what we will do next.

Now we return to equation (2.13). By adding and subtracting, the right-hand side of equation (2.13) becomes

$$\begin{aligned}&w^0_k + z - a_k - w^0_k - (z-a_k-w^0_k)(1+w^1_k)^{-1} \\ &\qquad + (z-a_k-w^0_k)\left[(1+w^1_k)^{-1}-(1+w^1_k+\varepsilon^2_k)^{-1}\right].\end{aligned}$$

Thus, we obtain

$$\begin{aligned}w^0_{k+1}(1+w^1_{k+1}+\varepsilon^2_{k+1}) &= w^0_k + (z-a_k-w^0_k)\left[1-(1-w^1_k)^{-1}\right] \\ &\quad + (z-a_k-w^0_k)\left[(1+w^1_k)^{-1}-(1+w^1_k+\varepsilon^2_k)^{-1}\right].\end{aligned}$$

Solving for $\varepsilon^2_{k+1}$ gives

$$\begin{aligned}\varepsilon^2_{k+1} &= \frac{w^0_k - w^0_{k+1} + (z-a_k-w^0_k)\left[1-(1+w^1_k)^{-1}\right] - w^0_{k+1}w^1_{k+1}}{w^0_{k+1}} \\ &\quad + \frac{z-a_k-w^0_k}{w^0_{k+1}}\left[(1+w^1_k)^{-1}-(1+w^1_k+\varepsilon^2_k)^{-1}\right].\end{aligned} \tag{2.17}$$

We claim that the first term on the right-hand side of (2.17) is $O(1/N^2)$. By adding and subtracting, this term can be broken into three fractions

$$\begin{aligned}&\frac{w_k^0 - w_{k+1}^0 + (z - a_k - w_k^0)w_k^1 - w_k^0 w_k^1}{w_{k+1}^0} \\ &+ \frac{(z - a_k - w_k^0)\left[1 - w_k^1 - (1 + w_k^1)^{-1}\right]}{w_{k+1}^0} + \frac{w_k^0 w_k^1 - w_{k+1}^0 w_{k+1}^1}{w_{k+1}^0}.\end{aligned} \tag{2.18}$$

Making use of (2.3)–(2.4) and (2.10)–(2.11), we have the following estimates for the last two fractions:

$$\begin{aligned}\frac{w_k^0 w_k^1 - w_{k+1}^0 w_{k+1}^1}{w_{k+1}^0} &= \frac{(w_k^0 - w_{k+1}^0)}{w_{k+1}^0} w_k^1 + (w_k^1 - w_{K+1}^1) \\ &= O\left(\frac{1}{N}\right) O\left(\frac{1}{N}\right) + O\left(\frac{1}{N^2}\right) = O\left(\frac{1}{N^2}\right)\end{aligned} \tag{2.19}$$

and

$$\frac{(z - a_k - w_k^0)}{w_{k+1}^0}\left[1 - w_k^1 - (1 + w_k)^{-1}\right] = O(1) O\left(\frac{1}{N^2}\right) = O\left(\frac{1}{N^2}\right). \tag{2.20}$$

For the first fraction in (2.18), we have

$$\begin{aligned}\frac{w_k^0 - w_{k+1}^0 + (z - a_k - 2w_k^0)w_k^1}{w_{k+1}^0} &= \frac{(z - a_k - 2w_k^0)}{w_{k+1}^0}\left[\frac{(w_k^0 - w_{k+1}^0)}{z - a_k - 2w_k^0} + w_k^1\right] \\ &= O(1)\left[\frac{w_k^0 - w_{k+1}^0}{z - a_k - 2w_k^0} + w_k^1\right].\end{aligned} \tag{2.21}$$

From (2.10) and (2.11), it also follows that the quantity inside the square brackets is equal to

$$\frac{-\sqrt{(z - a_k)^2 - 4b_k} + \sqrt{(z - a_{k+1})^2 - 4b_{k+1}}}{2\sqrt{(z - a_k)^2 - 4b_k}} + \frac{2(b_{k+1} - b_k) + (z - a_k)(a_{k+1} - a_k)}{2\left[(z - a_k)^2 - 4b_k\right]}.$$

Upon simplification, the last quantity becomes

$$\begin{aligned}&\frac{(a_k - a_{k+1})^2 + 2(z - a_k)(a_k - a_{k+1}) + 4(b_k - b_{k+1})}{2\sqrt{(z - a_k)^2 - 4b_k}\left[\sqrt{(z - a_k)^2 - 4b_k} + \sqrt{(z - a_{k+1})^2 - 4b_{k+1}}\right]} \\ &\qquad + \frac{2(b_{k+1} - b_k) + (z - a_k)(a_{k+1} - a_k)}{2\left[(z - a_k)^2 - 4b_k\right]},\end{aligned}$$

and it can be shown that it is equal to

$$\begin{aligned}&O\left(\frac{1}{N^2}\right) + \frac{(z - a_k)(a_k - a_{k+1}) + 2(b_k - b_{k+1})}{\sqrt{(z - a_k)^2 - 4b_k}\left[\sqrt{(z - a_k)^2 - 4b_k} + \sqrt{(z - a_{k+1})^2 - 4b_{k+1}}\right]} \\ &\qquad + \frac{2(b_{k+1} - b_k) + (z - a_k)(a_{k+1} - a_k)}{2\left[(z - a_k)^2 - 4b_k\right]}.\end{aligned}$$

Further calculation shows that this quantity is equal to

$$\begin{aligned}&O\left(\frac{1}{N^2}\right) + \left(\frac{2(b_k - b_k) + (z - a_k)(a_{k+1} - a_k)}{2\left[(z - a_k)^2 - 4b_k\right]}\right) \\ &\qquad \times \left(\frac{-\sqrt{(z - a_k)^2 - 4b_k} + \sqrt{(z - a_{k+1})^2 - 4b_{k+1}}}{\sqrt{(z - a_k)^2 - 4b_k} + \sqrt{(z - a_{k+1})^2 - 4b_{k+1}}}\right).\end{aligned}$$

Since each factor in the above product is $O(1/N)$, the calculation from (2.17) onwards gives

$$\frac{w_k^0 - w_{k+1}^0 + (z - a_k - 2w_k^0)w_k^1}{w_{k+1}^0} = O\left(\frac{1}{N^2}\right). \tag{2.22}$$

A combination of (2.14), (2.15), (2.16) and (2.18) shows that the first term on the right-hand side of (2.13) is indeed $O(1/N^2)$, thus proving our claim made earlier.

Following a similar argument as in the estimation of the first-order error term, we can now find constants $\delta \in (0,1)$, $L > 0$ and $N_1 > L$ such that $|\varepsilon_1^2| \leqslant L/N^2$ and

$$\left|\frac{z - a_k - w_k^0}{w_{k+1}^0}\right| \leqslant 1 - \delta,$$

$$\left|\frac{w_k^0 - w_{k+1}^0 + (z - a_k - w_k^0)[1 - (1 + w_k^1)^{-1}] - w_{k+1}^0 w_{k+1}^1}{w_{k+1}^0}\right| \leqslant \frac{\delta^2 L}{N^2},$$

$$\sup_{|s| \leqslant L/N^2} \frac{|(1 + w_k^1)^{-1} - (1 + w_k^1 + s)^{-1}|}{|s|} \leqslant 1 + \delta,$$

for all $k = 1, \cdots, n$ and $N > N_1$. On account of (2.13), it is easily verified that if $|\varepsilon_k^2| \leqslant L/N^2$ for some $k \geqslant 1$, then

$$|\varepsilon_{k+1}^2| \leqslant \frac{\delta^2 L}{N^2} + \frac{(1-\delta)(1+\delta)L}{N^2} = \frac{L}{N^2}.$$

By induction, we have $|\varepsilon_k^2| \leqslant L/N^2$ for all $k = 1, \cdots, n$ and $N > N_1$. This completes the proof of Lemma 2.1. □

On account of $\pi_0 = 1$ in (1.4), we can multiply (2.1) from $k = 1$ to $k = n$ and arrive at $\pi_n = w_1 \cdots w_n$. By taking logarithms on both sides of this equation and using Lemma 2.1, we obtain

$$\begin{aligned} \ln \pi_n &= \sum_{k=1}^{n} \ln w_k = \sum_{k=1}^{n} \ln w_k^0 + \sum_{k=1}^{n} \ln(1 + w_k^1) + O\left(\frac{1}{N}\right) \\ &= \sum_{k=1}^{n} \ln w_k^0 + \sum_{k=1}^{n} w_k^1 + O\left(\frac{1}{N}\right). \end{aligned} \tag{2.23}$$

To find an asymptotic formula for $\pi_n$, we only need to approximate $\sum_{k=1}^n \ln w_k^0$ and $\sum_{k=1}^n w_k^1$, respectively. Applying (2.3) and (2.4) to (2.10) and (2.11) gives

$$\frac{2w_k^0}{N^2} = T(y;t) + \frac{T_1(y;t)}{N} + O\left(\frac{1}{N^2}\right),$$

and

$$N w_k^1 = -\frac{f(t)h_0(t)}{\sqrt{y^2 + 4yf(t)}} - \frac{f(t)h_0(t)}{y + 4f(t)} + O\left(\frac{1}{N}\right),$$

where

$$T(y;t) := y + 2f(t) + \sqrt{y^2 + 4yf(t)}, \tag{2.24}$$

and

$$T_1(y;t) := f(t)h_1(t) + \frac{yf(t)h_1(t) + 2f(t)^2 h_0(t)}{\sqrt{y^2 + 4yf(t)}}. \tag{2.25}$$

Recall that $n/N = p$. It is easily seen from trapezoidal's rule that

$$\sum_{k=1}^{n} \ln w_k^0 = n \ln \frac{N^2}{2} + N \int_0^p \ln T(y;t)dt + \frac{1}{2} \ln \frac{T(y;p)}{T(y;0)} + \int_0^p \frac{T_1(y;t)}{T(y;t)}dt + O\left(\frac{1}{N}\right), \tag{2.26}$$

and

$$\sum_{k=1}^{n} w_k^1 = -\int_0^p \frac{f(t)h_0(t)}{\sqrt{y^2+4yf(t)}}dt - \int_0^p \frac{f(t)h_0(t)}{y+4f(t)}dt + O\left(\frac{1}{N}\right). \tag{2.27}$$

Finally, we have the following result.

**Theorem 2.2.** *Let $\pi_n(z)$ be the monic Racah polynomials with $\alpha + 1 = -N$. Assume $n/N = p$ is fixed in $(0, 1/2)$. Denote $y_p = -4p^2(p-1)^2/(2p-1)^2$. For any $y \in \mathbf{C} \setminus [y_p, 0]$, we have*

$$\pi_n(N^2 y) = \left(\frac{N^2}{2}\right)^n e^{Ng(y)+r(y)} \left[1 + O\left(\frac{1}{N}\right)\right], \tag{2.28}$$

*as $N \to \infty$, where the main function is given by*

$$g(y) := \int_0^p \ln T(y;t)dt, \tag{2.29}$$

*and the correction term is given by*

$$r(y) := \frac{1}{2} \ln \frac{T(y;p)}{T(y;0)} + \int_0^p \frac{T_1(y;t)}{T(y;t)}dt - \int_0^p \frac{f(t)h_0(t)}{\sqrt{y^2+4yf(t)}}dt - \int_0^p \frac{f(t)h_0(t)}{y+4f(t)}dt. \tag{2.30}$$

*Here, the functions $f(t), h_1(t), h_2(t), h_0(t), T(y;t), T_1(y;t)$ are defined in* (2.5)-(2.8) *and* (2.24)-(2.25)*, respectively.*

*Proof.* Use a combination of (2.23), (2.26) and (2.27). □

Next, we investigate the asymptotic behavior of $\pi_n(N^2 y)$ for $y$ in the oscillatory interval $(y_p, 0)$. Recall that $0 < p < 1/2$. By the definition in (2.5), the function $f(t)$ is monotonically increasing for $t \in [0, p]$. Moreover, $0 = -4f(0)$ and $y_p = -4p^2(p-1)^2/(2p-1)^2 = -4f(p)$. Thus, for any $y \in (y_p, 0)$, there exists a unique $t_y \in (0, p)$ such that $y = -4f(t_y)$. Note that the functions $T(y, t)$ and $T_1(y, t)$ in (2.24)-(2.25) have a branch cut for $y \in (-4f(t), 0)$. We introduce the one-side limit functions for $-4f(t) < y < 0$ (or equivalently, $t_y < t < p$):

$$T^{\pm}(y;t) := \lim_{\varepsilon \to 0^+} T(y \pm i\varepsilon;t) = y + 2f(t) \pm i\sqrt{-y[y+4f(t)]}, \tag{2.31}$$

and

$$T_1^{\pm}(y;t) := \lim_{\varepsilon \to 0^+} T_1(y \pm i\varepsilon;t) = f(t)h_1(t) \mp i\frac{yf(t)h_1(t) + 2f(t)^2 h_0(t)}{\sqrt{-y[y+4f(t)]}}. \tag{2.32}$$

We remark that $T(y;t)$ and $T_1(y;t)$, treated as functions in $y$, can be analytically continued to the interval $y_p < y < -4f(t)$. For convenience, we still define the one-side limit functions $T^{\pm}(y;t) = T(y;t)$ and $T_1^{\pm}(y;t) = T_1(y;t)$ for $y_p < y < -4f(t)$

(i.e., $0 < t < t_y$). Also, we define the functions

$$S^{\pm}(y;t) := \lim_{\varepsilon\to 0+} \sqrt{(y \pm i\varepsilon)[y \pm i\varepsilon + 4f(t)]}$$

$$= \begin{cases} \pm i\sqrt{-y[y+4f(t)]}, & -4f(t) < y < 0, \\ -\sqrt{(-y)[-y-4f(t)]}, & y_p < y < -4f(t). \end{cases} \tag{2.33}$$

It can be shown from (2.29)-(2.33) that

$$g^{\pm}(y) := \lim_{\varepsilon\to 0+} g(y \pm i\varepsilon; t) = \int_0^p \ln T^{\pm}(y;t)dt, \tag{2.34}$$

and

$$r^{\pm}(y) := \lim_{\varepsilon\to 0+} r(y \pm i\varepsilon; t)$$

$$= \frac{1}{2}\ln\frac{T^{\pm}(y;p)}{T^{\pm}(y;0)} + \int_0^p \frac{T_1^{\pm}(y;t)}{T^{\pm}(y;t)}dt - \int_0^p \frac{f(t)h_0(t)}{S^{\pm}(y;t)}dt - \Omega^{\pm}(y), \tag{2.35}$$

where

$$\Omega^{\pm}(y) = \text{P.V.}\int_0^p \frac{f(t)h_0(t)}{y+4f(t)}dt \pm i\pi\frac{f(t_y)h_0(t_y)}{4f'(t_y)}. \tag{2.36}$$

In (2.36), "P.V." denotes the Cauchy principal value and $f'$ denotes the derivative of $f$. In (2.35), the first three integrals are obtained by taking one-side limits, while the last term $\Omega^{\pm}(y)$ is derived from one-side limits of Stieltjes transform and Cauchy's theorem. Note that the last integral of $r(y)$ in (2.30) has a logarithmic singularity at $t = t_y$. We rewrite it as a Stieltjes integral

$$\Omega(y) := \int_0^p \frac{f(t)h_0(t)}{y+4f(t)}dt = \int_0^p \frac{F(y;t)}{t-t_y}dt, \ y \in \mathbf{C}\setminus[y_p, 0],$$

where $F(y;t) := f(t)h_0(t)(t-t_y)/[y+4f(t)]$. As a function of $t$, $F(y;t)$ can be continually defined at $t = t_y$. An application of Cauchy's theorem yields

$$\Omega^{\pm}(y) := \lim_{\varepsilon\to 0+} \Omega(y \pm i\varepsilon) = \text{P.V.}\int_0^p \frac{F(y;t)}{t-t_y}dt \pm i\pi \lim_{t\to t_y} F(y;t)$$

for $y \in (y_p, 0)$. By l'Hôpital's rule,

$$\lim_{t\to t_y} F(y;t) = \lim_{t\to t_y} \frac{f(t)h_0(t)(t-t_y)}{y+4f(t)} = f(t_y)h_0(t_y)\lim_{t\to t_y}\frac{t-t_y}{y+4f(t)} = \frac{f(t_y)h_0(t_y)}{4f'(t_y)}.$$

This gives (2.36).

Our second result is stated below. Its proof is based on the observation that the real part of an asymptotic approximant for orthogonal polynomials in the complex plane is half of the corresponding asymptotic approximant in the interval of orthogonality on the real line.

**Theorem 2.3.** *Let $\pi_n(z)$ be the monic Racah polynomials with $\alpha + 1 = -N$. Assume $n/N = p$ is fixed in $(0, 1/2)$. For $y \in (y_p, 0)$ with $y_p = -4p^2(p-1)^2/(2p-1)^2$, we have*

(2.37)

$$\pi_n(N^2y) = \left(\frac{N^2}{2}\right)^n \left\{e^{Ng^+(y)+r^+(y)}\left[1+O\left(\frac{1}{n}\right)\right] + e^{Ng^-(y)+r^-(y)}\left[1+O\left(\frac{1}{n}\right)\right]\right\},$$

*as $N \to \infty$, where $g^{\pm}(y)$ and $r^{\pm}(y)$ are defined as in (2.34) and (2.35), respectively.*

*Proof.* First, we note that the functions $g^{\pm}(y)$ and $r^{\pm}(y)$ can be analytically continued to a small neighborhood of any compact subset of $(y_p, 0)$ in the complex plane. In such a neighborhood, we further observe that the difference

$$g^+(y) - g^-(y) = \int_{t_y}^{p} \ln \frac{T^+(y;t)}{T^-(y;t)} dt$$

has positive real part if $y$ is in the upper-half plane and negative real part if $y$ is in the lower-half plane. Thus, the function $e^{N[g^+(y)-g^-(y)]}$ is exponentially large on the upper-half plane and exponentially small on the lower-half plane, which together with (2.28) implies that the asymptotic formula (2.37) is valid on both upper- and lower-half planes. By taking limits, this formula is also valid for $y \in (y_p, 0)$. □

## 3. Case II: $\gamma + 1 = -N$

In this section, we consider the case $\gamma + 1 = -N$ and $\alpha, \beta, \delta$ are fixed. It follows from (1.3) and (1.5) that

$$\begin{aligned} a_k = &- \frac{(k+\alpha+1)(k+\alpha+\beta+1)(k+\beta+\delta+1)(k-N)}{(2k+\alpha+\beta+1)(2k+\alpha+\beta+2)} \\ &- \frac{k(k+N+\alpha+\beta+1)(k+\alpha-\delta)(k+\beta)}{(2k+\alpha+\beta)(2k+\alpha+\beta+1)}; \\ b_k = &\frac{k(k+N+\alpha+\beta+1)(k+\alpha-\delta)(k+\beta)(k+\alpha)(k+\alpha+\beta)(k+\beta+\delta)(k-N-1)}{(2k+\alpha+\beta-1)(2k+\alpha+\beta)^2(2k+\alpha+\beta+1)}. \end{aligned}$$

Put $t := k/N$. We obtain for $0 < t < 1$,

$$\begin{aligned} \frac{a_k}{N^2} &= -\frac{t^2}{2} - \frac{t(2\alpha+2\beta+2) - (\alpha+\beta+2\delta+2)}{4N} + O\left(\frac{1}{N^2}\right); \\ \frac{b_k}{N^4} &= \frac{t^2(t^2-1)}{16} + \frac{(\alpha+\beta)t(t-1)(2t+1) - 2t^2}{16N} + O\left(\frac{1}{N^2}\right), \end{aligned} \tag{3.1}$$

and

$$\frac{a_{k+1}-a_k}{N} = -t + O\left(\frac{1}{N}\right); \quad \frac{b_{k+1}-b_k}{N^3} = \frac{t(2t^2-1)}{8} + O\left(\frac{1}{N}\right). \tag{3.2}$$

It is noted that $b_k < 0$ for large $k$ and $N$ such that $k < N$. Hence, Favard's theorem [3, §§1.4 & 1.5] is no longer applicable and it is expected that the zeros of $\pi_n(z)$ are not necessarily all real; see Figure 1.

As in the previous section, we introduce the ratios $w_k(z) = \pi_k(z)/\pi_{k-1}(z)$ for $k = 1, \cdots, n$. The corresponding difference equation remains the same:

$$w_{k+1}(z) = z - a_k - \frac{b_k}{w_k(z)}.$$

Let $z = N^2 y$. By using the same argument as in the proof of Lemma 2.1, we have

$$w_k = w_k^0[1 + w_k^1 + O(1/N^2)] \tag{3.3}$$

as $N \to \infty$, where the leading term is given by

$$w_k^0 = \frac{z - a_k + \sqrt{(z-a_k)^2 - 4b_k}}{2}$$

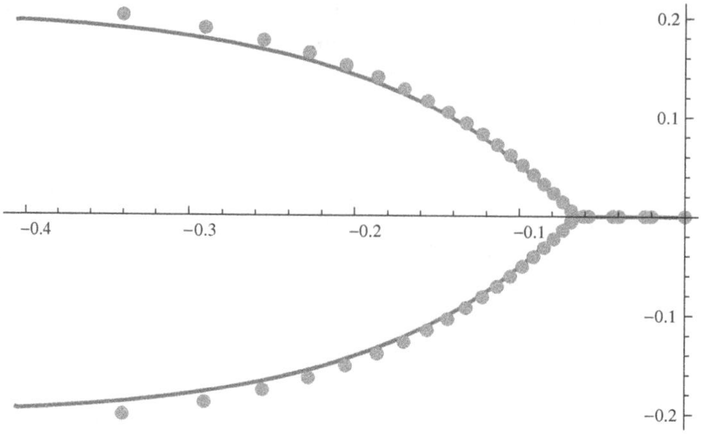


FIGURE 1. The zeros (red dots) of Racah polynomials $\pi_n(N^2y)$ with $N = 50$, $n = 45$, $p = n/N = 0.9$, $\alpha = 0.3$, $\beta = 0.8$, $\delta = -1.2$, $\gamma = -N-1 = -51$, and the branch cut $\Gamma_p$ (blue curves) with $p = 0.9$ of the main function $g(y)$ defined in (3.8).

and the first-order term by

$$w_k^1 = \frac{a_{k+1} - a_k}{2\sqrt{(z-a_k)^2 - 4b_k}} + \frac{2(b_{k+1} - b_k) + (z-a_k)(a_{k+1} - a_k)}{2[(z-a_k)^2 - 4b_k]}.$$

On account of (3.1) and (3.2), we rewrite the above two equations as

$$\frac{2w_k^0}{N^2} = T(y;t) + \frac{T_1(y;t)}{N} + O\left(\frac{1}{N^2}\right),$$

and

$$Nw_k^1 = -\frac{t}{2\sqrt{y^2 + t^2(y+1/4)}} - \frac{t(y+1/4)}{2y^2 + 2t^2(y+1/4)} + O\left(\frac{1}{N}\right),$$

where

$$T(y;t) := y + \frac{t^2}{2} + \sqrt{y^2 + yt^2 + \frac{t^2}{4}}, \tag{3.4}$$

and

(3.5)

$$T_1(y;t) := \frac{[2t(\alpha+\beta+1) - (\alpha+\beta+2\delta+2)]T(y;t) - (\alpha+\beta)t(t-1)(t+1/2) + t^2}{4\sqrt{y^2 + yt^2 + t^2/4}}.$$

Recall that $n/N = p$. It is easily seen from trapezoidal's rule that

(3.6)

$$\sum_{k=1}^{n} \ln w_k^0 = n \ln \frac{N^2}{2} + N \int_0^p \ln T(y;t)dt + \frac{1}{2} \ln \frac{T(y;p)}{T(y;0)} + \int_0^p \frac{T_1(y;t)}{T(y;t)} dt + O\left(\frac{1}{N}\right),$$

and

$$\sum_{k=1}^{n} w_k^1 = \frac{y - \sqrt{y^2 + p^2(y+1/4)}}{2(y+1/4)} - \frac{1}{4}\ln\frac{y^2 + p^2(y+1/4)}{y^2} + O\left(\frac{1}{N}\right). \tag{3.7}$$

Finally, we introduce the main function

$$g(y) := \int_0^p \ln T(y;t)dt, \tag{3.8}$$

and define the curve $\Gamma_p$ to be the set of complex numbers $y$ at which the real part of $g^+(y) - g^-(y)$ vanishes, where $g^\pm(y)$ are the one-sided limits of $g(y)$ on the curve $\Gamma_p$; see Figure 1.

**Theorem 3.1.** *Let $\pi_n(z)$ be the monic Racah polynomials with $\gamma = -N-1$. Assume $n/N = p$ is fixed in $(0,1)$. For any $y \in \mathbf{C} \setminus \bar{\Gamma}_p$, where $\bar{\Gamma}_p$ is the closure of $\Gamma_p$ in the complex plane, we have as $N \to \infty$,*

$$\pi_n(N^2 y) = \left(\frac{N^2}{2}\right)^n e^{Ng(y)+r(y)}\left[1 + O\left(\frac{1}{N}\right)\right], \tag{3.9}$$

*where the main function $g(y)$ is given in (3.8) and the correction term is given by*

$$\begin{aligned} r(y) := {}& \frac{1}{2}\ln\frac{T(y;p)}{T(y;0)} + \int_0^p \frac{T_1(y;t)}{T(y;t)}dt \\ &+ \frac{y - \sqrt{y^2 + p^2(y+1/4)}}{2(y+1/4)} - \frac{1}{4}\ln\frac{y^2 + p^2(y+1/4)}{y^2}. \end{aligned} \tag{3.10}$$

*Here, the functions $T(y;t)$ and $T_1(y;t)$ are defined in (3.4) and (3.5), respectively. Let $r^\pm(y)$ denote the one-sided limits of $r(y)$ on $\Gamma_p$. For any $y \in \Gamma_p$, we have as $N \to \infty$,*

$$\pi_n(N^2 y) = \left(\frac{N^2}{2}\right)^n \left\{e^{Ng^+(y)+r^+(y)}\left[1+O\left(\frac{1}{n}\right)\right] + e^{Ng^-(y)+r^-(y)}\left[1+O\left(\frac{1}{n}\right)\right]\right\}. \tag{3.11}$$

*Proof.* Just as in (2.23), we have

$$\ln \pi_n = \sum_{k=1}^{n} w_k = \sum_{k=1}^{n} \ln w_k^0 + \sum_{k=1}^{n} w_k^1 + O\left(\frac{1}{N}\right).$$

Substituting (3.6) and (3.7) in the above equation gives (3.9). Formula (3.11) follows from a similar argument as that in the proof of Theorem 2.3 and from the choice of the branch cut $\Gamma_p$. □

## 4. Case III: $\beta + \delta + 1 = -N$

In this section, we consider the case $\beta + \delta + 1 = -N$ and $\alpha, \delta$ are fixed. There are two subcases.

Subcase III.1: $\beta = -N - \delta - 1$ and $\alpha, \delta, \gamma$ are fixed.
Subcase III.2: $\delta = -N - \beta - 1$ and $\alpha, \beta, \gamma$ are fixed.

For the subcase III.1, we introduce two parameters $\tilde{\alpha} := \beta + \delta = -N - 1$ and $\tilde{\beta} = \alpha - \delta$. Note that the ${}_4F_3$ hypergeometric function in the definition of Racah polynomials (1.1) is symmetric with respect to the switch of parameters:

$$
{}_4F_3\left(\begin{matrix} -n, & n+\alpha+\beta+1, & -x, & x+\gamma+\delta+1 \\ & \alpha+1, & \beta+\delta+1, & \gamma+1 \end{matrix}\middle| 1\right)
$$
$$
={}_4F_3\left(\begin{matrix} -n, & n+\tilde{\alpha}+\tilde{\beta}+1, & -x, & x+\gamma+\delta+1 \\ & \tilde{\alpha}+1, & \tilde{\beta}+\delta+1, & \gamma+1 \end{matrix}\middle| 1\right).
$$

Thus, the subcase III.1 is the same as case I considered in Section 2.

For the subcase III.2, we introduce two parameters $\tilde{\gamma} := \beta+\delta = -N-1$ and $\tilde{\delta} = \gamma - \beta$. Note that the ${}_4F_3$ hypergeometric function in the definition of Racah polynomials (1.1) is symmetric with respect to the switch of parameters:

$$
{}_4F_3\left(\begin{matrix} -n, & n+\alpha+\beta+1, & -x, & x+\gamma+\delta+1 \\ & \alpha+1, & \beta+\delta+1, & \gamma+1 \end{matrix}\middle| 1\right)
$$
$$
={}_4F_3\left(\begin{matrix} -n, & n+\alpha+\beta+1, & -x, & x+\tilde{\gamma}+\tilde{\delta}+1 \\ & \alpha+1, & \beta+\tilde{\delta}+1, & \tilde{\gamma}+1 \end{matrix}\middle| 1\right).
$$

That is, the subcase III.2 is the same as case II considered in Section 3.

## Acknowledgment

We are very grateful to the anonymous referee for the careful reading and valuable suggestions which have helped to improve the presentation of this paper.

## References

[1] Richard Askey and James Wilson, *A set of orthogonal polynomials that generalize the Racah coefficients or* $6-j$ *symbols*, SIAM J. Math. Anal. **10** (1979), no. 5, 1008–1016, DOI 10.1137/0510092. MR541097

[2] Li-Chen Chen, Mourad E. H. Ismail, and Plamen Simeonov, *Asymptotics of Racah coefficients and polynomials*, J. Phys. A **32** (1999), no. 3, 537–553, DOI 10.1088/0305-4470/32/3/009. MR1670247

[3] T. S. Chihara, *An introduction to orthogonal polynomials*, Gordon and Breach Science Publishers, New York-London-Paris, 1978. Mathematics and its Applications, Vol. 13. MR0481884

[4] Mourad E. H. Ismail, *Classical and quantum orthogonal polynomials in one variable*, Encyclopedia of Mathematics and its Applications, vol. 98, Cambridge University Press, Cambridge, 2005. With two chapters by Walter Van Assche; With a foreword by Richard A. Askey. MR2191786

[5] Mourad E. H. Ismail and Plamen Simeonov, *Inequalities and asymptotics for a terminating* ${}_4F_3$ *series*, Illinois J. Math. **51** (2007), no. 3, 861–881. MR2379727

[6] Roelof Koekoek, Peter A. Lesky, and René F. Swarttouw, *Hypergeometric orthogonal polynomials and their q-analogues*, Springer Monographs in Mathematics, Springer-Verlag, Berlin, 2010. With a foreword by Tom H. Koornwinder. MR2656096

[7] X.-S. Wang and R. Wong, *Asymptotics of Racah polynomials with varying parameters*, J. Math. Anal. Appl. **436** (2016), no. 2, 1149–1164, DOI 10.1016/j.jmaa.2015.12.035. MR3447001

[8] James Arthur Wilson, *HYPERGEOMETRIC SERIES RECURRENCE RELATIONS AND SOME NEW ORTHOGONAL FUNCTIONS*, ProQuest LLC, Ann Arbor, MI, 1978. Thesis (Ph.D.)–The University of Wisconsin - Madison. MR2628179

Department of Mathematics, University of Louisiana at Lafayette, Lafayette, Louisiana 70503

Department of Mathematics, City University of Hong Kong, Tat Chee Avenue, Kowloon, Hong Kong

INTERNATIONAL JOURNAL OF
NUMERICAL ANALYSIS AND MODELING
Volume 15, Number 1-2, Pages 193–212



# ASYMPTOTICS OF ORTHOGONAL POLYNOMIALS

R. WONG

*In memory of Professor Benyu Guo*

**Abstract.** In this survey article, we present some recent results on the asymptotic behavior of four systems of orthogonal polynomials. These are Stieltjes-Wigert, Hahn, Racah and pseudo-Jacobi polynomials. In each case, the variable $z$ is allowed to be in any part of the complex plane. In some cases, asymptotic formulas are also given for their zeros.

**Key words.** Asymptotics, orthogonal polynomials, Riemann-Hilbert method, difference-equation techniques.

## 1. Introduction

Most of the asymptotic results of classical orthogonal polynomials (namely, Hermite $H_n(z)$, Laguerre $L_n^{(\alpha)}(z)$, and Jacobi $P_n^{(\alpha,\beta)}(z)$) can be found in the books of Szegö [45] and Erdélyi et al. [15]. Some of the asymptotic results of a few familiar discrete orthogonal polynomials (e.g., Charlier $C_n(z;a)$, Meixner $M_n(z;\beta,c)$, and Krawtchouk $K_n(z;p,N)$) have been summarized in a survey article of Wong [56]. With the new developments in asymptotic methods based on the Riemann-Hilbert approach [10, 11] and difference equations [48, 50, 51], asymptotic problems of some orthogonal polynomials, which have been considered to be more difficult to tackle, have also been resolved; for instance, discrete Chebyshev [33], Conrad-Flajolet [8], polynomials orthogonal with respect to Freud weights [27, 57], and Tricomi-Carlitz [30, 58]. The reasons for the asymptotic problems of the last few mentioned orthogonal polynomials being more difficult are : (i) they do not satisfy second-order linear ordinary differential equations, and (ii) they do not have integral representations to which one can apply the classical methods of steepest descent or stationary phase.

The purpose of this paper is to give a summary of asymptotic results obtained recently for four systems of orthogonal polynomials; they are Stieltjes-Wigert $S_n(z;q)$, Hahn $Q_n(z;\alpha,\beta,N)$, Racah $R_n(\lambda(x);\alpha,\beta,\gamma,\delta)$, and pseudo-Jacobi $P_n(z;a,b)$. The problem of finding asymptotic formulas for Stieltjes-Wigert polynomials has been around for some time. However, serious work began only at the beginning of this century. These polynomials do not have integral representations, and neither do they satisfy any second-order linear differential equation. Although they satisfy a three-term recurrence relation [6, p.174], the coefficients of the recurrence relation contain exponentially large terms of the form $q^{-n}, 0 < q < 1$. As a consequence, none of the existing methods for second-order difference equations can be applied. The Hahn polynomial was first introduced by Chebyshev in 1858. Despite its long history, there seems to be no literature on asymptotic results of this polynomial. A difficulty in dealing with asymptotic problems of this polynomial is that it has three free real parameters, whereas the other classical discrete orthogonal polynomials involve only two or less free parameters. This remark also applies to the

Received by the editors February 9, 2017 and, in revised form, April 5, 2017.
2000 *Mathematics Subject Classification.* 33C45, 41A60.

Racah polynomial, which has even one more free parameter. We pick Pseudo-Jacobi polynomials as our last example for presentation, since it also has a long history (over hundred years) and it is not well-known even to the researchers in the field of orthogonal polynomials until just recently. Another reason is that this example illustrates the asymptotic method developed for differential equations with a large parameter [38, Chapter 11], which can be applied to many applied disciplines, and deserves a larger audience.

## 2. Stieltjes-Wigert polynomials

Let $k > 0$ be a fixed number and

$$q = \exp\{-(2k^2)^{-1}\}. \tag{1}$$

Note that $0 < q < 1$. The $q$-shifted factorial is given by

$$(a;q)_0 = 1, \qquad (a;q)_n = \prod_{j=0}^{n-1}(1-aq^j), \qquad n = 1,2,\cdots.$$

The Stieltjes-Wigert polynomials

$$S_n(z;q) := \sum_{j=0}^{n} \frac{q^{j^2}}{(q;q)_j(q;q)_{n-j}}(-z)^j, \quad n = 0,1,2,\cdots, \tag{2}$$

are orthogonal with respect to the weight function

$$w(x) = k\pi^{-\frac{1}{2}} \exp\{-k^2 \log^2 x\} \tag{3}$$

for $0 < x < \infty$; see [26, (3.27.1)] and [39, (18.27.18)]. It is known that these polynomials belong to the indeterminate moment class and the weight function in (3) is not unique; see [7]. By changing the index $j$ to $n-j$ in the explicit expression given in (2), one can easily verify the symmetry relation

$$S_n(z;q) = (-zq^n)^n S_n\left(\frac{1}{zq^{2n}};q\right). \tag{4}$$

In some literatures, the variable $z$ in (2) is replaced by $q^{\frac{1}{2}}z$; see, for instance, [6], [45] and [52]. The notation for the Stieltjes-Wigert polynomials used in these literatures is

$$p_n(z) = (-1)^n q^{n/2+1/4}\sqrt{(q;q)_n} S_n(q^{\frac{1}{2}}z;q). \tag{5}$$

These polynomials arise in random walks and random matrix formulation of Chern-Simons theory on Seifert manifolds; see [4, 12].

The asymptotics of the Stieltjes-Wigert polynomials, as the degree tends to infinity, has been studied by several authors. First, Wigert [53] in 1923 proved that the polynomials have the limiting behavior

$$\lim_{n\to\infty}(-1)^n q^{-n/2} p_n(z) = \frac{q^{1/4}}{\sqrt{(q;q)_\infty}} \sum_{k=0}^{\infty}(-1)^k \frac{q^{k^2+k/2}}{(q;q)_k} z^k, \tag{6}$$

which can be put in terms of the $q$-Airy function (also known as the Ramanujan function)

$$A_q(z) = \sum_{k=0}^{\infty} \frac{q^{k^2}}{(q;q)_k}(-z)^k. \tag{7}$$

In terms of this function, Wigert's result can be stated as

$$\lim_{n\to\infty} S_n(z;q) = \frac{1}{(q;q)_\infty} A_q(z). \tag{8}$$

It has been shown in [52] that for large values of $n$, all zeros of $S_n(z;q)$ lie in the interval $(\frac{1}{4}, 4q^{-2n})$. Thus, it is natural to introduce a new scale

$$z = uq^{-tn} \tag{9}$$

with $u \in \mathbb{C}\backslash\{0\}$ and $t \in \mathbb{R}$. The values of $t = 0$ and $t = 2$ can be regarded as the turning points of $S_n(q^{-nt}u;q)$. In view of the symmetry relation in (4), one may restrict oneself to the case $t \geqslant 1$; see [47, (1.4)]. (However, we shall not make this restriction in our discussion here.) The case $t = 2$ has been studied by Ismail [20], and he proved that for $t = 2$

$$\lim_{n\to\infty} q^{n^2(t-1)}(-u)^{-n} S_n(uq^{-nt};q) = \frac{1}{(q;q)_\infty} A_q\left(\frac{q^{n(t-2)}}{u}\right) \tag{10}$$

uniformly for $u$ in compact subsets of $\mathbb{C} \setminus \{0\}$; see [20, Theorem 2.5]. This result can in fact be derived directly from Wigert's formula in (8) by using the symmetry relation mentioned in (4). In [21], Ismail and Zhang extended the validity of this result to $t \geqslant 2$. For $1 \leqslant t < 2$, Ismail and Zhang [21] gave asymptotic formulas for these polynomials in terms of the theta-type function (cf. also [17])

$$\Theta_q(z) = \sum_{k=-\infty}^{\infty} q^{k^2} z^k, \tag{11}$$

but in a very complicated manner. Their result was then simplified by Wang and Wong in [46]. For instance, when $1 \leqslant t < 2$, Wang and Wong proved that

$$\begin{aligned} S_n(uq^{-nt};q) = {}& \frac{(-u)^{n-m} q^{n^2(1-t)-m[n(2-t)-m]}}{(q;q)_n (q;q)_\infty} \\ & \times \left\{ \Theta_q\left(\frac{q^{2m-n(2-t)}}{-u}\right) + O(q^{n(l-\delta)}) \right\}, \end{aligned} \tag{12}$$

where $l = \frac{1}{2}(2-t), m = \lfloor nl \rfloor$ and $\delta > 0$ is any small number; see [46, Corollary 2]. However, none of these results is valid in a neighborhood of $t = 2$, one of the turning points. To address this issue, Wang and Wong in a second paper [47] presented a uniform asymptotic formula: for $z := uq^{-nt}$ with $t > 2(1-\delta), \delta > 0$, they showed that

$$S_n(z;q) = \frac{(-z)^n q^{n^2}}{(q;q)_n} [A_{q,n}(q^{-2n}/z) + r_n(z)], \tag{13}$$

where $r_n(z)$ is the remainder and $A_{q,n}(z)$ is the $q$-Airy polynomial obtained by truncating the infinite series in (7) at $k = n$, i.e.,

$$A_{q,n}(z) := \sum_{k=0}^{n} \frac{q^{k^2}}{(q;q)_k} (-z)^k.$$

This result, however, is not totally satisfactory, since the $q$-Airy polynomial is not a known special function and has not been well studied. (The situation is like that with the exponential polynomial $\sum_{k=0}^{n} z^k/k!$ , which is not as well-known as the exponential function $e^z$ itself.)

A more satisfactory solution was provided by Li and Wong [32] three years later. Their result is stated below.

THEOREM 2.1 *Let* $z := uq^{-nt}$ *with* $-\infty < t < 2, u \in \mathbb{C}$ *and* $|u| \leqslant R$, *where* $R > 0$ *is any fixed positive number. We have*

$$S_n(z;q) = \frac{1}{(q;q)_n}[A_q(z) + r_n(z)], \tag{14}$$

*where the remainder satisfies*

$$|r_n(z)| \leqslant \left[\frac{q^{n(1-\sigma)}}{1-q} + \frac{2}{1-q}\left(\frac{1}{2}\right)^{\lfloor n\sigma \rfloor}\right] A_q(-|z|) \tag{15}$$

*with* $\sigma = \max\{\frac{1}{2}, \frac{1}{2} + \frac{t}{4}\}$.

*Let* $z := uq^{-nt}$ *with* $0 < t < \infty, u \in \mathbb{C}$ *and* $|u| \geqslant 1/R$, *where* $R > 0$ *is any fixed number. We have*

$$S_n(z;q) = \frac{(-z)^n q^{n^2}}{(q;q)_n}\left[A_q(q^{-2n}/z) + r_n(z)\right], \tag{16}$$

*where the remainder satisfies*

$$|r_n(z)| \leqslant \left[\frac{q^{n(\delta-1)}}{1-q} + \frac{2}{1-q}\left(\frac{1}{2}\right)^{\lfloor n(2-\delta) \rfloor}\right] A_q(-q^{-2n}/|z|) \tag{17}$$

*with* $\delta = \min\{\frac{3}{2}, 1 + \frac{t}{4}\}$.

Since the quantities inside the square brackets in (15) and (17) are exponentially small, the two asymptotic formulas in (14) and (16) are quite satisfactory.

The above mentioned investigations all began with the explicit expression of $S_n(z;q)$ given in (2). In [52], Wang and Wong used a different method, namely, the Riemann-Hilbert approach, to construct a uniform asymptotic expression of the Stieltjes-Wigert polynomials $p_n(z)$ in terms of Airy functions. Unfortunately, the main result in [52] is not quite correct; it requires an additional condition. The additional condition is that the parameter $k$ in (1) should depend on the degree $n$ and tend to infinity as $n \to \infty$. More precisely, we need to impose the condition that there are two constants $c > 0$ and $\sigma > 0$ such that

$$\lim_{n\to\infty} \frac{k}{n^\sigma} = c; \tag{18}$$

see also Baik and Suidan [4]. For completeness, we state the amended result below.

First, we introduce some notations. Let $\gamma_n$ denote the leading coefficient of the polynomial $p_n(z)$ given in (5), and put $\pi_n(z) := p_n(z)/\gamma_n$. The Mhaskar-Rakhmanov-Saff (MRS) numbers $\alpha_n$ and $\beta_n$ for our present problem are given by

$$\begin{aligned}\alpha_n = {} & 2e^{(n+1/2)/k^2} - e^{(n+1/2)/2k^2} \\ & - 2e^{(n+1/2)/2k^2}\sqrt{e^{(n+1/2)/k^2} - e^{(n+1/2)/2k^2}},\end{aligned} \tag{19}$$

$$\begin{aligned}\beta_n = {} & 2e^{(n+1/2)/k^2} - e^{(n+1/2)/2k^2} \\ & + 2e^{(n+1/2)/2k^2}\sqrt{e^{(n+1/2)/k^2} - e^{(n+1/2)/2k^2}}.\end{aligned} \tag{20}$$

The probability density function supported on $[\alpha_n, \beta_n]$ is given by

$$\mu_n(x) = \frac{2k^2}{N\pi x} \arctan \frac{\sqrt{(x-\alpha_n)(\beta_n - x)}}{x + \sqrt{\alpha_n\beta_n}},$$

where $N = n + \frac{1}{2}$. Note that $\mu_n(x) \geqslant 0$ and

$$\int_{\alpha_n}^{\beta_n} \mu_n(x)dx = 1. \tag{21}$$

The logarithmic potential of $\mu_n(x)$, i.e., the $g$-function is given by

$$g(z) = \int_{\alpha_n}^{\beta_n} \log(z-s)\mu_n(s)ds. \tag{22}$$

A closely related function is

$$\begin{aligned}\phi_n(z) = \frac{k^2}{N}\int_{\beta_n}^{z} \frac{1}{\zeta}\Big\{&2\log\Big[\zeta+\sqrt{\alpha_n\beta_n}+\sqrt{(\zeta-\alpha_n)(\zeta-\beta_n)}\Big] \\ &-\log\Big[(\sqrt{\alpha_n}+\sqrt{\beta_n})^2\zeta\Big]\Big\}d\zeta\end{aligned} \tag{23}$$

for $z \in \mathbb{C}\setminus(-\infty,\beta_n]$; see [52, (3.30)-(3.31)]. The function

$$\zeta_n(z) := \left[\frac{3}{2}\phi_n(z)\right]^{2/3} \tag{24}$$

plays an important role in the theory of uniform asymptotic expansions. It arises in the Liouville transformation for differential equations [38, p. 398], as well as in the cubic transformation for integrals [55, p. 367]. Since $(\zeta_n)_+(x) = (\zeta_n)_-(x)$ for $x \in (\alpha_n, \beta_n)$, $\zeta_n(z)$ can be analytically continued to $\mathbb{C}\setminus(-\infty,\alpha_n]$. Finally, we introduce the Lagrange constant

$$l_n = 2\left(n+\frac{1}{2}\right)g(\beta_n) + \log w(\beta_n) - \log\frac{\beta_n-\alpha_n}{4}. \tag{25}$$

Note that the $g$-function in (22) has been explicitly evaluated in [52]; see (23) above and (3.7) and (3.37) in that paper.

THEOREM 2.2 *With $\pi_n(z), \zeta_n(z)$ and $l_n$ defined as above, we assume that the condition in (18) holds. Then, we have*

$$\begin{aligned}\pi_n(z) = \frac{\sqrt{\pi}e^{l_n/2}}{\sqrt{w(z)}}\Big\{&\left(n+\frac{1}{2}\right)^{1/6}\mathrm{Ai}\left(\left(n+\frac{1}{2}\right)^{2/3}\zeta_n\right)\boldsymbol{A}(z,n) \\ &-\left(n+\frac{1}{2}\right)^{-1/6}\mathrm{Ai}'\left(\left(n+\frac{1}{2}\right)^{2/3}\zeta_n\right)\boldsymbol{B}(z,n)\Big\},\end{aligned} \tag{26}$$

*where $\boldsymbol{A}(z,n)$ and $\boldsymbol{B}(z,n)$ are analytic functions of $z$ in $\mathbb{C}\setminus S_\delta$, $S_\delta = \{z : \frac{2\pi}{3} \leqslant \arg(z-(\alpha_n+\delta)) \leqslant \frac{4\pi}{3}\}$ and $\delta$ is any small positive number. Further, the asymptotic expansions*

$$\boldsymbol{A}(z,n) \sim \frac{\zeta_n^{1/4}(\beta_n-\alpha_n)^{1/2}}{[(z-\alpha_n)(z-\beta_n)]^{1/4}}\left[1+\sum_{k=1}^{\infty}\frac{A_k(z)}{n^k}\right], \tag{27}$$

$$\boldsymbol{B}(z,n) \sim \frac{[(z-\alpha_n)(z-\beta_n)]^{1/4}}{\zeta_n^{1/4}(\beta_n-\alpha_n)^{1/2}}\sum_{k=1}^{\infty}\frac{B_k(z)}{n^k} \tag{28}$$

*hold uniformly, and the coefficient functions $A_k(z)$ and $B_k(z)$ are analytic functions in $\mathbb{C}\setminus S_\delta$.*

## 3. Hahn polynomials

The Hahn polynomials are explicitly given by

$$\begin{aligned}Q_n(x;\alpha,\beta,N) &:= {}_3F_2(-n,-x,n+\alpha+\beta+1;-N,\alpha+1;1) \\ &= \sum_{k=0}^{n}\frac{(-n)_k(-x)_k(n+\alpha+\beta+1)_k}{(-N)_k(\alpha+1)_k k!},\end{aligned} \tag{29}$$

where $\alpha, \beta > -1$. These polynomials are orthogonal on the discrete set $\{0, 1, \cdots, N\}$ with respect to the weight function

$$\rho(x; \alpha, \beta, N) = \frac{(\alpha+1)_x}{x!} \frac{(\beta+1)_{N-x}}{(N-x)!}, \qquad x = 0, 1, \cdots, N; \tag{30}$$

that is,

$$\sum_{k=0}^{N} Q_n(k; \alpha, \beta, N) Q_m(k; \alpha, \beta, N) \rho(k; \alpha, \beta, N) = h_{N,n} \delta_{n,m}, \tag{31}$$

$n, m = 0, 1, \cdots, N$, where

$$h_{N,n} = \frac{(-1)^n (n+\alpha+\beta+1)_{N+1} (\beta+1)_n n!}{(2n+\alpha+\beta+1)(\alpha+1)_n (-N)_n N!}. \tag{32}$$

Formula (31) indicates that these polynomials are orthogonal on an interval which is unbounded as $n \to \infty$. To make the interval of orthogonality bounded, we introduce a rescaling. Let $X_N$ denote the set defined by

$$X_N := \left\{ x_{N,k} \right\}_{k=0}^{N-1}, \qquad \text{where } x_{N,k} := \frac{k+1/2}{N}. \tag{33}$$

The points $x_{N,k}$ are called *nodes*, and they all lie in the interval $(0, 1)$. Furthermore, we let

$$P_{N,n}(z) := Q_n(Nz - \frac{1}{2}; \alpha, \beta, N-1) \tag{34}$$

and

$$w(z) := N^{-\alpha-\beta} \alpha! \beta! \rho(Nz - 1/2; \alpha, \beta, N-1). \tag{35}$$

It is readily verified that the polynomials $P_{N,n}(z)$ are orthogonal on the nodes $x_{N,n}$ with respect to the weight $w_{N,k} := w(x_{N,k})$; i.e.,

$$\sum_{k=0}^{N-1} P_{N,n}(x_{N,k}) P_{N,m}(x_{N,k}) w_{N,k} = h^*_{N,n} \delta_{n,m}, \tag{36}$$

where $h^*_{N,n} = N^{-\alpha-\beta} \alpha! \beta! h_{N-1,n}$.

In 1989, Sharapudinov [43] considered the asymptotic behavior of $Q_n(x; \alpha, \beta, N)$ when the degree becomes large. His result is an asymptotic formula for $Q_n(x; \alpha, \beta, N)$ with $n = O(N^{1/2})$ as $n \to \infty$, which involves the Jacobi polynomial $P_n^{(\beta,\alpha)}(t)$, where $t$ is related to $x$ via the formula $x = (N-1)(1+t)/2$. More recently, Baik et al [3] used the Riemann-Hilbert method to investigate the asymptotics of discrete orthogonal polynomials with respect to a general weight function. Their results are very general, and it is difficult to use them to write out explicit formulas for specific polynomials. With regard to Hahn polynomials, they only considered the case of varying parameters, namely, $\alpha = NA$ and $\beta = NB$, where $A$ and $B$ are fixed positive numbers. Furthermore, their results are more local in nature; that is, one needs more asymptotic formulas to describe the behavior of these polynomials in the whole complex plane.

By using a modified version of the Riemann-Hilbert method, Lin and Wong [34] constructed globally uniform asymptotic formulas for the Hahn polynomials as $n \to \infty$. They considered the case when the parameters $\alpha$ and $\beta$ are fixed and the ratio $n/N$ is a constant $c \in (0, 1)$. To present their result, we need to recall some terminologies used in the Riemann-Hilbert method, including equilibrium measure, $g$-function, Lagrange constants, etc. In the existing literature, equilibrium measures are usually obtained by solving minimization problems of certain quadratic functions (see [3, 9]). In [34], the authors used a method introduced by Kuijlaars and Van Assche [29]. To state their result, we let

$$a := \frac{1}{2} - \frac{1}{2}\sqrt{1-c^2}, \qquad b := \frac{1}{2} + \frac{1}{2}\sqrt{1-c^2}, \tag{37}$$

where $c := n/N$. Define the function

$$\mu(x) = \begin{cases} \dfrac{2}{\pi c}\arcsin\left(\dfrac{c}{2\sqrt{x-x^2}}\right), & x \in (a,b), \\ \dfrac{1}{c}, & x \in [0,a] \cup [b,1]. \end{cases} \tag{38}$$

and note that $\mu(x) \geqslant 0$ and

$$\int_0^1 \mu(x)dx = 1; \tag{39}$$

that is, $\mu(x)$ is a probability density function. Also, note that

$$a + b = 1, \qquad a \cdot b = c^2/4. \tag{40}$$

The equilibrium measure associated with the Hahn polynomials is just the (indefinite) integral of $\mu(x)$. The $g$-function is the logarithmic potential of $\mu(x)$ defined by

$$g(z) := \int_0^1 \log(z-s)\mu(s)ds, \qquad z \in \mathbb{C} \setminus (-\infty, 1]. \tag{41}$$

An auxiliary function $\phi(z)$ is defined by

$$\phi(z) := l/2 - g(z), \qquad z \in \mathbb{C} \setminus (-\infty, 1], \tag{42}$$

where $l$ is the Lagrange constant given by

$$l := 2\int_0^1 \log|a-s| \cdot \mu(s)ds. \tag{43}$$

Moreover, we need a complex function $\nu(z)$ satisfying the requirement

$$\nu_\pm(x) = \pm\pi i\left(\mu(x) - \frac{1}{c}\right) \tag{44}$$

for $x \in (a,b)$, and the function is

$$\nu(z) := \frac{2}{c}\log\left[c/2 - \sqrt{(z-a)(z-b)}\right] - \frac{1}{c}\log(z - z^2), \tag{45}$$

where $\sqrt{(z-a)(z-b)}$ is analytic in $\mathbb{C} \setminus [a,b]$ and behaves like $z$ as $z \to \infty$, and where $\log(z-z^2)$ is an analytic function with branch cuts along $(-\infty, 0] \cup [1, +\infty)$.

Now, we are ready to introduce the auxiliary function

$$\tilde{\phi}(z) := \int_a^z \nu(s)ds, \qquad z \in \mathbb{C} \setminus (-\infty, 0] \cup [a, +\infty), \tag{46}$$

where the path of integration from $a$ to $z$ lies entirely in the region $z \in \mathbb{C}\setminus(-\infty, 0] \cup [a, +\infty)$, except for the initial point $a$. Similarly, we define

$$\phi^*(z) := \int_b^z \nu(s)ds, \qquad z \in \mathbb{C} \setminus (-\infty, b] \cup [1, +\infty), \tag{47}$$

where the path of integration from $b$ to $z$ lies entirely in the region $z \in \mathbb{C}\setminus(-\infty, b] \cup [1, +\infty)$, except for the initial point $b$. The connection between the $\tilde{\phi}$ - function ($\phi^*$ - function) and the $\phi$ - function in (42) is given by the formulas

$$\tilde{\phi}(z) = \phi(z) \pm \pi i \left(1 - \frac{1}{c} z\right) \tag{48}$$

and

$$\phi^*(z) = \phi(z) \pm \pi i \frac{1}{c}(1 - z) \tag{49}$$

for $z \in \mathbb{C}_\pm$.

The two functions $\tilde{\phi}(z)$ and $\phi^*(z)$ defined above play an important role in the construction of globally uniform asymptotic approximations given by Lin and Wong [34], since the arguments of the leading terms in our two asymptotics expansions are

$$\tilde{f}(z) = \left(-\frac{3}{2}\tilde{\phi}(z)\right)^{2/3}, \tag{50}$$

which is analytic in $\mathbb{C} \setminus (-\infty, 0] \cup [b, +\infty)$, and

$$f^*(z) = \left(-\frac{3}{2}\phi^*(z)\right)^{2/3}, \tag{51}$$

which is analytic in $\mathbb{C} \setminus (-\infty, a] \cup [1, +\infty)$.

Another important function used in our method is the $D$-function

$$D(z) := \frac{e^{Nz}\Gamma(Nz + 1/2)}{\sqrt{2\pi}(Nz)^{Nz}}, \qquad z \in \mathbb{C} \setminus (-\infty, 0]. \tag{52}$$

Two related functions are

$$D^*(z) := \frac{\sqrt{2\pi}e^{Nz}(-Nz)^{-Nz}}{\Gamma(-Nz + 1/2)}, \qquad z \in \mathbb{C} \setminus [0, +\infty), \tag{53}$$

and

$$\tilde{D}(z) := \begin{cases} D^*(z), & \operatorname{Re} z < x_0, \\ D^*(1 - z) & \operatorname{Re} z > x_0, \end{cases} \tag{54}$$

where $x_0$ is an arbitrary fixed point in the interval $(a, b)$ and $a, b$ are the numbers defined in (37).

To state the result in [34], we need to first introduce two more functions:

$$\begin{aligned} m(z) :=& \frac{(z-a)^{1/2} + (z-b)^{1/2}}{2(z-a)^{1/4}(z-b)^{1/4}} \left(\frac{z + c/2 + (z-a)^{1/2}(z-b)^{1/2}}{2z}\right)^{\alpha} \\ & \times \left(\frac{1 - z + c/2 - (z-a)^{1/2}(z-b)^{1/2}}{2(1-z)}\right)^{\beta} \end{aligned} \tag{55}$$

and

$$\begin{aligned} m^*(z) :=& \frac{(z-b)^{1/2} - (z-a)^{1/2}}{2(z-a)^{1/4}(z-b)^{1/4}} \left(\frac{z + c/2 - (z-a)^{1/2}(z-b)^{1/2}}{2z}\right)^{\alpha} \\ & \times \left(\frac{1 - z + c/2 + (z-a)^{1/2}(z-b)^{1/2}}{2(1-z)}\right)^{\beta}. \end{aligned} \tag{56}$$

Note that in view of (40), $m(z)$ is analytical in $\mathbb{C} \setminus [a, b]$. But, the function $m^*(z)$ has additional cuts $(-\infty, 0]$ and $[1, +\infty)$. Next, we divide the complex plane into the three regions I, II and III as shown in Figure 1, where $\delta$ is any positive number and $x_1$ is an arbitrary fixed number in $(0, a)$.

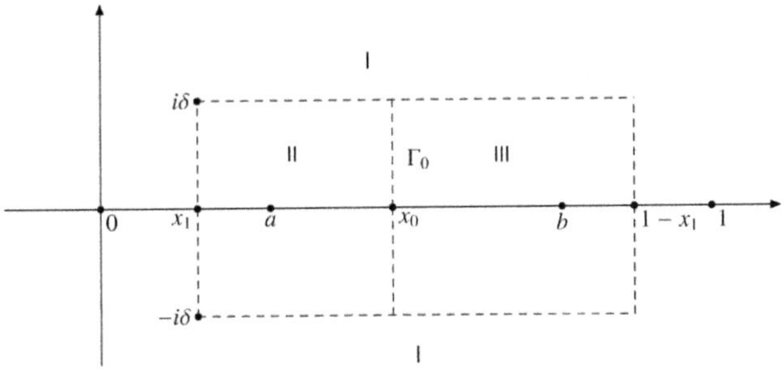


Figure 1: The regions I, II, III.

Recall the polynomial $P_{N,n}(z)$ defined in (34), and note that the leading coefficient of this polynomial is

$$k_{N,n} := \frac{N^n(n+\alpha+\beta+1)_n}{(\alpha+1)_n(-N+1)_n}. \tag{57}$$

Hence, the monic Hahn polynomials $\pi_{N,n}(z)$ are given

$$\pi_{N,n}(z) = \frac{1}{k_{N,n}} P_{N,n}(z). \tag{58}$$

Let $\Omega_+ = \{z \in \mathbb{C} : 0 < \text{Re } z < 1 \text{ and } 0 < \text{Im } z < \delta\}$ and $\Omega_- = \{z : \bar{z} \in \Omega_+\}$.

THEOREM 3.1. *Let* I, II *and* III *be the regions shown in Figure* 1, *and let $l$ denote the Lagrange constant given in* (43). *With $\phi(z)$ and $\tilde{D}(z)$ defined in* (42) *and* (53) – (54), *respectively, the asymptotic formula of the polynomial $\pi_{N,n}(z)$ is given by*

$$\pi_{N,n}(z) = e^{nl/2}\left\{\tilde{D}(z)e^{-n\phi(z)}m(z)\left[1+O\left(\frac{1}{n}\right)\right] + \delta(n)\right\} \tag{59}$$

*for $z \in \text{I}$, where $\delta(n) = 0$ for $z \in \text{I}\setminus\Omega_\pm$ and $\delta(n)$ is exponentially small in comparison with its leading term for $z \in \text{I}\cap\Omega_\pm$. More precisely, $\delta(n) = O(N^{\max(\alpha,\beta)}e^{n\phi(z)})$, where* Re $\phi(z)$ *is negative when* Re $z$ *is bounded away from the interval $(a,b)$.*

*Let $\tilde{f}(z)$ be defined as in* (50). *We have*

$$\pi_{N,n}(z) = (-1)^n\sqrt{\pi}e^{nl/2}\left\{\tilde{A}(z,n)\left[1+O\left(\frac{1}{n}\right)\right] + \tilde{B}(z,n)\left[1+O\left(\frac{1}{n}\right)\right]\right\} \tag{60}$$

*for $z \in \text{II}$, where*

$$\begin{aligned}\tilde{A}(z,n) =& \left[n^{2/3}\tilde{f}(z)\right]^{1/4}\left[m(z)+m^*(z)\right]\\ &\times\left\{\sin(N\pi z)\text{Ai}\left(n^{2/3}\tilde{f}(z)\right) + \cos(N\pi z)\text{Bi}\left(n^{2/3}\tilde{f}(z)\right)\right\}\end{aligned} \tag{61}$$

*and*

$$\begin{aligned}\tilde{B}(z,n) =& \left[n^{2/3}\tilde{f}(z)\right]^{-1/4}\left[m(z)-m^*(z)\right]\\ &\times\left\{\sin(N\pi z)\text{Ai}'\left(n^{2/3}\tilde{f}(z)\right) + \cos(N\pi z)\text{Bi}'\left(n^{2/3}\tilde{f}(z)\right)\right\}.\end{aligned} \tag{62}$$

*Similarly, with* $f^*(z)$ *defined in* (51),

$$\pi_{N,n}(z) = (-1)^N \sqrt{\pi} e^{nl/2} \Big\{ A^*(z,n)\Big[1+O\Big(\frac{1}{n}\Big)\Big] + B^*(z,n)\Big[1+O\Big(\frac{1}{n}\Big)\Big]\Big\} \tag{63}$$

*for* $z \in \mathrm{III}$, *where*

$$\begin{aligned} A^*(z,n) = {} & \Big[n^{2/3} f^*(z)\Big]^{1/4} \Big[m(z) - m^*(z)\Big] \\ & \times \Big\{ \cos(N\pi z)\mathrm{Bi}\Big(n^{2/3} f^*(z)\Big) - \sin(N\pi z)\mathrm{Ai}\Big(n^{2/3} f^*(z)\Big)\Big\} \end{aligned} \tag{64}$$

*and*

$$\begin{aligned} B^*(z,n) = {} & \Big[n^{2/3} f^*(z)\Big]^{-1/4} \Big[m(z) + m^*(z)\Big] \\ & \times \Big\{ \cos(N\pi z)\mathrm{Bi}'\Big(n^{2/3} f^*(z)\Big) - \sin(N\pi z)\mathrm{Ai}'\Big(n^{2/3} f^*(z)\Big)\Big\}. \end{aligned} \tag{65}$$

From (59), one can deduce asymptotic formulas for the Hahn polynomials $Q_n(x;\alpha, \beta, N-1)$ when $x$ is a fixed number. First, we note that with

$$x = Nz - 1/2, \tag{66}$$

it follows from (34) and (58) that

$$\begin{aligned} Q_n(x;\alpha,\beta,N-1) &= Q_n(Nz-1/2;\alpha,\beta,N-1) \\ &= \frac{N^n (n+\alpha+\beta+1)_n}{(\alpha+1)_n(-N+1)_n} \pi_{N,n}(z). \end{aligned} \tag{67}$$

The final result is summarized in the following

**Corollary** 3.2. *For fixed values of* $x$, *we have*

$$\begin{aligned} Q_n(x;\alpha,\beta,N-1) \sim {} & \frac{\Gamma(\alpha+1)\Gamma(N-n)}{\Gamma(N)\Gamma(-x)e^n n^{2x+2\alpha+2}} \\ & \times (1+c)^{n+N+\alpha+\beta+1/2} N^{x+n+\alpha+1} \end{aligned} \tag{68}$$

as $n \to \infty$.

Note that equation (68) is only an asymptotic equality, and is *not* an equation. So, even when the right-hand side of the equation vanishes when $x$ is a positive integer or zero, the polynomial on the left-hand side may not vanish. An error estimate for the approximation in (68) can be obtained by using the order estimates in (59).

## 4. Racah polynomials

Within the Askey scheme [39, p. 464] of hypergeometric orthogonal polynomials, Racah polynomials stay on the top of the hierarchy and they generalize all of the discrete hypergeometric orthogonal polynomials. These polynomials are named after Racah, because their orthogonal relation is equivalent to that of Racah coefficients or 6-$j$ symbols; see [2]. In [54], Wilson defined the Racah polynomials in terms of a $_4F_3$ hypergeometric function. Let $\lambda(x) := x(x+\gamma+\delta+1)$ and $N$ be a nonnegative integer. Define

(69)

$$R_n(\lambda(x);\alpha,\beta,\gamma,\delta) := {}_4F_3\left( \begin{matrix} -n, & n+\alpha+\beta+1, & -x, & x+\gamma+\delta+1 \\ & \alpha+1, & \beta+\delta+1, & \gamma+1 \end{matrix} \middle| 1 \right),$$

where $n = 0, \cdots, N$ and one of the following equalities is satisfied : $\alpha + 1 = -N$ or $\beta + \delta + 1 = -N$ or $\gamma + 1 = -N$. Let

$$\begin{aligned} A_n &:= -\frac{(n+\alpha+1)(n+\alpha+\beta+1)(n+\beta+\delta+1)(n+\gamma+1)}{(2n+\alpha+\beta+1)(2n+\alpha+\beta+2)}, \\ C_n &:= -\frac{n(n+\alpha+\beta-\gamma)(n+\alpha-\delta)(n+\beta)}{(2n+\alpha+\beta)(2n+\alpha+\beta+1)}. \end{aligned} \tag{70}$$

The Racah polynomials (69) satisfy the recurrence relation [26, (9.2.3)]

$$\lambda(x)R_n(\lambda(x)) = -A_n R_{n+1}(\lambda(x)) + (A_n + C_n)R_n(\lambda(x)) - C_n R_{n-1}((\lambda(x)).$$

This recurrence relation can be normalized as

$$\begin{aligned} \pi_{n+1}(z) &= (z - A_n - C_n)\pi_n(z) - A_{n-1}C_n\pi_{n-1}(z), \\ \pi_0(z) &= 1, \qquad \pi_1(z) = z - A_0, \end{aligned} \tag{71}$$

where

$$\pi_n(\lambda(x)) := \frac{(\alpha+1)_n(\beta+\delta+1)_n(\gamma+1)_n}{(n+\alpha+\beta+1)_n} R_n(\lambda(x)). \tag{72}$$

In addition to the variable $x$ and the degree $n$, the polynomials in (69) involve four free parameters. This inevitably makes the problem of deriving their asymptotic formulas much more complicated. By approximating the ratio of two shifted factorials in the hypergeometric representation (69), Chen, Ismail and Simeonov [5] obtained several asymptotic formulas in terms of hypergeometric function ${}_3F_2$ or ${}_2F_1$, when the parameters are fixed. Recently, Wang and Wong [49] studied the large-$n$ behavior of $\pi_n(z)$ with varying parameters $\alpha, \beta, \gamma, \delta$. More precisely, we set

$$\alpha + 1 = Na, \quad \beta = Nb, \quad \gamma + 1 = Nc, \quad \delta + 1 = Nd, \tag{73}$$

where either $a = -1$ or $b + d = -1$ or $c = -1$. For simplicity, we assume $A_n > 0$ and $C_n > 0$. From Favard's theorem, these conditions guarantee that the zeros of $\pi_n(z)$ are all real and simple; see [6, Sections 1.4 and 1.5]. Thus, we require some additional conditions:

1. when $a = -1$, we assume $b, c, d > 0$ and $b > c + 1$;
2. when $b + d = -1$, we assume $a, b, c > 0$ and $a + b + 1 < c$;
3. when $c = -1$, we assume $a, b, d > 0$ and $a + 1 < d$.

Let $n/N = p$ be a fixed number in (0,1). To present the results in [49], we need to introduce some notations. From (70) and (73), we have

$$\lim_{N\to\infty} \frac{A_n}{N^2} = A(p); \quad \lim_{N\to\infty} \frac{C_n}{N^2} = C(p),$$

where

$$\begin{aligned} A(t) &:= -\frac{(t+a)(t+a+b)(t+b+d)(t+c)}{(2t+a+b)^2}, \qquad 0 \leqslant t \leqslant p, \\ C(t) &:= -\frac{t(t+a+b-c)(t+a-d)(t+b)}{(2t+a+b)^2}, \qquad 0 \leqslant t \leqslant p. \end{aligned} \tag{74}$$

To use the asymptotic results from difference equation theory, we also need the concept of transition points (or turning points). These are the points where the characteristic roots of equation (71) coincide; see [50]. For $t \in [0, p]$, we define

$$y_\pm(t) := A(t) + C(t) \pm 2\sqrt{A(t)C(t)}; \tag{75}$$

These are transaction points when $t = p$. From (74), it is readily seen that $C(0) = 0$. Thus, $y_\pm(0) = A(0)$. For simplicity, in [49] Wang and Wong considered only the case when $y_+(x)$ is increasing and $y_-(t)$ is decreasing for $t \in [0,p]$. Following [49], we adopt the notations

$$\text{(76a)} \qquad \mathscr{A}_1(t) := t+a, \quad \mathscr{A}_2(t) := t+a+b, \quad \mathscr{A}_3(t) := t+b+d,$$

$$\text{(76b)} \qquad \mathscr{A}_4(t) := t+c, \quad \mathscr{C}_1(t) := t, \quad \mathscr{C}_2(t) := t+a+b-c,$$

$$\text{(76c)} \qquad \mathscr{C}_3(t) := t+a-d, \quad \mathscr{C}_4(t) := t+b, \quad \mathscr{D}(t) := 2t+a+b,$$

and

$$\text{(77a)} \quad \mathscr{A}^*(t) := \frac{4\mathscr{A}_1\mathscr{A}_2\mathscr{A}_3\mathscr{A}_4 - (\mathscr{A}_1\mathscr{A}_2\mathscr{A}_3 + \mathscr{A}_1\mathscr{A}_2\mathscr{A}_4 + \mathscr{A}_1\mathscr{A}_3\mathscr{A}_4 + \mathscr{A}_2\mathscr{A}_3\mathscr{A}_4)\mathscr{D}}{\mathscr{D}(t)^3}$$

$$\text{(77b)} \quad \mathscr{C}^*(t) := \frac{4\mathscr{C}_1\mathscr{C}_2\mathscr{C}_3\mathscr{C}_4 - (\mathscr{C}_1\mathscr{C}_2\mathscr{C}_3 + \mathscr{C}_1\mathscr{C}_2\mathscr{C}_4 + \mathscr{C}_1\mathscr{C}_3\mathscr{C}_4 + \mathscr{C}_2\mathscr{C}_3\mathscr{C}_4)\mathscr{D}}{\mathscr{D}(t)^3}$$

Furthermore, we put

$$\text{(78)} \quad S(y;t) := \sqrt{(y-A(t)-C(t))^2 - 4A(t)C(t)} = \sqrt{[y-y_-(t)][y-y_+(t)]},$$

$$\text{(79)} \quad T(y;t) := y - A(t) - C(t) + S(y;t) = \left[\sqrt{y-y_-(t)}\right] + \left[\sqrt{y-y_+(t)}\right]^2 \Big/ 2,$$

and

$$\text{(80)} \qquad G(y;t) := T(y;t)\left[A^*(t) + C^*(t)\right] + 2\left[A(t)C^*(t) + A^*(t)C(t)\right].$$

Despite all these notations, the functions in (76), (77) and (78)–(80) are simple and elementary. To state the results in [49], we also need the two integrals

$$\text{(81)} \qquad \Omega_0 = \int_0^p \left[\frac{A(t)-C(t)}{\mathscr{D}(t)S(y;t)} - \frac{2A(t)C(t)\mathscr{E}(t)}{S(y;t)T(y;t)}\right] dt$$

and

$$\text{(82)} \qquad \Omega_1 = \int_0^p \frac{G(y;t)}{2S(y;t)^2} dt.$$

We are now ready to state their main results.

**Theorem 4.1.** *Let $\pi_n(z)$ be the monic Racah polynomials satisfying the recurrence relation (71). Assume $n/N = p$ is a fixed number in $(0,1)$. Let $A(t), C(t)$ and $y_\pm(t)$ be defined as in (74) and (75). Also, let $\Omega_0(y)$ and $\Omega_1(y)$ denote, respectively, the integrals in (81) and (82). Assume that $y_+(t)$ is increasing and $y_-(t)$ is decreasing for $t \in [0,p]$. Then, for $y \in \mathbb{C} \setminus [y_-(p), y_+(p)]$, we have*

$$\text{(83)} \qquad \pi_n(N^2 y) = \left(\frac{N^2}{2}\right)^n e^{Ng(y)+r(y)}\left[1 + O\left(\frac{1}{n}\right)\right],$$

*where the main term $g(y)$ is given by*

$$\text{(84)} \qquad g(y) := \int_0^p \ln T(y;t)dt,$$

*and the correction term $r(y)$ is given by*

$$\text{(85)} \qquad r(y) := \frac{1}{2}\ln\frac{T(y;p)}{T(y;0)} + \Omega_0(y) + \Omega_1(y).$$

To derive the asymptotic formula of $\pi_n(N^2y)$ for $y$ in the interval of oscillation, i.e., $y \in [y_-(p), y_+(p)]$, Wang and Wong simply added the limits of the approximant in (83) as $y$ approaches the real axis from above and below. The reason behind their argument is based on the phenomenon that the real part of an asymptotic approximant for orthogonal polynomials in the complex plane is always half of the corresponding asymptotic approximant in the interval of orthogonality on the real line. This fact was apparently known to Heine [18, p. 175] and Szegö [45, p. 196, Theorem 8.21.9].

**Theorem 4.2.** *Let $\pi_n(z)$ be the monic Racah polynomials satisfying the recurrence relation* (71)*. Assume $n/N = p$ is a fixed number in* $(0,1)$*. Let $A(t), C(t)$ and $y_\pm(t)$ be defined as in* (74) *and* (75)*. We further assume that $y_+(t)$ is increasing and $y_-(t)$ is decreasing for $t \in [0,p]$. Then we have*

$$\pi_n(N^2y) = \left(\frac{N^2}{2}\right)^n \left\{ e^{Ng^+(y)+r^+(t)}\left[1+O\left(\frac{1}{n}\right)\right] + e^{Ng^-(y)+r^-(y)}\left[1+O\left(\frac{1}{n}\right)\right]\right\} \tag{86}$$

*for $y \in (y_-(p), A(0)) \cup (A(0), y_+(p))$.*

As a corollary, Wang and Wong [49] also gave an asymptotic formula for the Racah polynomials when the variable $z$ is fixed. The exact result is stated below.

**Corollary 4.1.** *Let $\pi_n(z)$ be the monic Racah polynomials satisfying the recurrence relation* (71)*. Assume $n/N = p$ is fixed in* $(0,1)$*. Let $A(t), C(t)$ and $y_\pm(t)$ be defined as in* (74) *and* (75)*. We further assume that $A(t) > C(t)$, $y_+(t)$ is increasing and $y_-(t)$ is decreasing for $t \in [0,p]$. Then as $n \to \infty$, we have for any fixed $z$,*

$$\begin{aligned}\pi_n(z) =& (-N^2)^n \exp\left[N\int_0^p \ln A(t)dt\right] \\ &\times \sqrt{\frac{ac(b+d)(2p+a+b)}{(p+a)(p+c)(p+b+d)(p+a+b)}}\left[1+O\left(\frac{1}{n}\right)\right].\end{aligned} \tag{87}$$

## 5. Pseudo-Jacobi polynomials

Pseudo-Jacobi (P-J) polynomials $P_n(z;a,b)$ are also known as the Romanovski-Routh polynomials. They are solutions to the second-order differential equation

$$(x^2+1)\frac{d^2y}{dx^2} + 2\left[(1+a)x+b\right]\frac{dy}{dx} - n(n+2a+1)y = 0. \tag{88}$$

These polynomials (of degree $n$) are orthogonal with respect to the weight function

$$w(x) = (x^2+1)^a e^{2b\arctan x}, \qquad x \in (-\infty,\infty), \tag{89}$$

only when the moments

$$\mu_n = \int_{-\infty}^{\infty} x^{2n} w(x)dx$$

exist; that is, only when $n < -a - 1/2$. If $a = -N-1$ with $N$ being a positive integer, then the orthogonal polynomials $P_n(x;a,b)$ are well-defined with respect to $w(x)$ only for $n = 0, \cdots, N$. Therefore, for fixed $a < -1$, we have only a finite number of orthogonal polynomials, in contrast to the classical Hermite, Laguerre and Jacobi polynomials.

The polynomials $P_n(x;a,b)$ were first discovered by Romanovski [41] in 1929 (in French) within the context of probability distribution functions in statistics. They also belong to a general family of Routh polynomials [42] introduced in 1884. For a historical account, see [35, 36]. Since they have not been discussed in standard books on orthogonal polynomials and special functions (e.g., [37, 45]), they are not known to most of the researchers in the field. In view of its close connection with the Jacobi polynomials of purely imaginary argument and complex parameters (see (91) below), in 1996 Lesky [31] (in German) coined the term "Pseudo-Jacobi polynomials". Later these polynomials are also included in the monograph [26]. It was probably the review article [40] that brought attention to the community of people working in orthogonal polynomials. This article discusses various applications of this polynomial ranging from quantum mechanics and quark physics to random matrix theory (see [16] and the references in [40]). For instance, they arise in the solutions of Schrödinger equations with certain non-central electric potential obtained by separation of variables in spherical coordinates; their role is similar to that of the associated Legendre functions; see [1] and [14].

The definition of monic P-J polynomials in terms of hypergeometric functions is given in [26], with parameters $N$ and $\nu$ related to the parameters $a$ and $b$ in (89); more precisely, with $a=-N-1$ and $b=\nu$, the monic P-J polynomials are defined by

$$P_n(x;a,b)=\frac{(-2i)^n(1+a+bi)_n}{(n+2a+1)_n}\,{}_2F_1\left(\begin{matrix}-n,n+2a+1\\1+a+bi\end{matrix};\frac{1-ix}{2}\right). \tag{90}$$

By comparing the definitions of the P-J polynomials and Jacobi polynomials $P_n^{(\alpha,\beta)}$, we have the relation

$$P_n(x;a,b)=(-i)^nP_n^{(a+bi,\,a-bi)}(ix). \tag{91}$$

This relation has also been used as the definition of P-J polynomials in [19, p. 508] and [23]. In view of the symmetry property of Jacobi polynomials, the P-J polynomials satisfy

$$P_n(z;a,-b)=(-1)^nP_n(-z;a,b); \tag{92}$$

see [23]. Since the coefficients of P-J polynomials are real, it also follows that

$$P_n(\overline{z};a,b)=\overline{P_n(z;a,b)}, \tag{93}$$

where an overbar means the complex conjugate. For other properties such as orthogonality, recurrence relations and Rodrigues formula, we refer to [26, p. 231-233].

Probably due to their close relation with Jacobi polynomials, recently P-J polynomials have attracted much attention [23, 24]. Properties of the zeros of P-J polynomials, including their asymptotics and bounds, have been studied in [23]. As a special case, the zeros of the pseudo-ultraspherical polynomials have been independently investigated in [13] almost at the same time. In [44], Song and Wong have investigated the asymptotic behavior of the P-J polynomials as the degree $n$ goes to infinity. Note that, for fixed $a$ or $a>-n$, there is no real-line orthogonality anymore; see [23, 28] for their orthogonality. For fixed parameters $a$ and $b$, in view of (90) the asymptotics of these polynomials can be obtained as a special case of that of the hypergeometric functions investigated in [22, 25]. In [44], Song and Wong consider the case when the parameters depend on the degree $n$,

$$a_n=-(An+A_0),\quad (A>1);\qquad b_n=Bn+B_0, \tag{94}$$

where $A, A_0, B, B_0$ are real constants. Their goal is to derive asymptotic formulas for $P_n(z;a_n,b_n)$ (with varying parameters) for $z$ in the whole complex plane. Their main tool is the well-developed method for differential equations with a large parameter as presented in Olver [38, Chapter 11]. Their other goal is to use this example as an illustration to show that this method is easy to understand and can be used in many problems in physical applications.

They first deal with the asymptotics when the variable $z$ lies outside the interval where the zeros are located. To do this, they let $\{x_{n,j}\}|_{j=1}^n$ be the zeros of

$P_n(x; a_n, b_n)$ and define the sequence of discrete measures $\lambda_n$ by

$$\lambda_n(f) \equiv \frac{1}{n}\sum_{j=1}^{n} f(x_{n,j}) = \int f(x)d\lambda_n(x),$$

where $f$ is any continuous function. They give a brief derivation of the result that there is a unique measure $\lambda$, which is the weak-$*$ limit of $\lambda_n$ as $n \to \infty$, and the measure $\lambda$ is supported on $[\gamma_-, \gamma_+]$ and given by

$$d\lambda = \frac{(A-1)\sqrt{(x-\gamma_-)(\gamma_+-x)}}{\pi(x^2+1)}dx, \tag{95}$$

where

$$\gamma_\pm = \frac{AB \pm \sqrt{(2A-1)[(A-1)^2+B^2]}}{(A-1)^2}. \tag{96}$$

This result was actually given earlier in [23]. To state the first main result in [44], we introduce the functions

$$\begin{aligned} \xi^*(z) = {} & 2(A-1)\log(\sqrt{z-\gamma_-}+\sqrt{z-\gamma_+}) - C_1 \\ & -(A+iB)\log(\sqrt{1+i\gamma_-}\sqrt{z-\gamma_+}+\sqrt{1+i\gamma_+}\sqrt{z-\gamma_-}) \\ & -(A-iB)\log(\sqrt{1-i\gamma_-}\sqrt{z-\gamma_+}+\sqrt{1-i\gamma_+}\sqrt{z-\gamma_-}) \end{aligned} \tag{97}$$

and

$$\begin{aligned} \chi^*(z) = {} & 2(A_0-1)\log(\sqrt{z-\gamma_-}+\sqrt{z-\gamma_+}) - C_2 \\ & -(A_0+iB_0)\log(\sqrt{1+i\gamma_-}\sqrt{z-\gamma_+}+\sqrt{1+i\gamma_+}\sqrt{z-\gamma_-}) \\ & -(A_0-iB_0)\log(\sqrt{1-i\gamma_-}\sqrt{z-\gamma_+}+\sqrt{1-i\gamma_+}\sqrt{z-\gamma_-}), \end{aligned} \tag{98}$$

where

$$\begin{aligned} C_1 = 2(A-1)\log 2 - \frac{1}{2}B\pi & - (A+iB)\log(\sqrt{1+i\gamma_-}+\sqrt{1+i\gamma_+}) \\ & - (A-iB)\log(\sqrt{1-i\gamma_-}+\sqrt{1-i\gamma_+}) \end{aligned} \tag{99}$$

and

$$\begin{aligned} C_2 = 2(A_0-1)\log 2 - \frac{1}{2}B_0\pi & - (A_0+iB_0)\log(\sqrt{1+i\gamma_-}+\sqrt{1+i\gamma_+}) \\ & - (A_0-iB_0)\log(\sqrt{1-i\gamma_-}+\sqrt{1-i\gamma_+}). \end{aligned} \tag{100}$$

In the last four equations, $A, A_0, B$ and $B_0$ are the constants in (94).

We also need the definition of a domain $D$ in three separate cases: (i) $\gamma_- \leqslant 0 \leqslant \gamma_+$, (ii) $0 < \gamma_- < \gamma_+$ and (iii) $\gamma_- < \gamma_+ < 0$. In case (i), $D$ denotes the set

$$D = \{z| \text{ Re } z \geqslant 0, \text{ Im } z \geqslant 0\} \setminus [0, \gamma_+], \qquad \gamma_- < 0 < \gamma_+, \tag{101}$$

where exclusion of the interval $[0, \gamma_+]$ is understood to mean the exclusion of a small neighborhood in the first quadrant containing the interval $[0, \gamma_+]$. In case (ii), we put

$$\xi(z) = \xi^*(z) + \frac{1}{2}(A+iB)\log(1+iz) + \frac{1}{2}(A-iB)\log(1-iz) \tag{102}$$

and let $\xi_0 = \xi(\gamma_+)$. Define a curve $L_1$ in the first quadrant of the $z$-plane by

$$\text{Re } \xi(z) = \xi_0, \text{ or equivalently Re}\left(\int_{\gamma_-}^{z} \frac{\sqrt{(t-\gamma_-)(t-\gamma_+)}}{1+t^2}dt\right) = 0, \tag{103}$$

which connects $\gamma_-$ and a point $z_1 = iy_1\,(0 < y_1 < 1)$ on the positive imaginary axis, and denote the region bounded by the axes and $L_1$ by $\Omega_1$; i.e.,

$$\begin{aligned}\Omega_1 = \{z|\ &\text{Re } z \geq 0,\ \text{Im } z \geq 0,\\ &\xi_0 \leqslant \text{Re } \xi(z) \leqslant \text{Re } \xi(0), \quad \text{Im } \xi(z_1) \leqslant \text{Im } \xi(z) \leqslant \text{Im } \xi(0)\}.\end{aligned} \tag{104}$$

The domain $D$ is now defined by

$$D = \{z|\ \text{Re } z \geqslant 0,\ \text{Im } z \geqslant 0\} \setminus ([\gamma_-, \gamma_+] \cup \Omega_1), \qquad 0 < \gamma_- < \gamma_+, \tag{105}$$

where small neighborhoods of the interval $[\gamma_-, \gamma_+]$ and the curve $L_1$ are again excluded. (For details; see Figure 2 in [44].) The final case (i.e., case (iii)) can be handled in a similar manner. We define a curve $L_2$ in the second quadrant of $z$-plane by

$$\text{Re } \xi(z) = \xi_0, \text{ or equivalently } \text{Re}\left(\int_{\gamma_+}^{z} \frac{\sqrt{(t-\gamma_-)(t-\gamma_+)}}{1+t^2}dt\right) = 0, \tag{106}$$

which connects $\gamma_+$ and a point $z_2 = iy_2\,(0 < y_2 < 1)$ on the positive imaginary axis, and denote the region bounded by the axes and $L_2$ by $\Omega_2$; i.e.,

$$\begin{aligned}\Omega_2 = \{z|\ &\text{Re } z \leqslant 0,\ \text{Im } z \geqslant 0,\\ &\xi_0 \leqslant \text{Re } \xi(z) \leqslant \xi(0), \quad 0 \leqslant \text{Im } \xi(z) \leqslant \text{Im } \xi(z_2)\}.\end{aligned} \tag{107}$$

The relevant domain $D$ is defined by

$$D = \{z|\ \text{Re } z \geqslant 0,\ \text{Im } z \geqslant 0\} \cup \Omega_2 \setminus [\gamma_+, \gamma_+ + \delta], \qquad \gamma_- < \gamma_+ < 0, \tag{108}$$

where a small neighborhood of $L_2$ can be included but a small neighborhood of the point $\gamma_+$ is excluded ($\delta > 0$); see Figure 3 in [44].

Note that the excluded region $\Omega_1$ in case (ii) is the mirror image of the region $\Omega_2$ in case (iii) with respect to the imaginary axis. Based on this observation and the symmetry properties (92) and (93) of the P-J polynomials, Song and Wong [44] proved that the domain $D$ defined as in (101), (105) and (108) for each of the three cases considered above can be used to cover the whole complex plane except a neighborhood of the critical interval $[\gamma_-, \gamma_+]$. Their first main result is stated below.

**Theorem 5.1.** *For $z \in D$ given by* (101), (105) *or* (108), *the* P-J *polynomials $P_n(z; a_n, b_n)$ with parameters in* (94) *have the uniform asymptotic expansion*

$$\begin{aligned}P_n(z; a_n, b_n) \sim\ & e^{b_n\pi/2}[(z-\gamma_-)(z-\gamma_+)]^{-1/4}\\ & \times e^{-n\xi^*(z)-\chi^*(z)} \sum_{s=0}^{\infty} \frac{(-1)^s w_s(z)}{n^s}\end{aligned} \tag{109}$$

*as $n \to \infty$, where $\xi^*(z)$ and $\chi^*(z)$ are given in* (97) *and* (98), *respectively, and the coefficients $w_s(z) (s \geqslant 1)$ can be determined by a recursive formula with $w_0(z) = 1$.*

The second main result in [44] is an asymptotic expansion for $P_n(z; a_n, b_n)$ which holds uniformly in a region that contains a turning point and a large part of the critical interval $[\gamma_-, \gamma_+]$. By symmetry, we can concentrate on the *turning point* $z = \gamma_+$. In the previous case, the function $\xi(z)$ in (102) is not well-defined near $\gamma_+$, which is a branch point. Following the method illustrated in [38, p. 426], Song and Wong modified the function $\xi(z)$ and defined

$$\zeta(z) := \left[\frac{3}{2}(\xi(z) - \xi_0)\right]^{2/3}, \qquad \xi_0 := \xi(\gamma_+). \tag{110}$$

To state their next result, we need some new notations. First, put

(111) $$n := N + \alpha \qquad \text{with } \alpha := -\frac{2A_0 - 1}{2(A-1)},$$

(112) $$\chi(z) := \chi^*(z) + \frac{1}{2}(A_0 + iB_0)\log(1+iz) + \frac{1}{2}(A_0 - iB_0)\log(1-iz),$$

where $\chi^*(z)$ is given in (98), and

(113) $$\Phi\left(\zeta(z)\right) := \zeta^{-1/2}(z)[\chi(z) - \chi_0], \qquad \chi_0 := \chi(\gamma_+).$$

Define

(114) $$\hat{\Phi}(\zeta) := \left[\Phi(\zeta) + \frac{2}{3}\alpha\zeta\right]$$

and

(115) $$\zeta_N := N^{2/3}\zeta + N^{-1/3}\hat{\Phi}(\zeta).$$

We now restrict $z$ to the domain

(116) $$D^* = \{z|\ \gamma_- + \delta \leqslant \text{Re } z < +\infty, \quad |\text{Im } z| \leqslant \delta_1\},$$

where $\delta, \delta_1$ are some positive numbers. Clearly, $D^*$ is a complex neighborhood of a semi-infinite real interval.

**Theorem 5.2.** *For $z \in D^*$, the Pseudo-Jacobi polynomials $P_n(z; a_n, b_n)$ with parameters in (94) have the uniform asymptotic expansion*

(117) $$\begin{aligned} P_n(z; a_n, b_n) &\sim C_n e^{b_n\pi/2}[w(z)]^{-1/2}\left[\frac{\zeta}{(z-\gamma_-)(z-\gamma_+)}\right]^{1/4} \\ &\times \left\{\text{Ai}(\zeta_N)\sum_{s=0}^{\infty}\frac{X_s(\zeta)}{N^s} + N^{-\frac{4}{3}}\text{Ai}'(\zeta_N)\sum_{s=0}^{\infty}\frac{Y_s(\zeta)}{N^s}\right\}\end{aligned}$$

*as $n \to \infty$, where $w(z)$ is given in (89), $\gamma_\pm$ are given in (96) and $\zeta(z)$ is given by (110). The constant $C_n$, independent of $z$, has an asymptotic series representation in powers of $N^{-1}$ with explicit expressions for the coefficients. The coefficient functions $X_s(\zeta)$ and $Y_s(\zeta)$ can be determined by recursive formulas with $X_0(\zeta) \equiv 1$.*

An immediate consequence of the above theorem is the following corollary.

**Corollary 5.1.** *For $z \in D^*$ defined in (116), the P-J polynomials $P_n(z; a_n, b_n)$ with parameters in (94) have the leading asymptotic behavior*

(118) $$\begin{aligned} P_n(z; a_n, b_n) &\sim 2\sqrt{\pi}N^{1/6}e^{-(n\xi_0+\chi_0)+b_n\pi/2}[w(z)]^{-1/2} \\ &\times \left[\frac{\zeta}{(z-\gamma_-)(z-\gamma_+)}\right]^{1/4}\text{Ai}(\zeta_N)\end{aligned}$$

*as $n \to \infty$, where $w(z)$ is given in (89), $\gamma_\pm$ are given in (96), $\zeta_N = N^{2/3}\zeta + N^{-1/3}\hat{\Phi}$ and $\zeta, \xi_0, \hat{\Phi}, \chi_0, N$ are given by (110), (114), (113) and (111).*

As a biproduct, Song and Wong [44] also obtained an asymptotic formula for the zeros of the P-J polynomials. To describe this result, we let the zeros be arranged in decreasing order:

(119) $$\gamma_- < x_{n,n} < x_{n,n-1} < \cdots < x_{n,2} < x_{n,1} < \gamma_+,$$

and introduce the functions

(120) $$f_i(z) = \frac{Q_i(z)}{(1+z^2)^2}, \qquad i = 0, 1,$$

where

$$(121)\quad \begin{aligned} Q_0(z) &= (A-1)^2z^2 - 2ABz + B^2 + 1 - 2A = (A-1)^2(z-\gamma_-)(z-\gamma_+), \\ Q_1(z) &= (A-1)(2A_0-1)z^2 - 2(A_0B + AB_0)z + 1 - A - 2A_0 + 2BB_0. \end{aligned}$$

**Corollary 5.2.** *Let the parameters $a_n$ and $b_n$ be given as in (94), and let the zeros of $P_n(z; a_n, b_n)$ be arranged as in (119). For each fixed $k$, we have*
(122)

$$x_{n,k} = \gamma_+ + c_0^{-2/3}\widetilde{a}_k N^{-2/3} - f_1(\gamma_+)c_0^{-2}N^{-1} - \frac{f_0''(\gamma_+)}{10c_0^{10/3}}\widetilde{a}_k^2 N^{-4/3} + O(N^{-5/3})$$

*as $n \to \infty$, where $N$ is given in (111), $\widetilde{a}_k$ is the $k$th negative zero of the Airy function $\mathrm{Ai}(x)$ and the $O$-symbol depends on $k$. The constant $\gamma_+$ is given in (96), and the constant $c_0$ is given by*

$$(123)\qquad c_0 = \frac{(A+1)\sqrt{\gamma_+ - \gamma_-}}{1+\gamma_+^2} > 0.$$

*The functions $f_i(z), i = 0, 1$, are given in (120).*

## References


[1] D. E. Alvarez-Castillo, Exactly Solvable Potentials and Romanovski Polynomials in Quantum Mechanics, Master's thesis, San Luis Potosi U., 2007; arXiv: 0808.1642.

[2] R. Askey and J. Wilson, A set of orthogonal polynomials that generalize the Racah coefficients or $6-j$ symbols, SIAM J. Math. Anal. 10 (1979), 1008-1016.

[3] J. Baik, T. Kriecherbauer, K. T.-R. McLaughlin and P. D. Miller, Discrete Orthogonal Polynomials: Asymptotics and Applications, Annals of Mathematics Studies, Vol. 164, Princeton University Press, Princeton, NJ, 2007.

[4] J. Baik and T. M. Suidan, Random matrix central limit theorems for nonintersecting random walks, Ann. Probab. 35 (2007), 1807-1834.

[5] L.-C. Chen, M. E. H. Ismail and P. Simeonov, Asymptotics of Racah coefficients and polynomials, J. Phys. A 32 (3) (1999), 537-553.

[6] T. S. Chihara, An Introduction to Orthogonal Polynomials, Gordon and Breach, New York, 1978.

[7] J. S. Christiansen, The moment problem associated with the Stieltjes-Wigert polynomials, J. Math. Anal. Appl. 277 (2003), 218-245.

[8] D. Dai, M. E. H. Ismail and X.-S. Wang, Plancherel-Rotach asymptotic expansion for some polynomials from indeterminate moment problems, Const. Approx. 40 (2014), 61-104.

[9] D. Dai and R. Wong, Global asymptotics of Krawtchouk polynomials - A Riemann-Hilbert approach, Chin. Ann. Math. Ser. B 28 (2007), 1-34.

[10] P. Deift, T. Kriecherbauer, K. T.-R. McLaughlin, S. Venakides and X. Zhou, Uniform asymptotics for polynomials orthogonal with respect to varying exponential weights and applications to universality questions in random matrix theory, Comm. Pure Appl. Math., 52 (1999), 1335-1425.

[11] P. Deift, T. Kriecherbauer, K. T.-R. McLaughlin, S. Venakides and X. Zhou, Strong asymptotics of orthogonal polynomials with respect to exponential weights, Comm. Pure Appl. Math. 52 (1999), 1491-1552.

[12] Y. Dolivet and M. Tierz, Chern-Simons matrix models and Stieltjes-Wigert polynomials, J. Math. Phys. 48 (2007) 0235207, 20pp.

[13] K. Driver and M. E. Muldoon, Zeros of pseudo-ultraspherical polynomials, Anal. Appl. 12 (2014), 563-581.

[14] R. Dutt, A. Gangopadhyaya and U. P. Sukhatme, Noncentral potentials and spherical harmonics using supersymmetry and shape invariance, Amer. J. Phys. 65 (1997), 400-403.

[15] A. Erdélyi, W. Magnus, F. Oberhettinger and F. G. Tricomi, Higher Transcendental Functions, Vol. II, McGraw - Hill, New York, NY 1953.

[16] P. J. Forrester, Log-gases and Random Matrices, (LMS - 34), Princeton University Press, 2010.

[17] W. K. Hayman, On the Zeros of a $q$-Bessel function, Contemporary Mathematics 382 (2005), 205-216.

[18] E. Heine, Handbuch der Kugelfunctionen, Vol. 1, 2nd edn., Berlin, 1878.

[19] M. E. H. Ismail, Classical and Quantum Orthogonal Polynomials in One Variable, Cambridge University Press, 2005.

[20] M. E. H. Ismail, Asymptotics of $q$-orthogonal polynomials and a $q$-Airy function, Int. Math. Res. Note 18 (2005), 1063-1088.

[21] M. E. H. Ismail and R.-M. Zhang, Chaotic and periodic asymptotics for $q$-orthogonal polynomials, Int. Math. Res. Note 2006 (2006) 83274, 1-33.

[22] D. S. Jones, Asymptotics of the hypergeometric function, Mathematical Methods in the Applied Sciences 24 (2001), 369-389.

[23] K. Jordaan and F. Toókos, Orthogonality and asymptotics of pseudo-Jacobi polynomials for non-classical parameters, J. Approx. Theory, 178 (2014), 1-12.

[24] M. A. Khan, A. H. Khan and S. M. Abbas, A note on pseudo Jacobi polynomials, Ain Shams Engineering Journal, 4 (2013), 127-131.

[25] S. F. Khwaja and A. B. Olde Daalhuis, Uniform asymptotic expansions for hypergeometric functions with large parameters IV, Anal. Appl. 12 (2014), 667-710.

[26] R. Koekoek, P. A. Lesky and R. F. Swarttouw, Hypergeometric Orthogonal Polynomials and Their $q$-Analogues, Springer Science & Business Media, 2010.

[27] T. Kriecherbauer and K. T.-R. McLaughlin, Strong asymptotics of polynomials orthogonal with respect to Freud weights, Int. Math. Res. Not. 6 (1999), 299-333.

[28] A. B. J. Kuijlaars, A. Martinez-Finkelshtein and R. Orive, Orthogonality of Jacobi polynomials with general parameters, Electronic Transactions on Numerical Analysis 19 (2005), 1-17.

[29] A. B. J. Kuijlaars and W. Van Assche, The asymptotic zero distribution of orthogonal polynomials with varying recurrence coefficients, J. Approx. Theory 99 (1999), 167-197.

[30] K. F. Lee and R. Wong, Uniform asymptotic expansions of the Tricomi-Carlitz polynomials, Proc. Amer. Math. Soc. 138 (2010), 2513-2519.

[31] P. A. Lesky, Endliche und unendliche systeme von kontinuierlichen klassischen orthogonalpolynomen, ZAMM - Journal of Appl. Math. and Mech. 76 (1996), 181-184.

[32] Y. T. Li and R. Wong, Global asymptotics of Stieltjes-Wigert polynomials, Anal. Appl. 11 (2013), 1350028 (12 pages).

[33] Y. Lin and R. Wong, Global asymptotics of the discrete Chebyshev polynomials, Asymptot. Anal. 82 (2013), 39-64.

[34] Y. Lin and R. Wong, Global asymptotics of the Hahn polynomials, Anal. Appl. 11 (2013), 1350018 (47 pages).

[35] M. Masjed-Jamei, F. Marcellán and E. J. Huertas, A finite class of orthogonal functions generated by Routh-Romanovski polynomials, Complex Variables and Elliptic Equations 59 (2014), 162-171.

[36] G. Natanson, Exact quantization of the Milson potential via Romanovski-Routh polynomials, arXiv: 1310.0796, 2013.

[37] A. F. Nikiforov and V. B. Uvarov, Special Functions of Mathematical Physics, Birkhäuser, Basel, 1988.

[38] F. W. J. Olver, Asymptotics and Special Functions, Academic Press, New York, 1974. Reprinted by A. K. Peters, Wellesley, 1997.

[39] F. W. J. Olver, D. W. Lozier, R. F. Boisvert and C. W. Clark, NIST Handbook of Mathematical Functions, Cambridge University Press, Cambridge, 2010.

[40] A. Raposo, H. Weber, D. E. Alvarez-Castillo and M. Kirchbach, Romanovski polynomials in selected physics problems, Open Physics 5 (2007), 253-284.

[41] V. Romanovski, Sur quelques classes nouvelles de polynomes orthogonaux, C. R. Acad. Sci. Paris 188 (1929), 1023-1025.

[42] E. J. Routh, On some properties of certain solutions of a differential equation of the second order, Proc. Lond. Math. Soc. 1 (1884), 245-262.

[43] I. I. Sharapudinov, Asymptotic properties of orthogonal Hahn polynomials in a discrete variable, Mat. Sb. 180 (1989), 1259-1277.

[44] Z. L. Song and R. Wong, Asymptotics of Pseudo-Jacobi Polynomials with Varying Parameters, Stud. Appl. Math., 139 (2017), 179-217.

[45] G. Szegö, Orthogonal Polynomials, 4th edn., AMS Colloquium Publications, Vol. 23, Amer. Math. Soc., Providence, R. I., 1975.

[46] X. S. Wang and R. Wong, Discrete analogues of Laplace's approximation, Asymptot. Anal. 54 (2007), 165-180.

[47] X. S. Wang and R. Wong, Uniform asymptotics of some $q$-orthogonal polynomials, J. Math. Anal. Appl. 364 (2010), 79-87.

[48] X. S. Wang and R. Wong, Asymptotics of orthogonal polynomials via recurrence relations, Anal. Appl. 10 (2012), 215-235.

[49] X.-S. Wang and R. Wong, Asymptotics of Racah polynomials with varying parameters, J. Math. Anal. Appl. 436 (2016), 1149-1164.

[50] Z. Wang and R. Wong, Asymptotic expansions for second-order linear difference equations with a turning point, Numer. Math. 94 (2003), 147-194.

[51] Z. Wang and R. Wong, Linear difference equations with transition points, Math. Comp. 74 (2005), 629-653.

[52] Z. Wang and R. Wong, Uniform asymptotics of the Stieltjes-Wigert polynomials via the Riemann-Hilbert approach, J. Math. Pures Appl. 85 (2006), 698-718.

[53] S. Wigert, Sur les polynômes orthogonaux el ĺapproximation des fonctions continues, Arkiv för matematik, astronomi och fysik 17 (1923) 15pp.

[54] J. A. Wilson, Hypergeometric series recurrence relations and some new orthogonal functions, PhD thesis, Univ. Wisconsin, Madison, 1978.

[55] R. Wong, Asymptotic Approximations of Integrals, Academic Press, Boston, 1989. (Reprinted by SIAM, Philadelphia, PA, 2001.)

[56] R. Wong, Orthogonal polynomials and their asymptotic behavior, Proceedings of the International Workshop on Special Functions, C. Dunkl, M. Ismail and R. Wong (eds.), World Scientific, 409-422, 2000.

[57] R. Wong and Y. Q. Zhao, Special functions, integral equations and Riemann-Hilbert problem, Proc. Amer. Math. Soc. 144 (2016), 4367-4380.

[58] X.-B. Wu, Y. Lin, S.-X. Xu and Y. Q. Zhao, Uniform asymptotics for discrete orthogonal polynomials on infinite nodes with an accumulation point, Anal. Appl. 14 (2016), 705-737.

Department of Mathematics, City University of Hong Kong, Kowloon, Hong Kong
*E-mail*: `rscwong@cityu.edu.hk`

Chin. Ann. Math. Ser. B
39(3), 2018, 553–596
DOI: 10.1007/s11401-018-0082-8

**Chinese Annals of Mathematics, Series B**


# Global Asymptotics of Orthogonal Polynomials Associated with a Generalized Freud Weight*

Zhi-Tao WEN[1] Roderick WONG[2] Shuai-Xia XU[3]

*(Dedicated to Professor Philippe G. Ciarlet on the occasion of his 80th birthday)*

**Abstract** In this paper, the authors consider the asymptotic behavior of the monic polynomials orthogonal with respect to the weight function $w(x) = |x|^{2\alpha}\mathrm{e}^{-(x^4+tx^2)}$, $x \in \mathbb{R}$, where $\alpha$ is a constant larger than $-\frac{1}{2}$ and $t$ is any real number. They consider this problem in three separate cases: (i) $c > -2$, (ii) $c = -2$, and (iii) $c < -2$, where $c := tN^{-\frac{1}{2}}$ is a constant, $N = n + \alpha$ and $n$ is the degree of the polynomial. In the first two cases, the support of the associated equilibrium measure $\mu_t$ is a single interval, whereas in the third case the support of $\mu_t$ consists of two intervals. In each case, globally uniform asymptotic expansions are obtained in several regions. These regions together cover the whole complex plane. The approach is based on a modified version of the steepest descent method for Riemann-Hilbert problems introduced by Deift and Zhou (1993).

**Keywords** Orthogonal polynomials, Globally uniform asymptotics, Riemann-Hilbert problems, The second Painlevé transcendent, Theta function

**2000 MR Subject Classification** 41A60, 30E15

## 1 Introduction

In a recent paper [8], Clarkson et al. made a detailed study of the monic orthogonal polynomials $\{S_n(x;t)\}_{n=0}^{\infty}$ associated with the generalized Freud weight

$$w(x;t) = |x|^{2\alpha}\mathrm{e}^{-(x^4+tx^2)}, \quad x \in \mathbb{R}, \tag{1.1}$$

where $\alpha > -\frac{1}{2}$ and $t \in \mathbb{R}$. In particular, they showed that these polynomials satisfy a differential-difference equation and a second-order linear differential equation. The coefficients in these equations involve the coefficient $\beta_n(t;\alpha)$ in the three-term recurrence relation

$$xS_n(x;t) = S_{n+1}(x;t) + \beta_n(t;\alpha)S_{n-1}(x;t), \tag{1.2}$$

where $S_{-1}(x;t) = 0$ and $S_0(x;t) = 1$. The recurrence coefficient $\beta_n(t;\alpha)$ satisfies the first discrete Painlevé equation $\mathrm{dP_I}$, and is related to solutions of the fourth Painlevé equation

Manuscript received September 28, 2017. Revised December 7, 2017.
[1]Department of Mathematics, Taiyuan University of Technology, Taiyuan 030024, China; Liu Bie Ju Centre for Mathematical Sciences, City University of Hong Kong, Hong Kong, China.
E-mail: wenzhitao@tyut.edu.cn
[2]Liu Bie Ju Centre for Mathematical Sciences, City University of Hong Kong, Hong Kong, China.
E-mail: rscwong@cityu.edu.hk
[3]Institut Franco-Chinois de l'Energie Nucléaire, Sun Yat-sen University, Guangzhou 510275, China.
E-mail: xushx3@mail.sysu.edu.cn
*This work was supported by the National Natural Science Foundation of China (Nos. 11771090, 11571376).

$\mathrm{P_{IV}}$. As a follow-up, in [7] Clarkson and Jordaan investigated the asymptotic behavior of the polynomials $S_n(x;t)$, as well as the recurrence coefficient $\beta_n(t;\alpha)$, when the degree $n$, or the parameter $t$, tends to infinity. However, with regard to the polynomial $S_n(x;t)$, what they gave is a solution $\widetilde{S}_n(x)$ to an approximate equation (see [7, (4.12)]), and not an asymptotic approximation to the polynomial $S_n(x;t)$.

In the present paper, we are mainly concerned with the problem of finding asymptotic formulas for the monic polynomials $S_n(x;t)$, as $n \to \infty$, for all values of $t$, whether $t$ is fixed or varying. This problem has actually already been studied by several people in special cases. For instance, when $\alpha = 0$ in (1.1) and the weight function is given in the form

$$w_N(x) = \mathrm{e}^{-N(\frac{g}{4}x^4 + \frac{h}{2}x^2)}, \quad g > 0, \tag{1.3}$$

asymptotic formulas (as $n \to \infty$) of the associated monic orthogonal polynomials have been given by Bleher and Its in the case $h < -2\sqrt{g}$ and $\varepsilon < \frac{n}{N} < \varepsilon^{-1}$, $\varepsilon > 0$. Their method makes use of a WKB formula (see [4, (7.5)]) from the differential equation theory. Without loss of generality, one can assume the constant $g$ in (1.3) to be equal to 1. To simplify the calculation, one may also take the parameter $N$ to be equal to the degree $n$. Thus, in [19] Wong and Zhang investigated the problem in the case when the weight function is given by $w(x) = \mathrm{e}^{-nV(x)}$ with

$$V(x) = \frac{x^4}{4} + \frac{h}{2}x^2$$

and $h = -2$. At about the same time, results for the same case (i.e., $h = -2$) but with a much more general $V(x)$ was obtained by Claeys et al. [5]. For fixed $t$, asymptotic formulas for the monic polynomials $S_n(x;t)$ in (1.2) can be written out directly from the results in [20]. Quite recently, in [2] and [3], Bertola and Tovbis obtained the asymptotics of the recurrence coefficients for the orthogonal polynomials associated with the weight (1.3), where $g \in \mathbb{C}$, $h = 1$ and the variable $x$ is on the rays $\arg x = \theta$ (for fixed $\theta$) in the complex plane. As a follow-up of the investigations in [2] and [3], one may also consider the case when the weight function has a singularity at the origin like the one in (1.1). But, we will leave this problem to a future investigation.

In this paper, we shall study the behavior of $S_n(x;t)$ as $n \to \infty$, for all values of $t$. Our approach is completely based on the nonlinear steepest-descent method for Riemann-Hilbert problems introduced by Deift and Zhou [12], and will not make use of any asymptotic result from differential equation theory. We divide our discussion into three cases: (i) $c > -2$, (ii) $c = -2$, and (iii) $c < -2$, where $c = tN^{-\frac{1}{2}}$ is a constant and $N = n + \alpha$. As we shall show, in the first two cases, the support of the associated equilibrium measure $\mu_t$ is a single interval, whereas in the third case the support of $\mu_t$ consists of two intervals. In all three cases, we will present infinite asymptotic expansions of $S_n(z;t)$ for $z$ in various different regions. These regions together cover the whole complex $z$-plane.

## 2 Riemann-Hilbert Problem

For notational convenience, from here on we will use $\pi_n(z)$ to denote the monic orthogonal polynomials in (1.2), and write $p_n(z) = \gamma_n \pi_n(z)$ for the orthogonal polynomials with respect to the weight in (1.1). Consider the following Riemann-Hilbert problem (RHP for short) for a $2 \times 2$ matrix-valued function $Y$:

$(Y_\mathrm{a})$ $Y(z)$ is analytic in $\mathbb{C} \setminus \mathbb{R}$;

($Y_{\rm b}$) $Y(z)$ takes continuous boundary values $Y_+(x)$ and $Y_-(x)$ such that

$$Y_+(x) = Y_-(x) \begin{pmatrix} 1 & w(x) \\ 0 & 1 \end{pmatrix}$$

for $x \in \mathbb{R} \setminus \{0\}$, where $w(x)$ is the weight function given by (1.1).

($Y_{\rm c}$) For $z \in \mathbb{C} \setminus \mathbb{R}$,

$$Y(z) = [I + O(1/z)] \begin{pmatrix} z^n & 0 \\ 0 & z^{-n} \end{pmatrix},$$

as $z \to \infty$.

($Y_{\rm d}$) For $z \in \mathbb{C} \setminus \mathbb{R}$,

$$Y(z) = \begin{cases} O\begin{pmatrix} 1 & |z|^{2\alpha} \\ 1 & |z|^{2\alpha} \end{pmatrix}, & \text{if } \alpha < 0, \\ O\begin{pmatrix} 1 & 1 \\ 1 & 1 \end{pmatrix}, & \text{if } \alpha > 0, \end{cases}$$

as $z \to 0$.

The following well-known theorem is due to Fokas, Its and Kitaev [15].

**Theorem 2.1** *The unique solution to the above RHP is given by*

$$Y(z) = \begin{pmatrix} \pi_n(z) & C(\pi_n w)(z) \\ c_n \pi_{n-1}(z) & c_n C(\pi_{n-1} w)(z) \end{pmatrix},$$

*where* $c_n = -2\pi \mathrm{i} \gamma_{n-1}^2$ *and*

$$C[f](z) := \frac{1}{2\pi \mathrm{i}} \int_{-\infty}^{\infty} \frac{f(\zeta)}{\zeta - z} \mathrm{d}\zeta, \quad z \in \mathbb{C} \setminus \mathbb{R},$$

*is the Cauchy transform of* $f$.

To asymptotically evaluate the solution of the RHP for $Y$, we will first follow the nonlinear steepest descent method introduced by Deift and Zhou [12], and further developed in [9–11]. This method consists of a sequence of transformations

$$Y \to U \to T \to V \to S,$$

which ultimately leads to a RHP that can be solved explicitly. The transformation $Y \to U$ is just a rescaling, and the transformation $U \to T$ is a normalization process. The transformation $T \to V$ involves a factorization of the jump matrix and deformation of contours so that $V$ can be approximated by an exact solution to a RHP for $n$ large. The final transformation $V \to S$ consists of some local analysis and the construction of parametrices. Since every step of the transformation is explicit and reversible, one can obtain the asymptotics of $Y$, and hence of $\pi_n(z)$, by a sequence of inverse transformations.

We define the first transformation $Y \to U$ by

$$U(z) := \begin{pmatrix} N^{-\frac{N}{4}} & 0 \\ 0 & N^{\frac{N}{4}} \end{pmatrix} Y(N^{\frac{1}{4}} z) \begin{pmatrix} N^{\frac{\alpha}{4}} & 0 \\ 0 & N^{-\frac{\alpha}{4}} \end{pmatrix}, \tag{2.1}$$

where $N = n + \alpha$. Then, U satisfies the following RHP:

($U_{\rm a}$) $U(z)$ is analytic in $\mathbb{C} \setminus \mathbb{R}$;

$(U_b)$ $U(z)$ takes continuous boundary values $U_+(x)$ and $U_-(x)$ such that

$$U_+(x) = U_-(x) \begin{pmatrix} 1 & |x|^{2\alpha} e^{-(Nx^4 + tx^2 N^{\frac{1}{2}})} \\ 0 & 1 \end{pmatrix}$$

for $x \in \mathbb{R} \setminus \{0\}$,

$(U_c)$ For $z \in \mathbb{C} \setminus \mathbb{R}$,

$$U(z) = [I + O(1/z)] \begin{pmatrix} z^n & 0 \\ 0 & z^{-n} \end{pmatrix}$$

as $z \to \infty$.

$(U_d)$ $U(z)$ has the same behavior as $Y(z)$ for $z$ near the origin.

From Theorem 2.1 and (2.1), it is easily seen that

$$U_{11}(z) = N^{-\frac{n}{4}} \pi_n(N^{\frac{1}{4}} z). \tag{2.2}$$

For convenience, we have scaled the variable $z$ by multiplying it by $N^{\frac{1}{4}} = (n+\alpha)^{\frac{1}{4}}$.

## 3 Normalization and Auxiliary Functions

In order to normalize the behavior of $U(z)$ at $z = \infty$, we first need to introduce some notations. Let $\alpha_t$ be a positive real number, and denote by $\mu_t(s)$ a probability density function supported on $[-\alpha_t, \alpha_t]$, i.e., $\mu_t(s) \geq 0$ and

$$\int_{-\alpha_t}^{\alpha_t} \mu_t(s)\, \mathrm{d}s = 1. \tag{3.1}$$

The explicit formulas for $\alpha_t$ and $\mu_t(s)$ will be determined later.

Next, we introduce the so-called $g$-function which is the logarithmic potential of $\mu_t(s)$; that is

$$g(z) = g_t(z) = \int_{-\alpha_t}^{\alpha_t} \log(z-s)\mu_t(s)\, \mathrm{d}s, \quad z \in \mathbb{C} \setminus (-\infty, \alpha_t], \tag{3.2}$$

where for each $s$ we view $\log(z-s)$ as an analytic function of the variable $z$ with a branch cut along $(-\infty, s]$.

Similarly, we also define

$$\widetilde{g}(z) = \widetilde{g}_t(z) = \int_{-\alpha_t}^{\alpha_t} \log(z-s)\mu_t(s)\, \mathrm{d}s, \quad z \in \mathbb{C} \setminus [-\alpha_t, \infty), \tag{3.3}$$

where for each $s$ we view $\log(z-s)$ as an analytic function of the variable $z$ with a branch cut along $[s, \infty)$.

From (3.1)–(3.2), it is easy to check that the $g$-function satisfies the jump conditions

$$g_+(x) - g_-(x) = 2\pi \mathrm{i}, \quad x < -\alpha_t, \tag{3.4}$$

and

$$g_+(x) - g_-(x) = 2\pi \mathrm{i} \int_x^{\alpha_t} \mu_t(s)\, \mathrm{d}s, \quad -\alpha_t < x < \alpha_t. \tag{3.5}$$

On account of (3.2) and (3.4), one readily sees that $e^{ng(z)}$ can be analytically extended to $\mathbb{C} \setminus [-\alpha_t, \alpha_t]$ and

$$e^{ng(z)} = z^n[1 + O(z^{-1})] \qquad \text{as } z \to \infty. \tag{3.6}$$

By adopting the convention that $\sigma_3$ denotes the Pauli matrix

$$\sigma_3 = \begin{pmatrix} 1 & 0 \\ 0 & -1 \end{pmatrix}, \tag{3.7}$$

we introduce the transformation

$$T(z) := e^{-\frac{1}{2}Nl_N\sigma_3}U(z)e^{-(Ng(z)-\frac{1}{2}Nl_N-\alpha\log z)\sigma_3}, \tag{3.8}$$

where $l_N$ is a constant to be determined. A straightforward calculation shows that $T(z)$ is the unique solution of the following RHP:

($T_a$) $T(z)$ is analytic in $\mathbb{C} \setminus \mathbb{R}$;

($T_b$) For $x \in \mathbb{R} \setminus \{0\}$,

$$T_+(x) = T_-(x) \begin{pmatrix} J_{11} & J_{12} \\ J_{21} & J_{22} \end{pmatrix},$$

where

$$J_{11}(x) = e^{-N(g_+(x)-g_-(x))+\alpha[(\log x)_+-(\log x)_-]},$$

$$J_{12}(x) = e^{N(g_+(x)+g_-(x)-l_N)-(Nx^4+tx^2N^{\frac{1}{2}})},$$

$$J_{22}(x) = e^{N(g_+(x)-g_-(x))-\alpha[(\log x)_+-(\log x)_-]},$$

$$J_{21}(x) = 0;$$

($T_c$) $T(z)$ behaves like the identity matrix at infinity:

$$T(z) = [I + O(1/z)] \qquad \text{as } z \to \infty,$$

for $z \in \mathbb{C} \setminus \mathbb{R}$;

($T_d$) $T(z)$ has the following behavior for $z \in \mathbb{C} \setminus \mathbb{R}$ as $z \to 0$

$$T(z) = \begin{cases} O\begin{pmatrix} |z|^\alpha & |z|^\alpha \\ |z|^\alpha & |z|^\alpha \end{pmatrix}, & \text{if } \alpha < 0, \\ O\begin{pmatrix} |z|^\alpha & |z|^{-\alpha} \\ |z|^\alpha & |z|^{-\alpha} \end{pmatrix}, & \text{if } \alpha > 0. \end{cases}$$

We proceed to seek for a probability density function $\mu_t(s)$ in (3.1) and a constant $l_N$ mentioned above so that $J_{12}$ in jump condition ($T_b$) becomes 1 for $x \in (-\alpha_t, \alpha_t)$. Thus, we set

$$N(g_+(x) + g_-(x) - l_N) - (Nx^4 + tx^2N^{\frac{1}{2}}) = 0 \tag{3.9}$$

for $x \in (-\alpha_t, \alpha_t)$. It follows from (3.9) by substituting $x = \alpha_t$ in (3.5) that

$$l_N = 2g(\alpha_t) - \alpha_t^4 - t\alpha_t^2N^{-\frac{1}{2}}. \tag{3.10}$$

Differentiating (3.9) yields

$$G_+(x) + G_-(x) = \frac{i}{\pi}(4x^3 + 2txN^{-\frac{1}{2}}), \quad x \in (-\alpha_t, \alpha_t), \tag{3.11}$$

where

$$G(z) := \frac{1}{\pi \mathrm{i}} \int_{-\alpha_t}^{\alpha_t} \frac{\mu_t(s)}{s-z} \mathrm{d}s, \quad z \in \mathbb{C} \setminus [-\alpha_t, \alpha_t]. \tag{3.12}$$

For convenience, we set

$$\widetilde{G}(z) := \frac{G(z)}{\sqrt{z^2 - \alpha_t^2}}, \tag{3.13}$$

where $\sqrt{z^2 - \alpha_t^2}$ is analytic in $\mathbb{C} \setminus [-\alpha_t, \alpha_t]$ and behaves like $z$ as $z \to \infty$. Since

$$\widetilde{G}_+(x) - \widetilde{G}_-(x) = \frac{4x^3 + 2txN^{-\frac{1}{2}}}{\pi \sqrt{\alpha_t^2 - x^2}}, \quad x \in (-\alpha_t, \alpha_t), \tag{3.14}$$

we can solve this scalar RHP to give

$$\widetilde{G}(z) = \frac{1}{2\pi^2 \mathrm{i}} \int_{-\alpha_t}^{\alpha_t} \frac{4s^3 + 2tsN^{-\frac{1}{2}}}{\sqrt{\alpha_t^2 - s^2}} \frac{\mathrm{d}s}{s-z},$$

or, equivalently,

$$G(z) = \frac{\sqrt{z^2 - \alpha_t^2}}{2\pi^2 \mathrm{i}} \int_{-\alpha_t}^{\alpha_t} \frac{4s^3 + 2tsN^{-\frac{1}{2}}}{\sqrt{\alpha_t^2 - s^2}} \frac{\mathrm{d}s}{s-z}. \tag{3.15}$$

From (3.1) and (3.12), it is easily seen that $G(z) \to 0$ and $zG(z) \to \frac{\mathrm{i}}{\pi}$ as $z \to \infty$. Hence, we have from (3.15)

$$\int_{-\alpha_t}^{\alpha_t} \frac{4s^3 + 2tsN^{-\frac{1}{2}}}{\sqrt{\alpha_t^2 - s^2}} \mathrm{d}s = 0 \tag{3.16}$$

and

$$\int_{-\alpha_t}^{\alpha_t} \frac{4s^4 + 2ts^2N^{-\frac{1}{2}}}{\sqrt{\alpha_t^2 - s^2}} \mathrm{d}s = 2\pi. \tag{3.17}$$

Since the integrand is an odd function, we find that (3.16) is trivially true. By a change of variable $s = \alpha_t \sin\theta$, it follows that

$$\alpha_t = \left( \frac{-tN^{-\frac{1}{2}} + \sqrt{t^2 N^{-1} + 12}}{3} \right)^{\frac{1}{2}}. \tag{3.18}$$

We next derive an explicit formula for the probability density function $\mu_t(s)$ in (3.1). From (3.12), one observes

$$\begin{aligned} G_+(x) &= \lim_{\varepsilon \to 0^+} G(x + \mathrm{i}\varepsilon) = \lim_{\varepsilon \to 0^+} \frac{1}{\pi \mathrm{i}} \int_{-\alpha_t}^{\alpha_t} \frac{(s-x) + \mathrm{i}\varepsilon}{(s-x)^2 + \varepsilon^2} \mu_t(s) \, \mathrm{d}s \\ &= \mu_t(x) + \frac{\mathrm{i}}{\pi} \mathrm{P.V.} \int_{-\alpha_t}^{\alpha_t} \frac{1}{x-s} \mu_t(s) \, \mathrm{d}s \end{aligned} \tag{3.19}$$

for $x \in (-\alpha_t, \alpha_t)$, where P.V. denotes the Cauchy principal value (see [1, p. 518]). Therefore,

$$\mu_t(x) = \mathrm{Re}\, G_+(x).$$

To evaluate $G(z)$ in (3.15), we first note that for any integer $m \geq 1$ and $z \in \mathbb{C} \setminus [-a, a]$ with $a > 0$, Lemma 1 in [20] gives

$$\int_{-a}^{a} \frac{s^m}{\sqrt{a^2 - s^2}} \frac{1}{s - z} \mathrm{d}s = \pi \Big( \sum_{i=0}^{[\frac{m-1}{2}]} a^{2\mathrm{i}} z^{m-1-2\mathrm{i}} A_i - \frac{z^m}{\sqrt{z^2 - a^2}} \Big), \tag{3.20}$$

where

$$A_0 = 1, \quad A_i = \prod_{j=1}^{i} \frac{2j - 1}{2j} \tag{3.21}$$

for $i \geq 1$, $[x]$ denotes the largest integer $\leq x$, and $\sqrt{z^2 - a^2}$ takes branch cut along $[-a, a]$ and behaves like $z$ as $z \to \infty$. It then follows from (3.15) and (3.20) that

$$\begin{aligned} G(z) &= \frac{\sqrt{z^2 - \alpha_t^2}}{2\pi^2 \mathrm{i}} \int_{-\alpha_t}^{\alpha_t} \frac{4s^3 + 2tsN^{-\frac{1}{2}}}{\sqrt{\alpha_t^2 - s^2}} \frac{\mathrm{d}s}{s - z} \\ &= \frac{\sqrt{z^2 - \alpha_t^2}}{\pi \mathrm{i}} \Big( 2z^2 + \alpha_t^2 + tN^{-\frac{1}{2}} - \frac{tN^{-\frac{1}{2}} z}{\sqrt{z^2 - \alpha_t^2}} - \frac{2z^3}{\sqrt{z^2 - \alpha_t^2}} \Big). \end{aligned} \tag{3.22}$$

A combination of (3.19) and (3.22) gives that

$$\mu_t(x) = \mathrm{Re}\ G_+(x) = \frac{\sqrt{\alpha_t^2 - x^2}}{\pi} (2x^2 + \alpha_t^2 + tN^{-\frac{1}{2}}) \tag{3.23}$$

for $x \in (-\alpha_t, \alpha_t)$, where $\alpha_t$ is shown in (3.18).

It is easy to see from (3.23) that $\mu_t(x)$ is not always larger than or equal to 0 for all $t \in \mathbb{R}$. Indeed, if $t < t_{cr} := -2\sqrt{N}$, then there exists $\delta > 0$ such that $\mu_t(x) < 0$ for $x \in (-\delta, \delta) \subset (-\alpha_t, \alpha_t)$, which contradicts our assumption that $\mu_t(x)$ is positive in the support. This fact shows that the support of $\mu_t(x)$ consists of more than one interval if $t < -2\sqrt{N}$. In the following, we divide our discussion into three cases: $c > -2$, $c = -2$ (i.e., the critical case) and $c < -2$, where $c = tN^{-\frac{1}{2}}$ is a constant. In the last case, the support of $\mu_t$ consists of two intervals.

### 3.1 Case for $c > -2$

In this subsection, we consider the first case. Thus, we have

$$\alpha_t < \alpha_{t_{cr}} = \sqrt{2}. \tag{3.24}$$

We first proceed to calculate $l_N$ in (3.10). By (3.1)–(3.2), the function $g(z) - \log(z + \alpha_t)$ is analytic in $\mathbb{C} \setminus [-\alpha_t, \alpha_t]$ and

$$g(z) - \log(z + \alpha_t) \to 0, \qquad \text{as } z \to \infty. \tag{3.25}$$

Let us define

$$H(z) := \frac{g(z) - \log(z + \alpha_t)}{\sqrt{z^2 - \alpha_t^2}}, \tag{3.26}$$

where $\sqrt{z^2 - \alpha_t^2}$ is analytic in $\mathbb{C} \setminus [-\alpha_t, \alpha_t]$ and behaves like $z$ as $z \to \infty$. It can be readily verified that for $x \in (-\alpha_t, \alpha_t)$,

$$H_+(x) - H_-(x) = \frac{g_+(x) + g_-(x) - 2\log(x + \alpha_t)}{i\sqrt{\alpha_t^2 - x^2}}.$$

By the Plemelj formula (see [1, p. 518]), we obtain

$$g(z) - \log(z + \alpha_t) = \frac{\sqrt{z^2 - \alpha_t^2}}{2\pi \mathrm{i}} \int_{-\alpha_t}^{\alpha_t} \frac{H_+(x) - H_-(x)}{x - z} \, \mathrm{d}x.$$

Recall that $t > -2\sqrt{N}$ and $\alpha_t$ in (3.18). From (3.9) it follows that

$$g(z) - \log(z + \alpha_t) = \frac{\sqrt{z^2 - \alpha_t^2}}{2\pi} \int_{-\alpha_t}^{\alpha_t} \frac{2\log(x + \alpha_t) - l_N - x^4 - tN^{-\frac{1}{2}}x^2}{\sqrt{\alpha_t^2 - x^2}(x - z)} \, \mathrm{d}x.$$

Now, let $z \to \infty$; on account of (3.25), we have

$$2\int_{-\alpha_t}^{\alpha_t} \frac{\log(x + \alpha_t)}{\sqrt{\alpha_t^2 - x^2}} \, \mathrm{d}x - l_N \int_{-\alpha_t}^{\alpha_t} \frac{\mathrm{d}x}{\sqrt{\alpha_t^2 - x^2}} - \int_{-\alpha_t}^{\alpha_t} \frac{x^4 + tN^{-\frac{1}{2}}x^2}{\sqrt{\alpha_t^2 - x^2}} \, \mathrm{d}x = 0. \tag{3.27}$$

Let the integrals on the left-hand side of (3.27) be denoted by $\mathrm{I}_1$, $\mathrm{I}_2$ and $\mathrm{I}_3$, respectively. It is easily seen that

$$\mathrm{I}_2 = \int_{-\alpha_t}^{\alpha_t} \frac{\mathrm{d}x}{\sqrt{\alpha_t^2 - x^2}} = \pi \tag{3.28}$$

and

$$\mathrm{I}_3 = \int_{-\alpha_t}^{\alpha_t} \frac{x^4 + tN^{-\frac{1}{2}}x^2}{\sqrt{\alpha_t^2 - x^2}} \, \mathrm{d}x = \frac{3}{8}\pi\alpha_t^4 + \frac{\pi}{2}\frac{t}{\sqrt{N}}\alpha_t^2. \tag{3.29}$$

To evaluate $\mathrm{I}_1$, we note that for any $0 \le a < b < \infty$,

$$\int_a^b \frac{\log s}{\sqrt{(s - a)(b - s)}} \, \mathrm{d}s = 2\pi \log \frac{\sqrt{a} + \sqrt{b}}{2} \tag{3.30}$$

(see [18, Lemma 1]). Hence, by taking $s = x + \alpha_t$, we obtain

$$\mathrm{I}_1 = \int_0^{2\alpha_t} \frac{\log s}{\sqrt{s(2\alpha_t - s)}} \, \mathrm{d}s = \pi \log \frac{\alpha_t}{2}. \tag{3.31}$$

Inserting (3.28)–(3.31) into (3.27) gives

$$l_N = 2\log\frac{\alpha_t}{2} - \frac{3}{8}\alpha_t^4 - \frac{t}{2\sqrt{N}}\alpha_t^2. \tag{3.32}$$

Let $\nu(z) := \sqrt{z^2 - \alpha_t^2}(2z^2 + \alpha_t^2 + tN^{-\frac{1}{2}})$, where $\sqrt{z^2 - \alpha_t^2}$ is analytic in $\mathbb{C} \setminus [-\alpha_t, \alpha_t]$ and behaves likes $z$ as $z \to \infty$. Clearly, $\nu_\pm(x) = \pm\pi \mathrm{i}\mu_t(x)$ for $x \in [-\alpha_t, \alpha_t]$. Define

$$\phi(z) := \int_{\alpha_t}^z \nu(\zeta) \, \mathrm{d}\zeta, \quad z \in \mathbb{C} \setminus (-\infty, \alpha_t], \tag{3.33}$$

where the path of integration from $\alpha_t$ to $z$ lies entirely in the region $z \in \mathbb{C} \setminus (-\infty, \alpha_t]$, except for the initial point $\alpha_t$.

Similarly, we define

$$\widetilde{\phi}(z) := \int_{-\alpha_t}^z \nu(\zeta) \, \mathrm{d}\zeta, \quad z \in \mathbb{C} \setminus [-\alpha_t, \infty), \tag{3.34}$$

where the path of integration from $-\alpha_t$ to $z$ lies entirely in the region $z \in \mathbb{C} \setminus [-\alpha_t, \infty)$, except for the initial point $-\alpha_t$.

The functions $\phi(z)$ and $\widetilde{\phi}(z)$ defined above will play an important role in our argument, and some of their properties are given below. For proofs of these results, see [20].

**Proposition 3.1** *Let* $\mathbb{C}_+ = \{z : \text{Im } z > 0\}$ *and* $\mathbb{C}_- = \{z : \text{Im } z < 0\}$. *We have*

$$\phi(z) = \widetilde{\phi}(z) \mp \pi \mathrm{i} \tag{3.35}$$

*for* $z \in \mathbb{C}_\pm$. *The mapping properties of* $\phi(z)$ *are given by*

$$\phi(x) > 0, \quad \arg \phi(x) = 0, \quad x \in (\alpha_t, \infty); \tag{3.36}$$

$$\phi(\alpha_t) = 0, \quad \phi_\pm(-\alpha_t) = \mp \pi \mathrm{i}, \quad \phi_\pm(0) = \mp \frac{\pi}{2}\mathrm{i}; \tag{3.37}$$

$$\text{Re } \phi_\pm(x) = 0, \quad \arg \phi_\pm(x) = \pm \frac{3}{2}\pi, \quad x \in (-\alpha_t, \alpha_t) \tag{3.38}$$

*and*

$$\phi(z) \sim \frac{z^4}{2} \tag{3.39}$$

*for* $z$ *large enough. Similarly, the mapping properties of* $\widetilde{\phi}(z)$ *are given by*

$$\widetilde{\phi}(x) > 0, \quad \arg \widetilde{\phi}(x) = 0, \quad x \in (-\infty, -\alpha_t); \tag{3.40}$$

$$\widetilde{\phi}(-\alpha_t) = 0, \quad \widetilde{\phi}_\pm(\alpha_t) = \pm \pi \mathrm{i}, \quad \widetilde{\phi}_\pm(0) = \pm \frac{\pi}{2}\mathrm{i}; \tag{3.41}$$

$$\text{Re } \widetilde{\phi}_\pm(x) = 0, \quad \arg \widetilde{\phi}_\pm(x) = \mp \frac{3}{2}\pi, \quad x \in (-\alpha_t, \alpha_t) \tag{3.42}$$

*and*

$$\widetilde{\phi}(z) \sim \frac{z^4}{2} \tag{3.43}$$

*for* $z$ *large enough.*

**Proposition 3.2** *With constant* $l_N$ *given in* (3.32) *and* $\widetilde{l}_N$ *defined by*

$$\widetilde{l}_N = l_N + 2\pi \mathrm{i}, \tag{3.44}$$

*the following connection formulas between the* $g$*-function* ($\widetilde{g}$*-function*) *and* $\phi$*-function* ($\widetilde{\phi}$*-function*) *hold*

$$g(z) + \phi(z) = \frac{1}{2}\Big(z^4 + \frac{tz^2}{\sqrt{N}} + l_N\Big), \tag{3.45}$$

$$\widetilde{g}(z) + \widetilde{\phi}(z) = \frac{1}{2}\Big(z^4 + \frac{tz^2}{\sqrt{N}} + \widetilde{l}_N\Big). \tag{3.46}$$

*Furthermore, we have*

$$\widetilde{g}(z) = \begin{cases} g(z), & z \in \mathbb{C}_+, \\ g(z) + 2\pi \mathrm{i}, & z \in \mathbb{C}_- \end{cases} \tag{3.47}$$

*and*

$$\widetilde{\phi}(z) = \begin{cases} \phi(z) + \pi \mathrm{i}, & z \in \mathbb{C}_+, \\ \phi(z) - \pi \mathrm{i}, & z \in \mathbb{C}_-. \end{cases} \tag{3.48}$$

### 3.2 Case for $c = -2$

Now, we consider the second case $t_{cr} = -2\sqrt{N}$. With

$$\alpha_t = \alpha_{t_{cr}} = \sqrt{2}, \tag{3.49}$$

we have from (3.23)

$$\mu_t(x) = \mu_{t_{cr}}(x) = \frac{2x^2\sqrt{2-x^2}}{\pi}. \tag{3.50}$$

With $\alpha_t = \sqrt{2}$ and $t = -2\sqrt{N}$, the formula in (3.32) gives

$$l_N = \frac{1}{2} - \log 2. \tag{3.51}$$

Let $\nu(z) := 2z^2\sqrt{z^2-2}$, where $\sqrt{z^2-2}$ is analytic in $\mathbb{C} \setminus [-\sqrt{2}, \sqrt{2}]$ and behaves likes $z$ as $z \to \infty$. Clearly, $(\nu)_\pm(x) = \pm\pi \mathrm{i}\mu_t(x)$ for $x \in [-\sqrt{2}, \sqrt{2}]$. Define

$$\phi(z) := \int_{\sqrt{2}}^{z} \nu(\zeta)\,\mathrm{d}\zeta, \quad z \in \mathbb{C} \setminus (-\infty, \sqrt{2}], \tag{3.52}$$

where the path of integration from $\sqrt{2}$ to $z$ lies entirely in the region $z \in \mathbb{C} \setminus (-\infty, \sqrt{2}]$, except for the initial point $\sqrt{2}$.

Similarly, we define

$$\widetilde{\phi}(z) := \int_{-\sqrt{2}}^{z} \nu(\zeta)\,\mathrm{d}\zeta, \quad z \in \mathbb{C} \setminus (-\sqrt{2}, \infty], \tag{3.53}$$

where the path of integration from $-\sqrt{2}$ to $z$ lies entirely in the region $z \in \mathbb{C} \setminus [-\sqrt{2}, \infty)$, except for the initial point $-\sqrt{2}$.

The functions $\phi(z)$ and $\widetilde{\phi}(z)$ defined above play the same role as $\phi(z)$ and $\widetilde{\phi}(z)$ defined in (3.33)–(3.34) for case (i). The following are some of their properties.

**Proposition 3.3** *Denoting $\mathbb{C}_+ = \{z : \mathrm{Im}\ z > 0\}$ and $\mathbb{C}_- = \{z : \mathrm{Im}\ z < 0\}$, we have*

$$\phi(z) = \widetilde{\phi}(z) \mp \pi \mathrm{i} \tag{3.54}$$

*for $z \in \mathbb{C}_\pm$. The mapping properties of $\phi(z)$ are given by*

$$\phi(x) > 0, \quad \arg\phi(x) = 0, \quad x \in (\sqrt{2}, \infty); \tag{3.55}$$

$$\phi(\sqrt{2}) = 0, \quad \phi_\pm(-\sqrt{2}) = \mp\pi \mathrm{i}, \quad \phi_\pm(0) = \mp\frac{\pi}{2}\mathrm{i}; \tag{3.56}$$

$$\mathrm{Re}\ \phi_\pm(x) = 0, \quad \arg\phi_\pm(x) = \pm\frac{3}{2}\pi, \quad x \in [0, \sqrt{2}); \tag{3.57}$$

$$\mathrm{Re}\ \phi(z) > 0, \quad \mathrm{Im}\ \phi(z) = -\frac{1}{2}\pi, \quad z \in (0, \infty \mathrm{i}). \tag{3.58}$$

*If $z \in \mathbb{C}_+$, we have*

$$\phi(z) \sim -\frac{\pi}{2}\mathrm{i} + \frac{2^{\frac{3}{2}}}{3} z^3 \mathrm{i} \tag{3.59}$$

*as $z \to 0$ and*

$$\phi(z) \sim \frac{z^4}{2} \tag{3.60}$$

*for $z$ large enough. Similarly, the mapping properties of $\widetilde{\phi}(z)$ are given by*

$$\widetilde{\phi}(x) > 0, \quad \arg \widetilde{\phi}(x) = 0, \quad x \in (-\infty, -\sqrt{2}); \tag{3.61}$$

$$\widetilde{\phi}(-\sqrt{2}) = 0, \quad \widetilde{\phi}_\pm(\sqrt{2}) = \pm\pi \mathrm{i}, \quad \widetilde{\phi}_\pm(0) = \pm\frac{\pi}{2}\mathrm{i}; \tag{3.62}$$

$$\mathrm{Re}\; \widetilde{\phi}_\pm(x) = 0, \quad \arg \widetilde{\phi}_\pm(x) = \mp\frac{3}{2}\pi, \quad x \in (-\sqrt{2}, 0]; \tag{3.63}$$

$$\mathrm{Re}\; \widetilde{\phi}(z) > 0, \quad \mathrm{Im}\; \widetilde{\phi}(z) = \frac{\pi}{2}, \quad z \in (0, \infty \mathrm{i}). \tag{3.64}$$

*If $z \in \mathbb{C}_+$, we have*

$$\widetilde{\phi}(z) \sim \frac{\pi}{2}\mathrm{i} + \frac{2^{\frac{3}{2}}}{3} z^3 \mathrm{i} \tag{3.65}$$

*as $z \to 0$ and*

$$\widetilde{\phi}(z) \sim \frac{z^4}{2} \tag{3.66}$$

*for $z$ large enough.*

**Proposition 3.4** *With constant $l_N$ given in* (3.51) *and $\widetilde{l}_N$ defined by*

$$\widetilde{l}_N = l_N + 2\pi \mathrm{i}, \tag{3.67}$$

*the following connection formulas between the $g$-function ($\widetilde{g}$-function) and $\phi$-function ($\widetilde{\phi}$-function) hold*

$$g(z) + \phi(z) = \frac{1}{2}(z^4 - 2z^2 + l_N), \tag{3.68}$$

$$\widetilde{g}(z) + \widetilde{\phi}(z) = \frac{1}{2}(z^4 - 2z^2 + \widetilde{l}_N) \tag{3.69}$$

*and*

$$\widetilde{g}(z) = \begin{cases} g(z), & z \in \mathbb{C}_+ \\ g(z) + 2\pi\mathrm{i}, & z \in \mathbb{C}_-. \end{cases} \tag{3.70}$$

The material in this and the previous section parallels that in [19, §3].

### 3.3 Case for $c < -2$

In this third case, the equilibrium measure will be defined on two intervals, and we denote them by $E := [-b, -a] \cup [a, b]$ with $0 < a < b$. To see that there are exactly two intervals and these intervals are symmetric, we refer to [4, p. 218] and [6, p. 603]. To normalize the behavior of $U(z)$ at $z = \infty$, we define as in (3.1) a probability density function $\mu_t(s)$ supported on $E$, i.e., $\mu_t(s) \geq 0$ and

$$\int_E \mu_t(s)\, \mathrm{d}s = 1. \tag{3.71}$$

The explicit formula for $a, b$, and $\mu_t(s)$ will be determined later.

Now, we repeat the analysis at the beginning of this section from (3.1) to (3.12) with the interval $[-\alpha_t, \alpha_t]$ replaced by $E$. With this minor change, we define

$$g(z) = \int_{-b}^{-a} \log(z-s)\mu_t(s)\,\mathrm{d}s + \int_a^b \log(z-s)\mu_t(s)\,\mathrm{d}s, \quad z \in \mathbb{C} \setminus (-\infty, b]. \tag{3.72}$$

From (3.71)–(3.72), it is easily checked that

$$g_+(x) - g_-(x) = \begin{cases} 2\pi\mathrm{i}, & x \in (-\infty, -b), \\ 2\pi\mathrm{i} \displaystyle\int_x^{-a} \mu_t(s)\,\mathrm{d}s + \pi\mathrm{i}, & x \in (-b, -a), \\ \pi\mathrm{i}, & x \in (-a, a), \\ 2\pi\mathrm{i} \displaystyle\int_x^b \mu_t(s)\,\mathrm{d}s, & x \in (a, b). \end{cases} \tag{3.73}$$

Also, on account of (3.72)–(3.73), one readily sees that $\mathrm{e}^{ng(z)}$ can be analytically extended to $\mathbb{C} \setminus [-b, b]$ and (3.6) holds. As in the previous cases, we introduce the transformation $T(z)$ given in (3.8), and obtain its associated RHP stated in ($T_\mathrm{a}$)-($T_\mathrm{d}$). To determine a probability density function $\mu_t(s)$ in (3.71) and a constant $l_N$ mentioned above so that the entry $J_{12}$ in jump condition ($T_\mathrm{b}$) becomes 1 for $x \in E$, we set

$$N(g_+(x) + g_-(x) - l_N) - (Nx^4 + tx^2 N^{\frac{1}{2}}) = 0 \tag{3.74}$$

for $x \in E$. By substituting $x = b$ in (3.73), it follows from (3.74) that

$$l_N = 2g(b) - b^4 - tb^2 N^{-\frac{1}{2}}. \tag{3.75}$$

Differentiating (3.74) yields

$$G_+(x) + G_-(x) = \frac{\mathrm{i}}{\pi}(4x^3 + 2txN^{-\frac{1}{2}}), \quad x \in E, \tag{3.76}$$

where

$$G(z) := \frac{1}{\pi\mathrm{i}} \int_E \frac{\mu_t(s)}{s-z}\,\mathrm{d}s, \quad z \in \mathbb{C} \setminus E. \tag{3.77}$$

Corresponding to (3.13), we set

$$\widetilde{G}(z) := \frac{G(z)}{\sqrt{(z^2-a^2)(z^2-b^2)}}, \tag{3.78}$$

where $\sqrt{(z^2-a^2)(z^2-b^2)}$ is analytic in $\mathbb{C} \setminus E$ and behaves like $z^2$ as $z \to \infty$ (see [9, p. 171]). Since

$$\widetilde{G}_+(x) - \widetilde{G}_-(x) = \frac{\mathrm{i}}{\pi} \frac{4x^3 + 2txN^{-\frac{1}{2}}}{(\sqrt{(x^2-a^2)(x^2-b^2)})_+}, \quad x \in E, \tag{3.79}$$

solving this scalar RHP gives

$$\widetilde{G}(z) = \frac{1}{2\pi^2} \int_E \frac{4s^3 + 2tsN^{-\frac{1}{2}}}{(\sqrt{(s^2-a^2)(s^2-b^2)})_+} \frac{\mathrm{d}s}{s-z},$$

or equivalently

$$G(z) = \frac{\sqrt{(z^2-a^2)(z^2-b^2)}}{2\pi^2} \int_E \frac{4s^3 + 2tsN^{-\frac{1}{2}}}{(\sqrt{(s^2-a^2)(s^2-b^2)})_+} \frac{\mathrm{d}s}{s-z}. \tag{3.80}$$

Recall (3.71) and (3.77). Since $G(z) \to 0$ and $zG(z) \to \frac{\mathrm{i}}{\pi}$ as $z \to \infty$, we have from (3.80)

$$\int_E \frac{4s^3 + 2tsN^{-\frac{1}{2}}}{(\sqrt{(s^2-a^2)(s^2-b^2)})_+} s^j\, \mathrm{d}s = 0 \tag{3.81}$$

for $j = 0, 1$, and

$$\int_E \frac{4s^3 + 2tsN^{-\frac{1}{2}}}{(\sqrt{(s^2-a^2)(s^2-b^2)})_+} s^2\, \mathrm{d}s = -2\pi\mathrm{i} \tag{3.82}$$

(see [9, p. 172]). Equations (3.81)–(3.82) are equivalent to

$$\int_a^b \frac{4s^3 + 2tsN^{-\frac{1}{2}}}{\sqrt{(s^2-a^2)(b^2-s^2)}}\, \mathrm{d}s = 0 \tag{3.83}$$

and

$$\int_a^b \frac{(4s^3 + 2tsN^{-\frac{1}{2}})s^2}{\sqrt{(s^2-a^2)(b^2-s^2)}}\, \mathrm{d}s = \pi. \tag{3.84}$$

Now we have two equations with two unknowns $a$ and $b$. By a change of variable

$$s = \frac{ab}{\sqrt{b^2\cos^2\theta + a^2\sin^2\theta}},$$

we obtain after a long and complicated calculation

$$a = \Big(-\frac{tN^{-\frac{1}{2}}+2}{2}\Big)^{\frac{1}{2}} \quad \text{and} \quad b = \Big(-\frac{tN^{-\frac{1}{2}}-2}{2}\Big)^{\frac{1}{2}} \tag{3.85}$$

(see [4, p. 218]). We next derive an explicit formula for the probability density function $\mu_t(s)$ in (3.71). It is well known that

$$\mu_t(x) = \mathrm{Re}\, G_+(x).$$

To evaluate $G(z)$ in (3.80), we need Lemma 1 in [20], which states that for $b > a > 0$, we have

$$\int_a^b \frac{s}{\sqrt{(b-s)(s-a)}} \frac{1}{s-z}\, \mathrm{d}s = \pi\Big(1 - \frac{z}{\sqrt{(z-a)(z-b)}}\Big). \tag{3.86}$$

Recall that

$$\int_a^b \frac{1}{\sqrt{(b-s)(s-a)}} \frac{1}{s-z}\, \mathrm{d}s = \frac{-\pi}{\sqrt{(z-a)(z-b)}}, \tag{3.87}$$

where $\sqrt{(z-a)(z-b)}$ takes branch cut along $[a,b]$ and behaves like $z$ as $z \to \infty$. It then follows from (3.80) and (3.86)–(3.87) that

$$G(z) = \frac{\sqrt{(z^2-a^2)(z^2-b^2)}}{2\pi^2} \int_E \frac{4s^3 + 2tsN^{-\frac{1}{2}}}{(\sqrt{(s^2-a^2)(s^2-b^2)})_+} \frac{\mathrm{d}s}{s-z}$$

$$= \frac{\sqrt{(z^2-a^2)(z^2-b^2)}}{2\pi\mathrm{i}}\left(4z - \frac{4z^3+2tN^{-\frac{1}{2}}z}{\sqrt{(z^2-a^2)(z^2-b^2)}}\right), \tag{3.88}$$

which gives

$$\mu_t(x) = \mathrm{Re}\ G_+(x) = \frac{2|x|}{\pi}\sqrt{(x^2-a^2)(b^2-x^2)} > 0 \tag{3.89}$$

for $x \in E$, where $a, b$ are given in (3.85).

Next, we proceed to calculate $l_N$ in (3.75). By (3.71)–(3.72), the function $g(z) - \frac{1}{2}(\log(z+b)+\log(z-b))$ is analytic in $\mathbb{C}\setminus E$ and

$$g(z) - \frac{1}{2}(\log(z+b)+\log(z-b)) \to 0 \quad \text{as } z \to \infty. \tag{3.90}$$

Let us define

$$H(z) := \frac{g(z) - \frac{1}{2}(\log(z+b)+\log(z-b))}{\sqrt{(z^2-a^2)(z^2-b^2)}}, \tag{3.91}$$

where $\sqrt{(z^2-a^2)(z^2-b^2)}$ is analytic in $\mathbb{C}\setminus E$ and behaves like $z^2$ as $z \to \infty$. It can be readily verified that for $x \in E$,

$$H_+(x) - H_-(x) = \frac{g_+(x)+g_-(x)-\log(b^2-x^2)}{(\sqrt{(x^2-a^2)(x^2-b^2)})_+}.$$

By the Plemelj formula (see [1, p. 518]), we obtain

$$g(z) - \frac{1}{2}(\log(z+b)+\log(z-b)) = \frac{\sqrt{(z^2-a^2)(z^2-b^2)}}{2\pi\mathrm{i}}\int_E \frac{H_+(x)-H_-(x)}{x-z}\,\mathrm{d}x.$$

From (3.74), it follows

$$\begin{aligned} &g(z) - \frac{1}{2}(\log(z+b)+\log(z-b)) \\ &= \frac{\sqrt{(z^2-a^2)(z^2-b^2)}}{2\pi\mathrm{i}}\int_E \frac{l_N + x^4 + tN^{-\frac{1}{2}}x^2 - \log(b^2-x^2)}{(\sqrt{(x^2-a^2)(x^2-b^2)})_+(x-z)}\mathrm{d}x, \end{aligned}$$

which is equivalent to

$$\begin{aligned} &g(z) - \frac{1}{2}(\log(z+b)+\log(z-b)) \\ &= \frac{\sqrt{(z^2-a^2)(z^2-b^2)}}{2\pi\mathrm{i}}\left(\int_a^b \frac{l_N + x^4 + tN^{-\frac{1}{2}}x^2}{\mathrm{i}\sqrt{(x^2-a^2)(b^2-x^2)}}\frac{2x}{x^2-z^2}\mathrm{d}x\right. \\ &\left. - 2\int_a^b \frac{x\log(b^2-x^2)}{\mathrm{i}\sqrt{(x^2-a^2)(b^2-x^2)}}\frac{\mathrm{d}x}{x^2-z^2}\right). \end{aligned}$$

Now, let $z \to \infty$; on account of (3.90), we have

$$\begin{aligned} &\int_a^b \frac{2xl_N}{\sqrt{(x^2-a^2)(b^2-x^2)}}\mathrm{d}x + \int_a^b \frac{(x^4+tN^{-\frac{1}{2}}x^2)2x}{\sqrt{(x^2-a^2)(b^2-x^2)}}\mathrm{d}x \\ &- 2\int_a^b \frac{x\log(b^2-x^2)}{\sqrt{(x^2-a^2)(b^2-x^2)}}\mathrm{d}x = 0. \end{aligned} \tag{3.92}$$

Let the integrals on the left-hand side of (3.92) be denoted by $\mathrm{I}_1$, $\mathrm{I}_2$ and $\mathrm{I}_3$, respectively. As in Subsections 3.1–3.2, we have

$$\mathrm{I}_1 = \int_a^b \frac{2x}{\sqrt{(x^2-a^2)(b^2-x^2)}}\,\mathrm{d}x = \pi, \tag{3.93}$$

$$\mathrm{I}_2 = \int_a^b \frac{(x^4 + tN^{-\frac{1}{2}}x^2)}{\sqrt{(x^2-a^2)(b^2-x^2)}} 2x\,\mathrm{d}x = \frac{\pi}{2} - \frac{\pi}{4}t^2N^{-1} \tag{3.94}$$

and

$$\int_a^b \frac{\log s}{\sqrt{(s-a)(b-s)}}\,\mathrm{d}s = 2\pi \log \frac{\sqrt{a}+\sqrt{b}}{2}.$$

Taking $s = b^2 - x^2$, we obtain

$$\mathrm{I}_3 = \int_0^{b^2-a^2} \frac{\log s}{\sqrt{s(b^2-a^2-s)}}\,\mathrm{d}s = \pi \log \frac{1}{2}. \tag{3.95}$$

Inserting (3.93)–(3.95) into (3.92) gives

$$l_N = -\log 2 - \frac{1}{2} + \frac{t^2N^{-1}}{4}. \tag{3.96}$$

As before, let $\nu(z) := 2z\sqrt{(z^2-a^2)(z^2-b^2)}$, where $\sqrt{(z^2-a^2)(z^2-b^2)}$ is analytic in $\mathbb{C}\setminus E$ and behaves likes $z^2$ as $z \to \infty$. Clearly, $\nu_\pm(x) = \pm\pi\mathrm{i}\mu_t(x)$ for $x \in E$. Define

$$\phi(z) := \int_b^z \nu(\zeta)\,\mathrm{d}\zeta, \quad z \in \mathbb{C}\setminus(-\infty, b], \tag{3.97}$$

where the path of integration from $b$ to $z$ lies entirely in the region $z \in \mathbb{C}\setminus(-\infty,b]$, except for the initial point $b$. Similarly, define

$$\widetilde{\phi}(z) := \int_{-b}^z \nu(\zeta)\,\mathrm{d}\zeta, \quad z \in \mathbb{C}\setminus[-b, \infty), \tag{3.98}$$

where the path of integration from $-b$ to $z$ lies entirely in the region $z \in \mathbb{C}\setminus[-b,\infty)$, except for the initial point $-b$. Moreover, we define

$$\overline{\phi}(z) := \int_a^z \nu(\zeta)\,\mathrm{d}\zeta, \quad z \in \mathbb{C}\setminus(-\infty,-a)\cup(a,\infty), \tag{3.99}$$

where the path of integration from $a$ to $z$ lies entirely in the region $z \in \mathbb{C}\setminus(-\infty,-a)\cup(a,\infty)$, except for the initial point $a$. The following two results correspond to Propositions 3.1–3.2 for case (i), and to Propositions 3.3–3.4 for case (ii).

**Proposition 3.5** *The mapping properties of $\phi(z)$ are given by*

$$\phi(x) > 0, \quad \arg\phi(x) = 0, \quad x \in (b,\infty); \tag{3.100}$$

$$\phi(b) = 0, \quad \phi_\pm(a) = \mp\frac{\pi}{2}\mathrm{i}, \quad \phi_\pm(-b) = \mp\pi\mathrm{i}; \tag{3.101}$$

$$\mathrm{Re}\ \phi_\pm(x) = 0, \quad \arg\phi_\pm(x) = \pm\frac{3}{2}\pi, \quad x \in (a,b) \tag{3.102}$$

*and*

$$\phi(z) \sim \frac{z^4}{2} \tag{3.103}$$

*for $z$ large enough. Similarly, the mapping properties of $\widetilde{\phi}(z)$ are given by*

$$\widetilde{\phi}(x) > 0, \quad \arg\widetilde{\phi}(x) = 0, \quad x \in (-\infty, -b); \tag{3.104}$$

$$\widetilde{\phi}(-b) = 0, \quad \widetilde{\phi}_\pm(-a) = \pm\frac{\pi}{2}\mathrm{i}, \quad \widetilde{\phi}_\pm(b) = \pm\pi\mathrm{i}; \tag{3.105}$$

$$\mathrm{Re}\,\widetilde{\phi}_\pm(x) = 0, \quad \arg\widetilde{\phi}_\pm(x) = \mp\frac{3}{2}\pi, \quad x \in (-b, -a) \tag{3.106}$$

*and*

$$\widetilde{\phi}(z) \sim \frac{z^4}{2} \tag{3.107}$$

*for $z$ large enough. Moreover, the mapping properties of $\overline{\phi}(z)$ are given by*

$$\overline{\phi}(x) > 0, \quad \arg\overline{\phi}(x) = 0, \quad x \in (-a, a); \tag{3.108}$$

$$\overline{\phi}(\pm a) = 0, \quad \overline{\phi}_\pm(-b) = \mp\frac{\pi}{2}\mathrm{i}, \quad \overline{\phi}_\pm(b) = \pm\frac{\pi}{2}\mathrm{i} \tag{3.109}$$

*and*

$$\overline{\phi}(z) \sim \frac{z^4}{2} \tag{3.110}$$

*for $z$ large enough.*

**Proposition 3.6** *The following connection formula holds between the g-function and $\phi$-function*

$$\frac{1}{2}\Big(z^4 + \frac{tz^2}{\sqrt{N}} + l_N\Big) = g(z) + \phi(z). \tag{3.111}$$

*Furthermore, we have for $z \in \mathbb{C}_\pm$,*

$$\phi(z) = \begin{cases} \widetilde{\phi}(z) \mp \pi\mathrm{i}, \\ \overline{\phi}(z) \mp \dfrac{1}{2}\pi\mathrm{i}. \end{cases} \tag{3.112}$$

## 4 Contour Deformation

In this section, we consider the problem of contour deformation in three separate cases. We give more details in the first case; for the other two cases, we only present the conclusions since the results can be obtained in the same manner as in the first case.

### 4.1 Case for $c > -2$

With the properties of $\phi(z)$ and $\widetilde{\phi}(z)$ established in Proposition 3.1, the jump matrix for $T(z)$ in condition $(T_\mathrm{b})$ can be written as follows:

$$T_+(x) = T_-(x)\begin{pmatrix} \mathrm{e}^{2N\phi_+(x)} & 1 \\ 0 & \mathrm{e}^{2N\phi_-(x)} \end{pmatrix} \tag{4.1}$$

for $x \in (0, \alpha_t)$,

$$T_+(x) = T_-(x)\begin{pmatrix} \mathrm{e}^{2N\widetilde{\phi}_+(x)} & 1 \\ 0 & \mathrm{e}^{2N\widetilde{\phi}_-(x)} \end{pmatrix} \tag{4.2}$$

for $x \in (-\alpha_t, 0)$,

$$T_+(x) = T_-(x) \begin{pmatrix} 1 & \mathrm{e}^{-2N\phi(x)} \\ 0 & 1 \end{pmatrix} \tag{4.3}$$

for $x \in (\alpha_t, \infty)$,

$$T_+(x) = T_-(x) \begin{pmatrix} 1 & \mathrm{e}^{-2N\widetilde{\phi}(x)} \\ 0 & 1 \end{pmatrix} \tag{4.4}$$

for $x \in (-\infty, -\alpha_t)$.

Note that for the jump matrix on $(0, \alpha_t)$, we have the following factorization:

$$\begin{pmatrix} \mathrm{e}^{2N\phi_+(x)} & 1 \\ 0 & \mathrm{e}^{2N\phi_-(x)} \end{pmatrix} = \begin{pmatrix} 1 & 0 \\ \mathrm{e}^{2N\phi_-(x)} & 1 \end{pmatrix} \begin{pmatrix} 0 & 1 \\ -1 & 0 \end{pmatrix} \begin{pmatrix} 1 & 0 \\ \mathrm{e}^{2N\phi_+(x)} & 1 \end{pmatrix}. \tag{4.5}$$

The first and the third matrices on the right-hand side of (4.5) have the analytic continuation

$$\begin{pmatrix} 1 & 0 \\ \mathrm{e}^{2N\phi(z)} & 1 \end{pmatrix}$$

on both sides of $(0, \alpha_t)$. When $x \in (-\alpha_t, 0)$, we have a similar result with $\phi(x)$ replaced by $\widetilde{\phi}(x)$. Based on the factorization in (4.5), we transform the RHP for $T(z)$ into a RHP for $V(z)$ given below, by opening a lens around $(-\alpha_t, \alpha_t)$ going through the origin (see Figure 1). The precise shape of the lens will be specified later; for now we choose it to be contained in the region of analyticity of $\phi(z)$ and $\widetilde{\phi}(z)$.

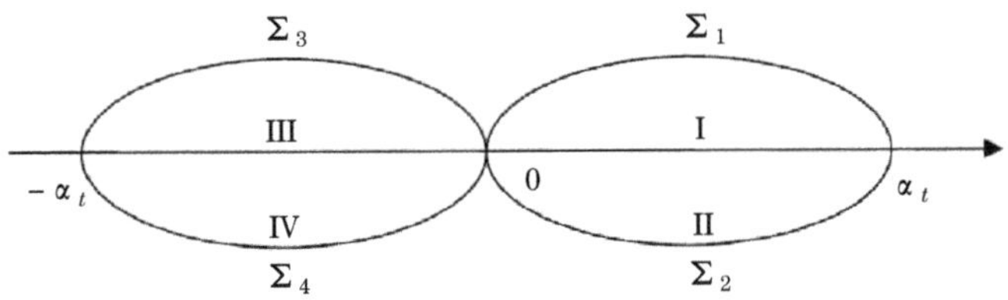


Figure 1 The lens-shaped contour Σ going through the origin.

Let $\Sigma = \bigcup_{i=1}^{4} \Sigma_i$ denote the lens-shaped contour shown in Figure 1. The second transformation $T \to V$ is then defined by

$$V(z) := \begin{cases} T(z) & \text{for } z \text{ outside the lens-shaped region,} \\ T(z) \begin{pmatrix} 1 & 0 \\ -\mathrm{e}^{2N\phi(z)} & 1 \end{pmatrix} & \text{for } z \in \mathrm{I}, \\ T(z) \begin{pmatrix} 1 & 0 \\ \mathrm{e}^{2N\phi(z)} & 1 \end{pmatrix} & \text{for } z \in \mathrm{II}, \\ T(z) \begin{pmatrix} 1 & 0 \\ -\mathrm{e}^{2N\widetilde{\phi}(z)} & 1 \end{pmatrix} & \text{for } z \in \mathrm{III}, \\ T(z) \begin{pmatrix} 1 & 0 \\ \mathrm{e}^{2N\widetilde{\phi}(z)} & 1 \end{pmatrix} & \text{for } z \in \mathrm{IV}, \end{cases} \tag{4.6}$$

where regions I, II, III, IV are also shown in Figure 1. Furthermore, we define the jump matrices

$$J_V(z) = \begin{pmatrix} 1 & 0 \\ e^{2N\phi(z)} & 1 \end{pmatrix}, \qquad z \in \Sigma_1 \cup \Sigma_2, \tag{4.7}$$

$$J_V(z) = \begin{pmatrix} 1 & 0 \\ e^{2N\widetilde{\phi}(z)} & 1 \end{pmatrix}, \qquad z \in \Sigma_3 \cup \Sigma_4, \tag{4.8}$$

$$J_V(x) = \begin{pmatrix} 1 & e^{-2N\widetilde{\phi}(x)} \\ 0 & 1 \end{pmatrix}, \qquad x < -\alpha_t, \tag{4.9}$$

$$J_V(x) = \begin{pmatrix} 1 & e^{-2N\phi(x)} \\ 0 & 1 \end{pmatrix}, \qquad x > \alpha_t, \tag{4.10}$$

$$J_V(x) = \begin{pmatrix} 0 & 1 \\ -1 & 0 \end{pmatrix}, \qquad x \in (-\alpha_t, 0) \cup (0, \alpha_t). \tag{4.11}$$

It is readily verified that $V$ satisfies the conditions of the following RHP:

($V_a$) $V(z)$ is analytic in $\mathbb{C} \setminus (\Sigma \cup \mathbb{R})$;

($V_b$) for $x \in \Sigma \cup \mathbb{R}$,

$$V_+(x) = V_-(x) J_V(x);$$

($V_c$) $V(z)$ behaves like the identity matrix at infinity:

$$V(z) = [I + O(1/z)] \quad \text{as } z \to \infty,$$

for $z \in \mathbb{C} \setminus (\Sigma \cup \mathbb{R})$;

($V_d$) if $\alpha < 0$,

$$V(z) = O\begin{pmatrix} |z|^\alpha & |z|^\alpha \\ |z|^\alpha & |z|^\alpha \end{pmatrix} \quad \text{as } z \to 0,\ z \in \mathbb{C} \setminus \Sigma \cup \mathbb{R};$$

if $\alpha > 0$,

$$V(z) = \begin{cases} O\begin{pmatrix} |z|^\alpha & |z|^{-\alpha} \\ |z|^\alpha & |z|^{-\alpha} \end{pmatrix} & \text{as } z \to 0 \text{ for } z \text{ outside the lens regions,} \\ O\begin{pmatrix} |z|^{-\alpha} & |z|^{-\alpha} \\ |z|^{-\alpha} & |z|^{-\alpha} \end{pmatrix} & \text{as } z \to 0 \text{ for } z \text{ inside the lens regions.} \end{cases}$$

From Proposition 3.1 and the behavior of $\phi(z)$ in (3.33) near $z = \alpha_t$, the lens-shaped regions can be chosen sufficiently small so that

$$\operatorname{Re} \phi(z) < 0 \tag{4.12}$$

for $z \in \mathrm{I} \cup \mathrm{II}$. Similarly, it can be shown that

$$\operatorname{Re} \widetilde{\phi}(z) < 0 \tag{4.13}$$

for $z \in \mathrm{III} \cup \mathrm{IV}$. These together with (3.36) and (3.40) imply that the jump matrix $J_V(z)$ tends exponentially to the identity matrix as $n \to \infty$, for $z \in (-\infty, -\alpha_t) \cup (\alpha_t, \infty) \cup \Sigma$. When

$z \in (-\alpha_t, 0) \cup (0, \alpha_t)$, $J_V(z)$ is the constant matrix given by (4.11). It is therefore natural to suggest that for large $n$, the solution of the RHP for $V(z)$ may behave asymptotically like the solution of the following RHP for $V_\infty(z)$:

($V_{\infty,\mathrm{a}}$) $V_\infty(z)$ is analytic in $\mathbb{C} \setminus [-\alpha_t, \alpha_t]$;

($V_{\infty,\mathrm{b}}$) for $x \in (-\alpha_t, 0) \cup (0, \alpha_t)$,

$$(V_\infty)_+(x) = (V_\infty)_-(x) \begin{pmatrix} 0 & 1 \\ -1 & 0 \end{pmatrix};$$

($V_{\infty,\mathrm{c}}$) $V_\infty(z)$ behaves like the identity matrix at infinity:

$$V_\infty(z) = [I + O(1/z)] \quad \text{as } z \to \infty,$$

for $z \in \mathbb{C} \setminus [-\alpha_t, \alpha_t]$.

The construction of $V_\infty(z)$ was done in [9–11], and we have

$$V_\infty(z) = \begin{pmatrix} \dfrac{\beta(z) + \beta(z)^{-1}}{2} & \dfrac{\beta(z) - \beta(z)^{-1}}{2\mathrm{i}} \\ \dfrac{\beta(z) - \beta(z)^{-1}}{-2\mathrm{i}} & \dfrac{\beta(z) + \beta(z)^{-1}}{2} \end{pmatrix}, \tag{4.14}$$

where

$$\beta(z) = \left(\frac{z - \alpha_t}{z + \alpha_t}\right)^{\frac{1}{4}} \tag{4.15}$$

with a branch cut along $[-\alpha_t, \alpha_t]$ and $\beta(z) \to 1$ as $z \to \infty$. It is worthwhile to point out that $V_\infty(z)$ has the factorization

$$V_\infty(z) = \frac{1}{2}\begin{pmatrix} 1 & -1 \\ -\mathrm{i} & -\mathrm{i} \end{pmatrix} \beta(z)^{-\sigma_3} \begin{pmatrix} 1 & \mathrm{i} \\ -1 & \mathrm{i} \end{pmatrix} = \frac{1}{2}\begin{pmatrix} 1 & -1 \\ i & \mathrm{i} \end{pmatrix} \beta(z)^{\sigma_3} \begin{pmatrix} 1 & -\mathrm{i} \\ -1 & -\mathrm{i} \end{pmatrix}, \tag{4.16}$$

where $\sigma_3$ is the Pauli matrix given in (3.7). By (4.6) and (4.12)–(4.13), it is clear that $T \sim V$ as $n \to \infty$. Hence, it follows that

$$T(z) \sim V_\infty(z) \quad \text{for } z \in \mathbb{C} \setminus [-\alpha_t, \alpha_t]. \tag{4.17}$$

On account of (3.8), we can work backwards to get

$$U(z) \sim \mathrm{e}^{\frac{1}{2}Nl_N\sigma_3} V_\infty(z) \mathrm{e}^{[N(g(z) - \frac{1}{2}l_N) - \alpha \log z]\sigma_3}, \tag{4.18}$$

where $l_N$ is given in (3.32).

Let $L(z)$ represent the right-hand side of (4.18). By (3.47), it is easily verified that

$$g(z) - \log z + \pi\mathrm{i} = \widetilde{g}(z) - \log(-z) \tag{4.19}$$

for $z \in \mathbb{C} \setminus \mathbb{R}$, where $\log(-z)$ is analytic in $z \in \mathbb{C} \setminus [0, \infty)$. Hence, we obtain from (3.44), (3.47) and (4.19) that

$$\begin{aligned} L(z) &= e^{\frac{1}{2}Nl_N\sigma_3} V_\infty(z) e^{[ng(z)+\alpha(g(z)-\log z)]\sigma_3} e^{-\frac{1}{2}Nl_N\sigma_3} \\ &= e^{-\alpha\pi i\sigma_3} e^{\frac{1}{2}N\widetilde{l}_N\sigma_3} V_\infty(z) e^{n\widetilde{g}(z)\sigma_3} e^{\alpha(g(z)-\log z+\pi i)\sigma_3} e^{-\frac{1}{2}N\widetilde{l}_N\sigma_3} \\ &= e^{-\alpha\pi i\sigma_3} e^{\frac{1}{2}N\widetilde{l}_N\sigma_3} V_\infty(z) e^{[N(\widetilde{g}(z)-\frac{1}{2}\widetilde{l}_N)-\alpha\log(-z)]\sigma_3} \end{aligned} \tag{4.20}$$

for $z \in \mathbb{C} \setminus \mathbb{R}$.

### 4.2 Case for $c = -2$

Here, we follow the arguments given in Subsection 4.1, and transform the RHP for $T(z)$ into the RHP for $V(z)$ by opening a lens passing through $-\sqrt{2}, 0$ and $\sqrt{2}$ as shown in Figure 1. By the same reasoning given in Subsection 4.1, it is natural to suggest that for large $n$, the solution of the RHP for $V(z)$ should behave asymptotically like the solution of the following RHP for $V_\infty(z)$:

($V_{\infty,\mathrm{a}}$) $V_\infty(z)$ is analytic in $\mathbb{C} \setminus [-\sqrt{2}, \sqrt{2}]$;

($V_{\infty,\mathrm{b}}$) for $x \in (-\sqrt{2}, 0) \cup (0, \sqrt{2})$,

$$(V_\infty)_+(x) = (V_\infty)_-(x) \begin{pmatrix} 0 & 1 \\ -1 & 0 \end{pmatrix};$$

($V_{\infty,\mathrm{c}}$) $V_\infty(z)$ behaves like the identity matrix at infinity:

$$V_\infty(z) = [I + O(1/z)] \quad \text{as } z \to \infty.$$

This problem can be solved explicitly, and its solution is given by

$$V_\infty(z) = \begin{pmatrix} \dfrac{\beta(z)+\beta(z)^{-1}}{2} & \dfrac{\beta(z)-\beta(z)^{-1}}{2i} \\ \dfrac{\beta(z)-\beta(z)^{-1}}{-2i} & \dfrac{\beta(z)+\beta(z)^{-1}}{2} \end{pmatrix}, \tag{4.21}$$

where

$$\beta(z) = \left(\frac{z-\sqrt{2}}{z+\sqrt{2}}\right)^{\frac{1}{4}} \tag{4.22}$$

with a branch cut along $[-\sqrt{2}, \sqrt{2}]$ and $\beta(z) \to 1$ as $z \to \infty$ (see (4.14)–(4.15)). Again, we note that $V_\infty(z)$ has the factorization given in (4.16).

### 4.3 Case for $c < -2$

With the properties of $\phi(z)$, $\widetilde{\phi}(z)$ and $\overline{\phi}(z)$ established in Proposition 3.6, the jump matrix for $T(z)$ in condition ($T_\mathrm{b}$) can be written as follows:

$$T_+(x) = T_-(x) \begin{pmatrix} e^{2N\phi_+(x)} & 1 \\ 0 & e^{2N\phi_-(x)} \end{pmatrix} \tag{4.23}$$

for $x \in (a, b)$,

$$T_+(x) = T_-(x) \begin{pmatrix} e^{2N\tilde{\phi}_+(x)} & 1 \\ 0 & e^{2N\tilde{\phi}_-(x)} \end{pmatrix} \tag{4.24}$$

for $x \in (-b, -a)$,

$$T_+(x) = T_-(x) \begin{pmatrix} 1 & e^{-2N\phi(x)} \\ 0 & 1 \end{pmatrix} \tag{4.25}$$

for $x \in (b, \infty)$,

$$T_+(x) = T_-(x) \begin{pmatrix} 1 & e^{-2N\tilde{\phi}(x)} \\ 0 & 1 \end{pmatrix} \tag{4.26}$$

for $x \in (-\infty, -b)$,

$$T_+(x) = T_-(x) \begin{pmatrix} e^{n\pi i + \alpha\pi i} & e^{-2N\bar{\phi}(x)} \\ 0 & e^{-n\pi i - \alpha\pi i} \end{pmatrix} \tag{4.27}$$

for $x \in (-a, 0)$,

$$T_+(x) = T_-(x) \begin{pmatrix} e^{-n\pi i - \alpha\pi i} & e^{-2N\bar{\phi}(x)} \\ 0 & e^{n\pi i + \alpha\pi i} \end{pmatrix} \tag{4.28}$$

for $x \in (0, a)$.

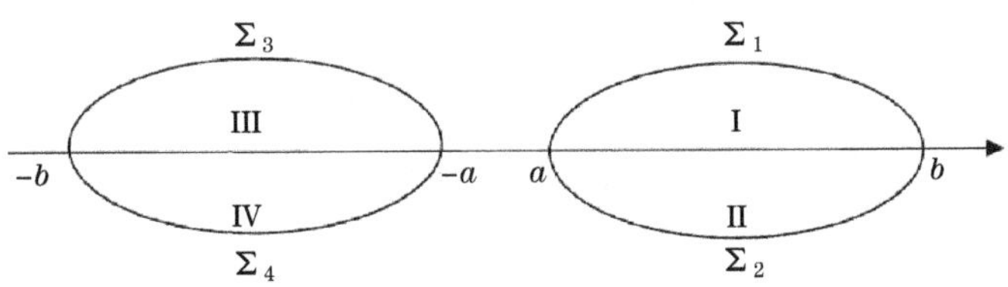


Figure 2 The lens-shaped contour $\Sigma = \Sigma_1 \cup \cdots \cup \Sigma_4$.

From (4.5), the transformation $T \to V$ is then defined by

$$V(z) := \begin{cases} T(z) & \text{for } z \text{ outside the lens}, \\ T(z) \begin{pmatrix} 1 & 0 \\ -e^{2N\phi(z)} & 1 \end{pmatrix} & \text{for } z \in \mathrm{I}, \\ T(z) \begin{pmatrix} 1 & 0 \\ e^{2N\phi(z)} & 1 \end{pmatrix} & \text{for } z \in \mathrm{II}, \\ T(z) \begin{pmatrix} 1 & 0 \\ -e^{2N\tilde{\phi}(z)} & 1 \end{pmatrix} & \text{for } z \in \mathrm{III}, \\ T(z) \begin{pmatrix} 1 & 0 \\ e^{2N\tilde{\phi}(z)} & 1 \end{pmatrix} & \text{for } z \in \mathrm{IV}. \end{cases} \tag{4.29}$$

Furthermore, we define the jump matrix

$$J_V(z) := \begin{cases} \begin{pmatrix} 1 & 0 \\ e^{2N\phi(z)} & 1 \end{pmatrix}, & z \in \Sigma_1 \cup \Sigma_2, \\ \begin{pmatrix} 1 & 0 \\ e^{2N\widetilde{\phi}(z)} & 1 \end{pmatrix}, & z \in \Sigma_3 \cup \Sigma_4, \\ \begin{pmatrix} 1 & e^{-2N\widetilde{\phi}(x)} \\ 0 & 1 \end{pmatrix}, & x < -b, \\ \begin{pmatrix} 1 & e^{-2N\phi(x)} \\ 0 & 1 \end{pmatrix}, & x > b, \\ \begin{pmatrix} 0 & 1 \\ -1 & 0 \end{pmatrix}, & x \in (-b,-a) \cup (a,b), \\ \begin{pmatrix} e^{n\pi i+\alpha\pi i} & e^{-2N\overline{\phi}(x)} \\ 0 & e^{-n\pi i-\alpha\pi i} \end{pmatrix}, & x \in (-a,0), \\ \begin{pmatrix} e^{-n\pi i-\alpha\pi i} & e^{-2N\overline{\phi}(x)} \\ 0 & e^{n\pi i+\alpha\pi i} \end{pmatrix}, & x \in (0,a). \end{cases} \tag{4.30}$$

Since $\overline{\phi}(x) > 0$ for $x \in (-a,a)$, $\widetilde{\phi}(x) > 0$ for $x < -b$ and $\phi(x) > 0$ for $x > b$, as in the previous cases we expect the solution of the RHP for $V(z)$ to behave asymptotically like the solution of the following RHP for $V_\infty(z)$:

($V_{\infty,\mathrm{a}}$) $V_\infty(z)$ is analytic in $\mathbb{C} \setminus [-b,b]$;

($V_{\infty,\mathrm{b}}$) for $x \in E$,

$$(V_\infty)_+(x) = (V_\infty)_-(x) \begin{pmatrix} 0 & 1 \\ -1 & 0 \end{pmatrix},$$

for $x \in (-a,0)$,

$$(V_\infty)_+(x) = (V_\infty)_-(x) \begin{pmatrix} e^{n\pi i+\alpha\pi i} & 0 \\ 0 & e^{-n\pi i-\alpha\pi i} \end{pmatrix},$$

and for $x \in (0,a)$,

$$(V_\infty)_+(x) = (V_\infty)_-(x) \begin{pmatrix} e^{-n\pi i-\alpha\pi i} & 0 \\ 0 & e^{n\pi i+\alpha\pi i} \end{pmatrix};$$

($V_{\infty,\mathrm{c}}$) for $z \in \mathbb{C} \setminus E$,

$$V_\infty(z) = [I + O(1/z)] \quad \text{as } z \to \infty.$$

We construct $V_\infty(z)$ in terms of the Szegö function $D$ associated with $e^{\alpha\pi i}$ on $E$, which is an analytic function on $\mathbb{C}\setminus[-b,b]$ satisfying $D_+(x)D_-(x) = 1$ for $x \in E$, and $D_+(x) = e^{-\pi i\alpha}D_-(x)$ for $x \in [-a,0]$ and $D_+(x) = e^{\pi i\alpha}D_-(x)$ for $x \in [0,a]$. We seek the function $D(z)$ in the form $D(z) = \exp(\widetilde{D})$. Then the problem is reduced to constructing a scalar function $\widetilde{D}(z)$, which is an analytic function on $\mathbb{C}\setminus E$ satisfying $\widetilde{D}_+(x)+\widetilde{D}_-(x) = 0$ for $x \in E$, and $\widetilde{D}_+(x) = \widetilde{D}_-(x)-\pi i\alpha$ for $x \in (-a,0)$ and $\widetilde{D}_+(x) = \widetilde{D}_-(x) + \pi i\alpha$ for $x \in (0,a)$. We can solve this scalar RHP by the Plemelj formula to yield

$$\widetilde{D}(z) = -\frac{\alpha\sqrt{(z^2-a^2)(z^2-b^2)}}{2}\left(\int_0^{a^2} \frac{1}{\sqrt{(a^2-x)(b^2-x)}}\frac{dx}{x-z^2}\right).$$

This gives

$$D(z) = \exp\Big[ -\frac{\alpha\sqrt{(z^2-a^2)(z^2-b^2)}}{2}\Big(\int_0^{a^2} \frac{1}{\sqrt{(a^2-x)(b^2-x)}}\frac{dx}{x-z^2}\Big)\Big] \tag{4.31}$$

and

$$D_\infty = \lim_{z\to\infty} D(z) = \Big(\frac{b+a}{b-a}\Big)^{\frac{\alpha}{2}}. \tag{4.32}$$

Therefore, the RHP for $V_\infty(z)$ can be transformed into the RHP for $\widetilde{V}_\infty(z)$ with the jump matrix $\begin{pmatrix} 0 & 1 \\ -1 & 0 \end{pmatrix}$ on $E$, via the formula

$$\widetilde{V}_\infty(z) = D_\infty^{-\sigma_3} V_\infty(z) D(z)^{\sigma_3}.$$

Thus, we have the RHP for $\widetilde{V}_\infty(z)$:

($\widetilde{V}_{\infty,\mathrm{a}}$) $\widetilde{V}_\infty(z)$ is analytic in $\mathbb{C}\setminus[-b,b]$;

($\widetilde{V}_{\infty,\mathrm{b}}$) for $x\in E$,

$$(\widetilde{V}_\infty)_+(x) = (\widetilde{V}_\infty)_-(x)\begin{pmatrix} 0 & 1 \\ -1 & 0 \end{pmatrix},$$

and for $x\in(-a,a)\setminus\{0\}$,

$$(\widetilde{V}_\infty)_+(x) = (\widetilde{V}_\infty)_-(x)\begin{pmatrix} e^{-n\pi i} & 0 \\ 0 & e^{n\pi i} \end{pmatrix};$$

($\widetilde{V}_{\infty,\mathrm{c}}$) for $z\in\mathbb{C}\setminus E$,

$$\widetilde{V}_\infty(z) = [I + O(1/z)] \quad \text{as } z\to\infty.$$

The RHP for $\widetilde{V}_\infty$ can be solved explicitly by

$$\widetilde{V}_\infty(z) = \begin{pmatrix} \dfrac{\widetilde{\gamma}(z)+\widetilde{\gamma}(z)^{-1}}{2} & \dfrac{\widetilde{\gamma}(z)-\widetilde{\gamma}(z)^{-1}}{-2i} \\ \dfrac{\widetilde{\gamma}(z)-\widetilde{\gamma}(z)^{-1}}{2i} & \dfrac{\widetilde{\gamma}(z)+\widetilde{\gamma}(z)^{-1}}{2} \end{pmatrix}, \tag{4.33}$$

where

$$\widetilde{\gamma}(z) = \Big(\frac{z+b}{z-b}\Big)^{\frac14}\Big(\frac{z-a}{z+a}\Big)^{\frac{(-1)^n}{4}} \tag{4.34}$$

with a branch cut along $E$ and $\widetilde{\gamma}(z)\to 1$ as $z\to\infty$. (This can actually be derived by using an elliptic theta function (see [16, pp. 15–16]). It is worthwhile to note that $\widetilde{V}_\infty(z)$ can be written as

$$\widetilde{V}_\infty(z) = \frac12\begin{pmatrix} 1 & -1 \\ -i & -i \end{pmatrix}\widetilde{\gamma}(z)^{\sigma_3}\begin{pmatrix} 1 & i \\ -1 & i \end{pmatrix} = \frac12\begin{pmatrix} 1 & -1 \\ i & i \end{pmatrix}\widetilde{\gamma}(z)^{-\sigma_3}\begin{pmatrix} 1 & -i \\ -1 & -i \end{pmatrix}, \tag{4.35}$$

where $\sigma_3$ is the Pauli matrix given in (3.7). Since $T\sim V$ as $n\to\infty$, it follows that

$$T(z) \sim D_\infty^{\sigma_3}\widetilde{V}_\infty(z)D(z)^{-\sigma_3} \quad \text{for } z\in\mathbb{C}\setminus E \tag{4.36}$$

(see (4.17)). On account of (3.8), we can work backwards to get

$$U(z) \sim e^{\frac12 N l_N\sigma_3} D_\infty^{\sigma_3}\widetilde{V}_\infty(z)D(z)^{-\sigma_3}e^{N(g(z)-\frac12 l_N-\alpha\log z)\sigma_3}, \tag{4.37}$$

where $l_N$ is given in (3.96).

# 5 Construction of the Parametrices

In this section, we will construct an approximation $U^*(z)$ to the solution of the RHP for $U(z)$ for large $n$. We again divide our discussion into three cases. For the first case, we bring in Bessel functions in the formation of a parametrix in a region containing the origin. For the second case, we introduce solutions associated with Painlevé II equation and construct the model RHP in the formation of a parametrix in a region containing the origin. For the third case, no special attention is required in the neighborhood of the origin, since it now lies outside the support of the equilibrium measure.

For regions outside the origin, we make use of Airy functions in all three cases. We give more details for the first case, and present only the conclusions for the two other cases.

## 5.1 Case for $c > -2$

Due to the singularity of $|x|^{2\alpha}$ in the weight function, special attention must be paid to the neighborhood of the origin, which we will discuss first. The arguments in this subsection parallel those in [20].

### 5.1.1 Parametrix in the neighborhood of the origin

Let $U_\delta$ be a small disk with center at 0 and radius $\delta > 0$. We seek a $2 \times 2$ matrix-valued function $P(z)$ in $U_\delta$, which has the same jumps as $U(z)$ and matches the behavior of $U(z)$ on the boundary $\partial U_\delta$ of the disk (see (4.18)). That is, we wish to find a $2 \times 2$ matrix-valued function that satisfies the following RHP:

($P_\mathrm{a}$) $P(z)$ is analytic in $U_\delta \setminus \mathbb{R}$;

($P_\mathrm{b}$) for $x \in (-\delta, \delta) \setminus \{0\}$,

$$P_+(x) = P_-(x) \begin{pmatrix} 1 & |x|^{2\alpha} e^{-(Nx^4 + t\sqrt{N}x^2)} \\ 0 & 1 \end{pmatrix};$$

($P_\mathrm{c}$) for $z \in \partial U_\delta \setminus \mathbb{R}$, $P(z)$ satisfies the matching condition

$$P(z) \sim e^{\frac{1}{2}Nl_N\sigma_3} V_\infty(z) e^{[N(g(z) - \frac{1}{2}l_N) - \alpha \log z]\sigma_3}$$

as $n \to \infty$.

Denote by

$$\widetilde{w}(z) := z^{2\alpha} e^{-(Nz^4 + t\sqrt{N}z^2)} \tag{5.1}$$

the analytic continuation of the function $|x|^{2\alpha} e^{-(Nx^4 + t\sqrt{N}x^2)}$ to the whole complex plane with a branch cut along the negative real-axis. It is easily seen that the jump matrix in ($P_\mathrm{b}$) has the following factorization:

$$\begin{pmatrix} 1 & |x|^{2\alpha} e^{-(Nx^4 + t\sqrt{N}x^2)} \\ 0 & 1 \end{pmatrix} = \begin{cases} \widetilde{w}(x)^{\frac{\sigma_3}{2}} \begin{pmatrix} 1 & 1 \\ 0 & 1 \end{pmatrix} \widetilde{w}(x)^{-\frac{\sigma_3}{2}} & \text{for } x > 0, \\ \widetilde{w}_-(x)^{\frac{\sigma_3}{2}} \begin{pmatrix} e^{2\pi\alpha i} & 1 \\ 0 & e^{-2\pi\alpha i} \end{pmatrix} \widetilde{w}_+(x)^{-\frac{\sigma_3}{2}} & \text{for } x < 0. \end{cases} \tag{5.2}$$

In order to solve the RHP for $P(z)$, we first transform it into the RHP for

$$\widetilde{P}(z) := P(z)\widetilde{w}(z)^{\frac{\sigma_3}{2}} = P(z) z^{\alpha\sigma_3} e^{-\frac{\sigma_3}{2}(Nz^4 + t\sqrt{N}z^2)}. \tag{5.3}$$

Clearly, the function on the right-hand side of $P(z)$ in (5.3) is the analytic continuation of the function $|x|^\alpha \mathrm{e}^{-\frac{Nx^4+t\sqrt{N}x^2}{2}}$ to the whole complex plane with a branch cut along the negative real-axis. Now, let us consider the RHP for $\widetilde{P}$:

($\widetilde{P}_{\mathrm{a}}$) $\widetilde{P}(z)$ is analytic in $U_\delta \setminus \mathbb{R}$;

($\widetilde{P}_{\mathrm{b}}$) for $x \in (-\delta, \delta) \setminus \{0\}$,

$$\widetilde{P}_+(x) = \widetilde{P}_-(x) \begin{cases} \begin{pmatrix} 1 & 1 \\ 0 & 1 \end{pmatrix} & \text{for } x \in (0, \delta), \\ \begin{pmatrix} \mathrm{e}^{2\pi\alpha \mathrm{i}} & 1 \\ 0 & \mathrm{e}^{-2\pi\alpha \mathrm{i}} \end{pmatrix} & \text{for } x \in (-\delta, 0); \end{cases}$$

($\widetilde{P}_{\mathrm{c}}$) $\widetilde{P}(z)$ satisfies the matching condition

$$\widetilde{P}(z) \sim \mathrm{e}^{\frac{1}{2}Nl_N\sigma_3} V_\infty(z) \mathrm{e}^{-N\phi(z)\sigma_3}$$

as $n \to \infty$ (see (3.45) and (4.18)).

To solve the above RHP for $\widetilde{P}(z)$, we first consider the matrix-valued function

$$\Phi_\alpha(\zeta) = \begin{cases} \zeta^{\frac{1}{2}} \begin{pmatrix} \sqrt{\pi} I_{\alpha+\frac{1}{2}}(\zeta \mathrm{e}^{-\frac{1}{2}\pi \mathrm{i}}) & -\dfrac{K_{\alpha+\frac{1}{2}}(\zeta \mathrm{e}^{-\frac{1}{2}\pi \mathrm{i}})}{\sqrt{\pi}} \\ -\mathrm{i}\sqrt{\pi} I_{\alpha-\frac{1}{2}}(\zeta \mathrm{e}^{-\frac{1}{2}\pi \mathrm{i}}) & -\dfrac{\mathrm{i}K_{\alpha-\frac{1}{2}}(\zeta \mathrm{e}^{-\frac{1}{2}\pi \mathrm{i}})}{\sqrt{\pi}} \end{pmatrix} \mathrm{e}^{\frac{1}{2}\pi\alpha \mathrm{i}\sigma_3}, & \operatorname{Im} \zeta > 0, \\ \zeta^{\frac{1}{2}} \begin{pmatrix} -\mathrm{i}\sqrt{\pi} I_{\alpha+\frac{1}{2}}(\zeta \mathrm{e}^{-\frac{1}{2}\pi \mathrm{i}}) & -\dfrac{\mathrm{i}K_{\alpha+\frac{1}{2}}(\zeta \mathrm{e}^{-\frac{1}{2}\pi \mathrm{i}})}{\sqrt{\pi}} \\ \sqrt{\pi} I_{\alpha-\frac{1}{2}}(\zeta \mathrm{e}^{-\frac{1}{2}\pi \mathrm{i}}) & -\dfrac{K_{\alpha-\frac{1}{2}}(\zeta \mathrm{e}^{-\frac{1}{2}\pi \mathrm{i}})}{\sqrt{\pi}} \end{pmatrix} \mathrm{e}^{-\frac{1}{2}\pi\alpha \mathrm{i}\sigma_3}, & \operatorname{Im} \zeta < 0, \end{cases} \tag{5.4}$$

where $\zeta^{\frac{1}{2}}$ takes the branch cut along the negative real-axis, and $I_\nu$, $K_\nu$ are modified Bessel functions defined in the complex plane with cut along the negative real-axis. Recall that

$$I_\nu(z\mathrm{e}^{m\pi \mathrm{i}}) = \mathrm{e}^{m\nu\pi \mathrm{i}} I_\nu(z),$$
$$K_\nu(z\mathrm{e}^{m\pi \mathrm{i}}) = \mathrm{e}^{-m\nu\pi \mathrm{i}} K_\nu(z) - \pi \mathrm{i} \sin(m\nu\pi) \csc(\nu\pi) I_\nu(z),$$

where $m$ is an integer. It is easily shown that

$$(\Phi_\alpha)_+(\zeta) = (\Phi_\alpha)_-(\zeta) \begin{cases} \begin{pmatrix} 1 & 1 \\ 0 & 1 \end{pmatrix} & \text{for } \zeta > 0, \\ \begin{pmatrix} \mathrm{e}^{2\pi\alpha \mathrm{i}} & 1 \\ 0 & \mathrm{e}^{-2\pi\alpha \mathrm{i}} \end{pmatrix} & \text{for } \zeta < 0. \end{cases} \tag{5.5}$$

Furthermore, from the well-known formulas

$$I_\nu(z) \sim \frac{1}{\sqrt{2\pi z}} \mathrm{e}^z, \quad K_\nu(z) \sim \sqrt{\frac{\pi}{2z}} \mathrm{e}^{-z} \tag{5.6}$$

as $z \to \infty$ in $|\arg z| < \frac{\pi}{2}$, we have

$$\Phi_\alpha(\zeta) \sim \begin{cases} \frac{1}{\sqrt{2}} \begin{pmatrix} 1 & -\mathrm{i} \\ -\mathrm{i} & 1 \end{pmatrix} \mathrm{e}^{(\frac{\pi}{4}i - \zeta i + \frac{1}{2}\pi\alpha \mathrm{i})\sigma_3}, & \mathrm{Im}\,\zeta > 0 \\ \frac{1}{\sqrt{2}} \begin{pmatrix} 1 & -\mathrm{i} \\ -\mathrm{i} & 1 \end{pmatrix} \mathrm{e}^{(\frac{\pi}{4}i - \zeta i + \frac{1}{2}\pi\alpha \mathrm{i})\sigma_3} \begin{pmatrix} 0 & -1 \\ 1 & 0 \end{pmatrix}, & \mathrm{Im}\,\zeta < 0 \end{cases} \tag{5.7}$$

as $\zeta \to \infty$. Next, we introduce the function

$$\eta_N(z) := \begin{cases} -\mathrm{i}\phi(z) + \mathrm{i}\phi_+(0), & \mathrm{Im}\, z > 0, \\ \mathrm{i}\phi(z) + \mathrm{i}\phi_+(0), & \mathrm{Im}\, z < 0, \end{cases} \tag{5.8}$$

which is analytic in a neighborhood of the origin by Proposition 3.1. In fact, it follows from (3.23) and (3.33) that

$$\eta_N(z) = \alpha_t(\alpha_t^2 + tN^{-\frac{1}{2}})z + O(z^2) \tag{5.9}$$

as $z \to 0$. A comparison of conditions $(\widetilde{P}_\mathrm{b})$ and $(\widetilde{P}_\mathrm{c})$ for $\widetilde{P}(z)$ with (5.5) and (5.7) shows that

$$\widetilde{P}(z) = \frac{1}{\sqrt{2}} \mathrm{e}^{\frac{1}{2}Nl_N\sigma_3} E_0(z) \begin{pmatrix} 1 & \mathrm{i} \\ \mathrm{i} & 1 \end{pmatrix} \Phi_\alpha[N\eta_N(z)], \tag{5.10}$$

where

$$E_0(z) = V_\infty(z)\mathrm{e}^{-(\frac{1}{2}\pi\alpha\mathrm{i} + N\phi_+(0) + \frac{1}{4}\pi\mathrm{i})\sigma_3} \tag{5.11}$$

for $z \in \mathbb{C}_+$ and

$$E_0(z) = V_\infty(z) \begin{pmatrix} 0 & 1 \\ -1 & 0 \end{pmatrix} \mathrm{e}^{-(\frac{1}{2}\pi\alpha\mathrm{i} + N\phi_+(0) + \frac{1}{4}\pi\mathrm{i})\sigma_3} \tag{5.12}$$

for $z \in \mathbb{C}_-$. The jump condition $(V_{\infty,\mathrm{b}})$ for $(V_\infty)$ implies that $E_0(z)$ is actually analytic in $U_\delta$. Indeed, with the aid of (4.14) and (5.11)–(5.12), we have the explicit formula

$$E_0(z) = \frac{1}{2} \begin{pmatrix} b(z) - ib(z)^{-1} & b(z) + ib(z)^{-1} \\ ib(z) - b(z)^{-1} & ib(z) + b(z)^{-1} \end{pmatrix} \mathrm{e}^{-(\frac{1}{2}\pi\alpha\mathrm{i} + N\phi_+(0))\sigma_3}, \tag{5.13}$$

where

$$b(z) = \left(\frac{\alpha_t - z}{\alpha_t + z}\right)^{\frac{1}{4}}$$

is analytic in $\mathbb{C} \setminus (-\infty, -\alpha_t] \cup [\alpha_t, \infty)$, where $\alpha_t$ is shown in (3.18). Finally, a combination of (5.3) and (5.10) gives

$$P(z) = \frac{1}{\sqrt{2}} \mathrm{e}^{\frac{1}{2}Nl_N\sigma_3} E_0(z) \begin{pmatrix} 1 & \mathrm{i} \\ \mathrm{i} & 1 \end{pmatrix} \Phi_\alpha[N\eta_N(z)] \mathrm{e}^{\frac{\sigma_3}{2}(Nz^4 + tz^2N^{\frac{1}{2}})} z^{-\alpha\sigma_3}. \tag{5.14}$$

### 5.1.2 Parametrices outside $U_\delta$

For $z$ outside the $\delta$-neighborhood $U_\delta$ of the origin, we will construct the parametrices by using Airy functions and elementary functions. To facilitate the following discussions, we divide the complex plane into four parts: $U_\delta$ and $\Omega_i$, $i = 1, 2, 3$, by the contours $\Gamma := \bigcup_{i=1}^{8} \Gamma_i$ (see Figure

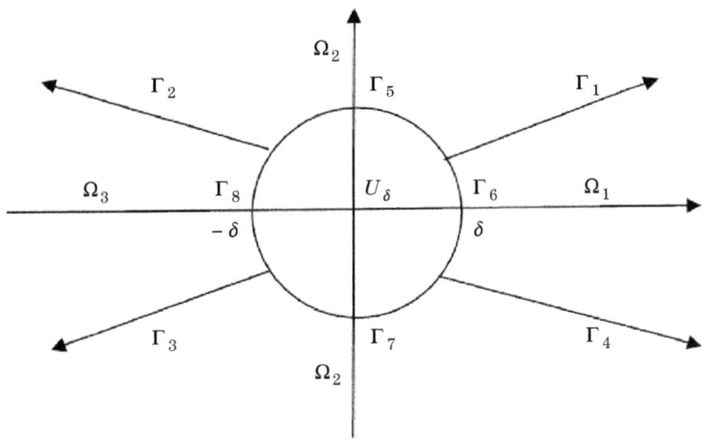


Figure 3 Contour $\Gamma_i$, $i = 1, \cdots, 8$ and the domains $\Omega_1, \Omega_2, \Omega_3$.

3). Note that $\partial U_\delta = \bigcup_{i=5}^{8} \Gamma_i$, and $\Gamma_3 \cup \Gamma_4$ is the complex conjugation of $\Gamma_1 \cup \Gamma_2$. The curve $\Gamma_1$ is chosen so that the function $\phi(z)$ is one-to-one in $\Omega_1 \cap \mathbb{C}_+$ and satisfies $0 < \arg\ \phi(z) < \frac{3}{2}\pi$ for $z \in \Omega_1 \cap \mathbb{C}_+$. Similarly, we choose $\Gamma_2$ such that $-\frac{3}{2}\pi < \arg\ \widetilde{\phi}(z) < 0$ for $z \in \Omega_3 \cap \mathbb{C}_+$.

In view of (4.16), we can rewrite (4.18) as

$$U(z) \sim \frac{1}{2} \mathrm{e}^{\frac{1}{2} N l_N \sigma_3} \begin{pmatrix} 1 & -1 \\ -\mathrm{i} & -\mathrm{i} \end{pmatrix} \beta(z)^{-\sigma_3} \begin{pmatrix} \mathrm{e}^{-N\phi(z)} & \mathrm{i}\mathrm{e}^{N\phi(z)} \\ -\mathrm{e}^{-N\phi(z)} & \mathrm{i}\mathrm{e}^{N\phi(z)} \end{pmatrix} \mathrm{e}^{[N(g(z)+\phi(z)-\frac{1}{2}l_N)-\alpha \log z]\sigma_3}$$

and from (3.45) we obtain

$$U(z) \sim \frac{1}{2} \mathrm{e}^{\frac{1}{2} N l_N \sigma_3} \begin{pmatrix} 1 & -1 \\ -\mathrm{i} & -\mathrm{i} \end{pmatrix} \beta(z)^{-\sigma_3} \begin{pmatrix} \mathrm{e}^{-N\phi(z)} & \mathrm{i}\mathrm{e}^{N\phi(z)} \\ -\mathrm{e}^{-N\phi(z)} & \mathrm{i}\mathrm{e}^{N\phi(z)} \end{pmatrix} \mathrm{e}^{[\frac{1}{2}Nz^4 + t\sqrt{N}z^2 - \alpha \log z]\sigma_3}. \tag{5.15}$$

To find an approximation to $U(z)$, we first look for a matrix which is asymptotic to

$$\begin{pmatrix} \mathrm{e}^{-N\phi(z)} & \mathrm{i}\mathrm{e}^{N\phi(z)} \\ -\mathrm{e}^{-N\phi(z)} & \mathrm{i}\mathrm{e}^{N\phi(z)} \end{pmatrix}. \tag{5.16}$$

From Proposition 3.1, it is clear that the function defined by

$$\xi_N = f_N(z) = \left(\frac{3}{2}\phi(z)\right)^{\frac{2}{3}} \tag{5.17}$$

is analytic in $\mathbb{C} \setminus (-\infty, -\alpha_t]$, where $\phi(z)$ is defined in (3.33) and depends on $N$. In particular, by the construction of $\Omega_1$, we have for $z \in \Omega_1 \cap \mathbb{C}_+$,

$$0 < \arg f_N(z) < \pi. \tag{5.18}$$

Also, for $z \neq \alpha_t$, $N^{\frac{2}{3}} \xi_N = N^{\frac{2}{3}} f_N(z) \to \infty$ as $n \to \infty$. From the asymptotic behavior of the Airy function (see [17, p. 392]), we have

$$\mathrm{Ai}(N^{\frac{2}{3}} \xi_N) \sim \frac{N^{-\frac{1}{6}}}{2\sqrt{\pi}} \xi_N^{-\frac{1}{4}} \mathrm{e}^{-\frac{2}{3} N \xi_N^{\frac{3}{2}}} = \frac{N^{-\frac{1}{6}}}{2\sqrt{\pi}} (f_N(z))^{-\frac{1}{4}} \mathrm{e}^{-N\phi(z)},$$

$$\mathrm{Ai}'(N^{\frac{2}{3}}\xi_N) \sim -\frac{N^{\frac{1}{6}}}{2\sqrt{\pi}}\xi_N^{\frac{1}{4}}\mathrm{e}^{-\frac{2}{3}N\xi_N^{\frac{3}{2}}} = -\frac{N^{\frac{1}{6}}}{2\sqrt{\pi}}(f_N(z))^{\frac{1}{4}}\mathrm{e}^{-N\phi(z)},$$

$$\mathrm{Ai}(N^{\frac{2}{3}}w^2\xi_N) = \mathrm{Ai}(N^{\frac{2}{3}}w^{-1}\xi_N) \sim \frac{N^{-\frac{1}{6}}}{2\sqrt{\pi}}\mathrm{e}^{\frac{\mathrm{i}\pi}{6}}(f_N(z))^{-\frac{1}{4}}\mathrm{e}^{N\phi(z)},$$

$$\mathrm{Ai}'(N^{\frac{2}{3}}w^2\xi_N) = \mathrm{Ai}'(N^{\frac{2}{3}}w^{-1}\xi_N) \sim -\frac{N^{\frac{1}{6}}}{2\sqrt{\pi}}\mathrm{e}^{-\frac{\mathrm{i}\pi}{6}}(f_N(z))^{\frac{1}{4}}\mathrm{e}^{N\phi(z)},$$

where $w = \mathrm{e}^{\frac{2\pi\mathrm{i}}{3}}$. It is then immediate that for $z \in \Omega_1 \cap \mathbb{C}_+$,

$$\begin{pmatrix} \mathrm{e}^{-N\phi(z)} & \mathrm{i}\mathrm{e}^{N\phi(z)} \\ -\mathrm{e}^{-N\phi(z)} & \mathrm{i}\mathrm{e}^{N\phi(z)} \end{pmatrix} \sim 2\sqrt{\pi}(N^{\frac{2}{3}}f_N(z))^{\frac{1}{4}\sigma_3}\begin{pmatrix} \mathrm{Ai}(N^{\frac{2}{3}}\xi_N) & -w^2\mathrm{Ai}(N^{\frac{2}{3}}w^2\xi_N) \\ \mathrm{Ai}'(N^{\frac{2}{3}}\xi_N) & -w\mathrm{Ai}'(N^{\frac{2}{3}}w^2\xi_N) \end{pmatrix}. \tag{5.19}$$

Similarly, for $z \in \Omega_1 \cap \mathbb{C}_-$, we also have $-\pi < \arg f_N(z) < 0$ and

$$\begin{pmatrix} \mathrm{e}^{-N\phi(z)} & \mathrm{i}\mathrm{e}^{N\phi(z)} \\ -\mathrm{e}^{-N\phi(z)} & \mathrm{i}\mathrm{e}^{N\phi(z)} \end{pmatrix} \sim 2\sqrt{\pi}(N^{\frac{2}{3}}f_N(z))^{\frac{1}{4}\sigma_3}\begin{pmatrix} \mathrm{Ai}(N^{\frac{2}{3}}\xi_N) & w\mathrm{Ai}(N^{\frac{2}{3}}w\xi_N) \\ \mathrm{Ai}'(N^{\frac{2}{3}}\xi_N) & w^2\mathrm{Ai}'(N^{\frac{2}{3}}w\xi_N) \end{pmatrix}. \tag{5.20}$$

For $z \in \Omega_3$, there is a result corresponding to (5.15) with $\phi(z)$ and $l_N$ replaced by $\widetilde{\phi}(z)$ and $\widetilde{l}_N$, respectively. Indeed, it follows from (3.46), (4.16) and (4.20) that

$$\begin{aligned} U(z) \sim \frac{1}{2}\mathrm{e}^{-\alpha\pi\mathrm{i}\sigma_3}\mathrm{e}^{\frac{1}{2}N\widetilde{l}_N\sigma_3}\begin{pmatrix} 1 & -1 \\ \mathrm{i} & \mathrm{i} \end{pmatrix}\beta(z)^{\sigma_3}\begin{pmatrix} \mathrm{e}^{-N\widetilde{\phi}(z)} & -\mathrm{i}\mathrm{e}^{N\widetilde{\phi}(z)} \\ -\mathrm{e}^{-N\widetilde{\phi}(z)} & -\mathrm{i}\mathrm{e}^{N\widetilde{\phi}(z)} \end{pmatrix} \\ \times \mathrm{e}^{[\frac{1}{2}Nz^4+t\sqrt{N}z^2-\alpha\log(-z)]\sigma_3}. \end{aligned} \tag{5.21}$$

Let

$$\widetilde{\xi}_N = \widetilde{f}_N(z) = \left(\frac{3}{2}\widetilde{\phi}(z)\right)^{\frac{2}{3}}, \tag{5.22}$$

which is analytic in $\mathbb{C} \setminus [\alpha_t, \infty)$. Also, note that

$$|\arg \widetilde{f}_N(z)| < \pi \tag{5.23}$$

for $z \in \Omega_3$. Hence, as before, it can be shown that the matrix

$$\begin{pmatrix} \mathrm{e}^{-N\widetilde{\phi}(z)} & -\mathrm{i}\mathrm{e}^{N\widetilde{\phi}(z)} \\ -\mathrm{e}^{-N\widetilde{\phi}(z)} & -\mathrm{i}\mathrm{e}^{N\widetilde{\phi}(z)} \end{pmatrix}$$

is the leading term in the asymptotic expansion of the matrices

$$2\sqrt{\pi}(N^{\frac{2}{3}}\widetilde{f}_N(z))^{\frac{1}{4}\sigma_3}\begin{pmatrix} \mathrm{Ai}(N^{\frac{2}{3}}\widetilde{\xi}_N) & -w\mathrm{Ai}(N^{\frac{2}{3}}w\widetilde{\xi}_N) \\ \mathrm{Ai}'(N^{\frac{2}{3}}\widetilde{\xi}_N) & -w^2\mathrm{Ai}'(N^{\frac{2}{3}}w\widetilde{\xi}_N) \end{pmatrix} \tag{5.24}$$

for $z \in \Omega_3 \in \mathbb{C}_+$, and

$$2\sqrt{\pi}(N^{\frac{2}{3}}\widetilde{f}_N(z))^{\frac{1}{4}\sigma_3}\begin{pmatrix} \mathrm{Ai}(N^{\frac{2}{3}}\widetilde{\xi}_N) & w^2\mathrm{Ai}(N^{\frac{2}{3}}w^2\widetilde{\xi}_N) \\ \mathrm{Ai}'(N^{\frac{2}{3}}\widetilde{\xi}_N) & w\mathrm{Ai}'(N^{\frac{2}{3}}w^2\widetilde{\xi}_N) \end{pmatrix} \tag{5.25}$$

for $z \in \Omega_3 \in \mathbb{C}_-$. Define the matrix function

$$Q(z) := \begin{cases} \begin{pmatrix} \mathrm{Ai}(N^{\frac{2}{3}}\xi_N) & -w^2\mathrm{Ai}(N^{\frac{2}{3}}w^2\xi_N) \\ \mathrm{Ai}'(N^{\frac{2}{3}}\xi_N) & -w\mathrm{Ai}'(N^{\frac{2}{3}}w^2\xi_N) \end{pmatrix}, & z \in \Omega_1 \cap \mathbb{C}_+, \\ \begin{pmatrix} \mathrm{Ai}(N^{\frac{2}{3}}\xi_N) & w\mathrm{Ai}(N^{\frac{2}{3}}w\xi_N) \\ \mathrm{Ai}'(N^{\frac{2}{3}}\xi_N) & w^2\mathrm{Ai}'(N^{\frac{2}{3}}w\xi_N) \end{pmatrix}, & z \in \Omega_1 \cap \mathbb{C}_-, \\ \begin{pmatrix} \mathrm{Ai}(N^{\frac{2}{3}}\widetilde{\xi}_N) & -w\mathrm{Ai}(N^{\frac{2}{3}}w\widetilde{\xi}_N) \\ \mathrm{Ai}'(N^{\frac{2}{3}}\widetilde{\xi}_N) & -w^2\mathrm{Ai}'(N^{\frac{2}{3}}w\widetilde{\xi}_N) \end{pmatrix}, & z \in \Omega_3 \cap \mathbb{C}_+, \\ \begin{pmatrix} \mathrm{Ai}(N^{\frac{2}{3}}\widetilde{\xi}_N) & w^2\mathrm{Ai}(N^{\frac{2}{3}}w^2\widetilde{\xi}_N) \\ \mathrm{Ai}'(N^{\frac{2}{3}}\widetilde{\xi}_N) & w\mathrm{Ai}'(N^{\frac{2}{3}}w^2\widetilde{\xi}_N) \end{pmatrix}, & z \in \Omega_3 \cap \mathbb{C}_-. \end{cases} \tag{5.26}$$

An appeal to the formulas

$$\mathrm{Ai}(z) + w\mathrm{Ai}(wz) + w^2\mathrm{Ai}(w^2z) = 0$$

and

$$\mathrm{Ai}'(z) + w^2\mathrm{Ai}'(wz) + w\mathrm{Ai}'(w^2z) = 0$$

shows that $Q(z)$ satisfies

$$Q_+(x) = Q_-(x)\begin{pmatrix} 1 & 1 \\ 0 & 1 \end{pmatrix}, \quad x \in \mathbb{R} \setminus [-\delta, \delta]. \tag{5.27}$$

The above heuristic argument now suggests that $U(z)$ is asymptotically approximated by

$$P^*(z) := \begin{cases} E(z)M(z), & z \in \Omega_1, \\ \widetilde{E}(z)\widetilde{M}(z), & z \in \Omega_3, \\ \mathrm{e}^{\frac{1}{2}Nl_N\sigma_3}V_\infty(z)\mathrm{e}^{(N(g(z)-\frac{1}{2}l_N)-\alpha\log z)\sigma_3}, & z \in \Omega_2, \end{cases} \tag{5.28}$$

where

$$E(z) := \sqrt{\pi}\mathrm{e}^{\frac{1}{2}Nl_N\sigma_3}\begin{pmatrix} 1 & -1 \\ -\mathrm{i} & -\mathrm{i} \end{pmatrix}\left(\frac{N^{\frac{1}{6}}\xi_N^{\frac{1}{4}}(z)}{\beta(z)}\right)^{\sigma_3},$$

$$M(z) := Q(z)\mathrm{e}^{[\frac{1}{2}(Nz^4+t\sqrt{N}z^2)-\alpha\log z]\sigma_3} \tag{5.29}$$

for $z \in \Omega_1$, and

$$\widetilde{E}(z) := \sqrt{\pi}\mathrm{e}^{-\alpha\pi\mathrm{i}\sigma_3}\mathrm{e}^{\frac{1}{2}N\widetilde{l}_N\sigma_3}\begin{pmatrix} 1 & -1 \\ \mathrm{i} & \mathrm{i} \end{pmatrix}(N^{\frac{1}{6}}\widetilde{\xi}_N^{\frac{-1}{4}}(z)\beta(z))^{\sigma_3},$$

$$\widetilde{M}(z) := Q(z)\mathrm{e}^{(\frac{1}{2}(Nz^4+t\sqrt{N}z^2)-\alpha\log(-z))\sigma_3} \tag{5.30}$$

for $z \in \Omega_3$. Note that the functions $\exp\{\frac{1}{2}(Nz^4+t\sqrt{N}z^2)-\alpha\log z\}$ and $\exp\{\frac{1}{2}(Nz^4+t\sqrt{N}z^2)-\alpha\log(-z)\}$ are analytic continuations of

$$(|x|^{2\alpha}\mathrm{e}^{-(Nx^4+tN^{\frac{1}{2}}x^2)})^{-\frac{1}{2}} = |x|^{-\alpha}\mathrm{e}^{\frac{1}{2}(Nx^4+tN^{\frac{1}{2}}x^2)}$$

to the cut planes $\mathbb{C} \setminus (-\infty, 0]$ and $\mathbb{C} \setminus [0, \infty)$, respectively. Hence, from (5.27) it follows that

$$M_+(x) = M_-(z)\begin{pmatrix} 1 & |x|^{2\alpha}\mathrm{e}^{-Nx^4-tN^{\frac{1}{2}}x^2} \\ 0 & 1 \end{pmatrix}, \quad x \in (\delta, \infty), \tag{5.31}$$

$$\widetilde{M}_+(x) = \widetilde{M}_-(z)\begin{pmatrix} 1 & |x|^{2\alpha}e^{-Nx^4-tN^{\frac{1}{2}}x^2} \\ 0 & 1 \end{pmatrix}, \quad x \in (-\infty, -\delta). \tag{5.32}$$

On the other hand, since $\frac{\xi_N^{\frac{1}{4}}(z)}{\beta(z)}$ and $\frac{\tilde{\xi}_N^{\frac{1}{4}}(z)}{\beta(z)}$ are analytic in $\mathbb{C}\setminus(-\infty,-\alpha_t]$ and $\mathbb{C}\setminus[\alpha_t,\infty)$, respectively, one easily sees that $E_+(x) = E_-(x)$ for $x\in(\delta,\infty)$ and $\widetilde{E}_+(x) = \widetilde{E}_-(x)$ for $x\in(-\infty,-\delta)$. From (5.28) and (5.31), it can be shown that

$$P_+^*(x) = P_-^*(x)\begin{pmatrix} 1 & |x|^{2\alpha}e^{-(Nx^4+tN^{\frac{1}{2}}x^2)} \\ 0 & 1 \end{pmatrix} \tag{5.33}$$

for all $x\in\mathbb{R}\setminus[-\delta,\delta]$. Furthermore, $P^*(z)$ has the same large $z$ behavior as $U(z)$ shown in $(U_c)$ (see [20, (5.38) and (5.39)]).

In summary, it is now natural to suggest that $U(z)$ is asymptotically approximated by

$$U^*(z) = \begin{cases} P(z), & z\in U_\delta, \\ P^*(z), & z\in\mathbb{C}\setminus U_\delta. \end{cases} \tag{5.34}$$

### 5.2 Case for $c = -2$

In this subsection, we first introduce the $\Psi_\alpha$ function associated with Painlevé II equation and then construct the model RHP to be used in the formation of a parametrix in a region containing the origin.

#### 5.2.1 $\Psi_\alpha$ functions for Painlevé II equation and model RHP

Let $\Psi_\alpha(\zeta,s)$ be a $2\times 2$ complex-valued matrix function. We consider the following linear differential equations:

$$\frac{\partial\Psi_\alpha}{\partial\zeta} = \begin{pmatrix} -4i\zeta^2 - i(s+2u^2) & 4\zeta u + 2iv + \dfrac{\alpha}{\zeta} \\ 4\zeta u - 2iv + \dfrac{\alpha}{\zeta} & 4i\zeta^2 + i(s+2u^2) \end{pmatrix}\Psi_\alpha \tag{5.35}$$

and

$$\frac{\partial\Psi_\alpha}{\partial s} = \begin{pmatrix} -i\zeta & u \\ u & i\zeta \end{pmatrix}\Psi_\alpha, \tag{5.36}$$

where $u$ and $v$ are functions of $s$. The compatibility condition for $\Psi_\alpha$ shows that $u(s)$ should satisfy the Painlevé equation

$$u'' = su + 2u^3 - \alpha \tag{5.37}$$

and $v(s) = u'(s)$. Now, we focus on equation (5.35) and view $u, v, s$ as parameters. In each sector

$$S_k := \left\{\zeta\in\mathbb{C}\;\middle|\;(k-2)\frac{\pi}{3} < \arg\zeta < k\frac{\pi}{3}\right\}$$

for $k = 1,\cdots,6$, there exists a unique solution $\Psi_{\alpha,k}$ of (5.35) such that

$$\Psi_{\alpha,k}(\zeta,s) \sim \Big(I + \sum_{j=1}^{\infty}\frac{\Phi_j}{\zeta^j}\Big)e^{-i(\frac{4}{3}\zeta^3+s\zeta)\sigma_3} = \Big(I + O\Big(\frac{1}{\zeta}\Big)\Big)e^{-i(\frac{4}{3}\zeta^3+s\zeta)\sigma_3} \tag{5.38}$$

as $\zeta \to \infty$ in each section $S_k$.

Since $\Psi_{\alpha,k}(\zeta, s)$ are solutions of the same linear differential equation, they are related by the so-called Stokes matrices $A_k$, i.e.,

$$\Psi_{\alpha,k+1}(\zeta, s) = \Psi_{\alpha,k}(\zeta, s)A_k, \quad k = 1, 2, \cdots, 5$$

and

$$\Psi_{\alpha,1}(\zeta, s) = \Psi_{\alpha,6}(\zeta, s)A_6.$$

Here, each $A_k$ is a triangular matrix of the form

$$A_k = \begin{pmatrix} 1 & 0 \\ a_k & 1 \end{pmatrix} \quad \text{for } k \text{ odd}$$

and

$$A_k = \begin{pmatrix} 1 & a_k \\ 0 & 1 \end{pmatrix} \quad \text{for } k \text{ even},$$

where $a_k$, $k = 1, 2, \cdots, 6$, are complex numbers satisfying

$$a_{k+3} = a_k, \quad k = 1, 2, 3, \quad \text{and} \quad a_1 + a_2 + a_3 + a_1a_2a_3 = -2\mathrm{i}\sin\pi\alpha. \tag{5.39}$$

The above properties can be found in [14, p. 164]. The entries $a_k$, which are also called Stokes multipliers, are in general dependent on the parameters $s, u$ and $v$. However, Flaschka and Newell [13] showed that if $u(s)$ is a solution of Painlevé II equation (5.37) and $v(s) = u'(s)$, then $a_k$ are constants. In particular, for the Hastings-Mcleod solution of Painlevé II equation, which satisfies the boundary conditions

$$u(s) \sim \frac{\alpha}{s} \quad \text{as } s \to \infty$$

and

$$u(s) \sim \sqrt{\frac{-s}{2}} \quad \text{as } s \to -\infty,$$

the choice is

$$a_1 = \mathrm{e}^{-\pi\mathrm{i}\alpha}, \quad a_2 = 0, \quad a_3 = -\mathrm{e}^{\pi\mathrm{i}\alpha}$$

(see [6, p. 604]). If we set

$$\widetilde{\Psi}(\zeta, s) := \begin{cases} \Psi_3(\zeta, s) & \text{for } \mathrm{Im}\,\zeta > 0, \\ \Psi_5(\zeta, s) & \text{for } \mathrm{Im}\,\zeta < 0, \end{cases} \tag{5.40}$$

then it follows from above equations that $\widetilde{\Psi}(\zeta, s)$ satisfies the following RHP:

($\widetilde{\Psi}_\mathrm{a}$) $\widetilde{\Psi}(\zeta, s)$ is analytic in $\mathbb{C} \setminus \mathbb{R}$;

($\widetilde{\Psi}_\mathrm{b}$) for $\zeta \in \mathbb{R}$,

$$\widetilde{\Psi}_+(\zeta, s) = \widetilde{\Psi}_-(\zeta, s) \begin{cases} \begin{pmatrix} 0 & -\mathrm{e}^{-\pi\mathrm{i}\alpha} \\ \mathrm{e}^{\pi\mathrm{i}\alpha} & 1 \end{pmatrix} & \text{for } \zeta < 0, \\ \begin{pmatrix} 0 & -\mathrm{e}^{\pi\mathrm{i}\alpha} \\ \mathrm{e}^{-\pi\mathrm{i}\alpha} & 1 \end{pmatrix} & \text{for } \zeta > 0; \end{cases}$$

($\widetilde{\Psi}_\mathrm{c}$) as $\zeta \to \infty$,

$$\widetilde{\Psi}(\zeta, s) = (I + O(\zeta^{-1}))\mathrm{e}^{-\mathrm{i}(\frac{4}{3}\zeta^3 + s\zeta)\sigma_3}.$$

Since all solutions to the second Painlevé equation are meromorphic functions with infinite number of poles, the RHP for $\widetilde{\Psi}$ shown above is solvable if and only if $s$ does not belong to the set of poles of $u(s)$.

With the above preparation, we are now ready to formulate a model RHP for our later use. Let $\theta \in \mathbb{R}$, and define

$$\overline{\Psi}(\zeta; s, \theta) := \begin{cases} e^{i\theta\sigma_3}\widetilde{\Psi}(\zeta, s)e^{i(\frac{4}{3}\zeta^3+s\zeta-\theta)\sigma_3} & \text{for Im } \zeta > 0, \\ e^{i\theta\sigma_3}\widetilde{\Psi}(\zeta, s)e^{i(\frac{4}{3}\zeta^3+s\zeta-\theta)\sigma_3}\begin{pmatrix} 0 & -1 \\ 1 & 0 \end{pmatrix} & \text{for Im } \zeta < 0. \end{cases} \tag{5.41}$$

It is then easily seen that $\overline{\Psi}(\zeta; s, \theta)$ is a solution of the following RHP:

($\overline{\Psi}_{\rm a}$) $\overline{\Psi}(\zeta; s, \theta)$ is analytic in $\mathbb{C} \setminus \mathbb{R}$;

($\overline{\Psi}_{\rm b}$) for $\zeta \in \mathbb{R}$,

$$\overline{\Psi}_+(\zeta; s, \theta) = \overline{\Psi}_-(\zeta; s, \theta) \begin{cases} \begin{pmatrix} e^{\pi i\alpha}e^{2i(\frac{4}{3}\zeta^3+s\zeta-\theta)} & 1 \\ 0 & e^{-\pi i\alpha}e^{-2i(\frac{4}{3}\zeta^3+s\zeta-\theta)} \end{pmatrix} & \text{for } \zeta < 0, \\ \begin{pmatrix} e^{-\pi i\alpha}e^{2i(\frac{4}{3}\zeta^3+s\zeta-\theta)} & 1 \\ 0 & e^{\pi i\alpha}e^{-2i(\frac{4}{3}\zeta^3+s\zeta-\theta)} \end{pmatrix} & \text{for } \zeta > 0, \end{cases}$$

where the real line is oriented from $-\infty$ to $\infty$, $\phi$ and $\widetilde{\phi}$ are defined in (3.52) and (3.53), respectively;

($\overline{\Psi}_{\rm c}$) as $\zeta \to \infty$,

$$\overline{\Psi}(\zeta; s, \theta) = \begin{cases} I + O(\zeta^{-1}), & \text{if Im } \zeta > 0, \\ (I + O(\zeta^{-1}))\begin{pmatrix} 0 & -1 \\ 1 & 0 \end{pmatrix}, & \text{if Im } \zeta < 0. \end{cases}$$

In the following subsection, we will show how $\overline{\Psi}(\zeta; s, \theta)$ is involved in the construction of a parametrix in the neighborhood of $z = 0$.

### 5.2.2 Parametrix in the neighborhood of the origin

Let $U_\delta$ be a domain in the complex plane containing the origin, the size of which will be determined later. We first wish to find a matrix-valued function $Q(z)$ that satisfies the following RHP:

($Q_{\rm a}$) $Q(z)$ is analytic in $U_\delta \setminus \mathbb{R}$;

($Q_{\rm b}$) for $x \in U_\delta \cap \mathbb{R}$,

$$Q_+(x) = Q_-(x) \begin{cases} \begin{pmatrix} e^{2N\phi_+} & 1 \\ 0 & e^{2N\phi_-} \end{pmatrix} & \text{for } x > 0, \\ \begin{pmatrix} e^{2N\widetilde{\phi}_+} & 1 \\ 0 & e^{2N\widetilde{\phi}_-} \end{pmatrix} & \text{for } x < 0, \end{cases}$$

where $\phi$ and $\widetilde{\phi}$ are defined in (3.52) and (3.53).

($Q_{\rm c}$) $Q(z)$ satisfies the matching condition

$$Q(z) \sim V_\infty(z)$$

as $n \to \infty$ for $z \in U_\delta \setminus \mathbb{R}$, where $V_\infty$ is the matrix given in (4.21)–(4.22).

A comparison of conditions $(\overline{\Psi}_\mathrm{b})$ and $(Q_\mathrm{b})$ invokes us to introduce the mapping

$$\mathrm{i}\Big(\frac{4}{3}\zeta^3 + s\zeta - \theta - \frac{\pi}{2}\alpha\Big) = \begin{cases} N\phi(z) & \text{for Im } z > 0, \\ -N\phi(z) & \text{for Im } z < 0. \end{cases} \tag{5.42}$$

By Proposition 3.3, we see that the right-hand side of (5.42) is analytic in

$$\mathbb{C} \setminus (-\infty, -\sqrt{2}] \cup [\sqrt{2}, \infty)$$

on account of $(\nu)_+ = -(\nu)_-$ for $x \in [-\sqrt{2}, \sqrt{2}]$, and can be written as

$$\mathrm{i}N\Big(-\frac{\pi}{2} + F(z)\Big),$$

where

$$F(z) := \int_0^z 2\zeta^2\sqrt{2-\zeta^2}\,\mathrm{d}\zeta \tag{5.43}$$

with branch cuts along $(-\infty, -\sqrt{2}] \cup [\sqrt{2}, \infty)$. It is clear that

$$F(0) = 0$$

and

$$F(z) \sim \frac{2}{3}\sqrt{2}z^3$$

as $z \to 0$. To make the mapping (5.42) one-to-one at least in a neighborhood of zero, it is natural to set

$$\zeta(z) = N^{\frac{1}{3}}\eta(z) = N^{\frac{1}{3}}\Big(\frac{3}{4}F(z)\Big)^{\frac{1}{3}}, \quad s = 0, \quad \theta = \frac{n\pi}{2}. \tag{5.44}$$

Since $s = 0 \in \mathbb{R}$, the RHP for $\widetilde{\Psi}(\zeta, s)$ is solvable (see the comment following $(\widetilde{\Psi}_\mathrm{c})$). Thus, $\overline{\Psi}(\zeta; s, \theta)$ is known. On account of $(\overline{\Psi}_c)$ and $(Q_c)$, we set

$$Q(z) = E_0(z)\overline{\Psi}\Big(N^{\frac{1}{3}}\eta(z); 0, \frac{n\pi}{2}\Big), \tag{5.45}$$

where

$$E_0 = \begin{cases} V_\infty(z), & \text{if Im } z > 0, \\ V_\infty(z)\begin{pmatrix} 0 & 1 \\ -1 & 0 \end{pmatrix}, & \text{if Im } z < 0. \end{cases} \tag{5.46}$$

The jump condition $(V_{\infty,\mathrm{b}})$ for $V_\infty$ shows that $E_0(z)$ is actually analytic in $\mathbb{C} \setminus (-\infty, -\sqrt{2}]$ $\cup[\sqrt{2}, \infty)$. In fact, it follows from (4.21) and (5.13) that we have the explicit formula

$$E_0(z) = \begin{pmatrix} \dfrac{\beta(z) + \beta(z)^{-1}}{2} & \dfrac{\beta(z) - \beta(z)^{-1}}{2\mathrm{i}} \\ \dfrac{\beta(z) - \beta(z)^{-1}}{-2\mathrm{i}} & \dfrac{\beta(z) + \beta(z)^{-1}}{2} \end{pmatrix}, \tag{5.47}$$

where

$$\beta(z) = \left(\frac{\sqrt{2}-z}{z+\sqrt{2}}\right)^{\frac{1}{4}} \tag{5.48}$$

is analytic in $\mathbb{C} \setminus (-\infty, -\sqrt{2}] \cup [\sqrt{2}, \infty)$.

In the literature, one may find the use of a "double-scaling" approach; that is, $c = -2+\rho N^{-\epsilon}$ for some constants $\rho$ and $\epsilon > 0$. In this case, the parametrix in the neighborhood of the origin can still be constructed in terms of the function $\Psi_\alpha(\zeta, s)$ for the model RHP associated with Painlevé II equation. However, the variable $s$ depends on the parameter $c$; in our case (i.e., $c = -2$), we have $s = 0$ (see [4, 6]).

### 5.2.3 Parametrices outside the origin

For $z$ outside the $\delta$-neighborhood $U_\delta$ of the origin, we will (as before) construct the parametrices by using Airy functions and elementary functions. To facilitate discussion, we divide the complex plane into four parts: $U_\delta$ and $\Omega_i$, $i = 1, 2, 3$ (see Figure 3).

Let $\beta(z)$ be given as in (5.48), and let $Q(z)$ be given as in (5.45) with $\phi(z)$ and $\widetilde{\phi}(z)$ in (3.52) and (3.53), respectively. As in Subsection 5.1, we expect $U(z)$ to be asymptotically approximated by

$$U^*(z) := \begin{cases} E(z)M(z), & z \in \Omega_1, \\ \widetilde{E}(z)\widetilde{M}(z), & z \in \Omega_3, \\ e^{\frac{1}{2}Nl_N\sigma_3} V_\infty(z) e^{(N(g(z)-\frac{1}{2}l_N)-\alpha \log z)\sigma_3}, & z \in \Omega_2, \\ e^{\frac{1}{2}Nl_N\sigma_3} Q(z) e^{(N(g(z)-\frac{1}{2}l_N)-\alpha \log z)\sigma_3}, & z \in U_\delta, \end{cases} \tag{5.49}$$

where

$$\begin{aligned} E(z) &:= \sqrt{\pi} e^{\frac{1}{2}Nl_N\sigma_3} \begin{pmatrix} 1 & -1 \\ -\mathrm{i} & -\mathrm{i} \end{pmatrix} \left(\frac{N^{\frac{1}{6}} \xi_N^{\frac{1}{4}}(z)}{\beta(z)}\right)^{\sigma_3}, \\ M(z) &:= Q(z) e^{(\frac{1}{2}(Nz^4 + t\sqrt{N}z^2) - \alpha \log z)\sigma_3}, \\ \xi_N &= f_N(z) = \left(\frac{3}{2}\phi(z)\right)^{\frac{2}{3}} \end{aligned} \tag{5.50}$$

for $z \in \Omega_1$, and

$$\begin{aligned} \widetilde{E}(z) &:= \sqrt{\pi} e^{\frac{1}{2}N\widetilde{l}_N\sigma_3} \begin{pmatrix} 1 & -1 \\ \mathrm{i} & \mathrm{i} \end{pmatrix} (N^{\frac{1}{6}} \widetilde{\xi}_N^{\frac{1}{4}}(z)\beta(z))^{\sigma_3}, \\ \widetilde{M}(z) &:= Q(z) e^{(\frac{1}{2}(Nz^4 + t\sqrt{N}z^2) - \alpha \log(-z))\sigma_3}, \\ \widetilde{\xi}_N &= \widetilde{f}_N(z) = \left(\frac{3}{2}\widetilde{\phi}(z)\right)^{\frac{2}{3}} \end{aligned} \tag{5.51}$$

for $z \in \Omega_3$, and $l_N$ and $\widetilde{l}_N$ are shown in (3.51) and (3.67). $V_\infty$ is the matrix given in (4.21)–(4.22). Note that $U^*(z)$ also satisfies the jump condition in $(U_{\rm b})$ for $U(z)$ in Section 2.

### 5.3 Case for $c < -2$

Note that unlike cases (i) and (ii), here no special attention is needed for the singularity of the weight function at the origin. The main reason is that the interval $(-a, a)$, and hence the origin, lies outside the support of the equilibrium measure. Therefore, the orthogonal polynomials

$$\pi_n(x) = S_n(x;t)$$

in (1.2) have no zero in a small neighborhood of the origin.

In this subsection, we will construct an approximation $U^*(z)$ to the solution of the RHP for $U(z)$ as $n \to \infty$, by using only Airy functions and elementary functions. To facilitate the following discussions, we divide the complex plane into four parts: $\Omega_i$, $i = 1, \cdots, 4$ (see Figure 4).

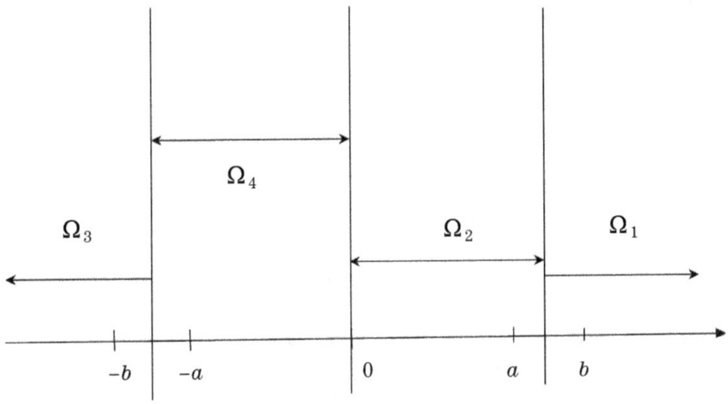


Figure 4 The domains $\Omega_1, \cdots, \Omega_4$.

In the view of (4.35), we can rewrite (4.37) as

$$\begin{aligned} U(z) \sim \frac{1}{2} \mathrm{e}^{\frac{1}{2} N l_N \sigma_3} D_\infty^{\sigma_3} \begin{pmatrix} 1 & -1 \\ -\mathrm{i} & -\mathrm{i} \end{pmatrix} \widetilde{\gamma}(z)^{\sigma_3} \begin{pmatrix} 1 & \mathrm{i} \\ -1 & \mathrm{i} \end{pmatrix} D^{-\sigma_3} \begin{pmatrix} 1 & \mathrm{i} \\ -1 & \mathrm{i} \end{pmatrix}^{-1} \\ \times \begin{pmatrix} \mathrm{e}^{-N\phi(z)} & \mathrm{i}\mathrm{e}^{N\phi(z)} \\ -\mathrm{e}^{-N\phi(z)} & \mathrm{i}\mathrm{e}^{N\phi(z)} \end{pmatrix} \mathrm{e}^{[N(g(z)+\phi(z)-\frac{1}{2}l_N)-\alpha \log z]\sigma_3}, \end{aligned}$$

and from (3.111) we obtain

$$\begin{aligned} U(z) \sim \frac{1}{2} \mathrm{e}^{\frac{1}{2} N l_N \sigma_3} D_\infty^{\sigma_3} \begin{pmatrix} 1 & -1 \\ -\mathrm{i} & -\mathrm{i} \end{pmatrix} \widetilde{\gamma}(z)^{\sigma_3} \begin{pmatrix} 1 & \mathrm{i} \\ -1 & \mathrm{i} \end{pmatrix} D^{-\sigma_3} \begin{pmatrix} 1 & \mathrm{i} \\ -1 & \mathrm{i} \end{pmatrix}^{-1} \\ \times \begin{pmatrix} \mathrm{e}^{-N\phi(z)} & \mathrm{i}\mathrm{e}^{N\phi(z)} \\ -\mathrm{e}^{-N\phi(z)} & \mathrm{i}\mathrm{e}^{N\phi(z)} \end{pmatrix} \mathrm{e}^{[\frac{1}{2}(Nz^4+t\sqrt{N}z^2)-\alpha \log z]\sigma_3}. \end{aligned} \tag{5.52}$$

To find an approximation to $U(z)$, we first look for a matrix which is asymptotic to

$$\begin{pmatrix} \mathrm{e}^{-N\phi(z)} & \mathrm{i}\mathrm{e}^{N\phi(z)} \\ -\mathrm{e}^{-N\phi(z)} & \mathrm{i}\mathrm{e}^{N\phi(z)} \end{pmatrix}. \tag{5.53}$$

From Proposition 3.5, it is clear that the function defined by

$$\xi_N = f_N(z) = \left(\frac{3}{2}\phi(z)\right)^{\frac{2}{3}} \tag{5.54}$$

is analytic in $\mathbb{C}\setminus(-\infty,a]$, where $\phi(z)$ is defined in (3.97) and depends on $N$. In particular, for $z\in\Omega_1\cap\mathbb{C}_+$, we have

$$0<\arg f_N(z)<\pi. \tag{5.55}$$

Also, for $z\neq b$, $N^{\frac{2}{3}}\xi_N=N^{\frac{2}{3}}f_N(z)\to\infty$ as $n\to\infty$. Now, as in Subsection 5.1.2, we have from the asymptotic behavior of the Airy function (see [17, p. 392]),

$$\begin{pmatrix} \mathrm{e}^{-N\phi(z)} & \mathrm{i}\mathrm{e}^{N\phi(z)} \\ -\mathrm{e}^{-N\phi(z)} & \mathrm{i}\mathrm{e}^{N\phi(z)} \end{pmatrix} \sim 2\sqrt{\pi}(N^{\frac{2}{3}}f_N(z))^{\frac{1}{4}\sigma_3} \begin{pmatrix} \mathrm{Ai}(N^{\frac{2}{3}}\xi_N) & -w^2\mathrm{Ai}(N^{\frac{2}{3}}w^2\xi_N) \\ \mathrm{Ai}'(N^{\frac{2}{3}}\xi_N) & -w\mathrm{Ai}'(N^{\frac{2}{3}}w^2\xi_N) \end{pmatrix} \tag{5.56}$$

for $z\in\Omega_1\cap\mathbb{C}_+$, where

$$w=\mathrm{e}^{\frac{2\pi\mathrm{i}}{3}}.$$

Also, we have

$$-\pi<\arg f_N(z)<0$$

and

$$\begin{pmatrix} \mathrm{e}^{-N\phi(z)} & \mathrm{i}\mathrm{e}^{N\phi(z)} \\ -\mathrm{e}^{-N\phi(z)} & \mathrm{i}\mathrm{e}^{N\phi(z)} \end{pmatrix} \sim 2\sqrt{\pi}(N^{\frac{2}{3}}f_N(z))^{\frac{1}{4}\sigma_3} \begin{pmatrix} \mathrm{Ai}(N^{\frac{2}{3}}\xi_N) & w\mathrm{Ai}(N^{\frac{2}{3}}w\xi_N) \\ \mathrm{Ai}'(N^{\frac{2}{3}}\xi_N) & w^2\mathrm{Ai}'(N^{\frac{2}{3}}w\xi_N) \end{pmatrix} \tag{5.57}$$

for $z\in\Omega_1\cap\mathbb{C}_-$. There is a result corresponding to (5.52) with $\phi(z)$ replaced by $\widetilde{\phi}(z)$ for $z\in\Omega_3$. Indeed, from (3.111) and (4.35), it follows that

$$\begin{aligned} U(z) \sim{}& \frac{1}{2}\mathrm{e}^{\frac{1}{2}N\widetilde{l}_N\sigma_3}\mathrm{e}^{-\alpha\pi\mathrm{i}\sigma_3}D_\infty^{\sigma_3}\begin{pmatrix} 1 & -1 \\ \mathrm{i} & \mathrm{i} \end{pmatrix}\widetilde{\gamma}(z)^{-\sigma_3}\begin{pmatrix} 1 & -\mathrm{i} \\ -1 & -\mathrm{i} \end{pmatrix}D^{-\sigma_3}\begin{pmatrix} 1 & -\mathrm{i} \\ -1 & -\mathrm{i} \end{pmatrix}^{-1} \\ &\times\begin{pmatrix} \mathrm{e}^{-N\widetilde{\phi}(z)} & -\mathrm{i}\mathrm{e}^{N\widetilde{\phi}(z)} \\ -\mathrm{e}^{-N\widetilde{\phi}(z)} & -\mathrm{i}\mathrm{e}^{N\widetilde{\phi}(z)} \end{pmatrix}\mathrm{e}^{[\frac{1}{2}(Nz^4+t\sqrt{N}z^2)-\alpha\log(-z)]\sigma_3}. \end{aligned} \tag{5.58}$$

Let

$$\widetilde{\xi}_N=\widetilde{f}_N(z)=\left(\frac{3}{2}\widetilde{\phi}(z)\right)^{\frac{2}{3}}, \tag{5.59}$$

which is analytic in $\mathbb{C}\setminus[-a,\infty)$. Also, note that

$$|\arg\widetilde{f}_N(z)|<\pi \tag{5.60}$$

for $z\in\Omega_3$. Hence, as before, it can be shown that the matrix

$$\begin{pmatrix} \mathrm{e}^{-N\widetilde{\phi}(z)} & -\mathrm{i}\mathrm{e}^{N\widetilde{\phi}(z)} \\ -\mathrm{e}^{-N\widetilde{\phi}(z)} & -\mathrm{i}\mathrm{e}^{N\widetilde{\phi}(z)} \end{pmatrix}$$

is the leading term in the asymptotic expansion of the matrices

$$2\sqrt{\pi}(N^{\frac{2}{3}}\widetilde{f}_N(z))^{\frac{1}{4}\sigma_3}\begin{pmatrix} \mathrm{Ai}(N^{\frac{2}{3}}\widetilde{\xi}_N) & -w\mathrm{Ai}(N^{\frac{2}{3}}w\widetilde{\xi}_N) \\ \mathrm{Ai}'(N^{\frac{2}{3}}\widetilde{\xi}_N) & -w^2\mathrm{Ai}'(N^{\frac{2}{3}}w\widetilde{\xi}_N) \end{pmatrix} \tag{5.61}$$

for $z \in \Omega_3 \cap \mathbb{C}_+$, and

$$2\sqrt{\pi}(N^{\frac{2}{3}}\widetilde{f}_N(z))^{\frac{1}{4}\sigma_3}\begin{pmatrix} \mathrm{Ai}(N^{\frac{2}{3}}\widetilde{\xi}_N) & w^2\mathrm{Ai}(N^{\frac{2}{3}}w^2\widetilde{\xi}_N) \\ \mathrm{Ai}'(N^{\frac{2}{3}}\widetilde{\xi}_N) & w\mathrm{Ai}'(N^{\frac{2}{3}}w^2\widetilde{\xi}_N) \end{pmatrix} \tag{5.62}$$

for $z \in \Omega_3 \cap \mathbb{C}_-$. For $z \in \Omega_4$ and $z \in \Omega_2$, we replace $\phi(z)$ and $\widetilde{\phi}(z)$ by $\overline{\phi}(z)$, and define

$$\overline{\xi}_N = \overline{f}_N(z) = \left(\frac{3}{2}\overline{\phi}(z)\right)^{\frac{2}{3}}, \tag{5.63}$$

which is analytic in $\mathbb{C} \setminus (-\infty, -b] \cup [b, \infty)$ and

$$|\arg \overline{f}_N(z)| < \pi$$

for $z \in \Omega_4$ and $z \in \Omega_2$. Thus, let us define the matrix function

$$Q(z) := \begin{cases} \begin{pmatrix} \mathrm{Ai}(N^{\frac{2}{3}}\xi_N) & -w^2\mathrm{Ai}(N^{\frac{2}{3}}w^2\xi_N) \\ \mathrm{Ai}'(N^{\frac{2}{3}}\xi_N) & -w\mathrm{Ai}'(N^{\frac{2}{3}}w^2\xi_N) \end{pmatrix}, & z \in \Omega_1 \cap \mathbb{C}_+, \\ \begin{pmatrix} \mathrm{Ai}(N^{\frac{2}{3}}\xi_N) & w\mathrm{Ai}(N^{\frac{2}{3}}w\xi_N) \\ \mathrm{Ai}'(N^{\frac{2}{3}}\xi_N) & w^2\mathrm{Ai}'(N^{\frac{2}{3}}w\xi_N) \end{pmatrix}, & z \in \Omega_1 \cap \mathbb{C}_-, \\ \begin{pmatrix} \mathrm{Ai}(N^{\frac{2}{3}}\widetilde{\xi}_N) & -w\mathrm{Ai}(N^{\frac{2}{3}}w\widetilde{\xi}_N) \\ \mathrm{Ai}'(N^{\frac{2}{3}}\widetilde{\xi}_N) & -w^2\mathrm{Ai}'(N^{\frac{2}{3}}w\widetilde{\xi}_N) \end{pmatrix}, & z \in \Omega_3 \cap \mathbb{C}_+, \\ \begin{pmatrix} \mathrm{Ai}(N^{\frac{2}{3}}\widetilde{\xi}_N) & w^2\mathrm{Ai}(N^{\frac{2}{3}}w^2\widetilde{\xi}_N) \\ \mathrm{Ai}'(N^{\frac{2}{3}}\widetilde{\xi}_N) & w\mathrm{Ai}'(N^{\frac{2}{3}}w^2\widetilde{\xi}_N) \end{pmatrix}, & z \in \Omega_3 \cap \mathbb{C}_-, \\ \begin{pmatrix} \mathrm{Ai}(N^{\frac{2}{3}}\overline{\xi}_N) & -w^2\mathrm{Ai}(N^{\frac{2}{3}}w^2\overline{\xi}_N) \\ \mathrm{Ai}'(N^{\frac{2}{3}}\overline{\xi}_N) & -w\mathrm{Ai}'(N^{\frac{2}{3}}w^2\overline{\xi}_N) \end{pmatrix}, & z \in \Omega_4 \cap \mathbb{C}_+, \\ \begin{pmatrix} \mathrm{Ai}(N^{\frac{2}{3}}\overline{\xi}_N) & w\mathrm{Ai}(N^{\frac{2}{3}}w\overline{\xi}_N) \\ \mathrm{Ai}'(N^{\frac{2}{3}}\overline{\xi}_N) & w^2\mathrm{Ai}'(N^{\frac{2}{3}}w\overline{\xi}_N) \end{pmatrix}, & z \in \Omega_4 \cap \mathbb{C}_-, \\ \begin{pmatrix} \mathrm{Ai}(N^{\frac{2}{3}}\overline{\xi}_N) & -w\mathrm{Ai}(N^{\frac{2}{3}}w\overline{\xi}_N) \\ \mathrm{Ai}'(N^{\frac{2}{3}}\overline{\xi}_N) & -w^2\mathrm{Ai}'(N^{\frac{2}{3}}w\overline{\xi}_N) \end{pmatrix}, & z \in \Omega_2 \cap \mathbb{C}_+, \\ \begin{pmatrix} \mathrm{Ai}(N^{\frac{2}{3}}\overline{\xi}_N) & w^2\mathrm{Ai}(N^{\frac{2}{3}}w^2\overline{\xi}_N) \\ \mathrm{Ai}'(N^{\frac{2}{3}}\overline{\xi}_N) & w\mathrm{Ai}'(N^{\frac{2}{3}}w^2\overline{\xi}_N) \end{pmatrix}, & z \in \Omega_2 \cap \mathbb{C}_-. \end{cases} \tag{5.64}$$

The identities

$$\mathrm{Ai}(z) + w\mathrm{Ai}(wz) + w^2\mathrm{Ai}(w^2z) = 0$$

and

$$\mathrm{Ai}'(z) + w^2\mathrm{Ai}'(wz) + w\mathrm{Ai}'(w^2z) = 0$$

show that $Q(z)$ satisfies

$$Q_+(x) = Q_-(x)\begin{pmatrix} 1 & 1 \\ 0 & 1 \end{pmatrix}, \quad x \in \mathbb{R}. \tag{5.65}$$

The above heuristic argument suggests that $U(z)$ is asymptotically approximated by

$$U^*(z) := \begin{cases} E(z)M(z), & z \in \Omega_1, \\ \widetilde{E}(z)\widetilde{M}(z), & z \in \Omega_3, \\ \overline{E}_1(z)\overline{M}_1(z), & z \in \Omega_4, \\ \overline{E}_2(z)\overline{M}_2(z), & z \in \Omega_2, \end{cases} \tag{5.66}$$

where

$$E(z) := \sqrt{\pi} e^{\frac{1}{2}Nl_N\sigma_3} D_\infty^{\sigma_3} \begin{pmatrix} 1 & -1 \\ -i & -i \end{pmatrix} \widetilde{\gamma}(z)^{\sigma_3} \begin{pmatrix} 1 & i \\ -1 & i \end{pmatrix}$$
$$\times D(z)^{-\sigma_3} \begin{pmatrix} 1 & i \\ -1 & i \end{pmatrix}^{-1} (N^{\frac{1}{6}} \xi_N^{\frac{1}{4}}(z))^{\sigma_3},$$
$$M(z) := Q(z) e^{[\frac{1}{2}(Nz^4 + t\sqrt{N}z^2) - \alpha \log z]\sigma_3}$$

for $z \in \Omega_1$, and

$$\widetilde{E}(z) := \sqrt{\pi} e^{\frac{1}{2}N\widetilde{l}_N\sigma_3} e^{-\alpha\pi i\sigma_3} D_\infty^{\sigma_3} \begin{pmatrix} 1 & -1 \\ i & i \end{pmatrix} \widetilde{\gamma}(z)^{-\sigma_3} \begin{pmatrix} 1 & -i \\ -1 & -i \end{pmatrix}$$
$$\times D(z)^{-\sigma_3} \begin{pmatrix} 1 & -i \\ -1 & -i \end{pmatrix}^{-1} (N^{\frac{1}{6}} \widetilde{\xi}_N^{\frac{1}{4}}(z))^{\sigma_3},$$
$$\widetilde{M}(z) := Q(z) e^{[\frac{1}{2}(Nz^4 + t\sqrt{N}z^2) - \alpha \log(-z)]\sigma_3}$$

for $z \in \Omega_3$, where $\widetilde{l}_N = l_N + 2\pi i$, and

$$\overline{E}_1(z) := \sqrt{\pi} e^{\frac{1}{2}N\widetilde{l}_N\sigma_3} e^{-\alpha\pi i\sigma_3} D_\infty^{\sigma_3} \begin{pmatrix} 1 & -1 \\ -i & -i \end{pmatrix} \widetilde{\gamma}(z)^{\sigma_3} \begin{pmatrix} 1 & i \\ -1 & i \end{pmatrix}$$
$$\times D(z)^{-\sigma_3} e^{\mp\frac{1}{2}\pi i N\sigma_3} \begin{pmatrix} 1 & i \\ -1 & i \end{pmatrix}^{-1} (N^{\frac{1}{6}} \overline{\xi}_N^{\frac{1}{4}}(z))^{\sigma_3},$$
$$\overline{M}_1(z) := Q(z) e^{[\frac{1}{2}(Nz^4 + t\sqrt{N}z^2) - \alpha \log(-z)]\sigma_3}$$

for $z \in \Omega_4 \cap \mathbb{C}_\pm$,

$$\overline{E}_2(z) := \sqrt{\pi} e^{\frac{1}{2}Nl_N\sigma_3} D_\infty^{\sigma_3} \begin{pmatrix} 1 & -1 \\ i & i \end{pmatrix} \widetilde{\gamma}(z)^{-\sigma_3} \begin{pmatrix} 1 & -i \\ -1 & -i \end{pmatrix}$$
$$\times D(z)^{-\sigma_3} e^{\pm\frac{1}{2}\pi i N\sigma_3} \begin{pmatrix} 1 & -i \\ -1 & -i \end{pmatrix}^{-1} (N^{\frac{1}{6}} \overline{\xi}_N^{\frac{1}{4}}(z))^{\sigma_3},$$
$$\overline{M}_2(z) := Q(z) e^{[\frac{1}{2}(Nz^4 + t\sqrt{N}z^2) - \alpha \log z]\sigma_3}$$

for $z \in \Omega_2 \cap \mathbb{C}_\pm$. From (5.66) it can be shown that

$$U_+^*(x) = U_-^*(x) \begin{pmatrix} 1 & |x|^{2\alpha} e^{-(Nx^4 + tN^{\frac{1}{2}}x^2)} \\ 0 & 1 \end{pmatrix} \tag{5.67}$$

for all $x \in \mathbb{R} \setminus \{0\}$. Furthermore, $U^*(z)$ has the same large $z$ behavior as $U(z)$ shown in $(U_c)$.

## 6 Uniform Asymptotic Expansions

In this section, we will present uniform asymptotic expansions of the monic polynomials $\pi_n(x)$ in Theorem 2.1. As in the previous sections, there are three cases to be considered, but in all three cases, we will use the same notations with different meaning.

### 6.1 Case for $c > -2$

To derive such expansions, we define the matrix-valued function

$$S(z) := \mathrm{e}^{-\frac{1}{2}Nl_N\sigma_3}U(z)U^*(z)^{-1}\mathrm{e}^{\frac{1}{2}Nl_N\sigma_3}. \tag{6.1}$$

Since $U(z)$ and $U^*(z)$ have the same jump condition on $(-\delta,\delta)\setminus\{0\}$, $S(z)$ is analytic in $U_\delta\setminus\{0\}$. In fact, 0 is a removable singularity of $S(z)$, (see [20, (6.2)–(6.9)]). Also, it is easily seen that $S(z)$ satisfies the following RHP:

($S_\mathrm{a}$) $S(z)$ is analytic for $z\in\mathbb{C}\setminus\Gamma$, where $\Gamma$ is the contour $\Gamma=\bigcup_{i=1}^{8}\Gamma_i$ shown in Figure 3.

($S_\mathrm{b}$) for $z\in\Gamma$,

$$S_+(z)=S_-(z)J(z),$$

where $J(z) := \mathrm{e}^{-\frac{1}{2}Nl_N\sigma_3}U_-^*(z)(U_+^*(z))^{-1}\mathrm{e}^{\frac{1}{2}Nl_N\sigma_3}$;

($S_\mathrm{c}$) for $z\in\mathbb{C}\setminus\Gamma$,

$$S(z)\to I$$

as $z\to\infty$.

To solve this problem, we first derive the asymptotic expansion of the jump matrix $J(z)$ as $n\to\infty$. This is done in exactly the same manner as in [20], and we have

$$J(z)\sim I+\sum_{k=1}^{\infty}\frac{J_k(z)}{N^k}, \tag{6.2}$$

where the coefficients $J_k(z)$ are explicitly given $2\times 2$ matrices; for details, see [20, p. 750].

An appeal to Theorem 3 in [20] shows that the solution of the RHP for $S$ has the asymptotic expansion

$$S(z)\sim I+\sum_{k=1}^{\infty}\frac{S_k(z)}{N^k} \tag{6.3}$$

as $n\to\infty$ uniformly for $z\in\mathbb{C}\setminus\Gamma$, where $N=n+\alpha$ and the coefficient functions $S_k(z)$ can be determined recursively by

$$S_k(z)=\frac{1}{2\pi\mathrm{i}}\int_\Gamma\sum_{j=1}^{k}(S_{k-j})_-(\zeta)J_j(\zeta)\frac{\mathrm{d}\zeta}{\zeta-z},\quad k=1,2,\cdots \tag{6.4}$$

for $z\in\mathbb{C}\setminus\Gamma$.

The proofs of the following theorems can be carried out along the same lines as given in [19–20].

**Theorem 6.1** *Let $\Omega_i$, $i=1,2,3$, and $U_\delta$ be the regions shown in Figure 3. With $N=n+\alpha$, $t>-2\sqrt{N}$, $l_N$ and $f_N(z)$ defined in (3.32) and (5.17), the asymptotic expansion of the polynomial $\pi_n(N^{\frac{1}{4}}z)$ is given by*

$$\begin{aligned}\pi_n(N^{\frac{1}{4}}z)&=\sqrt{\pi}N^{\frac{n}{4}}\mathrm{e}^{\frac{1}{2}(Nz^4+t\sqrt{N}z^2)+\frac{1}{2}Nl_N}z^{-\alpha}\\&\quad\times[\mathrm{Ai}(N^{\frac{2}{3}}f_N(z))\boldsymbol{A}(z,N)-\mathrm{Ai}'(N^{\frac{2}{3}}f_N(z))\boldsymbol{B}(z,N)],\end{aligned}\tag{6.5}$$

*where $\boldsymbol{A}(z,N)$ and $\boldsymbol{B}(z,N)$ are analytic in $\Omega_1$ and have asymptotic expansions*

$$\boldsymbol{A}(z,N) \sim \frac{N^{\frac{1}{6}} f_N^{\frac{1}{4}}(z)}{\beta(z)} \Big(1 + \sum_{k=1}^{\infty} \frac{S_{11}^{(k)}(z) - \mathrm{i}S_{12}^{(k)}(z)}{N^k}\Big) \tag{6.6}$$

*and*

$$\boldsymbol{B}(z,N) \sim \frac{\beta(z)}{N^{\frac{1}{6}} f_N^{\frac{1}{4}}(z)} \Big(1 + \sum_{k=1}^{\infty} \frac{S_{11}^{(k)}(z) + \mathrm{i}S_{12}^{(k)}(z)}{N^k}\Big) \tag{6.7}$$

*uniformly for $z \in \Omega_1$. In (6.6)–(6.7), the coefficient function $S_{ij}^{(k)}, i,j = 1,2$, refers to the element in the ith row and jth column of the matrix $S_k(z)$, which is given in (6.4). The function $\beta(z)$ is given in (4.15).*

*Similarly, with $\widetilde{f}_N(z)$ given in (5.22), we have*

$$\begin{aligned}\pi_n(N^{\frac{1}{4}}z) &= (-1)^n \sqrt{\pi} N^{\frac{n}{4}} \mathrm{e}^{\frac{1}{2}(Nz^4 + t\sqrt{N}z^2) + \frac{1}{2}Nl_N} (-z)^{-\alpha} \\ &\quad \times [\mathrm{Ai}(N^{\frac{2}{3}} \widetilde{f}_N(z)) \widetilde{\boldsymbol{A}}(z,N) - \mathrm{Ai}'(N^{\frac{2}{3}} \widetilde{f}_N(z)) \widetilde{\boldsymbol{B}}(z,N)],\end{aligned} \tag{6.8}$$

*where $\widetilde{\boldsymbol{A}}(z,N)$ and $\widetilde{\boldsymbol{B}}(z,N)$ are analytic in $\Omega_3$ and have asymptotic expansions*

$$\widetilde{\boldsymbol{A}}(z,N) \sim N^{\frac{1}{6}} \widetilde{f}_N^{\frac{1}{4}}(z) \beta(z) \Big(1 + \sum_{k=1}^{\infty} \frac{S_{11}^{(k)}(z) + \mathrm{i}S_{12}^{(k)}(z)}{N^k}\Big) \tag{6.9}$$

*and*

$$\widetilde{\boldsymbol{B}}(z,N) \sim \frac{1}{N^{\frac{1}{6}} \widetilde{f}_N^{\frac{1}{4}}(z) \beta(z)} \Big(1 + \sum_{k=1}^{\infty} \frac{S_{11}^{(k)}(z) - \mathrm{i}S_{12}^{(k)}(z)}{N^k}\Big) \tag{6.10}$$

*uniformly for $z \in \Omega_3$.*

*Let $\phi(z)$ be defined as in (3.33). We have*

$$\begin{aligned}\pi_n(N^{\frac{1}{4}}z) &= \frac{1}{2} N^{\frac{n}{4}} \mathrm{e}^{\frac{1}{2}(Nz^4 + t\sqrt{N}z^2) + \frac{1}{2}Nl_N - N\phi(z)} z^{-\alpha} \\ &\quad \times \Big[\Big(\Big(\frac{z-\alpha_t}{z+\alpha_t}\Big)^{\frac{1}{4}} + \Big(\frac{z-\alpha_t}{z+\alpha_t}\Big)^{-\frac{1}{4}}\Big) \boldsymbol{A}_1(z,N) \\ &\quad + \Big(\Big(\frac{z-\alpha_t}{z+\alpha_t}\Big)^{\frac{1}{4}} - \Big(\frac{z-\alpha_t}{z+\alpha_t}\Big)^{-\frac{1}{4}}\Big) \boldsymbol{B}_1(z,N)\Big],\end{aligned} \tag{6.11}$$

*where $\boldsymbol{A}_1(z,N)$ and $\boldsymbol{B}_1(z,N)$ are analytic functions of $z$ in $\Omega_2$. In addition, they have asymptotic expansions*

$$\boldsymbol{A}_1(z,N) \sim 1 + \sum_{k=1}^{\infty} \frac{S_{11}^{(k)}(z)}{N^k} \tag{6.12}$$

*and*

$$\boldsymbol{B}_1(z,N) \sim \mathrm{i} \sum_{k=1}^{\infty} \frac{S_{12}^{(k)}(z)}{N^k}, \tag{6.13}$$

*uniformly in $\Omega_2$.*

*Finally, when $z \in U_\delta$, we have*

$$\pi_n(N^{\frac{1}{4}}z) = \sqrt{\frac{\pi}{2}}N^{\frac{n+2}{4}}\mathrm{e}^{\frac{1}{2}(Nz^4+t\sqrt{N}z^2)+\frac{1}{2}Nl_N-\frac{1}{4}\pi\mathrm{i}}\eta_N(z)^{\frac{1}{2}}z^{-\alpha} \times [J_{\alpha+\frac{1}{2}}(N\eta_N(z))\boldsymbol{A}_0(z,N) + J_{\alpha-\frac{1}{2}}(N\eta_N(z))\boldsymbol{B}_0(z,N)], \tag{6.14}$$

*where $\eta_N(z)$ is defined in (5.8), $\boldsymbol{A}_0(z,N)$ and $\boldsymbol{B}_0(z,N)$ are analytic functions in $U_\delta$ with asymptotic expansions*

$$\boldsymbol{A}_0(z,N) \sim \sum_{k=0}^{\infty}\frac{A_{0,k}(z)}{N^k}, \quad \boldsymbol{B}_0(z,N) \sim \sum_{k=0}^{\infty}\frac{B_{0,k}(z)}{N^k}$$

*as $n \to \infty$. The leading coefficients are given by*

$$A_{0,0}(z) = (E_0)_{11}(z) + \mathrm{i}(E_0)_{12}(z) \quad \text{and} \quad B_{0,0}(z) = \mathrm{i}(E_0)_{11}(z) + (E_0)_{12}(z),$$

*where $E_0(z)$ is given in (5.13).*

### 6.2 Cases for $c \le -2$

Similarly, we have the following two theorems on global asymptotic expansions of the polynomials $\pi_n(x)$ in Theorem 2.1.

**Theorem 6.2** *Let $\Omega_i$, $i = 1,2,3$, and $U_\delta$ be the regions shown in Figure 3, with $N = n+\alpha$, $t = -2\sqrt{N}$, $l_N$ and $f_N(z)$ defined in (3.51) and (5.50). The asymptotic expansion of the polynomial $\pi_n(N^{\frac{1}{4}}z)$ is given*

$$\pi_n(N^{\frac{1}{4}}z) = \sqrt{\pi}N^{\frac{n}{4}}\mathrm{e}^{\frac{1}{2}N(z^4-2z^2)+\frac{1}{2}Nl_N}z^{-\alpha} \times [\mathrm{Ai}(N^{\frac{2}{3}}f_N(z))\boldsymbol{A}(z,N) - \mathrm{Ai}'(N^{\frac{2}{3}}f_N(z))\boldsymbol{B}(z,N)], \tag{6.15}$$

*where $\boldsymbol{A}(z,N)$ and $\boldsymbol{B}(z,N)$ are analytic in $\Omega_1$ and have asymptotic expansions*

$$\boldsymbol{A}(z,N) \sim \frac{N^{\frac{1}{6}}f_N^{\frac{1}{4}}(z)}{\beta(z)}\Big(1+\sum_{k=1}^{\infty}\frac{S_{11}^{(k)}(z) - \mathrm{i}S_{12}^{(k)}(z)}{N^k}\Big) \tag{6.16}$$

*and*

$$\boldsymbol{B}(z,N) \sim \frac{\beta(z)}{N^{\frac{1}{6}}f_N^{\frac{1}{4}}(z)}\Big(1+\sum_{k=1}^{\infty}\frac{S_{11}^{(k)}(z) + \mathrm{i}S_{12}^{(k)}(z)}{N^k}\Big) \tag{6.17}$$

*uniformly for $z \in \Omega_1$. In (6.16)–(6.17), the function $\beta(z)$ is given in (4.22). The coefficient function $S_{ij}^{(k)}, i,j = 1,2$, refers to the element in the $i$th row and $j$th column of the matrix $S_k(z)$, which is given in (6.4). In (6.4), the coefficients $J_k(z)$ are $2\times 2$ matrices explicitly given in* [19, p. 148].

*Let $\phi(z)$ be defined as in (3.52). We have*

$$\pi_n(N^{\frac{1}{4}}z) = \frac{1}{2}N^{\frac{n}{4}}\mathrm{e}^{\frac{1}{2}N(z^4-2z^2)+\frac{1}{2}Nl_N-N\phi(z)}z^{-\alpha} \times \Big[\Big(\Big(\frac{z-\sqrt{2}}{z+\sqrt{2}}\Big)^{\frac{1}{4}} + \Big(\frac{z-\sqrt{2}}{z+\sqrt{2}}\Big)^{-\frac{1}{4}}\Big)\boldsymbol{A}_1(z,N)$$

$$+\Big(\Big(\frac{z-\sqrt{2}}{z+\sqrt{2}}\Big)^{\frac{1}{4}}-\Big(\frac{z-\sqrt{2}}{z+\sqrt{2}}\Big)^{-\frac{1}{4}}\Big)\boldsymbol{B}_1(z,N)\Big],\tag{6.18}$$

*where $\boldsymbol{A}_1(z,N)$ and $\boldsymbol{B}_1(z,N)$ are analytic functions of $z$ in $\Omega_2$. In addition, they have asymptotic expansions*

$$\boldsymbol{A}_1(z,N)\sim 1+\sum_{k=1}^{\infty}\frac{S_{11}^{(k)}(z)}{N^k}\tag{6.19}$$

*and*

$$\boldsymbol{B}_1(z,N)\sim \mathrm{i}\sum_{k=1}^{\infty}\frac{S_{12}^{(k)}(z)}{N^k},\tag{6.20}$$

*uniformly in $\Omega_2$.*

*Let $(\psi_1(\zeta,s),\psi_2(\zeta,s))$ be the unique solution of equation (5.35) characterized by the asymptotic behavior*

$$\mathrm{e}^{\mathrm{i}(\frac{4}{3}\zeta^3+s\zeta)}\begin{pmatrix}\psi_1(\zeta,s)\\ \psi_2(\zeta,s)\end{pmatrix}=\begin{pmatrix}1\\0\end{pmatrix}+O(\zeta^{-1})$$

*as $\zeta\to\infty$ for $0<\arg\zeta<\pi$, and let $\zeta(z)$ be as shown in (5.44). We have*

$$\begin{aligned}\pi_n(N^{\frac{1}{4}}z)&=\sqrt{\pi}N^{\frac{n}{4}}\mathrm{e}^{\frac{1}{2}N(z^4-2z^2)+\frac{1}{2}Nl_N}z^{-\alpha}\mathrm{e}^{\frac{1}{2}\pi\mathrm{i}\alpha}\\&\quad\times\psi_1(N^{\frac{1}{3}}\eta(z),0)\boldsymbol{A}_1(z,N)+\psi_2(N^{\frac{1}{3}}\eta(z),0)\boldsymbol{B}_1(z,N))\end{aligned}\tag{6.21}$$

*as $n\to\infty$ uniformly for $z\in U_\delta$, where*

$$\boldsymbol{A}_1(z,N)\sim\sum_{k=0}^{\infty}\frac{A_k(z)}{N^{\frac{k}{3}}},\quad \boldsymbol{B}_1(z,N)\sim\sum_{k=0}^{\infty}\frac{B_k(z)}{N^{\frac{k}{3}}}$$

*with*

$$A_0(z)=\mathrm{e}^{\frac{\pi\mathrm{i}n}{2}}\Big(\mathrm{e}^{\frac{\pi\mathrm{i}}{4}}\Big(\frac{\sqrt{2}-z}{z+\sqrt{2}}\Big)^{\frac{1}{4}}+\mathrm{e}^{-\frac{\pi\mathrm{i}}{4}}\Big(\frac{\sqrt{2}-z}{z+\sqrt{2}}\Big)^{-\frac{1}{4}}\Big)$$

*and*

$$B_0(z)=\mathrm{e}^{-\frac{\pi\mathrm{i}n}{2}}\Big(\mathrm{e}^{-\frac{\pi\mathrm{i}}{4}}\Big(\frac{\sqrt{2}-z}{z+\sqrt{2}}\Big)^{\frac{1}{4}}+\mathrm{e}^{\frac{\pi\mathrm{i}}{4}}\Big(\frac{\sqrt{2}-z}{z+\sqrt{2}}\Big)^{-\frac{1}{4}}\Big).$$

*When $z\in\Omega_3$, a corresponding asymptotic expansion can be obtained by using the reflection formula $\pi_n^{(n)}(z)=(-1)^n\pi_n^{(n)}(-z)$.*

**Theorem 6.3** *Let $\Omega_i$, $i=1,\cdots,4$ be the regions shown in Figure 4, with $N=n+\alpha$, $t<-2\sqrt{N}$, $l_N$ and $f_N(z)$ defined in (3.96) and (5.54), respectively. The asymptotic expansion of the polynomial $\pi_n(N^{\frac{1}{4}}z)$ is given by*

$$\begin{aligned}\pi_n(N^{\frac{1}{4}}z)&=\sqrt{\pi}N^{\frac{n}{4}}\mathrm{e}^{\frac{1}{2}(Nz^4+t\sqrt{N}z^2)+\frac{1}{2}Nl_N}z^{-\alpha}\\&\quad\times[\mathrm{Ai}(N^{\frac{2}{3}}f_N(z))\boldsymbol{A}(z,N)-\mathrm{Ai}'(N^{\frac{2}{3}}f_N(z))\boldsymbol{B}(z,N)],\end{aligned}\tag{6.22}$$

*where $\boldsymbol{A}(z,N)$ and $\boldsymbol{B}(z,N)$ are analytic in $\Omega_1$ and have asymptotic expansions*

$$\boldsymbol{A}(z,N)\sim N^{\frac{1}{6}}f_N^{\frac{1}{4}}(z)\Big(\boldsymbol{W}_{11}(z)+\sum_{k=1}^{\infty}\frac{S_{11}^{(k)}(z)\,\boldsymbol{W}_{11}(z)-\mathrm{i}S_{12}^{(k)}(z)\,\boldsymbol{W}_{21}(z)}{N^k}\Big)\tag{6.23}$$

*and*

$$\boldsymbol{B}(z,N) \sim \frac{1}{N^{\frac{1}{6}} f_N^{\frac{1}{4}}(z)} \Big( \boldsymbol{W}_{12}(z) + \sum_{k=1}^{\infty} \frac{S_{11}^{(k)}(z)\boldsymbol{W}_{12}(z) - \mathrm{i} S_{12}^{(k)}(z)\boldsymbol{W}_{22}(z)}{N^k} \Big) \tag{6.24}$$

*uniformly for* $z \in \Omega_1$. *Furthermore,*

$$\boldsymbol{W} = \frac{1}{2} \begin{pmatrix} D_\infty D(z)^{-1} & D_\infty D(z) \\ D_\infty^{-1} D(z) & D_\infty^{-1} D(z)^{-1} \end{pmatrix} \begin{pmatrix} \widetilde{\gamma} + \widetilde{\gamma}^{-1} & \widetilde{\gamma} + \widetilde{\gamma}^{-1} \\ \widetilde{\gamma} - \widetilde{\gamma}^{-1} & -\widetilde{\gamma} + \widetilde{\gamma}^{-1} \end{pmatrix};$$

*the coefficient functions* $S_{ij}^{(k)}$ *and* $\boldsymbol{W}_{ij}$ *in* (6.23)–(6.24), $i, j = 1, 2$, *refer to the elements in the* $i$*th row and* $j$*th column of the matrices* $S_k(z)$ *and* $\boldsymbol{W}(z)$, *respectively. The function* $\widetilde{\gamma}(z)$ *is given in* (4.34). *Moreover,* $D(z)$ *and* $D_\infty$ *are given in* (4.31)–(4.32).

*Similarly, with* $\overline{f}_N(z)$ *given in* (5.63), *we have*

$$\begin{aligned} \pi_n(N^{\frac{1}{4}} z) &= \sqrt{\pi} N^{\frac{n}{4}} \mathrm{e}^{\frac{1}{2}(Nz^4 + t\sqrt{N}z^2) + \frac{1}{2} N l_N} z^{-\alpha} \\ &\quad \times [\mathrm{Ai}(N^{\frac{2}{3}} \overline{f}_N(z)) \overline{\boldsymbol{A}}(z,N) - \mathrm{Ai}'(N^{\frac{2}{3}} \overline{f}_N(z)) \overline{\boldsymbol{B}}(z,N)], \end{aligned} \tag{6.25}$$

*where* $\overline{\boldsymbol{A}}(z,N)$ *and* $\overline{\boldsymbol{B}}(z,N)$ *are analytic in* $\Omega_2$ *and have asymptotic expansions*

$$\overline{\boldsymbol{A}}(z,N) \sim N^{\frac{1}{6}} f_N^{\frac{1}{4}}(z) \Big( \overline{\boldsymbol{W}}_{11}(z) + \sum_{k=1}^{\infty} \frac{S_{11}^{(k)}(z)\overline{\boldsymbol{W}}_{11}(z) - \mathrm{i} S_{12}^{(k)}(z)\overline{\boldsymbol{W}}_{21}(z)}{N^k} \Big) \tag{6.26}$$

*and*

$$\overline{\boldsymbol{B}}(z,N) \sim \frac{1}{N^{\frac{1}{6}} f_N^{\frac{1}{4}}(z)} \Big( \overline{\boldsymbol{W}}_{12}(z) + \sum_{k=1}^{\infty} \frac{S_{11}^{(k)}(z)\overline{\boldsymbol{W}}_{12}(z) - \mathrm{i} S_{12}^{(k)}(z)\overline{\boldsymbol{W}}_{22}(z)}{N^k} \Big) \tag{6.27}$$

*uniformly for* $z \in \Omega_2$. *In* (6.26)–(6.27),

$$\overline{\boldsymbol{W}} = \frac{1}{2} \begin{pmatrix} D_\infty D(z)^{-1} \mathrm{e}^{\pm \frac{1}{2} N\pi \mathrm{i}} & D(z) D_\infty \mathrm{e}^{\mp \frac{1}{2} N\pi \mathrm{i}} \\ D_\infty^{-1} D(z)^{-1} \mathrm{e}^{\pm \frac{1}{2} N\pi \mathrm{i}} & D_\infty^{-1} D(z) \mathrm{e}^{\mp \frac{1}{2} N\pi \mathrm{i}} \end{pmatrix} \begin{pmatrix} \widetilde{\gamma} + \widetilde{\gamma}^{-1} & \widetilde{\gamma} + \widetilde{\gamma}^{-1} \\ \widetilde{\gamma}^{-1} - \widetilde{\gamma} & \widetilde{\gamma} - \widetilde{\gamma}^{-1} \end{pmatrix}.$$

*When* $z \in \Omega_3$ *and* $z \in \Omega_4$, *a corresponding asymptotic expansion can be obtained by using the reflection formula* $\pi_n^{(n)}(z) = (-1)^n \pi_n^{(n)}(-z)$.

By Theorem 2.1, there exists a $2 \times 2$ matrix $Y_1$ such that

$$Y(z) = \Big[ I + \frac{Y_1}{z} + O\Big(\frac{1}{z^2}\Big) \Big] z^{n\sigma_3} \quad \text{as } z \to \infty \tag{6.28}$$

(see condition $(Y_c)$). The recurrence coefficient $\beta_n(t;\alpha)$ in (1.2) is related to $Y_1$ by $\beta_n(t;\alpha) = (Y_1)_{12}(Y_1)_{21}$ (see, e.g., [6, (5.2)]). Thus, the asymptotics of the recurrence coefficient $\beta_n(t;\alpha)$ can also be derived by using the asymptotics of $Y$ as $n \to \infty$.

**Acknowledgement** The first author is grateful to Dan Dai, Yu Lin, Xiang-Sheng Wang and Lun Zhang for useful discussions.

## References

[1] Ablowitz, M. J. and Fokas, A. S., Complex variables: introduction and applications, Cambridge Texts in Applied Mathematics, Cambridge University Press, Cambridge, 1997.

[2] Bertola, M. and Tovbis, A., Asymptotics of orthogonal polynomials with complex varying quartic weight: Global structure, critical point behavior and the first Painlevé equation, *Constr. Approx.*, **41**(3), 2015, 529–587.

[3] Bertola, M. and Tovbis, A., On asymptotic regimes of orthogonal polynomials with complex varying quartic exponential weight, *SIGMA*, **12**(118), 2016, 50 pages.

[4] Bleher, P. and Its, A., Semiclassical asymptotics of orthogonal polynomials, Riemann-Hilbert problem, and universality in the matrix model, *Ann. Math.*, **150**(1), 1999, 185–266.

[5] Claeys, T. and Kuijlaars, A. B. J., Universality of the double scaling limit in random matrix models, *Comm. Pure Appl. Math.*, **59**(11), 2006, 1573–1603.

[6] Claeys, T., Kuijlaars, A. B. J. and Vanlessen, M., Multi-critical unitary random matrix ensembles and the general Painlevé II equation, *Ann. Math.*, **168**(2), 2008, 601–641.

[7] Clarkson, P. A. and Jordaan, K., Properties of generalized freud polynomials, *J. Approx. Theory*, **225**, 2018, 148–175.

[8] Clarkson, P. A., Jordaan, K. and Kelil, A., A generalized Freud weight, *Stud. Appl. Math.*, **136**(3), 2016, 288–320.

[9] Deift, P. A., Orthogonal polynomials and random matrices: A Riemann-Hilbert approach, Courant Lecture Notes in Mathematics, **3**, New York University, Courant Institute of Mathematical Sciences, New York, American Mathematical Society, Providence, RI, 1999.

[10] Deift, P. A., Kriecherbauer, T., McLaughlin, K. T.-R., et al., Uniform asymptotics for polynomials orthogonal with respect to varying exponential weights and applications to universality questions in random matrix theory, *Comm. Pure Appl. Math.,* **52**(11), 1999, 1335–1425.

[11] Deift, P. A., Kriecherbauer, T., McLaughlin, K. T.-R., et al., Strong asymptotics of orthogonal polynomials with respect to exponential weights, *Comm. Pure Appl. Math.*, **52**(12), 1999, 1491–1552.

[12] Deift, P. A. and Zhou, X., A steepest descent method for oscillatory Riemann-Hilbert problems, Asymptotics for the MKdV equation, *Ann. Math.*, **137**(2) , 1993, 295–368.

[13] Flaschka, H. and Newell, A. C., Monodromy- and spectrum-preserving deformations. I., *Comm. Math. Phys.*, **76**(1), 1980, 65–116.

[14] Fokas, A. S., Its, A. R., Kapaev, A. A. and Novokshenov, V. Y., Painlevé transcendents, the Riemann-Hilbert Approach, Mathematical Surveys and Monographs, **128**, American Mathematical Society, Providence, RI, 2006.

[15] Fokas, A. S., Its, A. R. and Kitaev, A. V., The isomonodromy approach to matrix models in 2D quantum gravity, *Comm. Math. Phys.*, **147**(2), 1992, 395–430.

[16] Mezzadri, F. and Mo, M. Y., On an average over the Gaussian unitary ensemble, *Int. Math. Res. Not. IMRN*, **18**, 2009, 3486–3515.

[17] Olver, F. W. J., Asymptotics and Special Functions, Academic Press, New York, London, 1974.

[18] Wang, Z. and Wong, R., Uniform asymptotics of the Stieltjes-Wigert polynomials via the Riemann-Hilbert approach, *J. Math. Pures Appl.*, **85**(5), 2006, 698–718.

[19] Wong, R. and Zhang, L., Global asymptotics for polynomials orthogonal with exponential quartic weight, *Asymptot. Anal.*, **64**(3–4), 2009, 125–154.

[20] Wong, R. and Zhang, L., Global asymptotics of orthogonal polynomials associated with $|x|^{2\alpha}e^{-Q(x)}$, *J. Approx. Theory*, **162**(4), 2010, 723–765.

PROCEEDINGS OF THE
AMERICAN MATHEMATICAL SOCIETY
Volume 147, Number 6, June 2019, Pages 2583–2597
https://doi.org/10.1090/proc/14405
Article electronically published on February 20, 2019

# ASYMPTOTICS OF THE ASSOCIATED POLLACZEK POLYNOMIALS

MIN-JIE LUO AND R. WONG

(Communicated by Mourad Ismail)

ABSTRACT. In this note, we investigate the large-$n$ behavior of the associated Pollaczek polynomials $P_n^\lambda(z;a,b,c)$. These polynomials involve four real parameters $\lambda$, $a$, $b$, and $c$, in addition to the complex variable $z$. Asymptotic formulas are derived for these polynomials, when $z$ lies in the complex plane bounded away from the interval of orthogonality $(-1,1)$, as well as in the interior of the interval of orthogonality. In the process of studying the asymptotic behavior of these polynomials when $z \in \mathbb{C}\setminus[-1,1]$, we found that the existing representations of $P_n^\lambda(z;a,b,c)$ do not provide useful information about their large-$n$ asymptotics. Here, we present a new representation in terms of the Gauss hypergeometric functions, from which the large-$n$ asymptotics for $z$ in $\mathbb{C}\setminus[-1,1]$ can be readily obtained. The asymptotic approximation in the interior of the interval of orthogonality is obtained by using asymptotic theory for difference equations.

## 1. INTRODUCTION

In a series of papers published in 1949 and 1950 in C. R. Acad. Sci. Paris, Pollaczek introduced certain families of orthogonal polynomials. Since these papers are very condensed and concise, you may also refer to his memoir [17], where Pollaczek provided more details about his technique. These orthogonal polynomials can be classified into two categories, depending on whether their intervals of orthogonality are finite or infinite; see [11, pp. 218–221]. In the present paper, we are only concerned with the first category, namely, those polynomials which are orthogonal over the finite interval $(-1,1)$. A representative class of polynomials in this category, which were studied by Pollaczek in [15], is now denoted by $P_n^\lambda(z;a,b)$ and can be defined by the three-term recurrence relation

$$\begin{aligned} nP_n^\lambda(z;a,b) &= 2\left[(n-1+\lambda+a)z+b\right]P_{n-1}^\lambda(z;a,b) \\ &\quad -(n+2\lambda-2)P_{n-2}^\lambda(z;a,b),\ n=1,2,\cdots, \end{aligned} \tag{1.1}$$

with initial conditions $P_{-1}^\lambda(z;a,b)=0$ and $P_0^\lambda(z;a,b)=1$; see [11, p. 219, Eq. (6)], [12, p. 147, Eq. (5.4.1)] and [13, p. 476, Eq. (18.35.2)]. In (1.1), $a$, $b$, and $\lambda$ are real parameters satisfying certain conditions. Usually, we write $z=\cos\theta$ with $\theta\in(0,\pi)$

Received by the editors August 7, 2018, and, in revised form, September 12, 2018, and September 19, 2018.
2010 *Mathematics Subject Classification.* Primary 33C45, 41A60; Secondary 33C05.
*Key words and phrases.* Asymptotics, difference equations, hypergeometric functions, Pollaczek polynomials.
The first author is the corresponding author.

so that the generating functions of $P_n^\lambda(\cos\theta; a, b)$ can be written as [13, p. 477, Eq. (18.35.7)]

$$\begin{aligned}\sum_{n=0}^{\infty} P_n^\lambda(\cos\theta; a, b)\, t^n &= G_{\mathrm{II}}(\cos\theta, t) \\ &:= \left(1 - te^{\mathrm{i}\theta}\right)^{-\lambda+\mathrm{i}\Phi}\left(1 - te^{-\mathrm{i}\theta}\right)^{-\lambda-\mathrm{i}\Phi}, \quad |t| < 1, \end{aligned} \tag{1.2}$$

where $\Phi \equiv \Phi(\theta) := (a\cos\theta + b)/\sin\theta$. The case when $\lambda = \frac{1}{2}$ in (1.1) gives another famous class of Pollaczek polynomials (see [3], [18], and [19]).

In the last paper [16] of his series, Pollaczek introduced a more general system of orthogonal polynomials $P_n^\lambda(z; a, b, c)$, which depends on four real parameters $a$, $b$, $c$, and $\lambda$. These polynomials are called *the associated Pollaczek polynomials* since they satisfy the same three-term recurrence relation as the Pollaczek polynomials $P_n^\lambda(z; a, b)$ just with $n$ replaced by $n + c$. That is, the associated Pollaczek polynomials $P_n^\lambda(z; a, b, c)$ satisfy

$$\begin{aligned}(n+c) P_n^\lambda(z; a, b, c) &= 2\left[(n - 1 + \lambda + a + c) z + b\right] P_{n-1}^\lambda(z; a, b, c) \\ &\quad - (n + 2\lambda + c - 2) P_{n-2}^\lambda(z; a, b, c), \quad n = 1, 2, \cdots, \end{aligned} \tag{1.3}$$

with initial conditions $P_{-1}^\lambda(z; a, b, c) = 0$, $P_0^\lambda(z; a, b, c) = 1$; see [16, p. 2254, Eq. (1)] and also [11, p. 220, Eq. (13)]. It was proved by Pollaczek that the generating function of $P_n^\lambda(z; a, b, c)$ is [16, p. 2254, Eq. (6)]

$$\begin{aligned}\sum_{n=0}^{\infty} P_n^\lambda(z; a, b, c)\, t^n &= G_{\mathrm{III}}(z, t) \\ &:= ct^{-c}(1 - \beta t)^A (1 - \alpha t)^B \int_0^t u^{c-1}(1 - \beta u)^{-A-1}(1 - \alpha u)^{-B-1} du, \end{aligned} \tag{1.4}$$

where

$$\alpha \equiv \alpha(z) := z + \sqrt{z^2 - 1}, \quad \beta \equiv \beta(z) := z - \sqrt{z^2 - 1} \tag{1.5}$$

and

$$A \equiv A(z) := -\lambda + \frac{az + b}{\sqrt{z^2 - 1}}, \quad B \equiv B(z) := -\lambda - \frac{az + b}{\sqrt{z^2 - 1}}. \tag{1.6}$$

Clearly, we have $\alpha(z)\beta(z) = 1$, $\alpha(z) + \beta(z) = 2z$, and $\alpha(z) - \beta(z) = 2\sqrt{z^2 - 1}$. A very precise description of these notation can be found in [5] and [12, pp. 144–146]. More importantly, we have $|\beta(z)| \le |\alpha(z)|$, and $|\alpha(z)| = |\beta(z)|$ if and only if $z \in [-1, 1]$; see [5, p. 1185, Lemma 2.5]. We need these properties in our later discussion. Derivation of (1.4) can also be found in [7, pp. 257–258]. The integral on the right-hand side of (1.4) cannot be expressed in terms of elementary functions, but can be represented in terms of a hypergeometric function of two variables. Recall that the Appell hypergeometric series $F_1$ is defined by [13, p. 413, Eq. (16.13.1)]

$$F_1[\alpha, \beta, \beta'; \gamma; z_1, z_2] := \sum_{m=0}^{\infty}\sum_{n=0}^{\infty} \frac{(\alpha)_{m+n}(\beta)_m(\beta')_n}{(\gamma)_{m+n}} \frac{z_1^m}{m!}\frac{z_2^n}{n!}, \quad \max\{|z_1|, |z_2|\} < 1.$$

When $\Re(\gamma) > \Re(\alpha) > 0$, $F_1$ has the single integral representation [13, p. 414, Eq. (16.15.1)]

$$\begin{aligned}&\frac{\Gamma(\alpha)\Gamma(\gamma-\alpha)}{\Gamma(\gamma)} F_1[\alpha,\beta,\beta';\gamma;z_1,z_2]\\ &\quad = \int_0^1 u^{\alpha-1}(1-u)^{\gamma-\alpha-1}(1-z_1u)^{-\beta}(1-z_2u)^{-\beta'}\,du,\\ &\qquad z_1\in\mathbb{C}\setminus[1,\infty),\ z_2\in\mathbb{C}\setminus[1,\infty).\end{aligned} \tag{1.7}$$

If we put $\alpha = c$, $\beta = 1 + A$, $\beta' = 1 + B$, $z_1 = \beta$, and $z_2 = \alpha$ in (1.7), then we have for $c > 0$,

$$G_{\mathrm{III}}(z,t) = (1-\beta t)^A (1-\alpha t)^B F_1[c;1+A,1+B;c+1;t\beta,t\alpha]. \tag{1.8}$$

Since the Appell hypergeometric function $F_1$ in (1.8) goes to 1 as $c \to 0$, we can obtain from (1.8) the generating function of $P_n^\lambda(z;a,b)$. Letting $z = \cos\theta$, $\theta \in (0,\pi)$, so that $\alpha(\cos\theta) = e^{i\theta}$, $\beta(\cos\theta) = e^{-i\theta}$, $A(\cos\theta) = -\lambda - i\Phi$, and $B(\cos\theta) = -\lambda + i\Phi$, we obtain

$$\begin{aligned}G_{\mathrm{III}}(z,t) &= \left(1-te^{-i\theta}\right)^{-\lambda-i\Phi}\left(1-te^{i\theta}\right)^{-\lambda+i\Phi}\\ &\quad\cdot F_1\left[c;1-\lambda-i\Phi,1-\lambda+i\Phi;c+1;te^{-i\theta},te^{i\theta}\right],\end{aligned} \tag{1.9}$$

where $\Phi$ is given in (1.2). Setting $a = b = 0$, $\lambda = \beta$, and $c = \gamma$ in (1.9) gives the generating function for the associated ultraspherical polynomials $C_n^\gamma(\cos\theta;\beta)$ [4, p. 730, Eq. (3.11)].

The large-$n$ asymptotics of the associated Pollaczek polynomials, when $z \in \mathbb{C}\setminus[-1,1]$, have been considered by Al-Salam and Ismail [6] under the frame of *seived orthogonal polynomials* by applying Darboux's method to (1.4). In fact, using the concept of the Hadamard integrals, Al-Salam and Ismail have established a general formula for the sieved associated Pollaczek polynomials of the second kind [6, p. 314, Theorem 3]. Letting $k = 1$, $r = 0$, and replacing $\lambda$ by $\lambda - 1$ in Theorem 3 of [6], we have

$$P_n^\lambda(z;a,b,c) \sim \frac{\Gamma(c+1)}{\Gamma(c-B)}\left(1-\beta^2\right)^A {}_2F_1\left[\begin{matrix}1+A,c\\ c-B\end{matrix};\beta^2\right]\beta^{-n}n^{-B-1} \tag{1.10}$$

as $n \to +\infty$, where $z$ is a fixed number in the cut plane $\in \mathbb{C}\setminus[-1,1]$. For $c = 0$, we have from [5, p. 1214, Eq. (6.11)]

$$P_n^\lambda(z;a,b) \sim \frac{1}{\Gamma(-B)}\left(1-\beta^2\right)^A \beta^{-n}n^{-B-1} \quad \text{as } n\to\infty.$$

The Pollaczek polynomials $P_n^\lambda(z;a,b)$ possess the following two widely used hypergeometric representations (see, for example, [11], [12], and [13]):

$$P_n^\lambda(\cos\theta;a,b) = e^{in\theta}\frac{(\lambda-i\Phi)_n}{n!}{}_2F_1\left[\begin{matrix}-n,\lambda+i\Phi\\ -n-\lambda+1+i\Phi\end{matrix};e^{-2i\theta}\right], \tag{1.11}$$

$$P_n^\lambda(\cos\theta;a,b) = e^{in\theta}\frac{(2\lambda)_n}{n!}{}_2F_1\left[\begin{matrix}-n,\lambda+i\Phi\\ 2\lambda\end{matrix};1-e^{-2i\theta}\right]. \tag{1.12}$$

If we allow $\theta$ to be complex, then (1.11) is particularly useful in deriving the large-$n$ asymptotics for the Pollaczek polynomials in the non-oscillatory region $\mathbb{C}\setminus[-1,1]$. Moreover, (1.11) can be easily obtained from the generating function (1.2). Also, it can be shown that (1.11) and (1.12) are closely related, by using a hypergeometric transformation of Pfaff [2, p. 79, Eq. (2.3.14)]. However, for the associated Pollaczek polynomials $P_n^\lambda(z;a,b,c)$, no such useful and easily obtainable representation can be found in the literature. A result of Pollaczek mentioned in [11, p. 220, Eq. (15)] shows that $P_n^\lambda(z;a,b,c)$ can be expressed as a ratio of cross products of Gauss hypergeometric functions; more precisely, we have

$$P_n^\lambda(z;a,b,c) := \frac{\mathsf{A}_{-1}\mathsf{B}_n - \mathsf{A}_n\mathsf{B}_{-1}}{\mathsf{A}_{-1}\mathsf{B}_0 - \mathsf{A}_0\mathsf{B}_{-1}}, \tag{1.13}$$

where

$$\mathsf{A}_n := \frac{\Gamma(2\lambda+c+n)}{\Gamma(c+n+1)} \mathrm{e}^{in\theta} {}_2F_1\left[\begin{matrix}-c-n, \lambda+\mathrm{i}\Phi \\ 2\lambda\end{matrix}; 1-\mathrm{e}^{-2\mathrm{i}\theta}\right]$$

and

$$\mathsf{B}_n := \mathrm{e}^{in\theta} {}_2F_1\left[\begin{matrix}1-2\lambda-c-n, 1-\lambda+\mathrm{i}\Phi \\ 2-2\lambda\end{matrix}; 1-\mathrm{e}^{-2\mathrm{i}\theta}\right].$$

In the above equations for $\mathsf{A}_n$ and $\mathsf{B}_n$, we have dropped the constants independent of $n$; this is just to make the expressions to appear as simple as possible; see our remark following (2.4). The same result (but in different notation) can also be found in [22, p. 303, Eq. (4)]. The representation (1.13) has several drawbacks. First, it is too complicated. Second, the Gauss hypergeometric functions appearing in (1.13) are not polynomials, and it is also not easy to verify that the right-hand side of equation (1.13) is a polynomial. Furthermore, since $n$ appears only in one of the first two parameters in ${}_2F_1$, letting $n \to \infty$ would not lead us to an asymptotic formula directly. In 1990, Wimp [22] developed a closed-form expression for $P_n^\lambda(z;a,b,c)$. Also, in a paper published in the same year, Hendriksen [9] found a formula with the same feature. Connections between these two results will be discussed in Section 2.

The main purpose of this paper is to find a neat and useful representation for the associated Pollaczek polynomials $P_n^\lambda(z;a,b,c)$, from which the asymptotic formula in (1.10) can be obtained by directly letting $n \to \infty$. The relationship between the Pollaczek polynomials and the Jacobi-Laurent polynomials is also discussed. Furthermore, we provide an asymptotic formula for $P_n^\lambda(z;a,b,c)$ in a region, containing the interval $(-1,1)$, where all the zeros lie.

## 2. Hypergeometric representation for the associated Pollaczek polynomials

In this section, we present a new representation for the associated Pollaczek polynomials $P_n^\lambda(z;a,b,c)$. Although this representation is new for $P_n^\lambda(z;a,b,c)$, a result for another polynomial of similar structure has appeared in the literature (see the discussion after the proof).

**Theorem 2.1.** *For $z \in \mathbb{C}$, the associated Pollaczek polynomial can be expressed as*

$$P_n^\lambda(z;a,b,c) = \frac{\Gamma(c+1)}{\Gamma(2\lambda+c-1)}\beta\left(\mathsf{F}_n\mathsf{G}_{-1} - \mathsf{F}_{-1}\mathsf{G}_n\right), \tag{2.1}$$

*where*

$$\mathsf{F}_n \equiv \mathsf{F}_n(\beta^2) := \frac{\Gamma(-B+n+c)}{\Gamma(n+c+1)}\alpha^n {}_2F_1\left[\begin{matrix}-n-c, -A \\ -n-c+1+B\end{matrix}; \beta^2\right]$$

*and*

$$\mathsf{G}_n \equiv \mathsf{G}_n(\beta^2) := \frac{\Gamma(2\lambda+c+n)}{\Gamma(-B+c+n+1)}\beta^n {}_2F_1\left[\begin{matrix}c+n+1, 1+A \\ -B+c+n+1\end{matrix}; \beta^2\right],$$

*where $\alpha$ and $\beta$, $A$ and $B$ are given in (1.5) and (1.6), respectively.*

*Proof.* Although the proof is quite long, its main idea is rather simple; it is actually the standard method used to find the explicit expression for associated orthogonal polynomials; see, for example, [9], [21], and [22]. We first construct two sequences $\mathfrak{f}_n$ and $\mathfrak{g}_n$ satisfying the same recurrence relation as $P_n^\lambda(z;a,b,c)$. We can write

$$P_n^\lambda(z;a,b,c) = \frac{\mathfrak{f}_{-1}\mathfrak{g}_n - \mathfrak{f}_n\mathfrak{g}_{-1}}{\mathfrak{f}_{-1}\mathfrak{g}_0 - \mathfrak{f}_0\mathfrak{g}_{-1}}, \tag{2.2}$$

provided that the denominator is not identically zero (i.e., $\mathfrak{f}_n$ and $\mathfrak{g}_n$ are linearly independent). Note that the quotient on the right-hand side of (2.2) is equal to 0 when $n=-1$, and equal to 1 when $n=0$; thus, this quotient satisfies the same initial-value conditions as $P_n^\lambda(z;a,b,c)$; see the line following equation (1.3). We then prove that the denominator in (2.2) is not identically zero and can be reduced to elementary functions.

Our starting point is the hypergeometric representation of the Pollaczek polynomial

$$P_n^\lambda(z;a,b) = \alpha^n \frac{(-B)_n}{n!} {}_2F_1\left[\begin{matrix}-n,-A\\-n+1+B\end{matrix};\beta^2\right]. \tag{2.3}$$

Since the Pochhammer symbol in front of the ${}_2F_1$–polynomial in (2.3) can be expressed in terms of gamma functions, replacing $n$ by $n+c$ in (2.3) gives

$$P_{n+c}^\lambda(z;a,b) = \alpha^{n+c}\frac{\Gamma(-B+n+c)}{\Gamma(-B)\Gamma(n+c+1)} {}_2F_1\left[\begin{matrix}-n-c,-A\\-n-c+1+B\end{matrix};\beta^2\right]. \tag{2.4}$$

Now, we consider the right-hand side of (2.2), and note that if $\mathfrak{f}_n$ and $\mathfrak{g}_n$ are replaced by their constant multiplies, say $\mathcal{C}\,\mathfrak{f}_n$ and $\mathcal{D}\,\mathfrak{g}_n$, the value of the quotient in (2.2) will remain the same, since the constants in the numerator and the denominator will cancel out. This obversion suggests that from now on we can omit the factors that are independent of $n$ in (2.4), and we define the first sequence $\mathfrak{f}_n$ by

$$\mathfrak{f}_n := \frac{\Gamma(-B+n+c)}{\Gamma(n+c+1)}\alpha^n {}_2F_1\left[\begin{matrix}-n-c,-A\\-n-c+1+B\end{matrix};\beta^2\right]. \tag{2.5}$$

To show that $\mathfrak{f}_n$ satisfies (1.3), we need the following relation:

$$\begin{aligned}{}_2F_1\left[\begin{matrix}\alpha,\beta\\\gamma\end{matrix};t\right] &= \left(1+\frac{-\alpha+\beta+1}{\gamma}t\right){}_2F_1\left[\begin{matrix}\alpha,\beta+1\\\gamma+1\end{matrix};t\right]\\ &\quad + \frac{(\alpha-\gamma-1)(\beta+1)}{\gamma(\gamma+1)}t\,{}_2F_1\left[\begin{matrix}\alpha,\beta+2\\\gamma+2\end{matrix};t\right];\end{aligned} \tag{2.6}$$

see [14, p. 376, Eq. (I)]; see also [9, p. 130, Eq. (2.10)].

Taking $\alpha=-A$, $\beta=-c-n$, $\gamma=-c+1+B-n$, and $t=\beta^2$ in (2.6) gives

$$\begin{aligned}{}_2F_1\left[\begin{matrix}-c-n,-A\\-c+1+B-n\end{matrix};\beta^2\right] &= \left(1+\frac{A-c-n+1}{B-c-n+1}\beta^2\right){}_2F_1\left[\begin{matrix}-c-n+1,-A\\-c+1+B-n+1\end{matrix};\beta^2\right]\\ &\quad - \frac{(2\lambda+c+n-2)(n+c-1)\beta^2}{(c-B+n-1)(c-B+n-2)}\,{}_2F_1\left[\begin{matrix}-c-n+2,-A\\-c+1+B-n+2\end{matrix};\beta^2\right].\end{aligned} \tag{2.7}$$

Multiplying (2.7) by $\alpha^n\Gamma(-B+n+c)/\Gamma(n+c+1)$ and using the definition of $\mathfrak{f}_n$ in (2.5) we obtain, $(n+c)\,\mathfrak{f}_n = \mathfrak{c}_n\mathfrak{f}_{n-1} - (2\lambda+c+n-2)\,\mathfrak{f}_{n-2}$, where $\mathfrak{c}_n := 2z(n+c-1) - \alpha B - \beta A$. Since

$$\alpha B + \beta A = -\lambda(\alpha+\beta) - 2\alpha\frac{az+b}{\alpha-\beta} + 2\beta\frac{az+b}{\alpha-\beta} = -2(\lambda z + az + b),$$

we conclude $\mathfrak{c}_n = 2\left[(\lambda+n+c+a-1)\,z+b\right]$. Thus, $\mathfrak{f}_n$ satisfies the recurrence relation (1.3) for the associated Pollaczek polynomials.

Next, we construct the second sequence $\mathfrak{g}_n$. First, we note that the hypergeometric part of $\mathfrak{f}_n$, denoted by

$$w_{1,n}(t) := {}_2F_1\left[\begin{matrix}-n-c,-A\\-n-c+1+B\end{matrix};t\right] \tag{2.8}$$

is a solution of the Gauss hypergeometric equation

$$t(1-t)\,w'' + [1-n-c+B-(1-A-n-c)\,t]\,w' - (-n-c)(-A)\,w = 0, \tag{2.9}$$

which has two linearly independent solutions. The first one is given in (2.8); the second one is given by (see [13, p. 395])

$$w_{2,n}(t) := t^{n+c-B}{}_2F_1\left[\begin{matrix}-B,2\lambda+c+n\\-B+c+n+1\end{matrix};t\right].$$

The latter one suggests that we can define the sequence $\mathfrak{g}_n$ by

$$\mathfrak{g}_n = \mathsf{K}_n\beta^{2n}{}_2F_1\left[\begin{matrix}-B,2\lambda+c+n\\-B+c+n+1\end{matrix};\beta^2\right], \tag{2.10}$$

where $\mathsf{K}_n$ is to be determined. As before, we also need $\mathfrak{g}_n$ to satisfy the recurrence relation (1.3).

This time we again need the result in (2.6). With $\alpha=-B$, $\beta=2\lambda+c+n-2$, $\gamma=-B+c+n-1$, and $t=\beta^2$, (2.6) becomes

$$\begin{aligned}{}_2F_1\left[\begin{matrix}-B,2\lambda+c+n-2\\-B+c+n-1\end{matrix};\beta^2\right] &= \left(1+\frac{A-c-n+1}{B-c-n+1}\beta^2\right){}_2F_1\left[\begin{matrix}-B,2\lambda+c+n-1\\-B+c+n\end{matrix};\beta^2\right]\\ &\quad -\frac{(c+n)(2\lambda+c+n-1)}{(-B+c+n)(-B+c+n-1)}\beta^2{}_2F_1\left[\begin{matrix}-B,2\lambda+c+n\\-B+c+n+1\end{matrix};\beta^2\right].\end{aligned} \tag{2.11}$$

Substituting (2.10) into (2.11) we get $(c+n)\,\mathfrak{g}_n = \mathfrak{s}_n\mathfrak{g}_{n-1} - \mathfrak{t}_n\mathfrak{g}_{n-2}$, where

$$\mathfrak{s}_n := \frac{\mathsf{K}_n}{\mathsf{K}_{n-1}}\left(1+\frac{A-c-n+1}{B-c-n+1}\beta^2\right)\frac{(-B+c+n)(-B+c+n-1)}{(2\lambda+c+n-1)},$$

$$\mathfrak{t}_n := \frac{\mathsf{K}_n}{\mathsf{K}_{n-2}}\frac{(-B+c+n)(-B+c+n-1)}{\alpha^2(2\lambda+c+n-1)}.$$

In order to have the sequence $\mathfrak{g}_n$ satisfying the recurrence relation (1.3), we must require that $\mathfrak{t}_n = (2\lambda+c+n-2)$. This choice gives the two-term recurrence relation

$$\frac{\mathsf{K}_n}{\mathsf{K}_{n-2}} = \alpha^2\frac{(2\lambda+c+n-1)(2\lambda+n+c-2)}{(-B+c+n)(-B+c+n-1)}. \tag{2.12}$$

A solution of (2.12) is given by $\mathsf{K}_n := \alpha^n\Gamma(2\lambda+c+n)/\Gamma(-B+c+n+1)$. Since $\mathsf{K}_n/\mathsf{K}_{n-1} = \alpha(2\lambda+c+n-1)/(-B+c+n)$, we immediately have

$$\mathfrak{s}_n = 2\left[(\lambda+n+c+a-1)\,z+b\right]. \tag{2.13}$$

With the choices made in (2.12) and (2.13), the sequence $\mathfrak{g}_n$ now also satisfies the recurrence relation (1.3).

If $\mathfrak{f}_n$ and $\mathfrak{g}_n$ are linearly independent solutions of the difference equation (1.3), which requires $\mathfrak{f}_{-1}\mathfrak{g}_0-\mathfrak{f}_0\mathfrak{g}_{-1}$ not identically zero, we can express the associated Pollaczek polynomials as a cross ratio of $\mathfrak{f}_n$ and $\mathfrak{g}_n$ as in (2.2). The linear independence is actually quite obvious in our case. But, we still need an explanation.

For convenience, we let $\mathfrak{f}_n = \alpha^n \hat{\mathfrak{f}}_n$ and $\mathfrak{g}_n = \beta^{-n}\hat{\mathfrak{g}}_n$ so that $\mathfrak{f}_{-1}\mathfrak{g}_0 - \mathfrak{f}_0\mathfrak{g}_{-1} = \beta(\hat{\mathfrak{f}}_{-1}\hat{\mathfrak{g}}_0 - \hat{\mathfrak{f}}_0\hat{\mathfrak{g}}_{-1})$. Equation (2.2) now can be rewritten as

$$P_n^{\lambda}(z;a,b,c) = \alpha^n \frac{\hat{\mathfrak{f}}_{-1}\hat{\mathfrak{g}}_n - \hat{\mathfrak{f}}_n\hat{\mathfrak{g}}_{-1}}{\hat{\mathfrak{f}}_{-1}\hat{\mathfrak{g}}_0 - \hat{\mathfrak{f}}_0\hat{\mathfrak{g}}_{-1}}.$$

Here we have made use of the result $\alpha\beta = 1$; see (1.5). We shall next show that

$$\Delta(\beta^2) := \hat{\mathfrak{f}}_{-1}\hat{\mathfrak{g}}_0 - \hat{\mathfrak{f}}_0\hat{\mathfrak{g}}_{-1} = -\frac{\Gamma(2\lambda+c-1)}{\Gamma(c+1)}\beta^{-2}\left(1-\beta^2\right)^{1-2\lambda}. \tag{2.14}$$

To this end, we write $\beta^2 = w$ in $\hat{\mathfrak{f}}_n$ and $\hat{\mathfrak{g}}_n$ so that $u_{1,n}(w) = \hat{\mathfrak{f}}_n(w)$ and $u_{2,n}(w) = z^{c-B}\hat{\mathfrak{g}}_n(w)$ are two linearly independent solutions to equation (2.9). It is known that the Wronskian $W(y_1,y_2)$ of any two linearly independently solutions of (2.9) is of the form

$$W(y_1,y_2) = D_n w^{n+c-1-B}(1-w)^{-2\lambda},$$

where $D_n$ is a non-zero constant depending on $a$, $b$, $c$, and $n$; see [13, p. 395]. We now establish the relationship between $\Delta(w)$ and $W(u_{1,0},u_{2,0})$.

Setting $\alpha = -A$, $\beta = -c$, $\gamma = -c+1+B$ in the identity [9, p. 128, Lemma 2.3]

$$\frac{\beta(\gamma-\alpha)}{\gamma}\,{}_2F_1\left[\begin{matrix}\alpha,\beta+1\\ \gamma+1\end{matrix};w\right] = \beta\,{}_2F_1\left[\begin{matrix}\alpha,\beta\\ \gamma\end{matrix};w\right] - (1-w)\frac{\mathrm{d}}{\mathrm{d}w}{}_2F_1\left[\begin{matrix}\alpha,\beta\\ \gamma\end{matrix};w\right]$$

gives the equation $(1-c-2\lambda)\hat{\mathfrak{f}}_{-1}(w) = -c\hat{\mathfrak{f}}_0(w) - (1-w)\hat{\mathfrak{f}}_0'(w)$, which is equivalent to

$$(1-w)u_{1,0}'(w) = (2\lambda+c-1)\hat{\mathfrak{f}}_{-1}(w) - c\hat{\mathfrak{f}}_0(w). \tag{2.15}$$

Setting $\alpha = -B$, $\beta = 2\lambda+c-1$ and $\gamma = -B+c$ in [9, p. 129, Lemma 2.4]

$$\gamma\,{}_2F_1\left[\begin{matrix}\alpha,\beta\\ \gamma\end{matrix};w\right] = (\gamma-\alpha w)\,{}_2F_1\left[\begin{matrix}\alpha,\beta+1\\ \gamma+1\end{matrix};w\right] + w(1-w)\frac{\mathrm{d}}{\mathrm{d}w}{}_2F_1\left[\begin{matrix}\alpha,\beta+1\\ \gamma+1\end{matrix};w\right]$$

gives $(2\lambda+c-1)w\hat{\mathfrak{g}}_{-1}(w) = (-B+c+Bw)\hat{\mathfrak{g}}_0(w) - w(1-w)\hat{\mathfrak{g}}_0'(w)$. Using the fact that $\hat{\mathfrak{g}}_0(w) = w^{-c+B}u_{2,0}(w)$, we obtain

$$-w^{-c+B}(1-w)u_{2,0}'(z) = c\hat{\mathfrak{g}}_0(w) - (2\lambda+c-1)\hat{\mathfrak{g}}_{-1}(w). \tag{2.16}$$

Since $W(u_{1,0},u_{2,0}) = u_{1,0}(w)u_{2,0}'(w) - u_{2,0}(w)u_{1,0}'(w)$, we have from (2.15) and (2.16)

$$\begin{aligned}
&-w^{-c+B}(1-w)W(u_{1,0},u_{2,0})\\
&\quad= -w^{-c+B}(1-w)u_{2,0}'(w)u_{1,0}(w) + w^{-c+B}u_{2,0}(w)\cdot(1-w)u_{1,0}'(w)\\
&\quad= \hat{\mathfrak{f}}_0(w)\left[c\hat{\mathfrak{g}}_0(w) - (2\lambda+c-1)\hat{\mathfrak{g}}_{-1}(w)\right] + \hat{\mathfrak{g}}_0(w)\left[(2\lambda+c-1)\hat{\mathfrak{f}}_{-1}(w) - c\hat{\mathfrak{f}}_0(w)\right]\\
&\quad= (2\lambda+c-1)\Delta(w).
\end{aligned}$$

On the other hand, we know in advance that the Wronskian must be of the form $W(u_{1,0},u_{2,0}) = Dw^{c-1-B}(1-w)^{-2\lambda}$, so

$$\Delta(w) = -\frac{D}{2\lambda+c-1}w^{-1}(1-w)^{1-2\lambda}. \tag{2.17}$$

The constant $D$ in $\Delta(w)$ is obtained from the coefficient of $w^{-1}$ in the Laurent series of

$$\begin{aligned}\Delta(w) &:= \hat{\mathfrak{f}}_{-1}\hat{\mathfrak{g}}_0 - \hat{\mathfrak{f}}_0\hat{\mathfrak{g}}_{-1}\\ &= \frac{\Gamma(2\lambda+c)\Gamma(-B-1+c)}{\Gamma(-B+c+1)\Gamma(c)}{}_2F_1\left[\begin{matrix}-B,2\lambda+c\\-B+c+1\end{matrix};w\right]{}_2F_1\left[\begin{matrix}1-c,-A\\2-c+B\end{matrix};w\right]\\ &\quad-\frac{1}{w}\frac{\Gamma(2\lambda+c-1)}{\Gamma(c+1)}{}_2F_1\left[\begin{matrix}-c,-A\\-c+1+B\end{matrix};w\right]{}_2F_1\left[\begin{matrix}-B,2\lambda+c-1\\-B+c\end{matrix};w\right],\end{aligned}$$

and we have

$$D=\frac{\Gamma(2\lambda+c)}{\Gamma(c+1)}. \tag{2.18}$$

Now, we can obtain (2.14) by substituting (2.18) into (2.17) and letting $w=\beta^2$ in the resulting equation.

Combining all the results together yields

$$\begin{aligned}P_n^\lambda(z;a,b,c) &= -\frac{\Gamma(c+1)}{\Gamma(2\lambda+c-1)}\alpha^n\beta^2\left(1-\beta^2\right)^{2\lambda-1}\left(\alpha\mathfrak{f}_{-1}\beta^n\mathfrak{g}_n-\beta^n\mathfrak{f}_n\alpha\mathfrak{g}_{-1}\right)\\ &= -\frac{\Gamma(c+1)}{\Gamma(2\lambda+c-1)}\beta\left(1-\beta^2\right)^{2\lambda-1}\left(\mathfrak{f}_{-1}\mathfrak{g}_n-\mathfrak{f}_n\mathfrak{g}_{-1}\right).\end{aligned} \tag{2.19}$$

Finally, we note that by using the Euler transformation [2, p. 68, Eq. (2.2.7)], $\mathfrak{g}_n$ can be rewritten as

$$\mathfrak{g}_n=\frac{\Gamma(2\lambda+c+n)}{\Gamma(-B+c+n+1)}\beta^n\left(1-\beta^2\right)^{1-2\lambda}{}_2F_1\left[\begin{matrix}c+n+1,1-B-2\lambda\\-B+c+n+1\end{matrix};\beta^2\right].$$

Introducing a pair of new sequences

$$\mathsf{F}_n:=\mathfrak{f}_n \text{ and } \mathsf{G}_n:=\frac{\Gamma(2\lambda+c+n)}{\Gamma(-B+c+n+1)}\beta^n{}_2F_1\left[\begin{matrix}c+n+1,1-B-2\lambda\\-B+c+n+1\end{matrix};\beta^2\right],$$

the factor $(1-\beta)^{2\lambda-1}$ on the right-hand side of (2.19) can be eliminated. This completes the proof. □

*Remark* 2.2. When $c=0$, we have $\mathsf{F}_{-1}=0$ and $\mathsf{G}_{-1}:=\alpha\Gamma(2\lambda-1)/\Gamma(-B)$ and (2.1) reduces to the Pollaczek polynomials of three parameters:

$$P_n^\lambda(z;a,b,0)=\frac{\beta}{\Gamma(2\lambda-1)}\mathsf{F}_n\mathsf{G}_{-1}=\frac{\mathsf{F}_n}{\Gamma(-B)}=P_n^\lambda(z;a,b).$$

Hendriksen [9] has applied this method to the *associated Jacobi-Laurent polynomials*. He considered the sequence $V_n(t;\mathsf{a},\mathsf{b},\mathsf{c})$ satisfying the recurrence relation

$$V_n(t;\mathsf{a},\mathsf{b},\mathsf{c})=(t-\mathfrak{A}_n)V_{n-1}(t;\mathsf{a},\mathsf{b},\mathsf{c})-\mathfrak{B}_n\,tV_{n-2}(t;\mathsf{a},\mathsf{b},\mathsf{c}),\ n=1,2,\cdots, \tag{2.20}$$

where

$$\mathfrak{B}_1:=-\frac{(\mathsf{b}+\mathsf{c})}{(-\mathsf{a}+\mathsf{c}+1)},\ \mathfrak{B}_n:=\frac{(n+\mathsf{c}-1)(\mathsf{b}-\mathsf{a}+n+\mathsf{c}-1)}{(-\mathsf{a}+n+\mathsf{c}-1)(-\mathsf{a}+n+\mathsf{c})},\ n=2,3,\cdots,$$

and

$$\mathfrak{A}_n:=-\frac{(\mathsf{b}+n+\mathsf{c}-1)}{(-\mathsf{a}+n+\mathsf{c})}\ n=1,2,\cdots.$$

The initial conditions are $V_{-1}(t; \mathsf{a}, \mathsf{b}, \mathsf{c}) = 0$ and $V_0(t; \mathsf{a}, \mathsf{b}, \mathsf{c}) = 1$. In terms of these polynomials, the associated Jacobi-Laurent polynomials $W_n(t; \mathsf{a}, \mathsf{b}, \mathsf{c})$ can be expressed as

$$W_{2n}(t; \mathsf{a}, \mathsf{b}, \mathsf{c}) = t^n V_{2n}(t; \mathsf{a}, \mathsf{b}, \mathsf{c}) \text{ and } W_{2n+1}(t; \mathsf{a}, \mathsf{b}, \mathsf{c}) = t^{-n-1} V_{2n+1}(t; \mathsf{a}, \mathsf{b}, \mathsf{c}).$$

Note that we have slightly modified the notation used by Hendriksen. For more details about the Jacobi-Laurent polynomials, the interested reader is referred to [8] and [10]. Here we shall focus only on the relationship between the polynomials $V_n(t; \mathsf{a}, \mathsf{b}, \mathsf{c})$ and the associated Pollaczek polynomials $P_n^\lambda(z; a, b, c)$.

When $\mathsf{c} = 0$, we have

$$V_n(t; \mathsf{a}, \mathsf{b}, 0) = \frac{(\mathsf{b})_n}{(-\mathsf{a}+1)_n} {}_2F_1\left[\begin{matrix}-n, -\mathsf{a}+1 \\ -\mathsf{b}-n+1\end{matrix}; t\right]; \tag{2.21}$$

see [9, p. 128]. Letting $\mathsf{a} = A+1$, $\mathsf{b} = -B$ and $t = \beta^2$ in (2.21) gives

$$V_n(\beta^2; A+1, -B, 0) = \frac{(-B)_n}{(-A)_n} {}_2F_1\left[\begin{matrix}-n, -A \\ -n+1+B\end{matrix}; \beta^2\right],$$

where $A$ and $B$ are defined in (1.6), and $\beta$ is given by (1.5). In view of the representation (2.3), we have

$$P_n^\lambda(z; a, b) = \alpha^n \frac{(-A)_n}{n!} V_n(\beta^2; A+1, -B, 0).$$

Since $P_n^\lambda(z; a, b)$ can be expressed in terms of $V_n(t; \mathsf{a}, \mathsf{b}; 0)$, it is natural to expect that $P_n^\lambda(z; a, b, c)$ can also be written in terms of $V_n(t; \mathsf{a}, \mathsf{b}, \mathsf{c})$. As in (2.10), we let

$$\mathsf{K}_n P_n^\lambda(z; a, b, c) = V_n(\beta^2; A+1, -B, c), \tag{2.22}$$

where $\mathsf{K}_n$ is to be determined. From the recurrence relation (2.20), we find that the sequence $V_n$ on the right-hand side of (2.22) satisfies

$$\begin{aligned} V_n(\beta^2; A+1, -B, c) &= \left(\beta^2 + \frac{n+c-1-B}{n+c-1-A}\right) V_{n-1}(\beta^2; A+1, -B, c) \\ &\quad - \frac{(n+c-1)(n+c+2\lambda-2)}{(n+c-A-2)(n+c-A-1)} \beta^2 V_{n-2}(\beta^2; A+1, -B, c), \end{aligned}$$

from which we obtain the recurrence relation for $\mathsf{K}_n P_n^\lambda(z; a, b, c)$:

$$\begin{aligned} \mathsf{K}_n P_n^\lambda(z; a, b, c) &= \left(\beta^2 + \frac{n+c-1-B}{n+c-1-A}\right) \mathsf{K}_{n-1} P_{n-1}^\lambda(z; a, b, c) \\ &\quad - \frac{(n+c-1)(n+c+2\lambda-2)}{(n+c-A-2)(n+c-A-1)} \beta^2 \mathsf{K}_{n-2} P_{n-2}^\lambda(z; a, b, c). \end{aligned}$$

The value of $\mathsf{K}_n$ can be determined by requiring the above recurrence relation to have the same form as (1.3), and the result is $\mathsf{K}_n = \beta^n (c+1)_n/(c-A)_n$. Thus,

$$P_n^\lambda(z; a, b, c) = \alpha^n \frac{(c-A)_n}{(c+1)_n} V_n(\beta^2; A+1, -B, c).$$

For the polynomials $V_n$, it has been proved by Hendriksen [9, p. 132] that

$$V_n(t;\mathsf{a},\mathsf{b},\mathsf{c}) = \frac{(\mathsf{b}+\mathsf{c})_n}{(-\mathsf{a}+\mathsf{c}+1)_n} {}_2F_1\left[\begin{matrix}-\mathsf{a}+1,-n-\mathsf{c}\\-\mathsf{b}-n-\mathsf{c}+1\end{matrix};t\right] {}_2F_1\left[\begin{matrix}\mathsf{c},\mathsf{a}\\\mathsf{b}+\mathsf{c}\end{matrix};t\right]$$

$$-\frac{(\mathsf{b}-\mathsf{a}+\mathsf{c})_{n+1}(\mathsf{c})_{n+1}}{(\mathsf{b}+\mathsf{c}-1)_{n+2}(-\mathsf{a}+\mathsf{c}+1)_n} t^{n+1} {}_2F_1\left[\begin{matrix}-\mathsf{a}+1,-\mathsf{c}+1\\-\mathsf{b}-\mathsf{c}+2\end{matrix};t\right] {}_2F_1\left[\begin{matrix}n+\mathsf{c}+1,\mathsf{a}\\\mathsf{b}+n+\mathsf{c}+1\end{matrix};t\right]. \tag{2.23}$$

By suitable choices of parameters, one can obtain the same result as given in our Theorem 2.1, after some calculation.

It is worth mentioning that the formula (2.23) is not the final form of Hendriksen's main result (see [9, p. 128, Theorem 2.1]). But this formula is surprisingly suitable for finding valuable information about the asymptotic behavior of $V_n$ for large $n$.

## 3. Asymptotics

The main task in this section is to establish an asymptotic formula for the associated Pollaczek polynomials in the oscillatory region by using the technique developed by Wang and Wong [20]. As a start, we present a simple proof of Al-Salam and Ismail's result stated in (1.10) by using Theorem 2.1.

Taking the limit $n\to\infty$ in the hypergeometric functions in $\mathsf{F}_n$ and $\mathsf{G}_n$, we have

$$\mathsf{F}_n \sim n^{-B-1}\alpha^n\left(1-\beta^2\right)^A \text{ and } \mathsf{G}_n \sim n^{2\lambda+B-1}\beta^n\left(1-\beta^2\right)^{-A-1},$$

as $n\to\infty$, $z\in\mathbb{C}\setminus[-1,1]$ $\left(|\beta|^2<1\right)$. Then, by Theorem 2.1, we have

$$P_n^\lambda(z;a,b,c) \sim \frac{\Gamma(c+1)}{\Gamma(2\lambda+c-1)}\beta\left(1-\beta^2\right)^A \mathsf{G}_{-1} n^{-B-1}\alpha^n$$

$$-\frac{\Gamma(c+1)}{\Gamma(2\lambda+c-1)}\beta\left(1-\beta^2\right)^{-A-1}\mathsf{F}_{-1} n^{2\lambda+B-1}\beta^n, \tag{3.1}$$

where

$$\mathsf{F}_{-1} = \frac{\Gamma(-B-1+c)}{\Gamma(c)}\alpha^{-1} {}_2F_1\left[\begin{matrix}1-c,-A\\2-c+B\end{matrix};\beta^2\right]$$

and

$$\mathsf{G}_{-1} = \frac{\Gamma(2\lambda+c-1)}{\Gamma(-B+c)}\beta^{-1} {}_2F_1\left[\begin{matrix}c,1+A\\c-B\end{matrix};\beta^2\right].$$

Since $|\beta|<|\alpha|$ when $z\in\mathbb{C}\setminus[-1,1]$, we can omit the second term in (3.1) to get the formula (1.10).

It is worth mentioning that as a direct consequence of (1.10), we can obtain the continued fraction $\chi(z)$ associated with the polynomials $P_n^\lambda(z;a,b,c)$. For short, we let $\mathfrak{p}_n(z) := P_n^\lambda(z;a,b,c)$, and denote by $\mathfrak{p}_n^*(z)$ the corresponding numerator polynomials. It is not difficult to see that

$$\mathfrak{p}_n^*(z) = \frac{2(\lambda+a+c)}{(c+1)} P_{n-1}^\lambda(z;a,b,c+1). \tag{3.2}$$

Then by using (1.10) and (3.2) we have

$$\chi(z) = \lim_{n\to\infty}\frac{\mathfrak{p}_n^*(z)}{\mathfrak{p}_n(z)} = \frac{2\beta(\lambda+a+c)}{c-B}\frac{{}_2F_1\left[\begin{matrix}1+A,c+1\\c-B+1\end{matrix};\beta^2\right]}{{}_2F_1\left[\begin{matrix}1+A,c\\c-B\end{matrix};\beta^2\right]}, \tag{3.3}$$

where $z \in \Omega := \{z \in \mathbb{C} \setminus [-1,1] \text{ and } \Re(B(z)) \neq 1, 2, \cdots\}$. This result can also be derived from a theorem of Al-Salam and Ismail [6, p. 316, Theorem 5] by taking $k = 1$, $r = 0$ and replacing $\lambda$ by $\lambda - 1$. For more detailed information about the continued fraction $\chi(z)$ and the numerator polynomials, the interested reader may refer to [12, Chapter 2].

The following two identities are used in the proof of Theorem 3.2.

**Lemma 3.1.** *When $\Re(\theta) \in (0, \pi)$ we have*

$$(1 - e^{2i\theta})^{-\lambda + i\Phi} e^{-in\theta} n^{i\Phi} = (2 \sin\theta)^{-\lambda} e^{(\pi/2-\theta)\Phi} e^{-i\Psi_n} \tag{3.4}$$

*and*

$$(1 - e^{-2i\theta})^{-\lambda - i\Phi} e^{in\theta} n^{-i\Phi} = (2 \sin\theta)^{-\lambda} e^{(\pi/2-\theta)\Phi} e^{i\Psi_n},$$

*where*

$$\Psi_n \equiv \Psi_n(\theta) := n\theta - \Phi \log(2n \sin\theta) - \lambda\left(\frac{\pi}{2} - \theta\right). \tag{3.5}$$

*Proof.* From the fact that $\sin\theta = \frac{1}{2i}(e^{i\theta} - e^{-i\theta})$ we can write

$$(1 - e^{2i\theta})^{-\lambda + i\Phi} = \left(2 \sin\theta\, e^{i(\theta - \pi/2)}\right)^{-\lambda + i\Phi}.$$

Since $\arg(\sin\theta) \in \left(-\frac{\pi}{2}, \frac{\pi}{2}\right)$ and $\arg\left(e^{i(\theta-\pi/2)}\right) \in \left(-\frac{\pi}{2}, \frac{\pi}{2}\right)$ for $\Re(\theta) \in (0, \pi)$, we have

$$\begin{aligned}(1 - e^{2i\theta})^{-\lambda + i\Phi} &= (2 \sin\theta)^{-\lambda + i\Phi} \left(e^{i(\theta - \pi/2)}\right)^{-\lambda + i\Phi} \\ &= (2 \sin\theta)^{-\lambda} e^{(\pi/2-\theta)\Phi} e^{i\Phi \log(2\sin\theta) + i\lambda(\pi/2-\theta)}.\end{aligned}$$

Then (3.4) follows immediately by noting that $e^{-in\theta} n^{i\Phi}$ can be expressed as $e^{-in\theta + i\Phi \log(n)}$. The second identity can be proved in the same manner. □

To simplify the notation, we define

$$\mathbf{F}_{\pm}(\theta) := \frac{1}{\Gamma(c + \lambda \pm i\Phi)} {}_2F_1\left[\begin{matrix}1 - \lambda \pm i\Phi, c \\ c + \lambda \pm i\Phi\end{matrix}; e^{\pm 2i\theta}\right], \tag{3.6}$$

where $a$, $b$, $c$, $\lambda$ are real and $\theta$ could be a complex number with $\Re(\theta) \in (0, \pi)$. The function $\Phi \equiv \Phi(\theta)$ is given in (1.2). For convenience, we write

$$\mathbf{F}_{+}(\theta) = |\mathbf{F}_{+}(\theta)| e^{i\psi_+} \text{ and } \mathbf{F}_{-}(\theta) = |\mathbf{F}_{-}(\theta)| e^{i\psi_-}. \tag{3.7}$$

In addition, if $\theta$ is real and satisfies $\theta \in (0, \pi)$, we have $\overline{\mathbf{F}_{+}(\theta)} = \mathbf{F}_{-}(\theta)$ and $\psi_+ = -\psi_-$, and functions $\mathbf{F}_{+}(\theta)$, $\mathbf{F}_{-}(\theta)$ can be expressed in a simpler manner:

$$\mathbf{F}_{+}(\theta) = |\mathbf{F}_{+}(\theta)| e^{-i\psi} \text{ and } \mathbf{F}_{-}(\theta) = |\mathbf{F}_{+}(\theta)| e^{i\psi}. \tag{3.8}$$

**Theorem 3.2.** *Let $\mathcal{N}$ be a small complex neighborhood of the interval $[-1+\delta, 1-\delta]$ $(0 < \delta < 1)$. For $z \in \mathcal{N}$, we have*

(3.9)
$$P_n^{\lambda}(z; a, b, c) \sim \frac{\Gamma(c+1)}{(2\sin\theta)^{\lambda}} e^{(\pi/2-\theta)\Phi} n^{\lambda-1} \left[|\mathbf{F}_{-}(\theta)| e^{i(\psi_- + \Psi_n)} + |\mathbf{F}_{+}(\theta)| e^{i(\psi_+ - \Psi_n)}\right],$$

*as $n \to \infty$, where $\theta \equiv \theta(z) := \arccos(z)$ with $\Re(\theta) \in (0, \pi)$. Functions $\mathbf{F}_{+}(\theta)$, $\mathbf{F}_{-}(\theta)$ and $\Psi_n$ are defined by (3.7) and (3.5), respectively.*

*In particular, setting $z = \cos\theta$ with $\theta \in (0, \pi)$, we obtain*

$$P_n^{\lambda}(z; a, b, c) \sim 2 |\mathbf{F}_{+}(\theta)| \frac{\Gamma(c+1)}{(2\sin\theta)^{\lambda}} e^{(\pi/2-\theta)\Phi} n^{\lambda-1} \cos(\psi + \Psi_n), \tag{3.10}$$

*as* $n \to \infty$.

*Proof.* First of all, we rewrite (1.3) as

$$P_{n+2}^{\lambda}(z;a,b,c) + r_1(n) P_{n+1}^{\lambda}(z;a,b,c) + r_2(n) P_n^{\lambda}(z;a,b,c) = 0, \tag{3.11}$$

where

$$r_1(n) := -\frac{2(n+1+\lambda+a+c)z+2b}{(n+2+c)} = -2z - \frac{2z(a+\lambda-1)+2b}{n} + \mathcal{O}\left(\frac{1}{n^2}\right)$$

and

$$r_2(n) := \frac{n+2\lambda+c}{n+2+c} = 1 + \frac{2(\lambda-1)}{n} + \mathcal{O}\left(\frac{1}{n^2}\right).$$

The characteristic equation of (3.11) is $\lambda^2 - 2\cos\theta\lambda + 1 = 0$, whose roots are $\lambda_1 = \mathrm{e}^{\mathrm{i}\theta}$ and $\lambda_2 = \mathrm{e}^{-\mathrm{i}\theta}$.

Recall that when $\lambda_1 \neq \lambda_2$, the second-order difference equation

$$P_{n+2}(z) + r_1(n) P_{n+1}(z) + r_2(n) P_n(z) = 0, \tag{3.12}$$

whose coefficients $r_1(n)$ and $r_1(n)$ possess asymptotic expansions of the form

$$r_1(n) \sim \sum_{j=0}^{\infty} \frac{a_j}{n^j}, \quad r_2(n) \sim \sum_{j=0}^{\infty} \frac{b_j}{n^j} \quad (b_0 \neq 0),$$

has two linearly independent solutions $y_1(n)$ and $y_2(n)$ of the form

$$y_i(n) \sim \lambda_i^n n^{\alpha_i} \sum_{r=0}^{\infty} \frac{c_i(r)}{n^r}, \quad i = 1, 2,$$

where $\alpha_i = (a_1\lambda_i + b_1)/(a_0\lambda_i + 2b_0)$ $(i = 1, 2)$ and $c_1(0) = c_2(0) = 1$ and the higher order coefficients can be determined recursively. The solutions $y_1(n)$ and $y_2(n)$ are also called *normal solutions* of (3.12). For a complete theory of the asymptotics of second-order difference equation of type (3.12), the interested reader is referred to the paper of Wong and Li [23].

In our case, $a_0 = -2\cos\theta$, $a_1 = -2\Phi\sin\theta - 2(\lambda-1)\cos\theta$, $b_0 = 1$, and $b_1 = 2(\lambda-1)$, and thus

$$\alpha_1 = -\mathrm{i}\Phi + \lambda - 1 \text{ and } \alpha_2 = \mathrm{i}\Phi + \lambda - 1.$$

The two normal solutions are

$$y_1(n) \sim \mathrm{e}^{\mathrm{i}n\theta} n^{-\mathrm{i}\Phi+\lambda-1} \sum_{r=0}^{\infty} \frac{\tilde{c}_1(r)}{n^r}, \quad y_2(n) \sim \mathrm{e}^{-\mathrm{i}n\theta} n^{\mathrm{i}\Phi+\lambda-1} \sum_{r=0}^{\infty} \frac{\tilde{c}_2(r)}{n^r},$$

where $\tilde{c}_1(0) = \tilde{c}_2(0) = 1$. So we have

$$P_n^{\lambda}(z;a,b,c) \sim C_1(\theta) \mathrm{e}^{\mathrm{i}n\theta} n^{-\mathrm{i}\Phi+\lambda-1} + C_2(\theta) \mathrm{e}^{-\mathrm{i}n\theta} n^{\mathrm{i}\Phi+\lambda-1}, \tag{3.13}$$

as $n \to \infty$, where $C_1(\theta)$ and $C_2(\theta)$ are functions to be determined. The relation (3.13) holds for $z \in \mathcal{N}$, which overlaps with the region $\mathbb{C} \setminus [-1, 1]$, where the asymptotic formula (1.10) holds. When $\Im(z) = \Im(\cos\theta) > 0$, we have $\Im(\theta) < 0$, and (3.13) gives

$$P_n^{\lambda}(z;a,b,c) \sim C_1(\theta) \mathrm{e}^{\mathrm{i}n\theta} n^{-\mathrm{i}\Phi+\lambda-1}.$$

Similarly, for $\Im(z) < 0$ (i.e., $\Im(\theta) > 0$), we have on account of (1.5) and (1.6)

$$P_n^{\lambda}(z;a,b,c) \sim C_2(\theta) \mathrm{e}^{-\mathrm{i}n\theta} n^{\mathrm{i}\Phi+\lambda-1}.$$

On the other hand, from (1.10), we have

$$P_n^\lambda(z;a,b,c) \sim \begin{cases} \dfrac{\Gamma(c+1)}{\Gamma(c+\lambda-\mathrm{i}\Phi)}\left(1-\mathrm{e}^{-2\mathrm{i}\theta}\right)^{-\lambda-\mathrm{i}\Phi} \\ \qquad \cdot {}_2F_1\left[\begin{matrix}1-\lambda-\mathrm{i}\Phi, c \\ c+\lambda-\mathrm{i}\Phi\end{matrix}; \mathrm{e}^{-2\mathrm{i}\theta}\right]\mathrm{e}^{\mathrm{i}n\theta}n^{-\mathrm{i}\Phi+\lambda-1}, & \Im(\theta)<0, \\ \dfrac{\Gamma(c+1)}{\Gamma(c+\lambda+\mathrm{i}\Phi)}\left(1-\mathrm{e}^{2\mathrm{i}\theta}\right)^{-\lambda+\mathrm{i}\Phi} \\ \qquad \cdot {}_2F_1\left[\begin{matrix}1-\lambda+\mathrm{i}\Phi, c \\ c+\lambda+\mathrm{i}\Phi\end{matrix}; \mathrm{e}^{2\mathrm{i}\theta}\right]\mathrm{e}^{-\mathrm{i}n\theta}n^{\mathrm{i}\Phi+\lambda-1}, & \Im(\theta)>0. \end{cases}$$

So by matching we have

$$\mathcal{C}_1(\theta) = \Gamma(c+1)\left(1-\mathrm{e}^{-2\mathrm{i}\theta}\right)^{-\lambda-\mathrm{i}\Phi}\mathbf{F}_-(\theta)$$

and

$$\mathcal{C}_2(\theta) = \Gamma(c+1)\left(1-\mathrm{e}^{2\mathrm{i}\theta}\right)^{-\lambda+\mathrm{i}\Phi}\mathbf{F}_+(\theta),$$

where $\mathbf{F}_+(\theta)$ and $\mathbf{F}_+(\theta)$ are defined in (3.6).

From Lemma 3.1, it follows that

$$\begin{aligned} P_n^\lambda(z;a,b,c) &\sim \Gamma(c+1)\,n^{\lambda-1}\Big[\left(1-\mathrm{e}^{-2\mathrm{i}\theta}\right)^{-\lambda-\mathrm{i}\Phi(\theta)}\mathrm{e}^{\mathrm{i}n\theta}n^{-\mathrm{i}\Phi}\left|\mathbf{F}_-(\theta)\right|\mathrm{e}^{\mathrm{i}\psi_-} \\ &\qquad + \left(1-\mathrm{e}^{2\mathrm{i}\theta}\right)^{-\lambda+\mathrm{i}\Phi(\theta)}\mathrm{e}^{-\mathrm{i}n\theta}n^{\mathrm{i}\Phi}\left|\mathbf{F}_+(\theta)\right|\mathrm{e}^{\mathrm{i}\psi_+}\Big] \\ &= \frac{\Gamma(c+1)}{(2\sin\theta)^\lambda}\mathrm{e}^{(\pi/2-\theta)\Phi}n^{\lambda-1}\Big[\left|\mathbf{F}_-(\theta)\right|\mathrm{e}^{\mathrm{i}(\psi_-+\Psi_n)} + \left|\mathbf{F}_+(\theta)\right|\mathrm{e}^{\mathrm{i}(\psi_+-\Psi_n)}\Big]. \end{aligned}$$

This completes the proof of (3.9). In view of (3.8), we obtain (3.10). □

*Remark* 3.3. To determine the weight function of a given sequence of orthogonal polynomials, we often resort to Nevai's theorem; see for example, [2, pp. 312–313] and [5, p. 1217, Theorem 6.4]. But in our case, the main condition of Nevai's theorem is not satisfied. Surprisingly, even so, Nevai's theorem still gives us the right weight function. In fact, Nevai's theorem provides us the following limit relation:

$$\limsup_n \left\{\psi'(x)\sqrt{1-x^2}\,\mathfrak{p}_n^2(x)/\lambda_n\right\} = \frac{2}{\pi} \text{ a.e. on } [-1,1], \tag{3.14}$$

where $\mathfrak{p}_n(x) := P_n^\lambda(x;a,b,c)$ as before and

$$\lambda_n = \frac{\lambda+a+c}{n+\lambda+a+c}\frac{\Gamma(c+1)\Gamma(2\lambda+c+n)}{\Gamma(2\lambda+c)\Gamma(c+n+1)} \sim (\lambda+a+c)\frac{\Gamma(c+1)}{\Gamma(2\lambda+c)}n^{2\lambda-2}, \tag{3.15}$$

as $n\to\infty$. Using (3.10), (3.14), and (3.15), our guess of the weight function is

$$\psi'(\cos\theta) = \frac{1}{\pi}\frac{(2\sin\theta)^{2\lambda-1}\mathrm{e}^{(2\theta-\pi)\Phi}}{\Gamma(2\lambda+c)\Gamma(c+1)}(\lambda+a+c)\left|\mathbf{F}_+(\theta)\right|^{-2},$$

where $\mathbf{F}_+(\theta)$ is given in (3.6). The orthogonality relation for the associated Pollaczek polynomials could be written as

(3.16)

$$\int_{-1}^1 P_m^\lambda(x;a,b,c)\,P_n^\lambda(x;a,b,c)\,w(x)\,\mathrm{d}x = \frac{2\pi[\Gamma(c+1)]^2\,\Gamma(n+2\lambda+c)}{2^{2\lambda}(n+\lambda+a+c)\Gamma(n+c+1)}\delta_{mn},$$

where

$$w(x) = \frac{\pi\Gamma(2\lambda+c)\Gamma(c+1)}{2^{2\lambda-1}(\lambda+a+c)}\psi'(x).$$

We know that the result (3.16) is correct, because Pollaczek gave it in his paper [16, p. 2254, Eq. (2)]. This result can also be obtained by taking $k=1$, $r=0$ and replacing $\lambda$ by $\lambda-1$ in [6, p. 319, Eq. (4.12)].

When $c=0$ and $\theta\in(0,\pi)$, $\mathbf{F}_+(\theta) = |\Gamma(\lambda+\mathrm{i}\Phi)|^{-1}\mathrm{e}^{-\mathrm{i}\psi}$. Theorem 3.2 provides an asymptotic formula for the Pollaczek polynomials $P_n^\lambda(\cos\theta;a,b)$, which corrects a result of Alhaidari et al. [1, p. 6, line 4]. In their paper, they state that the phase of $n^{\mathrm{i}\Phi}\mathrm{e}^{-\mathrm{i}n\theta}\left(1-\mathrm{e}^{2\mathrm{i}\theta}\right)^{-\lambda+\mathrm{i}\Phi}/\Gamma(\lambda+\mathrm{i}\Phi)$ is $\psi+n\theta-\lambda\left(\frac{\pi}{2}-\theta\right)-\Phi\log(2n)$, where $\psi$ is defined through $\Gamma(\lambda+\mathrm{i}\Phi)=|\Gamma(\lambda+\mathrm{i}\Phi)|\,\mathrm{e}^{\mathrm{i}\psi}$. The correct expression should be $\psi+n\theta-\lambda\left(\frac{\pi}{2}-\theta\right)-\Phi\log(2n\sin\theta)$.

**Corollary 3.4.** *For $\theta\in(0,\pi)$, we have*

$$P_n^\lambda(\cos\theta;a,b) \sim \frac{2\,\mathrm{e}^{(\pi/2-\theta)\Phi}}{|\Gamma(\lambda+\mathrm{i}\Phi)|\,(2\sin\theta)^\lambda}n^{\lambda-1}\cos(\psi+\Psi_n),$$

*as $n\to\infty$, where $\Gamma(\lambda+\mathrm{i}\Phi)=|\Gamma(\lambda+\mathrm{i}\Phi)|\,\mathrm{e}^{\mathrm{i}\psi}$ and $\Psi_n$ is defined in* (3.5).

## Acknowledgments

The first author is grateful to Professor Xiang-Sheng Wang, Professor Yu-Tian Li, and Dr. Xiao-Min Huang for useful discussions. The authors would also like to express their gratitude to Professor Mourad E. H. Ismail and the referee for valuable comments and suggestions.

## References

[1] A. D. Alhaidari, H. Bahlouli, and M. E. H. Ismail, *The Dirac-Coulomb problem: a mathematical revisit*, J. Phys. A **45** (2012), no. 36, 365204, 9, DOI 10.1088/1751-8113/45/36/365204. MR2967907

[2] George E. Andrews, Richard Askey, and Ranjan Roy, *Special functions*, Encyclopedia of Mathematics and its Applications, vol. 71, Cambridge University Press, Cambridge, 1999. MR1688958

[3] Rui Bo and R. Wong, *Asymptotic behavior of the Pollaczek polynomials and their zeros*, Stud. Appl. Math. **96** (1996), no. 3, 307–338, DOI 10.1002/sapm1996963307. MR1378865

[4] Joaquin Bustoz and Mourad E. H. Ismail, *The associated ultraspherical polynomials and their q-analogues*, Canad. J. Math. **34** (1982), no. 3, 718–736, DOI 10.4153/CJM-1982-049-6. MR663314

[5] Jairo A. Charris and Mourad E. H. Ismail, *On sieved orthogonal polynomials. V. Sieved Pollaczek polynomials*, SIAM J. Math. Anal. **18** (1987), no. 4, 1177–1218, DOI 10.1137/0518086. MR892496

[6] Nadhla A. Al-Salam and Mourad E. H. Ismail, *On sieved orthogonal polynomials. VIII. Sieved associated Pollaczek polynomials*, J. Approx. Theory **68** (1992), no. 3, 306–321, DOI 10.1016/0021-9045(92)90107-Y. MR1152221

[7] Jairo A. Charris, Mourad E. H. Ismail, and Sergio Monsalve, *On sieved orthogonal polynomials. X. General blocks of recurrence relations*, Pacific J. Math. **163** (1994), no. 2, 237–267. MR1262296

[8] E. Hendriksen and H. van Rossum, *Orthogonal Laurent polynomials*, Nederl. Akad. Wetensch. Indag. Math. **48** (1986), no. 1, 17–36. MR834317

[9] Erik Hendriksen, *Associated Jacobi-Laurent polynomials*, Extrapolation and rational approximation (Luminy, 1989), J. Comput. Appl. Math. **32** (1990), no. 1-2, 125–141, DOI 10.1016/0377-0427(90)90424-X. MR1091783

[10] Erik Hendriksen, *A weight function for the associated Jacobi-Laurent polynomials*, J. Comput. Appl. Math. **33** (1990), no. 2, 171–180, DOI 10.1016/0377-0427(90)90367-9. MR1090894

[11] Arthur Erdélyi, Wilhelm Magnus, Fritz Oberhettinger, and Francesco G. Tricomi, *Higher transcendental functions. Vols. I, II*, based, in part, on notes left by Harry Bateman, McGraw-Hill Book Company, Inc., New York-Toronto-London, 1953. MR0058756

[12] Mourad E. H. Ismail, *Classical and quantum orthogonal polynomials in one variable*, with two chapters by Walter Van Assche, with a foreword by Richard A. Askey, reprint of the 2005 original, Encyclopedia of Mathematics and its Applications, vol. 98, Cambridge University Press, Cambridge, 2009. MR2542683

[13] Frank W. J. Olver, Daniel W. Lozier, Ronald F. Boisvert, and Charles W. Clark (eds.), *NIST handbook of mathematical functions*, U.S. Department of Commerce, National Institute of Standards and Technology, Washington, DC; Cambridge University Press, Cambridge, 2010. With 1 CD-ROM (Windows, Macintosh and UNIX). MR2723248

[14] Henri Padé, *Recherches sur la convergence des développements en fractions continues d'une certaine catégorie de fonctions* (French), Ann. Sci. École Norm. Sup. (3) **24** (1907), 341–400. MR1509084

[15] Félix Pollaczek, *Systèmes de polynomes biorthogonaux qui généralisent les polynomes ultrasphériques* (French), C. R. Acad. Sci. Paris **228** (1949), 1998–2000. MR0031126

[16] Félix Pollaczek, *Sur une famille de polynômes orthogonaux à quatre paramètres* (French), C. R. Acad. Sci. Paris **230** (1950), 2254–2256. MR0035868

[17] Félix Pollaczek, *Sur une généralisation des polynomes de Jacobi* (French), Mémor. Sci. Math., no. 131, Gauthier-Villars, Paris, 1956. MR0075339

[18] G. Szegö, *On certain special sets of orthogonal polynomials*, Proc. Amer. Math. Soc. **1** (1950), 731–737, DOI 10.2307/2031975. MR0042546

[19] Gábor Szegő, *Orthogonal polynomials*, 4th ed., American Mathematical Society, Providence, R.I., 1975. American Mathematical Society, Colloquium Publications, Vol. XXIII. MR0372517

[20] X.-S. Wang and R. Wong, *Asymptotics of orthogonal polynomials via recurrence relations*, Anal. Appl. (Singap.) **10** (2012), no. 2, 215–235, DOI 10.1142/S0219530512500108. MR2903981

[21] Jet Wimp, *Explicit formulas for the associated Jacobi polynomials and some applications*, Canad. J. Math. **39** (1987), no. 4, 983–1000, DOI 10.4153/CJM-1987-050-4. MR915027

[22] Jet Wimp, *Pollaczek polynomials and Padé approximants: some closed-form expressions*, Extrapolation and rational approximation (Luminy, 1989), J. Comput. Appl. Math. **32** (1990), no. 1-2, 301–310, DOI 10.1016/0377-0427(90)90440-B. MR1091799

[23] R. Wong and H. Li, *Asymptotic expansions for second-order linear difference equations: Asymptotic methods in analysis and combinatorics*, J. Comput. Appl. Math. **41** (1992), no. 1-2, 65–94, DOI 10.1016/0377-0427(92)90239-T. MR1181710

Liu Bie Ju Centre for Mathematical Sciences, City University of Hong Kong, Hong Kong, China

*Email address*: `mathwinnie@live.com`

Liu Bie Ju Centre for Mathematical Sciences, City University of Hong Kong, Hong Kong, China

Received: 15 July 2019 | Revised: 8 November 2019 | Accepted: 11 November 2019
DOI: 10.1111/sapm.12292

ORIGINAL ARTICLE

STUDIES IN APPLIED MATHEMATICS WILEY

# Uniform asymptotics and zeros of the associated Pollaczek polynomials

**Xiao-Min Huang** | **R. Wong**

Liu Bie Ju Centre for Mathematical Sciences, City University of Hong Kong, Hong Kong, China

**Correspondence**
Xiao-Min Huang, Liu Bie Ju Centre for Mathematical Sciences, City University of Hong Kong, Hong Kong, China.
Email: huangxm37@mail2.sysu.edu.cn

**Abstract**

Using a difference-equation method in a previous paper, we study the associated Pollaczek polynomials $P_n^\lambda(x;a,b,c)$ defined by a three-term recurrence relation. Two asymptotic approximations are derived for these polynomials; one holds for $x = 1 + \frac{t}{n}$ with $-(a+b) < t$ and $(a+b) > 0$, and the other holds for $x = 1 + \frac{t}{n}$ with $t$ in a neighborhood of $t = -(a+b)$. An asymptotic formula is also provided for their largest zeros.

**KEYWORDS**
difference-equation method, turning point, uniform asymptotics, zeros

## 1 | INTRODUCTION

Recently, Luo and Wong[1] have obtained two asymptotic formulas for the associated Pollaczek polynomials $P_n^\lambda(z;a,b,c)$, one for $z \in \mathbb{C} \setminus [-1,1]$ and the other for $z = x \in (-1,1)$. These polynomials were introduced by Pollaczek[2] in 1950, and they can be defined by the recurrence relation

$$(n+c)P_n^\lambda(z;a,b,c) = 2[(n-1+\lambda+a+c)z+b]P_{n-1}^\lambda(z;a,b,c) \\ -(n+2\lambda+c-2)P_{n-2}^\lambda(z;a,b,c), \quad n = 1,2,\ldots, \tag{1}$$

with initial conditions $P_{-1}^\lambda(z;a,b,c) = 0$ and $P_0^\lambda(z;a,b,c) = 1$ (see Ref. 3, p. 220, eq. 13). The associated Pollaczek polynomials are orthogonal under either one of the following conditions:

1. $a > |b|,\ 2\lambda + c > 0,\ c \geq 0,$
2. $a > |b|,\ 2\lambda + c \geq 1,\ c \geq -1;$

(see Ref. 3, p. 219, eq. 13).

When $\lambda = \frac{1}{2}$ and $c = 0$, the above recurrence relation is also satisfied by the Pollaczek polynomials $P_n(z;a,b)$ discussed in Szegő.[4] The asymptotics of these polynomials was first investigated by Novikoff[5], and his results were later extended by Bo and Wong.[6] The method used by these authors is based on an integral representation derived from the generating function of these polynomials. For further results on these polynomials, see Wang and Zhao.[7]

The difficulty with the associated Pollaczek polynomials is that they do not have a generating function that can be expressed in terms of elementary functions. Indeed, their generating function is an integral which can be expressed as a hypergeometric function of two variables. As a consequence, Luo and Wong[1] had to turn to the recently developed asymptotic method for difference equations.

The purpose of this paper is to derive an asymptotic approximation for the largest zeros of the associated Pollaczek polynomials $P_n^\lambda(z;a,b,c)$ in their interval of orthogonality $(-1,1)$. More precisely, we show that for $x = 1+\frac{t}{n}$ and $-\infty < -M < t < 0$,

$$P_n(x) = K_n\mathcal{K}_1(x)\mathcal{A}_0(\eta(t))\left[\mathrm{Ai}\left(n^{\frac{1}{3}}\eta(t)\right) + O\left(\frac{1}{\sqrt{n}}\right)\right] \tag{2}$$

as $n \to \infty$, where $K_n$ is a constant, $\mathcal{K}_1(x)$ is an explicit function given in terms of an elementary function and a hypergeometric function, $\eta(t)$ and $\mathcal{A}_0(\eta(t))$ are analytic functions of $t \in (-\infty, 0)$. To achieve (2), we need to first derive an asymptotic formula for these polynomials when $x = 1+\frac{t}{n}$ and $t > -(a+b)$; see (32). Now, let $x_{vn}$, $v = 1, 2, \ldots$, denote the zeros of $P_n^\lambda(x;a,b,c)$, arranged in decreasing order. By using (2), we also establish the asymptotic expansion

$$x_{vn} = 1 - \frac{a+b}{n} + \frac{[2(a+b)]^{2/3}a_v}{2n^{4/3}} + O\left(\frac{1}{n^{5/3}}\right), \tag{3}$$

where $a_v$ is $v$th negative zero of the Airy function $\mathrm{Ai}(x)$. We should remark that the parameters $\lambda$ and $c$ in the polynomial $P_n^\lambda(x;a,b,c)$ appear only in the function $\mathcal{K}_1(x)$ in (2) and do not appear in the function $\eta(t)$, and that $\mathcal{K}_1(x)$ does not vanish when $x$ is close to 1. As a result, the expansion in (3) agrees with the formula given by Bo and Wong[6] for the zeros of the Pollaczek polynomials $P_n(x;a,b)$.

## 2 | EXPANSION IN THE NONOSCILLATORY REGION

To apply the theory developed in Wang and Wong[8-10] for difference equations, we first put the recurrence relation (1) in a standard form. For simplicity, from here on we use $P_n(z)$ to represent the associated Pollaczek polynomials; that is, $P_n(z) := P_n^\lambda(z;a,b,c)$. Thus, Equation (1) can be written as

$$\begin{aligned} P_{n+1}(x) = {} & \frac{2[x(n+c+\lambda+a)+b]}{n+c+1}P_n(x) \\ & - \frac{n+c+2\lambda-1}{n+c+1}P_{n-1}(x), \quad n = 0,1,2,\ldots. \end{aligned} \tag{4}$$

Now, define the sequence $\{K_n\}$ by

$$K_n := \frac{\Gamma\left(\frac{n+c}{2}+\lambda\right)}{\Gamma\left(\frac{n+c}{2}+1\right)} \sim \left(\frac{n}{2}\right)^{\lambda-1}. \tag{5}$$

Clearly,

$$\frac{K_n}{K_{n+1}} = \frac{\Gamma\left(\frac{n+c}{2}+\lambda\right)\Gamma\left(\frac{n+c+1}{2}+1\right)}{\Gamma(\frac{n+c}{2}+1)\Gamma\left(\frac{n+c+1}{2}+\lambda\right)} \sim 1+\frac{1-\lambda}{n}+\cdots. \tag{6}$$

Putting $\mathbb{P}_n(x) \equiv P_n(x)/K_n$, Equation (4) becomes

$$\mathbb{P}_{n+1}(x) - (A_n x + B_n)\mathbb{P}_n(x) + \mathbb{P}_{n-1}(x) = 0, \tag{7}$$

where

$$A_n := \frac{2(n+c+\lambda+a)}{n+c+1}\frac{K_n}{K_{n+1}} \sim \sum_{s=0}^{\infty}\frac{\alpha_s}{n^s} \tag{8}$$

and

$$B_n := \frac{2b}{n+c+1}\frac{K_n}{K_{n+1}} \sim \sum_{s=0}^{\infty}\frac{\beta_s}{n^s}. \tag{9}$$

Simple calculation gives

$$\alpha_0 = 2, \quad \alpha_1 = 2a, \quad \beta_0 = 0, \quad \beta_1 = 2b. \tag{10}$$

We also make use of the notation

$$\phi\left(x, \frac{1}{n}\right) := A_n x + B_n \sim \sum_{s=0}^{\infty}\left(\frac{\alpha_s}{n^s}x + \frac{\beta_s}{n^s}\right). \tag{11}$$

To investigate the uniform asymptotics of the solution to (7) when $x$ is near 1, we put

$$x = 1 + \frac{t}{n} \quad \text{with } t > \tau \quad \text{and } \tau = -(\mathrm{a}+\mathrm{b}), \tag{12}$$

and introduce the function

$$\Phi\left(t, \frac{1}{n}\right) := \phi(x, \frac{1}{n}) \sim \sum_{s=0}^{\infty}\left(\frac{\alpha_s t}{n^{s+1}} + \frac{\alpha_s+\beta_s}{n^s}\right) := \sum_{s=0}^{\infty}\frac{\bar{\alpha}_s t + \bar{\beta}_s}{n^s}, \tag{13}$$

where $\bar{\alpha}_s = \alpha_{s-1}$ and $\bar{\beta}_s = \alpha_s + \beta_s$. Defining $\alpha_{-1} = 0$, we have

$$\bar{\alpha}_0 = 0, \quad \bar{\beta}_0 = 2, \quad \bar{\alpha}_1 = 2, \quad \bar{\beta}_1 = 2(a+b) = -2\tau. \tag{14}$$

Under the orthogonality conditions of the associated Pollaczek polynomials, $a+b>0$; ie, $\bar{\beta}_1 > 0$. The *characteristic equation* associated with (7) is

$$\rho^2 - (\alpha_0 x + \beta_0)\rho + 1 = \rho^2 - 2x\rho + 1 = 0. \tag{15}$$

Its two roots are equal, when $x = \pm 1$. Here, we are concerned with only the case $x = 1$. Hence, we have a double root $\rho = 1$. With $\rho = x = 1$, this double root also satisfies the *auxiliary condition*

$$-(\alpha_1 + \beta_1)\rho + 0 = -2(a+b) \neq 0; \tag{16}$$

(cf. Ref. 11, eq. 1.6). Under this circumstance, the result in Wong and Li (Ref. 11, eq. 1.7) suggests that the two linearly independent solutions of (7) are of the form

$$\mathbb{P}_n(x) \sim \exp\left\{ n^{1/2}\xi(t) + \kappa \log n + \sum_{s=0}^{\infty} \frac{E_s(\xi(t))}{n^{s/2}} \right\}, \tag{17}$$

where $\xi(t)$ is an analytic function. Recall that we have put $x = 1 + \frac{t}{n}$ with $t \notin (-2n, \tau)$, where $\tau = -(a+b)$. [Here we should point out that the coefficients $a_s$ in (Ref. 11, eq. 1.2) are our $-(\alpha_s x + \beta_s)$ in (8) and (9). Also $\kappa$ in (17) is the constant $\alpha$ used in (Ref. 11, eq. 1.5).] The constant $\kappa$ in (17) can be determined by setting $t = 0$. This value corresponds to $x = 1$. The recurrence relation (7) is then independent of $x$, and becomes simply a second-order linear difference equation. The result in (Ref. 11, eq. 1.7) can now be applied directly, with the parameters $\gamma$ and $\alpha$ there being given by formula (2.16) and (2.17) in that paper. In the present situation, $\gamma = 2\sqrt{2(a+b)}$ and $\alpha = \kappa = \frac{1}{4}$. Note that eq. (2.17) in Ref. 11 shows that $\alpha$ (ie, $\kappa$ in our notation) is independent of $x$, whereas (2.16) in Ref. 11 shows that $\gamma$ depends on $x$. As we show later that our $\xi(t)$ in (27) indeed agree with the value $\gamma = 2\sqrt{2(a+b)}$, when $t = 0$; see (56) and (57).

Since $x$ is fixed, when we change $n$ to $n \pm 1$, we also change $t$ to

$$t_\pm = \left(1 \pm \frac{1}{n}\right) t. \tag{18}$$

Furthermore, $(n \pm 1)^{1/2}\xi(t_\pm) - n^{1/2}\xi(t)$ can be expanded into an asymptotic expansion of the form

$$(n \pm 1)^{1/2}\xi(t_\pm) - n^{1/2}\xi(t) \sim n^{-1/2} \sum_{s=0}^{\infty} \frac{\delta_s^\pm(t)}{n^s}. \tag{19}$$

It is readily verified that

$$\delta_0^\pm(t) = \pm\left(\frac{\xi(t)}{2} + t\xi'(t)\right) \tag{20}$$

and

$$\delta_1^\pm(t) = \frac{1}{2}\left[t^2\xi''(t) + t\xi'(t) - \frac{1}{4}\xi(t)\right]. \tag{21}$$

Note that there is no "±" sign on the right-hand side of (21).

For convenient, we denote by $D^j$ the derivative with respect to $t$; ie,

$$D^j E_s(\xi(t)) = \frac{d^j}{dt^j} E_s(\xi(t)), \quad s = 0, 1, 2, \ldots. \tag{22}$$

Substituting (13) and (17) into (7) and expanding $E_s(\xi(t_\pm))$ into Taylor series, we obtain from (19)-(22)

$$\begin{aligned}\sum_{s=0}^{\infty} \frac{\bar{\alpha}_s t + \bar{\beta}_s}{n^s} &= \exp\left\{ n^{-1/2} \sum_{s=0}^{\infty} \frac{\delta_s^+(t)}{n^s} + \kappa \log\left(1 + \frac{1}{n}\right) + \sum_{s=0}^{\infty}\left[\frac{E_s(\xi(t_+))}{(n+1)^{s/2}} - \frac{E_s(\xi(t))}{n^{s/2}}\right]\right\} \\ &+ \exp\left\{ n^{-1/2} \sum_{s=0}^{\infty} \frac{\delta_s^-(t)}{n^s} + \kappa \log\left(1 - \frac{1}{n}\right) + \sum_{s=0}^{\infty}\left[\frac{E_s(\xi(t_-))}{(n-1)^{s/2}} - \frac{E_s(\xi(t))}{n^{s/2}}\right]\right\}\end{aligned}$$

$$= \exp\left\{\frac{\delta_0^+(t)}{n^{1/2}} + \frac{tDE_0(\xi)}{n} + \frac{\kappa}{n} + \frac{\delta_1^+(\xi)}{n^{3/2}} + DE_1(\xi)\frac{t}{n^{3/2}} - \frac{1}{2}\frac{E_1(\xi)}{n^{3/2}}\right.$$

$$\left. + \; lower\, order\, terms\right\}$$

$$+ \exp\left\{\frac{\delta_0^-(t)}{n^{1/2}} - \frac{tDE_0(\xi)}{n} - \frac{\kappa}{n} + \frac{\delta_1^-(\xi)}{n^{3/2}} - DE_1(\xi)\frac{t}{n^{3/2}} + \frac{1}{2}\frac{E_1(\xi)}{n^{3/2}}\right.$$

$$\left. lower\, order\, terms\right\}. \tag{23}$$

Equating coefficients of like powers of $\frac{1}{n}$ on both sides of the equation gives

$$\bar{\alpha}_1 t + \bar{\beta}_1 = \frac{1}{2}\left(\delta_0^+(t)\right)^2 + \frac{1}{2}\left(\delta_0^-(t)\right)^2 = \left[\frac{1}{2}\xi(t) + t\xi'(t)\right]^2; \tag{24}$$

that is,

$$\xi'(t) + \frac{\xi(t)}{2t} = \pm\frac{\sqrt{2(t-\tau)}}{t}. \tag{25}$$

From (24), it is easily seen that if $\xi(t)$ is a solution of this equation, then so is $-\xi(t)$. These two solutions of (24) correspond to the two linearly independent solutions of (7). We now take the plus sign in (25). When $t > 0$, upon solving (25) we have

$$\begin{aligned}\xi(t) &= t^{-1/2}\int^t t^{-1/2}\sqrt{2(t-\tau)}dt \\ &= -\frac{\sqrt{2\tau}}{\sqrt{t}}\left[\log(\sqrt{t} + \sqrt{t-\tau}) - c_{-1}\right] + \sqrt{2(t-\tau)},\end{aligned} \tag{26}$$

where $c_{-1}$ is a constant to be determined. Using the same method, we can also derive an explicitly expansion for $\xi(t)$ for $\tau < t < 0$. Indeed, we have

$$\xi(t) = \begin{cases} -\frac{\sqrt{2\tau}}{\sqrt{t}}\left[\log(\sqrt{t} + \sqrt{t-\tau}) - c_{-1}\right] + \sqrt{2(t-\tau)}, & t > 0, \\ \frac{-\sqrt{2\tau}}{\sqrt{-t}}[\sin^{-1}\sqrt{\frac{t}{\tau}} - \tilde{c}_{-1}] + \sqrt{2(t-\tau)}, & \tau < t < 0, \end{cases} \tag{27}$$

where $\tilde{c}_{-1}$ is also a constant. In Section 3, it will be shown that we need to choose $c_{-1} = \log\sqrt{-\tau}$ for the purpose of matching and $\tilde{c}_{-1} = 0$ for the continuity of $\xi(t)$.

From (20) and (25), we can rewrite

$$\delta_0^\pm(t) = \pm\sqrt{2(t-\tau)}; \tag{28}$$

and from (21) and (25), we have

$$\delta_1^\pm(t) = -\frac{\xi(t)}{8} + \frac{t}{2}\xi'(t) + \frac{t^2}{2}\xi''(t) = -\frac{\sqrt{2(t-\tau)}}{4} + \frac{t}{2\sqrt{2(t-\tau)}}. \tag{29}$$

It can also be verified that

$$\delta_2^{\pm}(t) = \pm\left[\frac{\xi(t)}{16} - \frac{t}{8}\xi'(t) + \frac{t^2}{4}\xi''(t) + \frac{t^3}{6}\xi^{(3)}(t)\right]$$
$$= \pm\left[\frac{\sqrt{2(t-\tau)}}{8} - \frac{t}{6\sqrt{2(t-\tau)}} - \frac{t^2}{6(2(t-\tau))^{3/2}}\right].$$

In fact, by simple computation, it can be shown that in general, we have

$$\delta_j^{\pm}(t) = \frac{(\pm 1)^{j-1} Q_j(t)}{(2(t-\tau))^{j-\frac{1}{2}}}, \quad j = 1, 2 \ldots,$$

where $Q_j(t)$ is a polynomial of degree at most $j$.

Equating coefficients of like powers of $n^{-3/2}$ in (23), we obtain

$$\delta_1^{+}(t) + \delta_0^{+}(t)(tDE_0(\xi) + \kappa) = 0. \tag{30}$$

Recall that $\kappa = \frac{1}{4}$. Solving (30) gives

$$E_0(\xi) = \int_{\tau_0}^{t}\left[\frac{1}{4s} - \frac{1}{4(s-\tau)} - \frac{1}{4s}\right]ds = -\frac{1}{4}\log(t-\tau) \tag{31}$$

where $\tau_0 = \tau + 1, t > \tau$. Note that from (20) and (21), one readily see that $\delta_0^{+}(t)$ and $\delta_1^{+}(t)$ are odd with respect to $\xi(t)$. Hence from (30) it follows that $E_0(\xi)$ is an even function of $\xi$; that is $E_0(-\xi) = E_0(\xi)$. Since the asymptotic expansion of our polynomial $\mathbb{P}_n(x)$ is of the form (17) and both $\xi(t)$ and $-\xi(t)$ satisfy (24), we conclude that there are two explicit functions $C_1(x)$ and $C_1(x)$ which depend only on $x$ such that

$$\mathbb{P}_n(x) \sim C_1(x)\left(\frac{t-\tau}{n}\right)^{-1/4} \exp\left\{n^{1/2}\xi(t)\right\}$$
$$+ C_2(x)\left(\frac{t-\tau}{n}\right)^{-1/4} \exp\left\{-n^{1/2}\xi(t)\right\} \tag{32}$$

for $t > \tau$, where $\xi(t)$ is the explicit function given in (27). Note that $\xi(t)$ and $E_0(t)$ can be analytically continued to the non-oscillatory region $t \in \mathbb{C} \setminus [-2n, \tau]$, and that (32) holds in $t \in \mathbb{C} \setminus [-2n, \tau]$.

## 3 | MATCHING IN THE NONOSCILLATORY REGION

The result in (32) is valid in regions $t < -2n$ and $t > \tau$, ie, $x < -1$ and $x > 1 - \frac{a+b}{n}$. In particular, it is valid for $x \geq 1$. Our next step is to determine the coefficients $C_1(x)$ and $C_2(x)$ in (32) by means of the matching technique in Ref. 12. This can be achieved by matching our result (32) with an asymptotic formula recently obtained by Luo and Wong[1] for $x \geq 1$. To this end, we first quote some of their formulas. Define

$$\alpha \equiv \alpha(x) := x + \sqrt{x^2 - 1}, \qquad \beta \equiv \beta(x) := x - \sqrt{x^2 - 1}, \tag{33}$$

$$A \equiv A(x) := -\lambda + \frac{ax+b}{\sqrt{x^2-1}}, \qquad B \equiv B(x) := -\lambda - \frac{ax+b}{\sqrt{x^2-1}}. \tag{34}$$

They first show that for $z \in \mathbb{C}$, the associated Pollaczek polynomials $P_n(z)$ can be expressed as

$$P_n(z) = \frac{\Gamma(c+1)}{\Gamma(2\lambda+c-1)}\beta(\mathbf{F}_n\mathbf{G}_{-1} - \mathbf{F}_{-1}\mathbf{G}_n), \tag{35}$$

where

$$\mathbf{F}_n \equiv \mathbf{F}_n(\beta^2) := \frac{\Gamma(-B+n+c)}{\Gamma(n+1+c)}\alpha^n\,{}_2F_1\left[\begin{matrix}-n-c, -A\\ -n-c+1+B\end{matrix}; \beta^2\right] \tag{36}$$

and

$$\mathbf{G}_n \equiv \mathbf{G}_n(\beta^2) := \frac{\Gamma(2\lambda+c+n)}{\Gamma(-B+c+n+1)}\beta^n\,{}_2F_1\left[\begin{matrix}c+n+1,\ 1+A\\ -B+c+n+1\end{matrix}; \beta^2\right], \tag{37}$$

and ${}_2F_1[{a, b \atop c}; z]$ is the Gauss hypergeometric function. By just taking the limit $n \to \infty$ in the above hypergeometric functions, they deduce the asymptotic formula (Ref. 1, eq. 3.1)

$$\begin{aligned} P_n(z) \sim{}& \frac{\Gamma(c+1)}{\Gamma(2\lambda+c-1)}\beta(1-\beta^2)^A\mathbf{G}_{-1}n^{-B-1}\alpha^n \\ &- \frac{\Gamma(c+1)}{\Gamma(2\lambda+c-1)}\beta(1-\beta^2)^{-A-1}\mathbf{F}_{-1}n^{2\lambda+B-1}\beta^n \end{aligned} \tag{38}$$

for $z \in \mathbb{C} \setminus [-1, 1]$, where

$$\mathbf{F}_{-1} = \frac{\Gamma(-B-1+c)}{\Gamma(c)}\alpha^{-1}\,{}_2F_1\left[\begin{matrix}1-c, -A\\ 2-c+B\end{matrix}; \beta^2\right] \tag{39}$$

and

$$\mathbf{G}_{-1} = \frac{\Gamma(2\lambda+c-1)}{\Gamma(-B+c)}\beta^{-1}\,{}_2F_1\left[\begin{matrix}c,\ 1+A\\ c-B\end{matrix}; \beta^2\right]. \tag{40}$$

To match the two formulas (32) and (38), we recall that $x = 1 + \frac{t}{n}$. Thus, the functions $\alpha$, $\beta$, $A$, and $B$ in (33) and (34) all depend on $n$, and we need asymptotic formulas for these functions. Indeed, we have

$$\alpha(x) \sim 1 + \sqrt{\frac{2t}{n}} + \frac{t}{n}, \qquad \beta(x) \sim 1 - \sqrt{\frac{2t}{n}} + \frac{t}{n}, \tag{41}$$

$$1 - \beta^2 \sim 2\sqrt{\frac{2t}{n}}\left(1 - \sqrt{\frac{2t}{n}}\right), \tag{42}$$

$$A \sim (a+b)\sqrt{\frac{n}{2t}} - \lambda \sim -\tau\sqrt{\frac{n}{2t}} - \lambda, \tag{43}$$

and

$$B \sim -(a+b)\sqrt{\frac{n}{2t}} - \lambda \sim \tau\sqrt{\frac{n}{2t}} - \lambda. \tag{44}$$

By Stirling's formula, we also have

$$\begin{aligned}\Gamma(-B-1+c) &\sim \sqrt{\frac{2\pi}{c-1-B}}\left(\frac{c-1-B}{e}\right)^{c-1-B} \\ &\sim e^{c-1+\lambda}\left(-\frac{2\pi}{\tau}\sqrt{\frac{2t}{n}}\right)^{\frac{1}{2}}\left(-\frac{\tau}{e}\sqrt{\frac{n}{2t}}\right)^{-\tau\sqrt{\frac{n}{2t}}+c-1+\lambda}\end{aligned} \tag{45}$$

and

$$\frac{1}{\Gamma(-B+c)} \sim e^{-c-\lambda}\left(-\frac{2\pi}{\tau}\sqrt{\frac{2t}{n}}\right)^{-\frac{1}{2}}\left(-\frac{\tau}{e}\sqrt{\frac{n}{2t}}\right)^{\tau\sqrt{\frac{n}{2t}}-c-\lambda}. \tag{46}$$

For convenience, we introduce the notations

$$\mathcal{F} = \frac{\Gamma(c)}{\Gamma(-B-1+c)}\alpha\mathbf{F}_{-1} = {}_2F_1\left[\begin{matrix}1-c,\ -A \\ 2-c+B\end{matrix};\ \beta^2\right] \tag{47}$$

and

$$\mathcal{G} = \frac{\Gamma(-B+c)}{\Gamma(2\lambda+c-1)}\beta\mathbf{G}_{-1} = {}_2F_1\left[\begin{matrix}c,\ 1+A \\ c-B\end{matrix};\ \beta^2\right]. \tag{48}$$

Inserting (45) and (46) in (39) and (40), respectively, and applying the resulting expressions together with the asymptotic formulas in (41)-(44), we obtain from (38)

$$\begin{aligned}P_n(x) &\sim \Gamma(c+1)\left[2\sqrt{\frac{2t}{n}}\left(1-\sqrt{\frac{2t}{n}}\right)\right]^{-\tau\sqrt{\frac{n}{2t}}-\lambda} e^{-c-\lambda}\left(-\frac{2\pi}{\tau}\sqrt{\frac{2t}{n}}\right)^{-1/2} \\ &\times \left(-\frac{\tau}{e}\sqrt{\frac{n}{2t}}\right)^{\tau\sqrt{\frac{n}{2t}}-c-\lambda} n^{-\tau\sqrt{\frac{n}{2t}}+\lambda-1}\left(1+\sqrt{\frac{2t}{n}}+\frac{t}{n}\right)^n \mathcal{G} \\ &-\frac{\beta c}{\alpha\Gamma(2\lambda+c-1)}\left[2\sqrt{\frac{2t}{n}}\left(1-\sqrt{\frac{2t}{n}}\right)\right]^{\tau\sqrt{\frac{n}{2t}}+\lambda-1} e^{c-1+\lambda}\left(-\frac{2\pi}{\tau}\sqrt{\frac{2t}{n}}\right)^{1/2} \\ &\times \left(-\frac{\tau}{e}\sqrt{\frac{n}{2t}}\right)^{-\tau\sqrt{\frac{n}{2t}}+c-1+\lambda} n^{\tau\sqrt{\frac{n}{2t}}+\lambda-1}\left(1-\sqrt{\frac{2t}{n}}+\frac{t}{n}\right)^n \mathcal{F}.\end{aligned} \tag{49}$$

Note that in the first term of (49), the powers of 2, $-\tau$, $t$, $n$, $e$ are, respectively,

$$-\tau\sqrt{\frac{2n}{t}} - \lambda + \frac{c}{2} - \frac{3}{4}, \qquad \tau\sqrt{\frac{n}{2t}} - c - \lambda + \frac{1}{2},$$

$$-\tau\sqrt{\frac{n}{2t}} + \frac{c}{2} - \frac{1}{4}, \qquad -\frac{c}{2} + \frac{1}{4} + \lambda - 1, \qquad \sqrt{2nt} - \tau\sqrt{\frac{n}{2t}} + \tau.$$

Similarly, in the second term of (49), the powers of 2, $-\tau$, $t$, $n$, $e$ are, respectively,

$$\tau\sqrt{\frac{2n}{t}} + \lambda - \frac{c}{2} - \frac{1}{4}, \qquad -\tau\sqrt{\frac{n}{2t}} - \frac{3}{2} + c + \lambda,$$

$$\tau\sqrt{\frac{n}{2t}} - \frac{c}{2} + \frac{1}{4}, \qquad \frac{c}{2} - \frac{1}{4} + \lambda - 1, \qquad -\sqrt{2nt} + \tau\sqrt{\frac{n}{2t}} - \tau.$$

Thus, we can rewrite (49) as

$$\begin{aligned} P_n(x) &\sim 2^{-\lambda+\frac{c}{2}-\frac{3}{4}} \frac{(-\tau)^{-c-\lambda+\frac{1}{2}}}{\sqrt{\pi}} e^{\tau}\Gamma(c+1)\left(\frac{t}{n}\right)^{\frac{c}{2}-\frac{1}{4}} \\ &\times \left(2\sqrt{\frac{t}{-\tau}}\right)^{-\tau\sqrt{\frac{2n}{t}}} \exp\left(\sqrt{2nt} - \tau\sqrt{\frac{n}{2t}}\right) n^{\lambda-1}\mathcal{G} \\ &- 2^{\lambda-\frac{c}{2}-\frac{1}{4}}\sqrt{\pi}(-\tau)^{-\frac{3}{2}+c+\lambda} e^{-\tau} \frac{\beta c}{\alpha\Gamma(2\lambda+c-1)}\left(\frac{t}{n}\right)^{\frac{1}{4}-\frac{c}{2}} \\ &\times \left(2\sqrt{\frac{t}{-\tau}}\right)^{\tau\sqrt{\frac{2n}{t}}} \exp\left(-\sqrt{2nt} + \tau\sqrt{\frac{n}{2t}}\right) n^{\lambda-1}\mathcal{F}. \end{aligned} \tag{50}$$

On the other hand, the asymptotic formula in (38) is valid for all $z \in \mathbb{C} \setminus [-1, 1]$. Hence, Equation (50) holds for $0 < t < n$, and we may take $t = \sigma n$, $\sigma \in (0, \frac{1}{2})$, so that $(\frac{t-\tau}{n})^{-\frac{1}{4}} \sim (x-1)^{-\frac{1}{4}}$. Equation (32) then gives

$$\begin{aligned} \mathbb{P}_n(x) &\sim C_1(x)(x-1)^{-1/4}\exp\left\{ n^{1/2}\left[-\tau\sqrt{\frac{2}{t}}\left(\log(\sqrt{t} + \sqrt{t-\tau}) - c_{-1}\right) + \sqrt{2(t-\tau)}\right]\right\} \\ &+ C_2(x)(x-1)^{-1/4}\exp\left\{ n^{1/2}\left[\tau\sqrt{\frac{2}{t}}\left(\log(\sqrt{t} + \sqrt{t-\tau}) - c_{-1}\right) - \sqrt{2(t-\tau)}\right]\right. \end{aligned} \tag{51}$$

As $n \to \infty$, $t \to \infty$, (51) further simplifies to

$$\begin{aligned} \mathbb{P}_n(x) &\sim C_1(x)(x-1)^{-1/4}\left(2\sqrt{\frac{t}{-\tau}}\right)^{-\tau\sqrt{\frac{2n}{t}}} \exp\left\{\sqrt{2nt} - \tau\sqrt{\frac{n}{2t}}\right\} \\ &+ C_2(x)(x-1)^{-1/4}\left(2\sqrt{\frac{t}{-\tau}}\right)^{\tau\sqrt{\frac{2n}{t}}} \exp\left\{-\sqrt{2nt} + \tau\sqrt{\frac{n}{2t}}\right\}. \end{aligned} \tag{52}$$

For the purpose of matching (51) with (50), here we have chosen $c_{-1} = \log(\sqrt{-\tau})$; see the line following (27).

To match (50) against (52), we first recall the notation $\mathbb{P}_n(x) = P_n(x)/K_n$ and consider (5). The result is

$$C_1(x) = 2^{\frac{c}{2}-\frac{7}{4}} \frac{(-\tau)^{-c-\lambda+\frac{1}{2}}}{\sqrt{\pi}} e^{\tau}\Gamma(c+1)(x-1)^{\frac{c}{2}}\mathcal{G}, \tag{53}$$

$$C_2(x) = -2^{2\lambda-\frac{c}{2}-\frac{5}{4}}\sqrt{\pi}(-\tau)^{-\frac{3}{2}+c+\lambda}e^{-\tau}\frac{\beta c}{\alpha\Gamma(2\lambda+c-1)}(x-1)^{\frac{1}{2}-\frac{c}{2}}\mathcal{F}. \tag{54}$$

With $c_{-1} = \log(\sqrt{-\tau})$, we now rewrite (27) as

$$\xi(t) = \begin{cases} -\tau\sqrt{\frac{2}{t}}\log\frac{\sqrt{t}+\sqrt{t-\tau}}{\sqrt{-\tau}} + \sqrt{2(t-\tau)}, & t > 0, \\ \tau\sqrt{\frac{2}{-t}}\left[\sin^{-1}\sqrt{\frac{t}{\tau}} - \tilde{c}_{-1}\right] + \sqrt{2(t-\tau)}, & \tau < t < 0. \end{cases} \tag{55}$$

As $t \to 0^+$, we have

$$\xi(t) \sim -\tau\sqrt{\frac{2}{t}}\log\left(1+\sqrt{\frac{t}{-\tau}}\right) + \sqrt{-2\tau} \sim 2\sqrt{-2\tau}. \tag{56}$$

As $t \to 0^-$, we have

$$\xi(t) \sim -\tau\sqrt{\frac{2}{-t}}\left(\sqrt{\frac{t}{\tau}} - \tilde{c}_{-1}\right) + \sqrt{-2\tau} \sim 2\sqrt{-2\tau} + \tilde{c}_{-1}\tau\sqrt{\frac{2}{-t}}. \tag{57}$$

By the continuity of $\xi(t)$, we must choose $\tilde{c}_{-1} = 0$; see the line following (27).

The coefficients $C_1(x)$, $C_2(x)$ and the function $\xi(t)$ in (32) are now all determined.

## 4 | WORK OF LI AND WANG

We are now ready to derive the asymptotic formula given in (2). First, we recall the recurrence relation (7), where the coefficients $A_n$ and $B_n$ are assumed to have series representations given in (8) and (9) with general coefficients $\alpha_s$ and $\beta_s$. In Ref.,9 Wang and Wong have developed a turning point theory based on the characteristic equation

$$\lambda^2 - (\alpha_0 x + \beta_0)\lambda + 1 = 0. \tag{58}$$

The roots of this equation are given by

$$\lambda_\pm = \frac{1}{2}\left[\alpha_0 x + \beta_0 \pm \sqrt{(\alpha_0 x + \beta_0)^2 - 4}\right]; \tag{59}$$

they coincide when $x = x_\pm$, where

$$\alpha_0 x_\pm + \beta_0 = \pm 2. \tag{60}$$

The points $x_\pm$ are called the *turning points* of the polynomial satisfying (7) [see Ref. 9]. However, there is a restriction in the theory of Wang and Wong,[9] and the restriction is

$$\alpha_1 = 0 \quad \text{and} \quad \beta_1 = 0. \tag{61}$$

In our case, we have $\alpha_1 = 2a$, $\beta_1 = 2b$, and $(a + b) > 0$; see (10). Thus, we cannot apply their theory directly to our problem concerning the associated Pollaczek polynomials. Instead, we need to use a recent result of Li and Wang,[13] which gives asymptotic expansions that are valid in neighborhoods of critical values without the requirement $\alpha_1 = \beta_1 = 0$. Their idea is to modify the characteristic equation (58) and use, instead, the equation

$$\lambda^2 - (A_n x + B_n)\lambda + 1 = 0. \tag{62}$$

The roots of this equation are

$$\lambda_\pm = \frac{1}{2}\left[A_n x + B_n \pm \sqrt{(A_n x + B_n)^2 - 4}\right], \tag{63}$$

and they coincide when $x = x_\pm$, where

$$A_n x_\pm + B_n = \pm 2; \tag{64}$$

cf. (59) and (60). (Only the case $x_+$ is considered in Ref. 13; the analysis of $x_-$ is similar.) Substituting (8) and (9) into (64) gives

$$x_+ = \frac{2 - \beta_0}{\alpha_0} + \frac{\mu}{n} + O\left(\frac{1}{n^2}\right), \tag{65}$$

where

$$\mu = \frac{\alpha_0\beta_1 + 2\alpha_1 - \alpha_1\beta_0}{-\alpha_0^2}. \tag{66}$$

The difference between the definition of turning point introduced by Wang and Wong[9,10] and the modified definition used in Li and Wang[13] is that first one involves only the two leading coefficients $\alpha_0$ and $\beta_0$ (see Equation 60), whereas the second one involves $\alpha_0$, $\beta_0$, $\alpha_1$, and $\beta_1$; see (65)–(66).

Suggested by (65), Li and Wang[13] made a change of variable from $x$ to $t$ defined by

$$x = \frac{2 - \beta_0}{\alpha_0} + \frac{t}{n}. \tag{67}$$

The three-term recurrence relation (7) can now be written as

$$\mathbb{P}_{n+1}(x) - \left(A_n' t + B_n'\right)\mathbb{P}_n(x) + \mathbb{P}_{n-1}(x) = 0, \tag{68}$$

where

$$A_n' \sim n^{-1}\sum_{s=0}^{\infty}\frac{\alpha_s'}{n^s}, \qquad B_n' \sim \sum_{s=0}^{\infty}\frac{\beta_s'}{n^s}, \tag{69}$$

with the coefficients given by

$$\alpha_s' = \alpha_s, \qquad \text{and} \qquad \beta_s' = \beta_s + \frac{2-\beta_0}{\alpha_0}\alpha_s.$$

Furthermore, as in Ref.,9 Li and Wang[13] prove that Equation (7) has asymptotic solutions of the form

$$\mathbb{P}_n(x) \sim \chi\left(n^{\frac{1}{3}}\eta\right)\sum_{s=0}^{\infty}\frac{\mathcal{A}_s(\eta)}{n^{s/2}} + n^{-\frac{2}{3}}\chi'\left(n^{\frac{1}{3}}\eta\right)\sum_{s=0}^{\infty}\frac{\mathcal{B}_s(\eta)}{n^{s/2}}, \tag{70}$$

where $\chi$ is any solution of the Airy equation, $\eta(t)$ is an increasing function with $\eta(\mu) = 0$ and given by

$$\begin{cases} \frac{2}{3}(\eta(t))^{3/2} = -\sqrt{\alpha_0'}\left[\sqrt{t-\mu} + \frac{\mu}{\sqrt{-t}}\cos^{-1}\sqrt{\frac{t}{\mu}}\right], & \mu < t < 0, \\ \frac{2}{3}(-\eta(t))^{3/2} = \sqrt{\alpha_0'}\left[\sqrt{\mu-t} + \frac{\mu}{\sqrt{-t}}\cosh^{-1}\sqrt{\frac{t}{\mu}}\right], & t \le \mu, \end{cases} \tag{71}$$

for $\mu < 0$, or

$$\begin{cases} \frac{2}{3}(\eta(t))^{3/2} = \sqrt{\alpha_0'}\left[\sqrt{t-\mu} - \frac{\mu}{\sqrt{t}}\cosh^{-1}\sqrt{\frac{t}{\mu}}\right], & t \ge \mu, \\ \frac{2}{3}(-\eta(t))^{3/2} = \sqrt{\alpha_0'}\left[\sqrt{\mu-t} - \frac{\mu}{\sqrt{t}}\cos^{-1}\sqrt{\frac{t}{\mu}}\right], & 0 < t < \mu \end{cases} \tag{72}$$

for $\mu > 0$. Here, we have made the assumption that $\alpha_0' = \alpha_0 > 0$. If $\alpha_0 < 0$, the orthogonal polynomial $\mathbb{P}_n(-x)$ will satisfy the recurrence relation (7) with the coefficient $A_n$ replaced by $-A_n$. Hence, we may always assume $\alpha_0' = \alpha_0 > 0$. In Ref.,13 Li and Wang have also provided explicit formulas for the two leading coefficients

$$\mathcal{A}_0(\eta(t)) = t^{1/3}\left[\frac{\eta(t)}{\alpha_0' t + \beta_1'}\right]^{1/4}, \qquad \mathcal{B}_0(\eta(t)) = 0. \tag{73}$$

The proof of (70) is very much similar to that of the main theorem in (Ref. 9, p. 189).

## 5 | ASYMPTOTICS IN THE NEIGHBORHOOD OF $t = -(a+b)$

Now we turn to our specific case of the associated Pollaczek polynomials. In view of (10), Equation (67) becomes $x = 1 + \frac{t}{n}$, and the polynomials $\mathbb{P}_n(x)$ satisfy the recurrence relation (68) with specific values for $\alpha_s'$ and $\beta_s'$ in the coefficients $A_n'$ and $B_n'$. In particular, we have $\alpha_0' = 2$, $\alpha_1' = 2a$, $\beta_0' = 2$, and $\beta_1' = 2(a+b)$. Since they are solutions of (68), we can express them as a linear combination of two linearly independent asymptotic solutions of the form given in (70), one with $\chi$ replaced by Airy function Ai and the other with $\chi$ replaced by Bi. Taking just the leading terms in the expansions, we have

$$\mathbb{P}_n(x) = \mathcal{K}_1(x)\mathcal{A}_0(\eta(t))\left[\mathrm{Ai}\left(n^{\frac{1}{3}}\eta(t)\right) + O\left(\frac{1}{\sqrt{n}}\right)\right] + \mathcal{K}_2(x)\mathcal{A}_0(\eta(t))\left[\mathrm{Bi}\left(n^{\frac{1}{3}}\eta(t)\right) + O\left(\frac{1}{\sqrt{n}}\right)\right], \tag{74}$$

where the coefficients $\mathcal{K}_1(x)$ and $\mathcal{K}_2(x)$ are to be determined. In our case, a combination of (10), (12), and (66) gives

$$\mu = \tau < 0. \tag{75}$$

Hence, from (71) we have

$$\begin{cases} \frac{2}{3}(\eta(t))^{3/2} = \frac{-\sqrt{2}\tau}{\sqrt{-t}}\cos^{-1}\sqrt{\frac{t}{\tau}} - \sqrt{2(t-\tau)}, & \tau < t < 0, \\ \frac{2}{3}(-\eta(t))^{3/2} = \frac{\sqrt{2}\tau}{\sqrt{--t}}\cosh^{-1}\sqrt{\frac{t}{\tau}} + \sqrt{2(\tau - t)}, & t \le \tau. \end{cases} \tag{76}$$

Note that $\eta(t) > 0$ for $\tau < t < 0$ and $\eta(t) \sim (\frac{\sqrt{2}}{-\tau})^{2/3}(t-\tau)$ for $t \sim \tau$. Since $t < 0$, (73) yields

$$\mathcal{A}_0(\eta(t)) = e^{\frac{\pi i}{3}}(-t)^{1/3}\left[\frac{\eta(t)}{2(t-\tau)}\right]^{1/4}. \tag{77}$$

It is worth noting that $\mathcal{A}_0(\eta(t))$ is analytic and nonvanishing at $t = \tau$.

To determine $\mathcal{K}_1(x)$ and $\mathcal{K}_2(x)$ in (74), we recall the asymptotic formulas (Ref. 14, eqs. 9.7.5 and 9.7.7)

$$\mathrm{Ai}(z) \sim \frac{1}{2\sqrt{\pi}z^{1/4}}\exp\left(-\frac{2}{3}z^{3/2}\right), \qquad \mathrm{Bi}(z) \sim \frac{1}{\sqrt{\pi}z^{1/4}}\exp\left(\frac{2}{3}z^{3/2}\right), \quad \text{as} \quad z \to \infty.$$

On account of (76) and (77), we can rewrite (74) as

$$\mathbb{P}_n(x) \sim \mathcal{K}_1(x)\frac{e^{\frac{\pi i}{3}}(-t)^{1/3}}{2\sqrt{\pi}[2(t-\tau)]^{1/4}n^{1/12}}\exp\left\{-\frac{2}{3}\sqrt{n}(\eta(t))^{3/2}\right\} + \mathcal{K}_2(x)\frac{e^{\frac{\pi i}{3}}(-t)^{1/3}}{\sqrt{\pi}[2(t-\tau)]^{1/4}n^{1/12}}\exp\left\{\frac{2}{3}\sqrt{n}(\eta(t))^{3/2}\right\}, \quad \tau < t < 0. \tag{78}$$

Since $\cos^{-1}(z) = \frac{\pi}{2} - \sin^{-1}(z)$, from (55) and (76) we have $\xi(t) = \frac{-\tau\pi}{\sqrt{-2t}} - \frac{2}{3}(\eta(t))^{3/2}$ for $\tau < t < 0$. Comparing (78) with (32) gives

$$\begin{aligned} \mathcal{K}_1(x) &= \frac{2^{\frac{5}{4}}\sqrt{\pi}e^{-\frac{\pi i}{3}}}{(1-x)^{\frac{1}{3}}}\exp\left\{\frac{-\tau\pi}{\sqrt{2(1-x)}}\right\}C_1(x) \\ &= \frac{2^{\frac{c-1}{2}}e^{-\frac{\pi i}{3}}(-\tau)^{-c-\lambda+\frac{1}{2}}}{(1-x)^{\frac{1}{3}}}\Gamma(c+1)\exp\left\{\frac{-\tau\pi}{\sqrt{2(1-x)}} + \tau\right\}e^{\frac{c}{2}\pi i}(1-x)^{\frac{c}{2}}\mathcal{G} \end{aligned} \tag{79}$$

and

$$\mathcal{K}_2(x) = \frac{2^{\frac{1}{4}}\sqrt{\pi}\,e^{-\frac{\pi i}{3}}}{(1-x)^{\frac{1}{3}}} \exp\left\{\frac{\tau\pi}{\sqrt{2(1-x)}}\right\} C_2(x)$$
$$= -\frac{2^{2\lambda-\frac{c}{2}-1}\beta c\pi\, e^{-\frac{\pi i}{3}}(-\tau)^{-\frac{3}{2}+c+\lambda}}{\alpha\Gamma(2\lambda+c-1)(1-x)^{\frac{1}{3}}} \exp\left\{\frac{\tau\pi}{\sqrt{2(1-x)}} - \tau\right\} e^{(\frac{1}{2}-\frac{c}{2})\pi i}(1-x)^{\frac{1}{2}-\frac{c}{2}}\mathcal{F}. \quad (80)$$

From the asymptotic formulas in (91) and (92), it is easily seen that $(1-x)^{\frac{c}{2}}\mathcal{G}$ and $(1-x)^{\frac{1}{2}-\frac{c}{2}}\mathcal{F}$ are finite as $x \to 1$. Also, from (91) and Lemma 1, we have $(1-x)^{\frac{c}{2}}\mathcal{G} \neq 0$. Furthermore, from (77), (79)-(80), and (91)-(92), it follows that $\mathcal{A}_0(\eta(t))\mathcal{K}_1$ and $\mathcal{A}_0(\eta(t))\mathcal{K}_2$ are real and $\mathcal{A}_0(\eta(t))\mathcal{K}_1 \neq 0$.

Note that when $x = 1 + \frac{t}{n}$ for $-\infty < -M < t < 0$ and $n \to \infty$, $\exp\{\frac{-\tau\pi}{\sqrt{2(1-x)}}\}$ is exponentially large, whereas $\exp\{\frac{\tau\pi}{\sqrt{2(1-x)}}\}$ is exponentially small. Thus we have

$$\mathcal{A}_0(\eta(t))\mathcal{K}_1(x) \gg \mathcal{A}_0(\eta(t))\mathcal{K}_2(x). \quad (81)$$

In particular, when $\tau < t < 0$, we have $\cos^{-1}(\sqrt{\frac{t}{\tau}}) \in (0, \frac{\pi}{2})$. Hence, from (76) and (79), it can be easily verified that the exponential function in the first term of (78) is bigger than

$$\exp\left\{\frac{-\tau\pi}{\sqrt{2(1-x)}} + \tau\sqrt{\frac{2n}{t}} \times \frac{\pi}{2} + \sqrt{2n(t-\tau)}\right\} = \exp\{\sqrt{2n(t-\tau)}\}.$$

Since $x = 1 + \frac{t}{n}$, the first two terms inside the curly brackets on the left-hand side of the equality cancel out. This fact is used in reaching the last equality.

Similarly, the exponential function in the second term of (78) is less than

$$\exp\{-\sqrt{2n(t-\tau)}\}.$$

Thus, the first term in (78) is dominant.

When $-\infty < -M < t < \tau$, using the asymptotic formulas (Ref. 14, eqs. 9.7.9 and 9.7.11)

$$\mathrm{Ai}(-z) \sim \frac{1}{\sqrt{\pi}z^{1/4}}\cos\left(\frac{2}{3}z^{3/2} - \frac{\pi}{4}\right), \quad \mathrm{Bi}(-z) \sim -\frac{1}{\sqrt{\pi}z^{1/4}}\sin\left(\frac{2}{3}z^{3/2} - \frac{\pi}{4}\right) \quad \text{as } z \to \infty, \quad (82)$$

we have from (74)

$$\mathbb{P}_n(x) \sim \frac{\mathcal{A}_0(\eta(t))\mathcal{K}_1(x)}{\sqrt{\pi}[-\eta(t)]^{1/4}n^{1/12}}\cos\left(\frac{2}{3}\sqrt{n}(-\eta(t))^{3/2} - \frac{\pi}{4}\right)$$
$$- \frac{\mathcal{A}_0(\eta(t))\mathcal{K}_2(x)}{\sqrt{\pi}[-\eta(t)]^{1/4}n^{1/12}}\sin\left(\frac{2}{3}\sqrt{n}(-\eta(t))^{3/2} - \frac{\pi}{4}\right). \quad (83)$$

In view of (81), the first term of (83) is dominant. Thus, we have established the following asymptotic formula:

**Theorem 1.** *For $x = 1 + \frac{t}{n}$ and $-\infty < -M < t < 0$, we have*

$$P_n(x) = K_n\mathcal{K}_1(x)\mathcal{A}_0(\eta(t))\left[\mathrm{Ai}\left(n^{\frac{1}{3}}\eta(t)\right) + \epsilon_n(x)\right], \quad (84)$$

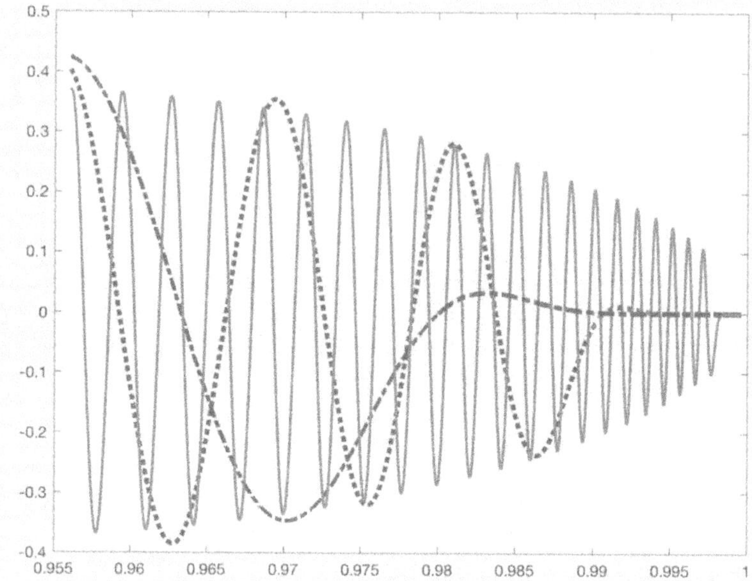


**FIGURE 1** Graphs of the error term

*with*

$$\epsilon_n(x) = \frac{P_n(x)}{K_n\mathcal{K}_1(x)\mathcal{A}_0(\eta(t))} - \mathrm{Ai}\left(n^{\frac{1}{3}}\eta(t)\right) = O\left(\frac{1}{\sqrt{n}}\right),$$

*as $n \to \infty$, where $K_n$ is the constant given in (5), $\mathcal{K}_1(x)$ is the function given in (79), $\eta(t)$ and $\mathcal{A}_0(\eta(t))$ are functions given in (76) and (77), respectively.*

Now let us fix the values of the parameters in the polynomials $P_n^{\lambda}(z; a, b, c)$:

$$a = 2, \qquad b = -1, \qquad c = 2.3, \qquad \lambda = 2.2. \tag{85}$$

By using Matlab, we have plotted in Figure 1 the curves of the error term $\epsilon_n(x)$ in (84) for $x \in [0.956, 0.9996]$. The dash-dot blue, dotted purple and solid red curves correspond to the cases $n = 50$, $n = 100$, and $n = 500$, respectively. These curves clearly indicate that the error term $\epsilon_n(x)$ gets smaller when the degree becomes bigger.

To derive an asymptotic formula for the zeros of the associated Pollaczek polynomials (ie, formula (3)), now we need to show that the function $\mathcal{K}_1(x)$ in (84) does not vanish for $x$ near 1. In view of (79), this is equivalent to showing that the function $(1-x)^{\frac{c}{2}}\mathcal{G}$ is nonvanishing for $x$ near 1, where $\mathcal{G}$ is the hypergeometric function given in (48). Let $x = \cos\theta$. Then, $x = 1 + \frac{t}{n}$ implies $\theta \sim \sqrt{\frac{-2t}{n}}$ as $n \to \infty$ with $-\infty < -M < t < \tau$. By (33) and (34), we also have $\beta(\cos\theta) = e^{-i\theta}$, $A(\cos\theta) = -\lambda - i\Phi$, and $B(\cos\theta) = -\lambda + i\Phi$ with

$$\Phi = \frac{a\cos\theta + b}{\sin\theta} \sim \frac{a+b}{\theta} \quad \text{as } \theta \to 0, \tag{86}$$

since $n \to \infty$. Equation (48) now gives

$$(1-x)^{\frac{c}{2}}\mathcal{G} = (1-\cos\theta)^{c/2}\,{}_2F_1\left[\begin{matrix} c, & 1-\lambda-i\Phi \\ & c+\lambda-i\Phi \end{matrix}; e^{-2i\theta}\right]. \tag{87}$$

Recall the transformation formula (Ref. 14, eq. 15.8.1)

$$ {}_2F_1\left[\begin{matrix}\alpha, & \beta \\ \gamma & \end{matrix}; z\right] = (1-z)^{-\alpha}\,{}_2F_1\left[\begin{matrix}\alpha, \gamma-\beta \\ \gamma \end{matrix}; \frac{z}{z-1}\right], \qquad |\arg(1-z)| < \pi. \tag{88}$$

With $\alpha = c$, $\beta = 1-\lambda-i\Phi$, $\gamma = c+\lambda-i\Phi$, and $z = e^{-2i\theta}$, it follows from (87) that

$$(1-x)^{\frac{c}{2}}\mathcal{G} = (1-\cos\theta)^{\frac{c}{2}}(1-e^{-2i\theta})^{-c}\,{}_2F_1\left[\begin{matrix} c, c+2\lambda-1 \\ c+\lambda-i\Phi \end{matrix}; -\frac{e^{-2i\theta}}{1-e^{-2i\theta}}\right]. \tag{89}$$

Recently, Farid Khwaja and Olde Daalhuis (Ref. 15, eq. 2.8) have shown that when $z$ is large, we have as $\eta \to \infty$ in $|\arg\eta| \le \pi/2$

$$\frac{\Gamma(\eta)}{\Gamma(\eta+\beta)}\,{}_2F_1\left[\begin{matrix}\alpha,\beta \\ \beta+\eta\end{matrix}; -z\right] \sim \zeta^{\beta} e^{-\alpha\zeta}\left(\frac{e^{\zeta}-1}{\zeta}\right)^{\alpha+\beta-1} U(\beta, \beta-\alpha+1, \eta\zeta)$$
$$+\,\zeta^{\beta}\left(1-e^{-\alpha\zeta}\left(\frac{e^{\zeta}-1}{\zeta}\right)^{\alpha+\beta-1}\right)U(\beta,\beta-\alpha+2,\eta\zeta), \tag{90}$$

uniformly for $|\zeta| \le \frac{1}{4}$ and $|\arg\zeta| \le \pi$, where $\zeta = \log(1+z^{-1})$ and the $U$-function is the confluent hypergeometric function (see Ref. 14, chap. 13). In the present case, $\alpha = c$, $\beta = c+2\lambda-1$, $\eta = 1-\lambda-i\Phi$, $z = \frac{e^{-2i\theta}}{1-e^{-2i\theta}}$, and $\zeta = \log(1+z^{-1}) = 2i\theta$. From (86), we have $\eta\zeta \sim 2(a+b)$ as $\theta \to 0$. Since $\zeta \to 0$ and $e^{-\alpha\zeta}(\frac{e^{\zeta}-1}{\zeta})^{\alpha+\beta-1} \sim 1$ as $\theta \to 0$, the second term on the right-hand side of (90) is asymptotically of lower order of magnitude than the first term. Thus, by (90), we have from (89)

$$(1-x)^{\frac{c}{2}}\mathcal{G} \sim \left(\frac{\theta^2}{2}\right)^{\frac{c}{2}}(2i\theta)^{-c}\frac{\Gamma(\eta+c+2\lambda-1)}{\Gamma(\eta)}\zeta^{c+2\lambda-1}U(c+2\lambda-1, 2\lambda, 2(a+b))$$
$$\sim 2^{-3c/2}e^{-\frac{c\pi i}{2}}[2(a+b)]^{c+2\lambda-1}U(c+2\lambda-1, 2\lambda, 2(a+b)). \tag{91}$$

Using the same argument, we also have from (47)

$$(1-x)^{\frac{1}{2}-\frac{c}{2}}\mathcal{F} \sim 2^{3(c-1)/2}e^{\left(\frac{3}{2}-\frac{c}{2}-2\lambda\right)\pi i}[2(a+b)]^{2-c-2\lambda}U(2-c-2\lambda, 2-2\lambda, -2(a+b)), \tag{92}$$

where $U(2-c-2\lambda, 2-2\lambda, -2(a+b))$ is assumed to have its principal value.

Now we prove that the value of $U$ in (91) is not zero under the orthogonality conditions of the associated Pollaczek polynomials. For convenience, we present the result as a lemma.

**Lemma 1.** *Under any one of the following two conditions:*

1. $a > |b|$, $2\lambda + c > 0$, $c \ge 0$;
2. $a > |b|$, $2\lambda + c \ge 1$, $c \ge -1$,

*we have* $U(c+2\lambda-1, 2\lambda, 2(a+b)) \ne 0$.

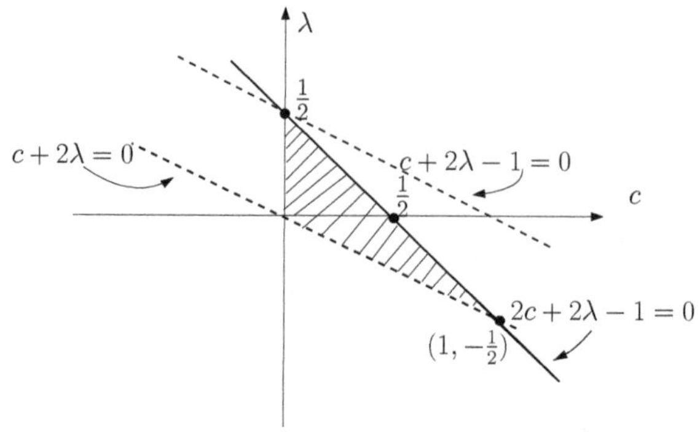


**FIGURE 2** The domain constrained by (95) and (97)

*Proof.* If condition (2) holds, then the conclusion follows directly from the formulas (Ref. 14, eqs. 13.4.4 and 13.6.3)

$$U(a, b, z) = \frac{1}{\Gamma(a)} \int_0^\infty e^{-zt} t^{a-1} (1+t)^{b-a-1} dt, \quad \text{Re } a > 0, \ |\arg z| < \frac{\pi}{2} \tag{93}$$

and

$$U(0, b, z) = 1. \tag{94}$$

If condition (1) holds and $c + 2\lambda - 1 \geq 0$, then from (93) and (94) we have that $U(c + 2\lambda - 1, 2\lambda, 2(a + b)) > 0$. Thus, we need consider only the case

$$-1 < c + 2\lambda - 1 < 0. \tag{95}$$

Recall the formula $U(a, a + 1, z) = z^{-a}$ (see Ref. 14, eq. 13.6.4). When $c = 0$, we have

$$U(c + 2\lambda - 1, 2\lambda, 2(a + b)) = U(2\lambda - 1, 2\lambda, 2(a + b)) = [2(a + b)]^{1-2\lambda} \neq 0. \tag{96}$$

Hence, we have further narrowed down our discussion to the case indicated in (95) plus the additional condition $c > 0$. In Ref. 16, Wimp has shown that if $a$ and $b$ are real and $2a - b > -1$, then $U(a, b, z)$ has no zero in the half-plane $|\arg z| \leq \frac{\pi}{2}$ (see also Ref. 14, sec. 13.9(ii), p. 331). Thus, if $2(c + 2\lambda - 1) - 2\lambda + 1 = 2c + 2\lambda - 1 > 0$, we again have $U(c + 2\lambda - 1, 2\lambda, 2(a + b)) \neq 0$. Thus, we may just restrict ourselves to the case

$$2c + 2\lambda - 1 \leq 0. \tag{97}$$

The domain in the $(c, \lambda)$-plane satisfying the conditions in (95) and (97) with $c > 0$ is shown in Figure 2, and is contained in the square

$$-1 < 2\lambda < 1, \qquad 0 < c < 1. \tag{98}$$

We now divide our discussion into two cases: $2\lambda = 0$ and $2\lambda \neq 0$. When $2\lambda = 0$, we use the result

$$U(a, b, z) = (a + 1)zU(a + 2, b + 2, z) + (z - b)U(a + 1, b + 1, z); \tag{99}$$

(see Ref. 14, eq. 13.3.14). In our case, this equation gives

$$\begin{aligned} U(c+2\lambda-1,2\lambda,2(a+b)) &= U(c-1,0,2(a+b)) \\ &= 2c(a+b)U(c+1,2,2(a+b)) + 2(a+b)U(c,1,2(a+b)) > 0, \end{aligned} \tag{100}$$

where the inequality follows from (93). When $2\lambda \neq 0$, we let $P(a,b)$ denote the number of positive zeros of $U(a,b,z)$, and $a$, $b$, and $a-b+1$ be nonintegers. From (Ref. 14, eq. 13.9.15), we have

$$P(a,b) = 0, \quad \text{if} \quad a+1 \geq b \quad \text{and} \quad b \leq 1. \tag{101}$$

In our present case, $a = c+2\lambda-1$, $b = 2\lambda$, $a-b+1 = c$. Thus, from (95) and (98), it follows that $U(c+2\lambda-1,2\lambda,2(a+b))$ is nonzero. This completes the proof of the lemma. ∎

## 6 | ZEROS

We are now ready to establish the asymptotics formula in (3) for the associated Pollaczek polynomials $P_n(x) := P_n^\lambda(x;a,b,c)$. To make the result more concise, we present it in the form of a theorem.

**Theorem 2.** *Let $x_{\upsilon n} = \cos\theta_{\upsilon n}$, $\upsilon = 1,2,\ldots,n$, denote the zeros of $P_n(x)$, arranged in decreasing order. For fixed $\upsilon$ and large $n$, we have*

$$x_{\upsilon n} = 1 - \frac{a+b}{n} + \frac{[2(a+b)]^{2/3}a_\upsilon}{2n^{4/3}} + O\left(\frac{1}{n^{5/3}}\right) \tag{102}$$

*and*

$$\theta_{\upsilon n} = \arccos x_{\upsilon n} = \sqrt{\frac{2(a+b)}{n}} - \frac{[2(a+b)]^{1/6}a_\upsilon}{2n^{5/6}} + O\left(\frac{1}{n^{7/6}}\right) \tag{103}$$

*as $n \to \infty$, where $a_\upsilon$ is $\upsilon$th negative zero of the Airy function* $\mathrm{Ai}(x)$.

*Proof.* We first return to Theorem 1 in Section 5. Since the coefficient function $\mathcal{K}_1(x)\mathcal{A}_0(\eta(t))$ in (84) is nonvanishing for $-\infty < t < 0$, the zeros of $P_n(x)$ must come from the zeros of the Airy function $\mathrm{Ai}(n^{1/3}\eta(t))$. From (76), we know that $\eta(t)$ is analytic and positive when $\tau < t < 0$, negative when $t < \tau$ and $\eta(\tau) = 0$. Now we recall the relationship $x = 1 + \frac{t}{n}$, and let $t_{\upsilon n}$, $\upsilon = 1,2,\ldots$ denote the values of $t$ corresponding to $x_{\upsilon n}$. Since $x_{\upsilon n}$ is decreasing in $\upsilon$, we can arrange $t_{\upsilon n}$ in the order

$$t_{nn} < \cdots < t_{1n} < \tau < 0. \tag{104}$$

Also since $|\mathrm{Ai}(n^{1/3}\eta(t))| \leq 1$ and $\mathcal{K}_1(x)\mathcal{A}_0(\eta(t))$ is nonvanishing for $-\infty < t < 0$, one can use (84) to prove that $t_{\upsilon n}$ is a root of

$$\mathrm{Ai}(n^{1/3}\eta(t)) + O\left(\frac{1}{\sqrt{n}}\right) = 0. \tag{105}$$

Let $a_\upsilon$ be the $\upsilon$th negative zero of the Airy function, $\rho$ be small enough so that $m = \min_{x\in[a_\upsilon-\rho,a_\upsilon+\rho]} |\mathrm{Ai}'(x)| > 0$, and $n$ be sufficiently large so that

$$O\left(\frac{1}{n}\right) \leq \frac{K}{n} < \min\{|\mathrm{Ai}(a_\upsilon-\rho)|, |\mathrm{Ai}(a_\upsilon+\rho)|\}.$$

In Ref. 17, Hethcote showed that in the interval $[a_\upsilon - \rho, a_\upsilon + \rho]$ there is a root $t_{\upsilon n}$ of (105) such that

$$|n^{1/3}\eta(t_{\upsilon n}) - a_\upsilon| \le \frac{K}{mn}, \tag{106}$$

from which it is readily seen that $\eta(t_{\upsilon n}) \to 0$ as $n \to \infty$. Since $\tau$ is the only zero of $\eta(t)$ in $(-\infty, 0)$, it follows that $\lim_{n\to\infty} t_{\upsilon n} = \tau$. This suggests that we put

$$t_{\upsilon n} = \tau + \delta_\upsilon(n), \tag{107}$$

where $\delta_\upsilon(n) < 0$ and $\lim_{n\to\infty} \delta_\upsilon(n) = 0$. Substituting (107) in (106) gives

$$\eta(\tau + \delta_\upsilon) = \frac{a_\upsilon}{n^{1/3}} + O\left(\frac{1}{n^{4/3}}\right). \tag{108}$$

Applying the identity $\cosh^{-1} x = \log[x + \sqrt{x^2 - 1}]$, $x > 1$, to the second formula in (76), we obtain

$$\begin{aligned}
\frac{2}{3}[-\eta(\tau+\delta_\upsilon)]^{3/2} &= -\sqrt{-2\tau}\left(1 + \frac{\delta_\upsilon}{\tau}\right)^{-1/2} \log\left[\sqrt{1 + \frac{\delta_\upsilon}{\tau}} + \sqrt{\frac{\delta_\upsilon}{\tau}}\right] + \sqrt{-2\delta_\upsilon} \\
&= -\sqrt{-2\tau}\left[1 - \frac{\delta_\upsilon}{2\tau} + \frac{3}{8}\left(\frac{\delta_\upsilon}{\tau}\right)^2 + \cdots\right]\left[\sqrt{\frac{\delta_\upsilon}{\tau}} - \frac{1}{6}\left(\frac{\delta_\upsilon}{\tau}\right)^{3/2}\right. \\
&\quad \left. + \frac{3}{40}\left(\frac{\delta_\upsilon}{\tau}\right)^{5/2} + \cdots\right] + \sqrt{-2\delta_\upsilon} \\
&= -\frac{2\sqrt{2}}{3\tau}(-\delta_\upsilon)^{3/2} - \frac{8\sqrt{2}}{15\tau^2}(-\delta_\upsilon)^{5/2} + \cdots.
\end{aligned} \tag{109}$$

Here, we have made use of the expansion

$$\log\left(x + \sqrt{1 + x^2}\right) = x - \frac{1}{6}x^3 + \frac{3}{40}x^5 - \cdots \quad \text{as} \quad x \to 0.$$

Coupling (108) and (109) yields

$$-\frac{2\sqrt{2}}{3\tau}(-\delta_\upsilon)^{3/2} - \frac{8\sqrt{2}}{15\tau^2}(-\delta_\upsilon)^{5/2} + \cdots = \frac{2}{3}\frac{(-a_\upsilon)^{3/2}}{n^{1/2}} + O\left(\frac{1}{n^{3/2}}\right). \tag{110}$$

Balancing the leading-order terms on both sides, we have

$$\delta_\upsilon = \delta_\upsilon(n) = \left(\frac{-\tau}{\sqrt{2}}\right)^{2/3} \frac{a_\upsilon}{n^{1/3}} + \sigma(n), \tag{111}$$

where $\sigma(n) = o(n^{-1/3})$. Inserting (111) into (110) gives $\sigma(n) = O(n^{-2/3})$. Therefore, it follows from (107) and (111) that

$$t_{\upsilon n} = -(a+b) + \left(\frac{a+b}{\sqrt{2}}\right)^{2/3} \frac{a_\upsilon}{n^{1/3}} + O(n^{-2/3}) \tag{112}$$

**TABLE 1** Numerical comparison between zeros $x_{\nu n}$ and approximate zeros $\bar{x}_{\nu n}$

| Degree $n$ | True values $x_{\upsilon n}$ | Asymptotic values $\bar{x}_{\upsilon n}$ | $x_{\upsilon n} - \bar{x}_{\upsilon n}$ | $n^{-5/3}$ |
|---|---|---|---|---|
| $n = 20$ | 0.92570314544 | 0.92570314544 | 0.00988648315016 | 0.00678604404149 |
| | 0.88245542221 | 0.89023381656222 | −0.00777839435222 | |
| $n = 50$ | 0.9704739385109 | 0.96992540057177 | 0.00054853793913 | 0.00147361259946 |
| | 0.9575100119778 | 0.96238555981294 | −0.00487554783514 | |
| $n = 100$ | 0.98588417960565 | 0.98600189256737 | −0.00011771296172 | 0.00046415888336 |
| | 0.98100422138714 | 0.98300970477931 | −0.00200548339217 | |

**TABLE 2** Numerical comparison between zeros $x_{n-\mu,n}$ and approximate zeros $\bar{x}_{n-\mu,n}$

| Degree $n$ | True values $x_{n-\mu,n}$ | Asymptotic values $\bar{x}_{n-\mu,n}$ | $x_{n-\mu,n} - \bar{x}_{n-\mu,n}$ | $n^{-5/3}$ |
|---|---|---|---|---|
| $n = 20$ | −0.82359998678523 | −0.77889579221118 | −0.04470419457405 | 0.00678604404149 |
| | −0.75827550720352 | −0.72568132866552 | −0.032594178538 | |
| $n = 50$ | −0.92517664857154 | −0.91904398870561 | −0.00613265986593 | 0.00147361259946 |
| | −0.90380353138620 | −0.90336048791477 | −0.00044304347143 | |
| $n = 100$ | −0.96286207139436 | −0.96168360140656 | −0.0011784699878 | 0.00046415888336 |
| | −0.95441485359861 | −0.95545959999308 | −0.00104474639447 | |

and

$$x_{\upsilon n} = 1 + \frac{t_{\upsilon n}}{n} = 1 - \frac{a+b}{n} + \frac{[2(a+b)]^{2/3} a_\upsilon}{2n^{4/3}} + O\left(\frac{1}{n^{5/3}}\right), \tag{113}$$

thus proving (102). Formula (103) is obtained directly from (102). This completes the proof of the theorem. ■

Let $\bar{x}_{\nu n}$, $\nu = 1, 2$, denote the three-term asymptotic approximant on the right-hand side of (102), and let the values of parameters be given in (85). Table 1 lists the values of the real zeros $x_{\nu n}$ versus the approximate zeros $\bar{x}_{\nu n}$. Table 2 gives the comparison of the real zeros $x_{n-\mu,n}$, $\mu = 1, 2$, and their approximate zeros $\bar{x}_{n-\mu,n}$.

Again with the values of the parameters given in (85), we have also plotted the curve of $P_n(x)$ for $x$ near 1 and $n = 50$ (see Figure 3). Note that $P_n(x)$ moves (nearly) vertically up once $x$ is away from its largest zero.

To conclude this paper, we wish to point out that the generating function of the Pallaczek polynomial used in Luo and Wong (Ref. 1, eq. 1.2) is slightly different from that in Bo and Wong (Ref. 6, eq. 1.1). In the present paper, we have followed the notations of Ref. 1, where the exponent function $\Phi(\theta)$ is defined to be $(a\cos\theta + b)/\sin\theta$ (see Ref. 1, eq. 1.2). The exponent function $h(\theta)$ in (Ref. 6, eq. 1.1) is defined by $h(\theta) = (a\cos\theta + b)/2\sin\theta$. Thus, our formula (103) agrees with the formula in Bo and Wong (Ref. 6, eq. 7.13). This is actually expected, since the zeros of the associated Pollaczek polynomials $P_n^\lambda(x; a, b, c)$ come from the zeros of the Airy function in (84) and the function $\eta(t)$ inside the argument of the Airy function is independent of the parameters $\lambda$ and $c$; see (76).

From (1), it is readily seen that

$$P_n^\lambda(z; a, b, c) = (-1)^n P_n^\lambda(-z; a, -b, c) \tag{114}$$

for all $n$ and $z$. From the orthogonality conditions, we have $a - b > 0$. Hence, by symmetry we have

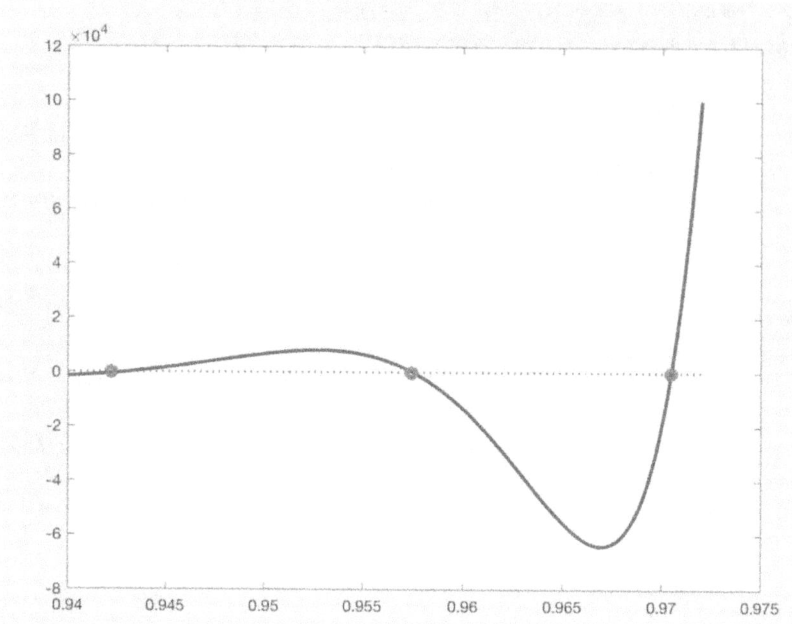


**FIGURE 3** Graph of $P_n(x)$ for $x \in [0.94, 0.972]$ and $n = 50$

**Theorem 3.** *Let $x_{n-\mu,n}$, $\mu = 1, 2, \ldots$ denote the zeros of $P_n(x)$, arranged in increasing order. For fixed $\mu$ and large n, we have*

$$x_{n-\mu,n} = -1 + \frac{a-b}{n} - \frac{[2(a-b)]^{2/3} a_\mu}{2n^{4/3}} + O\left(\frac{1}{n^{5/3}}\right) \tag{115}$$

*as $n \to \infty$, where $a_\mu$ is $\mu$th negative zero of the Airy function* Ai$(x)$.

## ACKNOWLEDGMENTS

We would like to thank the two referees for a careful reading of our paper and making some valuable suggestions. X.-M. Huang is grateful to Dr Dan Dai, Dr Wen-Gao Long, Dr Min-Jie Luo, and Prof Yu-Qiu Zhao for helpful discussions.

## ORCID

*Xiao-Min Huang* https://orcid.org/0000-0002-1939-2337

## REFERENCES

1. Luo M-J, Wong R. Asymptotics of the associated Pollaczek polynomials. *Proc Am Math Soc*. 2019;147:2583-2597.
2. Pollaczek F. Sur une famille de polynmes orthogonaux quatre paramtres (French). *C R Acad Sci Paris*. 1950;230:2254-2256.

3. Erdélyi A, Magnus W, Oberhettinger F, Tricomi F. *Higher Transcendental Functions*. Vol. 2. New York: McGraw-Hill; 1953.
4. Szegő G. *Orthogonal Polynomials*. 4th ed. Providence, RI: American Mathematical Society; 1975.
5. Novikoff AB. *On a Special System of Orthogonal Polynomials* [PhD thesis]. Stanford University; 1953.
6. Bo R, Wong R. Asymptotic behavior of the Pollaczek polynomials and their zeros. *Stud Appl Math*. 1996;96:307-338.
7. Wang H-Y, Zhao YQ. Uniform asymptotics and zeros of a system of orthogonal polynomials defined via a difference equation. *J Math Anal Appl*. 2010;369:453-472.
8. Wang Z, Wong R. Uniform asymptotic expansion of $J_\nu(\nu a)$ via a difference equation. *Numer Math*. 2002;91:147-193.
9. Wang Z, Wong R. Asymptotic expansions for second-order linear difference equations with a turning point. *Numer Math*. 2003;94:147-194.
10. Wang Z, Wong R. Linear difference equations with transition points. *Math Comput*. 2005;74:629-653.
11. Wong R, Li H. Asymptotic expansions for second-order linear difference equations. *J Comput Appl Math*. 1992;41:65-94.
12. Huang X-M, Lin Y, Zhao Y-Q. Asymptotics of Charlier polynomials via difference equation methods. arXiv:1901.06041.
13. Li Y-T, Wang H. Uniform asymptotics for second order linear difference equations with applications to Pollaczek polynomials. Preprint.
14. Olver F, Lozier D, Boisvert R, Clark C. *NIST Handbook of Mathematical Functions*. Cambridge: Cambridge University Press; 2010.
15. Farid Khwaja S, Olde Daalhuis AB. Uniform asymptotic expansions for hypergeometric functions with large parameters IV. *Anal Appl*. 2014;12:667-710.
16. Wimp J. On the zeros of a confluent hypergeometric function. *Proc Am Math Soc*. 1965;2(16):281-283.
17. Hethcote HW. *Asymptotic Approximations with Error Bounds for Zeros of Airy and Cylindrical Functions* [PhD thesis]. Ann Arbor: University of Michigan; 1968.


**How to cite this article:** Huang Xiao-Min, Wong R. Uniform asymptotics and zeros of the associated Pollaczek polynomials. *Stud Appl Math*. 2020;145:625–646. https://doi.org/10.1111/sapm.12292

Analysis and Applications, Vol. 20, No. 6 (2022) 1103–1146

DOI: 10.1142/S0219530522400012



# Recent advances in asymptotic analysis

R. Wong*
*Department of Mathematics, City University of Hong Kong*
*Tat Chee Avenue, Kowloon, Hong Kong*
*rodscwong@gmail.com*

Yu-Qiu Zhao
*Department of Mathematics, Sun Yat-sen University*
*GuangZhou 510275, P. R. China*

Received 13 January 2022
Accepted 19 April 2022
Published 20 June 2022

This is a survey paper on an old topic in classical analysis. We present some new developments in asymptotics in the last 50 years. We start with the classical method of Darboux and its generalizations, including a uniformity treatment which has a direct application to the Heisenberg polynomials. We then present the development of an asymptotic theory for difference equations, which is a major advancement since the work of Birkhoff and Trjitzinsky in 1933. A new method was introduced into this field in the 1990s, which is now known as the nonlinear steepest descent method or the Riemann–Hilbert approach. The advantage of this method is that it can be applied to orthogonal polynomials which do not satisfy any differential or difference equations neither do they have any integral representations. As an example, we mention the case of orthogonal polynomials with respect to the Freud weight. Finally, we show how the Wiener–Hopf technique can be used to derive asymptotic expansions for the solutions of an integral equation on a half line.

*Keywords*: Asymptotics; uniform asymptotic expansions; Darboux's method; second-order difference equations; Riemann–Hilbert approach; parametrix; Wiener–Hopf equations.

Mathematics Subject Classification 2020: 41-02, 41A60, 39A06, 45A05, 45E10

## 1. Introduction

Asymptotics is one of the oldest topics in classical analysis; see [5]. Examples from this topic include Stirling's formula [33, 79], prime number theorem [81], and Hardy–Ramanujan's formula [43]. Despite the fact that it is an old topic, new methods continue to be discovered. For instance, in the construction of error bounds for asymptotic expansions of the Stieltjes transform and the fractional integral transform, a method was introduced which makes use of distribution theory; see [61, 62].

*Corresponding author.

This method was later extended to derive asymptotic expansions of the generalized Mellin transform [101], which is a divergent integral but exists in the sense of generalized functions. In a study of the behavior of polynomials orthogonal with respect to the Freud weight $\exp(-|x|^\beta)$, $\beta > 0$, another new method was brought into the field of asymptotics, which is the Riemann–Hilbert approach introduced by Deift–Zhou [30] in their investigation of the behavior of solutions to the mKdV equation.

The development of an asymptotic theory for ordinary differential equations was quite successful in the last century. After Poincaré [75] introduced a proper definition for asymptotic expansion and established rigorously the results for irregular singularity at infinity, several people obtained asymptotic expansions for solutions of differential equations with a large parameter, including Schlesinger [78], Birkhoff [6], and Turrittin [85]. More important results were given by Langer [54–56] on uniform asymptotic solutions to turning-point problems, and his results were substantially extended by Olver [67–69] including extensions to unbounded domains of the independent variable [69] and explicit bounds for the error terms [70]. The corresponding development for difference equations, however, was not as successful as for differential equations. Formal theory of irregular linear difference equations was first given by Birkhoff [7], and rigorous analysis was later given by Birkhoff and Trjitzinsky [8]. The first paper appeared in 1930, and the second paper was published in 1933. For the remaining part of the century, there were essentially no significant results in this area of research. The first paper on turning-point problem for difference equation [86] did not appear until 2002; see also [98]. Subsequently, progress seems to have picked up, and results in this area are now nearly as complete as they are for differential equations; see [87, 88, 13, 57].

While there are now plenty of results on finding asymptotic solutions to differential equations and to difference equations, the same is not true for integral equations. The problem is not that difficult if the equation is either of Volterra type or Fredholm type with a difference kernel, since the equation can be reduced to an algebraic equation by taking either Laplace transforms or Fourier transforms, and asymptotic behavior of the solution can be obtained by taking their inverse transformations; see, e.g., [95; 4, p. 286]. The situation is quite different if the equation is of the form

$$u(t) = f(t) + \int_0^\infty k(t-\tau)u(\tau)d\tau. \tag{1.1}$$

As far as we are aware, there is only one paper on asymptotic solutions to such an equation [63]. Equation in (1.1) is known as of Wiener–Hopf type or renewal equation on half line. The famous Wiener–Hopf technique is to construct a contour-integral representation for the solution of (1.1), but to find the behavior of the solution is another story since the functions involved in the integrand are very complicated and contain Stieltjes and Hilbert transforms. A rather satisfactory solution to this problem was given only very recently; see [58].

The purpose of this survey is to bring people's attention to some of the developments in asymptotic analysis in the past 50 years. In Sec. 2, we begin with Darboux's method. Even this classical result has some very attractive extensions. We will first present a generalization that involves logarithmic singularities [102], and mention couple of examples to show that this generalization is indeed useful. Next, we give a result on the behavior of the coefficients in a Maclaurin series, whose generating function has two coalescing algebraic singularities on its circle of convergence, and provide as an application the asymptotic expansion of the Heisenberg polynomials as the degree of the polynomial goes to infinity; see [60, 103].

In Sec. 3, we briefly review the results that have been obtained for difference equations during the past 20 years. As examples, we present the asymptotic expansions of the Racah and the Wilson polynomials; these polynomials are listed at the top level of the Askey scheme [74, p. 464].

In Sec. 4, we introduce the Riemann–Hilbert approach, and show how to use it to derive the asymptotic behavior of the polynomials orthogonal with respect to the Freud weight $\exp(-|x|^{\beta})$, $\beta > 0$. If $\beta \geq 1$, then the problem was completely solved by Kriecherbauer and McLaughlin [50] in 1999. The difficulty of this problem occurs when $0 < \beta < 1$, in which case it is not known what is the dominant approximant in the asymptotic expansion of these polynomials; see [29, p. 62]. This problem was finally settled in 2016, and the answer is that the dominant approximant does not involve special functions such as Airy or Bessel; it involves two solutions to a pair of decoupled scalar linear integral equations.

In Sec. 5, we return to the integral equation (1.1). We first recall what is the Wiener–Hopf technique [92], the solution to Eq. (1.1) is then given by a contour integral. Thus, we reduce the problem to an asymptotic evaluation of a Fourier integral whose integrand involves Stieltjes and Hilbert transforms.

## 2. Darboux's Method

The following combinatorial problem has appeared in Whitworth [90].

*If there be $n$ straight lines in one plane, no three of which meet in a point, no two lines are parallel, there will be $n(n-1)/2$ points of intersection. Put the intersection points into groups of $n$, in each of which no three points lie in one of the straight lines. Find the number $g_n$ of such groups.*

The answer given in [90] is $g_n = \frac{1}{2}(n-1)!$ for $n = 3, 4, \ldots$. However, in 1951, Robinson [77] considered the problem and showed that the answer is wrong. He found the correspondence between $n$-point groups and $n$-polygons. By counting $g_n = \frac{1}{2}(n-1)!$, one misses the composite $n$-gons: those consist of several separate simple polygons with totally $n$ sides. Instead, $u_n = g_n/\left(\frac{1}{2}(n-1)!\right)$ satisfies the recurrence relation

$$u_{n+3} = u_{n+2} + \frac{1}{2n} u_n, \quad n = 1, 2, 3, \ldots, \tag{2.1}$$

with $u_1 = u_2 = 0$ and $u_3 = 1$. After confirming the existence of the number

$$b = \lim_{n\to\infty} \frac{u_n^2}{n} = 0.284098\ldots,$$

as noted by Robinson [77], $b$ assumes the role of an absolute geometric constant. Robinson further asked whether it is an algebraic number, which seems unlikely, whether it is expressible in terms of $\pi$ and $e$, or whether it is entirely new.

The answers came fast. It is stated that the constant is given by

$$b = 4e^{-3/2}/\pi;$$

cf. an editorial note [34]. Thirteen readers, including Harry Pollard, John Riordan, George Szekeres, Fritz Ursell and Morgan Ward, wrote to the editor about Robinson's constant. Several of them used the generating function

$$f(x) = -2 + 2(1-x)^{-1/2}\exp\left(-\frac{x^2+2x}{4}\right) = \sum_{n=3}^{\infty} \frac{u_n}{n} x^n, \tag{2.2}$$

which can readily be derived from (2.1) via the differential equation

$$2(1-x)f'(x) = x^2[f(x)+2], \quad f(0) = 0.$$

Mention had also been made of a theorem from Titchmarsh [82, p. 224]:

**Proposition 2.1.** *Let*

$$f(x) = \sum_{n=0}^{\infty} a_n x^n, \quad g(x) = \sum_{n=0}^{\infty} b_n x^n, \quad |x| < 1,$$

*where* $a_n \geq 0$, $b_n \geq 0$, *and* $\sum_{n=0}^{\infty} b_n$ *diverges. If* $\lim_{n\to\infty} a_n/b_n = C$, *then* $\lim_{x\to 1} f(x)/g(x) = C$.

To determine the behavior of $a_n$, it seems that the converse of Proposition 2.1 matters much. Yet Titchmarsh [82, p. 226] claimed that there is no general converse of the above proposition: *from the asymptotic behavior of* $f(x)$ *we cannot deduce that of* $a_n$.

Nevertheless, Darboux's method [24] demonstrates how to deduce the asymptotic behavior of $a_n$ from that of $f(x)$. Such problems arise in various mathematical fields, including number theory, combinatorics, and orthogonal polynomials. As a start, one may use the Cauchy integral formula

$$a_n = \frac{1}{2\pi i}\int_C f(z) z^{-n-1} dz,$$

where $C$ is a circle encircling the origin, but not any singularities of $f(z)$. Expanding the contour $C$ if possible, and applying the residue calculus if a pole is encountered, we obtain the leading contribution to the asymptotic behavior of $a_n$. This procedure fails, however, if an algebraic singularity is the closest to the origin, and the classic Darboux's method ([24], see also [80, 97]) becomes useful.

To describe briefly how this method works, we assume that

$$f(z) = \sum_{n=0}^{\infty} a_n z^n \tag{2.3}$$

is analytic in $|z| < 1$, with only one singularity on the circle $|z| = 1$, say, at $z = 1$, and that in a neighborhood of $z = 1$, $f(z)$ is of the form

$$f(z) = (1-z)^{\alpha} g(z), \tag{2.4}$$

where $g(z)$ is analytic at $z = 1$, and $\alpha \notin \mathbb{Z}$. To find the large-$n$ behavior of $a_n$, we expand

$$g(z) = \sum_{r=0}^{\infty} c_r (1-z)^r \tag{2.5}$$

at $z = 1$. The $m$th Darboux approximant of $f(x)$ is then defined by

$$f_m(z) = \sum_{r=0}^{m} c_r (1-z)^{\alpha+r},$$

which is also analytic in $|z| < 1$; cf. (2.3). Expanding $f_m(z)$ into a Maclaurin series

$$f_m(z) = \sum_{n=0}^{\infty} b_{mn} z^n,$$

it turns out $a_n \sim b_{mn}$ for finite $m$, as $n \to \infty$. Indeed, we can write $a_n = b_{m,n} + \varepsilon_m(n)$, and show that the error term

$$\varepsilon_m(n) = \frac{1}{2\pi i} \int_C [f(z) - f_m(z)] z^{-n-1} dz = o(n^{-N}),$$

where $\operatorname{Re}\alpha + m + 1 \leq N < \operatorname{Re}\alpha + m + 2$. The above error estimate has been proved rigorously; see, e.g., [97, Chap. 2.6]. Thus, we have the following result.

**Theorem 2.1.** *With $f(z)$ given in (2.3)–(2.5), assume that $\alpha \notin \mathbb{Z}$ is a complex constant, and $z = 1$ is the only singularity on the circle of convergence $|z| = 1$. For any $m \geq 0$, the Maclaurin coefficient of $f(z)$ has the asymptotic expansion*

$$a_n = \sum_{r=0}^{m} c_r \binom{\alpha+r}{n} (-1)^n + o(n^{-\operatorname{Re}\alpha - m - 1}) \tag{2.6}$$

*as $n \to \infty$.*

Formula (2.6) is referred to as a generalized asymptotic expansion in [97], which is more general than the usual Poincaré expansion; see [36]. The asymptotic nature of the expansion is rooted in

$$\binom{\alpha+r}{n}(-1)^n = \frac{\Gamma(n-\alpha-r)}{n!\,\Gamma(-\alpha-r)} \sim \frac{n^{-\alpha-1-r}}{\Gamma(-\alpha-r)}, \qquad r = 0, 1, \ldots \text{ as } n \to \infty.$$

When there are finite number of algebraic singularities on the circle of convergence, the final result can be extended by adding up the contributions from all singular points; see [80, p. 207, Theorem 8.4; 97, p. 119, Theorem 5].

Applying Theorem 2.1 to the generating function (2.2) (cf. [97, p. 120]), one finds that $u_n \sim b\sqrt{n}$ as $n \to \infty$, with $b = 4e^{-3/2}/\pi$; this is Robinson's constant. The theorem can also be used to solve a problem of Erlebach and Ruehr in counting Hamiltonian cycles for bipartite graphs; see [47]. Darboux's method plays an important role as well in determining orthogonal measures, especially those with explicit three-term recurrence relations; see [45].

We complete our discussion of the classical Darboux's method by mentioning the following interesting example investigated by Olver [71]. It is well known that the Legendre polynomials possess the generating function

$$f(z) = \left[(e^{i\theta} - z)(e^{-i\theta} - z)\right]^{-1/2} = \sum_{n=0}^{\infty} P_n(\cos\theta) z^n, \tag{2.7}$$

where the branches are chosen such that $(e^{\pm i\theta} - z)^{-\frac{1}{2}} \to e^{\mp\frac{1}{2}i\theta}$ as $z \to 0$. Now, applying Darboux's method yields

$$P_n(\cos\theta) \sim \left(\frac{2}{\sin\theta}\right)^{\frac{1}{2}} \sum_{k=0}^{\infty} \begin{pmatrix} -\frac{1}{2} \\ k \end{pmatrix} \begin{pmatrix} k - \frac{1}{2} \\ n \end{pmatrix} \frac{\cos\theta_{n,k}}{(2\sin\theta)^k}, \qquad n \to \infty,$$

where

$$\theta_{n,k} = \left(n - k + \frac{1}{2}\right)\theta + \left(n - \frac{1}{2}k - \frac{1}{4}\right)\pi.$$

On the other hand,

$$P_n(\cos\theta) = \left(\frac{1}{2\sin\theta}\right)^{\frac{1}{2}} \sum_{k=0}^{\infty} \begin{pmatrix} -\frac{1}{2} \\ k \end{pmatrix} \begin{pmatrix} k - \frac{1}{2} \\ n \end{pmatrix} \frac{\cos\theta_{n,k}}{(2\sin\theta)^k}$$

for $\frac{1}{6}\pi < \theta < \frac{5}{6}\pi$. A paradox now arises, since one has

$$P_n(\cos\theta) \sim 2P_n(\cos\theta), \quad \frac{1}{6}\pi < \theta < \frac{5}{6}\pi, \quad n \to \infty; \tag{2.8}$$

cf. Olver [71].

Darboux's method dates back to 1878. Although frequently used, it does not appear to have been extended until around 1970. In 1974, Wong and Wyman [102] have given a generalization of Darboux's method, which allows the generating function $f(z)$ in (2.3) to have logarithmic-type singularities on its circle of convergence. This paper was considered *an important early (and unduly neglected) reference* in analytic combinatorics; see [38, p. 438]. First, we quote a couple of examples of this type.

**Problem of Pólya [76, Exercise 8, p. 8 and p. 213].** Define $A_n$ by

$$\frac{z}{\ln(1+z)} = \sum_{n=0}^{\infty} A_n \frac{z^n}{n!}, \quad |z| < 1. \tag{2.9}$$

Show that

$$\frac{A_n}{n!} \sim (-1)^{n-1} \frac{1}{n \ln^2 n}, \quad \text{as } n \to \infty. \tag{2.10}$$

**Problem of Knuth [48].** What is the asymptotic behavior of the constant $l_n$ defined by the generating function

$$\frac{1}{\ln(1-z)} + \frac{1}{z} = \sum_{n=0}^{\infty} l_n z^n, \quad |z| < 1. \tag{2.11}$$

In [102], Wong and Wyman dealt with the general case

$$f(z) = (1-z)^{\alpha}[\ln(1-z)]^{\mu} g(z) = \sum_{n=0}^{\infty} a_n z^n, \quad |z| < 1, \tag{2.12}$$

where $\alpha$ and $\mu$ are complex constants, and $g(z)$ is analytic in $|z| \leq 1+\delta$ for some $\delta > 0$. Now, we briefly outline their treatment. Again the Cauchy integral formula applies, and one has, by deforming the contour

$$2\pi i a_n = \oint_{|z|=1+\delta_n} f(z) z^{-n-1} dz - \oint_{|z-1|=\delta_n} f(z) z^{-n-1} dz,$$

where the integration paths are counterclockwisely oriented, and $\delta_n = n^{-1/2}$. The first integral is of the order $O(e^{-c\sqrt{n}})$ for some positive constant $c$; the second integral alone contributes to the full asymptotic expansion. The approximants are "special" functions with integral representations

$$M(\alpha, \mu, n) = -\frac{1}{2\pi i} \int_{+\infty}^{(0+)} (-t)^{\alpha} [\ln(-t)]^{\mu} e^{-(n+1)t} dt, \quad n = 0, 1, 2, \ldots, \tag{2.13}$$

and have full asymptotic expansions of the form

$$M(\alpha, \mu, n) \sim \frac{(-\ln(n+1))^{\mu}}{(n+1)^{\alpha+1}} \sum_{k=0}^{\infty} \binom{\mu}{k} \frac{d^k}{d\alpha^k} \left[\frac{1}{\Gamma(-\alpha)}\right] \frac{1}{(-\ln(n+1))^k}.$$

Wong and Wyman further introduced the function

$$J_m = -\frac{1}{2\pi i} \oint_{|t|=\delta_n} (-t)^{\alpha+m} [\ln(-t)]^{\mu} P_m((n+1)t) e^{-(n+1)t} dt, \tag{2.14}$$

where $P_m(w)$ is a polynomial given by

$$P_m(w) = \frac{1}{m!} \frac{d^m}{dt^m} \left[ g(t+1) \exp\left\{ -\frac{1}{2} wt \left[ \frac{2(\ln(1+t)-t)}{t^2} \right] \right\} \right] \Bigg|_{t=0} := \sum_{s=0}^{m} p_s w^s,$$

so that

$$J_m \sim \sum_{s=0}^{m} p_s (n+1)^s M(\alpha+m+s,\mu,n) \quad \text{as } n\to\infty.$$

**Theorem 2.2 ([102]).** *If $f(z)$ is the function given in* (2.12), *then its Maclaurin coefficient $a_n$ has the asymptotic expansion*

$$a_n \sim \sum_{m=0}^{\infty} (-1)^m J_m,$$

*where $J_m$ is defined in* (2.14) *and has the asymptotic expansion*

$$J_m \sim \frac{(-\ln(n+1))^{\mu}}{(n+1)^{\alpha+m+1}} \sum_{k=0}^{\infty} \left\{ \binom{\mu}{k} \sum_{s=0}^{m} p_s \frac{d^k}{d\alpha^k}\left[\frac{1}{\Gamma(-\alpha-m-s)}\right]\right\} \frac{1}{(-\ln(n+1))^k}, \quad n\to\infty.$$

The above result includes Theorem 2.1 as a special case if $\mu = 0$, and can be extended to cases with several fixed singularities on the circle of convergence.

Now, it is readily seen that (2.10) follows accordingly, by applying Theorems 2.2 to (2.9). For the generating function (2.11), we take $\alpha = 0$, $\mu = -1$ and $g(z) \equiv 1$ in (2.12). Then, Theorem 2.2 applies and we have

$$l_n \sim \frac{1}{(n+1)\ln^2(n+1)} \quad \text{as } n\to\infty.$$

As a consistence check, it is worth noting that (2.10) and (2.11) are related in the straightforward manner: $A_n/n! = (-1)^{n-1} l_{n-1}$ for positive integers $n$.

When the singularities are free to move on the circle of convergence, Darboux's method will continue to work only if their essential configuration remains the same while the relative positions vary. However, this method breaks down when two or more singularities coalesce with each other. For example, the generating function (2.7) for the Legendre polynomials has a pair of algebraic singularities $e^{\pm i\theta}$. Darboux's method fails to apply when $e^{\pm i\theta} \to 1$ as $\theta \to 0^+$, and $e^{\pm i\theta} \to -1$ as $\theta \to \pi^-$; a difficulty arises. In the cases of coalescing singular points, the asymptotic expansion may involve transcendental functions instead of elementary ones.

In 1967, Fields [37] presented a uniform treatment of Darboux's method when two or three singularities coalesce. More precisely, he considered the case in which

$$f(z,\theta) = (1-z)^{-\lambda}\left[\left(e^{i\theta}-z\right)\left(e^{-i\theta}-z\right)\right]^{-\Delta} g(z,\theta) = \sum_{n=0}^{\infty} a_n(\theta) z^n, \tag{2.15}$$

where the Maclaurin expansion converges for $|z|<1$, $\lambda$ and $\Delta$ are bounded quantities, the branches of $(1-z)^{-\lambda}$ and $[(e^{i\theta}-z)(e^{-i\theta}-z)]^{-\Delta}$ are chosen such that each of them reduces to 1 at $z=0$, and $g(z,\theta)$ is analytic in $|z| \le e^{\eta}$ $(\eta>0)$, uniformly for $\theta \in [0,\pi]$. In [37], Fields first expressed $a_n(\theta)$ as a Cauchy integral, then made a change of variable and a rescaling, and finally obtained a generalized asymptotic expansion in the sense of Erdélyi and Wyman [36]. His results are uniform in certain $\theta$-intervals depending on $n$.

Despite the fact that Fields' results have achieved the so-called uniform reduction in the sense of Olver [73, p. 102], they are found to be too complicated for any practical application; see, e.g., [35, p. 167; 73, pp. 112–113; 97, p. 145]. For example, Olver commented that *It may be desirable to investigate whether any simplifications are feasible since the results in* [37] *are rather complicated to apply in their present form.*

A motivation for the uniform treatment of Darboux's method came from the uniform asymptotic expansions of integrals. As a point of information, we mention a few relevant references in this respect. For example, Chester *et al.* [19] first presented a uniform treatment of the steepest descent method, Bleistein [10] considered the problem of many nearby stationary points and algebraic singularities, Wong [97, Chap. VII] provided various types of coalescence of singular points, in particular uniform asymptotics of orthogonal polynomials, and Olde Daalhuis and Temme [66] introduced a class of rational functions based on which error terms can be estimated.

The work of Wong and Zhao [103] in 2005 seems to have responded to Olver's request mentioned above, i.e. to derive simpler forms of uniform asymptotic expansions when two or more algebraic singularities, on the circle of convergence, coalesce with each other as certain parameter approaches a critical value. To begin with, Wong and Zhao first considered the simplest case with two singularities, namely

$$f(z,\theta) = \left[(e^{i\theta} - z)(e^{-i\theta} - z)\right]^{-\alpha} g(z,\theta) = \sum_{n=0}^{\infty} a_n(\theta) z^n, \tag{2.16}$$

where $g(z,\theta)$ is analytic in $|z| \leq e^{\eta}$, $\eta > 0$, which is a special case of Fields; cf. (2.15) with $\lambda = 0$. Contribution to the large-$n$ behavior still comes from the singular points $e^{\pm i\theta}$, which are now subject to vary as $\theta \to 0^+$. The approximants in this case are

$$T_1(x) = \frac{1}{2\pi i} \int_{\Gamma_0} (s^2+1)^{-\alpha} e^{xs} ds, \quad T_2(x) = \frac{1}{2\pi i} \int_{\Gamma_0} s(s^2+1)^{-\alpha} e^{xs} ds,$$

where $\Gamma_0$ is a Hankel-type loop which starts and ends at $-\infty$, and encircles $s = \pm i$ in the positive sense. An asymptotic expansion of the form

$$a_n(\theta) = \theta^{1-2\alpha} T_1(n\theta) \sum_{k=0}^{m-1} \frac{\alpha_k(\theta)}{n^k} + \theta^{1-2\alpha} T_2(n\theta) \sum_{k=0}^{m-1} \frac{\beta_k(\theta)}{n^k} + \varepsilon(\theta, m)$$

was obtained for $m = 1, 2, 3, \ldots$, with the error term $\varepsilon(\theta, m)$ given explicitly and estimated, and the coefficients determined recursively as follows: Begin with an analytic function

$$h_0(s,\theta) := g(e^{-s\theta}, \theta) \left[\left(\frac{e^{-s\theta} - e^{i\theta}}{(-s-i)\theta}\right)\left(\frac{e^{-s\theta} - e^{-i\theta}}{(-s+i)\theta}\right)\right]^{-\alpha} \tag{2.17}$$

for $\mathrm{Re}\, s \geq -\eta/\theta$ and $|s \pm i| < 2\pi/\theta$, and make use of the iteration

$$\begin{cases} h_k(s,\theta) := \alpha_k(\theta) + s\beta_k(\theta) + (s^2+1)g_k(s,\theta), \\ h_{k+1}(s,\theta) = -\dfrac{1}{\theta}\left[(s^2+1)\dfrac{d}{ds} + 2(1-\alpha)s\right] g_k(s,\theta), \end{cases} \quad k = 0, 1, 2, \ldots. \tag{2.18}$$

Assuming analyticity of each $h_k(s,\theta)$ and $g_k(s,\theta)$ at $s = \pm i$, one determines the coefficients $\alpha_k(\theta)$ and $\beta_k(\theta)$. Furthermore, straightforward verification shows that

$$T_1(x) = \frac{\sqrt{\pi}}{\Gamma(\alpha)}\left(\frac{x}{2}\right)^{\alpha-\frac{1}{2}} J_{\alpha-\frac{1}{2}}(x),$$

and

$$T_2(x) = \frac{\sqrt{\pi}}{\Gamma(\alpha)}\left(\frac{x}{2}\right)^{\alpha-\frac{1}{2}}\left[\frac{2\alpha-1}{x}J_{\alpha-\frac{1}{2}}(x) - J_{\alpha+\frac{1}{2}}(x)\right] = \frac{\sqrt{\pi}}{\Gamma(\alpha)}\left(\frac{x}{2}\right)^{\alpha-\frac{1}{2}} J_{\alpha-\frac{3}{2}}(x).$$

Hence, we have the following theorem.

**Theorem 2.3 ([103]; see also [60, Corollary 1]).** *If $f(z,\theta)$ is the function given in (2.16). Then its Maclaurin coefficient has the asymptotic expansion*

$$a_n(\theta) \sim \frac{\sqrt{\pi}}{\Gamma(\alpha)}\left(\frac{n}{2\theta}\right)^{\alpha-\frac{1}{2}}\left[J_{\alpha-\frac{1}{2}}(n\theta)\sum_{k=0}^{\infty}\frac{\alpha_k(\theta)}{n^k} + J_{\alpha-\frac{3}{2}}(n\theta)\sum_{k=0}^{\infty}\frac{\beta_k(\theta)}{n^k}\right] \tag{2.19}$$

*as $n \to \infty$, holding uniformly for $\theta \in [0, \pi-\delta]$, $\delta > 0$, with coefficients $\alpha_k(\theta)$ and $\beta_k(\theta)$ given by (2.17) and (2.18). The leading coefficients are $\alpha_0(\theta) = \cos(\alpha\theta)\left(\frac{\sin\theta}{\theta}\right)^{-\alpha}$ and $\beta_0(\theta) = \sin(\alpha\theta)\left(\frac{\sin\theta}{\theta}\right)^{-\alpha}$.*

The above result can be extended to handle cases with many coalescing algebraic singularities. For instance, we have

$$f(z,\theta) = \left\{\prod_{k=1}^{q}(z_k(\theta)-z)^{-\alpha_k}\right\} g(z,\theta) = \sum_{n=0}^{\infty} a_n(\theta) z^n, \tag{2.20}$$

where $g(z,\theta)$ is analytic in $|z| \leq e^{\eta}$ with $\eta > 0$, $z_k(\theta) \equiv e^{i\theta s_k(\theta)}$ so that $z_k(\theta) \to 1$ as $\theta \to 0$. In this case, the approximants are the special functions

$$T_l(x) := \frac{1}{2\pi i}\int_{\Gamma_0} s^{l-1} e^{xs} \prod_{k=1}^{q}(s + is_k(\theta))^{-\alpha_k} ds \tag{2.21}$$

for $l = 1, 2, \ldots, q$, where $\Gamma_0$ is also a Hankel-type loop which starts and ends at $-\infty$, encircling all points $s = -is_k(\theta)$ in the positive direction.

**Theorem 2.4 ([103]).** *Let $f(z,\theta)$ be the function given in (2.20). If $z_k(\theta) = e^{is_k\theta}$, $s_k$ being real constants, then*

$$a_n(\theta) \sim \theta^{1-\alpha}\sum_{l=1}^{q} T_l(n\theta)\sum_{k=0}^{\infty}\frac{\beta_{k,l}(\theta)}{n^k} \tag{2.22}$$

*as* $n \to \infty$, *uniformly in* $\theta \in [0, \nu]$ *with* $\nu < \min_{1\le k\le q}\{\pi/|s_k|\}$, *where* $\alpha = \sum_{k=1}^{q} \alpha_k$, *and the coefficients* $\beta_{k,l}(\theta)$ *can be determined successively.*

Properties of the special functions $T_l(x)$ in (2.21) with $s_k(\theta) \equiv s_k$ are considered in [103]. For instance, from (2.21), we have

$$T_l(x) = \frac{d^{l-1}}{dx^{l-1}} T_1(x) \quad \text{for } l = 1, 2, \ldots, q,$$

and that $w = T_1(x)$ solves the differential equation

$$x\frac{d^q w}{dx^q} + \sum_{l=0}^{q-1}(d_l x + c_l)\frac{d^l w}{dx^l} = 0,$$

where the coefficients $c_l$ and $d_l$ are determined by

$$\prod_{k=1}^{q}(s + is_k) := s^q + \sum_{l=0}^{q-1} d_l s^l, \quad \sum_{k=1}^{q}(1-\alpha_k)\prod_{k'\neq k}(s + is_{k'}) := \sum_{l=0}^{q-1} c_l s^l.$$

Now, we check the special case (2.15) investigated by Fields [37]. Theorem 2.4 applies, and we may set

$$q = 3; \quad s_1 = 0, \ \alpha_1 = \lambda; \quad s_2 = 1, \ \alpha_2 = \Delta; \quad s_3 = -1, \ \alpha_3 = \Delta.$$

By expanding a part of the integrand in (2.21) into Laurent series, we find that in terms of the generalized hypergeometric functions, we have

$$T_l(x) = \frac{x^{\lambda+2\Delta-l}}{\Gamma(\lambda+2\Delta-l+1)}\,{}_1F_2\left(\Delta; \frac{\lambda+2\Delta-l+1}{2}, \frac{\lambda+2\Delta-l+2}{2}; -\frac{x^2}{4}\right)$$

for $l = 1, 2$, and 3; cf. [74, (16.2.1)].

We complete our discussion on uniform treatment of Darboux's method by listing several follow-up progresses. In 2007, Bai and Zhao [2] derived uniform asymptotics for the Jacobi polynomials via uniform treatment of Darboux's method, using not the results but the ideas discussed above. Later in 2013, Liu *et al.* [60] apply the uniform Darboux's method to analyze the Heisenberg polynomials. The asymptotic expansion obtained involves the Kummer function and its derivative.

We mention yet another situation, which may be termed a singular case. We take as an example the Pollaczek polynomials

$$(1 - ze^{i\theta})^{-\frac{1}{2}+ih(\theta)}(1 - ze^{-i\theta})^{-\frac{1}{2}-ih(\theta)} = \sum_{n=0}^{\infty} P_n(x; a, b) z^n, \tag{2.23}$$

each of the factors in the generating function reduces to 1 for $z = 0$, where $h(\theta) = \frac{a\cos\theta+b}{2\sin\theta}$, $a > |b|$. There are two singularities $z = e^{\pm i\theta}$ on the circle of convergence, coalescing with each other when $\theta \to 0$. Meanwhile the exponent $h(\theta) \sim \frac{a+b}{2\theta}$ as $\theta \to 0$, demonstrating a singular behavior. The reader is referred to [11] for an asymptotic analysis of these polynomials using integral methods. Such types of generating function with varying exponents, regular or singular, seem to be of interest.

## 3. Difference Equations

As remarked earlier in Sec. 1, the development of the asymptotic theory for difference equations took a halt in the 1930s, and did not make any progress until the turn of this century. It is interesting to note that in his lecture, given during the conference in honor of his 80th birthday, Frank Olver expressed that *In my view Birkhoff and Trjitzinsky* [8] *set back all research into the asymptotic solutions of difference equations for most of the 20th century.* Also, in 1985, Wimp and Zeilberger [94] made the interesting remark that *Once on the forefront of mathematical research in America, the asymptotics of the solutions of linear recurrence equations is now almost forgotten, especially by the people who need it most, namely combinatorics and computer scientists.* Later, in 1991, Wimp [93] also made the statement that *There are still vital matters to be resolved in asymptotic analysis. At least one widely quoted theory, the asymptotic theory of irregular difference equations expounded by G. D. Birkho and W. J. Trjitzinsky* [7, 8] *in the early 1930s, is vast in scope; but there is now substantial doubt that the theory is correct in all its particulars. The computations involved in the algebraic theory alone are truly mind-boggling.*

The above comments were certainly the driving force behind the work carried out by Wong and Li [99, 100] in the 1990s. In order to understand the problem better, they first studied in [99] the simple second-order difference equation

$$y(n+2) + a(n)y(n+1) + b(n)y(n) = 0,$$

where $a(n)$ and $b(n)$ have infinite expansions of the form

$$a(n) \sim \sum_{s=0}^{\infty} \frac{a_s}{n^s} \quad \text{and} \quad b(n) \sim \sum_{s=0}^{\infty} \frac{b_s}{n^s} \tag{3.1}$$

for large values of $n$, and $b_0 \neq 0$. Then in [100], they extended their investigation to include the more general equation

$$y(n+2) + n^p a(n)y(n+1) + n^q b(n)y(n) = 0, \tag{3.2}$$

where $p$ and $q$ are integers, and $a(n)$ and $b(n)$ are as given in (3.1) with the leading coefficients $a_0$ and $b_0$ being nonzero.

The basic technique adopted in [99, 100] is essentially the method of successive approximations, which is customarily used in the asymptotic theory of differential equations. The ultimate goal of Wong's investigation was to develop a turning point theory for the three-term recurrence relation

$$y_{n+1} = (a_n x + b_n)y_n - c_n y_{n-1}, \quad n = 1, 2, \ldots, \tag{3.3}$$

where $a_n$, $b_n$ and $c_n$ are constants. If $x$ is a fixed number, then the recurrence relation is equivalent to the second-order linear difference equation (3.2). The importance of developing of a turning point theory for the recurrence relation (3.3) lies in the fact that many special functions of mathematical physics (Bessel functions, parabolic cylinder functions, Legendre functions, etc.) satisfy such an equation. In fact, any sequence of orthogonal polynomials satisfies an equation of the form (3.3); see [80, p. 42]. This statement, of course, applies to classical orthogonal polynomials such as Hermite $H_n(x)$, Laguerre $L_n^{(\alpha)}(x)$ and Jacobi $P_n^{(\alpha,\beta)}(x)$; more importantly, it includes all those that do not satisfy any second-order linear differential equations, e.g., Charlier $C_n^{(a)}(x)$, Meixner $m_n(x;\beta,c)$, Pollaczek $P_n(x;a,b)$, Meixner–Pollaczek $M_n(x;\delta,\eta)$ and Krawtchouk polynomials $K_n^N(x;p,q)$.

While a turning-point theory for second-order differential equations has been satisfactorily developed from 1930 to 1960 by Langer [54], Cherry [18], Olver [69] and others, not much progress has been made in the development of a corresponding theory for the recurrence equation (3.3). A possible explanation is that difference equations are considerably more complicated to analyze than differential equations; see a remark made by Iserles [44, p. 743].

To illustrate the difficulty in hand, we consider the Bessel function $J_\nu(\nu x)$, which has been used as a typical example in some of the most profound research in asymptotic analysis; see, e.g., [19, 67, 68]. The differential equation approach proceeds with the equation

$$\frac{d^2w}{dx^2} = \left\{\nu^2\frac{1-x^2}{x^2} - \frac{1}{4x^2}\right\}w, \tag{3.4}$$

which is satisfied by $x^{\frac{1}{2}}J_\nu(\nu x)$. From (3.4), it is clear that $x=\pm 1$ are two turning points, i.e. zeros of the coefficients function multiplied by the large parameter $\nu^2$. An application of the transformations

$$\frac{2}{3}\zeta^{\frac{3}{2}} = \ln\frac{1+(1-x^2)^{\frac{1}{2}}}{x} - (1-x^2)^{\frac{1}{2}} \tag{3.5}$$

and

$$W = \hat{f}^{\frac{1}{4}}w, \quad \hat{f} = \frac{1-x^2}{x^2\zeta} \tag{3.6}$$

gives

$$\frac{d^2W}{d\zeta^2} = \{\nu^2\zeta + \psi(\zeta)\}W,$$

where

$$\psi(\zeta) = \frac{5}{16\zeta^2} + \frac{\zeta x^2(x^2+4)}{4(x^2-1)}. \tag{3.7}$$

The basic approximation equation is

$$\frac{d^2W}{d\zeta^2} = \nu^2\zeta W; \tag{3.8}$$

two linearly independent solutions are the Airy functions $\mathrm{Ai}(\nu^{\frac{2}{3}}\zeta)$ and $\mathrm{Bi}(\nu^{\frac{2}{3}}\zeta)$. After identifying $J_\nu(\nu x)$ and $\mathrm{Ai}(\nu^{\frac{2}{3}}\zeta)$ and matching their behavior as $x\to\infty$, one obtains the expansion

$$J_\nu(\nu x) \sim \left(\frac{4\zeta}{1-x^2}\right)^{\frac{1}{4}} \left\{ \frac{\mathrm{Ai}(\nu^{\frac{2}{3}}\zeta)}{\nu^{\frac{1}{3}}} \sum_{s=0}^{\infty} \frac{A_s(\zeta)}{\nu^{2s}} + \frac{\mathrm{Ai}'(\nu^{\frac{2}{3}}\zeta)}{\nu^{\frac{5}{3}}} \sum_{s=0}^{\infty} \frac{B_s(\zeta)}{\nu^{2s}} \right\} \tag{3.9}$$

as $\nu \to \infty$, uniformly with respect to $x \geq 0$. The coefficients $A_s(\zeta)$ and $B_s(\zeta)$ are defined recursively by $A_0(\zeta) = 1$,

$$B_s(\zeta) = \frac{1}{2\zeta^{\frac{1}{2}}} \int_0^\zeta \{\psi(v)A_s(v) - A_s''(v)\} \frac{dv}{v^{\frac{1}{2}}} \tag{3.10}$$

and

$$A_{s+1}(\zeta) = -\frac{1}{2}B_s'(\zeta) + \frac{1}{2}\int \psi(\zeta)B_s(\zeta)d\zeta + \text{constants}. \tag{3.11}$$

From a difference-equation point of view, it is natural to suggest that the same result can be obtained directly from the three-term recurrence equation

$$J_{\nu+1}(x) - \frac{2\nu}{x}J_\nu(x) + J_{\nu-1}(x) = 0. \tag{3.12}$$

This problem turns out to be considerably more difficult to tackle than what we would have anticipated. To start with the very form of expansion (3.9) immediately raises the following questions:

(i) Since the Airy function $\mathrm{Ai}(x)$ does not satisfy any second-order difference equation, how is it going to arise from the three-term recurrence relation (3.12)?
(ii) Since we cannot change the discrete variable $n$ (or $\nu$) to a new independent variable, like what we have done with the Langer transformation (3.5), how are we going to produce the function $\zeta$ in (3.9) (cf. (3.5))?
(iii) Even if we can derive an asymptotic expansion similar to that in (3.9), can the coefficients be determined recursively like (3.10)–(3.11)?
(iv) Can the region of validity of our expansion be as large as the one for (3.9), i.e. the whole positive real axis $x \geq 0$; or is it just a finite interval $0 \leq x \leq M$ like in the case of integral approach (see [19, p. 610])?

Answers to these questions are given in [86].

In 1967, Dingle and Morgan [31, 32] made an attempt to study second-order difference equations of the form

$$f_{n+\omega}(x) + f_{n-\omega}(x) = 2\sigma(n,x)f_n(x), \tag{3.13}$$

where $\sigma(n,x)$ is a slowly varying function of the independent variable $n$. For convenience, they expressed $\sigma(n,x)$ as

$$\sigma(n,x) = a(n,x) + \omega^2 b(n,x) + \omega^4 c(n,x) + \cdots \tag{3.14}$$

with the understanding that $b(n,x)$ is two "order" smaller than $a(n,x)$, and that $c(n,x)$ is four "order" smaller than $a(n,x)$ and so on; see [32, Eq. (35)]. In (3.14),

$\omega$ is used as an "ordering" parameter which picks out terms of comparable magnitude. These authors called the zeros $n_0(x)$ of the equation

$$a(n, x) = 1 \tag{3.15}$$

turning points, and suggested that near $a = 1$ (i.e. in a neighborhood of $n_0(x)$), two linearly independent asymptotic expansions are given by

$$f_n^{(1)}(x) = J_{n+F(x)-n_0(x)}(F(x)) \tag{3.16}$$

and

$$f_n^{(2)}(x) = Y_{n+F(x)-n_0(x)}(F(x)), \tag{3.17}$$

where

$$F(x) = \left(\frac{da(n,x)}{dn}\right)^{-1}_{n=n_0(x)}. \tag{3.18}$$

In the case of (3.12), $\omega = 1$ and $\sigma(n, x) = n/x$. Thus, in (3.14) we just take $a(n, x) = \sigma(n, x) = n/x$ and $b(n, x) = c(n, x) = \cdots = 0$. The turning point (in the sense of Dingle and Morgan) is at $n = n_0(x) = x$, and the function in (3.18) is given by $F(x) = x$. As a result, the asymptotic solution $f_n^{(1)}(x)$ in (3.16) is just the function $J_n(x)$ itself, which is what we wish to approximate. In any case, the arguments in [31, 32] are too sketchy and non-rigorous. Nevertheless, in view of (3.9), the result of Dingle and Morgan is most likely correct.

Another paper that is relevant to our discussion here is by Costin and Costin [21]. These authors have studied the asymptotic behavior of the solutions of recurrence relations of the form

$$a_2(k\varepsilon, \varepsilon)y_{k+2} + a_1(k\varepsilon, \varepsilon)y_{k+1} + a_0(k\varepsilon, \varepsilon)y_k = 0 \tag{3.19}$$

as $\varepsilon \to 0^+$, where the coefficients $a_i(x, \varepsilon)$, $i = 0, 1, 2$, are $C^\infty$-functions in $x$ and $\varepsilon$ in some domain $I \times [0, \varepsilon_0]$, $I$ being a compact interval. Furthermore, they assume that for any $n \geq 0$

$$a_i(x, \varepsilon) = \sum_{s=0}^{n} a_{i,s}(x)\varepsilon^s + O(\varepsilon^{n+1}), \quad i = 0, 1, 2, \tag{3.20}$$

where $a_{i,s}(x)$ are $C^\infty$-functions in $x$ and the $O$-symbol is uniform with respect to $x$. Let $\lambda_1(x)$ and $\lambda_2(x)$ be the roots of the characteristic polynomial

$$a_2(x, 0)\lambda^2 + a_1(x, 0)\lambda + a_0(x, 0) = 0, \tag{3.21}$$

and assume that $\lambda_1(0) = \lambda_2(0)$ but $\lambda_1(x) \neq \lambda_2(x)$ when $x \neq 0$. Fix $\frac{1}{3} < \alpha < \frac{1}{2}$. One of the major results in [21] is that for $k < \varepsilon^{-\alpha}$, Eq. (3.19) has two solutions of the form

$$\exp\left\{F_\pm\left(k\varepsilon^{\frac{1}{3}}, \varepsilon^{\frac{1}{3}}\right)\right\},$$

where $\exp\{F_{\pm}(x,0)\} = \mathrm{Ai}(\Theta x) \pm \mathrm{Bi}(\Theta x)$ and $\Theta$ is an explicitly given constant in terms of $a_i(0,0)$ and $D_x a_i(0,0)$. A particular solution has the behavior

$$y_{k,\varepsilon} \sim \mathrm{Ai}(\Theta k \varepsilon^{\frac{1}{3}})\left[1 + \varepsilon^{\frac{1}{3}} A_1(\Theta k \varepsilon^{\frac{1}{3}}) + \cdots\right] \tag{3.22}$$

for large $k < \varepsilon^{-\alpha}$. A comparison of Eqs. (3.12) and (3.19) readily shows that the results of Costin and Costin cannot be used to derive a uniform asymptotic expansion for $J_\nu(\nu x)$ such as the one given in (3.9).

We are now ready to present a summary of the investigation carried out by Wong and his collaborators in the last 20 years. Returning to Eq. (3.3), we first define a sequence $\{K_n\}$ recursively by $K_{n+1}/K_{n-1} = c_n$, with $K_0$ and $K_1$ depending on the particular sequence $\{y_n\}$ satisfying (3.3). Then we put $A_n = a_n K_n/K_{n+1}$, $B_n = b_n K_n/K_{n+1}$ and $x_n = y_n/K_n$, so that (3.3) becomes

$$x_{n+1} - (A_n x + B_n)x_n + x_{n-1} = 0. \tag{3.23}$$

The coefficients $A_n$ and $B_n$ are assumed to have asymptotic expansions of the form

$$A_n \sim n^{-\theta} \sum_{s=0}^{\infty} \frac{\alpha_s}{n^s} \quad \text{and} \quad B_n \sim \sum_{s=0}^{\infty} \frac{\beta_s}{n^s}, \tag{3.24}$$

where $\theta$ is a real number and $\alpha_0 \neq 0$.

Let $\tau_0$ be a constant and put $\nu := n + \tau_0$. Clearly, the expansions in (3.24) can be recast in the form

$$A_n \sim \nu^{-\theta} \sum_{s=0}^{\infty} \frac{\alpha'_s}{\nu^s} \quad \text{and} \quad B_n \sim \sum_{s=0}^{\infty} \frac{\beta'_s}{\nu^s}. \tag{3.25}$$

In (3.23), we now let $x = \nu^\theta t$ and $x_n = \lambda^n$. Substituting (3.25) into (3.23) and letting $n \to \infty$ (and hence $\nu \to \infty$), we obtain the characteristic equation

$$\lambda^2 - (\alpha'_0 t + \beta'_0)\lambda + 1 = 0. \tag{3.26}$$

The roots of this equation are given by

$$\lambda = \frac{1}{2}\left[(\alpha'_0 t + \beta'_0) \pm \sqrt{(\alpha'_0 t + \beta'_0)^2 - 4}\right], \tag{3.27}$$

and they coincide when $t = t_{\pm}$, where

$$\alpha'_0 t_{\pm} + \beta'_0 = \pm 2. \tag{3.28}$$

The values $t_{\pm}$ play an important role in the asymptotic theory of the three-term recurrence relation (3.23), and they correspond to the transition points (i.e. turning points and poles) occurring in differential equations; [72, p. 362]. For this reason, we shall also call them *transition points*. Since $t_+$ and $t_-$ have different values, we may restrict ourselves to just consider the case $t = t_+$. In terms of the exponent $\theta$ in (3.25) and transition point $t_+$, we have three cases to consider; namely, (i) $\theta \neq 0$

and $t_+ \neq 0$; (ii) $\theta \neq 0$ and $t_+ = 0$; and (iii) $\theta = 0$. In the following, we present the main result in each of these three cases in the form of a theorem.

**Case (i).** In this case, we assume $\theta > 0$. The analysis for the case $\theta < 0$ is very similar; for an important example with $\theta = -1$, see [86]. Since the discussion of the two cases $t_+ > 0$ and $t_+ < 0$ are similar, we shall consider only the case $t_+ > 0$. Furthermore, to make the presentation simpler, we also assume that $t_- < 0$. These two assumptions together are equivalent to the condition $|\beta_0| < 2$. Before stating our first theorem, we need to impose one more condition; that is,

$$\alpha_1' = \beta_1' = 0. \tag{3.29}$$

It is interesting to note that this condition holds in most of the classical cases. In fact, in [31] Dingle and Morgan have assumed that $\alpha_{2s+1} = \beta_{2s+1} = 0$ for $s = 0, 1, 2, \ldots$. For a result without this assumption, we refer the reader to [87]. Now, we define the function

$$\begin{cases} \dfrac{2}{3}[\zeta(t)]^{\frac{3}{2}} = \alpha_0' t^{\frac{1}{\theta}} \displaystyle\int_{t_+}^{t} \frac{s^{-1/\theta} ds}{\sqrt{(\alpha_0' s + \beta_0')^2 - 4}} - \ln \frac{\alpha_0' t + \beta_0' + \sqrt{(\alpha_0' t + \beta_0')^2 - 4}}{2}, \\ \hfill t \geq t_+, \\ \dfrac{2}{3}[-\zeta(t)]^{\frac{3}{2}} = -\alpha_0' t^{\frac{1}{\theta}} \displaystyle\int_{t}^{t_+} \frac{s^{-1/\theta} ds}{\sqrt{4 - (\alpha_0' s + \beta_0')^2}} + \arccos\left(\frac{\alpha_0' t + \beta_0'}{2}\right), \quad t < t_+, \end{cases} \tag{3.30}$$

which plays the role of the Langer transform for differential equations; cf. (3.5).

**Theorem 3.1.** *Assume that the coefficients $A_n$ and $B_n$ in the recurrence relation* (3.23) *have asymptotic expansions of the form given in* (3.24) *with $\theta \neq 0$ and $|\beta_0| < 2$. (See the statement just before* (3.29).*) Let $\zeta(t)$ be defined as in* (3.30), *and recall the number $\nu := n + \tau_0$ in the asymptotic expansions in* (3.25). *Then, with $x = \nu^\theta t$, Eq.* (3.23) *has a pair of solutions $P_n(x)$ and $Q_n(x)$ given by*

$$P_n(\nu^\theta t) \sim \left(\frac{4\zeta}{(\alpha_0' t + \beta_0')^2 - 4}\right)^{\frac{1}{2}} \left[\mathrm{Ai}(\nu^{\frac{2}{3}}\zeta) \sum_{s=0}^{\infty} \frac{\widetilde{A}_s(\zeta)}{\nu^{s-\frac{1}{6}}} + \mathrm{Ai}'(\nu^{\frac{2}{3}}\zeta) \sum_{s=0}^{\infty} \frac{\widetilde{B}_s(\zeta)}{\nu^{s+\frac{1}{6}}}\right]$$

*and*

$$Q_n(\nu^\theta t) \sim \left(\frac{4\zeta}{(\alpha_0' t + \beta_0')^2 - 4}\right)^{\frac{1}{2}} \left[\mathrm{Bi}(\nu^{\frac{2}{3}}\zeta) \sum_{s=0}^{\infty} \frac{\widetilde{A}_s(\zeta)}{\nu^{s-\frac{1}{6}}} + \mathrm{Bi}'(\nu^{\frac{2}{3}}\zeta) \sum_{s=0}^{\infty} \frac{\widetilde{B}_s(\zeta)}{\nu^{s+\frac{1}{6}}}\right],$$

*where the coefficients $\widetilde{A}_s(\zeta)$ and $\widetilde{B}_s(\zeta)$ are determined successively from some recursive formulas, beginning with $\widetilde{A}_0(\zeta) = 1$ and $\widetilde{B}_0(\zeta) = 0$.*

**Case (ii).** Historically, this case was dealt with nearly 10 years after Case (iii) was settled. One of the reasons is that it was not clear what exactly the dominant approximant should be, although the example of Laguerre polynomials suggested that it most likely is a Bessel function. Since $\theta \neq 0$ in this case, we may let $\tau_0 := -\alpha_1/(\alpha_0\theta)$ and $N := n + \tau_0$. Then, from (3.23) and (3.24), we have

$$A_n x + B_n \sim n^{-\theta} \sum_{s=0}^{\infty} \frac{\alpha_s}{n^s} x + \sum_{s=0}^{\infty} \frac{\beta_s}{n^s} := \sum_{s=0}^{\infty} \frac{\alpha_s' t + \beta_s'}{N^s}, \tag{3.31}$$

where $x := N^\theta t$. A simple calculation gives

$$\alpha_0' = \alpha_0, \quad \alpha_1' = 0, \;\; \beta_0' = \beta_0, \quad \beta_1' = \beta_1, \;\; \beta_2' = \beta_2 + \beta_1 \tau_0. \tag{3.32}$$

If $\beta_0 = 2$ (or $-2$), from (3.28) it follows that one of the transition points is zero. Without loss of generality, we assume that $t_+ = 0 < t_-$ and $\beta_0 = 2$, $\alpha_0 < 0$. (For other cases, see Remark 3.1.) As in Case (i), we assume that $\beta_1' = \beta_1 = 0$ so that

$$\alpha_1' t_+ + \beta_1' = 0. \tag{3.33}$$

(In most of the classical cases, $\beta_1' = \beta_1 = 0$; see the statement following (3.29).)

With $\alpha_0' < 0$ and $\beta_0 = 2$, we now define the function

$$\pm\,\zeta(t) = \arccos\left(\frac{\alpha_0' t + \beta_0'}{2}\right) + \alpha_0' t^{\frac{1}{\theta}} \int_a^t \frac{\phi^{-1/\theta}}{\sqrt{4 - (\alpha_0'\phi + \beta_0')^2}} d\phi \tag{3.34}$$

for $t < t_-$, where "+" sign is taken for $\theta < 2$ and "−" sign is taken for $\theta > 2$. The sign is chosen so that $\zeta(t)$ is positive for $t \in (0, t_-)$. The choice of the lower limit of integration is just for the purpose of convergence of the integral. For instance, we can choose $a = 0$ if $\theta < 0$ or $\theta > 2$ and

$$a = \begin{cases} -\infty, & t < 0, \\ t_-, & 0 \le t \le t_- \end{cases} \qquad \text{if } 0 < \theta < 2. \tag{3.35}$$

With this choice, it can be shown that $\zeta(0) = 0$.

**Theorem 3.2.** *Assume that the coefficients $A_n$ and $B_n$ in the recurrence relation (3.23) have asymptotic expansions given in (3.24) with $\theta \neq 0, 2$ and $\beta_0 = 2$. Let $t_+ = 0$ be a transition point defined in (3.28), and the function $\zeta(t)$ be given as in (3.34) and (3.35). Then, (3.23) has a pair of linearly independent solutions*

$$P_n(N^\theta t) \sim N^{\frac{1}{2}} \left(\frac{4\zeta^2}{4 - (\alpha_0' t + \beta_0')^2}\right)^{\frac{1}{4}} \left[J_\nu(N\zeta) \sum_{s=0}^{\infty} \frac{\widetilde{A}_s(\zeta)}{N^s} + J_{\nu+1}(N\zeta) \sum_{s=0}^{\infty} \frac{\widetilde{B}_s(\zeta)}{N^s}\right] \tag{3.36}$$

*and*

$$Q_n(N^\theta t) \sim N^{\frac{1}{2}} \left(\frac{4\zeta^2}{4 - (\alpha_0' t + \beta_0')^2}\right)^{\frac{1}{4}} \left[W_\nu(N\zeta) \sum_{s=0}^{\infty} \frac{\widetilde{A}_s(\zeta)}{N^s} + W_{\nu+1}(N\zeta) \sum_{s=0}^{\infty} \frac{\widetilde{B}_s(\zeta)}{N^s}\right]$$

*for* $-\infty < t \le t_- - \delta$, $\delta$ *being an arbitrary positive constant. Here,* $W_\nu(x) := Y_\nu(x) - iJ_\nu(x)$, $x = N^\theta t$, $N = n + \tau_0$, $\tau_0 = -\alpha_1/(\alpha_0\theta)$ *and* $\nu$ *is given by*

$$\nu = \sqrt{\frac{1 + 4\beta_2'}{(\theta - 2)^2}}. \tag{3.37}$$

*The coefficients* $\widetilde{A}_s(\zeta)$ *and* $\widetilde{B}_s(\zeta)$ *are determined successively from some recursive formulas, beginning with* $\widetilde{A}_0(\zeta) = 1$ *and* $\widetilde{B}_0(\zeta) = 0$.

In the above theorem, we have used the functions $J_\nu(x)$ and $W_\nu(x) := Y_\nu(x) - iJ_\nu(x)$ as two linearly independent solutions of the Bessel equation. This is because $J_\nu(x)$ and $H_\nu^{(1)}(x)$ form a numerically satisfactory pair when $0 < \arg x < \pi$; see [74, §10.2(iii)]. Note that $i\zeta(t)$ is negative for $t < 0$; thus $J_\nu(N\zeta)$ is exponentially large and $H_\nu^{(1)}(N\zeta)$ is exponentially small as $N \to \infty$.

**Remark 3.1.** For the case $\alpha_0 > 0$ and $\beta_0 = -2$, the two transition points $t_\pm$ satisfy $t_- = 0 < t_+$. Note that we again have one transition point at the origin and the other being positive. Put $\mathcal{P}_n(x) := (-1)^n P_n(-x)$. Theorem 3.2 then applies to $\mathcal{P}_n(x)$. If the two transition points satisfy $t_- < 0 = t_+$, we set $x := -N^\theta t$, instead of $x := N^\theta t$. Theorem 3.2 again holds with $P_n(N^\theta t)$ replaced by $P_n(-N^\theta t)$ (in the case $\alpha_0 > 0$, $\beta_0 = 2$) or $(-1)^n P_n(-N^\theta t)$ (in the case $\alpha_0 < 0$, $\beta_0 = -2$).

**Remark 3.2.** In Theorem 3.2, we assume $\theta \ne 0, 2$. The situation with $\theta = 0$ was studied in [88], and will be discussed in Case (iii). The result in that case holds regardless whether the transition point is at 0 or not. $\theta = 2$ is exceptional in our present case (i.e. Case (ii)), since $\theta - 2$ appears in the denominator of (3.37). However, $\theta = 2$ and one of the transition points being zero is exactly the case of Wilson polynomials. For the results in that case, we refer to [57].

The presentation of Case (ii) here is based on the work of Cao and Li [13].

**Case (iii).** We now consider the case $\theta = 0$. Here, there is no need to make the scale change $x = \nu^\theta t$ in Case (i) or $x = N^\theta t$ in Case (ii); cf. (3.25) and (3.31). Also, we will not be concerned with whether the transition point $t_+(= x_+)$ in (3.28) is at the origin or not. However, we still need to make the assumptions $\alpha_1 = \beta_1 = 0$; see (3.29) and (3.33). Let

$$\tau_0 := -\frac{\alpha_3 x_+ + \beta_3}{2(\alpha_2 x_+ + \beta_2)}, \qquad N := n + \tau_0, \tag{3.38}$$

and define

$$\nu = \left(\alpha_2' x_+ + \beta_2' + \frac{1}{4}\right)^{1/2}, \tag{3.39}$$

$$\zeta^{\frac{1}{2}}(x) = \operatorname{arccosh}\left(\frac{\alpha_0 x + \beta_0}{2}\right). \tag{3.40}$$

Under the assumption in (3.29), it is easily verified that $\alpha_2' = \alpha_2$ and $\beta_2' = \beta_2$; see (3.32).

**Theorem 3.3.** *Assume that the coefficients $A_n$ and $B_n$ in the recurrence relation (3.23) have asymptotic expansions given in (3.24) with $\theta = 0$. Let $x_\pm (= t_\pm)$ be the transition points defined by (3.28). Then, Eq. (3.23) has a pair of linearly independent solutions*

$$P_n(x) \sim \left(\frac{4\zeta}{(\alpha_0 x + \beta_0)^2 - 4}\right)^{\frac{1}{4}} \times \left[N^{\frac{1}{2}} I_\nu(N\zeta^{\frac{1}{2}}) \sum_{s=0}^{\infty} \frac{A_s(\zeta)}{N^s} + N^{\frac{1}{2}} \zeta^{\frac{1}{2}} I_{\nu-1}(N\zeta^{\frac{1}{2}}) \sum_{s=0}^{\infty} \frac{B_s(\zeta)}{N^s}\right] \tag{3.41}$$

*and*

$$Q_n(x) \sim \left(\frac{4\zeta}{(\alpha_0 x + \beta_0)^2 - 4}\right)^{\frac{1}{4}} \times \left[N^{\frac{1}{2}} K_\nu(N\zeta^{\frac{1}{2}}) \sum_{s=0}^{\infty} \frac{A_s(\zeta)}{N^s} - N^{\frac{1}{2}} \zeta^{\frac{1}{2}} K_{\nu-1}(N\zeta^{\frac{1}{2}}) \sum_{s=0}^{\infty} \frac{B_s(\zeta)}{N^s}\right] \tag{3.42}$$

*as $n \to \infty$, for $x_- + \delta \le x < \infty$, $\delta$ being an arbitrary positive constant, where $N = n + \tau_0$, $\tau_0$ is given in (3.38), $\nu$ is given in (3.39) and $\zeta(x)$ is defined by (3.40). The coefficients $A_s(\zeta)$ and $B_s(\zeta)$ can be determined successively for any given $A_0(\zeta)$ and $B_0(\zeta)$.*

## 4. Riemann–Hilbert Approach

A significant connection between orthogonal polynomials and matrix-valued Riemann–Hilbert problems was established in the work of Fokas, Its and Kitaev [40]. Assuming that there is a sequence of polynomials $\{p_n\}$, orthonormal with respect to the weight function $d\alpha(\zeta) = w(\zeta)d\zeta$ supported on a curve $\Gamma$, the formulation of Fokas, Its and Kitaev is the following Riemann–Hilbert problem for a $2 \times 2$ matrix-valued function $Y(z)$:

($Y_1$) $Y(z)$ is analytic in $\mathbb{C}\backslash\Gamma$.

($Y_2$) The jump condition on $\Gamma$ is

$$Y_+(\zeta) = Y_-(\zeta) \begin{pmatrix} 1 & w(\zeta) \\ 0 & 1 \end{pmatrix}, \quad \zeta \in \Gamma, \tag{4.1}$$

where $Y_\pm(\zeta)$ denotes the limits of $Y(z)$ as $z \to \zeta \in \Gamma$ from the left and right of the oriented curve $\Gamma$, respectively.

($Y_3$) The behavior at infinity is

$$Y(z) = \left(I + O\left(\frac{1}{z}\right)\right) \begin{pmatrix} z^n & 0 \\ 0 & z^{-n} \end{pmatrix}, \quad z \to \infty. \tag{4.2}$$

The significance lies in that the unique solution to the above Riemann–Hilbert problem can be determined explicitly in terms of the orthogonal polynomials, that is,

$$Y(z) = \begin{pmatrix} \pi_n(z) & \dfrac{1}{2\pi i}\displaystyle\int_\Gamma \frac{\pi_n(\zeta)w(\zeta)d\zeta}{\zeta - z} \\ -2\pi i\gamma_{n-1}^2\pi_{n-1}(z) & -\gamma_{n-1}^2\displaystyle\int_\Gamma \frac{\pi_{n-1}(\zeta)w(\zeta)d\zeta}{\zeta - z} \end{pmatrix}, \quad z \in \mathbb{C}\setminus\Gamma, \tag{4.3}$$

where $\pi_n(z)$ is the monic polynomial corresponding to $p_n(z)$, and $\gamma_n$ is the leading coefficient of $p_n(z)$ such that $p_n(z) = \gamma_n\pi_n(z)$. The uniqueness of the solution $Y(z)$ can be justified by using Liouville's theorem, with asymptotic behavior appropriately preassigned to each critical point such as an endpoint or a possible point of intersection of $\Gamma$, etc.

To extract asymptotics from (4.3), we may use a steepest descent type method for oscillatory Riemann–Hilbert problems introduced by Deift and Zhou [30]; see [29] for a brief historical account of how this steepest descent method, also termed the Deift–Zhou method or Riemann–Hilbert approach, was further developed. This new approach to study asymptotic questions for orthogonal polynomials is rigorous, essentially global, and has proved to be very powerful. The pioneering works in which the nonlinear steepest descent method was actually applied to orthogonal polynomials are by Deift *et al.* [27] on orthogonal polynomials with varying exponential weight, with an application to a rigorous proof of the universality in random matrix theory, and by Bleher and Its [9] on semi-classical asymptotics with applications in random matrix theory; see also [28].

There are many more publications afterwards. Here, we mention only a very few. In [52], Kuijlaars *et al.* studied the polynomials orthogonal with respect to the modified Jacobi weight. This provides a situation where one seems to have no choice but to use the Riemann–Hilbert approach, since the orthogonal measure is the only property known. Discrete orthogonal polynomials are investigated in Johansson [46] and Baik *et al.* [3]. The orthogonal polynomials with respect to Freud weights are considered in Kriecherbauer and McLaughlin [50]. Wong and his coauthors have made attempts to bring in a global treatment to the Riemann–Hilbert approach; see [89] and a brief account in [104], in which all zeros of the polynomials belong to one single domain of uniformity.

There have been excellent introductions to the Riemann–Hilbert approach. The reader is referred to the books of Deift [26] and Fokas *et al.* [39], and the paper of Kuijlaars [51], for a better understanding of this method; see also [104, 110] for some more recent progress.

The asymptotic analysis of the Riemann–Hilbert problem consists of the following successive transformations, starting from the Riemann–Hilbert problem for $Y$.

(i) The first transformation, if applicable, is a re-scaling of the variable to make the equilibrium measure compactly supported.

(ii) The second transformation is a normalization of the behavior at infinity, involving a function $g$ which is the log transformation of the equilibrium measure, and turning the problem into an oscillatory Riemann–Hilbert problem.

(iii) The next transformation is the central piece of the analysis, which involves a factorization of the jump matrix and a deformation of the contour; oscillatory terms are transformed into exponentially decaying terms and may be neglected along the deformed contour, which reminds us of Debye's classical method of steepest descent for integrals.

(iv) The last transformation combines together the parametrices at all critical points such as the endpoints of the contour, and edges of the support of the equilibrium measure. In many circumstances technical difficulties may arise in the construction of such a parametrix, that is, an explicitly given local solution to the modified Riemann–Hilbert problem in a neighborhood of the critical point.

Now, we distinguish two types of parametrices. For the first type, an example is the frequently used parametrix involving the Airy functions [26, 41, 83], usually constructed in a neighborhood of the soft edge of the spectrum. An application of it to the random matrix theory is to describe the soft edge behavior by using the Airy kernel

$$\mathbb{A}(x,y)=\frac{\mathrm{Ai}(x)\mathrm{Ai}'(y)-\mathrm{Ai}(y)\mathrm{Ai}'(x)}{x-y}.$$

While the Bessel kernel

$$\mathbb{J}_\alpha(x,y)=\frac{J_\alpha(\sqrt{x})\sqrt{y}J_\alpha'(\sqrt{y})-J_\alpha(\sqrt{y})\sqrt{x}J_\alpha'(\sqrt{x})}{2(x-y)},$$

resulted from a Bessel parametrix, is often used to characterize the behavior at the hard edge of the spectrum; cf. [41, 52, 84]. The list of these parametrices is much longer, including special functions such as the confluent hypergeometric functions, the parabolic cylinder functions, the Painlevé functions, and more general Painlevé-type functions. There is a quite vast literature on these more transcendental parametrices; see, e.g., [39, 106, 107] and the references therein.

A common feature of the above parametrices, is that the special function involved satisfies a certain differential equation, linear or nonlinear. The reason for this phenomenon is that each of these parametrices can be transformed, by making use of a certain local conformal mapping, into a model Riemann–Hilbert problem, say $P(\zeta)$, with constant jump.

In the constant jump case, $P(\zeta)$ and $P'(\zeta)$ share the same jump condition. Hence all jumps vanish for $A(\zeta):=P'(\zeta)\{P(\zeta)\}^{-1}$. Then $A(\zeta)$ is analytic in $\mathbb{C}$, with only isolated singularities of known types. This would give a differential equation $P'(\zeta)=A(\zeta)P(\zeta)$.

However, there are still paramatrices of another type, in which the model problems could not be turned into the ones with constant jumps. In what follows, we briefly describe three examples of this type. These examples deal with polynomials orthogonal with respect to a logarithmic weight, a Bessel-function weight, and the Freud weight, separately.

**Example 4.1 (Logarithmic weight).** In 2018, Conway and Deift [20] derived the asymptotics of the recurrence coefficients for the polynomials orthogonal with respect to the logarithmic weight $w(x)dx$, where

$$w(x) = \ln \frac{2k}{1-x}, \quad x \in [-1,1], \tag{4.4}$$

and $k > 1$ so that the measure is strictly positive on its support.

The formulation in (4.1) and (4.2) fits well with $\Gamma = [-1,1]$, the weight $w$ specified in (4.4), and $Y(z)$ having at most weak singularities at the endpoints $z = \pm 1$. Following a now standard series of transformations, we change the original Riemann–Hilbert problem for $Y$ into a modified problem for $Q$, of which the jumps on curves adjacent to $z = 1$ are

$$J_Q(s) = \begin{cases} \begin{pmatrix} 1 & 0 \\ \dfrac{F^2}{w}\phi^{-2n} & 1 \end{pmatrix}, & s \in \Sigma_1 \cup \Sigma_2, \\ \begin{pmatrix} 0 & 1 \\ -1 & 0 \end{pmatrix}, & s \in (-1,1), \\ \begin{pmatrix} 1 & 0 \\ \left( \dfrac{F^2}{w_+} + \dfrac{F^2}{w_-} \right) \phi^{-2n} & 1 \end{pmatrix}, & s \in (1,+\infty); \end{cases} \tag{4.5}$$

cf. [20]. Here, $(\ )_\pm$ denote the boundary values (cf. (4.1)), all curves are in a rightward direction, $\Sigma_1$ and $\Sigma_2$ are, respectively, the upper and lower boundaries of the lens opening from $(-1,1)$, $w(z) = \ln \frac{2k}{1-z}$ is analytic in $\mathbb{C} \setminus [1,+\infty)$, the mapping $\phi(z)$ and the Szegő function $F(z)$ (cf. [52]) defined by

$$\phi(z) = z + \sqrt{z^2-1}, \quad F(z) = \exp\left( \frac{(z^2-1)^{1/2}}{2\pi i} \int_{-1}^{1} \frac{\ln w(s)}{(s^2-1)_+^{1/2}} \frac{ds}{s-z} \right)$$

are analytic in $\mathbb{C} \setminus [-1,1]$. The new difficulty is that no explicit local solution is known near the logarithmic singularity $x = 1$ of the weight (4.4), since the jump matrices in (4.5) are unlikely to be simultaneously turned into constant ones.

Fortunately, for the above problem with logarithmic weight, no local solution is needed. The strategy in [20] is to compare the above modified Riemann–Hilbert problem with the corresponding modified Riemann–Hilbert problem for the Legendre case. The comparison is not to examine the quotient, but to estimate the difference of solutions to these two Riemann–Hilbert problems. The estimates given in [20] seem to allow an effective comparison of two Riemann–Hilbert problems on the same contour in the general case.

**Example 4.2 (Bessel function weight).** In 2016, Deaño *et al.* [25] investigated the asymptotics of the polynomials $P_n(z)$ that are orthogonal with respect to the weight function $J_\nu(x)$ on $[0,\infty)$, where $J_\nu$ is the Bessel function of order $\nu \geq 0$. The Bessel function is oscillatory with an amplitude that decays like $O(x^{-1/2})$ as $x \to +\infty$, and therefore the moments $\int_0^\infty x^j J_\nu(x)dx$ do not exist. However, the polynomials $P_n$ can be defined via a regularization of the weight with an exponential factor; see [1].

$$\int_0^\infty P_n(x;s)x^j J_v(x)e^{-sx}dx = 0, \quad j = 0, 1, \ldots, n-1, \tag{4.6}$$

and then

$$P_n(x) = \lim_{s\to 0^+} P_n(x;s). \tag{4.7}$$

From numerical experiments, Asheim and Huybrechs [1] observe that the zeros of $P_n(z)$ seem to cluster along the vertical line $\mathrm{Re}\, z = \nu\pi/2$. When $0 \leq \nu \leq 1/2$, Deaño *et al.* [25] derive large-$n$ asymptotic behavior of $P_n(in\pi z)$ by using the Riemann–Hilbert approach, and give a rigorous proof of the observation in this case. More precisely, they have shown that for the scaled zeros, that is, dividing the imaginary parts of the zeros by $n$ while keeping the real parts fixed, the limiting curve is a vertical line segment of $\mathrm{Re}\, z = \nu\pi/2$. They further conjecture, based also on numerical evidence, that there is a limiting curve for the scaled zeros that differs from the vertical line segment when $\nu > 1/2$, partially because in this case the method in [25] to construct a local parametrix at the origin fails: this difficulty may very well be related to the different behavior of the zeros. Yet a very recent research [109] shows that for $\nu > 1/2$, the limiting curve for the scaled zeros is still a vertical segment of $\mathrm{Re}\, z = \nu\pi/2$.

The most technical part of [25, (3.50)–(3.51), Proposition 3.16] is the analysis of the local parametrix at the origin, with jump matrices involving the Bessel functions in a complicated way. Since by no means one can transform the jumps into constant matrices, an integral operator argument was brought in to show that the modified local Riemann–Hilbert problem has a solution for $n$ large enough. The leading order behavior of the parametrix has not yet been explicitly given, though it does not affect the determination of zero distribution.

In 1982, Wong [96] considered quadrature formulas for oscillatory integral transforms with Fourier kernel and Bessel kernels. Also, in [22], Dai *et al.* have proposed a study of the asymptotics of orthogonal polynomials whose weight function is closely related to the Bessel weight stated above.

**Example 4.3 (Freud weight).** In 1999, Kriecherbauer and McLaughlin [50] used the Riemann–Hilbert approach to study the strong asymptotics of the polynomials orthogonal with respect to the Freud weight

$$w_\beta(x)dx = e^{-\kappa_\beta |x|^\beta} dx, \quad x \in \mathbb{R}, \quad \kappa_\beta = \frac{\Gamma\left(\frac{\beta}{2}\right)\Gamma\left(\frac{1}{2}\right)}{\Gamma\left(\frac{\beta+1}{2}\right)}, \quad \beta > 0. \tag{4.8}$$

Although the Riemann–Hilbert approach works, an obstacle arises, that is, to find the explicit local solution in a neighborhood of the origin.

When $0 < \beta < 1$, the problem is reduced to solving a model Riemann–Hilbert problem for a certain $2 \times 2$ matrix-valued function $L$:

$$L : \mathbb{C}\backslash\mathbb{R} \to \mathbb{C}^{2\times 2} \text{ is analytic,}$$

$$L_+(s) = L_-(s)v_L(s) \quad \text{for } s \in \mathbb{R}, \tag{4.9}$$

$$L(s) = O(\ln|s|) \quad \text{as } s \to 0, \tag{4.10}$$

$$L(s) \to I \quad \text{for } s \to \infty, \tag{4.11}$$

where the jump matrix is given by

$$v_L(s) = \begin{pmatrix} 1 & -\eta_L(s) \\ \eta_L(-s) & 1 \end{pmatrix}, \quad s \in \mathbb{R},$$

with

$$\eta_L(s) = 2i(-1)^{n+1} e^{-|s|^\beta \sin(\pi\beta/2)} \sin\left(|s|^\beta \cos\frac{\pi\beta}{2}\right) \mathbf{1}_{[0,\infty)}(s), \quad s \in \mathbb{R},$$

$\mathbf{1}_{[0,\infty)}(s)$ denoting the indicator function of the set $[0, \infty)$, and $n$ being the polynomial degree; see [50, (6.41)–(6.43)].

In [50], the existence and uniqueness of $L(s)$ were established in the Wiener class. However, as commented in Deift *et al.* [29, p. 62, Remark 3.6].

*For $0 < \beta < 1$ the leading order behavior of the solution to the model problem at the origin has not been determined explicitly. It is defined through a Riemann–Hilbert problem which only depends on $\beta$ and on the parity of $n$.*

The reader is referred to [50, 105] and the references there for polynomials orthogonal with respect to weights similar to the Freud weight. For example, Chen and Ismail [17] considered the Freud-like orthogonal polynomials arising from a recurrence relation related to the indeterminate moment problems; see also [23]. In fact, the polynomials orthogonal with respect to the Bessel weight in [25]; cf. (4.6) and (4.7), provides another example of Freud-type weights. Expressing $J_\nu(x)$ as a linear combination of $K_\nu(\pm ix)$, one encounters a varying exponential weight of Freud-type $e^{-n\pi|x|}$ with a potential function $\pi|x|$. The jump (4.9) with constraint (4.11), cannot be transformed into a constant jump. Further, to solve the model problem, one may either represent them in terms of known special functions, or alternatively use new special functions to serve the same purpose.

It is well known that the Riemann–Hilbert problems are closely related to singular integral equations; see, e.g., [64]. Assume

$$Y(z) = \frac{1}{2\pi i}\int_\Gamma \frac{\Phi(\zeta)d\zeta}{\zeta - z}, \quad z \notin \Gamma, \tag{4.12}$$

where $\Phi(\zeta)$ for $\zeta \in \Gamma$ is the new matrix-valued unknown function. From the Plemelj formula, we have

$$Y_\pm(\zeta) = \pm\frac{1}{2}\Phi(\zeta) + \frac{1}{2\pi i}\fint_\Gamma \frac{\Phi(\tau)d\tau}{\tau - \zeta}, \quad \zeta \in \Gamma,$$

where the bar indicates that the integral is a Cauchy principal value; cf. (5.41). Substituting the above formulas into the jump condition (4.1), we obtain a homogeneous singular integral equation

$$\Phi(\zeta) - \frac{1}{2\pi i}\fint_\Gamma \frac{\Phi(\tau)d\tau}{\tau - \zeta}\begin{pmatrix} 0 & w(\zeta) \\ 0 & 0 \end{pmatrix} = 0 \tag{4.13}$$

with index 0. Finding asymptotic results from (4.13) for specific $\Gamma$ or more general integral equations is an important topic worth exploring.

The rest of this section will be devoted to the discussion of special functions determined by integral equations. Returning to Example 4.3, we examine the model problem for $L$. To facilitate our discussion, instead of the behavior (4.10) of $L(s)$ at the origin, we assume

$$L(s) = O(\epsilon_\beta(s)) \quad \text{as } s \to 0, \quad \epsilon_\beta(s) := \begin{cases} 1, & 1/2 < \beta < 1, \\ \ln|s|, & \beta = 1/2, \\ |s|^{\beta - 1/2}, & 0 < \beta < 1/2. \end{cases} \tag{4.14}$$

The objective is to introduce a pair of special functions, $u_\beta$ and $v_\beta$, as solutions of two scalar linear integral equations. Based on these functions, the solution to the model problem $L$ of Kriecherbauer and McLaughlin [50] can readily be constructed. More precisely, we assume that $u_\beta$ and $v_\beta$ solve, respectively, the following integral equations:

$$u(x) = 1 + \int_0^\infty \frac{K(t)u(t)dt}{t + x}, \quad x \in (0, \infty), \tag{4.15}$$

$$v(x) = 1 - \int_0^\infty \frac{K(t)v(t)dt}{t + x}, \quad x \in (0, \infty), \tag{4.16}$$

where

$$K(t) = \frac{1}{2\pi i}\left[\exp\left(e^{\left(\frac{\pi}{2} + \frac{\beta\pi}{2}\right)i}t^\beta\right) - \exp\left(e^{-\left(\frac{\pi}{2} + \frac{\beta\pi}{2}\right)i}t^\beta\right)\right] \tag{4.17}$$

$$= \frac{1}{\pi}\exp\left(-t^\beta \sin\frac{\pi\beta}{2}\right)\sin\left(t^\beta \cos\frac{\pi\beta}{2}\right). \tag{4.18}$$

Note that $\eta_L(t) = 2\pi i(-1)^{n+1}K(t)$ for $t \in [0, +\infty)$.

First, we formally derive the integral equations from the model problem for $L$ satisfying (4.9), (4.11) and (4.14). From the jump condition (4.9) and the behavior in (4.11) and (4.14), it is readily seen that the $(1,1)$-entry $L_{11}(s)$ is analytic in $\mathbb{C} \setminus (-\infty, 0]$, such that

$$
\begin{aligned}
&(L_{11})_+(s) - (L_{11})_-(s) = \eta_L(-s)L_{12}(s), \quad s \in (-\infty, 0),\\
&L_{11}(s) = 1 + o(1), \quad s \to \infty,\\
&L_{11}(s) = O(\epsilon_\beta(s)), \quad s \to 0.
\end{aligned}
$$

Hence, in view of the behavior of $L_{12}$ obtained from (4.11) and (4.14), we have from the Plemelj formula

$$
\begin{aligned}
L_{11}(s) &= 1 + \frac{1}{2\pi i}\int_{-\infty}^{0} \frac{\eta_L(-\tau)L_{12}(\tau)d\tau}{\tau - s}\\
&= 1 + \frac{1}{2\pi i}\int_{0}^{\infty} \frac{\eta_L(\tau)L_{12}(-\tau)d\tau}{-\tau - s}
\end{aligned} \tag{4.19}
$$

for $s \in \mathbb{C}\setminus(-\infty, 0]$, especially for $s \in (0, \infty)$. Similarly, it is also seen from (4.9) that $L_{12}(s)$ is analytic in $\mathbb{C}\setminus[0,\infty)$, and solves the scalar Riemann–Hilbert problem

$$
\begin{aligned}
&(L_{12})_+(s) - (L_{12})_-(s) = -\eta_L(s)L_{11}(s), \quad s \in (0, \infty),\\
&L_{12}(s) = o(1), \quad s \to \infty,\\
&L_{12}(s) = O(\epsilon_\beta(s)), \quad s \to 0.
\end{aligned}
$$

Hence, we have

$$
L_{12}(s) = \frac{1}{2\pi i}\int_{0}^{\infty} \frac{-\eta_L(\tau)L_{11}(\tau)d\tau}{\tau - s} \quad \text{for } s \in \mathbb{C}\setminus[0,\infty), \tag{4.20}
$$

especially for $s \in (-\infty, 0)$. For $x \in (0, +\infty)$, now we define

$$
u_\beta(x) = L_{11}(x) + (-1)^n L_{12}(-x), \quad v_\beta(x) = L_{11}(x) - (-1)^n L_{12}(-x), \tag{4.21}
$$

$n$ being the integer appearing in the definition of $\eta_L(s)$. From (4.19)–(4.21), it is readily verified that $u_\beta$ and $v_\beta$ solve the integral equations (4.15) and (4.16), respectively.

It is worth noting that similar coupled scalar integral equations have been derived in [25], although the equations in that paper were used only to construct a contraction mapping. Also, it is mentioned in Fokas *et al.* [39, p. 161] that a function $u(x)$ can be parametrized via the solution of a linear integral equation. This is exactly what we are doing: We are rigorously defining a pair of new special functions, using the above integral equations, to construct a parametrix.

We now outline the investigation carried out in Wong and Zhao [105]. To begin with, we have the following lemma.

**Lemma 4.1.** *The operator $T$ is a compact operator on $L^2[0,+\infty)$, where*

$$(Tu)(x) = \int_0^\infty \frac{K(t)u(t)dt}{t+x}. \tag{4.22}$$

The reason is that $\frac{K(t)}{t+x} \in L^2([0,+\infty) \times [0,+\infty))$, with $K(t)$ given by (4.18). Thus, $T$ is a Hilbert–Schmidt operator, and hence is compact; cf. e.g. [108, p. 277].

The next step is to show that the operators $I \pm T$ have trivial null space by using a vanishing lemma technique. To this aim, we need to know more about the analytic structure of the solutions to (4.15) and (4.16). It is readily seen that $(Tu)(z)$ in (4.22) is a Stieltjes transform, and is hence analytic in the cut plane $|\arg z| < \pi$. For later use, we need the following behavior of $(Tu)(z)$ at the origin:

**Lemma 4.2.** *For $u \in L^2[0,+\infty)$, we have $(Tu)(z) = O(\epsilon_\beta(z))$ for $|\arg z| \le \pi$ as $z \to 0$, where $\epsilon_\beta(z)$ is given by (4.14).*

A description of the null space is as follows.

**Lemma 4.3.** *Assume that $u \in L^2[0,+\infty)$ and $u - Tu = 0$ (or $u + Tu = 0$) in $L^2[0,+\infty)$. Then, $u \equiv 0$ for $x \in [0,\infty)$ in $L^2[0,+\infty)$.*

To conclude, we have the unique existence of $u_\beta(x)$ and $v_\beta(x)$ in $L^2$ sense.

**Theorem 4.1 ([105]).** *There exist unique solutions $u(x)$ and $v(x)$ to the integral equations (4.15) and (4.16), respectively, such that $u(x) - 1 \in L^2[0,+\infty)$ and $v(x) - 1 \in L^2[0,+\infty)$.*

Further analysis of the integral equations (4.15) and (4.16) show that $u_\beta(x)$ and $v_\beta(x)$ are in fact bounded for all $\beta \in (0,1)$.

**Theorem 4.2 ([105]).** *Let $u_\beta(x)$ and $v_\beta(x)$ be solutions of the integral equations in (4.15) and (4.16), respectively. Then*

$$u_\beta(x), v_\beta(x) \in L^\infty[0,\infty) \quad \textit{and} \quad u_\beta(x), v_\beta(x) \in C[0,\infty).$$

Since $u_\beta(x)$ and $v_\beta(x)$ solve the integral equations (4.15) and (4.16), with $u_\beta - 1, v_\beta - 1 \in L^2[0,\infty)$, we can deduce that $u_\beta(z), v_\beta(z) = 1 + O(1/z)$ as $z \to \infty$ for $|\arg z| < 3\pi/2$ and are of the order $O(\epsilon_\beta(z))$ as $z \to 0$ for $|\arg z| \le \pi$. Thus, we have the following theorem.

**Theorem 4.3 ([105]).** *The piece-wise analytic function*

$$L(s) = \begin{cases} \begin{pmatrix} L_\beta(s) & (-1)^n U_\beta(se^{-\pi i}) \\ (-1)^n U_\beta(s) & L_\beta(se^{-\pi i}) \end{pmatrix}, & 0 < \arg s < \pi; \\ \begin{pmatrix} L_\beta(s) & (-1)^n U_\beta(se^{\pi i}) \\ (-1)^n U_\beta(s) & L_\beta(se^{\pi i}) \end{pmatrix}, & -\pi < \arg s < 0 \end{cases} \tag{4.23}$$

*solves the Riemann–Hilbert problem* (4.9), (4.11) *and* (4.14) *for* $L$, *where*

$$L_\beta(z) = \frac{1}{2}(u_\beta(z) + v_\beta(z)), \quad U_\beta(z) = \frac{1}{2}(u_\beta(z) - v_\beta(z)) \tag{4.24}$$

*for* $\arg z \in (-\infty, \infty)$, $z \neq 0$.

Regarding $u_\beta(z)$ and $v_\beta(z)$ as a pair of special functions, a very preliminary analysis has been carried out in [105]. For example, for analytic continuation, applying the Plemelj formula to the integral equations (4.15)–(4.16), we have

$$u(ze^{-\pi i}) - u(ze^{\pi i}) = 2\pi i K(z)u(z), \quad v(ze^{-\pi i}) - v(ze^{\pi i}) = -2\pi i K(z)v(z), \tag{4.25}$$

initially for real and positive $z$, and then for complex $z$ with $\arg z \in (-\infty, \infty)$; cf. (4.17).

Also, treating the integral equations as Stieltjes transform and applying Theorem 4.1, we readily obtain the asymptotic approximations

$$u_\beta(z) \sim 1 + \sum_{k=1}^{\infty} \frac{c_k}{z^k} \quad \text{and} \quad v_\beta(z) \sim 1 + \sum_{k=1}^{\infty} \frac{d_k}{z^k} \quad \text{as } z \to +\infty, \quad |\arg z| < \frac{3\pi}{2}, \tag{4.26}$$

where

$$c_k = (-1)^{k-1} \int_0^\infty K(t)t^{k-1}u_\beta(t)dt \quad \text{and} \quad d_k = (-1)^k \int_0^\infty K(t)t^{k-1}v_\beta(t)dt \tag{4.27}$$

for $k = 1, 2, \ldots$.

The Stokes phenomenon is also addressed in [105]. Assume that the functions $u_\beta(z)$ and $v_\beta(z)$ solve, respectively, (4.15) and (4.16), such that $u_\beta - 1$, $v_\beta - 1 \in L^2[0, \infty)$. Then the Stokes lines for both functions are

$$\arg z = \pm\left\{\frac{\pi}{2}(3 - \alpha) + 2(l - 1)\pi\right\},$$

where $\alpha = \frac{1}{\beta} \in (4l - 3, 4l + 1]$, $l = 1, 2, \ldots$.

Much remains to be done. We propose a thorough investigation of the analytic and asymptotic properties of the functions $u_\beta$ and $v_\beta$, such as their zeros, modulus function [74, §10.18], kernel [53], and differential or difference equations, etc. Here, we list a few.

(i) Determine the coefficients $c_k$ and $d_k$ in the expansions at infinity given by (4.26) and (4.27), which are now expressed in terms of $u_\beta$ and $v_\beta$. It would be interesting and challenging to decode the coefficients from (4.15) and (4.16) in an explicit manner.

(ii) As a refinement of Theorem 4.2, one might reasonably expect that the behaviors of $u_\beta$ and $v_\beta$ at the origin are of the form $\sum_{k=0} \tilde{c}_k z^k + \sum_{k=1} \tilde{d}_k z^{\beta k}$.

A natural problem is to determine the small-$z$ asymptotic formula and to evaluate explicitly the coefficients $\tilde{c}_k$ and $\tilde{d}_k$, at least the leading coefficients such as $k = 0$ and $k = 1$.

(iii) The eigenvalues of the compact operator $T$ given by (4.22), that is, find $\lambda$ such that the null space of $\lambda I - T$ is nontrivial in $L^2[0, +\infty)$.

(iv) It has been conjectured in [105] that the function $u_\beta(x)$ is monotonically decreasing in $x \in [0, \infty)$, and also that the initial value $u_\beta(0)$ is decreasing in $\beta \in (0, 1)$. Corresponding conjectures can be made for the function $v_\beta(x)$, such as it is increasing in $x \in [0, \infty)$, and also that the initial value $v_\beta(0)$ is increasing in $\beta \in (0, 1)$. It is also of interest to consider the complete monotonicity of these special functions.

## 5. Wiener–Hopf Equation

As far as we are aware, the first paper that addresses asymptotic solutions to an integral equation of the form in (1.1) is by Muki and Sternberg [63] in 1970. But, their approach assumes that the solution behaves like $t^{-\delta}/M$ as $t \to +\infty$, where $\delta$ is a positive number and $M$ is a constant. An attempt was made by Li and Wong [59] in 1994 to remove such an assumption and to derive the asymptotic solution directly from the integral equation, but their attempt was not successful. Later, it was realized that Eq. (1.1) is known as of Wiener–Hopf type. This naturally led Li and Wong [58] to use the famous Wiener–Hopf technique [92]. Here, we first give an outline of this technique, leading to an integral representation of the solution. For convenience we reproduce (1.1) below

$$u(t) = f(t) + \int_0^\infty k(t-\tau)u(\tau)d\tau, \quad t > 0. \tag{5.1}$$

The following notations will be used throughout this section:

$$u_+(t) = \begin{cases} u(t), & t > 0, \\ 0, & t < 0 \end{cases} \tag{5.2}$$

and

$$u_-(t) = \begin{cases} 0, & t > 0, \\ f(t) + \displaystyle\int_0^\infty k(t-\tau)u(\tau)d\tau, & t < 0. \end{cases} \tag{5.3}$$

Since we are only concerned with $t > 0$ in Eq. (5.1), in most cases the function $f(t)$ is identically zero for $t$ negative. Equation (5.1) can now be written as

$$u_+(t) + u_-(t) = f(t) + \int_{-\infty}^\infty k(t-\tau)u_+(\tau)d\tau, \quad -\infty < t < \infty. \tag{5.4}$$

Taking Fourier transforms on both sides of this equation gives

$$\left[1 - \sqrt{2\pi}K(\lambda)\right]U_+(\lambda) = F(\lambda) - U_-(\lambda), \tag{5.5}$$

where

$$K(\lambda) = \frac{1}{\sqrt{2\pi}} \int_{-\infty}^{\infty} k(t) e^{i\lambda t} dt, \quad F(\lambda) = \frac{1}{\sqrt{2\pi}} \int_{-\infty}^{\infty} f(t) e^{i\lambda t} dt,$$
$$U_{\pm}(\lambda) = \frac{1}{\sqrt{2\pi}} \int_{-\infty}^{\infty} u_{\pm}(t) e^{i\lambda t} dt, \tag{5.6}$$

and $\lambda = \sigma + i\tau$. In (5.5), we note that $U_+(\lambda)$ and $U_-(\lambda)$ are unknown functions, and that $K(\lambda)$ and $F(\lambda)$ are known functions. To solve this functional equation involves a factorization of the function $1 - \sqrt{2\pi} K(\lambda)$, which is the crux of the idea in the Wiener–Hopf technique.

Let $L(\lambda)$ be an analytic function of $\lambda = \sigma + i\tau$ in the strip $\tau_- < \tau < \tau_+$ such that $|L(\sigma + i\tau)| \le C|\sigma|^{-p}$, $p > 0$ and $C > 0$, as $|\sigma| \to \infty$. This inequality holds uniformly for $\tau$ in the strip $\tau_- + \varepsilon \le \tau \le \tau_+ - \varepsilon$, $\varepsilon > 0$. For $\tau_- < c < \tau < d < \tau_+$, we have by Cauchy's theorem

$$L(\lambda) = \frac{1}{2\pi i} \int_{\mathcal{C}} \frac{L(\zeta)}{\zeta - \lambda} d\zeta,$$

where $\mathcal{C}$, a positively oriented contour, is the boundary of the rectangle whose four corners are at $-R + ic$, $R + ic$, $R + id$ and $-R + id$. Letting $R \to +\infty$ yields the decomposition

$$L(\lambda) = L_+(\lambda) - L_-(\lambda), \tag{5.7}$$

where

$$L_+(\lambda) = \frac{1}{2\pi i} \int_{-\infty+ic}^{\infty+ic} \frac{L(\zeta)}{\zeta - \lambda} d\zeta, \quad L_-(\lambda) = \frac{1}{2\pi i} \int_{-\infty+id}^{\infty+id} \frac{L(\zeta)}{\zeta - \lambda} d\zeta, \tag{5.8}$$

and $\tau_- < c < d < \tau_+$. It is easily seen that $L_+(\lambda)$ is analytic for all $\tau > c$ and $L_-(\lambda)$ is analytic for all $\tau < d$.

If $\ln K(\lambda)$ satisfies the conditions imposed on the function $L(\lambda)$ above, then we immediate have

$$K(\lambda) = \frac{K_+(\lambda)}{K_-(\lambda)}, \tag{5.9}$$

where

$$K_+(\lambda) = \exp\left\{ \frac{1}{2\pi i} \int_{-\infty+ic}^{\infty+ic} \frac{\ln K(\zeta)}{\zeta - \lambda} d\zeta \right\} \tag{5.10}$$

and

$$K_-(\lambda) = \exp\left\{ \frac{1}{2\pi i} \int_{-\infty+id}^{\infty+id} \frac{\ln K(\zeta)}{\zeta - \lambda} d\zeta \right\}. \tag{5.11}$$

It is not difficult to see that $K_+(\lambda)$ and $K_-(\lambda)$ are analytic, bounded and nonzero in the half-planes $\tau > \tau_-$ and $\tau < \tau_+$, respectively.

Returning to (5.5), we assume that $1-\sqrt{2\pi}K(\lambda)$ has zeros $a_1, \ldots, a_m$. As in (5.9), we can find $K_+(\lambda)$ and $K_-(\lambda)$, which are analytic and zero-free in the respective half-plane $\tau > \tau_-$ and $\tau < \tau_+$, such that

$$1-\sqrt{2\pi}K(\lambda) = \frac{K_+(\lambda)}{K_-(\lambda)}(\lambda - a_1)(\lambda - a_2)\cdots(\lambda - a_m). \tag{5.12}$$

Equation (5.5) may now be written as

$$U_+(\lambda)K_+(\lambda)(\lambda - a_1)(\lambda - a_2)\cdots(\lambda - a_m) = K_-(\lambda)F(\lambda) - K_-(\lambda)U_-(\lambda). \tag{5.13}$$

As in (5.7), the term $K_-(\lambda)F(\lambda)$ can be decomposed into the form

$$K_-(\lambda)F(\lambda) = C_+(\lambda) + C_-(\lambda), \tag{5.14}$$

where $C_+(\lambda)$ and $C_-(\lambda)$ are analytic in $\tau > \tau_-$ and $\tau < \tau_+$, respectively. With (5.14), we rearrange (5.13) so that we can define a function $E(\lambda)$ by

$$\begin{aligned} E(\lambda) &= U_+(\lambda)K_+(\lambda)(\lambda - a_1)(\lambda - a_2)\cdots(\lambda - a_m) - C_+(\lambda) \\ &= C_-(\lambda) - K_-(\lambda)U_-(\lambda). \end{aligned} \tag{5.15}$$

This equation defines $E(\lambda)$ only in the strip $\tau_- < \tau < \tau_+$. But, by the first equality $E(\lambda)$ is actually defined and analytic in $\tau > \tau_-$, and by the second equality $E(\lambda)$ is defined and analytic in $\tau < \tau_+$. Hence, using analytic continuation, we can define $E(\lambda)$ in the whole $\lambda$-plane. Suppose that it can be shown

$$|U_+(\lambda)K_+(\lambda)(\lambda - a_1)(\lambda - a_2)\cdots(\lambda - a_m) - C_+(\lambda)| \le |\lambda|^p, \quad \text{as } \lambda \to \infty \text{ in } \tau > \tau_-,$$

$$|C_-(\lambda) - K_-(\lambda)U_-(\lambda)| \le |\lambda|^q, \quad \text{as } \lambda \to \infty \text{ in } \tau < \tau_+.$$

Then, by the extended form of the Liouville theorem, $E(\lambda)$ is a polynomial $P(\lambda)$ of degree less than or equal to the integer part of $\min\{p, q\}$, i.e.

$$\begin{aligned} U_+(\lambda)K_+(\lambda)(\lambda - a_1)(\lambda - a_2)\cdots(\lambda - a_m) - C_+(\lambda) &= P(\lambda), \\ C_-(\lambda) - K_-(\lambda)U_-(\lambda) &= P(\lambda). \end{aligned} \tag{5.16}$$

To determine the degrees $p$ and $q$, we need to know the behaviors of $U_+(\lambda)$ and $U_-(\lambda)$. In special cases, these can be found from the behavior of $u(t)$. For example, if $u(t) \sim t^{-1/2}$ as $t \to 0^+$, then by the Abelian theorem $U_+(\lambda) \sim \lambda^{-\frac{1}{2}}$ as $\lambda \to \infty$ in the upper half-plane $\tau > 0$; see [65, p. 36 and 69]. In equations in (5.16), $U_+(\lambda)$ and $U_-(\lambda)$ are determined only within an arbitrary polynomial $P(\lambda)$; that is, within a finite number of arbitrary constants, which must be determined by some other methods.

By using the analytic properties of the function $U_+(\lambda)$ and Cauchy's theorem, we first derive the integral representation

$$U_+(\lambda) = \frac{1}{2\pi i}\int_{-\infty+i\alpha}^{\infty+i\alpha} \frac{U_+(\zeta)}{\zeta - \lambda} d\zeta, \tag{5.17}$$

where $\alpha$ is a real number bigger than $\tau_-$ and $\lambda$ is any point in the upper half-plane $\tau > \alpha$. The denominator of the above Cauchy integral can be written as

$$\frac{1}{i(\zeta - \lambda)} = \int_0^\infty e^{-i(\zeta-\lambda)t} dt, \tag{5.18}$$

where $\lambda$ is any point in the half-plane $\tau > \alpha$. Substituting (5.18) into (5.17) and interchanging the order of integration give

$$U_+(\lambda) = \frac{1}{\sqrt{2\pi}} \int_0^\infty e^{i\lambda t} \left[ \frac{1}{\sqrt{2\pi}} \int_{-\infty+i\alpha}^{\infty+i\alpha} U_+(\zeta) e^{-it\zeta} d\zeta \right] dt.$$

Comparing the integral on the right side of this equation with the Fourier transforms given in (5.6), we have

$$u(t) = \frac{1}{\sqrt{2\pi}} \int_{-\infty+i\alpha}^{\infty+i\alpha} U_+(\zeta) e^{-it\zeta} d\zeta, \quad t > 0. \tag{5.19}$$

On account of (5.16), the solution of (5.1) has the integral representation

$$u(t) = \frac{1}{\sqrt{2\pi}} \int_{-\infty+i\alpha}^{\infty+i\alpha} \frac{P(\lambda) + C_+(\lambda)}{K_+(\lambda)(\lambda - a_1)(\lambda - a_2)\cdots(\lambda - a_m)} e^{-it\lambda} d\lambda, \quad t > 0, \tag{5.20}$$

for any $\alpha$, $\tau_- < \alpha < \tau_+$.

A major assumption in the preceding analysis is that the kernel function $k(t)$ in Eq. (5.1) is exponentially decaying at infinity. This enables one to factor

$$\frac{1 - \sqrt{2\pi} K(\lambda)}{(\lambda - a_1)(\lambda - a_2)\cdots(\lambda - a_m)}$$

into two components $K_+(\lambda)$ and $K_-(\lambda)$, one analytic in an upper half-plane, the other analytic in a lower half-plane, and the two half-planes overlapping in an infinite strip parallel to the real axis.

If the kernel is not exponentially decaying, e.g., the Cauchy density function

$$k(t) = \frac{1}{\pi} \frac{1}{1 + t^2},$$

then one would not to be able to decompose $\left[1 - \sqrt{2\pi} K(\lambda)\right] / (\lambda - a_1)(\lambda - a_2)\cdots(\lambda - a_m)$ into two factors analytic in two overlapping half-planes. A natural way to extend the method presented above is to introduce an exponential function, such as $e^{-\varepsilon|t|}$, in the kernel. This idea has been used by Carlson and Heins [15] and by Carrier [16], but the arguments in both papers are only formal. Rigorous justifications were given by Widom [91] for the case $k \in L^2(\mathbb{R})$, and by Krein for $k \in L^1(\mathbb{R})$. (For a brief introduction to the work of Krein [49] and Gohberg and Krein [42], see [4, Chap. 20].) Here, we give a quick review of the argument given in Widom [91].

Let $\{\varepsilon_n\}$ be a monotonically decreasing sequence tending to zero as $n \to \infty$. Define

$$\tilde{e}_n(t) = \begin{cases} e^{-\varepsilon_n t}, & t > 0, \\ 0, & t < 0. \end{cases}$$

We call $u(t)$ a solution of (5.1), if for any $\eta > 0$ it satisfies

$$\lim_{n\to\infty} \left\| e^{-\eta|t|} \int_{-\infty}^{\infty} k(t-\tau)\tilde{e}_n(\tau)u(\tau)d\tau - e^{-\eta|t|}\left[u(t) - f(t)\right] \right\|_2 = 0. \qquad (5.21)$$

In (5.21), $e^{-\eta|t|}$ is inserted for some technical reasons. If such a solution does exist, then it is well known that there will exist a subsequence $\{\varepsilon_{n_k}\}$ such that $\varepsilon_{n_k} \to 0$ as $k \to \infty$ and

$$\lim_{k\to\infty} \int_0^{\infty} k(t-\tau)\tilde{e}_{n_k}(\tau)u(\tau)d\tau = u(t) - f(t) \qquad (5.22)$$

holds almost everywhere; see [12, Proposition 0.1.10]. To make this argument work, Widom first proved that the functional equation

$$U_-(\zeta) = F(\zeta) + \left[\sqrt{2\pi}K(\zeta) - 1\right]U_+(\zeta), \quad \zeta \in \mathbb{R}, \qquad (5.23)$$

holds almost everywhere; cf. (5.5). Next, he introduced the function

$$\Psi(\zeta) := \frac{\left[1 - \sqrt{2\pi}K(\zeta)\right](\zeta+i)^{\frac{n}{2}}(\zeta-i)^{\frac{n}{2}}}{(\zeta-\alpha_1)^{m_1}(\zeta-\alpha_2)^{m_2}\cdots(\zeta-\alpha_l)^{m_l}} \left(\frac{\zeta-i}{\zeta+i}\right)^k, \qquad (5.24)$$

which has no zero on the real line. Here, $\alpha_j$ for $1 \le j \le l$ is a zero of $1 - \sqrt{2\pi}K(\zeta)$ with multiplicity $m_j$, and $n = m_1 + \cdots + m_l$. To obtain a factorization formula like the one in (5.12), he set

$$\chi(\lambda) := \frac{1}{2\pi i} \int_{-\infty}^{\infty} \frac{\ln \Psi(\zeta)}{\zeta - \lambda} d\zeta; \qquad (5.25)$$

cf. (5.10) and (5.11). Note that this integral defines an analytic function in $\mathbb{C}\backslash\mathbb{R}$. Put

$$\begin{aligned} \phi_+(\lambda) &:= (\lambda+i)^{\frac{n}{2}-k}e^{-\chi_+(\lambda)}, \quad &\operatorname{Im}\lambda > 0, \\ \phi_-(\lambda) &:= (\lambda-i)^{-\frac{n}{2}-k}e^{-\chi_-(\lambda)}, \quad &\operatorname{Im}\lambda < 0, \end{aligned} \qquad (5.26)$$

where

$$\chi_+(\lambda) = \chi(\lambda) \quad \text{in } \operatorname{Im}\lambda > 0, \quad \chi_-(\lambda) = \chi(\lambda) \quad \text{in } \operatorname{Im}\lambda < 0$$

with boundary values

$$\begin{aligned} \chi_+(\zeta) &= \lim_{\varepsilon\to 0^+} \chi(\zeta + i\varepsilon) = \frac{1}{2}\ln\Psi(\zeta) + \frac{1}{2\pi i} \int_{-\infty}^{\infty} \frac{\ln\Psi(\xi)}{\xi - \zeta} d\xi, \\ \chi_-(\zeta) &= \lim_{\varepsilon\to 0^+} \chi(\zeta - i\varepsilon) = -\frac{1}{2}\ln\Psi(\zeta) + \frac{1}{2\pi i} \int_{-\infty}^{\infty} \frac{\ln\Psi(\xi)}{\xi - \zeta} d\xi \end{aligned} \qquad (5.27)$$

for $\zeta \in \mathbb{R}$, where the bar indicates that the integral is a Cauchy principal value. It is now readily seen that

$$1 - \sqrt{2\pi}K(\zeta) = \frac{\phi_-(\zeta)}{\phi_+(\zeta)}(\zeta - \alpha_1)^{m_1}(\zeta - \alpha_2)^{m_2} \cdots (\zeta - \alpha_l)^{m_l}; \tag{5.28}$$

cf. (5.12). Applying (5.28) to the functional equation (5.23) yields

$$\frac{U_+(\zeta)}{\phi_+(\zeta)}(\zeta - \alpha_1)^{m_1}(\zeta - \alpha_2)^{m_2} \cdots (\zeta - \alpha_l)^{m_l} = -\frac{U_-(\zeta)}{\phi_-(\zeta)} + \frac{F(\zeta)}{\phi_-(\zeta)}. \tag{5.29}$$

To obtain an analogue of (5.14), Widom set

$$H(\lambda) = \frac{1}{2\pi i}\int_{-\infty}^{\infty} \frac{F(\zeta)}{\phi_-(\zeta)}\frac{d\zeta}{\zeta - \lambda} \tag{5.30}$$

with

$$H_+(\lambda) = H(\lambda) \quad \text{in } \operatorname{Im}\lambda > 0, \quad H_-(\lambda) = H(\lambda) \quad \text{in } \operatorname{Im}\lambda < 0$$

and

$$\begin{aligned} H_+(\zeta) &= \frac{1}{2}\frac{F(\zeta)}{\phi_-(\zeta)} + \frac{1}{2\pi i} \fint_{-\infty}^{\infty} \frac{F(\xi)}{\phi_-(\xi)}\frac{d\xi}{\xi - \zeta}, \\ H_-(\zeta) &= -\frac{1}{2}\frac{F(\zeta)}{\phi_-(\zeta)} + \frac{1}{2\pi i} \fint_{-\infty}^{\infty} \frac{F(\xi)}{\phi_-(\xi)}\frac{d\xi}{\xi - \zeta} \end{aligned} \tag{5.31}$$

for $\zeta \in \mathbb{R}$; cf. (5.27). From (5.31), it follows

$$\frac{F(\zeta)}{\phi_-(\zeta)} = H_+(\zeta) - H_-(\zeta). \tag{5.32}$$

Coupling (5.29) and (5.32) yields

$$\frac{U_+(\zeta)}{\phi_+(\zeta)}(\zeta - \alpha_1)^{m_1}(\zeta - \alpha_2)^{m_2} \cdots (\zeta - \alpha_l)^{m_l} - H_+(\zeta) = -\frac{U_-(\zeta)}{\phi_-(\zeta)} - H_-(\zeta). \tag{5.33}$$

The left-hand side of (5.33) defines an analytic function on the upper half-plane $\mathbb{C}_+$, and the right-hand side defines an analytic function on the lower half-plane $\mathbb{C}_-$; these two functions agree on the real line $\mathbb{R}$. By using a theorem of Carleman [14, p. 40], Widom showed that these two functions are analytic continuations of each other and together represent an entire function $P(\lambda)$. Furthermore, by using an estimate of the function $\phi_-(\lambda)$ in (5.26), Widom concluded that $P(\lambda)$ is in fact a polynomial of degree less than $[\frac{n}{2} + k]$. From (5.33), we have

$$U_+(\lambda) = \frac{[P(\lambda) + H_+(\lambda)]\phi_+(\lambda)}{(\lambda - \alpha_1)^{m_1}(\lambda - \alpha_2)^{m_2} \cdots (\lambda - \alpha_l)^{m_l}}. \tag{5.34}$$

By inversion, we obtain

$$u(t) = \frac{1}{\sqrt{2\pi}}\int_{-\infty+i\gamma}^{\infty+i\gamma} \frac{[P(\lambda) + H_+(\lambda)]\phi_+(\lambda)}{(\lambda - \alpha_1)^{m_1}(\lambda - \alpha_2)^{m_2} \cdots (\lambda - \alpha_l)^{m_l}} e^{-it\lambda} d\lambda, \quad t > 0, \tag{5.35}$$

for any $\gamma > 0$; cf. (5.19). In his final step, Widom established the limit in (5.21), thus proving that $u(t)$ is a solution of (5.1).

For illustration purpose, it suffices to consider just the special case in which all zeros of $1 - \sqrt{2\pi}K(\lambda)$ are simple; i.e. $m_1 = \cdots = m_l = 1$, and $K(\lambda)$ and $F(\lambda)$ are both smooth, except possibly for a finite number of points in $\mathbb{R}\backslash\{\alpha_1, \ldots, \alpha_l\}$. By Cauchy's residue theorem, the integral in (5.35) can be expressed as

$$u(t) = \frac{1}{\sqrt{2\pi}} \fint_{-\infty}^{\infty} \frac{[P(\lambda) + H_+(\lambda)]\,\phi_+(\lambda)}{(\lambda - \alpha_1)(\lambda - \alpha_2)\cdots(\lambda - \alpha_l)} e^{-it\lambda} d\lambda - i\sqrt{\frac{\pi}{2}} \sum_{j=1}^{l} G_j(\alpha_j) e^{-it\alpha_j}, \tag{5.36}$$

where

$$G_j(\lambda) = \frac{[P(\lambda) + H_+(\lambda)]\phi_+(\lambda)(\lambda - \alpha_j)}{(\lambda - \alpha_1)(\lambda - \alpha_2)\cdots(\lambda - \alpha_l)}. \tag{5.37}$$

Note that here we have made use of the formula

$$\frac{1}{\sqrt{2\pi}} \fint_{-\infty}^{\infty} \frac{e^{-it\xi}}{\xi} d\xi = -\sqrt{\frac{\pi}{2}} i. \tag{5.38}$$

Finally, choose $l$ cut-off functions $\eta_j(\lambda)$, $j = 1, \ldots, l$, such that $\eta_j(\lambda)$ is equal to 1 for $\lambda$ near $\alpha_j$, $j = 1, \ldots, l$, and $\operatorname{supp} \eta_j \cap \operatorname{supp} \eta_i = \emptyset$ for each $j \neq i$. Let $\eta_0 = 1 - \sum_{j=1}^{l} \eta_j$, the integral in (5.36) can now be written as

$$\begin{aligned} u(t) = &\frac{1}{\sqrt{2\pi}} \sum_{j=1}^{l} e^{-it\alpha_j} \int_{-\infty}^{\infty} \frac{G_j(\xi)\eta_j(\xi) - G_j(\alpha_j)}{\xi - \alpha_j} e^{-it(\xi - \alpha_j)} d\xi \\ &+ \frac{1}{\sqrt{2\pi}} \int_{-\infty}^{\infty} \frac{[P(\xi) + H_+(\xi)]\phi_+(\xi)\eta_0(\xi)}{(\xi - \alpha_1)(\xi - \alpha_2)\cdots(\xi - \alpha_l)} e^{-it\xi} d\xi \\ &- \sqrt{2\pi} i \sum_{j=1}^{l} G_j(\alpha_j) e^{-it\alpha_j}. \end{aligned} \tag{5.39}$$

Asymptotic behavior of the solution $u(t)$ to Eq. (5.1) can be obtained from the Fourier integrals in (5.39); see [97, Theorem 1, p. 199 and Example 11, p. 234]. But, the amount of work to derive just the leading term in the asymptotic expansion of $u(t)$ is still tremendous. This is because the integrands in (5.39) involve functions $\phi_+(\lambda)$ and $H_+(\lambda)$, which are given in terms of Hilbert transforms; see (5.26), (5.27) and (5.31). Fortunately, asymptotic expansions of Hilbert transforms can be obtained via corresponding results for the Stieltjes transform

$$S_\varphi(z) = \int_0^{\infty} \frac{\varphi(t)}{t + z} dt, \quad |\arg z| < \pi, \tag{5.40}$$

and the one-sided Hilbert transform

$$H_\varphi^+(x) = \fint_0^{\infty} \frac{\varphi(t)}{t - x} dt, \quad x \in \mathbb{R}^+, \tag{5.41}$$

where again the bar indicates that the integral is a Cauchy principal value. In deriving the asymptotic expansions of these transforms, we also run into the Mellin transform of $\varphi$, which is defined by

$$M[\varphi; z] = \int_0^\infty t^{z-1}\varphi(t)dt, \quad \operatorname{Re} z > 0. \tag{5.42}$$

We assume that $\varphi \in L^1(\mathbb{R}^+)$ has an asymptotic expansion of the form

$$\varphi(t) \sim \sum_{s=0}^{\infty}\sum_{i=0}^{s} \alpha_{s,i} t^{-s-\theta} \ln^i t, \quad t \to \infty, \tag{5.43}$$

for fixed $\theta \in (0, 1]$. Let

$$\varphi_n(t) = \varphi(t) - \sum_{s=0}^{n-1}\sum_{i=0}^{s} \alpha_{s,i} t^{-s-\theta} \ln^i t. \tag{5.44}$$

For convenience, we also introduce a new notation. Recall the Pochhammer symbol $(\theta)_n = \theta(\theta+1)\cdots(\theta+n-1)$. Since

$$(-1)^n \frac{d^n}{dt^n}\left(\frac{t^{-\theta}}{(\theta)_n}\right) = t^{-n-\theta}$$

for any $\theta \in \mathbb{R}$ and $n \in \mathbb{N}$ with $(\theta)_n \neq 0$, we have

$$(-1)^{n+i} \frac{d^n}{dt^n}\left\{\frac{d^i}{d\theta^i}\left(\frac{t^{-\theta}}{(\theta)_n}\right)\right\} = t^{-n-\theta} \ln^i t \tag{5.45}$$

for any $i \in \mathbb{N}$. Let

$$H_{\theta,n,i}(t) = (-1)^{n+i} \frac{d^i}{d\theta^i}\left(\frac{t^{-\theta}}{(\theta)_n}\right) = \sum_{j=0}^{i} C^j_{\theta,n,i} t^{-\theta} \ln^j t, \tag{5.46}$$

where

$$C^j_{\theta,n,i} = (-1)^{n+i+j} \binom{i}{j} \frac{d^{i-j}}{d\theta^{i-j}}\left(\frac{1}{(\theta)_n}\right). \tag{5.47}$$

Note that from (5.46), we have

$$H^{(n)}_{\theta,n,i}(t) = t^{-n-\theta} \ln^i t.$$

**Theorem 5.1.** *Suppose that $\varphi$ satisfies (5.43) with $\theta \in (0,1)$. Then for any $z \in \mathbb{C}$ with $|\arg z| < \pi$ and $n \in \mathbb{N}^+$, we have*

$$S_\varphi(z) = \sum_{s=0}^{n-1}\sum_{i=0}^{s}\sum_{j=0}^{i} (-1)^j \pi \alpha_{s,i} C^j_{\theta,s,i} \frac{d^j}{d\theta^j}\left[\frac{(\theta)_s}{\sin\theta\pi} z^{-\theta}\right] z^{-s}$$
$$- \sum_{s=1}^{n} (s-1)! c_s z^{-s} + \frac{(-1)^n}{z^n} \int_0^\infty \frac{\tau^n \varphi_n(\tau)}{\tau + z} d\tau, \tag{5.48}$$

*where $c_s = \frac{(-1)^s}{(s-1)!} M[\varphi; s]$ and $M[\varphi; s]$ is the Mellin transform of $\varphi$.*

To state the next result, we need another notation, namely

$$\gamma_{k,s} = \int_0^\infty \frac{\ln^k t}{(t+1)^{s+2}} dt, \tag{5.49}$$

where $s,\ k \in \mathbb{N}$. Direct calculation gives

$$\gamma_{0,s} = \frac{1}{s+1}, \quad \gamma_{1,s} = -\frac{1}{s+1}\sum_{i=1}^{s}\frac{1}{i}, \tag{5.50}$$

$$\gamma_{2,s} = \frac{4}{s+1}\sum_{i=1}^{\infty}\frac{(-1)^{i-1}}{i^2} - \frac{2}{s+1}\sum_{i=1}^{s}\int_0^\infty \frac{\ln t}{(t+1)^{i+1}} dt \tag{5.51}$$

and

$$\begin{aligned}\gamma_{k,s} &= \frac{[1+(-1)^k]k!}{s+1}\sum_{i=1}^{\infty}\frac{(-1)^{i-1}}{i^k} - \frac{k}{s+1}\sum_{i=1}^{s}\int_0^\infty \frac{\ln^{k-1} t}{(t+1)^{i+1}} dt \\ &= \frac{[1+(-1)^k]k!}{s+1}\sum_{i=1}^{\infty}\frac{(-1)^{i-1}}{i^k} - \frac{k}{s+1}\sum_{i=1}^{s}\gamma_{k-1,i-1},\end{aligned} \tag{5.52}$$

empty sum being understood to be zero.

**Theorem 5.2.** *Suppose that $\varphi$ satisfies (5.43) with $\theta = 1$. Then for any $z \in \mathbb{C}$ with $|\arg z| < \pi$ and $n \in \mathbb{N}^+$, we have*

$$\begin{aligned}S_\varphi(z) &= \sum_{s=0}^{n-1}\sum_{i=0}^{s}\sum_{j=0}^{i}\sum_{l=0}^{j+1}\frac{(s+1)!\alpha_{s,i}C_{1,s,i}^{j}\binom{j+1}{l}\gamma_{j+1-l,s}}{j+1} z^{-s-1}\ln^l z \\ &\quad - \sum_{s=1}^{n}(s-1)!c_s^* z^{-s} + \frac{(-1)^n}{z^n}\int_0^\infty \frac{\tau^n \varphi_n(\tau)}{\tau+z} d\tau,\end{aligned} \tag{5.53}$$

*where* $c_s^* = \varphi_{s,s}(1) - \int_0^1 \varphi_{s-1,s-1}(t)dt$, $\varphi_{s,s}(t) = \frac{(-1)^s}{(s-1)!}\int_t^\infty (\tau-t)^{s-1}\varphi_s(\tau)d\tau$ *for* $s \in \mathbb{N}^+$, *and* $\varphi_{0,0}(t) = \varphi(t)$.

In view of the well-known formula of Plemelj

$$H_\varphi^+(x) = \frac{1}{2}\lim_{\varepsilon\to 0^+}\int_0^\infty \left[\frac{1}{t-x+i\varepsilon} - \frac{1}{t-x-i\varepsilon}\right]\varphi(t)dt, \tag{5.54}$$

the results of the last two theorems also give the asymptotic expansions of the one-side Hilbert transform of $\varphi$.

**Theorem 5.3.** *Suppose that $\varphi$ satisfies* (5.43) *with $\theta \in (0,1)$. Then for any $n \in \mathbb{N}^+$, we have*

$$H_\varphi^+(x) = \sum_{s=0}^{n-1}\sum_{i=0}^{s}\sum_{j=0}^{i}(-1)^{s+j}\pi\alpha_{s,i}C^j_{\theta,s,i}\frac{d^j}{d\theta^j}\left[(\theta)_s\cot(\theta\pi)x^{-\theta}\right]x^{-s}$$

$$-\sum_{s=1}^{n}(-1)^s(s-1)!c_s x^{-s} + \frac{1}{x^n}\fint_0^\infty \frac{\tau^n\varphi_n(\tau)}{\tau - x}d\tau, \tag{5.55}$$

*where* $c_s = \frac{(-1)^s}{(s-1)!}M[\varphi; s]$.

**Theorem 5.4.** *Suppose that $\varphi$ satisfies* (5.43) *with $\theta = 1$. Then for any $n \in \mathbb{N}^+$, we have*

$$H_\varphi^+(x) = \sum_{s=0}^{n-1}\sum_{i=0}^{s}\sum_{j=0}^{i}\sum_{l=0}^{j+1}\frac{(-1)^{s+1}(s+1)!\alpha_{s,i}C^j_{1,s,i}\binom{j+1}{l}\gamma_{j+1-l,s}}{2(j+1)}$$

$$\times\left[(\ln x + i\pi)^l + (\ln x - i\pi)^l\right]x^{-s-1}$$

$$-\sum_{s=1}^{n}(-1)^s(s-1)!c_s^* x^{-s} + \frac{1}{x^n}\fint_0^\infty \frac{\tau^n\varphi_n(\tau)}{\tau - x}d\tau, \tag{5.56}$$

*where* $c_s^* = \varphi_{s,s}(1) - \int_0^1 \varphi_{s-1,s-1}(t)dt$, $\varphi_{s,s}(t) = \frac{(-1)^s}{(s-1)!}\int_t^\infty (\tau - t)^{s-1}\varphi_s(\tau)d\tau$ *for* $s \in \mathbb{N}^+$, *and* $\varphi_{0,0}(t) = \varphi(t)$.

To obtain the asymptotic behavior of the solution to Eq. (5.1), one would have to make use of the results in the above four theorems. For examples, see [58].

## Acknowledgment

The second author was supported in part by the National Natural Science Foundation of China under Grant Numbers 11571375 and 11971489.

## References

[1] A. Asheim and D. Huybrechs, Complex Gaussian quadrature for oscillatory integral transforms, *IMA J. Numer. Anal.* **33** (2013) 1322–1341.

[2] X.-X. Bai and Y.-Q. Zhao, A uniform asymptotic expansion for Jacobi polynomials via uniform treatment of Darboux's method, *J. Approx. Theory* **148** (2007) 1–11.

[3] J. Baik, T. Kriecherbauer, K. T.-R. McLaughlin and P. D. Miller, *Discrete Orthogonal Polynomials. Asymptotics and Applications*, Annals of Mathematics Studies, Vol. 164 (Princeton University Press, Princeton, NJ, 2007).

[4] R. Beals and R. Wong, *Explorations in Complex Functions* (Springer, Cham, 2020).

[5] G. Birkhoff (ed.), *A Source Book in Classical Analysis* (Harvard University Press, Cambridge, MA, 1973). Assisted by U. Merzbach.

[6] G. D. Birkhoff, On the asymptotic character of the solutions of certain linear differential equations containing a parameter, *Trans. Amer. Math. Soc.* **9** (1908) 219–231.

[7] G. D. Birkhoff, Formal theory of irregular linear difference equations, *Acta Math.* **54** (1930) 205–246.

[8] G. D. Birkhoff and W. J. Trjitzinsky, Analytic theory of singular difference equations, *Acta Math.* **60** (1933) 1–89.

[9] P. Bleher and A. Its, Semiclassical asymptotics of orthogonal polynomials, Riemann–Hilbert problem, and universality in the matrix model, *Ann. Math.* **150** (1999) 185–266.

[10] N. Bleistein, Uniform asymptotic expansions of integrals with many nearby stationary points and algebraic singularities, *J. Math. Mech.* **17** (1967) 533–560.

[11] R. Bo and R. Wong, Asymptotic behavior of the Pollaczek polynomials and their zeros, *Stud. Appl. Math.* **96** (1996) 307–338.

[12] P. L. Butzer and R. J. Nessel, *Fourier Analysis and Approximation I* (Academic Press, New York, 1971).

[13] L.-H. Cao and Y.-T. Li, Linear difference equations with a transition point at the origin, *Anal. Appl.* **12** (2014) 75–106.

[14] T. Carleman, *L'intégrale de Fourier et Questions Qui s'y rattachent* (Uppsala, 1944).

[15] J. F. Carlson and A. E. Heins, The reflection of an electromagnetic plane wave by an infinite set of plates I, *Quart. Appl. Math.* **4** (1947) 313–329.

[16] G. F. Carrier, Sound transmission from a tube with flow, *Quart. Appl. Math.* **13** (1956) 457–461.

[17] Y. Chen and M. Ismail, Some indeterminate moment problems and Freud-like weights, *Constr. Approx.* **14** (1998) 439–458.

[18] T. M. Cherry, Uniform asymptotic expansions, *J. London Math. Soc.* **24** (1949) 121–130.

[19] C. Chester, B. Friedman and F. Ursell, An extension of the method of steepest descents, *Proc. Cambridge Philos. Soc.* **53** (1957) 599–611.

[20] T. O. Conway and P. Deift, Asymptotics of polynomials orthogonal with respect to a logarithmic weight, *SIGMA Symmetry Integr. Geom. Methods Appl.* **14** (2018) 056.

[21] O. Costin and R. Costin, Rigorous WKB for finite-order linear recurrence relations with smooth coefficients, *SIAM J. Math. Anal.* **27** (1996) 110–134.

[22] D. Dai, W.-Y. Hu and X.-S. Wang, Uniform asymptotics of orthogonal polynomials arising from coherent states, *SIGMA Symmetry Integr. Geom. Methods Appl.* **11** (2015) 070.

[23] D. Dai, M. E. H. Ismail and X.-S. Wang, Plancherel–Rotach asymptotic expansion for some polynomials from indeterminate moment problems, *Constr. Approx.* **40** (2014) 61–104.

[24] G. Darboux, Mémoire sur l'approximation des fonctions de très grands nombres, et sur une classe étendue de développements en serie, *J. Math. Pures. Appl.* **4** (1878) 5–56.

[25] A. Deaño, A. B. J. Kuijlaars and P. Román, Asymptotic behavior and zero distribution of polynomials orthogonal with respect to Bessel functions, *Constr. Approx.* **43** (2016) 153–196.

[26] P. Deift, *Orthogonal Polynomials and Random Matrices: A Riemann–Hilbert Approach*, Courant Lecture Notes, Vol. 3 (American Mathematical Society, 1999).

[27] P. Deift, T. Kriecherbauer, K. T.-R. McLaughlin, S. Venakides and X. Zhou, Uniform asymptotics for polynomials orthogonal with respect to varying exponential weights and applications to universality questions in random matrix theory, *Comm. Pure Appl. Math.* **52** (1999) 1335–1425.

[28] P. Deift, T. Kriecherbauer, K. T.-R. McLaughlin, S. Venakides and X. Zhou, Strong asymptotics of orthogonal polynomials with respect to exponential weights, *Comm. Pure Appl. Math.* **52** (1999) 1491–1552.

[29] P. Deift, T. Kriecherbauer, K. T.-R. McLaughlin, S. Venakides and X. Zhou, A Riemann–Hilbert approach to asymptotic questions for orthogonal polynomials, *J. Comput. Appl. Math.* **133** (2001) 47–63.

[30] P. Deift and X. Zhou, A steepest descent method for oscillatory Riemann–Hilbert problems. Asymptotics for the MKdV equation, *Ann. Math.* **137** (1993) 295–368.

[31] R. B. Dingle and G. J. Morgan, WKB methods for difference equations. I, *Appl. Sci. Res.* **18** (1967) 221–237.

[32] R. B. Dingle and G. J. Morgan, WKB methods for difference equations. II, *Appl. Sci. Res.* **18** (1967) 238–245.

[33] J. Dutka, The early history of the factorial function, *Arch. History Exact Sci.* **43** (1991) 225–249.

[34] Editorial note, Robinson's constant, *Amer. Math. Mon.* **59** (1952) 296–297.

[35] A. Erdélyi, Uniform asymptotic expansion of integrals, in *Analytic Methods in Mathematical Physics*, eds. R. P. Gilbert, R. G. Newton (Gordon and Breach, New York, 1970), pp. 149–168.

[36] A. Erdélyi and M. Wyman, The asymptotic evaluation of certain integrals, *Arch. Ration. Mech. Anal.* **14** (1963) 217–260.

[37] J. L. Fields, A uniform treatment of Darboux's method, *Arch. Ration. Mech. Anal.* **27** (1967) 289–305.

[38] P. Flajolet and R. Sedgewick, *Analytic Combinatorics* (Cambridge University Press, Cambridge, 2009).

[39] A. S. Fokas, A. R. Its, A. A. Kapaev and V. Y. Novokshenov, *Painlevé Transcendents. The Riemann–Hilbert Approach*, Mathematical Surveys and Monographs, Vol. 128 (American Mathematical Society, Providence, RI, 2006).

[40] A. S. Fokas, A. R. Its and A. V. Kitaev, The isomonodromy approach to matrix models in $2D$ quantum gravity, *Comm. Math. Phys.* **147** (1992) 395–430.

[41] P. J. Forrester, The spectrum edge of random matrix ensembles, *Nucl. Phys. B* **402** (1993) 709–728.

[42] I. C. Gohberg and M. G. Krein, Systems of integral equations on a half line with kernels depending on the difference of arguments, *Uspehi Mat. Nauk* **13** (1958) 3–72; *Amer. Math. Soc. Transl. Ser.* (2) **14** (1960) 217–287.

[43] G. H. Hardy and S. Ramanujan, Asymptotic formulæ in combinatory analysis, *Proc. London Math. Soc.* (2) **17** (1918) 75–115.

[44] A. Iserles, Stephen Smale: The mathematician who broke the dimension barrier by Steve Batterson, *SIAM Rev.* **42** (2000) 739–745.

[45] M. E. H. Ismail, *Classical and Quantum Orthogonal Polynomials in One Variable*, Encyclopedia of Mathematics and its Applications, Vol. 98 (Cambridge University Press, Cambridge, 2005).

[46] K. Johansson, Discrete orthogonal polynomial ensembles and the Plancherel measure, *Ann. Math.* **153** (2001) 259–296.

[47] D. E. Knuth, Asymptotic behavior of a sequence, *SIAM Rev.* **22** (1980) 101–102.

[48] D. E. Knuth, Communication between D. E. Knuth and F. W. J. Olver (2004).

[49] M. G. Krein, Integral equations on a half-line with a kernel depending on the difference of the arguments, *Uspehi Mat. Nauk* **13** (1958) 3–120; *Amer. Math. Soc. Transl. Ser.* (2) **22** (1962) 163–288.

[50] T. Kriecherbauer and K. T.-R. McLaughlin, Strong asymptotics of polynomials orthogonal with respect to Freud weights, *Int. Math. Res. Not.* **6** (1999) 299–333.

[51] A. B. J. Kuijlaars, Riemann–Hilbert analysis for orthogonal polynomials, *Orthogonal Polynomials and Special Functions*, Lecture Notes in Mathematics, Vol. 1817 (Springer, Berlin, 2003), pp. 167–210.

[52] A. B. J. Kuijlaars, K. T.-R. McLaughlin, W. Van Assche and M. Vanlessen, The Riemann–Hilbert approach to strong asymptotics for orthogonal polynomials on $[-1,1]$, *Adv. Math.* **188** (2004) 337–398.

[53] A. B. J. Kuijlaars and M. Vanlessen, Universality for eigenvalue correlations from the modified Jacobi unitary ensemble, *Int. Math. Res. Not.* **2002** (2002) 1575–1600.

[54] R. E. Langer, On the asymptotic solutions of ordinary differential equations, with an application to the Bessel functions of large order, *Trans. Amer. Math. Soc.* **33** (1931) 23–64.

[55] R. E. Langer, On the asymptotic solutions of differential equations, with an application to the Bessel functions of large complex order, *Trans. Amer. Math. Soc.* **34** (1932) 447–480.

[56] R. E. Langer, The asymptotic solutions of ordinary linear differential equations of the second order, with special reference to a turning point, *Trans. Amer. Math. Soc.* **67** (1949) 461–490.

[57] Y.-T. Li, X.-S. Wang and R. Wong, Asymptotics of the Wilson polynomials, *Anal. Appl.* **18** (2020) 237–270.

[58] K. Li and R. Wong, Asymptotic expansions for Wiener–Hopf equations, *Anal. Appl.* **19** (2021) 1059–1092.

[59] X. Li and R. Wong, Error bounds for asymptotic expansions of Laplace convolutions, *SIAM J. Math. Anal.* **25** (1994) 1537–1553.

[60] S.-Y. Liu, R. Wong and Y.-Q. Zhao, Uniform treatment of Darboux's method and the Heisenberg polynomials, *Proc. Amer. Math. Soc.* **141** (2013) 2683–2691.

[61] J. P. McClure and R. Wong, Explicit error terms for asymptotic expansions of Stieltjes transforms, *J. Inst. Math. Appl.* **22** (1978) 129–145.

[62] J. P. McClure and R. Wong, Exact remainders for asymptotic expansions of fractional integrals, *J. Inst. Math. Appl.* **24** (1979) 139–147.

[63] R. Muki and E. Sternberg, Note on an asymptotic property of solutions of a class of Fredholm integral equations, *Quart. Appl. Math.* **28** (1970) 277–281.

[64] N. I. Muskhelishvili, *Singular Integral Equations* (Groningen, Noordhoff, 1953). Reprinted by (Dover Publications, New York, 1992).

[65] B. Noble, *Methods Based on the Wiener–Hopf Technique for the Solution of Partial Differential Equations* (Pergamon Press Inc., New York, 1958).

[66] A. B. Olde Daalhuis and N. M. Temme, Uniform Airy-type expansions of integrals, *SIAM J. Math. Anal.* **25** (1994) 304–321.

[67] F. W. J. Olver, The asymptotic solution of linear differential equations of the second order for large values of a parameter, *Phil. Trans. R. Soc. Lond. A* **247** (1954) 307–327.

[68] F. W. J. Olver, The asymptotic expansion of Bessel functions of large order, *Phil. Trans. R. Soc. Lond. A* **247** (1954) 328–368.

[69] F. W. J. Olver, The asymptotic solution of linear differential equations of the second order in a domain containing one transition point, *Phil. Trans. R. Soc. Lond. A* **249** (1956) 65–97.

[70] F. W. J. Olver, Error bounds for the Liouville–Green (or WKB) approximation, *Math. Proc. Cambridge Philos. Soc.* **57** (1961) 790–810.

[71] F. W. J. Olver, A paradox in asymptotics, *SIAM J. Math. Anal.* **1** (1970) 533–534.

[72] F. W. J. Olver, *Asymptotics and Special Functions* (Academic Press, New York, 1974). Reprinted by (A. K. Peters, Ltd., Wellesley, 1997).

[73] F. W. J. Olver, Unsolved problems in the asymptotic estimation of special functions, in *Theory and Application of Special Functions*, ed. R. Askey (Academic Press, New York, 1975), pp. 99–142.

[74] F. Olver, D. Lozier, R. Boisvert and C. Clark, *NIST Handbook of Mathematical Functions* (Cambridge University Press, Cambridge, 2010).

[75] H. Poincaré, Sur les intégrales irrégulières des équations linéaires, *Acta Math.* **8** (1886) 295–344.

[76] G. Pólya, *Mathematics and Plausible Reasoning, Volume 1: Induction and Analogy in Mathematics* (Princeton University Press, Princeton, NJ, 1954).

[77] R. Robinson, A new absolute geometric constant? *Amer. Math. Mon.* **58** (1951) 462–469.

[78] L. Schlesinger, Über asymptotische darstellungen der lösungen linearer differentialsysteme als funktionen eines parameters, *Math. Ann.* **63** (1907) 277–300.

[79] J. Stirling, *Methodus Differentialis* (Springer, London, 1730).

[80] G. Szegő, *Orthogonal Polynomials,* 4th edn. (American Mathematical Society, Providence, RI, 1975).

[81] E. C. Titchmarsh, *The Theory of the Riemann Zeta-Function* (Clarendon Press, Oxford, 1951).

[82] E. C. Titchmarsh, *The Theory of Functions* (Oxford University Press, Oxford, 1958), reprint of the 2nd edn.

[83] C. A. Tracy and H. Widom, Level-spacing distributions and the Airy kernel, *Comm. Math. Phys.* **159** (1994) 151–174.

[84] C. A. Tracy and H. Widom, Level spacing distributions and the Bessel kernel, *Comm. Math. Phys.* **161** (1994) 289–309.

[85] H. L. Turrittin, Asymptotic solutions of certain ordinary differential equations associated with multiple roots of the characteristic equation, *Amer. J. Math.* **66** (1936) 364–376.

[86] Z. Wang and R. Wong, Uniform asymptotic expansion of $J_\nu(\nu a)$ via a difference equation, *Numer. Math.* **91** (2002) 147–193.

[87] Z. Wang and R. Wong, Asymptotic expansions for second-order linear difference equations with a turning point, *Numer. Math.* **94** (2003) 147–194.

[88] Z. Wang and R. Wong, Linear difference equations with transition points, *Math. Comp.* **74** (2005) 629–653.

[89] Z. Wang and R. Wong, Bessel-type asymptotic expansions via the Riemann–Hilbert approach, *Proc. R. Soc. Lond. A* **461** (2005) 2839–2856.

[90] W. A. Whitworth, *Choice and Chance*, 5th edn. (George Bell, London, 1901).

[91] H. Widom, Equations of Wiener–Hopf type, *Illinois J. Math.* **2** (1958) 261–270.

[92] N. Wiener and E. Hopf, Über eine klasse singulärer integral gleichungen, *Sitz. Ber. Preuss. Akad. Wiss.* **30** (1931) 696–706.

[93] J. Wimp, Review of R. Wong's book 'Asymptotic approximations of integrals', *Math. Comp.* **56** (1991) 388–394.

[94] J. Wimp and D. Zeilberger, Recurrenting the asymptotics of linear recurrences, *J. Math. Anal. Appl.* **111** (1985) 162–176.

[95] J. S. W. Wong and R. Wong, Asymptotic solutions of linear Volterra integral equations with singular kernels, *Trans. Amer. Math. Soc.* **189** (1974) 185–200.

[96] R. Wong, Quadrature formulas for oscillatory integral transforms, *Numer. Math.* **39** (1982) 351–360.

[97] R. Wong, *Asymptotic Approximations of Integrals* (Academic Press, Boston, 1989). Reprinted by (SIAM, Philadelphia, PA, 2001).

[98] R. Wong, Asymptotics of linear recurrences, *Anal. Appl.* **12** (2014) 463–484.

[99] R. Wong and H. Li, Asymptotic expansions for second-order linear difference equations, *J. Comput. Appl. Math.* **41** (1992) 65–94.

[100] R. Wong and H. Li, Asymptotic expansions for second-order linear difference equations II, *Stud. Appl. Math.* **87** (1992) 289–324.

[101] R. Wong and J. P. McClure, Generalized Mellin convolutions and their asymptotic expansions, *Canad. J. Math.* **36** (1984) 924–960.

[102] R. Wong and M. Wyman, The method of Darboux, *J. Approximation Theory* **10** (1974) 159–171.

[103] R. Wong and Y.-Q. Zhao, On a uniform treatment of Darboux's method, *Constr. Approx.* **21** (2005) 225–255.

[104] R. Wong and Y.-Q. Zhao, Asymptotics of orthogonal polynomials via the Riemann–Hilbert approach, *Acta Math. Sci. Ser. B* **29** (2009) 1005–1034.

[105] R. Wong and Y.-Q. Zhao, Special functions, integral equations and a Riemann–Hilbert problem, *Proc. Amer. Math. Soc.* **144** (2016) 4367–4380.

[106] S.-X. Xu, D. Dai and Y.-Q. Zhao, Critical edge behavior and the Bessel to Airy transition in the singularly perturbed Laguerre unitary ensemble, *Comm. Math. Phys.* **332** (2014) 1257–1296.

[107] S.-X. Xu and Y.-Q. Zhao, Gap probability of the circular unitary ensemble with a Fisher–Hartwig singularity and the coupled Painlevé V system, *Comm. Math. Phys.* **377** (2020) 1545–1596.

[108] K. Yosida, *Functional Analysis* (Springer-Verlag, Berlin, 1995), reprint of the 6th edn.

[109] C.-R. Zhao, X.-B. Wu and Y.-Q. Zhao, Asymptotic zero distribution of orthogonal polynomials with respect to Bessel functions via Riemann–Hilbert approach, preprint (2021).

[110] Y.-Q. Zhao, Uniform asymptotics for orthogonal polynomials via the Riemann–Hilbert approach, *Appl. Anal.* **85** (2006) 1165–1176.

*Proceedings of the Royal Society of Edinburgh*, **153**, 417–440, *2023*
DOI:10.1017/prm.2021.90

# Error bounds for the asymptotic expansions of the Hermite polynomials

**Wei Shi**
College of Science, Huazhong Agricultural University, Wuhan 430070, P. R. China

**Gergő Nemes**
Alfréd Rényi Institute of Mathematics, Reáltanoda utca 13-15, Budapest H-1053, Hungary

**Xiang-Sheng Wang**
Department of Mathematics, University of Louisiana at Lafayette, Lafayette, LA 70503, USA (xswang@louisiana.edu)

**Roderick Wong**
Department of Mathematics, City University of Hong Kong, Tat Chee Avenue, Kowloon, Hong Kong

(Received 2 September 2021; accepted 22 December 2021)

In this paper, we present explicit and computable error bounds for the asymptotic expansions of the Hermite polynomials with Plancherel–Rotach scale. Three cases, depending on whether the scaled variable lies in the outer or oscillatory interval, or it is the turning point, are considered separately. We introduce the 'branch cut' technique to express the error terms as integrals on the contour taken as the one-sided limit of curves approaching the branch cut. This new technique enables us to derive simple error bounds in terms of elementary functions. We also provide recursive procedures for the computation of the coefficients appearing in the asymptotic expansions.

*Keywords:* Error bounds; asymptotic expansions; Hermite polynomials

2020 *Mathematics subject classification:* Primary: 41A60; 33C45

## 1. Introduction

Plancherel–Rotach asymptotic expansions for the orthogonal polynomials as the polynomial degree tends to infinity have been studied extensively in the literature. There are many standard asymptotic techniques including the steepest descent method for integrals [**12**], the WKB method for differential equations [**8**], the Deift–Zhou method for Riemann–Hilbert problems [**4**, **5**], asymptotic theory of difference equations [**13**] and Darboux's method [**14**]. The error term for the truncated asymptotic expansion is usually of the same order of magnitude as the first

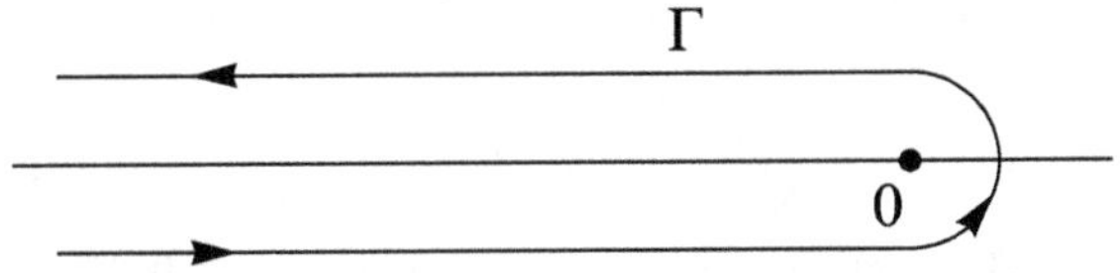


Figure 1. The contour Γ with counter-clockwise orientation.

neglected term. In other words, the ratio of the error term and the first omitted term is bounded by a constant independent of the polynomial degree. The existence of such a bound can be proved by standard 'soft' analysis. However, as far as we know, the quantitative information for this error bound is unknown even for the classical orthogonal polynomials. One may expect a rather complicated 'hard' analysis on finding explicit and computable expressions for error bounds. Various reasons why we need error bounds for asymptotic expansions are discussed in Ref. [**9**]. As mentioned in Ref. [**11**], the upper bounds of the error term obtained from standard asymptotic techniques are difficult to compute, and thus may not be realistic. In this paper, we will derive explicit and computable error bounds for the asymptotic expansions of the Hermite polynomials.

The main challenge is to find a convenient integral representation of the error term on an appropriate contour so that the error bound is computable. Berry and Howls [**2**] proposed the so-called 'adjacent saddles' method which was further developed in [**1, 3**]. The key idea of their method is to express the error term as an integral over the 'adjacent contours' passing through the 'adjacent saddles'. However, the technique of 'adjacent saddles' cannot be applied directly to the Hermite polynomials and many other orthogonal polynomials because the integrand (or the phase function) of the integral representation for the Hermite polynomials has a branch point in the complex plane. It is thus difficult to express the error term as an integral over the 'adjacent contours'. Moreover, for the turning point case, when two saddle points coincide with each other, there does not exist any 'adjacent saddle'. To resolve these two difficulties, we introduce a new 'branch cut' technique which deforms the contour of integration for the error term to the branch cut of the phase function. More precisely, the contour is defined as the limit of the curves approaching one side of the branch cut. By virtue of the 'branch cut' technique, we are able to find simple error bounds in terms of elementary functions.

The Hermite polynomials can be expressed as contour integrals:

$$H_n(x) = \frac{n!}{2\pi i}\int_\Gamma e^{2xt-t^2}t^{-n-1}dt,$$

where Γ is a counter-clockwisely oriented contour encircling the negative real line (cf. [**7**, § 18.10(iii)]). To be more specific, we may choose

$$\Gamma = \{\lambda - i\delta : \lambda < 0\} \cup \{\delta e^{i\theta} : -\pi/2 \leqslant \theta \leqslant \pi/2\} \cup \{\lambda + i\delta : \lambda < 0\},$$

where $\delta > 0$ is any fixed positive number; see figure 1.

After introducing the Plancherel–Rotach scale, we rewrite the integral representation as

$$H_n(\sqrt{N}x) = \frac{n!}{2\pi i N^{n/2}} \int_\Gamma e^{-N[t^2-2xt+(\ln t)/2]} t^{-1/2} dt, \tag{1.1}$$

where $N = 2n+1$. By symmetry, we may assume that $x \geqslant 0$. We shall denote the phase function in the above integral by

$$f(t;x) = t^2 - 2xt + \frac{1}{2}\ln t. \tag{1.2}$$

The zeros of $f'(t;x)$ are called saddle points and can be calculated as

$$t_\pm = \frac{x \pm \sqrt{x^2-1}}{2} = \begin{cases} e^{\pm\beta}/2, & x = \cosh\beta > 1,\ \beta > 0; \\ e^{\pm i\alpha}/2, & x = \cos\alpha \in [0,1),\ \alpha \in (0,\pi/2]; \\ 1/2, & x = 1. \end{cases}$$

Note that the two saddle points coincide ($t_+ = t_-$) at the turning point $x = 1$. For simplicity, we shall drop the dependence of $f$ on $x$. There are three cases to be considered.

Case I: $x = \cosh\beta > 1$ with $\beta > 0$. This is called the outer interval.

Case II: $x = \cos\alpha \in [0,\ 1)$ with $\alpha \in (0,\ \pi/2]$. This is called the oscillatory interval.

Case III: $x = 1$. This is called the turning point.

We will consider these three cases separately. We mention here that an asymptotic expansion with error bounds for the slightly differently scaled $H_n(\sqrt{N+1}\cos\alpha)$ was given earlier by van Veen [**10**].

The rest of the paper is organized as follows. In § 2–4, we derive the asymptotic expansions with error bounds for Case I, II and III, respectively. The main results are stated at the end of each section. In § 5, we demonstrate the accuracy of the error bounds through numerical examples. Recursive procedures for the computation of the coefficients appearing in the asymptotic expansions are given in appendix A.

## 2. Case I: $x = \cosh\beta > 1$ with $\beta > 0$

The saddle points are $t_\pm = e^{\pm\beta}/2$ with $0 < t_- < t_+$. Moreover, $f''(t_-) = 2(1 - e^{2\beta}) < 0$ and $f''(t_+) = 2(1 - e^{-2\beta}) > 0$. We shall deform the contour of integration $\Gamma$ to the path of steepest descent passing through the saddle point $t_-$. To describe this steepest descent contour, we shall introduce the analytic function

$$w(z) = [f(z) - f(t_-)]^{1/2}, \tag{2.1}$$

where the branch of the square root function is chosen so that

$$\begin{aligned} w(z) &= -i(z-t_-)\sqrt{-f''(t_-)/2} + \mathcal{O}((z-t_-)^2) \\ &= -i(z-t_-)\sqrt{e^{2\beta}-1} + \mathcal{O}((z-t_-)^2), \end{aligned}$$

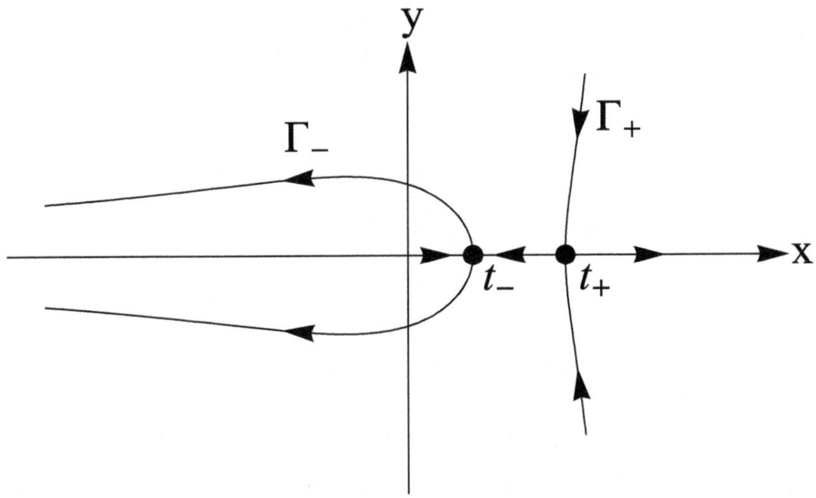


Figure 2. The contours $\Gamma_\pm$. The arrows indicate the increasing direction of the phase function $f$.

for $z$ in a small complex neighbourhood of $t_-$. Let $z = re^{i\theta}$ with $r > 0$ and $\theta \in (-\pi, \pi)$. We investigate the equation $\mathrm{Im}[f(z) - f(t_-)] = \mathrm{Im}\, f(z) = 0$; namely,

$$r^2 \sin(2\theta) - 2r \cosh\beta \sin\theta + \frac{\theta}{2} = 0. \tag{2.2}$$

Clearly, $\theta = 0$ satisfies the above equation. By symmetry, we can restrict $\theta \in (0, \pi)$ and find two solutions

$$r_-(\theta) = \frac{\theta/(2\sin\theta)}{\cosh\beta + \sqrt{\cosh^2\beta - \theta/\tan\theta}}, \quad \theta \in (0, \pi),$$

$$r_+(\theta) = \frac{\cosh\beta + \sqrt{\cosh^2\beta - \theta/\tan\theta}}{2\cos\theta}, \quad \theta \in (0, \pi/2). \tag{2.3}$$

It is readily seen that $r_\pm(\theta) \to t_\pm = e^{\pm\beta}/2$ as $\theta \to 0$. For consistency, we define $r_\pm(0) = t_\pm$. Consequently, we can regard $r_-(\theta)$ as a symmetric function on $(-\pi, \pi)$ and $r_+(\theta)$ a symmetric function on $(-\pi/2, \pi/2)$. Define two contours

$$\Gamma_- = \{r_-(\theta)e^{i\theta} : \theta \in (-\pi, \pi)\}, \quad \Gamma_+ = \{r_+(\theta)e^{i\theta} : \theta \in (-\pi/2, \pi/2)\}.$$

It follows that the solution of $\mathrm{Im}\, f(z) = 0$ is the union $\Gamma_- \cup \Gamma_+ \cup \mathbb{R}^+$, where $\mathbb{R}^+$ is the positive real line. The contours $\Gamma_\pm$ are depicted in figure 2.

Since $f'(z)$ has no zeros other than $t_\pm$, it follows from the asymptotic behaviour $f(z) \sim z^2$ as $z \to \infty$ that $\Gamma_-$ is a steepest descent contour passing through $t_-$, while $\Gamma_+$ is a steepest ascent contour passing through $t_+$. Moreover, the solutions of $f(z) - f(t_-) = 0$ are real and positive. Consider $f(t) = t^2 - 2t\cosh\beta + (\ln t)/2$ for $t > 0$. Since $f(t) \to -\infty$ as $t \to 0^+$ and $f(t) \to \infty$ as $t \to \infty$, we obtain from

$f'(t_\pm) = 0$ that the equation $f(t) - f(t_-) = 0$ has exactly two positive roots: one is $t_-$; and the other, denoted by $t_c$, is larger than $t_+$. Hence, we can choose $[t_c, \infty)$ as a branch cut and extend the definition of $w(z)$ on $\mathbb{C} \setminus (-\infty, 0] \cup [t_c, \infty)$ by analytic continuation.

Now, we deform the contour of integration from $\Gamma$ to the steepest descent contour $\Gamma_-$ with counter-clockwise orientation and then make a change of variable $u = w(t)$:

$$\begin{aligned} H_n(\sqrt{N}\cosh\beta) &= \frac{n! e^{-Nf(t_-)}}{2\pi i N^{n/2}} \int_{\Gamma_-} e^{-N[f(t)-f(t_-)]} t^{-1/2} dt \\ &= \frac{n! e^{-Nf(t_-)}}{2\pi i N^{n/2}} \int_{\mathbb{R}} e^{-Nu^2} \frac{[w^{-1}(u)]^{-1/2}}{w'(w^{-1}(u))} du. \end{aligned} \tag{2.4}$$

Recall that $\Gamma_+$ is the steepest ascent contour passing through $t_+$. We also denote by $\Gamma_0$ a clockwise oriented contour encircling $(-\infty, 0]$; namely,

$$\Gamma_0 = \{\lambda + i\delta : \lambda < 0\} \cup \{\delta e^{i\theta} : -\pi/2 \leqslant \theta \leqslant \pi/2\} \cup \{\lambda - i\delta : \lambda < 0\},$$

where $\delta > 0$ is a small positive number such that $\delta < t_c$. In fact, $\Gamma_0$ is identical to $\Gamma$ but has the opposite orientation. Since $|w(z)|/|z| \to 1$ as $z \to \infty$, we have from the Cauchy integral formula

$$\frac{[w^{-1}(u)]^{-1/2}}{w'(w^{-1}(u))} = \frac{1}{2\pi i} \int_{\Gamma_0 \cup \Gamma_+} \frac{z^{-1/2}}{w(z) - u} dz$$

for any $u \in \mathbb{R}$. Substituting this expression into (2.4) gives a double integral representation

$$H_n(\sqrt{N}\cosh\beta) = \frac{n! e^{-Nf(t_-)}}{(2\pi i)^2 N^{n/2}} \int_{\mathbb{R}} e^{-Nu^2} \int_{\Gamma_0 \cup \Gamma_+} \frac{z^{-1/2}}{w(z) - u} dz du.$$

For any $p \geqslant 0$, it follows from

$$\frac{1}{w(z) - u} = \sum_{k=1}^{2p} \frac{u^{k-1}}{w(z)^k} + \frac{u^{2p}}{w(z)^{2p}[w(z) - u]}$$

and

$$\int_{\mathbb{R}} e^{-Nu^2} u^{k-1} du = \begin{cases} 0, & k = 2j, \\ \Gamma(j + 1/2)/N^{j+1/2}, & k = 2j + 1 \end{cases} \tag{2.5}$$

that

$$\begin{aligned} H_n(\sqrt{N}\cosh\beta) &= \frac{n! e^{-Nf(t_-)}}{(2\pi i)^2 N^{n/2}} \left\{ \sum_{k=1}^{2p} \int_{\mathbb{R}} e^{-Nu^2} u^{k-1} du \int_{\Gamma_0 \cup \Gamma_+} \frac{z^{-1/2}}{w(z)^k} dz \right. \\ &\quad \left. + \int_{\mathbb{R}} e^{-Nu^2} u^{2p} \int_{\Gamma_0 \cup \Gamma_+} \frac{z^{-1/2}}{w(z)^{2p}[w(z) - u]} dz du \right\} \\ &= \frac{n! e^{-Nf(t_-)}}{2\sqrt{\pi} \sinh\beta N^{(n+1)/2}} \left\{ \sum_{j=0}^{p-1} \frac{A_j(\coth\beta)}{N^j} + \varepsilon_p(N, \beta) \right\}, \end{aligned} \tag{2.6}$$

where

$$A_j(\coth\beta) = -\frac{\sqrt{\sinh\beta}\,\Gamma(j+1/2)}{2\pi^{3/2}}\int_{\Gamma_0\cup\Gamma_+}\frac{z^{-1/2}}{w(z)^{2j+1}}dz, \tag{2.7}$$

$$\varepsilon_p(N,\beta) = -\frac{\sqrt{N\sinh\beta}}{2\pi^{3/2}}\int_{\mathbb{R}} e^{-Nu^2}u^{2p}\int_{\Gamma_0\cup\Gamma_+}\frac{z^{-1/2}}{w(z)^{2p}[w(z)-u]}dzdu. \tag{2.8}$$

We show in appendix A.1 that the coefficients $A_j(\coth\beta)$ are polynomials in $\coth\beta$ of degree $3j$ with rational coefficients, and can be calculated using a recursive formula.

To find an upper bound for the remainder $\varepsilon_p(N,\beta)$, we need to estimate the integral

$$c_p(u) = \int_{\Gamma_0\cup\Gamma_+}\frac{z^{-1/2}}{w(z)^{2p}[w(z)-u]}dz, \quad u\in\mathbb{R}. \tag{2.9}$$

If there exists $C_p > 0$ such that $|c_p(u)| \leqslant C_p$ for all $u\in\mathbb{R}$, then we obtain from (2.5) with $k = 2p+1$ and (2.8) that

$$|\varepsilon_p(N,\beta)| \leqslant \frac{\sqrt{\sinh\beta}}{2\pi^{3/2}}\frac{C_p\Gamma(p+1/2)}{N^p}. \tag{2.10}$$

First, we estimate the integral

$$I_0 = \int_{\Gamma_0}\frac{z^{-1/2}}{w(z)^{2p}[w(z)-u]}dz.$$

Recall that the contour $\Gamma_0$ encircles the negative real line having a distance $\delta > 0$ from it. By analyticity of the integrand, the value $I_0$ does not change if we let $\delta\to 0^+$. Hence,

$$|I_0| \leqslant \int_0^{+\infty}\frac{s^{-1/2}}{|w(se^{i\pi})|^{2p}|w(se^{i\pi})-u|}ds + \int_0^{+\infty}\frac{s^{-1/2}}{|w(se^{-i\pi})|^{2p}|w(se^{-i\pi})-u|}ds, \tag{2.11}$$

where $w(se^{\pm i\pi})$ is defined as the limit of $w(se^{i\theta})$ as $\theta\to\pm\pi$. Note from (1.2) and (2.1) that

$$w(se^{\pm i\pi})^2 = s^2 + 2s\cosh\beta + (\ln s)/2 - f(t_-) \pm i\pi/2.$$

Thus, we obtain

$$|w(se^{\pm i\pi})| = \{[s^2 + 2s\cosh\beta + (\ln s)/2 - f(t_-)]^2 + (\pi/2)^2\}^{1/4} =: R(s), \tag{2.12}$$

where the last equality defines $R(s)$, the modulus of $w(se^{\pm i\pi})$. Let $\theta_\pm(s)$ be the phase of $w(se^{\pm i\pi})$. We then have

$$R(s)^2\cos[2\theta_\pm(s)] = \mathrm{Re}[w(se^{\pm i\pi})^2] = s^2 + 2s\cosh\beta + (\ln s)/2 - f(t_-),$$

which implies that

$$\begin{aligned}|w(se^{\pm i\pi})-u| &\geqslant |\mathrm{Im}[w(se^{\pm i\pi})]| = R(s)|\sin[\theta_\pm(s)]| = R(s)\sqrt{\frac{1-\cos[2\theta_\pm(s)]}{2}}\\ &= \sqrt{\frac{R(s)^2-[s^2+2s\cosh\beta+(\ln s)/2-f(t_-)]}{2}}\\ &= \sqrt{\frac{(\pi/2)^2/2}{R(s)^2+[s^2+2s\cosh\beta+(\ln s)/2-f(t_-)]}}\\ &\geqslant \sqrt{\frac{(\pi/2)^2}{4R(s)^2}} = \frac{\pi}{4R(s)}.\end{aligned}$$

Assume that $p \geqslant 1$. Substituting this inequality into (2.11) yields

$$|I_0| \leqslant \frac{8}{\pi}\int_0^{+\infty}\frac{s^{-1/2}}{R(s)^{2p-1}}ds.$$

To estimate the integral on the right-hand side of the above inequality, we note from (2.12) that $R(s) \geqslant \sqrt{\pi/2}$ for $s \in [0,\, 1]$ and $R(s) \geqslant \sqrt{2s\cosh\beta}$ for $s \geqslant 1 > t_-$. Thus,

$$\begin{aligned}|I_0| &\leqslant \frac{8}{\pi}\int_0^1\frac{s^{-1/2}}{(\pi/2)^{p-1/2}}ds + \frac{8}{\pi}\int_1^{+\infty}\frac{s^{-1/2}}{(2s\cosh\beta)^{p-1/2}}ds\\ &= \frac{2^{p+7/2}}{\pi^{p+1/2}} + \frac{1}{\pi(p-1)2^{p-7/2}(\cosh\beta)^{p-1/2}},\end{aligned} \tag{2.13}$$

where we have assumed $p \geqslant 2$ to ensure convergence of the second integral.

It remains to estimate the integral

$$I_c = \int_{\Gamma_+}\frac{z^{-1/2}}{w(z)^{2p}[w(z)-u]}dz.$$

For $z \in \Gamma_+$, we have $\mathrm{Im}\ f(z) = 0$ and $\mathrm{Re}[f(z)-f(t_-)] < f(t_+)-f(t_-) < 0$. It then follows from (2.1) that $w(z)$ is purely imaginary and $|w(z)-u| \geqslant |w(z)|$ for $z \in \Gamma_+$ and $u \in \mathbb{R}$. Now, we parameterize $\Gamma_+$ as $z = z_+(\theta) = r_+(\theta)e^{i\theta}$ with $\theta \in (-\pi/2,\, \pi/2)$, where $r_+(\theta)$ is given in (2.3). Since $dz/d\theta = [r'_+(\theta)+ir_+(\theta)]e^{i\theta}$ and $|w(z_+(\theta))|^2 = |f(z_+(\theta))-f(t_-)| = f(t_-)-f(z_+(\theta))$, we deduce

$$\begin{aligned}|I_c| &\leqslant \int_{-\pi/2}^{\pi/2}\frac{r_+(\theta)^{-1/2}\sqrt{[r'_+(\theta)]^2+[r_+(\theta)]^2}}{|w(z_+(\theta))|^{2p+1}}d\theta\\ &= \int_0^{\pi/2}\frac{2r_+(\theta)^{-1/2}\sqrt{[r'_+(\theta)]^2+[r_+(\theta)]^2}}{[f(t_-)-f(z_+(\theta))]^{p+1/2}}d\theta.\end{aligned}$$

As both $\theta/\tan\theta$ and $\cos\theta$ are decreasing functions for $\theta \in (0,\, \pi/2)$, it follows from (2.3) that $r_+(\theta)$ is an increasing function for $\theta \in (0,\, \pi/2)$; namely, $r'_+(\theta) \geqslant 0$. To estimate $r_+'(\theta)$, we apply an implicit differentiation on (2.2) and obtain

$$r_+'(\theta) = \frac{-2r_+(\theta)^2\cos(2\theta) + 2r_+(\theta)\cosh\beta\cos\theta - 1/2}{2r_+(\theta)\sin(2\theta) - 2\cosh\beta\sin\theta}.$$

A simple calculation together with (2.3) gives

$$\begin{aligned} 2r_+(\theta)\sin(2\theta) - 2\cosh\beta\sin\theta &= 2\sin\theta[2r_+(\theta)\cos\theta - \cosh\beta] \\ &= 2\sin\theta\sqrt{\cosh^2\beta - \theta/\tan\theta} \end{aligned}$$

and

$$\begin{aligned} &2r_+(\theta)\cosh\beta\cos\theta - 2r_+(\theta)^2\cos^2\theta \\ &= \cosh\beta(\cosh\beta + \sqrt{\cosh^2\beta - \theta/\tan\theta}) - \frac{(\cosh\beta + \sqrt{\cosh^2\beta - \theta/\tan\theta})^2}{2} \\ &= \frac{\theta}{2\tan\theta}. \end{aligned} \tag{2.14}$$

Consequently,

$$\begin{aligned} 0 \leqslant r_+'(\theta) &= \frac{2r_+(\theta)^2\sin^2\theta + \theta/(2\tan\theta) - 1/2}{2\sin\theta\sqrt{\cosh^2\beta - \theta/\tan\theta}} \\ &\leqslant \frac{r_+(\theta)^2\sin\theta}{\sqrt{\cosh^2\beta - \theta/\tan\theta}} \leqslant \frac{r_+(\theta)\tan\theta}{\tanh\beta}, \end{aligned}$$

and

$$|I_c| \leqslant \frac{2}{\tanh\beta}\int_0^{\pi/2} \frac{r_+(\theta)^{1/2}/\cos\theta}{[f(t_-) - f(z_+(\theta))]^{p+1/2}}d\theta. \tag{2.15}$$

Let $\theta_0 = \arccos(1/4)$. If $\theta \in (0, \theta_0)$, then $\cos\theta > 1/4$, $f(t_-) - f(z_+(\theta)) > f(t_-) - f(t_+) = \sinh(2\beta)/2 - \beta$ and $r_+(\theta) < 4\cosh\beta$. Hence,

$$\int_0^{\theta_0} \frac{r_+(\theta)^{1/2}/\cos\theta}{[f(t_-) - f(z_+(\theta))]^{p+1/2}}d\theta \leqslant \frac{2\pi(4\cosh\beta)^{1/2}}{[\sinh(2\beta)/2 - \beta]^{p+1/2}}. \tag{2.16}$$

If $\theta \in (\theta_0, \pi/2)$, then $r_+(\theta) > 2e^\beta = 4t_+ > 2\cosh\beta > 2$. This together with (2.14) implies that

$$\begin{aligned} -f(z_+(\theta)) &= -r_+(\theta)^2\cos(2\theta) + 2r_+(\theta)\cosh\beta\cos\theta - \frac{1}{2}\ln r_+(\theta) \\ &= r_+(\theta)^2 + \frac{\theta}{2\tan\theta} - \frac{1}{2}\ln r_+(\theta) \\ &\geqslant \frac{4}{5}r_+(\theta)^2 + \frac{1}{5}r_+(\theta)^2 - \frac{1}{2}\ln r_+(\theta) \geqslant \frac{4}{5}r_+(\theta)^2, \end{aligned}$$

and

$$-f(t_-) = \frac{e^{-2\beta}}{4} + \beta 2 + \frac{1+\ln 2}{2} \leqslant \frac{e^{2\beta}}{4} + \frac{19}{20} \leqslant t_+^2 + \frac{19}{80} r_+(\theta)^2 \leqslant \frac{3}{10} r_+(\theta)^2.$$

Coupling the above two inequalities gives

$$f(t_-) - f(z_+(\theta)) \geqslant \frac{1}{2} r_+(\theta)^2.$$

Therefore,

$$\begin{aligned}\int_{\theta_0}^{\pi/2} \frac{r_+(\theta)^{1/2}/\cos\theta}{[f(t_-) - f(z_+(\theta))]^{p+1/2}} d\theta &\leqslant \int_{\theta_0}^{\pi/2} \frac{r_+(\theta)^{1/2} \cdot r_+(\theta)/t_+}{r_+(\theta)^{2p+1}/2^{p+1/2}} d\theta \\ = \frac{2^{p+1/2}}{t_+} \int_{\theta_0}^{\pi/2} r_+(\theta)^{-2p+1/2} d\theta &\leqslant \frac{2^{p+1/2}}{t_+} \int_{\pi/3}^{\pi/2} (4t_+)^{-2p+1/2} d\theta = \frac{2^{-3p+1/2}\pi}{3 t_+^{2p+1/2}},\end{aligned} \tag{2.17}$$

where we have assumed $p \geqslant 1$. Substituting (2.16) and (2.17) into (2.15) yields

$$|I_c| \leqslant \frac{2}{\tanh\beta} \left\{ \frac{4\pi\sqrt{\cosh\beta}}{[\sinh(2\beta)/2 - \beta]^{p+1/2}} + \frac{2^{-p+1}\pi}{3e^{(2p+1/2)\beta}} \right\}, \quad p \geqslant 1. \tag{2.18}$$

A combination of (3.5), (2.13) and (2.18) gives

$$\begin{aligned}|c_p(u)| \leqslant |I_0| + |I_c| &\leqslant \frac{2^{p+7/2}}{\pi^{p+1/2}} + \frac{1}{\pi(p-1)2^{p-7/2}(\cosh\beta)^{p-1/2}} \\ &+ \frac{2}{\tanh\beta} \left\{ \frac{4\pi\sqrt{\cosh\beta}}{[\sinh(2\beta)/2 - \beta]^{p+1/2}} + \frac{2^{-p+1}\pi}{3e^{(2p+1/2)\beta}} \right\},\end{aligned}$$

for $p \geqslant 2$. Let $C_p$ be the number on the right-hand side of the above inequality. The error bound (2.10) holds for $p \geqslant 2$. Note from (2.6) that

$$\varepsilon_p(N,\beta) = \frac{A_p(\coth\beta)}{N^p} + \varepsilon_{p+1}(N,\beta).$$

Hence, we may improve the error bound (2.10) to

$$|\varepsilon_p(N,\beta)| \leqslant \frac{|A_p(\coth\beta)|}{N^p} + \frac{\sqrt{\sinh\beta}}{2\pi^{3/2}} \frac{C_{p+1}\Gamma(p+3/2)}{N^{p+1}} = \frac{\widetilde{C}_p}{N^p},$$

for $p \geqslant 1$, where

$$\widetilde{C}_p = |A_p(\coth\beta)| + \frac{\sqrt{\sinh\beta}}{2\pi^{3/2}} \frac{C_{p+1}\Gamma(p+3/2)}{N}.$$

Using the explicit value

$$-f(t_-) = \frac{e^{-2\beta}}{4} + \frac{\beta}{2} + \frac{1+\ln 2}{2}$$

in (2.6), we can summarize our results as follows.

THEOREM 2.1. *Let $\beta > 0$ and $N = 2n+1 \geqslant 1$. Then for any $p \geqslant 1$, we have*

$$H_n(\sqrt{N}\cosh\beta) = \frac{2^n n! e^{N(e^{-2\beta}+2\beta+2)/4}}{\sqrt{2\pi\sinh\beta}N^{(n+1)/2}}\left\{\sum_{j=0}^{p-1}\frac{A_j(\coth\beta)}{N^j} + \varepsilon_p(N,\beta)\right\}, \tag{2.19}$$

*where*

$$|\varepsilon_p(N,\beta)| \leqslant \frac{\widetilde{C}_p}{N^p}, \tag{2.20}$$

*with*

$$\widetilde{C}_p = |A_p(\coth\beta)| + \frac{\sqrt{\sinh\beta}}{2\pi^{3/2}}\frac{C_{p+1}\Gamma(p+3/2)}{N},$$

*and*

$$\begin{aligned} C_{p+1} &= \frac{2^{p+9/2}}{\pi^{p+3/2}} + \frac{1}{\pi p 2^{p-5/2}(\cosh\beta)^{p+1/2}} \\ &\quad + \frac{2}{\tanh\beta}\left\{\frac{4\pi\sqrt{\cosh\beta}}{[\sinh(2\beta)/2-\beta]^{p+3/2}} + \frac{2^{-p}\pi}{3e^{(2p+5/2)\beta}}\right\}. \end{aligned}$$

*In particular, $\widetilde{C}_p \to |A_p(\coth\beta)|$ as $N \to +\infty$; namely, the error bound is close to the absolute value of the first neglected term when $N$ is large. The coefficients $A_j(\coth\beta)$ are polynomials in $\coth\beta$ of degree $3j$ with rational coefficients, and these polynomials can be calculated using a recursive formula given in appendix A.1.*

## 3. Case II: $x = \cos\alpha \in [0, 1)$ with $\alpha \in (0, \pi/2]$

The saddle points are $t_\pm = e^{\pm i\alpha}/2$, and we have

$$f''(t_\pm) = 2 - 2e^{\mp 2i\alpha} = 4\sin\alpha e^{\pm i(\pi/2-\alpha)}.$$

Now, we define an analytic function $w(z)$ via

$$w(z) = [f(z) - f(t_+)]^{1/2}, \tag{3.1}$$

where the branch of the square root function is chosen so that

$$w(z) = \sqrt{2\sin\alpha}e^{i(\pi/4-\alpha/2)}(z-t_+) + \mathcal{O}((z-t_+)^2)$$

for $z$ in a small complex neighbourhood of $t_+$. The steepest descent contour passing through $t_+$ is described via the equation $\mathrm{Im}\,[f(z) - f(t_+)] = 0$. By using polar coordinates $z = re^{i\theta}$, this equation becomes

$$r^2 \sin(2\theta) - 2r\cos\alpha\sin\theta + \frac{\theta-\theta_0}{2} = 0,$$

with

$$\theta_0 = \alpha - \sin(2\alpha)/2 \in (0, \alpha).$$

We are only interested in solutions in the upper half-plane; namely, $\theta \in (0, \pi)$. By the quadratic formula, there are exactly two solutions:

$$r_{d+}(\theta) = \begin{cases} \dfrac{(\theta-\theta_0)/(2\sin\theta)}{\cos\alpha + \sqrt{\cos^2\alpha - (\theta-\theta_0)/\tan\theta}}, & \theta \in [\alpha, \pi), \\[2ex] \dfrac{\cos\alpha + \sqrt{\cos^2\alpha - (\theta-\theta_0)/\tan\theta}}{2\cos\theta}, & \theta \in (0, \alpha], \end{cases}$$

and

$$r_{a+}(\theta) = \begin{cases} \dfrac{\cos\alpha + \sqrt{\cos^2\alpha - (\theta-\theta_0)/\tan\theta}}{2\cos\theta}, & \theta \in [\alpha, \pi/2), \\[2ex] \dfrac{\cos\alpha - \sqrt{\cos^2\alpha - (\theta-\theta_0)/\tan\theta}}{2\cos\theta}, & \theta \in (\theta_0, \alpha], \end{cases}$$

where $r_{d+}(\theta)$ and $r_{a+}(\theta)$ are defined as piecewise functions so that they are differentiable at $\theta = \alpha$. This is because the function $\sqrt{\cos^2\alpha - (\theta-\theta_0)/\tan\theta}$ is not differentiable at $\theta = \alpha$. For any $\theta$ in a small real neighbourhood of $\alpha$, we have, by Taylor expansion, $\cos^2\alpha - (\theta-\theta_0)/\tan\theta = (\theta-\alpha)^2 + \mathcal{O}((\theta-\alpha)^3)$ and $\sqrt{\cos^2\alpha - (\theta-\theta_0)/\tan\theta} = |\theta-\alpha| + \mathcal{O}((\theta-\alpha)^2)$. Hence,

$$\begin{aligned} r_{d+}(\theta) &= \frac{\cos\alpha - (\theta-\alpha)}{2[\cos\alpha - (\theta-\alpha)\sin\alpha]} + \mathcal{O}((\theta-\alpha)^2) \\ &= \frac{1}{2} - \frac{1-\sin\alpha}{2\cos\alpha}(\theta-\alpha) + \mathcal{O}((\theta-\alpha)^2), \\ r_{a+}(\theta) &= \frac{\cos\alpha + (\theta-\alpha)}{2[\cos\alpha - (\theta-\alpha)\sin\alpha]} + \mathcal{O}((\theta-\alpha)^2) \\ &= \frac{1}{2} + \frac{1+\sin\alpha}{2\cos\alpha}(\theta-\alpha) + \mathcal{O}((\theta-\alpha)^2). \end{aligned}$$

In particular, we have

$$r_{d+}(\alpha) = r_{a+}(\alpha) = \frac{1}{2}, \quad r'_{d+}(\alpha) = -\frac{1}{2}\tan\left(\frac{\pi}{4} - \frac{\alpha}{2}\right), \; r'_{a+}(\alpha) = \frac{1}{2}\cot\left(\frac{\pi}{4} - \frac{\alpha}{2}\right).$$

Now, we define two contours

$$\Gamma_{d+} = \{r_{d+}(\theta)e^{i\theta} : \theta \in (0, \pi)\}, \quad \Gamma_{a+} = \{r_{a+}(\theta)e^{i\theta} : \theta \in (\theta_0, \pi/2)\}.$$

Recall that $-Nf(t)$ is the phase function in the integral representation of $H_n(\sqrt{N}\cos\alpha)$. It is readily seen from the asymptotic behaviours $f(z) \sim z^2$ as

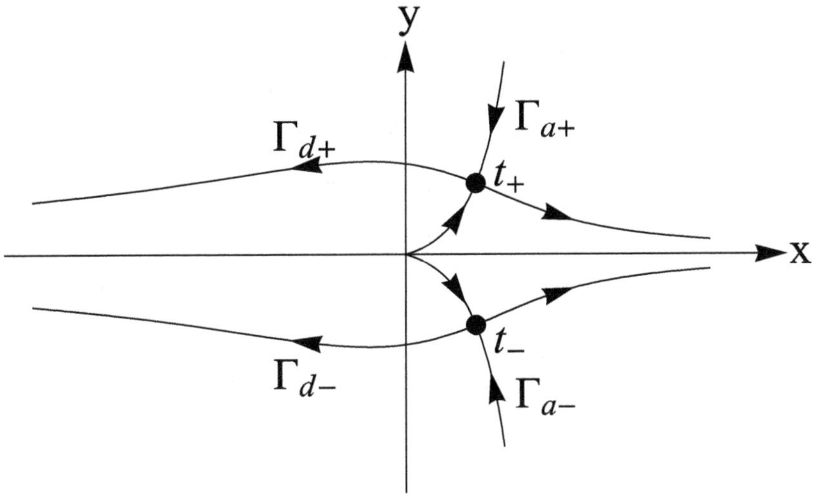


Figure 3. The contours $\Gamma_{d\pm}$ and $\Gamma_{a\pm}$. The arrows indicate the increasing direction of the phase function $f$.

$z \to \infty$ and $2f(z) \sim \ln z$ as $z \to 0$ that $\Gamma_{d+}$ is a steepest descent contour and $\Gamma_{a+}$ is a steepest ascent contour passing through the saddle point $t_+ = e^{i\alpha}/2$. Moreover, the only zero of $f(z) - f(t_+)$ in the upper half-plane is $t_+$. Hence, $w(z)$ defined in (3.1) is analytic for all Im $z > 0$ and positive $z > 0$. For $z = se^{i\pi}$ with $s > 0$, we define

$$w(se^{i\pi}) = \lim_{\theta \to \pi^-} w(se^{i\theta}).$$

We can define in a similar manner the contours $\Gamma_{d-}$ and $\Gamma_{a-}$ which, together with $\Gamma_{d+}$ and $\Gamma_{a+}$, are illustrated in figure 3.

Now, we return to the integral representation of $H_n(\sqrt{N}\cos\alpha)$ in (1.1) and deform the contour of integration to the steepest descent contours $\Gamma_{d+} \cup \Gamma_{d-}$ oriented with increasing parameter $\theta$ (i.e., counter-clockwise direction):

$$H_n(\sqrt{N}\cos\alpha) = \frac{n!}{2\pi i N^{n/2}} \int_{\Gamma_{d+}} e^{-Nf(t)} t^{-1/2} dt + \frac{n!}{2\pi i N^{n/2}} \int_{\Gamma_{d-}} e^{-Nf(t)} t^{-1/2} dt =: I_+ + I_-,$$

where the last equality defines the two integrals $I_+$ and $I_-$, respectively. Since the contours $\Gamma_{d+}$ and $\Gamma_{d-}$ are symmetric with respect to the real axis, have opposite orientations, and $f(t) = \overline{f(\overline{t})}$, the integrals $I_+$ and $I_-$ are complex conjugates of each other. Therefore, it suffices to study the integral $I_+$. First, we introduce the change of variable $u = w(t)$:

$$I_+ = -\frac{n! e^{-Nf(t_+)}}{2\pi i N^{n/2}} \int_{\mathbb{R}} e^{-Nu^2} \frac{[w^{-1}(u)]^{-1/2}}{w'(w^{-1}(u))} du.$$

Since $w(z)$ is analytic in the upper half-plane and $|w(z)|/|z| \to 1$ as $z \to \infty$, we can infer from the Cauchy integral formula that

$$\frac{[w^{-1}(u)]^{-1/2}}{w'(w^{-1}(u))} = \frac{1}{2\pi i} \int_{\Gamma_0^+ \cup \mathbb{R}^+} \frac{z^{-1/2}}{w(z) - u} dz,$$

where $\mathbb{R}^+$ is the positive real line and $\Gamma_0^+$ is the negative real line with argument $\pi$:

$$\Gamma_0^+ = \lim_{\delta \to 0^+} \{\lambda + i\delta : \lambda < 0\} = \{se^{i\pi} : s > 0\}.$$

Both $\mathbb{R}^+$ and $\Gamma_0^+$ are oriented from left to right. Now, we have the following double integral representation:

$$I_+ = -\frac{n! e^{-Nf(t_+)}}{(2\pi i)^2 N^{n/2}} \int_{\mathbb{R}} e^{-Nu^2} \int_{\Gamma_0^+ \cup \mathbb{R}^+} \frac{z^{-1/2}}{w(z) - u} dz du.$$

For any $p \geqslant 0$, we use the identities (2.5) and

$$\frac{1}{w(z) - u} = \sum_{k=1}^{2p} \frac{u^{k-1}}{w(z)^k} + \frac{u^{2p}}{w(z)^{2p}[w(z) - u]}$$

to derive

$$I_+ = \frac{n! e^{-Nf(t_+)}}{2\sqrt{\pi \sin\alpha} N^{(n+1)/2}} \left\{ \sum_{j=0}^{p-1} \frac{A_j(i\cot\alpha)}{N^j} e^{\pi i/4} + \widetilde{\varepsilon}_p(N, \alpha) \right\}, \tag{3.2}$$

where

$$A_j(i\cot\alpha) = e^{-\pi i/4} \frac{\sqrt{\sin\alpha}\,\Gamma(j + 1/2)}{2\pi^{3/2}} \int_{\Gamma_0^+ \cup \mathbb{R}^+} \frac{z^{-1/2}}{w(z)^{2j+1}} dz, \tag{3.3}$$

$$\widetilde{\varepsilon}_p(N, \alpha) = \frac{\sqrt{N \sin\alpha}}{2\pi^{3/2}} \int_{\mathbb{R}} e^{-Nu^2} u^{2p} \int_{\Gamma_0^+ \cup \mathbb{R}^+} \frac{z^{-1/2}}{w(z)^{2p}[w(z) - u]} dz du. \tag{3.4}$$

The coefficients $A_j(i\cot\alpha)$ are polynomials in $i\cot\alpha$ and these polynomials are identical to those appearing in the expansion (2.19) (see appendix A.1 for details).

To find an upper bound for the remainder $\widetilde{\varepsilon}_p(N, \alpha)$, we need to estimate the integral

$$c_p(u) = \int_{\Gamma_0^+ \cup \mathbb{R}^+} \frac{z^{-1/2}}{w(z)^{2p}[w(z) - u]} dz, \quad u \in \mathbb{R}. \tag{3.5}$$

By expressing $w(z)$ in polar coordinates, one can easily show that

$$|w(z) - u| \geqslant |\text{Im } w(z)| \geqslant \frac{|\text{Im } w(z)^2|}{2|w(z)|}.$$

Note from (3.1) that

$$|\text{Im } w(z)^2| = |\text{Im } f(z) - \theta_0/2| = \begin{cases} \theta_0/2, & z \in \mathbb{R}^+, \\ (\pi - \theta_0)/2, & z \in \Gamma_0^+. \end{cases}$$

Assume that $p \geqslant 1$. Substituting the above two formulas into (2.9) yields

$$|c_p(u)| \leqslant \frac{4}{\theta_0} \int_0^{+\infty} \frac{s^{-1/2}}{|w(s)|^{2p-1}} ds + \frac{4}{\pi - \theta_0} \int_0^{+\infty} \frac{s^{-1/2}}{|w(se^{i\pi})|^{2p-1}} ds.$$

For $s \in (0, 2)$, we have

$$|w(s)| \geqslant \sqrt{|\text{Im } w(s)^2|} = \sqrt{\theta_0/2},$$
$$|w(se^{i\pi})| \geqslant \sqrt{|\text{Im } w(se^{i\pi})^2|} = \sqrt{(\pi - \theta_0)/2},$$

and for $s > 2$, we have

$$|w(s)|^2 \geqslant |\text{Re } w(s)^2| = s^2 - 2s\cos\alpha + \frac{\ln s}{2} + \frac{\cos^2\alpha + \ln 2 + 1/2}{2} \geqslant (s-1)^2,$$
$$|w(se^{i\pi})|^2 \geqslant |\text{Re } w(se^{i\pi})^2| = s^2 + 2s\cos\alpha + \frac{\ln s}{2} + \frac{\cos^2\alpha + \ln 2 + 1/2}{2} \geqslant s^2.$$

Accordingly,

$$\begin{aligned}|c_p(u)| &\leqslant \frac{4}{\theta_0}\left\{\int_0^2 \frac{s^{-1/2}}{(\theta_0/2)^{p-1/2}} ds + \int_2^{+\infty} \frac{s^{-1/2}}{(s-1)^{2p-1}} ds\right\} \\ &\quad + \frac{4}{\pi-\theta_0}\left\{\int_0^2 \frac{s^{-1/2}}{[(\pi-\theta_0)/2]^{p-1/2}} ds + \int_2^{+\infty} \frac{s^{-1/2}}{s^{2p-1}} ds\right\} \\ &\leqslant \frac{4}{\theta_0}\left\{\frac{2^{3/2}}{(\theta_0/2)^{p-1/2}} + \frac{1}{2p-3/2}\right\} \\ &\quad + \frac{4}{\pi-\theta_0}\left\{\frac{2^{3/2}}{[(\pi-\theta_0)/2]^{p-1/2}} + \frac{2^{3/2-2p}}{2p-3/2}\right\}.\end{aligned}$$

Denote the right-hand side of the above inequality by $C_p$. It then follows from (2.5), (3.4) and (2.9) that

$$|\widetilde{\varepsilon}_p(N,\alpha)| \leqslant \frac{\sqrt{\sin\alpha}}{2\pi^{3/2}} \frac{C_p\Gamma(p+1/2)}{N^p},$$

provided $p \geqslant 1$. We define

$$\varepsilon_p(N,\alpha) := \text{Re}(\widetilde{\varepsilon}_p(N,\alpha)e^{-iN\theta_0/2}),$$

and observe that

$$\begin{aligned}|\varepsilon_p(N,\alpha)| &\leqslant \left|\text{Re}\left(A_p(i\cot\alpha)e^{-i(\theta_0 N-\pi/2)/2}\right)\right| \frac{1}{N^p} + |\text{Re}\widetilde{\varepsilon}_{p+1}(N,\alpha)| \\ &\leqslant \left|\text{Re}\left(A_p(i\cot\alpha)e^{-i(\theta_0 N-\pi/2)/2}\right)\right| \frac{1}{N^p} + |\widetilde{\varepsilon}_{p+1}(N,\alpha)|\end{aligned}$$

for any $p \geqslant 0$. Using the explicit value

$$-f(t_+) = \frac{\cos(2\alpha)}{4} + \frac{1+\ln 2}{2} - i\frac{\theta_0}{2}$$

in (3.2), and the fact that $H_n(\sqrt{N}\cos\alpha) = I_+ + I_- = I_+ + \overline{I_+} = 2\text{Re } I_+$, we can summarize our results as follows.

THEOREM 3.1. *Let* $\alpha \in (0, \pi/2]$, $N = 2n+1 \geqslant 1$ *and* $\theta_0 = \alpha - \sin(2\alpha)/2$. *Then for any* $p \geqslant 0$, *we have*

$$H_n(\sqrt{N}\cos\alpha) = \frac{2^{n+1}n!e^{N(\cos(2\alpha)+2)/4}}{\sqrt{2\pi\sin\alpha}N^{(n+1)/2}} \times \left\{\sum_{j=0}^{p-1}\frac{\mathrm{Re}\left(A_j(i\cot\alpha)e^{-i(\theta_0 N-\pi/2)/2}\right)}{N^j} + \varepsilon_p(N,\alpha)\right\}$$

$$= \frac{2^{n+1}n!e^{N(\cos(2\alpha)+2)/4}}{\sqrt{2\pi\sin\alpha}N^{(n+1)/2}}\left\{\cos\left(\frac{\theta_0 N}{2}-\frac{\pi}{4}\right)\sum_{j=0}^{p-1}\frac{\mathrm{Re}(A_j(i\cot\alpha))}{N^j}\right.$$

$$\left. + \sin\left(\frac{\theta_0 N}{2}-\frac{\pi}{4}\right)\sum_{j=1}^{p-1}\frac{\mathrm{Im}(A_j(i\cot\alpha))}{N^j} + \varepsilon_p(N,\alpha)\right\},$$

*where*

$$|\varepsilon_p(N,\alpha)| \leqslant \frac{\widetilde{C}_p}{N^p}, \tag{3.6}$$

*with*

$$\widetilde{C}_p = \left|\mathrm{Re}\left(A_p(i\cot\alpha)e^{-i(\theta_0 N-\pi/2)/2}\right)\right| + \frac{\sqrt{\sin\alpha}}{2\pi^{3/2}}\frac{C_{p+1}\Gamma(p+3/2)}{N},$$

*and*

$$C_{p+1} = \frac{4}{\theta_0}\left\{\frac{2^{3/2}}{(\theta_0/2)^{p+1/2}} + \frac{1}{2p+1/2}\right\} + \frac{4}{\pi-\theta_0}\left\{\frac{2^{3/2}}{[(\pi-\theta_0)/2]^{p+1/2}} + \frac{2^{-1/2-2p}}{2p+1/2}\right\}.$$

*In particular,* $\widetilde{C}_p \to |\mathrm{Re}(A_p(i\cot\alpha)e^{-i(\theta_0 N-\pi/2)/2})|$ *as* $N \to +\infty$; *namely, the error bound is close to the absolute value of the first omitted term for large* $N$. *The coefficients* $A_j(i\cot\alpha)$ *are polynomials in* $i\cot\alpha$ *of degree* $3j$ *with rational coefficients, and these polynomials can be calculated using a recursive formula given in appendix* A.1.

## 4. Case III: $x = 1$

The saddle points coincide $t_s = t_\pm = 1/2$. We have $f(t_s) = -3/4 - (\ln 2)/2$, $f'(t_s) = f''(t_s) = 0$ and $f'''(t_s) = 8$. We introduce the analytic function

$$w(z) = [f(z) - f(t_s)]^{1/3}, \tag{4.1}$$

with the branch of the cube root being chosen so that

$$w(z) = (4/3)^{1/3}(z-t_s)e^{-2i\pi/3} + \mathcal{O}((z-t_s)^2) \tag{4.2}$$

for $z$ in a small complex neighbourhood of $t_s$. We shall extend the definition of $w(z)$ by analytic continuation to $\mathbb{C}\setminus(-\infty, 0]$. First, we investigate the equation

$\mathrm{Im}[f(z) - f(t_s)] = \mathrm{Im}\, f(z) = 0$, which in polar coordinates $z = re^{i\theta}$ is

$$r^2 \sin(2\theta) - 2r \sin\theta + \theta/2 = 0. \tag{4.3}$$

This equation is clearly satisfied by all $z > 0$. There are two further solutions:

$$r_+(\theta) = \begin{cases} \dfrac{\theta/(2\sin\theta)}{1+\sqrt{1-\theta\cot\theta}}, & \theta \in (0,\pi), \\ \dfrac{1}{2}, & \theta = 0, \\ \dfrac{1+\sqrt{1-\theta\cot\theta}}{2\cos\theta}, & \theta \in (-\pi/2, 0), \end{cases}$$

and

$$r_-(\theta) = \begin{cases} \dfrac{1+\sqrt{1-\theta\cot\theta}}{2\cos\theta}, & \theta \in (0,\pi/2), \\ \dfrac{1}{2}, & \theta = 0, \\ \dfrac{\theta/(2\sin\theta)}{1+\sqrt{1-\theta\cot\theta}}, & \theta \in (-\pi, 0), \end{cases}$$

where we have defined $r_\pm(\theta)$ as piecewise functions so that they are differentiable at $\theta = 0$. Note that since $1 - \theta\cot\theta = \theta^2/3 + \mathcal{O}(\theta^3)$ for small $|\theta| > 0$, we have

$$\sqrt{1-\theta\cot\theta} = \frac{|\theta|}{\sqrt{3}} + \mathcal{O}(\theta^2)$$

and

$$r_\pm(\theta) = \frac{1}{2} \mp \frac{\theta}{2\sqrt{3}} + \mathcal{O}(\theta^2)$$

for small values of $\theta$. Especially,

$$r_\pm(0) = \frac{1}{2}, \quad r'_\pm(0) = \mp\frac{1}{2\sqrt{3}}. \tag{4.4}$$

Now, we introduce two contours

$$\Gamma_1 = \{r_+(\theta)e^{i\theta} : \theta \in (-\pi/2, \pi)\}, \quad \Gamma_2 = \{r_-(\theta)e^{i\theta} : \theta \in (-\pi, \pi/2)\},$$

which we illustrate in figure 4.

We have the following lemma.

Lemma 4.1. *Let $z_+(\theta) = r_+(\theta)e^{i\theta}$ for $\theta \in (-\pi/2, \pi)$; the function $f(z_+(\theta)) - f(t_s)$ increases from $-\infty$ to $+\infty$ as $\theta$ increases from $-\pi/2$ to $\pi$. Let $z_-(\theta) = r_-(\theta)e^{i\theta}$ for $\theta \in (-\pi, \pi/2)$; the function $f(z_-(\theta)) - f(t_s)$ decreases from $+\infty$ to $-\infty$ as $\theta$ increases from $-\pi$ to $\pi/2$.*

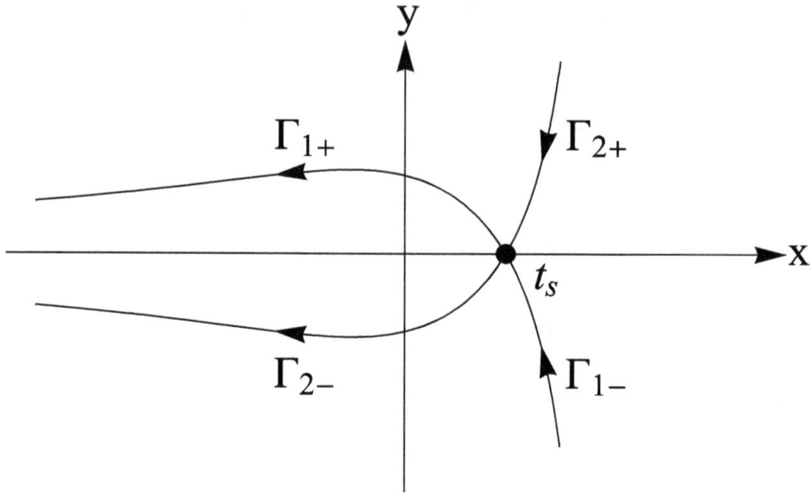


Figure 4. The contours $\Gamma_1$ and $\Gamma_2$. The arrows indicate the increasing direction of the phase function $f$.

*Proof.* For small $\theta$, we have from $f'(t_s) = f''(t_s) = 0$, $f'''(t_s) = 8$ and (4.4) that

$$z_+(\theta) - t_s = [r'_+(0) + ir_+(0)]\theta + \mathcal{O}(\theta^2) = \frac{e^{2i\pi/3}}{\sqrt{3}}\theta + \mathcal{O}(\theta^2),$$

and

$$f(z_+(\theta)) - f(t_s) = \frac{4}{3}(z_+(\theta) - t_s)^3 + \mathcal{O}((z_+(\theta) - t_s)^4) = \frac{4}{9\sqrt{3}}\theta^3 + \mathcal{O}(\theta^4).$$

Hence, the function $f(z_+(\theta)) - f(t_s)$ is increasing for small $\theta$. We claim that $f(z_+(\theta)) - f(t_s)$ is increasing for all $\theta \in (-\pi/2, \pi)$. Assume to the contrary that

$$0 = \frac{d}{d\theta} f(z_+(\theta)) = f'(z_+(\theta))[r'_+(\theta) + ir_+(\theta)]e^{i\theta}$$

for some $\theta \in (-\pi/2, 0) \cup (0, \pi)$. Since $f'(z) \neq 0$ for any $z \neq t_s$, we have $r'_+(\theta) = r_+(\theta) = 0$. This contradicts to the equation (4.3). Hence, the function $f(z_+(\theta)) - f(t_s)$ is increasing for all $\theta \in (-\pi/2, \pi)$. Note that $r_+(\theta) \to \infty$ as $\theta$ tends to $-\pi/2$ from the right or $\pi$ from the left. It is readily seen that $f(z_+(\theta)) - f(t_s) = r_+(\theta)^2 \cos(2\theta) - 2r_+(\theta)\cos\theta + (1/2)\ln r_+(\theta) - f(t_s)$ increases from $-\infty$ to $+\infty$ as $\theta$ increases from $-\pi/2$ to $\pi$.

The second statement of the lemma follows in a similar manner. □

Since the solutions of Im $f(z) = 0$ are the union $\Gamma_1 \cup \Gamma_2 \cup \mathbb{R}^+$, the equation $f(z) - f(t_s) = 0$ has a unique solution $z = t_s$ in $\mathbb{C} \setminus (-\infty, 0]$. This implies that $w(z)$ defined in (4.1) can be analytically continued to $\mathbb{C} \setminus (-\infty, 0]$. Moreover, it follows from (4.2) that $w(z_+(\theta))$ increases from $-\infty$ to $\infty$ as $\theta$ increases from $-\pi/2$ to $\pi$.

Next, we let $\Gamma_{1+}$ be the portion of $\Gamma_1$ in the upper half-plane and $\Gamma_{2-}$ be the portion of $\Gamma_2$ in the lower half-plane; namely,

$$\Gamma_{1+} = \{r_+(\theta)e^{i\theta} : \theta \in [0, \pi)\}, \quad \Gamma_{2-} = \{r_-(\theta)e^{i\theta} : \theta \in (-\pi, 0]\}.$$

Both $\Gamma_{1+}$ and $\Gamma_{2-}$ are oriented with increasing $\theta$ (counter clockwise direction). We now deform the contour of integration for $H_n(\sqrt{N})$ in (1.1) from $\Gamma$ to $\Gamma_{1+} \cup \Gamma_{2-}$:

$$\begin{aligned} H_n(\sqrt{N}) &= \frac{n!}{2\pi i N^{n/2}} \int_{\Gamma_{1+}} e^{-Nf(t)} t^{-1/2} dt \\ &\quad + \frac{n!}{2\pi i N^{n/2}} \int_{\Gamma_{2-}} e^{-Nf(t)} t^{-1/2} dt =: I_+ + I_-, \end{aligned}$$

where the last equality defines the two integrals $I_+$ and $I_-$, respectively. Since the contours $\Gamma_{1+}$ and $\Gamma_{2-}$ are symmetric with respect to the real axis, have opposite orientations, and $f(t) = \overline{f(\overline{t})}$, the integrals $I_+$ and $I_-$ are complex conjugates of one another. Thus, it is sufficient to study the integral $I_+$. First, we introduce the change of variable $u = w(t)$:

$$I_+ = \frac{n! e^{-Nf(t_s)}}{2\pi i N^{n/2}} \int_0^{+\infty} e^{-Nu^3} \frac{[w^{-1}(u)]^{-1/2}}{w'(w^{-1}(u))} du. \tag{4.5}$$

We denote by $\Gamma_0^{\pm}$ the negative real lines with arguments $\pm\pi$ and orientation from left to right. For convenience, we denote by $\Gamma_*$ the union of $\Gamma_0^+$ oriented to the right and $\Gamma_0^-$ oriented to the left. Since $w(z)$ is analytic in $\mathbb{C} \setminus (-\infty, 0]$ and $|w(z)|/|z|^{2/3} \to 1$ as $z \to \infty$, we can assert from the Cauchy integral formula that

$$\frac{[w^{-1}(u)]^{-1/2}}{w'(w^{-1}(u))} = \frac{1}{2\pi i} \int_{\Gamma_*} \frac{z^{-1/2}}{w(z) - u} dz.$$

Substituting this into (4.5) gives

$$I_+ = \frac{n! e^{-Nf(t_s)}}{(2\pi i)^2 N^{n/2}} \int_0^{+\infty} e^{-Nu^3} \int_{\Gamma_*} \frac{z^{-1/2}}{w(z) - u} dz du.$$

For any $p \geqslant 1$, we use the identities

$$\frac{1}{w(z) - u} = \sum_{j=1}^{p-1} \frac{u^{j-1}}{w(z)^j} + \frac{u^{p-1}}{w(z)^{p-1}[w(z) - u]}$$

and

$$\int_0^{+\infty} e^{-Nu^3} u^{j-1} du = \frac{\Gamma(j/3)}{3N^{j/3}} \tag{4.6}$$

to find

$$I_+ = \frac{2^{1/2} n! e^{-Nf(t_s)}}{6\pi N^{n/2}} \left\{ \sum_{j=1}^{p-1} \left(\frac{3}{4}\right)^{j/3} D_j e^{2\pi i j/3 - \pi i/2} \frac{\Gamma(j/3)}{N^{j/3}} + \widetilde{\varepsilon}_p(N) \right\}, \tag{4.7}$$

where

$$D_j = \left(\frac{4}{3}\right)^{j/3} \frac{e^{-2\pi ij/3}}{2^{3/2}\pi i} \int_{\Gamma_*} \frac{z^{-1/2}}{w(z)^j} dz, \tag{4.8}$$

$$\widetilde{\varepsilon}_p(N) = -\frac{3}{2^{3/2}\pi} \int_{\mathbb{R}} e^{-Nu^3} u^{p-1} \int_{\Gamma_*} \frac{z^{-1/2}}{w(z)^{p-1}[w(z)-u]} dz du. \tag{4.9}$$

It is shown in appendix A.2 that the coefficients $D_j$ are rational numbers which can be calculated via a recurrence relation.

To find an upper bound for the remainder $\widetilde{\varepsilon}_p(N)$, we need to estimate the integral

$$c_p(u) = \int_{\Gamma_*} \frac{z^{-1/2}}{w(z)^{p-1}[w(z)-u]} dz, \quad u \in \mathbb{R}. \tag{4.10}$$

By expressing $w(z)$ in polar coordinates, one can easily show that

$$|w(z) - u| \geqslant |\text{Im } w(z)| \geqslant \frac{|\text{Im } w(z)^3|}{3|w(z)|^2}.$$

We also note from (4.1) that

$$|\text{Im } w(z)^3| = |\text{Im } f(z)| = \pi/2,$$

for $z = se^{\pm i\pi} \in \Gamma_*$. Assume that $p \geqslant 4$. Substituting the above two formulas into (4.10) yields

$$|c_p(u)| \leqslant \frac{6}{\pi} \int_0^{+\infty} \frac{s^{-1/2}}{|w(se^{i\pi})|^{p-3}} ds + \frac{6}{\pi} \int_0^{+\infty} \frac{s^{-1/2}}{|w(se^{-i\pi})|^{p-3}} ds.$$

For $z = se^{\pm i\pi} \in \Gamma_*$ with $s = |z| \in (0, 1)$, we have

$$|w(z)| \geqslant |\text{Im } w(z)^3|^{1/3} = (\pi/2)^{1/3}.$$

For $z = se^{\pm i\pi} \in \Gamma_*$ with $s = |z| > 1$, we have

$$|w(z)|^3 \geqslant |\text{Re } w(z)^3| = s^2 + 2s + \frac{\ln s}{2} + \frac{\ln 2 + 3/2}{2} \geqslant s^2.$$

Consequently,

$$\begin{aligned} |c_p(u)| &\leqslant \frac{12}{\pi} \left\{ \int_0^1 s^{-1/2} (\pi/2)^{p/3-1} ds + \int_1^{+\infty} \frac{s^{-1/2}}{s^{2p/3-2}} ds \right\} \\ &\leqslant \frac{12}{\pi} \left\{ \frac{2}{(\pi/2)^{p/3-1}} + \frac{6}{4p-15} \right\}. \end{aligned}$$

Denote the right-hand side of the above inequality by $C_p$. It then follows from (4.6), (4.9) and (4.10) that

$$|\widetilde{\varepsilon}_p(N)| \leqslant \frac{1}{2^{3/2}\pi}\frac{C_p\Gamma(p/3)}{N^{p/3}},$$

provided $p \geqslant 4$. We define

$$\varepsilon_p(N) := \text{Re } \widetilde{\varepsilon}_p(N),$$

and observe that

$$\begin{aligned}|\varepsilon_p(N)| &\leqslant \left|\left(\frac{3}{4}\right)^{p/3} D_p \sin\left(\frac{2\pi p}{3}\right)\right| \frac{\Gamma(p/3)}{N^{p/3}} + |\text{Re}\widetilde{\varepsilon}_{p+1}(N)| \\ &\leqslant \left|\left(\frac{3}{4}\right)^{p/3} D_p \sin\left(\frac{2\pi p}{3}\right)\right| \frac{\Gamma(p/3)}{N^{p/3}} + |\widetilde{\varepsilon}_{p+1}(N)|\end{aligned}$$

for any $p \geqslant 3$. Using the explicit value $-f(t_s) = 3/4 + (\ln 2)/2$ in (4.7), and the fact that $H_n(\sqrt{N}) = I_+ + I_- = I_+ + \overline{I_+} = 2\text{Re } I_+$, we can summarize our results as follows.

THEOREM 4.2. *Denote $N = 2n + 1 \geqslant 1$. Then for any $p \geqslant 3$, we have*

$$H_n(\sqrt{N}) = \frac{2^{n+1}n!e^{3N/4}}{3\pi N^{n/2}}\left\{\sum_{j=1}^{p-1}\left(\frac{3}{4}\right)^{j/3} D_j \sin\left(\frac{2\pi j}{3}\right)\frac{\Gamma(j/3)}{N^{j/3}} + \varepsilon_p(N)\right\},$$

*where*

$$|\varepsilon_p(N)| \leqslant \frac{\widetilde{C}_p}{N^{p/3}}, \tag{4.11}$$

*with*

$$\widetilde{C}_p = \left|\left(\frac{3}{4}\right)^{p/3} D_p \sin\left(\frac{2\pi p}{3}\right)\right| \Gamma(p/3) + \frac{1}{2^{3/2}\pi}\frac{C_{p+1}\Gamma((p+1)/3)}{N^{1/3}},$$

*and*

$$C_{p+1} = \frac{12}{\pi}\left\{\frac{2}{(\pi/2)^{(p-2)/3}} + \frac{6}{4p-11}\right\}.$$

*In particular, if $p$ is not divisible by 3, $\widetilde{C}_p \to \left|(3/4)^{p/3} D_p \sin(2\pi p/3)\right|\Gamma(p/3)$ as $N \to +\infty$; namely, the error bound is close to the absolute value of the first neglected term when $N$ is large. The coefficients $D_j$ are rational numbers and can be calculated using a recurrence relation given in appendix* A.2.

## 5. Numerical examples

In tables 1, 2 and 3, we present some numerical results that demonstrate the accuracy of our error bounds given in theorems 2.1, 3.1 and 4.2, respectively. We can infer from the tables that the bounds are rather realistic, that is, they do not seriously overestimate the actual error.

Table 1. *Bounds for* $|\varepsilon_p(N,\,\beta)|$ *with various* $N = 2n+1$, $\beta$ *and* $p$, *using (2.20)*

| | | |
|---|---|---|
| Values of $n$, $\beta$ and $p$ | $n = 50,\ \beta = 1,\ p = 1$ | $n = 50,\ \beta = 1,\ p = 3$ |
| Numerical value of $\lvert\varepsilon_p(N,\,\beta)\rvert$ | $0.0985 \times 10^{-2}$ | $0.0543 \times 10^{-5}$ |
| The bound (2.20) for $\lvert\varepsilon_p(N,\,\beta)\rvert$ | $0.1917 \times 10^{-2}$ | $0.1704 \times 10^{-5}$ |
| Values of $n$, $\beta$ and $p$ | $n = 50,\ \beta = 4,\ p = 1$ | $n = 50,\ \beta = 4,\ p = 3$ |
| Numerical value of $\lvert\varepsilon_p(N,\,\beta)\rvert$ | $0.8229 \times 10^{-3}$ | $0.1879 \times 10^{-7}$ |
| The bound (2.20) for $\lvert\varepsilon_p(N,\,\beta)\rvert$ | $0.9811 \times 10^{-3}$ | $0.7375 \times 10^{-7}$ |
| Values of $n$, $\beta$ and $p$ | $n = 100,\ \beta = 1,\ p = 1$ | $n = 100,\ \beta = 1,\ p = 3$ |
| Numerical value of $\lvert\varepsilon_p(N,\,\beta)\rvert$ | $0.5004 \times 10^{-3}$ | $0.0701 \times 10^{-6}$ |
| The bound (2.20) for $\lvert\varepsilon_p(N,\,\beta)\rvert$ | $0.7357 \times 10^{-3}$ | $0.1441 \times 10^{-6}$ |
| Values of $n$, $\beta$ and $p$ | $n = 100,\ \beta = 4,\ p = 1$ | $n = 100,\ \beta = 4,\ p = 3$ |
| Numerical value of $\lvert\varepsilon_p(N,\,\beta)\rvert$ | $0.4134 \times 10^{-3}$ | $0.2383 \times 10^{-8}$ |
| The bound (2.20) for $\lvert\varepsilon_p(N,\,\beta)\rvert$ | $0.4533 \times 10^{-3}$ | $0.5887 \times 10^{-8}$ |

Table 2. *Bounds for* $|\varepsilon_p(N,\,\alpha)|$ *with various* $N = 2n+1$, $\alpha$ *and* $p$, *using (3.6)*

| | | |
|---|---|---|
| Values of $n$, $\alpha$ and $p$ | $n = 50,\ \alpha = \pi/4,\ p = 1$ | $n = 50,\ \alpha = \pi/4,\ p = 3$ |
| Numerical value of $\lvert\varepsilon_p(N,\,\alpha)\rvert$ | $0.0406 \times 10^{-1}$ | $0.0103 \times 10^{-3}$ |
| The bound (3.6) for $\lvert\varepsilon_p(N,\,\alpha)\rvert$ | $0.1147 \times 10^{-1}$ | $0.3159 \times 10^{-3}$ |
| Values of $n$, $\alpha$ and $p$ | $n = 50,\ \alpha = \pi/3,\ p = 1$ | $n = 50,\ \alpha = \pi/3,\ p = 3$ |
| Numerical value of $\lvert\varepsilon_p(N,\,\alpha)\rvert$ | $0.1551 \times 10^{-2}$ | $0.0119 \times 10^{-4}$ |
| The bound (3.6) for $\lvert\varepsilon_p(N,\,\alpha)\rvert$ | $0.2779 \times 10^{-2}$ | $0.1192 \times 10^{-4}$ |
| Values of $n$, $\alpha$ and $p$ | $n = 100,\ \alpha = \pi/4,\ p = 1$ | $n = 100,\ \alpha = \pi/4,\ p = 3$ |
| Numerical value of $\lvert\varepsilon_p(N,\,\alpha)\rvert$ | $0.0689 \times 10^{-2}$ | $0.0066 \times 10^{-4}$ |
| The bound (3.6) for $\lvert\varepsilon_p(N,\,\alpha)\rvert$ | $0.2494 \times 10^{-2}$ | $0.2001 \times 10^{-4}$ |
| Values of $n$, $\alpha$ and $p$ | $n = 100,\ \alpha = \pi/3,\ p = 1$ | $n = 100,\ \alpha = \pi/3,\ p = 3$ |
| Numerical value of $\lvert\varepsilon_p(N,\,\alpha)\rvert$ | $0.0932 \times 10^{-2}$ | $0.1532 \times 10^{-6}$ |
| The bound (3.6) for $\lvert\varepsilon_p(N,\,\alpha)\rvert$ | $0.1249 \times 10^{-2}$ | $0.8399 \times 10^{-6}$ |

Table 3. *Bounds for* $|\varepsilon_p(N)|$ *with various* $N = 2n+1$ *and* $p$, *using (4.11)*

| | | | |
|---|---|---|---|
| Values of $n$ and $p$ | $n = 50,\ p = 4$ | $n = 50,\ p = 7$ | $n = 50,\ p = 10$ |
| Numerical value of $\lvert\varepsilon_p(N)\rvert$ | $0.2148 \times 10^{-3}$ | $0.0300 \times 10^{-5}$ | $0.0538 \times 10^{-7}$ |
| The bound (4.11) for $\lvert\varepsilon_p(N)\rvert$ | $0.6615 \times 10^{-3}$ | $0.4092 \times 10^{-5}$ | $0.6667 \times 10^{-7}$ |
| Values of $n$ and $p$ | $n = 100,\ p = 4$ | $n = 100,\ p = 7$ | $n = 100,\ p = 10$ |
| Numerical value of $\lvert\varepsilon_p(N)\rvert$ | $0.0835 \times 10^{-3}$ | $0.0601 \times 10^{-6}$ | $0.0523 \times 10^{-8}$ |
| The bound (4.11) for $\lvert\varepsilon_p(N)\rvert$ | $0.2253 \times 10^{-3}$ | $0.6658 \times 10^{-6}$ | $0.5438 \times 10^{-8}$ |

**Acknowledgments**

We are very grateful to the anonymous referee for his/her careful reading and valuable suggestions which have helped to improve the presentation of this paper. The second author was supported by the JSPS Postdoctoral Research Fellowship No. P21020.

## Appendix A. The computation of the coefficients

### Appendix A.1. The coefficients $A_j$

Our starting point is the integral representation (2.7). We first reverse the orientation of the path $\Gamma_0 \cup \Gamma_+$ and thereby remove the minus sign in front of the integral. Next, by an appeal to Cauchy's theorem, the contour of integration can be shrunk into a small, positively oriented circle around $t_-$. After performing the change of variable $u = ze^{\beta}$, we arrive at

$$A_j(\coth\beta) = \sqrt{\frac{e^{-2\beta}-1}{\pi}}\frac{\Gamma(j+1/2)}{2\pi i}\oint_{(1/2+)}\frac{(2u)^{-1/2}}{(w(ue^{-\beta})^2)^{j+1/2}}du. \tag{A.1}$$

The contour of integration is a small loop surrounding $1/2$ in the positive sense. We would like to obtain a recursive scheme for the evaluation of the integrals in (A.1). To this end, we employ a method of Lauwerier [**6**]. The method requires the power series expansions about $1/2$ of the functions appearing in the integrand of (A.1), namely

$$w(ue^{-\beta})^2 = \left(u-\tfrac{1}{2}\right)^2\left((e^{-2\beta}-1)+\sum_{k=1}^{\infty}\frac{(-2)^{k+1}}{k+2}\left(u-\tfrac{1}{2}\right)^k\right)$$

and

$$(2u)^{-1/2} = \sum_{k=0}^{\infty}\frac{(-1)^k}{2^k}\binom{2k}{k}\left(u-\tfrac{1}{2}\right)^k,$$

respectively. An application of Lauwerier's method [**6**, eqs. (8) and (9)] (with the slightly different notation $P_j(t) := (-1)^j q_j(-t/2)$) then yields

$$A_j(\coth\beta) = \frac{(-1)^j}{\sqrt{\pi}}\left(\frac{1+\coth\beta}{2}\right)^j\int_0^{+\infty}e^{-s}s^{j-1/2}P_{2j}((1+\coth\beta)s)ds, \tag{A.2}$$

where $P_0(t) = 1$ and

$$P_j(t) = \frac{1}{2^j}\binom{2j}{j} - \sum_{k=1}^{j}\frac{2^k}{k+2}\int_0^t P_{j-k}(s)ds \tag{A.3}$$

for $j \geqslant 1$. It is readily seen from (A.3) that $P_j(t)$ is a polynomial in $t$ of degree $j$ with rational coefficients. With the substitution $t = \coth\beta$, (A.2) reduces to the more compact form

$$A_j(t) = \frac{(-1)^j}{\sqrt{\pi}}\left(\frac{1+t}{2}\right)^j\int_0^{+\infty}e^{-s}s^{j-1/2}P_{2j}((1+t)s)ds.$$

The $A_j(t)$ is the product of $(1+t)^j$ and a polynomial in $t$ of degree $2j$, and has rational coefficients. In particular, $A_0(t) = 1$ and

$$A_1(t) = \frac{(1+t)(1+5t-5t^2)}{24},$$

$$A_2(t) = \frac{(1+t)^2(-143+298t+231t^2-770t^3+385t^4)}{1152},$$

$$A_3(t) = \frac{(1+t)^3(-6187-240549t+750468t^2-334565t^3-1021020t^4+1276275t^5-425425t^6)}{414720}.$$

Consider now the right-hand side of (3.3). We deform the contour of integration into a small, positively oriented circle around $t_+$, employ the identity $\sqrt{2}\sin\alpha = e^{-\pi i/4}e^{-i\alpha/2}\sqrt{e^{2i\alpha}-1}$, and perform the change of variable $u = ze^{-i\alpha}$, to obtain the equivalent form

$$\sqrt{\frac{e^{2i\alpha}-1}{\pi}}\frac{\Gamma(j+1/2)}{2\pi i}\oint_{(1/2+)}\frac{(2u)^{-1/2}}{(w(ue^{i\alpha})^2)^{j+1/2}}du. \tag{A.4}$$

Since $\cos\alpha = \cosh(-i\alpha)$, we see that (A.1) is identical to (A.4) when $\beta$ is replaced by $-i\alpha$. This confirms that the right-hand side of (3.3) is indeed $A_j(\coth(-i\alpha)) = A_j(i\cot\alpha)$.

### Appendix A.2. The coefficients $D_j$

We proceed similarly as in the case of the coefficients $A_j$. By Cauchy's theorem, we can deform the contour of integration in (4.8) into a small, positively oriented loop contour around the saddle point $t_s = 1/2$, to obtain

$$D_j = \left(\frac{4}{3}\right)^{j/3}\frac{e^{-2\pi ij/3}}{2^{3/2}\pi i}\oint_{(1/2+)}\frac{z^{-1/2}}{w(z)^j}dz = \left(\frac{4}{3}\right)^{j/3}\frac{1}{2\pi i}\oint_{(1/2+)}\frac{(2z)^{-1/2}}{(w^3(z))^{j/3}}dz. \tag{A.5}$$

The square and cube roots in the second integral are defined to be positive for positive real $z$ and are defined by continuity elsewhere. To apply Lauwerier's method, we expand the functions in the rightmost integral in (A.5) into Taylor series about 1/2:

$$w^3(z) = \left(z-\tfrac{1}{2}\right)^3\sum_{k=0}^{\infty}\frac{(-2)^{k+2}}{k+3}\left(z-\tfrac{1}{2}\right)^k, \quad (2z)^{-1/2} = \sum_{k=0}^{\infty}\frac{(-1)^k}{2^k}\binom{2k}{k}\left(z-\tfrac{1}{2}\right)^k.$$

Lauwerier's method [**6**, eqs. (8) and (9)] (with the notation $Q_j(t) := (-1)^{j-1}q_{j-1}(3t/4)$) then gives

$$D_j = \frac{(-1)^{j-1}}{\Gamma(j/3)}\int_0^{+\infty}e^{-s}s^{j/3-1}Q_j(s)ds, \tag{A.6}$$

where $Q_1(t) = 1$ and

$$Q_{j+1}(t) = \frac{1}{2^j}\binom{2j}{j} - 3\sum_{k=0}^{j-1}\frac{2^{k+1}}{k+4}\int_0^t Q_{j-k}(s)ds$$

for $j \geqslant 1$. It is easy to verify that $Q_j(t)$ is a polynomial in $t$ of degree $j-1$ with rational coefficients. When integrating $Q_j(s)$ term-by-term in (A.6), the factor $\Gamma(j/3)$ in the denominator cancels, showing that $D_j$ is always a rational number. In particular,

$$D_1 = 1, \quad D_2 = 0, \; D_3 = -\frac{3}{20}, \; D_4 = \frac{1}{6}, \; D_5 = -\frac{9}{70},$$
$$D_6 = \frac{3}{40}, \; D_7 = -\frac{199}{7200}, \; D_8 = -\frac{3}{700}.$$

## References


1 T. Bennett, C. J. Howls, G. Nemes and A. B. Olde Daalhuis. Globally exact asymptotics for integrals with arbitrary order saddles. *SIAM J. Math. Anal.* **50** (2018), 2144–2177.

2 M. V. Berry and C. J. Howls. Hyperasymptotics for integrals with saddles. *Proc. Roy. Soc. London Ser. A* **434** (1991), 657–675.

3 W. G. C. Boyd. Error bounds for the method of steepest descents. *Proc. Roy. Soc. London Ser. A* **440** (1993), 493–518.

4 P. Deift, T. Kriecherbauer, K. T.-R. McLaughlin, S. Venakides and X. Zhou. Strong asymptotics of orthogonal polynomials with respect to exponential weights. *Comm. Pure Appl. Math.* **52** (1999), 1491–1552.

5 P. Deift and X. Zhou. A steepest descent method for oscillatory Riemann–Hilbert problems. Asymptotics for the MKdV equation. *Ann. Math.* **137** (1993), 295–368.

6 H. A. Lauwerier. The calculation of the coefficients of certain asymptotic series by means of linear recurrent relations. *Appl. Sci. Res. B* **2** (1952), 77–84.

7 *NIST Digital Library of Mathematical Functions*, http://dlmf.nist.gov/, Release 1.1.3 of 2021-09-15, F. W. J. Olver, A. B. Olde Daalhuis, D. W. Lozier, B. I. Schneider, R. F. Boisvert, C. W. Clark, B. R. Miller, B. V. Saunders, H. S. Cohl, and M. A. McClain, eds..

8 F. W. J. Olver. *Asymptotics and Special Functions* (Academic Press, New York, 1974).

9 F. W. J. Olver. Asymptotic approximations and error bounds. *SIAM Rev.* **22** (1980), 188–203.

10 S. C. van Veen. Asymptotische Entwicklung und Nullstellenabschätzung der Hermiteschen Funktionen. *Math. Annal.* **105** (1931), 408–436.

11 R. Wong. Error bounds for asymptotic expansions of integrals. *SIAM Rev.* **22** (1980), 401–435.

12 R. Wong. *Asymptotic Approximations of Integrals* (Academic Press, Boston, 1989).

13 R. Wong. Asymptotics of linear recurrences. *Anal. Appl.* **12** (2014), 463–484.

14 R. Wong and Y.-Q. Zhao. On a uniform treatment of Darboux's method. *Constr. Approx.* **21** (2005), 225–255.

# Permissions

**Journal papers included in this collection are reprinted with permission from the following publishers:**

**American Mathematical Society**

Author(s): K. F. Lee & R. Wong
Article title: Uniform asymptotic expansions of the Tricomi-Carlitz polynomials
Journal title: Proc. Amer. Math. Soc.
Volume/Page(s): ***138(2010), 2513–2519***
Year of Publication: © 2010 American Mathematical Society

Author(s): S.-Y. Liu & R. Wong & Y.-Q. Zhao
Article title: Uniform treatment of Darboux's method and the Heisenberg polynomials
Journal title: Proc. Amer. Math. Soc.
Volume/Page(s): ***141(2013), 2683–2691***
Year of Publication: © 2013 American Mathematical Society

Author(s): K. F. Lee & R. Wong
Article title: Asymptotic expansion of the modified Lommel polynomials $h_{n,\nu}(x)$ and their zeros
Journal title: Proc. Amer. Math. Soc.
Volume/Page(s): ***142(2014), 3953–3964***
Year of Publication: © 2014 American Mathematical Society

Author(s): R. Wong & Y.-Q. Zhao
Article title: Special functions, integral equations and a Riemann-Hilbert problem
Journal title: Proc. Amer. Math. Soc.
Volume/Page(s): ***144(2016), 4367–4380***
Year of Publication: © 2016 American Mathematical Society

Author(s): X.-S. Wang & R. Wong
Article title: Asymptotics of Racah polynomials with fixed parameters
Journal title: Proc. Amer. Math. Soc.
Volume/Page(s): ***146(2018), 1083–1096***
Year of Publication: © 2018 American Mathematical Society

Author(s): M.-J. Luo & R. Wong
Article title: Asymptotics of the associated Pollaczek polynomials
Journal title: Proc. Amer. Math. Soc.
Volume/Page(s): ***147(2019), 2583–2597***
Year of Publication: © 2019 American Mathematical Society

**Cambridge University Press**

Author(s): W. Shi & G. Nemes & X.-S. Wang & R. Wong
Article title: Error bounds for the asymptotic expansions of the Hermite polynomials
Journal title: Proc. Roy. Soc. Edinburgh Sect. A
Volume/Page(s): ***153(2023), 417–440***
Year of Publication: © 2023 Cambridge University Press

**Elsevier**

Author(s): K. F. Lee & R. Wong
Article title: Asymptotic expansion of the Tricomi-Carlitz polynomials and their zeros
Journal title: J. Comput. Appl. Math.
Volume/Page(s): ***265(2014), 220–242***
Year of Publication: © 2014 Elsevier

Author(s): X.-S. Wang & R. Wong
Article title: Asymptotics of Racah polynomials with varying parameters
Journal title: J. Math. Anal. Appl.
Volume/Page(s): ***436(2016), 1149–1164***
Year of Publication: © 2016 Elsevier

**Institute for Scientific Computing and Information**

Author(s): R. Wong
Article title: Asymptotics of orthogonal polynomials
Journal title: Int. J. Numer. Anal. Model.
Volume/Page(s): ***15(2018), 193–212***
Year of Publication: © 2018 Institute for Scientific Computing and Information

**IOS Press**

Author(s): Y. Lin & R. Wong
Article title: Global asymptotics of the discrete Chebyshev polynomials
Journal title: Asymptot. Anal.
Volume/Page(s): ***82(2013), 39–64***
Year of Publication: © 2013 The Authors

Author(s): C. Ou & R. Wong
Article title: A novel generalized solution expansion for the Lagerstrom model
Journal title: Asymptot. Anal.
Volume/Page(s): ***88(2014), 111–123***
Year of Publication: © 2014 The Authors

**John Wiley & Sons, Inc**

Author(s): Z. Wang & R. Wong
Article title: Uniform asymptotics for orthogonal polynomials with exponential weights — the Riemann-Hilbert approach
Journal title: Stud. Appl. Math.
Volume/Page(s): ***115(2005), 139–155***
Year of Publication: © 2005 John Wiley & Sons, Inc

Author(s): X. H. Jiang & R. Wong
Article title: Asymptotic analysis of a perturbed periodic solution for the KdV equation
Journal title: Stud. Appl. Math.
Volume/Page(s): ***116(2006), 21–33***
Year of Publication: © 2006 John Wiley & Sons, Inc

Author(s): R. Wong & H. Y. Zhang
Article title: Asymptotic solutions of a fourth order differential equation
Journal title: Stud. Appl. Math.
Volume/Page(s): ***118(2007), 133–152***
Year of Publication: © 2007 John Wiley & Sons, Inc

Author(s): J. Wang & W.-Y. Qiu & R. Wong
Article title: Global asymptotics for Meixner-Pollaczek polynomials with a varying parameter
Journal title: Stud. Appl. Math.
Volume/Page(s): ***130(2013), 345–392***
Year of Publication: © 2013 John Wiley & Sons, Inc

Author(s): J. H. Pan & R. Wong
Article title: Asymptotics of the discrete Chebyshev polynomials
Journal title: Stud. Appl. Math.
Volume/Page(s): ***132(2014), 13–49***
Year of Publication: © 2014 John Wiley & Sons, Inc

Author(s): Z. Song & R. Wong
Article title: Asymptotics of pseudo-Jacobi polynomials with varying parameters
Journal title: Stud. Appl. Math.
Volume/Page(s): ***139(2017), 179–217***
Year of Publication: © 2017 John Wiley & Sons, Inc

Author(s): X.-M. Huang & R. Wong
Article title: Uniform asymptotics and zeros of the associated Pollaczek polynomials
Journal title: Stud. Appl. Math.
Volume/Page(s): ***145(2020), 625–646***
Year of Publication: © 2020 John Wiley & Sons, Inc

**Springer Nature**

Author(s): Z.-T. Wen & R. Wong & S.-X. Xu
Article title: Global asymptotics of orthogonal polynomials associated with a generalized Freud weight
Journal title: Chinese Ann. Math. Ser. B
Volume/Page(s): ***39(2018), 553–596***
Year of Publication: © 2018 Springer Nature

**The Royal Society**

Author(s): R. Wong & Y.-Q. Zhao
Article title: Uniform asymptotic expansion of the Jacobi polynomials in a complex domain
Journal title: Proc. R. Soc. Lond. Ser. A Math. Phys. Eng. Sci.
Volume/Page(s): ***460(2004), 2569–2586***
Year of Publication: © 2004 The Royal Society

Author(s): Z. Wang & R. Wong
Article title: Bessel-type asymptotic expansions via the Riemann-Hilbert approach
Journal title: Proc. R. Soc. Lond. Ser. A Math. Phys. Eng. Sci.
Volume/Page(s): ***461(2005), 2839–2856***
Year of Publication: © 2005 The Royal Society

**World Scientific Publishing**

Author(s): Y. Lin & R. Wong
Article title: Global asymptotics of the Szegő-Askey polynomials
Journal title: Anal. Appl. (Singap.)
Volume/Page(s): ***12(2014), 727–746***
Year of Publication: © 2014 World Scientific Publishing

Author(s): R. Wong & Y.-Q. Zhao
Article title: Recent advances in asymptotic analysis
Journal title: Anal. Appl. (Singap.)
Volume/Page(s): ***20(2022), 1103-1146***
Year of Publication: © 2022 World Scientific Publishing

www.ingramcontent.com/pod-product-compliance
Lightning Source LLC
LaVergne TN
LVHW081935200926
848484LV00009B/2437
* 9 7 8 9 8 1 1 2 9 0 8 5 5 *